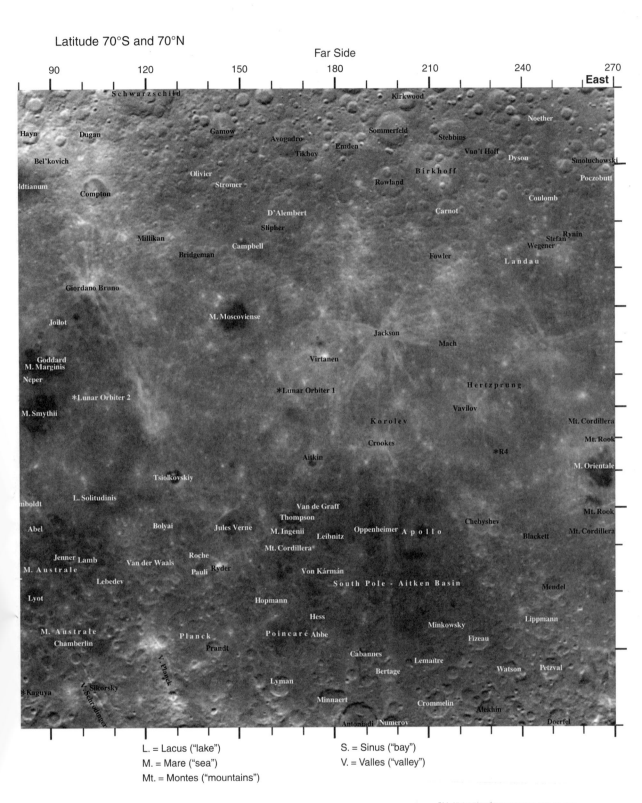

The New Moon

Water, Exploration, and Future Habitation

Explore Earth's closest neighbor, the Moon, in this fascinating and timely book and discover what we should expect from this seemingly familiar, but strange, new frontier. What startling discoveries are being uncovered on the Moon? What will these tell us about our place in the Universe? How can exploring the Moon benefit development on Earth?

Discover the role of the Moon in Earth's past and present; read about the lunar environment and how it could be made more habitable for humans; consider whether continued exploration of the Moon is justified; and view rare Apollo-era photos and film stills.

This is a complete story of the human lunar experience, presenting many interesting but little-known and significant events in lunar science for the first time. It will appeal to anyone wanting to know more about the stunning discoveries being uncovered about the Moon.

ARLIN CROTTS is Professor of Astronomy at Columbia University and has won numerous awards for his work. Having observed objects as distant as ten billion light years and as close as the Moon, he finds the problems of lunar science particularly intriguing.

The New Moon

Water, Exploration, and Future Habitation

Arlin Crotts
Columbia University, New York

CAMBRIDGE
UNIVERSITY PRESS

32 Avenue of the Americas, New York, NY 10013–2473, USA

Cambridge University Press is part of the University of Cambridge.

It furthers the University's mission by disseminating knowledge in the pursuit of education, learning, and research at the highest international levels of excellence.

www.cambridge.org
Information on this title: www.cambridge.org/9780521762243

© Arlin Crotts 2014

This publication is in copyright. Subject to statutory exception and to the provisions of relevant collective licensing agreements, no reproduction of any part may take place without the written permission of Cambridge University Press.

First published 2014

Printed in the United States of America

A catalog record for this publication is available from the British Library.

Library of Congress Cataloging in Publication Data
Crotts, Arlin.
The new moon : water, exploration, and future habitation / Arlin Crotts, Columbia University, New York.
   pages   cm
Includes index.
1. Moon.   I. Title.
QB581.C89   2013
523.3–dc23   2013030434

ISBN 978-0-521-76224-3 Hardback

Cambridge University Press has no responsibility for the persistence or accuracy of URLs for external or third-party Internet Web sites referred to in this publication and does not guarantee that any content on such Web sites is, or will remain, accurate or appropriate.

Notes on photo credits: all images from the Lunar Reconnaissance Camera are used courtesy of NASA, Goddard Space Flight Center, and Arizona State University.

Contents

	Preface	*page* vii
	Acknowledgements	ix
1	The Importance of the Moon	1
2	First Steps	18
3	Moon/Mars	58
4	An International Flotilla	126
5	The Moon Rises from the Ashes	158
6	Moons Past	174
7	The Pull of the Far Side	211
8	Water in a Land of False Seas	227
9	Inconstant Moon	261
10	Moonlighting	298
11	Lunar Living Room	331
12	Lunar Power	368
13	Stepping Stone	392
14	Return to Earth	432
	Glossary	463
	Appendix A Von Braun et al. Space and Lunar Exploration Issues	477
	Appendix B Topics in Transient Phenomena on the Moon	480
	Index	495

Preface

There is a new Moon. Many of the most dramatic recent discoveries in planetary science are lunar. They transform our understanding of the Moon and lunar exploration's prospects for exciting work there. Yet in many minds the Moon is an old story. NASA sent a dozen men to its surface in 1969–1972, and most people do not remember those events personally. Only in the past few years has lunar exploration accelerated again, and many, including policy makers deciding about the space program, do not realize how rapidly our knowledge of the Moon is changing.

This ignorance is unfortunate, since from 2004 to 2010 we were headed to the Moon but changed our minds. Some of these recent discoveries about the Moon might have changed our minds back again. By "our" in this case I refer to the United States, since many other nations' space programs still see the Moon as an essential objective.

This book's purpose is to provide information that the reader might need to decide if the Moon is worthy of our exploration and what we might do there. What has affected our decisions to explore the Moon (or not) and what must we consider in the future, at least until the mid-21st century? What place could lunar exploration play in space exploration, and in the human future in general? There is politics involved, and international relations. Humans will significantly influence Earth's environment over this time, soon to our detriment. This will affect our economics. Human technology and communications will change, and lunar exploration might play a natural role in this, all by mid-century.

Because of recent political decisions in the United States we do not know where we are headed beyond low Earth orbit (and by "we" here I could mean the whole world). The Obama administration has suggested a mission to an asteroid by the mid-2020s and then on to Mars, but this decision is scheduled to be reconsidered in a few years. None of the political, budgetary, and programmatic considerations tilting towards the asteroids and Mars change the fact that the Moon is 100 times closer on average. This is qualitatively decisive, since round-trip light-travel times to the Moon allow remote control of robots in real time, whereas on Mars or an asteroid the delays change this drastically. Also, the environment of the Moon is not as harsh as sometimes advertised; the implications of this must be considered. I want to present these facts and let the reader decide what to do about them. Is the Moon interesting enough?

Humans face challenges in the coming decades, and to many of these the Moon holds no answer. War, poverty, strife, and environmental degradation are problems for which we must look to ourselves for solutions, not to the Moon. Still, the Moon has great potential. Perhaps if we had continued the course set in 1961 but abandoned in 1972, more tools would be available now and in time to defeat obstacles to continued civilized existence. Perhaps that was a vain hope.

But the Moon offers hope. It can be a productive, comfortable, economically viable place, and depending on available modes of transportation, can benefit directly and host many people. Possibly within the lifetimes of many people today, we will see ourselves, relatives, and neighbor's children live and work on the Moon. It is like a new continent, larger than North America or Africa, nearly the area of Asia, with real possibilities for economically viable and comfortable lives for people who live there.

Like Columbus, starting with three ships, then a few more, followed by dozens and hundreds within a few generations, we have started (except this time with robots) the human race's journey to this new land. With six ships, and twelve men, we have just scratched its surface. It is our choice as to whether we continue. Like the Ming emperor, we might burn the fleets of Zheng He (Chung Ho) and journey no more across the great oceans, or like Leif Ericsson and his clan, we might establish a toehold in this new world, only to see it sink into obscurity, an archeological curiosity, no more than an amusing tale in the Icelandic Sagas.

In the lexicon of human symbols, the Moon is among the most potent, but it is much more than a symbol; it is a whole world, the one most familiar to us, save Earth, but surprisingly unknown, and still potentially concealing surprises. In this book we discuss little the Moon of old, of historical record, the poet, the shaman, or the artist. Instead, we will discuss the Moon of the scientist, astronaut, and engineer, and hopefully the Moon in humanity's future. It is for us to choose.

This book was not written to dwell on the politics of The Return to The Moon, although one cannot speak of humankind and the Moon without touching on NASA and its initiatives. The Moon arose 4.5 billion years before the creation of NASA, and humanity's reach for the Moon was set in motion at least 100,000 years ago.

The Moon is a world, a physical environment in which people can be comfortable and productive someday. This book's purpose is to describe how this world could become familiar to us. I have no crystal ball to reveal this, but the journey may have now begun in earnest. I invite readers to consider if this should be our coming destination.

Acknowledgements

The author greatly appreciates the current and former astronauts, scientists, engineers, aerospace workers, and others who took the time to talk with him. He thanks Paola, Caleb, and Steve for their patience in hearing and tolerating the impassioned version that then spewed forth.

Chapter 1
The Importance of the Moon

If God meant us to explore space, He would have given us a Moon.
– Krafft A. Ehricke (1917–1984), aerospace scientist from
pre–World War II Germany to 1980s United States[1]

Krafft Ehricke not only helped pioneer some of the earliest modern, liquid-propellant rockets, but also lived to develop workhorse boosters for the space age and concepts for lunar mining and planetary exploration now in the works. He envisioned the Moon as a stepping-stone, a role it played in several ways throughout humanity's development starting long ago. He had a clever way of stating the profoundly obvious.

Ehricke's life-span saw astounding human achievements: harnessing amazing new energy sources, traveling hundreds of times faster than ever before, probing scales millions of times larger and thousands of times smaller than imagined before, and transforming the Moon and planets from dreamlands to mapped worlds. We once ascribed romantic notions to the Moon; now we see how alien worlds differ from Earth and distant worlds of our imagination.

To many the Moon is the most fascinating, beloved pearl in the sky. It rules the night (half of the time)[2] with more personality than the day's Sun, transforming throughout each month, coming and going with regularity rare among celestial objects, which seemingly never change, or planets, which perform irregular loops and whirls.[3] Only the Sun and the Moon perform their regular, regal glide across the sky. We can perceive features on the Moon with the naked eye – the Man in the Moon (or woman, rabbit, buffalo, dragon) – something we cannot do with the Sun and most celestial objects. Whereas the Sun is like molten gold, the Moon is silver, engraved with patterns for the mind to explore.[4] However, until lately, these patterns beckoned from distances too great to cross.

The Moon and the Sun are a pair. Almost the same in apparent angular size, they seem to chase each other across the sky. This is visible near the full Moon, when the Sun is setting and the Moon rising (or vice versa) on opposite sides of the sky. (Ask Eudora Welty.[5]) The Sun is manifestly important. Without it we would freeze, starve, and die. The naïve human mind will ascribe similar importance to the equally sized disk of the Moon but suspect that the Moon's significance is more subtle, covert.

Science confirms mythology's suspicion that objects looming large in the sky affect Earth. We are immersed in the expanse of our Milky Way Galaxy, from which our Sun was born and that produces the starstuff of heavier elements forming planets such as Earth and creatures like us. Less obvious objects extensive in the sky will have their huge impact, such as the Andromeda Galaxy M31 (which modern-day urbanites rarely see because of light pollution). Someday, several billion years from now, M31, now appearing only a few times larger in angle than the Moon, will grow to fill Earth's sky and engulf our galaxy. In the future when most stars more massive than the Sun have died, the great nebula from Andromeda will shred our galaxy, plowing many billion times the Sun's mass in loose gas back into stars and reawakening stellar nurseries that set our skies aglow billions of years ago, before the Sun's birth. This will likely not destroy the Sun or Earth: if the Sun is not ejected in this great collision, our graves will come to rest in this new Andromeda galaxy.

The Moon in its time was no less important to Earth, and loomed even larger in the sky, as we shall see. Although the Moon rules the night, it does not shed much light; what power does it exert? Modern humans think that the Moon is less important. Or is it? This is the question we explore in this chapter.

We must discuss the real Moon: the one we know and the one just offshore in the sea of our ignorance. The Moon is amusing in myth and lore[6] and endlessly discussed, misguidedly, in astrology. The Moon is not a goddess; it is a five billion cubic mile sphere of rock and magma orbiting over our heads at three times the speed of sound. It is not green cheese but largely olivine, a green silicate mineral at thousands of degrees, like the mantle deep within Earth. The Moon is no fairy tale; its reality is as comforting as realizing that we orbit a four million solar mass black hole at our Galaxy's center. Although the Moon's appearance in literature is boundless (lovers swoon beneath the Moon while they spoon in June – but not here), we will be more serious (see Figure 1.1).

Some lovers actually do swoon beneath the Moon, with regularity. Many crab species spawn only at specific lunar phases,[7] which makes sense because the highest tides occur near the new and full Moon. And some species not spawning along the shore reproduce in keeping with the lunar phase;[8] this occurs even in cattle.[9] Humans, arising in East Africa far from the sea, seemingly retain vestigial ties to the Moon. Women's menstrual cycles peak in period at 29.1 days, nearly the same as a lunation (despite the typical menstrual periods often misquoted as four weeks).[10] Those with lunar-like 29.5-day menstruation tend to menstruate near the full Moon and ovulate near the new Moon. No lunar phase dependence is apparent in fertility in artificial in vitro fertilization of women as this is determined by when the embryo is transferred.[11] The reason for a lunar period in human fertility is not settled science, and some – Isaac Asimov, for example – have argued it is pure coincidence. As fundamental as reproduction is to behavior, humans show little lunar effect – dependence of human behavior (crime, conception, insanity) on lunar phase, although it may affect human sleep.[12]

The tides, dominated by the Moon (30% solar, 70% lunar), determine when some species mate, spawn, and die, with strong influence on life along the shore. Tides

The Importance of the Moon

Figure 1.1. **Moons of imagination**. (a) From a 1920s Parisian postcard, used in *Drawing-Collage* by Joan Miró, August 8, 1933. (b) Futuristic scene to inspire Soviet youth, from *Tekhnika Molodezhi* (*Youth's Technology* magazine) August 1953. Compare this to space hardware envisioned by Von Braun and Bonestell in 1952 *Collier's* magazine (Figure 3.1). (c) *Lunar Jim* and Rover from the Canadian children's TV cartoon series, on Blue Moon L22 (See Box 3.4, courtesy LJ Productions 2003 Ltd).

The New Moon

Figure 1.2. **What a difference a quarter day makes**. The tide's affect on the Saint John, New Brunswick waterfront, on the Bay of Fundy in maritime Canada. These two photographs were taken about 6.5 hours apart. Note the water level rising about two stories, or 7 m, as evidenced by disappearing walls along the wharfs. The cruise ship rises by a similar amount. (Photographs by the author.)

can be impressive where tidal period (12.5 hours) resonates with the time for waves to traverse a body of water (depending on its size, shape, and depth profile). The Bay of Fundy in southeastern Canada, with the right resonant frequency, has tides ranging up to 17 meters (56 feet)[13] (see Figure 1.2). In flat intertidal zones the incoming tide can overtake a fleeing person. People living in such areas learn to respect the importance of the Moon.

As dangerous and powerful as the tides can be now, they were once a world-dominating force. The reason for these stronger tides in the past is, ultimately, the tides themselves. The bulge raised on Earth by the Moon's tidal force is the one way rotational angular momentum from the spinning Earth can be transferred, in this case into orbital motion of the revolving Moon (see Figure 1.3). Thus Earth's days grow longer, and the month does, too. As the Earth-Moon distance increases with

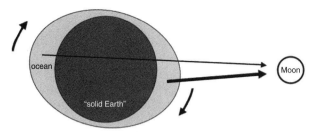

Figure 1.3. **The days (and months) grow longer**. The Moon pulls the ocean closest to it away from Earth, raising a tidal bulge beneath the Moon. Simultaneously, it pulls Earth away from ocean on the planet's Far Side, resulting in a second tidal bulge opposite the Moon. Add more complexity: tidal bulges rise with a slight delay. Because Earth rotates faster than the Moon revolves around it, the Moon tugs on the closer bulge in the direction slowing Earth's rotation; the bulge farthest from the Moon pulls to speed up Earth's rotation. These effects do not cancel; the bulge closer to the Moon interacts more strongly and so wins out. Energy is robbed from Earth's rotation, lengthening the day. Where does the energy go? Mainly into the Moon's orbit, forcing it to move away from Earth, lengthening the month.

time, the tidal force falls off. This happens slowly, with the day lengthening only about 0.00003 second annually and the Moon creeping outward by 3.8 centimeters per year (1.5 inches, or a ten-billionth part of the distance to Earth).[14] The tidal energy lost in Earth's oceans and interior is about 3 terawatts, equivalent to 15% of the output from all of humanity's power plants, 6% of the heat flow from Earth's interior, and 4% of the rate energy is stored by all photosynthesis. Over billions of years, the distance from Earth to the Moon has changed radically, as has the power of the tides.[15]

The Moon's uneven pull produces tides with a force varying as the reciprocal of the cube of the distance. If the Moon was significantly closer in the past, the tides were much greater. At today's rate of drift away from Earth, the Moon would have been twice as close 4.5 billion years ago, and the tidal force eight times greater. Accounting for accelerated tidal interaction at closer distances and retracing the Moon's orbital evolution, one would predict that less than two billion years ago the Moon had merged with Earth.[16] The geological record obviously rules out such drastic recent conditions. When, if ever, did such terrifying effects occur? Earth's oceans are ancient, with minerals indicating their formation at least 4 billion and perhaps 4.4 billion years ago,[17] only 150 million years after Earth's birth.

How do we reconcile such seemingly paradoxical facts – the inferred age of the Moon, with its purported merger with Earth being much younger than our planet? We actually measure the ancient length of the day and month with the musically named rock type, the "rhythmites" (see Figure 1.4). Strong tides wash material into sedimentary layers, and variations in composition of these sediments over daily, monthly, or annual cycles can highlight the passage of these individual time intervals,[18] as can daily, monthly, and yearly coral growth rings.[19] As opposed to 365.25 days per year, 29.5 days per month, and 12.4 months per year, ancient stripe patterns in rhythmites imply more days per year, more months per year, and slightly more days per month.[20] (Actually, we should use the lunar day – now

Figure 1.4. **Rock 'n' Rhythmites**. (a) Varved shale sediment samples showing rhythmite structure. The scale at upper left shows 1-centimeter reference squares. (b) Rhythmites in Precambrian shale deposits, 900 million years old, in Big Cottonwood Canyon, Utah. (Photographs courtesy of Dr. Marjorie Chan, University of Utah.)

24.83 hours – during which the Moon returns to the same position east-west in Earth's sky.) With an astrophysical model for how angular momentum is transferred from Earth to the Moon, we can infer the Moon's orbital radius and the length of the day and month. Comparing rhythmites over the past 2.5 billion years produces the odd result that the Moon is receding faster now than in the past, by about a factor of two.

Does this theory work? Can the recession of the Moon from Earth be faster now than when the Moon was closer? Yes, because like in the Bay of Fundy, the match between the basin's resonant frequency and the frequency of the tides is also crucial for the oceans.[21] Generally, as Earth's rotation speeds up, this match worsens, slowing the angular momentum transfer to the Moon.[22] Furthermore, tidal friction depends critically on the depth of the oceans and shape of the continents dividing them, both radically different long ago.[23] These can change radically over several hundred million years, with the oceans merging occasionally into one, roughly 0.25, 1.1, and 1.7 billion years ago. In addition, we might detect ancient Earth thickly covered with ice, not liquid ocean, in part because of the cooler Sun in the first billion years. Changes in the upper mantle might also play a role. These effects are not all incorporated in models of the tidal evolution of the Earth-Moon system but would influence changes in the length of the day and month.

With tides varying so strongly with distance, if the Moon was significantly closer in the past, the tides were much greater. When the Moon first formed, as we will soon discuss, it was perhaps five to ten times closer than it is now (implying tidal forces hundreds of times greater). Whereas the range of ocean tides currently averages about 0.5 meter (and 0.2 meter over continents), there were once tides hundreds of meters high. Although no landmass familiar to us existed so long ago, such ocean tides would sweep every few hours over lands as large as subcontinents. Depending on when life arose on Earth, and when tides were hugely larger, tides may have promoted the first DNA.[24] This controversial theory is in keeping with the idea that DNA polymerization is aided by periodic concentration such as that experienced in this large intertidal zone, as well as the observation that, unlike many competing molecules, DNA tolerates high salt concentrations that would accompany its polymerase chain reaction between tides. Perhaps the Moon aided the start of life on Earth.

However lunar tides affected Earth's early oceans, they have produced a stabilizing influence on the tilt of Earth's rotational axis since then. The Moon pulls on Earth's equatorial bulge to keep Earth's rotational axis from shifting as a result of other planets' gravitation. Earth's axis tilts from its orbital plane (23.5° "the obliquity"), produces our seasons, and oscillates within a narrow range of 21°–25°. Without the Moon, the obliquity would vary wildly; indeed, Earth will lose some stability as the Moon continues to draw away.[25] Without the Moon, the flip-flopping obliquity would produce huge temperature fluctuations promoting harsher seasons and ice ages. This effect causes radical changes in the axial tilt and seasons of Mars;[26] without the Moon, Earth would suffer a similar fate.

The earlier in time one views the Earth-Moon system, the more perilously close the Moon was to Earth. How did the Moon form? With little evidence that its early orbit was elliptical, it was likely not captured gravitationally as an intact body.[27] Instead, the prime scientific conclusion from the Apollo program is that the Moon formed in a catastrophic glancing blow to the proto-Earth by a smaller planet: the Giant Impact hypothesis. (See Chapter 5.) This collision ripped much of the crust and mantle from proto-Earth and flung it into space. The smaller planet's core, about 10% of Earth's, plunged into Earth's center. In essence, we live on this new heavier and denser planet. The collision's debris that did not escape or fall to Earth coalesced into the Moon. The original planet that would have been Earth was destroyed.

We might never know what the third planet from the Sun was like before the formation of the Moon. Much depends on when this "Big Whack" occurred in the narrow interval between the oldest meteorites (4.567 billion years ago[28]) and the oldest Earth minerals (4.4 billion years). In most models Earth had largely accumulated about 10–15 million years after the first meteorites, whereas the Moon's heavy-element isotopes indicate it came along in 30–55 million years after these meteorites, probably enough time for the proto-Earth to have formed a crust and a solid (although deformably plastic) mantle and even to accumulate a significant atmosphere (first primordial hydrogen, then water vapor, on the way to forming an ocean). All this changed when it was hit by another planet roughly the size of Mars.

We do not know exactly how the Big Whack occurred, but conditions thought sufficient to make the Moon are a proto-Earth of a mass about 95% of today's planet struck at a 45° angle at 11–15 kilometers per second by another planet roughly 13%

of Earth's mass. The crust and much of the mantle of the proto-Earth were scalped by a shock wave that circled the planet for several hours at half orbital speed, raising surface temperatures to 7,000K (6,727°C or 12,140°F) – more than 1,000°C hotter than the Sun. Most of the mantle was re-melted, and 20% was vaporized. A mixture of rock, magma, and about 25% vapor at 3,000K average temperature was flung into orbit or beyond Earth's influence altogether, while the rest fell back to Earth, with 1.2% of the total mass coalescing into the Moon.[29] The Moon could not form closer than the distance at which Earth's tidal forces would tear it apart (the Roche limit, at 3 Earth radii) but, in simulations, it tends to form just beyond, at 3.5 Earth radii.[30] We can measure where the Moon was shortly after this. The Moon's shape appears to incorporate a "fossil bulge" in equilibrium with a rotational period of 3.5 days (not the current 29.5 days).[31] This close to Earth, the Moon would almost certainly settle into turning the same face toward Earth (hence, the rotation period equals the period of revolution), implying an orbit at 14.5 Earth radii when the Moon solidified, 25% of its radius now.

The Giant Impact hypothesis predicts huge tides from the Moon soon after formation, looming in Earth's sky, larger in angle than one's fist held at arm's length. It was invisible from Earth, however, in the immediate post-impact environment. For 1,000 years, the terrestrial atmosphere we usually think of as gas (hydrogen, water vapor) was replaced by a thick silicate cloud – rock vapor at roughly 2,000°C. Two million years passed before a solid surface began to re-form. Perhaps water vapor spewing from this magma ocean filled the atmosphere with further gases, particularly carbon dioxide, leading to runaway greenhouse warming that partially replaced the heat radiated from magma and rock.[32] Meanwhile, lunar tides were an important heat source. A sharp transition from this superheated state occurred several million years later when water began raining from the sky and oceans replaced the steam atmosphere. Instead of water vapor, carbon dioxide, at 100 times the density of Earth's whole current atmosphere, filled the air at about 200°C.[33] Over tens of millions of years, carbon dioxide incorporated into oceans and rock, and temperatures dropped further. (The first limestone appeared hundreds of millions of years later, indicating oceans acidified by dissolved carbon dioxide.) For roughly the next 500 million years Earth alternated between a water-ice-covered world (about –50°C) and one punctuated with sudden heating by lesser asteroidal and cometary impacts, swinging between fire and ice, and awaited the coming of life.

Within several million years post–Big Whack, Earth's surface returned to a light-element, atmospherically dominated phase in closer analogy to Venus or early Mars, but having changed forever. The violent disruption and mixing of Earth's crust and mantle permanently affected rock's distribution in the crust. The heat shortly after the giant impact was so great that the surface was probably more a roiling layer of mantle-like material than the crust we know today, but eventually lighter materials rose and heavy materials subducted until embryonic continents appeared. Earth's crust is unique compared to Mars or Venus (or Mercury or the Moon) in having two distinct terranes (continental and oceanic crust) and the terranes' evolution via plate tectonics.[34] Earth's surface elevation separates into two distinct zones: one centered just above sea level corresponding to continents and another 4 kilometers down corresponding to oceans. They are truly different: oceanic basin crust (two-thirds of the total) is only 3–10 kilometers thick and dense, compared to continents, which are 30–50 kilometers thick and composed of lighter material.[35] The continents are old, up to several billion years, riding above plate tectonic processing of crust, whereas oceans are young, less than 200 million years, with older ocean basins having been subducted into the interior. Earth's "twin" planet, Venus, has no such dichotomy, with crust a relatively uniform slab 40–70 kilometers thick. Venus and Mars show the first hints of the rifting and subduction dominating Earth's crust, with its "rings of fire" of volcanoes and earthquakes stretching around the world.

Much of what would have become Earth's continental crust (and on the Mars-sized impactor) is now on the Moon. Earth's average crustal density, 2.9 grams per cubic centimeter, is much lower than the rest of Earth, 5.51 g/cm^3, but close to the entire Moon's, 3.34 g/cm^3. Earth's moonless twin, Venus, has an intermediate value of 5.20 g/cm^3. Although the Moon probably contains a small ferrous core, it is high in the silicates and incompatible elements seen in Earth's crust and ferromagnesian mixtures in the mantle (plus the similar, primitive mineral pyroxene). If one could spread the volume of the Moon over Earth's surface, it would cover it to a depth of 43 kilometers, almost exactly the "missing" volume of Earth's crust compared to Venus and the amount required to supplement Earth's oceanic crust to continental thickness. Although more than just density defines "crust,"[36] in a sense the reason our planet has plate tectonics rests in the Moon. We explore this conjecture in Chapter 6.

Comparing Earth and Venus illustrates essential differences that lunar formation's effects provide us on Earth. It is more than crust deep. Earth rotates 250 times faster than Venus, as a consequence of the Giant Impact's glancing blow. For a planet to produce a strong magnetic field, it must rotate rapidly and contain a conducting fluid. Earth's strong magnetic field not only protects life from intense solar wind radiation but also keeps the wind from eroding gas from the upper atmosphere, explaining why Earth has so much more water than Venus (by about 10,000 times). Water lubricates plate tectonic motion and dissolves carbon dioxide and sequesters it at the ocean bottom. If all carbon dioxide disappeared, however, ice would cover Earth. Plate tectonics restore volcanically a small fraction of the CO_2 (about 1/10,000th) by subducting ocean crust into the mantle and baking out the water and CO_2, where roughly 97% of the resulting CO_2 is taken up by life, leaving the rest in the atmosphere. Together, water, carbon dioxide, life, and plate tectonics regulate Earth's thermostat, and the basic configuration that enables this is tied intimately to the Moon's formation.[37]

The speed of the Moon's withdrawal from Earth, several centimeters per year, is typical of drift velocities between continents in plate tectonics – a curiously appropriate coincidence. The surface area of the Moon is almost as large as that of Asia and larger than that of Africa. The Moon is Earth's unexplored continent. What mysteries does it conceal? The sixteenth century had the Americas, the eighteenth had Australia, and the twentieth had Antarctica. The Moon is the strangest of all, likely concealing mysteries we cannot anticipate. When we explore the Moon, we will gain the wherewithal to explore the remaining Solar System. Humans can then transition from Earth's denizens to inhabitants of the Universe. Slowly, the Moon is pulling away from Earth, and it is not coming back. We are going to have to go get it.

When humans first roamed the African plains, surely they looked up and wondered about the Moon. Imagine an adult hominid awake at night, listening for predators, gazing over the moonlit landscape. Sometimes the Moon is bright, other times a faint crescent, sometimes absent from the sky altogether: How can one not notice it and its effects on the nighttime environment's appearance? Anticipating the Moon and its light's effects tonight, tomorrow, or next week was vital for survival for early humans. Certainly people thought about it. As the human mind has developed, the Moon has served in several ways as a bridge from Earth to the Uinverse.

Whereas the Sun is responsible for two basic units of time in our life, the day and the year, the Moon keeps its own time, the month – obvious to any human who can see its phases. The Moon is the sky's dedicated timekeeper. To the mind, its subtle changes over the course of 29.5 days are a convenient bridge between the short day and the 365.24-day year. Many societies employ the Moon to pace their rituals. It gives us the month and the week, which are just long enough to allow people to see things grow older, to bridge between cyclical and permanent changes in our lives.

The Moon is the only object in space that appreciably blocks the Sun. (Venus and Mercury occasionally transit the Sun, blocking only 0.0025% and 0.00008% of sunlight reaching Earth, respectively.) The Sun's size in the sky ranges from 31.6 to 32.7 arc minutes (or 32.7/60ths of a degree), whereas the Moon ranges from 29.9 to 33.9 arc minutes, roughly half the width of one's little finger at arm's length. Thus, we are treated to the astounding phenomenon of the total solar eclipse (see Figure 1.5). In earlier historic times, a ruler could impress his/her subjects no more effectively than by having the court astronomers predict when the next solar eclipse would occur; seemingly nobody could be better tapped into the power of the gods than one able to proclaim when the Sun would disappear from the sky. This close match in sizes makes the event more dramatic and arises because the Moon is 400 times closer to Earth than the Sun is, while the Sun is 400 times larger than the Moon. This fearful delight is peculiar to our time in the life of Earth; in the future there will be no more total solar eclipses. The Moon is edging away from Earth by about 3.8 centimeters per year. After receding from Earth another 27,000 kilometers in the next few hundred million years, the Moon will no longer block the Sun to totality.

One can trace people thinking of the Moon as a thing in itself, separate from humanity and its gods, to ancient Greece, during the third to fourth centuries BCE. Aristarchus of Samos noted circa 320 BCE that Earth's shadow cast on the Moon during a lunar eclipse showed that the Moon is several times smaller than Earth (see Figure 1.6). Eratosthenes determined Earth's diameter a few decades later by finding how a solar shadow changes angle from vertical as one's vantage point changes from Earth's equator toward the pole. Together with Aristarchus' measurement, this implied a lunar diameter of slightly less than 4,000 kilometers. Because the

The Importance of the Moon

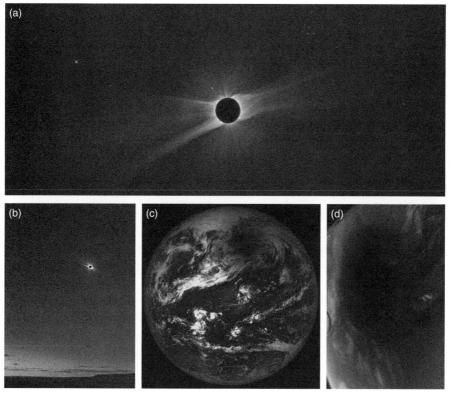

Figure 1.5. **The sun goes dark, as does a bit of Earth.** (a) Closeup view of solar eclipse (August 1, 2008, in Mongolia), with Mercury at upper left, Praesepe star cluster at upper right. (b) Appearance of this eclipse in a clear sky, which turns dark blue, as does the Sun's occulted disk. A ring of solar chromosphere joins in tendrils to a wispy corona. Venus is at upper left. (c) Earth seen from Moon by *LRO* during annular solar eclipse of May 22, 2012, showing the Pacific with an eclipsed area over the Aleutian Islands at upper right. (d) A total solar eclipse has profound effects on a patch of Earth, between Cyprus, center right, and southwestern Turkey (*top*) in deep shadow, seen from *ISS* on March 29, 2006. The air cools, and animals retire as if for night. (2008 photos by Miloslav Druckmüller, NASA photo E192199689L by LROC, ISS012E21343 by Commander Bill McArthur and Flight Engineer Valery Tokarev, *ISS* Expedition 12.)

Moon subtends one-half degree, it is a little more than 100 times distant, or about 380,000 kilometers from Earth. Aristarchus made another measurement that could have, in principle, extended our knowledge of the Moon's distance to a measurement of the Earth-Sun distance, but in reality this failed. For 18 centuries the accepted distance to the Sun was too small by 20 times. Nonetheless, Aristarchus used the Moon to show that the Sun is many millions of kilometers away and larger than Earth.[38]

Nineteen centuries later, with European intellectual leaders in conflict over whether Earth revolves around the Sun or vice versa, the one thing on which all these debaters agreed was that the Moon revolves around Earth. In 1609–1610, Galileo Galilei used the astronomical telescope to revolutionize humanity's view

11

The New Moon

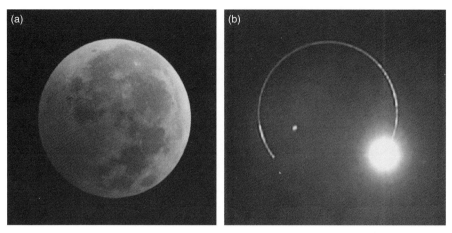

Figure 1.6. **Lit by all Earth's sunrises and sunsets**. (a) A lunar eclipse is visible from a whole hemisphere of Earth simultaneously (from South Miami, Florida, December 21, 2010). Earth's angular size is four times the Sun's seen from the Moon, so the Sun's disk is easily blocked entirely. Nonetheless, the Moon in total eclipse is often illuminated nonuniformly and acquires a reddish hue from the Moon receiving light refracted around Earth by all Earth's sunsets and sunrises simultaneously. (b) This is seen for Earth imaged from spacecraft *Kaguya* orbiting the Moon near its north pole during the February 10, 2009, lunar eclipse. The Sun begins to peek from behind Earth, but Earth is ringed by light refracting through its atmosphere. Aristarchus used the Earth's shadow being four times larger than the Moon to determine the Moon's size. (See http://space.jaxa.jp/movie/20090218_kaguya_movie01_j.html for this scene in video.) (Images courtesy of Jan Kratochvil and JAXA/NHK.)

of the universe with several observations, in important ways involving the Moon. Most obviously, when he turned his telescope toward the lunar surface, he found it marked by mountains, valleys, plains, and strange craters, many of these features resembling familiar ones on Earth. The Moon is not some idealized heavenly body but a landscape with structures in many cases similar to Earth's. In addition, as he observed Venus over time, Galileo noticed it undergoing the full range of phases (new, quarter, gibbous, full), just as the Moon does during a month, as a result of changing illumination from the Sun. These phases change in concert with Venus's angular diameter, although one would expect this diameter to stay constant for a planet orbiting Earth. This observation and the lunar analogy more than any other analogy were death knells for the Earth-centered view of what we now call the Solar System. Dramatically, when Galileo observed Jupiter, he noted it was orbited by four "stars" best understood in analogy with our Moon orbiting Earth. These were the first moons (of hundreds now known) around other worlds, proving the geocentrists wrong in thinking Earth the unique center of celestial motion.

A half-century later, Isaac Newton mathematically described the motions of celestial objects by making an intellectual leap to the Moon. This was not because of the apocryphal hit on his head by an apple but rather resulted from thinking of an object falling to Earth attracted by its gravitation, and then looking at the Moon as a radically more distant object acting under the same force. Both motions are described by adopting the proposition that gravitation's strength falls proportionally to the inverse of the square of the distance between the gravitating objects. Only the Moon could extend the laws on Earth to objects in the heavens. Newton's theory, in addition to Galileo's work, led to the first application of astrophysics, originally to the Moon but soon to the entire universe and its contents. In the intervening centuries, the Moon has been exploited in many ground-breaking ways to explore the Universe (such as in 1946 America, when Project Diana showed that radar beams could penetrate the Earth's ionosphere and measure distances to objects in the Solar System, starting with the Moon).

As humanity ponders expanding beyond Earth's immediate environs, the Moon looms at least 100 times closer than any other major body in the Solar System.[39] This book describes this harsh but welcoming world and our potential interaction with it. We must learn how to live in a planetary environment different from that to which we adapted – the first of many that will draw us across the Solar System. The Moon as the first step in this journey will determine the future in surprising ways.

The Moon was crucial in differentiating Earth from planets such as Venus when preparing for the emergence and maintenance of life. In the language of ancient mythology, the Moon is the part of our world sacrificially banished to allow Earth to become the nurturing place we know. The Moon plays a significant role in how life procreates in the ocean and interacts via the tides along continental margins. Intelligent life on Earth has looked toward the Moon to regulate its concept of time and change, and has depended on the Moon as the mental stepping-stone to explore how the wider universe works. In the past few decades we have done an amazing thing: we have translated this mental leap into space into a substantial one, with our robotic probes and spacecraft with human crews. As Krafft Ehricke encapsulated, the Moon is the essential rung in the practical ascent from Earth to the larger Solar System, and we have made our first reconnaissance of what exploring it is like. The Moon is a complex, surprising world that does not obey our preconceived notions of it. In exploring this strange, new outpost of Earth, we will encounter and learn new things, allowing us to further ascend this ladder and to look back to Earth and deliver what we have gained. The Moon is ours if we are wise and capable enough to win her back again.

Notes

1. *Krafft Ehricke's Extraterrestrial Imperative* by Krafft Ehricke and Marsha Freeman, 2009, Apogee: Burlington, Ontario, 89. Ehricke had different versions of this saying (see "Krafft Ehricke's Moon: Extraterrestrial Imperative" by Marsha Freeman, in *Lunar Settlements*, ed. Haym Benaroya, 2010, Boca Raton: Taylor & Francis, 19).
2. Augustus de Morgan (1806–1871) wrote in "The Irishman's Astronomy":

    "Long life to the moon, for a dear noble cratur
    Which serves us for a lamplight all night in the dark
    While the sun only shines in the day, which by natur
    Wants no light at all, as ye all remark."

This echoes Shakespeare's 1597 reverie on night's sky (Juliet's soliloquy, *Romeo and Juliet*, Act 3, Scene 2):

"Come, gentle night, come, loving, black-browed night,
Give me my Romeo. And when I shall die,
Take him and cut him out in little stars,
And he will make the face of heaven so fine
That all the world will be in love with night
And pay no worship to the garish sun."

Night is lugubrious but, to humans, can be more compelling. Its ruler is the Moon.

3. *Romeo and Juliet* addresses this, too (Act 2, Scene 2):

"ROMEO: Lady, by yonder blessed moon I swear,
          That tips with silver all these fruit-tree tops –
 JULIET: O, swear not by the moon, the inconstant moon,
          That monthly changes in her circled orb,
          Lest that thy love prove likewise variable."

4. The Incas thought of gold as the sweat of the Sun, and silver the tears of the Moon. (*Sweat of the Sun and Tears of the Moon: Gold and Silver in Pre-Columbian Art* by André Emmerich, 1965, Seattle: University of Washington Press).

5. Welty describes her early impressions of the Moon from her grandfather's Ohio farm:

"At around age six, perhaps, I was standing by myself in our front yard waiting for supper, just at that hour in a late summer day when the sun is already below the horizon and the risen full moon in the visible sky stops being chalky and begins to take on light. There comes the moment, and I saw it then, when the moon goes from flat to round. For the first time it met my eyes as a globe. The word 'moon' came into my mouth as though fed to me out of a silver spoon. Held in my mouth the moon became a word. It had the roundness of a Concord grape Grandpa took off his vine and gave to me to suck out of its skin and swallow whole, in Ohio. This love did not prevent me from living for years in foolish error about the moon. The new moon just appearing in the west was the rising moon to me. The new should be rising. And in early childhood the sun and moon, those opposite reigning powers, I just as easily assumed rose in east and west respectively in their opposite sides of the sky, and like partners in a reel they advanced, sun from the east, moon from the West, crossed over (when I wasn't looking) and went down on the other side. My father couldn't have known I believed that when, bending behind me and guiding my shoulder, he positioned me at our telescope in the front yard, with careful adjustment of the focus, brought the moon close to me." (*One Writer's Beginnings*, 1984, Cambridge, MA: Harvard University Press, 10 – reprinted with permission). I wonder if Welty felt forming the word "Moon" as if how a wolf might howl. Although no evidence indicates that wolves howl in connection with lunar phase (*The Wolf Almanac* by Robert H. Busch, 1998, Guilford, CT: Lyons, 50), the notion of howling at the Moon is ingrained in human cultures, particularly European–American.

6. A charming old study of the Moon in myth: *Moon Lore* by Rev. Timothy Harley, 1885, London; Swan Sonnenschein, Le Bas & Lowry; a recent photo essay in *Moon: A Tribute to Our Nearest Neighbor* by Scott Montgomery, 2009, Scoresby, Australia: Five Mile Press. Not all such myths are ancient: rumors spread in Iran of Ayatollah Khomeini's face appearing across the Moon in 1979. Libyan schoolchildren were told the same about Muammar Gaddafi.

7. According to "Marching to a Different Drummer: Crabs Synchronize Reproduction to a 14-Month Lunar-Tidal Cycle" by Martin W. Skov, Richard G. Hartnoll, Renison K. Ruwa, Jude P. Shunula, Marco Vannini & Stefano Cannicci, 2005, *Ecology*, 86, 1164, most crab species spawn according to a cycle whereby new or full Moon aligns with the Moon's closest approach to Earth (perigee) in its orbit, every 7 months (actually every 14 distinguishing between full and new).

8. "Lunar Cycles and Reproductive Activity in Reef Fishes with Particular Attention to Rabbitfishes" by Akihiro Takemura, Sigeo Nakamura, Young J. Park & Kazunori Takano, 2005, *Export Fish and Fisheries*, 5, 317.

9. "Effect of Season of the Year and Phase of the Moon on Puberty and on the Occurrence of Oestrus and Conception on Dairy Heifers Reared on High Planes of Nutrition" by J. H. B. Roy, Catherine M. Gilles, M. W. Perfitt & I. J. F. Stobo, 1980, *Animal Production*, 13, 31; *Reproduction in Cattle* by A. R. Peters & P. J. H. Ball, 1987, Wellington, New Zealand: Butterworths, 42.

10. "Lunar and Menstrual Phase Locking" by Winnifred B. Cutler, 1980, *American Journal of Obstetrics and Gynecology*, 137, 834; "Lunar Influences on the Reproductive Cycle in Women" by Winnifred B. Cutler, Wolfgang M. Schleidt, Erika Freidmann, George Preti & Robert Stine, 1987, *Human Biology*, 59, 6.

    More personally, and poetic (*Of Woman Born: Motherhood as Experience and Institution* by Adrienne Rich, 1976, New York: Norton, 78): "The ocean, whose tides respond, like women's menses, to the pull of the moon, the ocean which corresponds to the amniotic fluid in which human life begins, the ocean on whose surface vessels (personified as female) can ride but in whose depth sailors meet their death and monsters conceal themselves ... it is unstable and threatening as the earth is not; it spawns new life daily, yet swallows up lives; it is changeable like the moon, unregulated, yet indestructible and eternal."

11. "Do Lunar Phases Affect Conception Rates in Assisted Reproduction?" by Sangeeta Das, Susanna Dodd, David I. Lewis-Jones, Foram M. Patel, Andrew J. Drakeley, Charles R. Kingsland & Rafet Gazvani, 2005, *Journal of Assisted Reproduction and Genetics*, 22, 15, DOI: 106.1007/s10815–005–0815-y.

12. "Evidence that the Lunar Cycle Influences Human Sleep" by Christian Cajochen, Songül Altanay-Ekici, Mirjam Münch, Sylvia Frey, Vera Knoblauch & Anna Wirz-Justice, 2013, *Current Biology*, 23, 1.

13. We present nonmetrical measures for each kind of quantity, then present the conversion (here, meters to feet) in the first such instance. Thereafter, we stick to metric units throughout this book.

14. "Lunar Laser Ranging: A Continuing Legacy of the Apollo Program" by J. O. Dickey, et al. 1994, *Science*, 265, 482.

15. "Effects of Tidal Dissipation in the Oceans on the Moon's Orbit and the Earth's Rotation" by Kurt Lambeck, 1975, *Journal of Geophysical Research*, 80, 291.

16. "Earth's Precambrian Rotation and the Evolving Lunar Orbit: Implications of Tidal Rhythmite Data for Palæogeophysics" by George E. Williams, in *The Precambrian Earth: Tempos and Events*, eds. P. G. Eriksson, et al., 2004, Elsevier: Amsterdam, 472.

17. "Evidence from Detrital Zircons for the Existence of Continental Crust and Oceans on the Earth 4.4 Gyr Ago" by Simon A. Wilde, John W. Valley, William H. Peck & Colin M. Graham, 2001, *Nature*, 409, 175. There is a range of opinion about when the oceans formed; also see "Prime Controls on Archaean–Palaeoproterozoic Sedimentation: Change over Time" by P. G. Eriksson, S. Banerjee, O. Catuneanu, S. Sarkar, A. J. Bumby & M. N. Mtimkulu, 2007, *Gondwana Research*, 12, 550.

18. Such laminated deposits, usually of alternating laminations of coarser and finer silt, clay, or sand, are not always tied with certainty to specific time intervals. If not, these usually striped rocks are called simply varved deposits, not rhythmites.

19. A drawback of coral growth rings is that they extend no more than about 450 million years into the past, but they nonetheless show that there were about 410 days per year and 31 days per month some 400 million years ago. ("Length of the Year during the Silurian and Devonian Periods: New Values" by S. J. Mazzullo, 1971, *Bulletin of Geological Society of America*, 82, 1085.)

20. From 620-million-year-old South American rhythmites, one finds 13.1 ± 0.1 months per year, 400 ± 7 days per year, and 30.5 ± 0.5 days per month, implying a day of 21.9 ± 0.4 hours and a lunar orbital radius $96.5 \pm 0.5\%$ of current values ("Geological Constraints on the Precambrian History of Earth's Rotation and the Moon's Orbit" by George A. Williams, 2000, *Review of Geophysics*, 38, 1). Similar analyses of 2.45-billion-year-old Western Australian deposits indicate a lunar orbital radius of $90.6 \pm 2.9\%$ of the current, for a lunar recessional rate 1.8 billion years ago of only 1.24 ± 0.71 cm per year versus 3.82 ± 0.07 cm per year currently. Older samples (3.2 billion years old) indicate larger changes, with the number of days per month only about three-quarters of the current ("Quantifying the Oldest Tidal Record: The 3.2 Ga Moodies Group, Barberton Greenstone Belt, South Africa" by Kenneth A. Eriksson & Edward L. Simpson, 2000, *Geology*, 28, 831).

21. "Secular Effects of Oceanic Tidal Dissipation on the Moon's Orbit and the Earth's Rotation" by K. S. Hansen, 1982, *Reviews of Geophysics and Space Physics*, 20, 457; "Tides and the Evolution of the Earth–Moon System" by D. J. Webb, 1982, *Geophysical Journal of the Royal Astronomical Society*, 70, 261; and "Lunar Orbital Evolution: A Synthesis of Recent Results" by Bruce G. Bills & Richard D. Ray, 1999, *Geophysical Research Letters*, 26, 2045.

22. "A Stochastic Model of the Earth–Moon Tidal Evolution Accounting for Cyclic Variations of Resonant Properties of the Ocean: An Asymptotic Solution" by B. A. Kagan & N.B Maslova, 1994, *Earth, Moon and Planets*, 66, 173.

23. "The Supercontinent Medley: Recent Views" by A. V. Sankaran, 2003, *Current Science*, 85, 1121.

24. "Fast Tidal Cycling and the Origin of Life" by Richard Lathe, 2004, *Icarus*, 168, 18, c.f., "Comment on the Paper 'Fast Tidal Cycling and the Origin of Life' by Richard Lathe" by P. Varga, K. R. Rybicki & C. Denis, 2006, *Icarus*, 180, 274; and "Early Tides: Response to Varga et al." by Richard Lathe, 2006, *Icarus*, 180, 277.
25. "Comments on the Long-Term Stability of the Earth's Obliquity" by William R. Ward, 1982, *Icarus*, 50, 444; "Obliquity Variations of a Moonless Earth" by Jack J. Lissauer, Jason W. Barnes & John E. Chambers, 2012, *Icarus*, 217, 77.
26. "The Chaotic Obliquity of Mars" by Jihad Touma & Jack Wisdom, 1993, *Science*, 259, 1294.
27. "Quantifying the Oldest Tidal Record: The 3.2 Ga Moodies Group, Barberton Greenstone Belt, South Africa" by Kenneth A. Eriksson & Edward L. Simpson, 2000, *Geology*, 28, 831.
28. The age of high-temperature (calcium and aluminum-rich) inclusions in primitive, undifferentiated meteorites (chondrites) has been measured at 4.5672 ± 0.0006 billion years, on the basis of radioactive decay of uranium isotopes (^{235}U and ^{238}U) into lead (^{207}Pb and ^{206}Pb) ("Lead Isotopic Ages of Chondrules and Calcium–Aluminum–Rich Inclusions" by Y. Amelin, A. N. Krot, I. D. Hutcheon & A. A. Ulyanov, 2002, *Science*, 297, 1678). In comparison the main sequence age of the Sun is about 4.52 billion years.
29. "Simulations of a Late Lunar-Forming Impact" by Robin Canup, 2004, *Icarus*, 168, 433.
30. "Lunar Accretion from an Impact-Generated Disk" by E. Kokubo, R. M. Canup & S. Ida, in *Origin of the Earth and Moon*, eds. R. M. Canup & K. Righter, 2000, Tucson: University of Arizona Press, 145.
31. "Evidence for a Past High-Eccentricity Lunar Orbit" by I. Garrick-Bethell, J. Wisdom & M. Zuber, 2006, *Science*, 313, 652.
32. "Emergence of a Habitable Planet" by Kevin Zahnle, Nick Arndt, Charles Cockell, Alex Halliday, Euan Nisbet, Franck Selsis & Norman H. Sleep, 2007, *Space Science Review*, 129, 35.
33. We use the unit of "atmosphere," roughly the atmospheric pressure of Earth's atmosphere now at sea level: 1 atmosphere = 1.01 bar = 101 kilopascal = 760 torr = 760 mm of mercury.
34. Knowledge about the Moon is essential to understanding the Earth but directly applicable to interpreting data from Mercury, in many ways similar to the Moon. (See Chapter 8.)
35. The composition of the Earth's continental crust is "felsic" – largely silicate minerals such as feldspar, quartz, and mica – with an average density about 2.8 times water's at standard temperature and pressure (hence about 2.8 grams per cubic centimeter), versus the ocean basin crust's "mafic" composition: more iron/magnesium, often darker minerals including basalt, gabbro, and so forth, averaging about 3.2 grams per cubic centimeter. A primary upper mantle mineral is olivine, a dark greenish magnesium-iron silicate, typically with a density of about 3.3 grams per cubic centimeter. The continental crust contains a much higher fraction of incompatible elements (potassium, phosphorus, thorium, uranium, rare Earth elements, etc.) that would not dissolve in the liquid magma ocean of Earth's early outer layer or incorporate easily into its major minerals.
36. The Mohorovičić discontinuity (or Moho) is conventionally taken as the crust/mantle boundary. It marks a discontinuity of about 15% in earthquake wave (P-wave) speed but perhaps not an actual compositional boundary.
37. One could argue that the Moon is as essential to life on Earth as is the Sun. Without sunlight our planet could support deep-Earth bacteria and whole ecosystems at deep-ocean hydrothermal vents. Earth without the Moon would be more Venusian and probably lifeless. Note that although Venus is only 72% as far from the Sun as Earth, receiving 1.91 times as much light, it absorbs less light than Earth because it so reflective. The reflectivity or albedo of Venus (actually, its cloud tops) is a fairly constant 75–76%, whereas Earth's average albedo ranges between 30% and 40%, meaning that Venus gets 75–100% as much energy from the Sun as Earth. Earth's albedo changes radically over time; ice and snow albedos are typical about 85%, green forest about 14%, cloudless ocean about 25%, and their shares of the surface rise and fall over the æons.
38. Aristarchus realized that a lunar eclipse is an opportunity to see the disk of the Earth projected to the distance of the Moon. Particularly, measuring the angular diameter of the umbral eclipse (the totally shadowed region) and of the penumbral eclipse (partially shadowed region – a difficult measurement to make in practice) and averaging the two diameters produces the angular diameter of the Earth as seen as the distance of the Moon. Because this diameter is about 2° (about 1/30 of a radian), the Moon is 30 times further than the 12,756 km diameter of Earth, or 383,000 km.

Aristarchus utilized this measurement (unsuccessfully) seeking the distance from Earth to the Sun. His idea was that when the Moon was exactly at first quarter (or last), the Earth–Moon–Sun angle would be exactly 90°, so one could find the Earth–Sun–Moon angle (call it θ) by measuring the Moon–Earth–Sun angle (90°− θ),

The Importance of the Moon

apparent in the sky as the angular distance between the Sun and Moon. Aristarchus tried measuring this and found $90° - \theta = 87°$. In actuality it is $89.85°$. The distance to the Sun (D) in terms of the distance to the Moon (d) is $D = d/\sin\theta \approx d/\theta$, with θ measured in radians (not degrees). The Sun is 400 times farther than the Moon; Aristarchus calculated that the Sun was 20 times farther. The Sun is also about 400 times larger than the Moon (hence, we get a nearly perfect solar eclipses), but Aristarchus calculated that the Sun was 20 times bigger than the Moon, hence about 5 times bigger than Earth (off by a factor of 20).

Not until 1630 did Johannes Kepler make a better measurement of the distance to the Sun (incorrect by about 30%), whereas Jean Richer and Giovanni Cassini in 1672 measured it to within 5%. From the transit of Venus on the 1769 Cook expedition to the South Pacific, Charles Green, Daniel Solander, and Thomas Hornsby measured it to within 1%. Today we know the Sun's distance to within two trillionths of a percent.

39. The Moon is a major body, 1.2% of Earth's mass, 13th most massive object in the Solar System, between Jupiter's moons Io and Europa, and more massive than Pluto but lighter than Mercury.

Chapter 2
First Steps

> That's one small step for (a) man, one giant leap for Mankind.
> – Neil Armstrong, July 20, 1969

With these words, Armstrong concluded the competition that had gripped the world for years between the two most powerful nations – the United States and the Soviet Union – on Earth and now the Moon. Six and a half hours earlier, he and Edwin "Buzz" Aldrin[1] set their 8-ton spacecraft carefully onto the lunar surface with fewer than 45 seconds of fuel left,[2] fulfilling a promise made eight years before by their martyred president. Stepping onto the Moon, they reached the apex of a perilous journey that in several days would fulfill the president's second promise – to return them safely to Earth.[3] Their success engaged the work of hundreds of thousands of people and the attentions of billions of people. It was a cosmic and geopolitical culmination that, with the following lunar landings, bears on any future human exploration of space, as we will see.

Armstrong had much on his mind, but his terseness also reflected his characteristic lack of self-absorption. [BOX 2.1] He was selected in part for his steely, split-second ability to make the right choice concerning balky, expensive, and complex flying machines.[4] He had made the right choices in combat, as a test pilot, in Earth's orbit, in training for lunar landing, and in the landing that day at Tranquility Base, or else he would not have lived to step onto the Moon. He was not easily distracted from the mission. [BOX 2.2]

To those of us watching on Earth, there had been no other event like this. This achievement was intended to eclipse the flight of Yuri Gagarin as first man into outer space in 1961, but its significance seemed more like that of hundreds of thousands of years ago when humans, not yet Homo sapiens, first left the cradle continent of Africa. Wernher von Braun compared it to the first creature leaving the ocean for dry land. Just as early humans adapted to the cold and seasons far from the equator, humans of our era would learn to adapt to greater, new, and alien environs. Armstrong became the first creature from our planet to disturb the soil of another world. Many still feel honored to have witnessed this. Nothing else matches it.

Under these conditions it is understandable that Armstrong seems to have aspirated a syllable of his momentous statement, that little *a* before *man* – hardly a

Box 2.1: Gagarin's Giant Leap

Armstrong's brevity of self-expression contrasts, for instance, with Yuri Gagarin's composition on his first flight into outer space on April 12, 1961 (translated):

Dear Friends, known and unknown to me, my dear compatriots and all people of the world. In the next few minutes a mighty spaceship will carry me off into the distant spaces of the universe. What can I say to you during these last minutes before the start? All my life now appears as a single beautiful moment to me. All I have done and lived for has been done and lived for this moment. It is difficult for me to analyze my feelings, now that the hour of trial for which we have prepared so long and passionately, is so near. It's hardly worth talking about the feelings I experienced when I was asked to make this first space flight in history. Joy? No, it was not only joy. Pride? No, it was not only pride. I was immensely happy to be the first in outer space, to meet nature face to face in this unusual single-handed encounter. Could I possibly have dreamed of more? Then I thought of the tremendous responsibility I had taken on: to be the first to accomplish what generations of people dreamed of: to be the first to pave the way for humanity to outer space. Can you name a more complex task than the one I am undertaking? This is a responsibility, not to one, not to many, and not to a collective group. This is a responsibility to all the Soviet people, to all of humanity, to its present and future. I know I have to summon all my will power to carry out my assignment to the best of my ability. I understand the importance of my mission and shall do all I can to fulfill the assignment for the Communist party and the Soviet people. Only a few minutes are left before the start. I am saying goodbye to you, dear friends, as people always say goodbye to each other when leaving on a long journey.

Gagarin is remembered as an avid poetry reader ("Elena Gagarina Remembers her Father" by Andrea Rose, April 11, 2011, *Voices*, http://blog.britishcouncil.org/2011/04/11/gagarin_orbit1/).

Box 2.2: Armstrong and the Right Stuff

Armstrong was a Korean War jet fighter pilot in 1951–1952, flying 78 combat missions. On September 3, 1951, his F9F-2 was hit by anti-aircraft fire, a meter of one wing was severed by a cable strung across the valley on its rapidly descending path. Armstrong piloted the severely damaged plane 500 kilometers to a friendly airfield, then ejected because the plane could not be landed.

On March 22, 1956, Armstrong co-piloted a B-29 bomber carrying a D558–2 Skyrocket plane (flown by Jack McKay). As he prepared to drop the Skyrocket for flight, one of the B-29's engines lost its governor and spun out of control. Armstrong and pilot Stan Butchart delayed dropping the D558–2 only long enough for its safe launch but just before the engine exploded. Propeller fragments slashed the B-29 (including where the D558–2 had been), disabling two more engines, severing one set of control cables, and damaging the second. Armstrong landed the B-29 safely with one working engine.

In command of *Gemini 8* in September 1965, Armstrong, along with David Scott, docked with an Agena target vehicle in Earth's orbit. When the two craft started spinning out of control, Armstrong undocked from the Agena, only to find that the spin was accelerating, revealing that its cause was a misfiring thruster on their own Gemini spacecraft. Before the two astronauts could lose consciousness, with spin approaching one revolution per second, Armstrong deactivated the stuck thruster by engaging the reentry control system, thus aborting the mission safely.

On May 6, 1968, Armstrong was rehearsing lunar landings flying the Lunar Landing Research Vehicle *LLRV-1* (or *Flying Bedstead*) near Houston, when its attitude control

Box 2.2: (continued)

system failed. He tried to rescue the *LLRV-1* but ejected about one second before the ejection rocket would propel him into the ground, fatally. Two Lunar Landing Training Vehicles (LLTV, essentially the LLRV) later crashed with no fatalities, fortunately. Video of Armstrong's LLRV crash is available on the Internet, as are other LLRV videos.

Apollo 11's landing was very tense as Armstrong was forced to manually steer Lunar Module (LM) *Eagle* past an unanticipated field of boulders large enough to endanger the LM (Figure 2.1, central panel). Data, unavailable in 1969, alter assessments of how critical these issues became. First, new estimates indicate that *Eagle* had 45 seconds of fuel left, not 30 as first thought. This is not comforting; the mission was nearing an abort with 20 seconds of fuel left. Aborting near the surface was riskier than they likely realized. A 1966 NASA report described how a LM descent stage nearly spent of propellant but with a significant amount onboard might explode hitting the ground after an abort. With

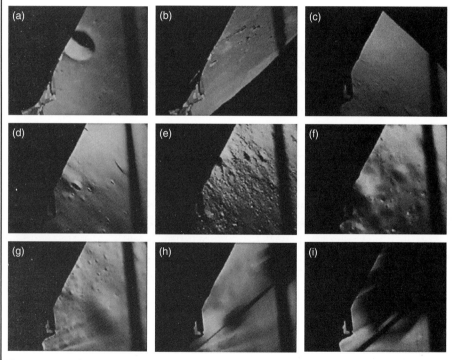

Figure 2.1. *Eagle* **lands**. View out *Apollo 11* LM window recorded by the 16-mm data acquisition camera (DAC).[5] LM Pilot Edwin "Buzz" Aldrin operates the Apollo Guidance Computer (suffering data overloads) while Commander Neil Armstrong adjusts to landmark positions. Up to 13 minutes before landing (a), *Eagle* flies engine forward and face down, then (b) yaws 180°. The Moon leaves the view, then (c) reappears with its skyline horizontal. Nine minutes to landing, Armstrong sees *Eagle* pass crater Maskelyne W 3 seconds early, overshooting its target. He sees crater West (d, top of picture) strewn with boulder hazards (e). Flying manually with less than 1 minute of propellant, he finds smooth ground (f), sees dust blown by exhaust from *Eagle* (g), then its shadow (h). Contact probes touch (i). "The *Eagle* has landed." (Courtesy of NASA.)

> **Box 2.2: (continued)**
>
> 5 seconds between when abort is initiated and the LM's ascent motor is engaged (during which and for sometime thereafter it could crash), even longer intervals existed in which an exploding descent stage might engulf the ascent stage in fire and debris, fratricidally. If Armstrong had less effectively located a boulder-free site, he might have entered this danger zone. Unfortunately, there is no evidence that Mission Control took this report into account in 1969. (See *Hazards Associated with a LEM Abort Near the Lunar Surface* by Charles Teixeira, June 24, 1966, NASA Program Apollo Working Paper #1203, Manned Spaceflight Center, Houston.)
>
> Other problems arose when the astronauts forgot to switch off one radar unit before switching on their downward radar, overloading the LM computer. Mission Control guidance officer Steve Bales successfully diagnosed the problem and allowed the final seconds of the landing to proceed. Less well known are computer overloads because of an undetected software error. Also, the LM engine throttle responded poorly to Armstrong's controls. An account 45 years later by an engineer on *Eagle*'s guidance computer ("Tales from the Lunar Module Guidance Computer" by Don Eyles, 2004, *27th Guidance and Control Conference*, American Astronautical Society, paper 04–064) details five alarms requiring a restart of the computer as a result of this software error, alone demanding 13% of computing capacity and causing overloads, a condition unknown in simulations. Five alarms in four minutes just before landing were alleviated when Armstrong took manual control. Eyles details another unseen error on *Apollo 11* and *12* causing the engine throttle to vary by 25% several times per second because of a nearly unstable feedback loop. Had this loop actually been implemented as intended, it could have precluded manual landing of the LM.

phoneme. NASA claimed that the *a* was lost in transmission because of static, but the press questioned that explanation.[6] Some people think he ran *for* and *a* together: *fra*.[7] Alternatively, in 2006 a computer programmer claimed to have found the missing *a* – a 35-millisecond rush of air recorded during Armstrong's quote, explaining this as consistent with Armstrong's tendency to clip vowels.[8] Armstrong admitted that there is little time for that *a* in the recordings,[9] but in the end it matters little; one could imagine that in the moment Armstrong refers to an example of man, as in "Behold Man" ("Ecce homo" – John 19:5).[10]] He had more important things to consider. [BOX 2.3]

Armstrong had been relatively silent descending the ladder from the "front porch" of the lander's crew cabin to the surface level. After pulling the release to deploy the platform carrying equipment and the camera to record his descent (Figure 2.2), three minutes before reaching the ground, Armstrong spoke only a few seconds to check his radio link.[11] He had described the scene out the lander window[12] but had been hurried along by Mission Control with a massive checklist of post-landing and pre-moonwalk preparations.

NASA mission designers presented Armstrong with a rapid series of tasks for this most significant moment: taking a quick lunar soil sample and inspecting the spacecraft, just in case something went wrong. To be safe, Aldrin watched from the LM window, and Armstrong was filmed by a second camera through this same

> **Box 2.3: One Small Quote for a Man**
>
> How did Armstrong's saying originate? NASA officials avow setting only basic guidelines for what he should say. Julian Scheer, assistant administrator for public affairs during Apollo, required official policy never control public statements of flight crews (Memo: Julian Scheer to George M. Low, Apollo spacecraft program manager, March 12, 1969, and response, March 18). Paul Dembling, then NASA's general counsel, said in a 2010 interview for this book that NASA insisted Armstrong only avoid anything possibly construed to claim U.S. sovereignty or control over the Moon. The plaque left at Tranquility Base so stated: "Here Men From The Planet Earth First Set Foot Upon the Moon, July 1969 A.D. We Came in Peace For All Mankind." (Dembling admitted making a political point of adding Richard Nixon's signature.) Armstrong recalls:
>
> > I thought about it after landing, and because we had a lot of other things to do, it was not something that I really concentrated on but just something that was kind of passing around subliminally or in the background. But it, you know, was a pretty simple statement, talking about stepping off something. Why, it wasn't a very complex thing. It was what it was. ("Oral History Transcript: Neil A. Armstrong" interview by Stephen E. Ambrose & Douglas Brinkley, September 19, 2001, *Oral History Project*, NASA Johnson Space Center, 81.)
>
> Authors note a memo from Willis H. Shapley, associate deputy administrator, to George Mueller, associate director for manned space flight, on symbolic public themes: "the forward step of all mankind." ("Symbolic Items for the First Lunar Landing," April 19, 1969.) A July 2, 1969, memo from Shapley to Mueller on the topic explicitly requires including flight crew operations director Deke Slayton and other astronauts, so perhaps Armstrong heard of this "step for all mankind."
>
> *Apollo 12* commander Pete Conrad had his own opinion. His first words after stepping on the Moon were "Whoopie! Man, that may have been a small one for Neil, but that's a long one for me!" Conrad need not take careful first steps, so he descended straight from the ladder's bottom rung. He explains that he said this to win a bet with a reporter (Oriana Fallaci, it seems) about NASA telling astronauts what to say when stepping on the Moon ("That May Have Been a Small One for Neil…" by Eric M. Jones, 1995, *Apollo 12 Lunar Surface Journal*).

window (Figure 2.3). These provide some of the few photos of Armstrong on the surface, because Aldrin was instructed to leave the second still camera in the LM.

In many science fiction movies at this point, the extraterrestrial monster would appear, or astronauts might fall victim to unanticipated natural hazards. Armstrong and Aldrin were busy and cautious in these first few minutes on the surface but relatively secure against external, non-engineering threats. Why was this?

In the decade before *Apollo 11*, many robots flew past, impacted, orbited, and landed on the Moon, variously photographing, probing, and disturbing it. Armstrong carefully examined the lunar dust beyond the LM's landing pad, commenting on its consistency and mechanical properties to assure that it would support him.[13] Several years prior, one theory predicted he would sink deeply into the dust. Or hypothetically, the dust's consistency might make him slip and fall onto his back like a turtle, or dust clouds kicked up by his boots might stick to his suit and blind him.[14] Armstrong was not surprised by the lunar surface's

First Steps

Figure 2.2. **Before one small step**. Armstrong trains for *Apollo 11* with a mockup Modularized Equipment Stowage Assembly (MESA, in foreground) and Lunar Module. MESA carried the camera recording the first step on the Moon; Armstrong deployed it by pulling a cable while descending the LM ladder. Compare to Figure 2.3. (NASA photo S69–31044.)

characteristics, however, because in the past three years several Surveyor robotic landers had analyzed, bounced on, manipulated, and excavated lunar soil. A decade earlier, scientists inferred and science fiction writers imagined stronger rains of meteorites than later found. A micrometeorite could have penetrated Armstrong's suit in less time than his silent *a*, and a major leak would have disabled him faster than his finishing his famous first sentence. Although he knew he was relatively safe from these hazards, he was judiciously cautious.

Standing on the landing pad below the ladder, Armstrong extended his leg and toed the soil of this alien world in a carefully choreographed step. The dust was loose but so fine and compact that it resisted further packing, almost like a solid, like cement not yet set. He made careful note of its resistance and consistency for Houston[15] then walked carefully away from the pad and started work. The momentous event passed.[16]

Easy to forget between the victories of *Sputnik* and Gagarin and the Soviet lunar program's defeat by Apollo are lunar exploration milestones first reached by Soviet craft. In 1959, *Luna 1* was first to fly by the Moon (two months before *Pioneer 4*), *Luna 2* was the first to hit it (31 months ahead of *Ranger 4*), and *Luna 3* photographed its Far Side. In 1966 *Luna 9* was the first soft lander (four months before *Surveyor 1*), and *Luna 10* the first lunar orbiter (four months before *Lunar Orbiter 1*). In 1968 robotic *Zond 5* was first to circle the Moon and return to Earth (three months prior to *Apollo 8* with its human crew), and *Lunokhod 1* in 1970 was the first lunar rover (8 months before *Apollo 15*, again with humans). At first reaching the Moon was exceedingly difficult: 7 of the first 10 Soviet missions failed (although the world did not know this then), as did 11 of the 12 first U.S. missions, initiated by the U.S. Army, USAF, then NASA.[17]

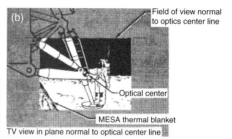

Figure 2.3. **One small step, two views**. Armstrong steps onto the Moon at 9:56:15 PM CDT, July 20, 1969 (Houston time – with several seconds' ambiguity in official time), recorded by (a) Westinghouse Apollo Lunar Television Camera SSTV video (on the MESA on *Eagle*'s right leg) via NASA's Honeysuckle Creek tracking station in Australia, as seen by hundreds of millions of TV viewers at the time. The LM front leg is at left, and Armstrong is the vague figure just below right of center. (b) Anticipated view of first step from MESA camera; compare to Figure 2.2. (Figure 3.3 from *Apollo 11 Lunar Surface Operations Plan, Final*, June 27, 1969, 18.) (c) View from *Eagle*'s ascent stage window, provided by the DAC. Reflections of Aldrin and the cabin interior are seen in this video. Armstrong's helmet and backpack are at lower left, and the LM shadow dominates the remainder of the image. Compare to Figure 2.2. (Courtesy of NASA.)

Luna 3's Far Side images in October 1959 were shocking, showing few dark maria (versus 32% of the Near Side's area): one small patch dubbed Mare Moscoviense, and a large, dark crater, Tsiolkovsky (see Figure 2.4). The remaining areas looked heavily cratered like the south central Near Side highlands.[18] Furthermore, sophisticated craft orienting themselves and returning dozens of pictures (with quality suffering from bad radio reception) exceeded U.S. expectations of the Soviets, which was highlighted by immediate dissemination of Far Side images to the Western television media. (First U.S. spacecraft pictures of the Far Side come from Lunar Orbiter in 1966.) The CIA "borrowed" overnight a working copy of *Luna 3* on tour in Mexico to understand how the probe could deliver such data and how large Soviet booster throw weights might be.

The United States soon initiated a series of missions with impressive goals: to place scientific instruments on the Moon and take closeup lunar images from space. After delay, NASA picked a new booster (an Atlas intercontinental ballistic

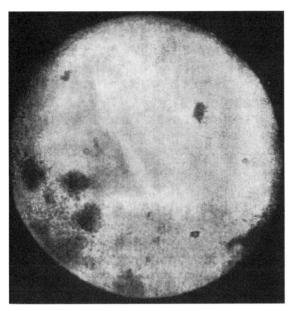

Figure 2.4. **Hidden moon revealed**. *Luna 3*, launched just two years after *Sputnik 1*, returned the first photographs of the lunar Far Side, here from 65,000 km away on October 7, 1959. *Luna 3* recorded 75% of the Far Side, surprising scientists with its paucity of maria. This image is centered 60° from the central Near Side. Far Side Mare Moscoviense is at upper right and crater Tsiolkovskiy at lower right, with marginal maria Smythii and Marginus at lower left.

missile – ICBM – topped by an ex-military Agena-B stage) to launch a new probe: Ranger, designed by the Jet Propulsion Laboratory (JPL) near Pasadena, recently transferred from the U.S. Army to NASA. Ranger's Block 1 design embodied a multipurpose spacecraft for the Moon and planets. In 1961 Block 1 probes *Ranger 1* and *2* orbited Earth, but their upper stages misfired, and few useful data resulted before the craft reentered Earth's atmosphere. (Still, a Block 1 as *Mariner 2* flew successfully past Venus in 1962. Rangers used solar panels for electrical power, superfluous on short lunar flights but failing occasionally.) Block 2 was innovative: it would decelerate above the Moon to 50 meters per second then release a 0.64-meter-wide sphere containing a seismograph. This sphere would bounce across the surface to a stop and fire a bullet through the balsawood exterior, causing fluid cushioning the seismograph to drain as the instrument settled into good mechanical contact for detecting moonquakes. The sphere was sterilized against bio-contamination. In 1962 *Ranger 3* and *5* missed the Moon entirely. *Ranger 4* lost power and control soon after launch but was tracked to a Far Side lunar impact. Regrettably, little science derived from Block 2. In 1964–1965, Block 3 succeeded at its simpler but important task: photographing the Moon on scales relevant to human lunar landings – NASA's new goal. These missions would take pictures – some rivaling in resolution (better than 0.4 meter) the best images from lunar orbit today – and transmit them until impact. *Ranger 6* followed a proper trajectory, but its camera failed. *Ranger 7*, *8*, and *9* succeeded, together taking 17,000 images approaching impacts at Mare

Cognitum, Mare Tranquillitatis, and crater Alphonsus, respectively. *Ranger 8* scouted sites for *Surveyor 5* and *Apollo 11*'s landings several years later.

The trove of unprecedentedly detailed images from Ranger sparked immediate scientific debate. Detailed features in Mare Cognitum were seemingly indistinguishable from Tranquillitatis. Regarding the structure of the lunar surface and regolith (the impact-pulverized lunar soil), Ranger posed more questions than it answered. Scientists noticed many small, rimless "dimple craters" – many aligned in chains across the surface. University of Arizona professor Gerard Kuiper argued that these were sinkhole-like collapses into underground chambers, perhaps evidence of "frothy" lunar lava in contact with the vacuum, which might pose a hazard for future landers. Eugene Shoemaker of the U.S. Geological Survey argued instead that these craters' impactors were thrown from larger craters, aligned in "rays" by impact splatter, their rims eroded away by myriad tiny impacts. Although Kuiper's idea may explain an interesting minority of cases (see Figure 6.25, ignoring the froth idea), Shoemaker won the argument generally on the basis of later data. Intriguingly, precise tracking of Ranger trajectories allowed scientists to discover that the Moon's geometric center is offset from its gravitational center by several kilometers (see Chapter 7) and allowed the first precise measure of the lunar diameter. More to the point, Ranger showed that flat areas where Apollo might land appeared to be common, at least in the maria. It would be possible to put people on the surface without crashing on rocks. However, would Apollo sink into the surface?

In August 1964 the Soviet Union secretly approved its own human lunar landing program, so preparation by robotic craft became paramount for both nations. Crucially, the planned Soviet human landing craft would be guided by a radio beacon onboard a robotic scout already on the surface. The year 1966 marked for both American and Soviet programs the first orbits of the Moon, and the first craft from both nations to make a controlled lunar landing. A modified Soviet ICBM R-7, the 8K78 (Molniya), and a new spacecraft design, the Ye-6, flew in 1963–1965 in the guise of five unnamed Cosmos launches as well as *Luna 4–8*; all ten of which failed. Ye-6 was redesigned and finally, on February 3, 1966, using a sequence similar to Ranger Block 2, resulted in *Luna 9* firing to a stop 250 meters above the Moon, releasing its 1.1-meter-long ovoid, which inflated its double airbag and bounced to a halt on the surface. Its four clamshell petals opened, and *Luna 9* righted itself to expose its camera, antennae, and several scientific instruments – the first human-built device operating on another world. It did not sink into the lunar dust and disappear. Six months earlier *Zond 3* had flown past the Moon and finished the Far Side atlas started by *Luna 3*, and on April 3, 1966, *Luna 10* became the first artificial satellite of another world. After many failures over several years, the Soviet lunar robotic exploration program was back on track. *Luna 13*'s spheroid landed successfully, and three of four lunar satellites succeeded (*Luna 11, 12*, and *14* in August 1966–April 1968).

NASA was close behind, following rapidly on Ranger's successes with two sophisticated designs: Lunar Orbiter and Surveyor. Surveyor became the first U.S. craft to land on another world, not with airbags or balsawood spheres, but with a tripod of legs and footpads, controlled by landing rockets like Apollo's lunar module. Surveyor would eject its main engine 7–11 kilometers above the Moon, then land softly with its vernier rockets. It would provide crucial data about the lunar surface in dress rehearsal for human landings (despite the project beginning two years before Kennedy's Apollo commitment), but first NASA needed another Atlas upper stage, the Centaur, ultimately more reliable than Agena but delayed by competing demands of the Defense department.

Surveyor originated as a scientific project in its own right but was curtailed after congressional friction regarding cost overruns to largely an Apollo-support role, ultimately costing $426,000,000 ($3 billion in 2012 U.S. currency). Nonetheless, Surveyors collected 87,000 images and voluminous other data on largely successful missions. On June 2, 1966, *Surveyor 1* landed in Oceanus Procellarum and conducted experiments on the regolith. *Surveyor 2* suffered a bad midcourse correction approaching the Moon, crashing near crater Copernicus. *Surveyor 3* also landed in Procellarum and, with its new robotic sampler arm, could manipulate the lunar surface. *Surveyor 4* vanished on final approach to Sinus Medii. *Surveyor 5* landed in Tranquillitatis near the future Apollo site, its sampler arm replaced by an alpha particle scattering instrument to measure the composition of heavy elements in the soil. *Surveyor 6* successfully returned to Sinus Medii. To better probe regolith properties, *Surveyor 6* fired its engines and jumped 4 meters off the surface, landing 3 meters away (the first craft to launch from another world, if only for a few seconds). *Surveyor 7* landed in rough terrain on the flank of crater Tycho, and conducted a complete study with both sampler arm and alpha particle probe. Surveyor's scientific results were extensive and sometimes surprising. In addition to measuring the regolith's mechanical strength (strong enough for Apollo), Surveyor detected dust levitating above the horizon at sunrise and sunset as a result of electrostatic repulsion caused when solar ultraviolet blasts electrons from lunar rocks (via the photoelectric effect). In Procellarum at crater Flamsteed, *Surveyor 1* found amazingly thin regolith, indicating volcanic outflows a billion years old or less. Even after shutdown, *Surveyor 3* expanded our knowledge of lunar space weathering when some of its parts were returned to Earth by the *Apollo 12* astronauts two years later.

Lunar Orbiter was the only robotic program initiated specifically for Apollo preparation. With JPL absorbed in Ranger and Surveyor, Langley Space Center put the craft on debugged Atlas/Agena boosters, contracting Boeing and Kodak to orbit the Moon with a camera much like those in spy satellites. (A competing Lockheed proposal literally would have sent to lunar orbit a downward-looking surveillance system once used by the U.S. Air Force.) The camera could resolve features down to 0.5 meter to see rocks that might befoul an Apollo lunar landing. Lunar Orbiter finished its primary goals ahead of schedule. At first it was not a science project, but after the candidate Apollo lunar landing sites and other objectives were covered in three flights, *Lunar Orbiter 4* and *5* were devoted primarily to science targets. The project's five lunar satellites, all successful, made a nested sequence of lunar maps ranging from a whole global atlas (at about 100 meters resolution) to finer pictures of selected science targets and candidate Apollo sites with details less than 1 meter. The 1,654 images delivered were huge in format, about 100 times more pixels than in a TV image. The system would take a photograph (panning to ameliorate spacecraft motion), develop it onboard, scan it, and relay the scan to Earth. Lunar Orbiter also carried micrometeorite impact detectors and found a rate half that near Earth. Strange fluctuations in Lunar Orbiter's trajectory around the Moon led to the discovery of lunar *mascons* (mass concentrations; see Chapter 7), and maps of the Moon's gravitational field by *Lunar Orbiter 3, 4*, and *5* showed it to be much more lumpy than Earth's. Preliminary correction for these effects helped *Apollo 11* land, but they still contributed to *Eagle* falling several kilometers beyond its planned error ellipse.

The cost of American and Soviet robotic lunar missions before *Apollo 11* was roughly $10 billion (in 2012 U.S. currency).[19] While showing that humans could land safely on the Moon, they also performed the first investigation of the lunar surface in situ and foreshadowed many issues discussed in this book: the odd division of lunar regions in origin, elevation, and roughness; odd behaviors of dust in unanticipated ways; the wide range of lunar surface ages; and geology dominated by impacts but punctuated by volcanism, sometimes bizarre.

Impressive successes by *Luna 9–14*, Surveyor, and Lunar Orbiter set the path for new human missions, but for both the United States and Soviet Union, this brought catastrophe. The Soviet space program's central figure, rocket designer Sergey Korolev, died in January 1966 during surgery to remove a colon tumor. His heart, weakened by years of imprisonment in the gulag, failed. In the West, his identity was revealed for the first time to NASA and von Braun; in the Soviet Union his absence become apparent in the disorganization of the Moon program. At Cape Kennedy in January 1967, haste in design and testing of Apollo's command module caused the deaths of astronauts Virgil "Gus" Grissom (second American in space), Edward White (first American to walk in space), and Roger Chaffee (on his first space flight mission) in the conflagration of Apollo/Saturn 204 (a.k.a. Apollo 1). Three months later, the new Soyuz capsule, having failed all three test launches and killing one ground crew member, was launched into orbit nonetheless with cosmonaut Vladimir Komorov. He died as multiple failures slammed his reentering capsule into the ground at 40 meters per second. Delays and expense mounted in both programs.

The United States completed Mercury's one-person spacecraft program, with flights in 1961–1963. Kennedy's Moon initiative ushered in two-person Gemini flights in 1965–1966, sending 10 missions into orbit to perfect techniques necessary for human lunar landings. After recovering from the Apollo 1 disaster, Project Apollo fulfilled that dream. Between 1968 and 1972, Apollo reigned triumphant, placing 12 astronauts on the Moon and 9 three-person crews in lunar orbit, as well as two missions in Earth's orbit. No program before or since has sent humans to another world beyond Earth (photographs spanning Apollo are seen in Figures 2.5–2.26).

By mid-1968 the momentum of lunar exploration turned squarely to the United States with resumption of Apollo, where it stayed. In mid-October, following a dozen successful crewless flights of Apollo spacecraft and/or Saturn rockets

First Steps

Figure 2.5. **Rising from the ashes**. *Apollo 7* launches toward Earth's orbit on its Saturn I-B on October 11, 1968, with the Vehicle Assembly Building (VAB) and other Kennedy Space Center structures in background, seen from an Airborne Lightweight Optical Tracking System (ALOTS) C-135 aircraft, flying at 10 km altitude (35,000 feet) and about 50 km away. *Apollo 7* was assembled and launched at Launch Complex 34 (where the three Apollo AS-204 astronauts died 20 months before), 11 km from the VAB. Later Apollo rockets were assembled at the VAB and launched from Launch Complex 39's pads, 5 km from the VAB. (Photo KSC-68PC-187 courtesy NASA.)

(including the mighty Saturn V), Walter "Wally" Schirra, Donn Eisele, and Walter Cunningham rode a smaller Saturn 1B into Earth's orbit on *Apollo 7* to successfully test the Command/Service Module (CSM), which would later carry astronauts into lunar orbit (Figure 2.5). A similar Earth-orbital test of the entire Saturn V and both Apollo crafts (including the lander) was slated for *Apollo 8*, but NASA decided boldly to send a CSM with Frank Borman, Jim Lovell, and William Anders into lunar orbit (reasons and consequences discussed in note 30). On December 21–27, 1968 (Figure 2.6), the first humans left Earth and its orbit to visit, but not land, on another world (Figure 2.7). A test of all Apollo/Saturn V hardware was conducted in Earth's orbit on March 3–13, 1969, with James McDivitt, David Scott, and Russell "Rusty" Schweickart on *Apollo 9* (Figure 2.8). *Apollo 10*, with Thomas P. Stafford, John W. Young, and Eugene Cernan, tested all but an actual landing in lunar orbit down to only 15 kilometers altitude, on May 18–26, 1969 (Figures 2.9 and 2.10).

Apollo's Lunar Module was a true spacecraft – the first to carry humans only in space outside Earth's atmosphere – hence, more economical than designs incorporating aerodynamics. Grumman Aircraft Engineering Corporation built the LM with considerable payload margin, allowing long stays and a robust science return. The LM was further upgraded for J missions (*Apollo 15–17*), with larger

The New Moon

Figure 2.6. **Promise and threat**. *Apollo 8* launches, 7:51 AM EST, December 21, 1968: first Saturn V with crew, first humans to the Moon. This is the moment at which the Soviet KGB threatened the rocket would explode (Chapter 3). Compare its height (111 m) to the I-B (68 m) with same upper stage (Figure 2.5). (Courtesy of NASA.)

durations and science payloads. Unfortunately, the LM was delivered a few months late. This produced an amazing coup, however, by requiring *Apollo 8* to fly to the Moon sans LM, while *Apollo 9* tested the LM in Earth's orbit, rather than vice versa. This brave but necessary swap in Apollo's manifest doomed the Soviets to fail as the first to circumnavigate the Moon, despite being nearly ready. With the success of *Apollo 8*, they abandoned that program (while still pursuing a separate landing mission).

Apollo 7–10 set the stage for human lunar landings, and two months later *Apollo 11* was headed to the Moon (Figure 2.11), where Armstrong and Aldrin landed on its

Figure 2.7. **In the beginning**. Earthrise from lunar orbit by *Apollo 8*'s William Anders, December 24, 1968. Borman, Lovell, and Anders were the first humans to see Earth from another world. Here the Moon is brownish-grey, and Earth brilliant blue. *Apollo 8* probably preempted by weeks or months the Soviet Zond attempt to send two cosmonauts into lunar orbit, which they soon disavowed despite continued, secret efforts through 1972. (Photo AS8–14–2383, courtesy of NASA.)

Figure 2.8. *Spider* **and** *Gumdrop*. *Apollo 9* CM Pilot David Scott in the hatch of CM (*Gumdrop*) on March 6, 1969, photographed by LM Pilot Russell L. Schweickart from the porch of the LM *Spider* seen in foreground. The Service Module extends behind the CM, in orbit around Earth in the background. Commander James McDivitt is in the LM. This was the first test of the LM and Command/Service Module together; a crewless LM flew on *Apollo 5*. (Photo AS09–20–3064, courtesy of NASA.)

Figure 2.9. **Full-up test**. One of the first exterior photographs of a spacecraft in lunar orbit: *Apollo 10* LM *Snoopy* carrying Commander Thomas Stafford and LM Pilot Gene Cernan, taken through the window of CM *Charlie Brown* on May 22, 1969, by CM Pilot John Young. Stafford and Cernan flew the LM to within 14.4 km of the Moon's surface. (Photo AS10–34–5091, courtesy of NASA.)

Figure 2.10. **Barnstorming the Moon**. *Apollo 10* CM *Charlie Brown* over mountains on the Moon, taken from LM *Snoopy*. The mountain at upper left was unofficially dubbed Mount Marilyn by *Apollo 8* astronaut Jim Lovell, after his wife. The feature's official name is Secchi Theta in Montes Secchi of south Mare Tranqillitatis. (Photo AS10–28–4165, courtesy NASA.)

First Steps

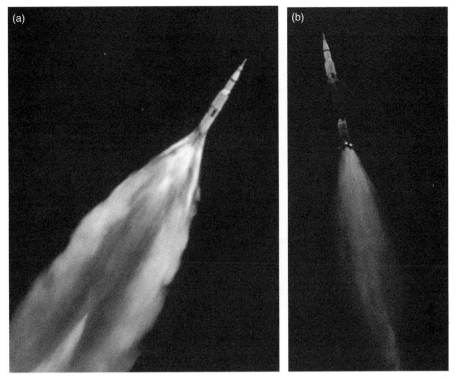

Figure 2.11. **Fire shakes the sky**. Launch of *Apollo 11*, up to first-stage S-IC booster separation, 2.5 minutes after launch, as filmed by ALOTS on an Air Force EC-135N aircraft. At separation, *Apollo 11* was 63 km high and 89 km downrange. (NASA images KSC-69PC-413 and S69–39958.)

surface, while Michael Collins orbited above (Figures 2.1 and 2.3). Walking on the Moon for 2.5 hours, they collected 21.6 kilograms (47.7 pounds) of samples and deployed a solar wind particle collector, the first successful lunar seismograph, and a laser reflector to measure the Earth-Moon distance accurate to centimeters (Figure 2.12). Although the seismograph failed after one month, it recorded the first moonquakes and showed the Moon to be relatively quiet seismically. The laser reflector still performs valuable science.

The year 1969 was dismal for the Soviet Moon program, with all six Luna attempts lost and test flight disasters in the human lunar effort. Nothing worked. *Luna 15* launched three days before *Apollo 11* to upstage it by returning a Soviet lunar sample; instead the Luna robot hit a mountain on its lunar landing approach. In contrast *Apollo 11*'s success was followed that year by another human landing with even greater science return.

Apollo 12 (November 14–24, 1969) put Richard Gordon in lunar orbit, and Charles "Pete" Conrad and Alan Bean in a pinpoint landing beside *Surveyor 3* (Figure 2.13), despite being struck twice by lightning during launch from Cape Kennedy. [Box 3.5] This first, better-equipped H mission more thoroughly probed the mare, deploying the first large science instrument package, the Apollo Lunar Science Equipment Package (ALSEP). Whereas *Apollo 11* had strayed several kilometers outside its

Figure 2.12. **A leap for science**. Aldrin, lunar seismometer (PSEP: Passive Seismic Experiment Package), American flag, and LM *Eagle*, photographed by Armstrong. With 83 minutes thus far outside, Aldrin has just set up and leveled the seismometer, one of two experiments deployed. (The Laser Ranging Retroreflector [LRRR] still functions.) In 33 minutes he and Armstrong would be back inside the LM, having completed *Apollo 11*'s lunar EVA. (Photo AS11–40–5948, courtesy of NASA.)

landing area as a result of lunar gravity's lumpiness, *Apollo 12* landed only 163 meters from *Surveyor 3* (Figure 2.14). Hence, future missions could target more challenging sites, even in highlands rubble or recent craters. The astronauts made two walks[20] lasting 7.75 hours total. Unfortunately, the mission quickly became a public relations anticlimax after Bean accidentally damaged the TV camera, pointing it at the Sun. (Also, several magazines of still photographs were left unintentionally on the Moon.) Scientifically, the mission succeeded. ALSEP contained a second seismometer and experiments to study ionized atmospheric components, the Moon's magnetic field, and the charged-particle solar wind. Although *Apollo 12* landed opposite *Apollo 11* along the lunar Near Side equator in similar-looking mare, its 34 kilograms of samples revealed a surprisingly different composition of superficially similar, largely volcanic rocks. Immediately scientists saw that *Apollo 12*'s site, more than *Apollo 11*'s, contained volcanic basalts rather than impact-formed breccias,[21] with basalts sufficiently diverse to indicate many different flows, younger than at *Apollo 11*. The abundant element titanium was more rare at *Apollo 12*. Minerals proved higher in radioactive elements and in potassium (K), rare Earth elements (REE), and phosphorus (P) – dubbed KREEP[22] – distinguishing them from similar-looking, eastern Near Side maria.

Figure 2.13. **One small step at a time**. *Apollo 12* LM Pilot Al Bean begins his 3-minute descent from LM *Intrepid*, as photographed by Commander Pete Conrad, on November 19, 1969, on the first of their two 4-hour lunar surface EVAs. Note the ladder's nearly 1-meter final step. The MESA (holding the TV camera that later broke after removed from MESA) is at lower left. (Photo AS12–46–6726, courtesy of NASA.)

We will see that KREEP is fundamental to understanding the Moon's origin and evolution. Furthermore, Bean and Conrad visited *Surveyor 3* and brought back pieces, showing how materials fare with two years of lunar environmental exposure. Bacteria found in these artifacts initiated controversy regarding their survival on the Moon (see Chapter 11). *Apollo 12* presaged scientific productivity from later landings and allowed mission planners to rely on landing at specific lunar features rather than broad landing zones, another scientific enhancement.

After *Apollo 12*, lunar missions were designed to emphasize further science. Crews were trained thoroughly in techniques of photography and field geology (Figure 2.15).[23] Most were commanded by veteran Apollo astronauts, which proved immediately fortuitous with the unlucky flight of *Apollo 13*, crewed by James Lovell (of *Apollo 8*) and rookies Leonard "Jack" Swigert and Fred Haise. On April 13, 1970 (a Monday), more than halfway to the Moon, a temperature sensor electrically shorted causing a main oxygen tank to explode and blow a quadrant out of the Service Module from which the mission draws power, fuel, oxygen, and water (Figure 2.16). By heroic coordination of innovation, quick action, and brave luck, the three astronauts returned

Figure 2.14. **Aliens from Earth.** *Apollo 12* Commander Pete Conrad reaching for *Surveyor 3*'s TV camera (later returned to Earth), photographed by Bean on November 20, 1969, 150 minutes into their second lunar EVA. Surveyor's sample scoop arm protrudes in front of Conrad. *Intrepid* rests 180 m in the background. *Surveyor 3* is lightly dusted as a result of *Intrepid*'s arrival the previous day, landing closer than intended. (Photo AS12–48–7133, courtesy of NASA.)

Figure 2.15. **Field trip.** In preflight training for *Apollo 16*, Commander John Young (*left*) simulates using a telephoto lens while wearing mock lunar EVA equipment, accompanied by his LM Pilot Charlie Duke during geology field simulations near Taos, New Mexico. All Apollo lunar astronauts, especially on later missions, received extensive geology education and field training. *Apollo 17* LM Pilot Jack Schmitt was already a professional geologist. (Courtesy of NASA.)

Figure 2.16. **Forget not peril**. The damaged Service Module of *Apollo 13*'s disabled *Odyssey*, as photographed by CM Pilot Jack Swigert about 4 hours before the end of the mission on April 17, 1970, and 82 hours after a large explosion on April 13 (Houston time) tore one quadrant side panel from the spacecraft, as seen here. About 90 minutes before splashdown, Swigert, Commander Jim Lovell, and LM Pilot Fred Haise transferred from LM *Aquarius* into the CM, after surviving in the LM more than three days, including a swing around the Moon. The three spacecraft reentered Earth's atmosphere over the South Pacific. (Photo AS13–59–8500, courtesy of NASA.)

to Earth alive and well in their crippled spacecraft, having swung in a "free-return" single orbit around the Moon (Figure 2.17).[24] Public interest was raised by this outer-space drama, but the program was delayed by accident investigation and remediation until *Apollo 14* nine months later.

For the Soviets 1970 continued 1969's misfortune with two more failed Lunas, but by autumn their luck changed. On September 24, *Luna 16* became the first robot to return a sample to Earth, from Mare Fecunditatis 1,000 kilometers east of *Apollo 11* near the limb of the Moon.[25] *Luna 16* showed that regolith can hold significant amounts of melt glasses from the heat of meteoroid impacts. The sample had still different titanium abundance, intermediate between *Apollo 11* and *12*. Every site looked different. Two months later *Luna 17* performed another amazing technical feat, placing the first wheeled roving vehicle on another world, *Lunokhod 1*, in the mare feature Sinus Iridium in a largely unexplored lunar quadrant 1,300 kilometers north of *Apollo 12*. In 11 months *Lunokhod 1* traveled 10.5 kilometers over the lunar surface, delivering insight into regolith's mechanical properties. Although deprived of lunar science instrumentation except for an X-ray spectrometer for mineral composition, it set the stage for successful rovers of the future.

The Fra Mauro site chosen for *Apollo 13* was so crucial that *Apollo 14* (January 31–February 9, 1971) was retargeted there. The Fra Mauro formation is thought to have formed peripherally to the impact excavating the giant Mare Imbrium basin

The New Moon

Figure 2.17. **Sweet home.** *Apollo 13*'s crew returns on April 17, 1971, in *Odyssey* to the South Pacific 6 km from recovery ship *U.S.S. Iwo Jima*, as photographed from a recovery aircraft (with structure partially obscuring the lower left corner). After almost 6 days in space, most of it being spent in a cold spacecraft forced to shut down, the three men were met with rejoicing by the recovery crew and the world. (NASA photo KSC–70PC–121.)

(1,100 kilometers across). Imbrium so dominates that if *Apollo 14* could tell its age, it would establish minimum ages to many other Near Side features splattered by its ejecta. Nearby Cone crater (300 meters wide) excavates to the Fra Mauro bedrock, exposing Imbrium-aged material. With rookies Stuart Roosa in orbit overhead and Edgar Mitchell in the LM, the first American in space Alan Shepard set the lander down 53 meters from the target next to Cone crater. (This was Shepard's first space mission since his 1961 Mercury flight, enabled by his recovery from an inner ear condition.)

Shepard and Mitchell collected 42 kilograms of Fra Mauro samples, allowing Imbrium to be dated, although time prevented them from reaching Cone crater (Figure 2.18). Many rock samples were fused by the Imbrium event 3.85 billion years ago, although detailed conditions were more complex (see Chapter 6). ALSEP included another working seismometer (*Apollo 11*'s died), remote-controlled mortars and thumpers generating local seismic signals, another copy of the atmospheric composition instrument (one channel having failed on *Apollo 12*), and another retroreflector for Earth-Moon laser ranging (Figure 2.19).

Although three Apollo missions (18–20) were cancelled, in 1971–1972 the last three lunar landings returned voluminous results. These J missions exploited larger LM payload capacity to deliver more experiments to the Moon, more samples to Earth, and more lunar surface exploration time. Each mission carried a two-person Lunar Roving Vehicle (LRV) allowing astronauts greater lunar travel range (Figure 2.20). This was designed to alleviate the stress on the *Apollo 14* astronauts encountered near Cone crater (which we discuss in Chapter 11), but on *Apollo 15*

Figure 2.18. **Pulling the weight**. *Apollo 14* Commander Al Shepard with the Modularized Equipment Transporter (MET), a two-wheeled, ricksaw-like equipment handcart to aid in the 3.5-km walking traverse and sample collection over 9 hours and two EVAs, as photographed by LM Pilot Ed Mitchell on February 6, 1971. Here on the second EVA, Shepard is preparing a core sample. The MET carries cameras, hand tools, trenching shovel, collection bags, core tubes, sample containers, and occasionally experiments such as geophones and seismic thumpers. The MET, only used on *Apollo 14*, was covered by a thermal blanket between EVAs. (NASA photo AS14–68–9404.)

(July 26–August 7, 1971) this plan may have gone tragically wrong (see Figure 2.21). Leaving this unpleasantness for Chapter 11, we must appreciate the intriguing results from this mission, which put Al Worden in orbit with Dave Scott and Jim Irwin on the banks of the Hadley Rille lava flow valley, next to the Apennine mountains upthrust by the Imbrium impact. ALSEP's instruments are too numerous to inventory here but included the third laser retroreflector, completing the triad needed to measure distance and orientation of the Moon. *Apollo 15* deployed the first of three heat flow experiments to measure temperature below the Moon's surface (crucial to Chapter 8's topic) and heat leaking from the interior. In orbit the CSM was also loaded with instruments to probe the Moon and its atmosphere, including a free-flying subsatellite making its own measurements (Figure 2.22). These kept Worden busy with his experiments and gave him opportunity to walk in deep space (between Earth and the Moon) to retrieve the large-format film cassette used to map much of the Moon in detail.

Apollo 15's results are profound. Among 77 kilograms of samples (approaching the total from all three previous landings) were rocks of nearly primordial age, over

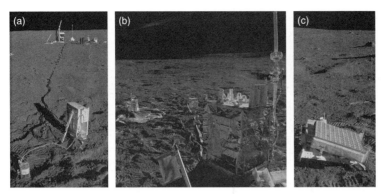

Figure 2.19. **Science aplenty**. Three hours of *Apollo 14*'s first EVA involved deploying scientific equipment, as photographed by LM Pilot Al Mitchell on February 5, 1971. After *Apollo 11*, each LM bore a diverse Apollo Lunar Surface Experiments Package (ALSEP). (a) Atmospheric sensors, Suprathermal Ion Detector Experiment (SIDE), and Cold Cathode Ion Gauge (CCIG), cabled to the main ALSEP cluster in distance (for isolation from contaminants), powered by radioactive decay in the Radioisotope Thermoelectric Generator (RTG), the black object, top middle-right. (b) Central Station (and Lunar Dust Detector), at right, and 58 cm long antenna pointed at Earth; at left, the Passive Seismic Experiment (PSE) seismometer. The small box, center, is the Charged Particle Lunar Environment Experiment (CPLEE) for electrons and ions. (Not shown: Active Seismic Experiment, a seismic "thumper"/geophone array to probe the local subsurface, and a Lunar Portable Magnetometer.) (c) One Lunar Ranging Retroreflectors (LRRR) of *Apollo 11, 14,* and *15* (like those on *Lunokhod 1* and *2*), the only instruments operating after all ALSEPs were turned off September 30, 1977. (Photos AS14–67–9372, AS14–67–9384 and AS14–67–9386, courtesy of NASA.)

4 billion years old. These consisted of deep "magma ocean" material formed soon after the giant impact along with the Moon and transformed Earth. Furthermore the astronauts harvested collections of green glass beads produced when magma from the deep interior exploded as lava "fire fountains" when gas dissolved within hit the vacuum and expanded violently, propelling sprays of solidifying lava droplets raining down as spherules. Its green color betrays primitive silicates (olivine and pyroxene) perhaps expected from deep lunar mantle. Thus the mission recovered two channels into the Moon's heart, with surprising results from the fire fountain's gas, the presence of water, becoming apparent 36 years later (see Chapter 8). *Apollo 15* also revealed great detail about behavior of lunar lava flows such as those forming Hadley Rille (Figure 2.20). Orbital data taken by Worden are unique and still used today, including some 3,000 of the highest-resolution images taken from lunar orbit.

Apollo 16 (April 16–27, 1972) put Thomas "Ken" Mattingly in orbit and John Young and Charles Duke on the surface to expand the insights from previous missions (Figures 2.15 and 2.23). This landing – in a flat zone in the Descartes highlands of the south-central Near Side – was the sole landing in the lunar highlands, which covers 80% of the Moon's surface and is much rougher to land on than maria. Scientists hoped that samples would include ancient crust, perhaps 4.4 billion years old, in analogy to Earth's primordial surface. Found instead in 96 kilograms of samples were many breccias: impact-melted agglomerations with characteristics ages much younger than ancient crust, set by their impact date. Guesses from photographic

First Steps

Figure 2.20. **You can see forever**. *Apollo 15* Commander Dave Scott working at the Lunar Roving Vehicle (LRV), overlooking Hadley Rille valley extending 15 km into the distance, with the toe of the Apennine Front on the right, as photographed by LM Pilot Jim Irwin on July 31, 1971, at the first station of their first of three roughly 7-hour EVAs. *Apollo 15* was the first J-series mission, with the LM made 1.2 tonne heavier with more propellant capacity, more payload (including the LRV), and modifications to extend lunar stays to 75 hours, allowing more work to be done on the Moon. Compare this view to Figures 6.6 and 6.7. (Mosaic of photos AS15–85–11450 and AS15–85–11451, courtesy of NASA.)

reconnaissance missed the mark, and rocks at Descartes proved younger than suspected (sometimes younger even than Imbrium), seemingly blown there by distant impacts. Furthermore, major features appeared impact generated, not volcanic as previously suspected in some cases. Much of the geological lessons of *Apollo 16* came from surprises of missed expectation. Still, some of the oldest lunar samples, about 4.4 billion years old, came from *Apollo 16*, although their origin is less clear than desired and their quantity radically less than anticipated.[26] ALSEP was varied and successful, with local seismic experiments more complete than those of *Apollo 15*, extending understanding of the regolith and shallow subsurface structure. Although the heat flow experiment was broken during deployment, other experiments performed satisfactorily.

The grand finale, *Apollo 17* (December 7–19, 1972) landed Gene Cernan and geologist Harrison "Jack" Schmitt in the Taurus–Littrow valley, with Ron Evans in orbit (Figures 2.24 and 2.25). Sitting below the 2 kilometer-tall Montes Taurus uplifted by the Mare Serenitatis impact and plausibly composed of ancient highlands crust, the site might probe this and likely younger, volcanic Littrow formations in the valley (Figure 2.26). It might date Serenitatis itself as well as a young chain of craters perhaps from ejecta in a ray from distant Tycho crater. Scientists thought the 100-meter Shorty crater might possibly be volcanic, but it proved an impact feature that nonetheless excavated orange-black fire fountain glass: "orange soil" rich in iron and titanium. Now it appears no late impact excavated pre-

Figure 2.21. **Heroic sacrifice**. Jim Irwin digs a trench to sample unexposed soil, on *Apollo 15*'s EVA 2 near the ALSEP site, as photographed by Scott on August 1, 1971. (Note that the antenna atop his backpack broke on EVA 1 and has been taped on.) At left is the gnomon photographic reference device. Scott and Irwin fully exploited J-series capabilities, collecting nearly as many samples (77.3 kg) as *Apollo 11–14* combined (98.2 kg). Tragically both men suffered severe dehydration on the Moon and developed heart problems – Irwin chronically. He later died after three heart attacks. (Photo AS15–92–12424, courtesy of NASA.)

Serenitatis bedrock, making the ancient crust an elusive, unattained goal. From 110 kilograms of samples, rocks likely tied to the Serenitatis impact seemed nearly identical in age to Imbrium, sampled earlier, perhaps only several tens of millions of years younger than Imbrium. Tycho, if relevant, corresponds to relatively fresh samples only about 100 million years old.

Overshadowed by Apollo, Soviet lunar robots maintained momentum while their human Moon project crashed and burned. *Luna 20* (February 14–22, 1972) returned 55 grams from the Apollonius highlands between maria Fecunditatis and Crisium, similar in composition to 101 grams from *Luna 16*, 130 kilometers due south. Whereas *Luna 16*'s 39-kilogram return capsule landed in Kazakhstan without incident, *Luna 20* landed in a river during a snowstorm. Fortunately ice and a midstream island kept the capsule from sinking. On January 15, 1973, *Luna 21* carried Lunokhod 2 to the Moon on a 113-day mission covering 37 kilometers – an extraterrestrial robotic land speed record – and returning many data including 80,000 pictures (see Chapter 10 for more on the Lunokhods).

Figure 2.22. **Science onboard**. *Apollo 15*'s Command/Service Module *Endeavor* and its Scientific Instrument Module (SIM) bay, photographed by Irwin from LM *Falcon*, August 2, 1971. This side of the Service Module is opposite that in Figure 2.12. The SIM flew on *Apollo 15–17*, holding many instruments, as well as others extended on a long boom, including on *Endeavor* three cameras, a laser altimeter, and alpha-particle, gamma-ray, and mass spectrometers. One-third of the way on the return from the Moon to Earth, each CM Pilot (in this case Al Wordon) would spacewalk from the CM to the SIM to retrieve exposed photographic film canisters. *Upper left*: *Apollo 15* and *16* released a lunar subsatellite (shown to scale) with three instruments to study charged particles and magnetic and gravitational fields. (Photos AS15–88–11972 and AS15–96–13068, courtesy of NASA.)

Luna 16, 20, and *24* were the only successful robotic sample-return missions[27] until *Genesis* delivered solar wind particles to Earth in 2004 (plus *Stardust* from Comet Wild 2 in 2006 and *Hayabusa* from asteroid Itokawa in 2010). Indeed, Luna samples are the only ones brought via robot from an extraterrestrial planet-like world. A Luna sample return capsule would escape the Moon and achieve trans-Earth injection in a single rocket burn. This required leaving from a narrow strip on the Eastern limb, near maria Crisium and Fecunditatis. *Luna 24* added to this transect by landing on August 18, 1976, in southern Crisium, a couple of hundred kilometers north of *Luna 20*, and returning 170 grams of maria regolith to Earth. While *Luna 24* reported mildly surprising results for mare regolith mineral composition, this sample also contained some of the earliest evidence of hydrated material, possibly water on the Moon (see Chapter 8).

The August 1976 *Luna 24* was the last Soviet lunar mission, following secret cancellation of the human Moon program in March 1976. *Luna 25* was ready to carry

Figure 2.23. **Dust world**. *Apollo 16* Commander Jim Young and LRV covered with dust, on EVA 3, on April 23, 1972, as photographed by Charlie Duke. Young's originally white suit is now largely gray. On both *Apollo 16* and *17*, the dust fender on an LRV wheel was damaged, for different reasons, causing dust to spray the vehicle. Astronauts occasionally lost their balance, falling onto regolith. Ubiquitous dust caused problems, such as befouling sample container seals, allowing them to become contaminated by leaking air. (Photo AS16–116–18718, courtesy of NASA.)

an enhanced Lunokhod 3 to the Moon but, in 1977, was cancelled despite the popularity of previous Lunokhods with the Soviet people and glowing official propaganda. Since the fall of the Soviet Union 14 years later, Russia has sent no spacecraft to the Moon but plans to launch Luna–Resurs and Luna–Grunt starting in 2016, both an orbiter and lander (one which may be renamed *Luna 25*). On paper projects began in the 1970s and 1980s for a Soviet Moon base, but no metal was cut. Academy of Sciences president and chief rocketry theoretician Mstislav Keldysh and Space Research Institute director Roald Sagdeev proposed to Soviet leaders sending probes to Venus to avoid direct competition with the United States on the Moon. In the mid-1980s several robotic lunar probes were planned, but these died with the Soviet Union.

When *Apollo 11*'s *Eagle* landed on the Moon, hundreds of millions were watching and billions soon knew the outcome. The common reaction was a feeling of human accomplishment, even in Moscow, like that felt across the world when Gagarin first orbited Earth. This reaction was strengthened not by surprise, as in *Sputnik*, but by fulfillment of a promise that seemed eight years earlier to be

Figure 2.24. **Rock hound**. *Apollo 17* LM Pilot Harrison "Jack" Schmitt on the third EVA removing the gnomon after sampling operations at the Station 6 boulder (Tracy's rock after Cernan's daughter), as photographed by Commander Gene Cernan on December 31, 1972. A few meters to Schmitt's right one can see dust sampling marks made by Cernan. East Massif protrudes above the horizon at left. Schmitt and Cernan returned 110.5 kg of samples, more than any other mission, including many vital to our later discussions. (Photo AS17–140–21496, courtesy of NASA.)

unrealistic, or at least daringly inspired. Most people saw it as something we, humankind, accomplished, not they, the United States. Several hours after the landing, the Soviet newspaper *Pravda* published the story on the front page, and Soviet TV that day showed the first moonwalks multiple times (but not live). (The story stayed out of the papers in China, North Korea, and North Vietnam.) The world, even the Soviet Union, felt better about itself for several days.

Ultimately the Soviet human lunar landing program could only have outstripped Apollo following another major setback between Apollo 1 and *Apollo 11*, but not for Soviet's lack of trying. They built the most powerful rocket that they ever produced (the N1), as well as a spacecraft to land one person on the Moon (the *Lunniy Korabl* – lunar craft or LK) but never tested it beyond Earth's orbit and never with a cosmonaut. All four N1 launches failed (1969–1972), including an explosion two weeks before *Apollo 11*, destroying the launch tower and delaying the program for two years. In the Soviet Union these efforts were not commonly known, or their complete failure, when the program was cancelled quietly in 1974.[28] It occurred in

Figure 2.25. **Last on the Moon**. *Apollo 17* Commander Cernan with flag and LRV, starting EVA 3, the last Apollo moonwalk, photographed on December 13, 1972, by Schmitt. Along with patriotic symbols, one can sense pride in human achievement and skilled capability that garnered a powerful science return. Cernan said later that day that the prime thing "about Apollo is that it has opened for us – 'for us' being the world – a challenge of the future. The door is now cracked, but the promise of the future lies in the young people, not just in America, but the young people all over the world learning to live and work together." (Photo AS17–140–21389, courtesy of NASA.)

parallel to the famed *Vostok* that first placed humans in space in 1961, followed by the first multi-person (three) craft (Voskhod) in 1964, the three-person Soyuz in 1967, the first space station *Salyut* in 1971, as well as many covert military-crewed vehicles. Their human lunar program was quietly airbrushed from history, and many Western media assumed that Apollo only competed with itself.[29]

Despite the deception, the Soviet lunar landing scenario was designed in scale and complexity comparable to an Apollo landing. Three stages of the N1 rocket would place two cosmonauts and their lunar vessels in Earth's orbit, and the fourth stage would send them toward the Moon (versus stage three for a Saturn V). The fifth stage, Block D, would place both spacecraft into lunar orbit, and one of the two cosmonauts would spacewalk from the orbital craft into the LK lander, which would be sent toward the lunar surface by Block D. (In contrast, two of three Apollo astronauts would transfer through a pressurized tunnel connecting the

First Steps

Figure 2.26. **A world remains**. Taurus-Littrow valley with *Apollo 17* LM *Challenger* 3 km away (small dot, center) and Sculptured Hills 8 km in background, photographed by Schmitt on December 13, 1972. This evokes the Moon's scale compared to our human place amid its isolation. Cernan closed: "I'd like to just (say) what I believe history will record, that America's challenge of today has forged man's destiny of tomorrow. And, as we leave the Moon at Taurus-Littrow, we leave as we came and, God willing, as we shall return with peace and hope for all mankind. Godspeed." (Photo AS17–139–21204, courtesy of NASA.)

CSM into the LM.) The LK would power itself in the final landing approach, having shed the Block D, guided by the radio beacon from an earlier robotic Luna lander. Having arrived safely, the lone cosmonaut would exit the depressurized LK in his space suit, plant a flag, collect samples, deploy several experiments, enter the LK, and blast off within a day or less. The same LK engine would carry the lunar explorer back to the orbiting Soyuz-like spacecraft, leaving LK's landing tripod behind. (In contrast, the Apollo LM with its two astronauts had three times the mass of the LK, and separate descent and ascent stages, each with their own main engine.) After the cosmonaut spacewalked back to join his comrade in the orbiter, the LK would be jettisoned and the orbiter's engine would send the two toward Earth, for direct entry into Earth's atmosphere and recovery on Soviet territory. Although their mission profiles were similar, compared to the LK, the LM was a more capable scientific and exploration vehicle, with the later model carrying hundreds of pounds of scientific equipment, a Lunar Rover to drive many kilometers across the surface, and supplies to sustain the two astronauts for several days. The 800 pounds of lunar samples returned by Apollo have sustained lunar geology ever since.

Soviet rocket programs developed under the direction of several design bureaus, the most prominent coming closest to challenging Apollo's lead, under Sergey Korolev's leadership until his death in January 1966. By 1968 Soviet space planners seriously believed they might send cosmonauts to circumnavigate the Moon in 1968–1969. In September 1968 the crewless *Zond-5* spacecraft flew around the Moon and was recovered successfully from the Indian Ocean. With one more good test flight, designers felt the Zond series craft (otherwise known as the L1 or 7K-L1) could carry cosmonauts in a single loop around the Moon ("free return trajectory") and back safely. After circumnavigating the Moon, the crewless *Zond-6* crashed on landing in Kazakhstan on November 17, 1968, and forced a delay in the program. In part because the CIA thought that Zond might succeed and because the LM was not quite ready for an Earth orbital test flight, *Apollo 8*'s mission became that of the CSM alone, in lunar orbit. At the competition's cusp, as the Soviets saw their chances slip away to fly humans around the Moon before Apollo, the Soviet KGB sunk to trying "dirty tricks," desperately attempting to delay NASA with threats to sabotage *Apollo 8*'s launch (as well as *Apollo 11*'s).[30] Before crewless *Zond-7* was launched in August 1969, however, *Apollo 8* orbited the Moon in December 1968, followed by *Apollo 10* in May 1969 and *Apollo 11*'s landing – a total of nine astronauts around the Moon and two on its surface before *Zond-7* – and six Apollo lunar flights still to go. Zond was never designed to take cosmonauts to the Moon's surface. On January 1, 1969, five days after *Apollo 8*, the Soviet Union announced plans to construct a space station in low Earth orbit, simultaneously disparaging Apollo as a unilateral stunt. Much American press echoed the notion that there had never been a real Moon Race.[31] Nonetheless, the Soviet Moon effort continued.

Opportunity for cosmonauts setting foot on the Moon was years away, and in the end, the human lunar Soviet program proved fruitless. Although the reasons for this are many, the important ones are few. Paradoxically, unlike their command-economy Soviet counterparts who nurtured a vigorous competition between rocket design groups – in the end proving to be unhealthy – the Americans engaged in a top-down, single-minded, single-architecture effort to reach the Moon. Five different Soviet lunar spacecraft, most of them orbiters, were initiated. Two separate boosters, the Proton by Vladimir Chelomei's design bureau OKB-52 and the N1 under Korolev himself heading Design Bureau Number One (OKB-1), proved unreliable. The Proton rocket, originally conceived as a giant ICBM, was instead born into a miserable childhood in space exploration, with most of its lunar payloads falling short of the Moon between first attempts in 1965 and the last Soviet lunar launch in 1976.

In contrast, the N1 booster, 105 meters tall (nearly as large as the Saturn-V and more powerful at ignition), never left Earth's atmosphere. Most failures (first on February 21, 1969) arose from its cluster of 30 rocket engines in its first stage; therein lay the problem: making a reliable engine when its success rate is raised mathematically to the thirtieth power. Even a 98% successful engine in such a cluster would produce a booster failing almost half of the time. N1 incorporated a complex engine firing control system to shut down failing engines, but failures of this system itself proved catastrophic. Exploding first-stage engines in the second N1 launch on July 3, 1969, destroyed the launch pad (one of two for the N1), shattered windows 50 kilometers away, and doomed Soviet challenges to Apollo before the last astronaut walked on the Moon in December 1972. The fourth N1 failure on November 23, 1972, occurred in the first stage as the second stage prepared to ignite. A fifth N1 launch might have worked, but the project was cancelled and thus ended the Soviets' flights of human lunar exploration, as well as Korolev's OKB-1 group itself soon thereafter (forming NPO Energia). All Soviet human Moon programs were doomed by lack of reliable boosters of sufficient power.

Given how close the Soviet moon program seemed to be in 1968, one wonders if they might have passed the United States had Armstrong used up that 45 seconds of fuel and aborted the mission or crashed. There would have been inquests, investigations, and long delays. NASA had planned three Apollo lunar landing missions in 1969. Seventeen months remained "before this decade was out" as President Kennedy promised (technically, 1970 being part of the 1960s decade). There was little time for the Soviets to simultaneously fix the N1 rocket, test the LK lunar lander, and surpass the Americans, but they would have tried. The lunar orbits of *Apollo 8* and the landing of *Apollo 11* were the sharpest tipping points in the Cold War. Perhaps if designers of the N1 had built a larger rocket engine, decreasing the number of engines, catastrophic failures could have been reduced. Of the 32 Saturn rocket launches (13 Saturn Vs and 19 of the smaller Saturn I and Saturn 1B), only *Apollo 6* and *13* suffered unplanned main engine shutdowns (none of the five massive first-stage F-1 engines, but smaller J-2s of which five power the Saturn V second stage). The unmanned test *Apollo 6* partially succeeded even after losing two J-2s; *Apollo 13* proceeded unhindered with one J-2 extinguished (unrelated to the famous explosion onboard the Apollo spacecraft 56 hours later). No Saturn rocket ever exploded.

Difficulties revolved around increasing haste in the Soviet program after Krushchev decided in August 1964 to accept Apollo's challenge. The designers would abbreviate testing of booster components on the ground for sake of flying them, only to have them fail catastrophically, often inexplicably, given destruction of diagnostic forensic evidence. Saturn was sometimes rushed; the first full-up Saturn V test on *Apollo 4* skipped some component flight tests that were usually considered prudent. It was not just luck, however. Excluding Apollo 1's horrendous demise, Apollo spacecraft suffered far fewer system failures than Soyuz, as did corresponding boosters compared to their Soviet counterparts. The Soviet human Moon program's tremendously duplicated effort meant resources were too diffused: first with Korolev's N1 rocket and L3 space vehicles (ultimately the LK lunar lander and LOK lunar ferry orbiter) competing with Chelomei's LK-1 lunar orbiter on the Proton, before the LK-1 was replaced by a shrunken, one or two cosmonaut Soyuz, the L1 (or 7K-L1). Additionally, work on intermediate steps of these four different spacecraft (versus two for Apollo) competed with well-known spacecraft for cosmonauts in Earth's orbit: Vostok, Voskhod, and Soyuz. If they had started earlier and pursued one program with deliberation, the Soviets might easily own later space race victories, especially sending humans into lunar orbit. Until 1965 they achieved all milestones first, then things fell apart. Attention soon focused on Soviet space stations.[32]

The Soviet Politburo itself interfered, demanding launches in time for random May Day celebrations and Communist Party Congresses. Cosmonauts and ground crew died in part as a result of this; for instance, Vladimir Komarov died in the crash of *Soyuz 1* on April 23, 1967, as a result of flying too soon at Soviet leader Leonid Brezhnev's behest. Politicians setting goals years in advance can provide leadership, but micromanaging on daily or monthly bases is usually counterproductive. Also the ouster of Khrushchev in favor of Brezhnev in October 1964 rearranged the priority of Korolev's program over Chelomei's. Too many decisions were made precipitously and arbitrarily as much on the basis of personality as national interest or engineering realities. As with the United States, national interest was perceived in terms of ICBM technology more than science or exploration, except for when it embarrassed one's Cold War adversary.

Thus the Soviets lost the Space Race and the United States won on the quality of their management styles. The Soviets interfered too much from the top and decided issues too much on the basis of changing personalities. It is ironic that Soviet efforts allowed too much diverse competition, whereas NASA organization had a much more singular command structure. Does this decide who had the better political system? Hardly, because Americans have produced big, technological boondoggles before and since. Occasionally the United States can pull it together. It did not help that the Soviet program was secret; the United States had no graceful exit strategy except to win.

We discuss economic benefits from NASA in Chapters 3 and 14, but the case for Soviet space exploration is less clear. However, important spin-offs were less technological. Soviet space generated international prestige and internal pride. Furthermore the United States and Soviet Union engaged in a Space Race as proxy competition for military conflict, and only a few brave cosmonauts died. Unlike the Olympics, "kitchen debates" (à la Nixon/Khrushchev), or chess matches (à la Spassky/Fischer), the Moon Race went straight to the issue of who had better rocket technology and superior rapid innovation. During the Cold War the rival superpowers could demonstrate their prowess and measure one another without engaging in battle, probably averting at least one catastrophic war. Expenditure on national defense is dozens of times greater than for space exploration. Society never spends as much as rationally justified for preventing waste of human potential caused by deprivation of children from the education and nutrition they need. Space exploration may be an indicator more than a motive force of a healthy, technological society. To young people who decide to make their careers in science and engineering, the message is received. If society is intent on progress, young people will join the effort. If society is not forward-looking, they will invest their efforts in moneymaking and security. This way the world signals to its youth its interest in expanding the frontier. Unfortunately, the late Soviet program's history is more troublesome.

The Soviets put the first satellite and first human in space; for many the first person on the Moon being American was less crucial. It was dismissed with official spin: Soviets would do good science with robotic spacecraft, at lower costs and less the quality of a stunt. Apollo did not send the Soviet Union into apprehension and self-examination as *Sputnik* had America. Gagarin is more important than Armstrong to a Russian. The Soviet reaction to Apollo was largely limited to the politicians, design bureaus, and space workers themselves. The Soviet human space program launched several classes of space stations into Earth's orbit (the more civilian *Salyut* and *Mir*, as well as purely military *Almaz* space stations) and continued a robust response to the United States on many fronts. Apollo taught the Soviets not to underestimate an insecure United States' ability to mobilize a technological offensive, no matter how seemingly implausible, such as Reagan's Star Wars. The Soviet Union again rose to this challenge, only to fail in obscurity. One year after beginning construction of the huge space station *Mir*, they built the largest successful Soviet rocket, *Energia*, and used it in 1987 (despite Gorbachev's objections) to launch a confrontational response to Reagan's threat: *Polyus* (*Skif*), crew-capable and fitted for armament and stealth technology, an 88-ton monster orbital weapons platform that misfired and fell into the Pacific Ocean.[33] They launched *Energia* once more as a booster for the sole flight of Space Shuttle-like *Buran* the next year (after making two complete and three partial shuttles for flight), then terminated all these programs except *Mir*, ending the last gasp of Cold War space competition. These projects cost the equivalent of tens of billions of dollars, money better spent building roads or farm equipment, and constituted a forlorn mini-space race that the greater world ignored. One can only imagine the hysteria and intense competition with the United States resulting had *Polyus* actually succeeded in becoming a weapons platform. Coming just prior to the fall of the Berlin Wall and Soviet Union, these projects (in addition to huge military spending in general) provide an object lesson of how a nation can mount space efforts beyond its capability. The United States and Russia cooperated on *Mir*, and along with Europe and Japan, now cooperate on the *International Space Station*, but *Polyus*, *Energia*, and *Buran* came to naught and drained the Soviet Union at its most vulnerable.

One hears statements along the line of "Russians succeed in space because they stick to half-century-old designs and improve them, not always abandoning working systems like the Americans," but the *Polyus/Energia/Buran* era demonstrates this was not always the case, even relatively recently (and even ignoring the failed lunar efforts). This truism about Russian space technology may hold as a result of this profligate and unsuccessful spending, enriching the design bureaus as the Soviet economy was collapsing near the end of the Soviet Union: no more blank checks. It is a hard-learned lesson, not quickly forgotten.

The full impact of the Moon Race on the Russian psyche was delayed until August 1989, when the Soviet newspaper *Isvestia* revealed that there had been a Soviet program to land a human on the Moon after all. Russians then had much to distract them just prior to the collapse of the Soviet Union, but once in a while the issue crops up again. Russian cinema, remembering the Space Race, is cynical about its accomplishments.[34] During the nationalistic bluster of 2007's competing Arctic claims, one heard unintentional irony in the boast that a Russian flag under the ice of the North Pole was "like placing a flag on the Moon; this is really a massive scientific achievement."[35] The Moon landings still bother some Russians, but not as much as Americans may think. One cannot see an impact of the expensive, covert Soviet lunar program on the society that sponsored it, despite pride in the larger Soviet/Russian space program.

The Moon landings had a different impact on Americans, discussed in Chapter 3. Most are proud of this achievement. Despite a deepening war in Vietnam and riots in American cities, there was perverse pride even among those who would use it to shame their fellow citizens: "If we can send a man to the Moon, why can't we do something about (*fill in the blank*)?" (see Box 3.3). Although many dismissed the Space Race as just a proxy for warfare, during the Cold War this was to be lauded: better the death of a few brave astronauts and cosmonauts than hundreds of millions by alternative means, a war postponed for a dozen years and ultimately made obsolete. Perhaps the billions of rubles and years of work that went into the secret Soviet Moon program diverted efforts of the same personnel to make better ICBMs. In the United States, extra funds from Apollo funneled into aerospace probably strengthened the companies that armed the Cold War and accelerated the miniaturization of computers and other electronics. Fortunately, many new post-*Sputnik* American aerospace engineers were entering the workforce.

In private many Soviet space leaders recall bitterly how close their program came to reaching the Moon and how it should not have been cancelled even as the Americans succeeded.[36] This is not often exorcised in public, but rather airbrushed from history. Consequentially, the Russian populace disowns the aspiration, consistent with the Politburo disavowal in 1969. One unfortunate result is that highly valuable scientific contributions of Soviet lunar exploration are commonly discounted. Despite Apollo returning 1,000 times the sample mass of Luna probes, Soviet missions explored unique areas, made important discoveries that the Americans missed, and developed techniques mastered by no others. They should not be ignored, but we encounter examples where this unfortunately transpires (Chapter 8). Partially this results from Cold War antagonism and American pride over their lunar program's success, but the Soviets also brought it on themselves. It should be reversed.

Although Apollo succeeded, its own ending was abrupt and ultimately damaging. How this happened, what led to it, and what differs now in planning future exploration is crucial to what we must consider next.

Notes

1. Aldrin legally changed his first name to "Buzz" in 1988.
2. *Apollo by the Numbers: A Statistical Reference* by Richard W. Orloff, 2001, NASA pub. 2000–4029. They thought they had 20 seconds left.
3. In the event Armstrong and Aldrin were stranded on the Moon, White House speech writer William Safire had the tragically macabre duty of composing for President Nixon a eulogy for the doomed astronauts. The world would have heard the following in the case of that alternate reality:

   "In Event of Moon Disaster, Memo to H.R. Haldeman from William Safire, July 18, 1969: 'Fate has ordained that the men who went to the Moon to explore in peace will stay on the Moon to rest in peace. These brave men, Neil Armstrong and Edwin Aldrin, know that there is no hope for their recovery. But they also know that there is hope for mankind in their sacrifice. These two men are laying down their lives in mankind's most noble goal: the search for truth and understanding. They will be mourned by their families and friends; they will be mourned by the nation; they will be mourned by the people of the world; they will be mourned by a Mother Earth that dared send two of her sons into the unknown. In their exploration, they stirred the people of the world to feel as one; in their sacrifice, they bind more tightly the brotherhood of man. In ancient days, men looked at the stars and saw their heroes in the constellations. In modern times, we do much the same, but our heroes are epic men of flesh and blood. Others will follow, and surely find their way home. Man's search will not be denied. But these men were the first, and they will remain the foremost in our hearts. For every human being who looks up at the Moon in the nights to come will know that there is some corner of another world that is forever mankind.' Prior to the President's Statement: The President should telephone each of the widows-to-be. After the President's statement, at the point when NASA ends communications with the men: A clergyman should adopt the same procedure as a burial at sea, commending their souls to 'the deepest of the deep,' concluding with the Lord's Prayer" (U.S. National Archives and Records Administration).

4. "Neil Armstrong makes decisions slowly and well. As Borman gulps decisions, Armstrong savors them – rolling them around on his tongue like a fine wine and swallowing at the very last moment. (He had twenty seconds of fuel remaining when he landed on the moon.) Neil is a classy guy, and I can't offhand think of a better choice for the first man on the moon" (*Carrying the Fire: An Astronaut's Journeys* by Michael Collins, 1974, New York: Farrar, Straus & Giroux, 60).

    Regarding Armstrong's terseness, perhaps this impeded him in dealing with the press, e.g., his famously uninspiring press conference on August 12, 1969.

5. A useful video animation of this sequence, "Apollo 11" by Andrew Chaikin, July 19, 2009, *Google Earth*, http://www.youtube.com/watch?v=6R3j1NU2nQM. See also note 15 to this chapter.
6. "Armstrong Adds an 'a' to Historical Quotation," *New York Times*, July 31, 1969.
7. *International Atlas of Lunar Exploration* by Phillip J. Stooke, 2007, Cambridge: Cambridge University Press, 211.
8. Peter Shann Ford paraphrasing Armstrong, in an interview by Melissa Block for *All Things Considered*, October 2, 2006: "Programmer Finds Moon Landing's Missing 'A.'"
9. *First Man: The Life of Neil A. Armstrong* by James R. Hansen (2005, New York: Simon and Schuster): Armstrong said, "It doesn't sound like there was time for the word to be there. On the other hand, I didn't intentionally make an inane statement, and ... certainly the 'a' was intended, because that's the only way the statement makes any sense. So I would hope that history would grant me leeway for dropping the syllable and understand that it was certainly intended, even if it wasn't said ... although it might actually have been."
10. This occurs in literature if not common speech, e.g., in Arthur Koestler's 1941 *Darkness at Noon* (London: Macmillan), prisoner Rubastov admits to his inquisitors to putting commonly human over Soviet goals: "I plead guilty for having rated the question of guilt and innocence higher than that of utility and harmfulness. Finally, I plead guilty to having placed the idea of man above the idea of mankind."
11. When possible the *Apollo 11* astronauts often described their lunar environs – Aldrin more vociferously than Armstrong. The following dialog came 8 minutes after landing, a time of "housekeeping" and checklists. Armstrong first detailed his delay in landing.

    From *Apollo 11*'s mission transcript (Charles Duke is Capsule Communicator at the Manned Spacecraft Center in Houston – now NASA Johnson Space Center):

ARMSTRONG:	Roger. Stay for T2. We thank you.
DUKE:	Roger, sir.
ARMSTRONG:	Tranquility Base, Houston. We recommend you exit P12. Over.
ARMSTRONG:	Hey, Houston, that may have seemed like a very long final phase. The Auto targeting was taking us right into a football-field-sized crater, with a large number of big boulders and rocks for about one or two crater diameters around it, and it required us going in P66 and flying manually over the rock field to find a reasonably good area.
DUKE:	Roger. We copy. It was beautiful from here, Tranquility. Over.
ALDRIN:	We'll get to the details of what's around here (later), but it looks like a collection of just about every variety of shape, angularity, granularity, about every variety of rock you could find. The

	color is... Well, it varies pretty much depending on how you're looking relative to the zero-phase point. There doesn't appear to be too much of a general color at all.
	However, it looks as though some of the rocks and boulders, of which there are quite a few in the near area... It looks as though they're going to have some interesting colors to them. Over.
DUKE:	Roger. Copy. Sounds good to us, Tranquility. We'll let you press on through the simulated countdown, and we'll talk to you later. Over.
ARMSTRONG:	Roger.
ALDRIN:	Okay. This one-sixth g is just like the airplane.

(Aldrin refers to the KC-135 Vomit Comet airplane that could fly on a parabolic trajectory to provide weightlessness about 30 seconds at a time, or 1/6 g for a similar interval, several dozen parabolae per flight.

The zero-phase point arises in one's vision opposite the light source – the Sun – where lunar dust appears highly reflective and almost shadow free. See Chapter 6.)

12. Early NASA documents call the LM lander the Lunar Excursion Module (LEM). It was felt that in discussing Apollo with Congress, it would be best not to suggest that astronauts were using the taxpayers' money on excursions.

13. From *Lunar Sourcebook, A User's Guide to the Moon* (henceforth *LSB*) edited by Grant H. Heiken, David T. Vanuman & Bevan M. French, 1991, Cambridge: Cambridge University Press, 476: "In spite of this evidence, some scientists were still concerned about the bearing capacity of the lunar surface and recommended that the astronauts have snowshoes and that the LM be equipped with a long radio antenna in case it disappeared into the soft dust at touchdown (Cooper 1969, 1970). Just a few months before the flight of *Apollo 11*, it was seriously suggested that, when Astronaut Armstrong took his first steps on the lunar surface, the soil would jump onto his spacesuit because of the electrostatic attraction. The resulting coating of soil covering his suit was expected to be so thick that he would not be able to see and might not be able to move. Then, if he were able to stagger back into the LM, there was concern that the highly reduced soil would burst into flames when the cabin was re-pressurized with pure oxygen" (*Apollo on the Moon* by H. S. F. Cooper, Jr., 1969, Dial Press: New York, and *Moon Rocks*, 1970, Dial Press: New York).

14. Farouk El-Baz (Lunar Science Planning supervisor at ATT/Bellcomm, prime contractor for Apollo landing planning) on November 29, 2009, recalling Gold's lunar dust hypotheses (e.g., "Ranger Moon Pictures: Implications" by T. Gold, 1964, *Science*, 145, 1046), said of working with Tommy Gold: "It was not a problem, really. You could stand there discussing the Moon with a very definite idea of what to expect and then Tommy would make you look at it in a new way. Even though he was often wrong, you need people like that. I loved Tommy Gold in that way."

15. "I'm at the foot of the ladder.

The LEM footpads are only depressed in the surface about 1 or 2 inches, although the surface appears to be very, very fine-grained, as you get close to it, it's almost like a powder; down there, it's very fine ... I'm going to step off the LM now.

That's one small step for (a) man. One giant leap for mankind.

As the... the surface is fine and powdery. I can – I can pick it up loosely with my toe. It does adhere in fine layers like powdered charcoal to the sole and sides of my boots. I only go in a small fraction of an inch. Maybe an eighth of an inch, but I can see the footprints of my boots and the treads."

16. Regarding the first step, Armstrong put it best: "It was special and memorable but it was only instantaneous because there was work to do." ("An Audience with Neil Armstrong," interview by Alex Malley, August 24, 2011, *The Bottom Line*, Certified Practicing Accountants of Australia, http://thebottomline.cpaaustralia.com.au/?__utma=1.1378473257.1337973142.1337973142.1337973142.1. This was Armstrong's last official interview.)

17. The first U.S. attempt, the USAF/NASA *Pioneer 1*, reached 30% of the distance to the Moon, where it found the micrometeorite impact rate in deep space much smaller than anticipated. Eight Pioneer missions, on three different kinds of launchers, were attempted and largely failed. Dispute exists whether some first failed Soviet missions should really be counted as attempts.

18. Patrick Moore in *Mission to the Planets* (1990, New York: Norton: p. 26–27), describes the scientific reaction to the first Far Side images. Moore had extrapolated from Near Side features that perhaps the Far Side had fewer maria. Their poor radio reception and shadow-free lighting conditions made these images difficult to interpret. Researchers soon identified craters and mountains similar to Near Side highlands features, but as well as the

stark paucity of maria (and the anomalously bright crater Giordano Bruno). A few modest scientific papers resulted from *Luna 3* images; a fuller understanding of the mare distribution would await more data (see Chapter 7).

19. Much robotic lunar exploration history is summarized in *Lunar Exploration: Human Pioneers and Robotic Explorers* by P. Ulivi & D. M. Harland, 2004, Chichester UK: Praxis; *Surveyor: Lunar Exploration Program, The NASA Mission Reports* by R. Godwin, 2004, Burlington, Ontario: Apogee; *Destination Moon: A History of the Lunar Orbiter Program* by B. K. Byers, 1977 (NASA TM-X3487); *Lunar Impact: A History of Project Ranger* by R. G. Hall, 1977 (NASA SP-4210); *Soviet and Russian Lunar Exploration* by B. Harvey, 2007, Chichester UK: Praxis; *Unmanned Space Project Management: Surveyor and Lunar Orbiter* by E. H. Kloman, 1972 (NASA SP-4901). A good summary of the Soviet human program of lunar exploration is two books by A.A. Siddiqi: *Challenge to Apollo: The Soviet Union and the Space Race*, 2000 (NASA SP-2000–4408), and *The Soviet Space Race with Apollo*, 2003, Gainesville: University of Florida Press.
20. Moonwalks are EVA (extravehicular activity), i.e., leaving the spacecraft during a mission.
21. *Terranes* refers to distinct regions of geological characteristics, whereas *terrain* is the up-and-down variation of landscape's topography.
22. See Chapter 6.
23. See *To a Rocky Moon: A Geologist's View of Lunar Exploration* by Don E. Wilhelms, 1993, Tucson: University of Arizona Press; *The U.S. Geological Survey, Branch of Astrogeology – A Chronology of Activities from Conception through the End of Project Apollo, 1960–1973* by Gerald R. Schaber, 2005, USGS Open File Report 2005–1190; "Training Apollo Astronauts in Lunar Orbital Observations and Photography" by Farouk El-Baz, in *Analogs for Planetary Exploration*, eds. W. B. Garry & J. E. Bleacher, 2011, Geological Society of America Special Paper 483, 49; "Geologic Field Training of the *Apollo* Astronauts and Implications for Future Manned Exploration" by Gary E. Lofgren, Friedrich Horz & Dean Eppler (ibid., 33).
24. Consequently Lovell, Swigert, and Haise hold the record for humans traveling farthest from Earth.
25. Directions are reversed for a person standing on the lunar surface compared to someone looking up at the Moon from Earth: east becomes west (although north is still north, of course).
26. "Sample 67215: An Anomalous Ferroan Anorthosite" by I. S. McCallum, J. M. Schwartz, F. Camara & M. Norman, 2002, *Lunar and Planetary Science Conference*, 33, 1830 and citations therein.
27. In the late 1950s both the United States and Soviet Union seriously considered exploding nuclear bombs on the Moon, as a demonstration and means of producing incandescent clouds to measure spectral composition of the lunar surface and even to expel samples toward Earth to be collected as meteorites (Project A119 in the United States). After accidentally contaminating areas in the South Pacific with plutonium from launch failures of nuclear-armed missiles, the United States abandoned sending such payloads to the Moon. The Partial Test Ban Treaty in 1963 ended further plans. In an unpublished 1958 memo, the young Carl Sagan first suggested this approach then later warned it might contaminate the Moon with radioactivity sufficient to corrupt lunar age measurements using abundance ratios of radioisotopes, later a key result of Apollo ("Possible Contribution of Lunar Nuclear Weapons Detonations to the Solution of Some Problems in Planetary Astronomy" by Carl Sagan, 1958; "Radiological Contamination of the Moon by Nuclear Weapons Detonations" by Carl Sagan, 1958). Fortunately, both countries abandoned the lunar nuclear explosion idea. As discussed in Chapter 6, some 77 lunar rocks have been found on Earth naturally as meteorites blasted from the Moon by large impacts. Unfortunately, the nuclear explosion idea was resurrected in late 1969 by lunar seismologist Gary V. Latham as a way of producing a moonquake to probe the lunar interior but was quickly squelched by public disapproval and concerns about lunar radioactive contamination. Previously, some had seriously suggested using lunar missile bases to launch retaliatory attacks on Earth. In retrospect this seems a lunar expansion of the arms race that would rapidly continue to escalate.
28. The heady hopes in late 1968 that the Soviet program might beat Apollo into lunar orbit were crushed by *Apollo 8*[30]. As late as early 1969 some Soviet program leaders expected unrealistically that an Apollo landing failure would set NASA back again, allowing Soviet landings to succeed. For example, in a May 8, 1969, meeting of Soviet space program leaders, this was considered seriously (Mark Wade, 2008, http://www.astronautix.com). Nonetheless the Kremlin disagreed, and on May 16, 1969, Soviet space leader Keldysh revealed the new "party line" at a press conference on *Venera 5*'s Venus landing. When asked, he revealed that Russia would only use robot probes in lunar endeavors and not risk lives. Several months earlier priority went to a robotic lunar sample return probe to launch before *Apollo 11*.

29. "It turned out there never had been a race to the Moon," said Walter Cronkite, July 20, 1974, CBS News TV Special *Fifth Anniversary – Apollo in Retrospect*; see "Russians Finally Admit They Lost Race to Moon" by John Noble Wilford, December 18, 1989, *New York Times*. Also "According to some observers in Washington and some American scientists, the Russians may never have had a high priority goal and timetable for a lunar landing," John Noble Wilford, "Soviet Apparently Drops Plan to Put Man on Moon," *New York Times*, October 26, 1969. One can find similar quotes from American (and British) press before and after 1969 supporting the one-sided Moon Race truism, but later revelations of Soviet lunar efforts show how wrong these were. *Soviet Life*'s August 1969 edition contains 65 pages on the Soviet space program, including four pages exclusively on the Moon, without mention of any human lunar plans.

30. A few days before *Apollo 8*'s launch on December 18, 1968, Kennedy Space Center's Security Office received a letter claiming a plot afoot to destroy its Saturn V rocket on launch. NASA received many such crank communications, but unknown to NASA this was different: from the KGB, the Soviet security agency.

    The letter, supposedly written by M. Miller and mailed from Atlanta with no return address, described a conversation overheard at Miami International Airport between a "tall husky fellow" and another, "a Texan from his accent," describing technical aspects of *Apollo 8*'s launch and its sabotage to explode like a "lovely Christmas cracker when it takes off." The letter's author expressed concern for Apollo and how sinister the discussion had been.

    Although the letter had no effect on the launch, years later its sender materialized. The FBI held a press conference on March 3, 1980, outing former KGB agent Rudolph Albert Herrmann who claims to have sent the letter, among other espionage activities ("Double Agent Revealed by FBI," *New York Times*, March 4, 1980). Simultaneously, Kennedy Space Center security released the letter (Associated Press, March 3, 1980), without actually identifying Herrmann as the sender. The FBI claimed the letter was transmitted from the Soviet Union by the KGB and only transcribed by Herrmann. Constructions in the letter, "Kennedy Space Centre" and "Christmas cracker" (an English tradition), bespeak an author tutored in British language schools as might befit KGB Moscow, not the southeastern United States.

    Note that news stories published in Moscow at this time (December 13, 1968) emphasize the dangerous nature of *Apollo 8*'s planned flight.

    From my research of these events, primarily via Freedom of Information Act requests and interviews with former journalists, NASA employees, and their relatives, records indicate Kennedy received a second KGB letter threatening the destruction of *Apollo 11*. Unfortunately, by the time of the FBI investigation (May 1979), the *Apollo 11* letter had been misfiled.

    Although the Apollo sabotage letters seem to be late acts of desperation, Herrmann's presence in the United States was anything but. Born Ludek Zemenek, a Czeck, he was recruited by the KGB and given the identity of an East German (who actually died in World War II Soviet Union). He immigrated to East Germany then West Germany in 1957 with an East German wife (also an intelligence agent) and child. After four years they moved to Canada, providing undercover support to the KGB in Ottawa and establishing his cover as a photographer/film producer. Six years later they moved to the New York area, where one of Herrmann's first tasks was the Apollo sabotage letters. Several years later Herrmann enlisted his own son, completely westernized, in serving the KGB (*Silent Warfare: Understanding The World of Intelligence*, 3rd ed., by Abram N. Shulsky & Gary J. Schmitt, 2002, Dulles, VA: Brassey's Inc.). In 1977 the FBI caught Herrmann, turning him to a double agent, and in 1980 the FBI established new identities for him and his family. Apollo suffered no actual sabotage. (In 2007, a NASA subcontractor's employee cut wires in a computer bound for the *International Space Station*, but the flaw was detected on the ground and would have posed no danger to astronauts in flight. This act appears connected with no international intrigues.)

31. Walter Cronkite, *Fifth Anniversary*. Nonetheless the flight crew had already been selected for L1's circumlunar flight in December 1968: Nikolai Rukavishnikov and Valeri Bykovsky. Later Oleg Makarov and Aleksei Leonov replaced them.

32. Anecdotally, consider the difficult career course of cosmonaut Alexei Leonov after the early Soviet human space program. On March 18, 1965, Leonov became the first human to perform an EVA. He was then selected to command the first Soviet flight around the Moon in 1968, which was canceled. He was then chosen as first cosmonaut to land on the Moon, but that mission was also canceled. In 1971 he was to command the *Soyuz 11* mission to the *Salyut 1* space station but was replaced by his backup when a fellow crew member was suspected of becoming ill. He was selected to command the next Soyuz, but that was canceled after the three *Soyuz 11*

cosmonauts died on reentry. No successful missions to Salyut space stations took place until 1975. Instead Leonov's second and last mission was the Apollo-Soyuz mission in July 1975. Leonov became the first cosmonaut to greet American astronauts in space. After that he retired.

33. "Soviet Star Wars: The Launch that Saved the World from Orbiting Laser Battle Stations" by Dwayne A. Day & Robert G. Kennedy III, January 2010, *Air and Space*, based on "The 'Star Wars' That Never Happened: The True Story of the Polyus (Skif-DM) Space-based Laser Battle Stations" (Parts 1–3) by Konstantin Lantratov, 2007, *Quest: History of Spaceflight Quarterly*, 14, #1, 5; #2, 5; #3, 11. Also "Unknown Polyus" by Yuri Kornilov, 1992, *Zemlya I Vselennaya*, 4, 18, reprinted in English in *Science and Technology, Central Eurasia: Space*, JPRS Reports USP-93–001, March 25, 1993.

34. There are two recent treatments of early Soviet space accomplishments. *First on the Moon* (*Pervye Na Lune – Первые на Луне*), directed by Aleksey Fedorchenko, won the Venice Film Festival Horizons Documentary Award in 2005. This "mockumentary" excavates the issue by portraying the purported landing, expunged from history, of a Soviet cosmonaut on the Moon in 1938. It pokes fun at Stalinist hero worship, with ironic reference to Nazis and "The Great War," gazing wistfully at the real early Soviet space program. It shows the trials and exploits of the hero and his comrades before the mission and follows his years-long odyssey home after returning from the Moon to land in Chile. In one flash-forward scene at the end, the twenty-first-century film crew carefully designs and constructs a model of the 1938 rocket, moves it onto a frozen lake for launch, then watches it explode on the pad (as we zoom in on the burning miniature crew capsule).

More brutal to the Soviet space legacy is *Paper Soldier* (*Bumaznyj soldat – Бумажный солдат*), garnering awards in 2008 portraying events around Yuri Gagarin's launch in 1961, then his death in a 1968 plane crash. The main characters are Danya and Nina, both doctors, he being a flight surgeon to the Vostok cosmonauts. Their marriage is ending, and he is haunted by mortal risks he conspires to impose on the cosmonauts. We see the character Valentin Bondarenko, in real life dying when an alcohol swab combusted in his oxygen-rich test capsule, igniting his space suit, two weeks before Gagarin's launch. The film's most common sentiment is "maybe this flight will improve things" – but in a milieu evoked by oppressive rain, slush, and mud. One sequence centers on the burning of Tyuratam's prison camp to make way for Baikonur Cosmodrome. Danya tries to be optimistic; recalling that Stalin died only eight years before, he predicts in eight more years the Soviets will have a Moon base, then lunar factories staffed by hundreds, missions to Mars, and grand space liners to the Moon, all by the year 2000. Then they will press on to Jupiter's moons. The men in *Paper Soldier* are depressingly unrealistic, and women suffering and stoic. Things end badly.

In comparison, Hollywood films like *Apollo 13*, *The Right Stuff*, or *Space Cowboys* are buoyant.

35. "This may sound grandiloquent but for me this is like placing a flag on the moon, this is really a massive scientific achievement," Sergei Balyasnikov, spokesman for Russia's Arctic and Antarctic Institute, told Reuters ("Russian Sub Plants Flag under North Pole," August 2, 2007).

36. See summary in *Soviet and Russian Lunar Exploration* by Brian Harvey, 2007, Chichester, UK: Springer Praxis, 293–294.

Chapter 3
Moon/Mars

> There can be no thought of finishing, for aiming at the stars, both literally and figuratively, is the work of generations, but no matter how much progress one makes there is always the thrill of just beginning.
> – Robert H. Goddard (letter to writer H.G. Wells, 1932)[1]

Long before the Cold War and the Space Race, people dreamed of traveling to places beyond Earth, like islands in a mythical sea, and they persist now that the Cold War is over. Dreams change, however, with the centuries' achievements and the day's technology and society. An odd mixture of myth, politics, and scientific knowledge sets the goals. More than five decades after the first robotic Moon missions, we have seen enough cycles of interest and dispassion about space exploration to see why we sometimes advance and at other times fail. To grasp the difference between the times in which Apollo was born versus later thrusts into space (and to understand why some succeed), momentarily let us return to the beginning.

Humans have dreamed of space flight for thousands of years. In second century AD Syria, Lucian of Samosata wrote[2] of a sailing ship blown hundreds of kilometers skyward to an inhabited, cultivated, and luminescent island: the Moon. After war between the kings of the Moon and of the Sun, the ship and sailors return home to the Mediterranean. Before rocketry's importance was realized, imaginative means were proposed to reach the Moon. In eleventh-century Persia, Firdausí wrote of King Kai-Kaus who "fetched four vigorous eagles and bound them firmly to the throne" to ride them to the Moon,[3] much like the hero in Lucian's other space fiction (*Icaro-Menippus*), whereas in 1630 the protagonist of Johannes Kepler's *Somnium* is transported by demons. In Francis Godwin's 1638 *The Man in the Moone*, the traveler exploits a flock of magic geese to find himself on the Moon with its human inhabitants (Christians, no less).[4] In 1657 Cyrano de Bergerac's hero reached the Moon propelled by rockets (fireworks, actually).[5] In 1870 Edward Hale described a "brick moon" to be built in Earth's orbit as a navigational aid for shipping. In 1865–1870, Jules Verne[6] blasted his lunar explorers from a giant cannon on Florida's west coast (not Canaveral due east), whereupon like Apollo they experienced weightlessness and the cold of space shadowed from sunlight, and worried over an orbit to round the Moon and return them to Earth (by attempted use of their rockets). They splashed down in the Pacific to be rescued by the U.S. Navy.

Although much of twentieth century rocketry and space exploration developed in Russia, Germany, the United States, and to a degree Britain, much early inspiration was French. Jules Verne (1828–1905) grew in influence by inspiring the first science fiction film, *A Trip to the Moon*, by George Méliès in 1902, contemporary to H.G. Wells' novel *The First Men in the Moon* (1901).[7] Russian rocketry theorist Konstantin Tsiolkovsky mocked Verne's cannon (which would squash its occupants flat via rapid acceleration), advocating rocket propulsion in his seminal 1903 treatise.[8]

Perhaps rocketry began around 400 BC with a toy made by Archytas of Tarentum, who suspended a bird-shaped steam rocket on wires and let it fly around the room. A different steam-rocket device was made by Hero of Alexandria in the first century AD. Whereas gunpowder rockets were used nine centuries ago in China (if not a millennium before),[9] work on liquid-fuel rocketry first took off in 1921 with Robert Goddard's research, following his *A Method of Reaching Extreme Altitudes* in 1919, including discussions of reaching the Moon.[10] His liquid-fuel rocket construction starting in 1923 (with the first launch success in 1926) attracted the attention of Hermann Oberth,[11] whose private publication in 1923 of *By Rocket into Planetary Space* largely initiated growth of rocketry research in Germany and encouraged Tsiolkovsky to popularize ideas of space exploration. This culminated in the popular moon-trip film *Cosmic Voyage*[12] made in part by Tsiolkovsky and released in 1936 shortly after his death. In the Soviet Union the first liquid-fuel rocket was launched in 1933. In the Great Purge of 1938, however, the head engineers in these efforts were imprisoned in the Gulag until 1945, postponing Soviet rocketry advancement.

Oberth also popularized space exploration (working in 1929 with Fritz Lang on his own popular film,[13] inspiration for many early German rocketeers). He mentored Berlin's League for Spaceship Travel (*VfR: Verein für Raumschiffahrt*, along with founder Johannes Winkler) and built his own liquid-fuel rocket in 1929. In the United States, Goddard's liquid-propellant rocketry ran out of fuel in 1941, after introducing gyroscopic internal guidance to aim rockets, reaching altitudes of several kilometers, whereas the VfR enjoyed increasingly successful launches from 1931 to 1934, having attracted support from the German army in 1929. Goddard died in 1945, seeing many of his ideas incorporated into the German V-2.

Germany's Aggregat rocket program used liquid propellants in potent weaponry for the first time, directed by Wernher von Braun, one of the original (and youngest) VfR members. Starting in 1933 and expanding in 1937, they soon launched an A-4 rocket (renamed V-2 by Nazi leaders: *Vergeltungswaffe*-2 for "vengeance weapon"), with a top altitude of more than 100 kilometers and 1-metric-tonne warhead capacity, becoming the first artificial object to reach outer space in 1942.[14]

The Aggregat-4 project ended in March 1945 after 3,225 V-2 launches, killing 7,000 victims downrange (and some 20,000 workers who perished in the A-4 labor camps). The Soviet Union captured most of the A-4 production sites, but the project's personnel and technology had dispersed in early 1945, accelerating as the Allies overran Germany. Although some VfR members went to the Soviet Union, most A-4 rocket scientists surrendered to the West in 1945. (VfR member Willy Ley escaped

Nazi anti-Semitism for the United States in 1935.) Hundreds of freight car loads of V-2 rockets and parts left for the United States, filling 16 Liberty cargo ships,[15] whereas in 1946 the Soviet Union built a V-2 factory near the Caspian Sea, teaming often newly released Russian engineers with captive Germans. By 1949 the Soviets were building improved, single-stage V-2s, separate from the displaced Germans' work, whereas the United States promptly began launching German-made V-2s, adding upper stages developed by Pasadena's Jet Propulsion Laboratory in 1947 to increase their range. In 1949 over White Sands, New Mexico, a two-stage V-2/WAC Corporal (Bumper) reached nearly 400 kilometers altitude and more than five times the speed of sound.

Goddard broke rocket altitude records during most of 1930s, never receiving U.S. government funding – only foundation grants. In 1947 the Soviets started work on intercontinental ballistic missiles (ICBM), but President Truman, advised by Vannevar Bush,[16] cancelled the U.S. ICBM effort that same year. Aeronautical and weaponry firms continued ICBM research without U.S. federal funding. By the early 1950s American leaders realized that launching a few ICBMs might cripple entire nations in less than an hour. Truman started the Redstone project for an intermediate-range (about 300 kilometers) liquid-fuel missile in 1950–1951.

The Cold War was underway. Soviets exploded its first atomic bomb in 1949. By 1950 Americans were shocked to find their pilots falling over Korea to Soviet-built Migs alarmingly competitive with U.S. fighter jets. Maintaining advantage in rocketry, to loft nuclear warheads across the planet in 45 minutes, became a national security imperative. Near World War II's close, Sergei Korolev emerged from prison to lead rocket design and construction that Krushchev would use a decade later to affordably counter American advantages in jet bombers and nuclear warheads.[17]

In this context people began detailing plans to explore space. Several developments encouraged these ambitions. Technological advances fascinated Americans. The atomic bomb had ended World War II early, for "only" $2 billion (one of the most costly government research program heretofore, but cheap versus $300 billion from the United States for the war). Airline travel was rapidly improving in quality and accessibility, as was television. In October 1947 Chuck Yeager achieved human supersonic flight (a premier accomplishment of the newly minted U.S. Air Force).[18] Oldsmobile adopted the White Sands V-2 as its new sedan's marketing logo. In December 1948 defense secretary James Forrestal introduced the necessity of an "earth satellite vehicle program" largely for surveillance of the Soviet military.

Outer space grew to a fad in the United States in the early 1950s. Several popular science fiction films and radio shows[19] promoted notions of alien life and space travel, building on fiction from the past few decades. Children playing space man were becoming commonplace. The 1946–1947 spate of UFO reports was not explained away to the satisfaction of many and was followed by further UFO mania. Despite the passing of notorious "canali" on Mars, having inspired imaginings of Edgar Rice Burroughs' Martians and the *War of the Worlds* Martian invasion scare of 1938, one could often hear statements in the early 1950s on the probability of simple life forms on Mars.[20]

In December 1949, the Gallup organization found that 70% of respondents did not think America would send people to the Moon in 50 years, whereas only 15% thought that it would.[21] Scientists and writers began to bring an air of reality to popular imaginings about outer space. In 1949 former VfR member Willy Ley wrote a guide for young people, *The Conquest of Space* (with Chesley Bonestell, New York: Viking), designed to realistically portray the Solar System, enhanced by Chesley Bonestell's marvelous paintings of spacecraft and planetary scenes.[22] Working on V-2s at White Sands, Wernher von Braun also articulated space flight's potential (the passion for which the Gestapo jailed him for several weeks in March 1944, feeling him insufficiently devoted to the V-2). In 1948 he summarized his plans in the book *Das Marsprojekt* (*The Mars Project*).[23] Although the U.S. government gave space exploration low priority,[24] von Braun articulated a scientific and engineering proposal for an entire space program, culminating in humans landing on Mars – the first time most practical issues were assessed. By 1957, another Gallup poll showed Americans typically thought humans would reach the Moon in about 20 years; this change was well underway by 1955.[25] This changed perception was largely accomplished by von Braun, Ley, and their collaborators, including Walt Disney, as much as by rocket flights, and this story is worth special attention.

In October 1951 at New York's Hayden Planetarium, a conference on space travel held by 250 aerospace experts was venue to the idea of merging these discussions with von Braun's into a popular publication. Ley, von Braun, as well as noted astronomer Fred Whipple and other experts and artists (particularly Bonestell) presented eight articles, 82 pages total, in *Collier's* magazine, over two years starting in March.[26] The series "Man Will Conquer Space Soon" unveiled a graphically detailed plan for how the United States could, indeed must, embark on a space program in the following decade, starting with simple satellites to carry instruments and animal subjects, followed by large, multistage rockets to loft humans into space, leading to a space station and crew in Earth's orbit, then expeditions to the Moon and finally to Mars. Each issue sold about three million copies and was likely read by millions of people more.

The von Braun and company's series laid out problems to be encountered in exploration of outer space and the Moon. Most are still major concerns today. Beyond the high velocities and great distances involved, the authors anticipated and presented with skill many concerns that must be considered in developing space flight and planetary exploration. We detail many of these throughout the book as well as in Appendix A. Von Braun, physicist Joseph Kaplan, and physician Heinz Haber suggested plausible solutions to most of these, and reader responses were largely positive.

Von Braun and company's plans are radically oversold but presented on a scale to capture people's interest and imagination. The lunar landing proposal's book version (*Conquest of the Moon*, 1953, New York: Viking) details the expedition, for example, 8 tables with 265 entries describing the spacecraft and mission parameters. The projects would be huge, far exceeding any ever seen even now in reality: 50 men in three spacecraft to the Moon for 60 days (see Figure 3.1) and 10 spacecraft to Mars with 70 men for 970 days, including 50 men on its surface for 400 days. (The major

The New Moon

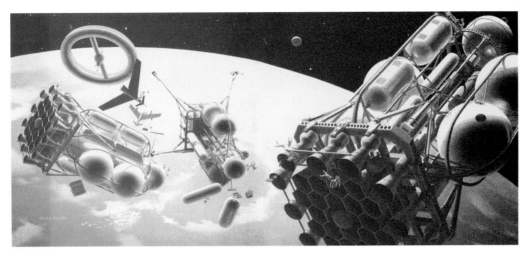

Figure 3.1. **The importance of dreams**. "Assembly of the Moon Ships 1,075 miles above the Hawaiian Islands" by Chesley Bonestell appeared in the October 18, 1952, *Collier's* magazine, presenting Wernher von Braun's vision of a realistic lunar expedition. Another scene on the magazine's cover surprised people: the craft landing on the Moon, with no fins or nose cones, no aerodynamics at all – a true spaceship. Realism is relative: this mission would land 50 people in three craft (foreground) on the Moon for six weeks, for an estimated $500 million. In reality Apollo landed 12 men for up to three days for $26 billion (several thousand times more costly per person-day). Apollo's LM is about as large as one of 30 engine bells below each of three landers. (Note the tethered astronauts for scale.) Also note other major elements of von Braun's space program plan: a space station (ring at upper left) and reusable space shuttles (two below right of space station). He also envisioned a fleet of 10 spacecraft carrying 70 people to Mars. (Used with permission of Bonestell LLC.)

error in the series is the erroneous density value for the Martian atmosphere – one-tenth of Earth's, or about 15 times the actual value – in reality, von Braun's airplane-like Mars landers would crash.) The space station would be 250 feet (76 meters) in diameter; the booster 265 feet tall, 65 feet across, 7,000 ton mass, with propellant totaling 6,100 tons and 36 tons payload. In comparison the current *International Space Station* (*ISS*) is 108 meters long; the Saturn V was 3,300 tons. The quoted price tag was highly underestimated: $4 billion total for boosters, orbital spaceships, and space station, and $500 million for the Moon expedition. The 1952 March 22 *Collier's* (p. 23) cites "$54 billion for rearmament since the Korean War began," so von Braun's vision was cheap. The Apollo program eventually cost $26 billion. Von Braun's price tags were flagrantly overoptimistic.

One could never fault von Braun's vision as lacking grandeur, and it and Bonestell's art became iconic, with many copies and derivatives. Another contemporary influence was Arthur C. Clarke and the British Interplanetary Society, although they produced more boosterism than rocket boosters. Von Braun and company's series attracted Walt Disney, who aired three programs educating the public to engineering, physics, and human aspects of space flight: *Man in Space* (March 8, 1955), *Man and the Moon* (December 28, 1955), and *Mars and Beyond*

(December 4, 1957, two months after *Sputnik 1*). Each was viewed by about one-quarter of the U.S. population. A feature film based partially on the series and produced by George Pal, *Conquest of Space* (1955), faired less well. (An analogous, successful film *Road to the Stars – Doroga k zvezdam* – was released by Soviet director Pavel Klushantsev in 1957.)

By 1960, outer space was oversold, not so much by von Braun as tagalong productions from Hollywood and beyond. We see in Chapter 13 that there are no other Earth-like environments in the Solar System. (The only world with surface atmospheric pressure even within a factor of ten of Earth's is Saturn's moon Titan at temperature −180°C and dominated by nitrogen and methane.) Some imagined we would move en masse to the Moon. Although the Moon is most easily adapted for human habitation (Chapter 13), mass entertainment offered humanoid aliens from Earth-like planets, more plausible if one accepted decades of UFO fads. In 1964 comics author Chester Gould moved the popular, grittily urban strip *Dick Tracy* to the Moon, but by 1969 his Moon Valley theme was extinguished and its lunar alien characters terminated in the 1970s. *Star Trek* rose in 1966 to die in 1969 (then reincarnated a dozen times beginning in the 1970s). Many Americans suffered a poor understanding of the relative difficulty of different goals in space flight. The distinction between flying to Earth's orbit versus the Moon, or Mars versus the stars, was often blurred despite the huge increment in effort separating each step. (Note the ambiguity even in Gagarin's flowery language [Box 2.1].) As knowledgeable an author as Willy Ley played on this, teasing readers in 1964 with prospects of seeing the first crew to depart for another star.[27] Von Braun maintained these distinctions in public discourse, despite his wildly optimistic *Collier's* series budgets (and despite the name of his 1960 feature length biography by Columbia Pictures, *I Aim at the Stars*). Even today these distinctions are blurred by technologists, who should know better.[28]

As films and magazines prepared the populace for space, the major action occurred out of public view in Soviet and U.S. military and scientific laboratories. In the early 1950s the Soviet Union found itself surrounded by American forward bases equipped with nuclear-armed, heavy bombers such as the B-47 (with a 6,000 kilometer range), and then the B-52 (more than 15,000 kilometers range). The Soviets had no such bases nor could they afford to match the United States' bomber investment.[29] Especially after Nikita Khrushchev's rise to first secretary of the Communist Party in 1953, long-range rockets with nuclear warheads were seen as affordable counterbalances to the American threat. That year Korolev's OKB-1 design bureau began work on the R-7 rocket, later modified to launch Sputniks in 1957 and becoming fully operational as an ICBM in 1959, with a 12,000-kilometer range and 5 tonnes payload. Because the Soviet Union could then respond to an attack ten times faster than with bombers, the United States was mortally concerned.

In 1953 the United States had already launched the Redstone rocket (which went into military service in 1958 and launched the first two Americans into space in 1961), but it had only a 300-kilometer range. Within a year the U.S. Atomic Energy Commission rapidly developed miniature nuclear warhead technology, and in early 1954 studies of ICBMs led by John von Neumann and the RAND Corporation showed that larger missiles could soon accommodate nuclear explosives. In response, in 1955–1956 several programs were begun to meet the Soviet ICBM threat: the Titan and Atlas long-range missile programs and the shorter-ranged Jupiter, Thor, and Polaris missiles (a different one for each U.S. military service). The first two took many years to reach fruition. First launched in 1957, the Atlas became operational in 1959. The Titan II entered as an ICBM in 1963, replacing the Atlas (which later became a satellite-launching workhorse). In the meantime von Braun and his team in Huntsville, Alabama, began to adapt the Redstone into a vehicle to deliver a small payload into Earth's orbit. Adding three more small stages, they produced the Juno 1 rocket, which launched the first U.S. satellite, *Explorer 1*, on January 31, 1958. Additionally, civilian scientists with the Navy produced the Vanguard rocket, launching the second U.S. satellite *Vanguard 1* a few weeks later, following the stunning failure of the Vanguard TV3 launch attempt on December 6, 1957. The United States had an impressive stable of rockets soon after *Sputnik 1* was launched.

By 1960 the United States had one class of operational ICBMs (with a larger one soon to arrive) and several kinds of medium/intermediate nuclear-armed rockets, amounting to hundreds of launch vehicles. It also had many rockets capable of placing satellites in orbit (variants of Vanguard, Juno, Atlas, and Thor/Delta). In contrast the Soviet Union never fielded more than about 10 operational copies of the R-7 ICBM, as a result of its difficult maintenance. Nonetheless modified R-7s (Sputnik, Vostok, and Soyuz models) formed the backbone of early Soviet space exploration. Liquid-fuel rockets found their greatest application in space, not war.

There had been no missile gap except in perception, despite the issue's importance during the 1960 U.S. Presidential election. Any gap was largely eliminated after Americans awoke from their technological stupor in the early 1950s. Nonetheless the Eisenhower administration had played too clever a game regarding the development of rocketry and satellites that resulted in one of history's monumental propaganda backfires. At the same time, Khrushchev played on the shock and uncertainty.[30]

Americans had been secure in their mastery of technology before Sputnik, particularly in aerospace. For individuals born after 1960 it might be hard to appreciate the visceral change regarding people's perceptions of air transportation around this time. True, commercial airlines had been operational for decades but utilized propeller planes at fairly low altitude, slow and rumbling in flight. In comparison jetliners, introduced in large numbers only around 1958,[31] were sleek and smoothly powerful. There was nothing so impressive as gazing up to a clear, blue sky and seeing a silver streak, many miles high, whisking a hundred people or more to their distant destination. Humans seemed halfway to outer space already. Americans would get there.

In a sense nobody should have been surprised when an artificial moon orbited Earth in 1957. From July 1, 1957, to December 31, 1958, the International Geophysical Year (IGY) probed Earth's atmosphere, exosphere, and interior in a scientific effort rich in international cooperation. In 1954 the council initiating the IGY adopted a resolution giving priority to launching an artificial satellite to map Earth, and the Soviet Academy of Science (and Radio Moscow) announced in response that it would. The United States (and National Academy of Sciences) revealed in 1955 its intention to conduct a satellite launch, which would become Project Vanguard. The Soviet announcement missed the American newspapers but was no secret. As the Sputnik launch approached in 1957, the Soviet press increasingly talked of satellites. In some sense the IGY's results would not be purely peaceful; to successfully target hardened sites in the Soviet Union or United States with ICBMs, better coordinates on Earth's surface, as the IGY promoted, would be needed, as well as better knowledge of how rockets exit and reenter the atmosphere.

Along with NASA's birth in 1958, Congress created post-Sputnik educational scholarships, loan programs, curriculum reforms (1958 National Defense Education Act), and the Advanced Research Projects Agency (1958 Defense Reorganization Act), later DARPA, which is largely responsible for the creation of the Internet (hence, the top-level Internet domain .arpa – now Address and Routing Parameter Area, but originally ARPANET). NASA's investment, and later DARPA's, greatly encouraged developing integrated circuit and multiprocessor technologies forming the basis of today's electronics-based economy.

Satellites also permitted unhindered surveillance of missiles and other military sites. This was much less a threat to the United States than the Soviet Union, which made great effort to hide sensitive sites.[32] Indeed, to confirm the location of Baikonur Cosmodrome, the Soviet space program's main launch site, CIA pilot Francis Gary Powers penetrated into Soviet airspace on May 1, 1960, the day his U-2 spy plane was shot down, ending American aerial surveillance of the Soviet Union (although previous U-2 missions had photographed the site as early as 1955). In contrast some American launch sites could be found using common road maps. Eisenhower appreciated that surveillance satellite overflight would provide needed intelligence to ultimately help stabilize the arms race by allowing each side to better count its adversary's missile stockpiles.[33] Unfortunately for the Soviet Union, this knowledge would come at their temporary disadvantage. By allowing a Soviet satellite launch first and by not objecting to a violation of U.S. airspace (almost unavoidable by a Soviet satellite), Eisenhower would establish precedent for unhindered satellite surveillance. He was not surprised when the Soviet Union orbited a satellite. What surprised him, to his political peril, was the American public's reaction.

The Powers U2 incident not only made necessary spy satellites but weakened Khrushchev with Kremlin hard-liners for being insufficiently aggressive toward the United States. On February 25, 1956, Khrushchev had denounced Stalin at the Communist Party congress, offending Mao and widening the Sino-Soviet rift. Under pressure, addressing Western ambassadors at a Polish embassy reception on November 18, 1956, in Moscow, Khrushchev uttered his famous "Whether you like it or not, history is on our side. We will bury you."[34] Over several years U.S.-Soviet relations became most dangerous: the Bay of Pigs invasion on April 17, 1961, the rise of the Berlin Wall starting August 13, 1961, with its standoff between Soviet and U.S. tanks in October 1961, then the Cuban Missile Crisis (or Caribbean Crisis) in October 1962. Khrushchev was found lacking; on October 14, 1964, the Presidium and Central Committee each finally voted to accept his "voluntary" retirement.

The orbit of *Sputnik 1* on October 4, 1957, shocked many Americans. They demanded and tortured themselves with how a relatively backward, troubled Soviet Union could beat the United States of America into outer space. The Soviets had announced the first successful ICBM launch only six weeks before.

The nuclear security implications were manifest, especially a month later when *Sputnik 2* orbited with a mass of one-half tonne and its canine passenger. U.S. Senate Majority Leader Lyndon B. Johnson recalled looking skyward after hearing *Sputnik 1*'s announcement on the radio at his Texas ranch:

> I felt uneasy and apprehensive. In the open West, you learn to live with the sky. It is a part of your life. But now, somehow, in some new way, the sky seemed almost alien. I also remember the profound shock of realizing that it might be possible for another nation to achieve technological superiority over this great country of ours.[35]

This galvanized Johnson and others to accelerate the U.S. space program. From this came NASA, the Advanced Research Projects Agency, and the Defense Education Act (hence, many "Sputnik babies") in 1958. Fairly or not, Sputnik left Republicans open to charges of a "missile gap" from Eisenhower's inattentive leadership. Kennedy and Johnson beat Nixon, Eisenhower's vice president, in 1960's election.

Americans' worry expanded beyond Sputnik and missile counts. Soviets won new space firsts and continued to trumpet them.[36] Popular magazines enticed youth with future Soviet space quests (Figure 1.1), and Soviet propaganda broadcasts promised comrades on the Moon by fiftieth anniversary celebrations of the Russian Revolution in 1967.[37] As demoralizing as Soviet successes were to Americans, they more profoundly affected opinions in other countries, to America's detriment.[38] Outside the United States, people thought the United States was losing its grip on future leadership.

Mass media added to perceptions of the Soviets as supermen. High-circulation *Look* magazine (January 5, 1960, 45) predicted Russians landing a man on the Moon by 1968, but the United States doing so by 1969. *Life* magazine (August 24, 1962, 25) predicted a Russian on the Moon by 1965, an American by 1968–1970. However, by 1960 U.S. public interest in space was growing rapidly, with popular articles on space increasing exponentially.[39] Hardly anything was so impressive as an issue of *National Geographic* or *Life* filled with Chesley Bonestell's space art or better yet actual pictures from space. *Life*'s content measures public interest well, like Internet Web site hit counts today. By 1962 astronauts' photos on the cover of *Life* were nearly as common as movie stars'.[40]

Into this challenging landscape strode the administration of John F. Kennedy, quickly finding itself humbled by events: the April 12, 1961, orbital flight of Soviet Air Force Senior Lieutenant Yuri Gagarin, first human in space, followed quickly by the Bay of Pigs fiasco in Cuba. As late as mid-1959, the United States planned to put a man in space on a suborbital flight in 1960 and in orbit by 1961. (See note 36.) It failed. Three weeks after Gagarin, it launched U.S. Navy Commander Alan Shepard into space, but only on a 15-minute suborbital flight. Khrushchev purportedly referred to Shepard's mission as a "flea hop." John Glenn would not orbit the Earth for another 10 months. United States prestige continued downward.

Kennedy faced several challenges with space exploration. If the United States' position over the Soviet Union was to be restored, NASA needed a prominent goal in which it could prevail – when you are losing the race, try moving the finish line. This required something beyond the capacity of current Soviet rockets, which also required the United States to develop its own heavy-lift boosters. The Saturn I rocket series was approaching reality, essentially a cluster of eight Redstones fused to a Jupiter rocket, with eight new, more powerful engines. Kennedy, Johnson, and NASA leadership considered building a space station or sending men around the Moon, but these might be done first by the Soviets. Kennedy chose a goal that NASA might achieve that would require the Soviets to design completely new rockets and spacecraft.[41] NASA had already tested the giant F-1 rocket engine (each as powerful as all eight engines on the Saturn I). This offered a large, powerful booster sufficient for spacecraft needed for Moon landings, soon to become the Saturn V.

Kennedy also needed to convince the American people of this quest; their support was questionable.[42] On May 25, 1961, with only 15 minutes logged by Americans in outer space, Kennedy delivered one of his best orations before a special joint session of Congress:

> This nation should commit itself to achieving the goal, before this decade is out, of landing a man on the Moon and returning him safely to the Earth. No single space project in this period will be more impressive to mankind, or more important for the long-range exploration of space; and none will be so difficult or expensive to accomplish. [BOX 3.1]

Box 3.1: Throwing Down the Gauntlet

Continuing Kennedy asked Congress for several space priorities, some seeming curious today:

We propose to accelerate the development of the appropriate lunar space craft. We propose to develop alternate liquid and solid fuel boosters, much larger than any now being developed, until certain which is superior. We propose additional funds for other engine development and for unmanned explorations – explorations which are particularly important for one purpose which this nation will never overlook: the survival of the man who first makes this daring flight. But in a very real sense, it will not be one man going to the Moon – if we make this judgment affirmatively, it will be an entire nation. For all of us must work to put him there.

Secondly, an additional 23 million dollars, together with 7 million dollars already available, will accelerate development of the Rover nuclear rocket. This gives promise of some day providing a means for even more exciting and ambitious exploration of space, perhaps beyond the moon, perhaps to the very end of the Solar System itself.

Third, an additional 50 million dollars will make the most of our present leadership, by accelerating the use of space satellites for world-wide communications.

Fourth, an additional 75 million dollars – of which 53 million dollars is for the Weather Bureau – will help give us at the earliest possible time a satellite system for world-wide weather observation.

Let it be clear – and this is a judgment which the Members of the Congress must finally make – let it be clear that I am asking the Congress and the country to accept a firm commitment to a new course of action, a course which will last for many years and carry very heavy costs: 531 million dollars in fiscal '62 – an estimated seven to nine billion dollars additional over the next five years. If we are to go only half way, or reduce our sights in the face of difficulty, in my judgment it would be better not to go at all.

Now this is a choice which this country must make, and I am confident that under the leadership of the Space Committees of the Congress, and the Appropriating Committees, that you will consider the matter carefully.

It is a most important decision that we make as a nation. But all of you have lived through the last four years and have seen the significance of space and the adventures in space, and no one can predict with certainty what the ultimate meaning will be of mastery of space.

He delivered a second, powerful speech on space at Rice University on September 12, 1962. [BOX 3.2] There he emphasized economic and scientific benefits to also sell the program to Americans.

> **Box 3.2: JFK at Rice**
>
> In short, our leadership in science and in industry, our hopes for peace and security, our obligations to ourselves as well as others, all require us to make this effort, to solve these mysteries, to solve them for the good of all men, and to become the world's leading space-faring nation.
>
> We set sail on this new sea because there is new knowledge to be gained, and new rights to be won, and they must be won and used for the progress of all people. For space science, like nuclear science and all technology, has no conscience of its own. Whether it will become a force for good or ill depends on man, and only if the United States occupies a position of pre-eminence can we help decide whether this new ocean will be a sea of peace or a new terrifying theater of war. I do not say that we should or will go unprotected against the hostile misuse of space any more than we go unprotected against the hostile use of land or sea, but I do say that space can be explored and mastered without feeding the fires of war, without repeating the mistakes that man has made in extending his writ around this globe of ours.
>
> There is no strife, no prejudice, no national conflict in outer space as yet. Its hazards are hostile to us all. Its conquest deserves the best of all mankind, and its opportunity for peaceful cooperation may never come again. But why, some say, the moon? Why choose this as our goal? And they may well ask why climb the highest mountain? Why, 35 years ago, fly the Atlantic? Why does Rice play Texas?
>
> We choose to go to the Moon. We choose to go to the Moon in this decade and do the other things, not because they are easy, but because they are hard, because that goal will serve to organize and measure the best of our energies and skills, because that challenge is one that we are willing to accept, one we are unwilling to postpone, and one which we intend to win, and the others, too.
>
> It is for these reasons that I regard the decision last year to shift our efforts in space from low to high gear as among the most important decisions that will be made during my incumbency in the office of the Presidency.
>
> In the last 24 hours we have seen facilities now being created for the greatest and most complex exploration in man's history. We have felt the ground shake and the air shattered by the testing of a Saturn C-1 booster rocket, many times as powerful as the Atlas that launched John Glenn, generating power equivalent to 10,000 automobiles with their accelerators on the floor. We have seen the site where five F-1 rocket engines, each one as powerful as all eight engines of the Saturn combined, will be clustered together to make the advanced Saturn missile, assembled in a new building to be built at Cape Canaveral as tall as a 48 story structure, as wide as a city block, and as long as two lengths of this field.
>
> Within these last 19 months at least 45 satellites have circled the Earth. Some 40 of them were "made in the United States of America" and they were far more sophisticated and supplied far more knowledge to the people of the world than those of the Soviet Union.
>
> The Mariner spacecraft now on its way to Venus is the most intricate instrument in the history of space science. The accuracy of that shot is comparable to firing a missile from Cape Canaveral and dropping it in this stadium between the 40-yard lines.
>
> Transit satellites are helping our ships at sea to steer a safer course. Tiros satellites have given us unprecedented warnings of hurricanes and storms, and will do the same for forest fires and icebergs.

On the strength of Kennedy's announcement, NASA started work on a rocket incorporating five giant F-1 engines and the development of another large engine, the J-2, with the ability to restart in a vacuum and burn liquid hydrogen, the most efficient fuel practical. A new plan for lunar landing missions meant that a single rocket would take the command craft and lander to lunar orbit, where they would separate then meet up again after the landing (Lunar orbit rendezvous: LOR). LOR was consistent with a crew of two on the lunar surface, starting with the new rocket, the C-5 or Saturn V – five times more powerful than any rocket in existence (and with an energy release approaching that of the Hiroshima atomic bomb) – to deliver 120 tonnes to low Earth orbit or 8 tonnes to lunar landing.[43]

Despite Apollo's importance to Kennedy, he had considered raising American prestige via other alternatives, even on Earth (large-scale irrigation projects in poorer countries, for instance), and continued to fret over Apollo's rising costs.[44] Nonetheless, despite the massive Kennedy federal tax cut of 1963–1964, he let NASA's budget double in 1963 and increase 63% in 1964. However, before his death, he weighed an alternative joint American-Soviet lunar program.[45] Khrushchev initially rejected the proposal because of Kremlin politics, but his son reveals the Soviet premier's later warming to a joint program.[46] However, no joint approach resulted.

President Johnson supported Apollo at least as strongly as Kennedy and had from the beginning. For the astronaut and the worker on the ground in Houston, Canaveral, or dozens of other locations feeding into Apollo, funding was little worry: all systems go. "One nice thing about Apollo was that no one ever told us we were running the price up too high," says Michael Collins, astronaut on *Apollo 11* and *Gemini 10*.[47] This extended well beyond Texan Johnson's enthusiasm for the massive space center in Houston. He, like Kennedy, fretted over NASA's large budget but continued his support, regularly singing NASA's praises.[48] Despite reservations behind closed doors about costs, Johnson shepherded Apollo through increasingly fraught politics and budgets plagued by the growing Vietnam War, during his tenure seeing astronauts orbit the Moon and final preparations for lunar landings.[49] Congress was harder to convince; NASA's budget declined first in fiscal 1966, with more cuts to follow.

As a funding priority, most Americans never supported Apollo (Figure 3.2). Various polls asked similar questions from 1961 to 1972 about whether the Moon program or space exploration was worth the expense, and the answer was nearly invariably negative.[50] Only one poll, in October 1965, gave a slightly positive response. By May 1972, with one Apollo lunar landing left, the situation was worse: 58% wanted to reduce or eliminate all space exploration, and only 12% wanted to increase funding, despite recent NASA budget cuts. The only positive aspect to the polls is that most people found watching space events on television a memorable and positive experience (once forgetting about the money).

For a brief moment Americans were united in shock and mourning at losing three astronauts to the Apollo/Saturn-204 fire (Apollo 1). Virgil "Gus" Grissom had been the second American in space, Edward H. White the first American to walk in space, and Roger B. Chafee a rookie. Their deaths were horrible. Americans remembered, briefly, what heroes these people are, that it is not just about money and politics. As months of investigation and congressional inquiries wore on, however, it became grimly apparent that mistakes had been made, and blame was assigned. Relations between NASA and Congress, especially a particular few of its members, were

"I suppose the Government knows all it wants to about us"

Figure 3.2. **The importance of dreams, too**. Kenneth Mahood's editorial cartoon (June 2, 1966, date of *Surveyor 1*'s lunar landing) portrays parents of a family living in poverty, gazing at the Moon, and commenting on their neglect by the government (*The Times* of London). Another editorial cartoon ("American Know-How" by Hugh Haynie, July 17, 1969, *Courier-Journal*, Louisville, Kentucky) portrays an impoverished child holding his toy rocket and gazing at the Moon through a broken window and torn curtains, one day after *Apollo 11*'s launch.

severely strained. Throughout 1967–1968, in a series of amazing engineering successes, Apollo regained momentum, but many Americans had turned away.

Compared to more recent and favorable public opinion (Chapter 14), the negative view of space exploration in the mid-1960s is surprising. Its decline near the end of Apollo is not. The 18 months prior to *Apollo 11* saw the heaviest American casualties in Vietnam, the murder of Robert Kennedy, Martin Luther King, Jr.'s assassination and ensuing riots, bloodshed at the Democratic Party Convention in Chicago, a giant oil spill closing Santa Barbara harbor, the polluted Cuyahoga River catching fire,[51] and violent protests on many U.S. college campuses. On the nine days *Apollo 11* was in flight, 190 U.S. soldiers died in Vietnam. To many Americans, the country seemed ready to fly apart; they could not concentrate on Apollo. [BOX 3.3] This began the era of "If we can send a man to the Moon, why can't we do something about (*fill in the blank*)?"

What did we think of the people heading to space and the Moon? American culture was changing. Astronauts' images ranged from clean-cut superhero to slightly naughty (and spoiled?) all-American, without the subtleties and conflicts found later in Tom Wolfe's book *The Right Stuff* (1979). *Life* magazine paid handsomely for exclusive rights to stories of astronauts and their families, partially

Box 3.3: When the Moon Is Not Enough

From President Nixon's inaugural address, January 20, 1969: "We have found ourselves rich in goods, but ragged in spirit; reaching with magnificent precision for the Moon, but falling into raucous discord on Earth. We are caught in war, wanting peace. We are torn by divisions, wanting unity."

The "If we can send a man to the Moon ..." argument was popularized by President Richard Nixon himself, on August 8, 1969, in his Radio and Television Address to the Nation on Domestic Programs: New Federalism: "Poverty is not only a state of income. It is also a state of mind, a state of health. Poverty must be conquered without sacrificing the will to work, for if we take the route of the permanent handout, the American character will itself be impoverished."

In my recent trip around the world, I visited countries in all stages of economic development; countries with different social systems, different economic systems, different political systems. In all of them, however, I found that one event had caught the imagination of the people and lifted their spirits almost beyond measure: the trip of *Apollo 11* to the moon and back. On that historic day, when the astronauts set foot on the moon, the spirit of Apollo truly swept through this world. It was a spirit of peace and brotherhood and adventure, a spirit that thrilled to the knowledge that man had dreamed the impossible, dared the impossible, and done the impossible.

Abolishing poverty, putting an end to dependency – like reaching the moon a generation ago – may seem to be impossible. But in the spirit of Apollo, we can lift our sights and marshal our best efforts. We can resolve to make this the year not that we reached the goal, but that we turned the corner – turned the corner from a dismal cycle of dependency toward a new birth of independence; from despair toward hope; from an ominously mounting impotence of government toward a new effectiveness of government, and toward a full opportunity for every American to share the bounty of this rich land.

By no means was Nixon's optimistic slant the only one; consider Gil Scott-Heron's "Whitey on the Moon" released in 1969 by the Last Poets and in 1972 by Scott-Heron himself:

> A rat done bit my sister Nell. (with Whitey on the moon)
> Her face and arms began to swell. (and Whitey's on the moon)
> I can't pay no doctor bill. (but Whitey's on the moon)
> Ten years from now I'll be payin' still. (while Whitey's on the moon)
> The man jus' upped my rent las' night. ('cause Whitey's on the moon)
> Which ends (after a fusillade of grievances):
> Y'know I jus' 'bout had my fill (of Whitey on the moon)
> I think I'll sen' these doctor bills, (to Whitey on the moon)

(*Now and Then* by Gil Scott-Heron, 2000, Edinburgh: Ganongate, 21)

A day before the launch of *Apollo 11*, the Reverend Ralph Abernathy came to Kennedy Space Center with several hundred people to protest poverty in America while NASA pursued space exploration. Meeting with NASA Administrator Thomas O. Paine he called on America to "feed the hungry, clothe the naked, tend the sick, and house the homeless." Paine told Abernathy: "If we could solve the problems of poverty by not pushing the button to launch men to the Moon tomorrow, then we would not push that button" ("Memo for Record" by Thomas O. Paine, July 17, 1969) (see Figure 3.2).

offsetting the scarcity of life insurance for their profession. Astronauts appeared intimately tied to The Establishment. Almost all Mercury, Gemini, and Apollo astronauts were military pilots, with attitudes increasingly unpopular as opposition to the Vietnam War grew. Some authors, such as Norman Mailer for *Life* in 1969 and in *Of a Fire on the Moon* (1971), capitalized on anti-military antipathy and disinterest toward Apollo.[52] Contemporary songs do the same, such as David Bowie's "Space Oddity" (1969) about Major Tom, or Barry McGuire's 1965 "Eve of Destruction" ("You may leave here for four days in space, but when you return, it's the same old place.") Film and music often have lasting, visceral effects on our feelings about human activities. Only recently have the music and film industries seen real space exploration (not science fiction) as acceptably cool again – *The Right Stuff* (1983), *Apollo 13* (1995), *From the Earth to the Moon* (1998), and so on – even though Apollo inspired as many feature films as the space shuttle.[53]

Box 3.4: Moon of Picture and Song

Bowie released "Space Oddity" five days before *Apollo 11*, and the BBC used it in its mission coverage. In Britain it became his first hit, introduced him to America, and grew into standard musical commentary on space flight:

> Ground control to major tom
> Take your protein pills and put your helmet on
> (Ten, nine, eight, seven, six, five,
> Four, three, two, one, liftoff)
> Ground control to major tom
> Commencing countdown, engines on
> Check ignition and may god's love be with you.

Before his space walk, ground control congratulates Major Tom on his newfound fame. Unlike real lunar missions, however, events go very wrong, and Tom is powerless to affect them:

> Though I'm past one hundred thousand miles
> I'm feeling very still
> And I think my spaceship knows which way to go
> Tell me wife I love her very much (she knows)
> Ground control to major tom
> Your circuits dead, there's something wrong
> Can you hear me, major tom?

And things do not improve (contrasting with events of Apollo 13 two years later):

> Can you hear me, major tom?
> Can you. . . .
> Here. . . Am I sitting in a tin can
> Far above the moon
> Planet earth is blue
> And there's nothing I can do.

Established in America, Bowie heaped cynicism on Tom with his 1980 "Ashes to Ashes" and 1995 "Hallo Spaceboy." Peter Schilling in Germany counters with "Major Tom (Coming Home)" (1983): "Far beneath the ship the world is mourning. They don't realize he's alive.

Box 3.4: (continued)

No one understands, but Major Tom sees now the life commands. This is my home. I'm coming home." Not a hit in the United States or England, it reached number 14 worldwide, especially in Europe. Real-life analog Sergei Krikalev, left 10 months on *Mir* by the Soviet Union's fall, is celebrated in "Casiopea" by Cuban songwriter Silvio Rodriguez. "Space Oddity" is echoed by Elton John/Bernie Taupin's "Rocket Man" (1972, recalling Ray Bradbury's story) but offset by Sun Ra and his afro-cosmic force from Saturn, an early artist of New Age space music, which slides into space rock, for example, Pink Floyd. Space themes extend into new classical, for instance, Meredith Monk's "Astronaut Anthem."

The film much of *2001, A Space Odyssey* (1968) is set, confidently, on the Moon, inspiring the British TV series set, *Space 1999* (1975–1977). Duncan Jones released the film *Moon* (2009), timed like "Space Oddity" to *Apollo 11*, but for its fortieth anniversary. Perhaps Jones (born Zowie Bowie) took a lesson from his father, leveraging off the lunar landing for his first hit. As in the song, space supplies emotional isolation: a lone astronaut mining Helium-3 on a bleak Moon many decades in the future, at the mercy of corporate interests. Another Hollywood film about the race for Helium-3 (Doug Liman, director) is on hold in 2014. The ABC-TV film *Plymouth* (1991) shows lunar settlers overcoming collectively the lunar environment and intimidation by big business. Ron Underwood's lunar sci-fi comedy *The Adventures of Pluto Nash* (2002), starring Eddie Murphy, was a box-office flop. Farcical *Iron Sky* (2012) by Finnish director Timo Vuorensola portrays Earth's invasion (with a Sarah Palin-like U.S. president) in 2018 by Nazis waiting on the Far Side since 1945, mining Helium-3. Chinese cinema contributes *Love in Space* (2010) with a subplot on a space station and the Moon; optimistic and syrupy, the lovers planting a red flag on a heart-shaped Moon. In 2009 ABC cancelled the near future drama *Defying Gravity* after eight episodes and minimal promotion, despite success by a similar, award-winning BBC mockumentary *Space Odyssey: Voyage to the Planets* (2004). American space drama is most popular set in deep fantasy or the far future: *Star Trek, Stargate, Firefly, Farscape, Battlestar Galactica*, and so forth, but new films released in 2013 – *Gravity, Europa Report*, and *Stranded* (on the Moon) – offer more realism. Recent TV movies about near future space travel (*Meteor, Impact, Earthstorm*, etc.) are grossly unrealistic about the Moon. Aliens and Apollo are further confused in people's minds by the films *Apollo 18* and *Transformers: Dark of the Moon. Capricorn One* (1977), about a Mars mission faked by NASA, has sets and props so like Apollo's that implications of faked Moon landings are inescapable. Director John Moore plans a remake of this film in 2014.

Moon hoax themes filter into R.E.M.'s "Man on the Moon" (1992): "If you believed they put a man on the moon ... If you believe there's nothing up my sleeve, then nothing is cool." Even straight pop used Apollo in dark metaphor e.g., Jonathan King's (1965) "Everyone's Gone to the Moon": "Streets full of people, all alone; Roads full of houses, never home; Church full of singing, out of tune; Everyone's gone to the moon." At least the Police compare being in love to "Walking on the Moon" (1979), as does Kanye West in 2009. As time passed, American music's mention of space or the Moon grew abstracted. Still, in 2014 Ziggy Marley sings "I Don't Wanna Live On Mars."

Popular but iconoclastic is Canadian songwriter Neil Young's apocalyptic vision: "Look at Mother Nature on the run in the 1970s. I was lying in a burned-out basement with the full moon in my eyes" – title song on his 1970 album *After the Gold Rush*: "Well, I dreamed I saw the silver space ships flying in the yellow haze of the sun; there were children crying and colors flying all around the chosen ones. All in a dream, all in a dream

> **Box 3.4: (continued)**
>
> the loading had begun. They were flying Mother Nature's silver seed to a new home in the sun." Young updates this to "look at Mother Nature on the run in the 21st century." Contrast this to Jackson Browne's "Before the Deluge" (1974) from *Late for the Sky*.
>
> Is there hope? The Canadian popular, Canadian kids' show *Lunar Jim* portrays a small, multiethnic community solving problems together creatively on a lunar surface rather like Arizona with craters ("Blue Moon L22," not quite our Moon – see Fig. 1.1). There are pets, robots, and aliens and lots of spacecraft, rovers, space suits, and other gadgets. Jim ends episodes gazing skyward through his telescope, and his looks are reminiscent of George Remi's cartoon hero Tintin, whose *Objectif Lune* and *On a Marche Sur La Lune* (*Destination Moon* and *We Walked on the Moon*) inspired young space enthusiasts in 1950s and 1960s Europe.
>
> Why do no songs extol astronauts' heroism, as about the days of sail? (There are some songs by or about cosmonauts, for example, "Do you know what kind of guy he was!" – "Знаете, каким он парнем был!" – about Gagarin. Not all of these are "state-sponsored," apparently.) Chanteys of whaling and exploration (sung in unison while pulling sail) are echoed in recent years, respecting bravery and sacrifice – Canadian Gordon Lightfoot's (1975) "The Wreck of the *Edmund Fitzgerald*" or, my favorite, "Northwest Passage" by another Canadian, Stan Rogers (1981):
>
>> Westward from the Davis Straight 'tis there 'twas said to lie
>> The sea route to the Orient for which so many died
>> Seeking gold and glory, leaving weathered broken bones
>> And a long forgotten lonely cairn of stones
>> (chorus) Ah, for just one time I would take the Northwest Passage
>> To find the hand of Franklin reaching for the Beaufort Sea
>> Tracing one warm line through a land so wide and savage
>> And make a Northwest Passage to the sea.
>
> Regarding how Americans perceived Apollo, they were proud but exhausted, disappointed, even cynical, best summarized by Paul Simon ("American Tune," 1973): "We come on the ship they call the Mayflower / We come on the ship that sailed the moon / We come in the age's most uncertain hour / and sing an American tune / But it's all right, it's all right / You can't be forever blessed / Still, tomorrow's going to be another working day / And I'm trying to get some rest."
>
> The late 1960s and 1970s were cynical times, and astronauts had no time for ballads (Figure 11.7). Mailer might be right about the split between astronauts and the rest of us. When common people with time for song head into space, this could change; and maybe Americans will grow less cynical.

Unfortunate timing forced Apollo's heroic effort to swim upstream against the late 1960s and 1970s counterculture. The "right stuff" was exactly the wrong stuff in an age of missile mutual assured destruction and air war over Hanoi.[54] The subsequent absence of lunar exploration for two decades produced more unfortunate knock-on effects.

When Nixon replaced Johnson as president, Apollo lost its most avidly influential backer. As Nixon's term began, six months before *Apollo 11*, it was unclear how many more lunar landings would be funded. On June 10, 1969, Nixon cancelled the

U.S. Air Force human space program (*Gemini-2/MOL*: the *Manned Orbiting Laboratory*, a surveillance station largely superceded by automated satellites). What of NASA? Enough Saturn V rockets were already built to loft missions through Apollo 20, as were most of the required spacecraft. Nixon seemed sincerely pleased with *Apollo 11*'s landing and his role in it, including his televised phone call from the White House to Armstrong and Aldrin on the Moon. (Onboard the *U.S.S. Hornet* recovery ship with the returning astronauts, he called the week of *Apollo 11*'s flight "the greatest since Creation.") Nixon attended *Apollo 12*'s launch on November 14, 1969, and was impressed, despite lightning striking the rocket post-launch temporarily disabling the command module.[BOX 3.5] (see Figure 3.3.)

Box 3.5: *Apollo 12*'s **One-Minute Brush with Catastrophe**

Apollo 12's planned lift-off time occurred during a thunderstorm, but conditions were judged acceptable for launch. President Richard Nixon waited in the VIP viewing stand 6 kilometers from the pad. Not only was a current president a very rare attendee, Nixon's approval of NASA funding was uncertain. Launch Operations Director Walter Kapryan has denied that Nixon's presence influenced the launch decision. The lightning caused temporary electrical failures that could easily have caused a mission abort. Never again would a launch proceed under such conditions. (See *Analysis of Apollo 12 Lightning Incident* by R. Godfrey, E.R. Mathews & James A. McDivitt, 1970, NASA Pub. MSC-0154, and *Moonport: A History of Apollo Launch Facilities and Operations* by Charles D. Benson & William B. Faherty, 1978, NASA Pub. SP-4204.)

The tension onboard *Apollo 12* after the second lightning hit is evident in communications between its crew (Conrad, Gordon, and Bean) and Capcom Carr 37 to 112 seconds into the flight.

000:00:31 CONRAD:	Looks good.
000:00:33 CONRAD:	Roll's complete.
000:00:33 BEAN:	This thing moves, doesn't it?
000:00:34 CARR:	Roger, Pete.
000:00:37 GORDON:	What the hell was that?
000:00:38 CONRAD:	Huh?
000:00:39 GORDON:	I lost a whole bunch of stuff; I don't know ...
000:00:40 CONRAD:	Turn off the buses.

(Public Affairs Office Announcement: 40 seconds.)

000:00:42 CARR:	Mark.
000:00:43 CARR:	One Bravo.
000:00:43 CONRAD:	Roger. We had a whole bunch of buses drop out.
000:00:44 CONRAD:	Roger. We [garble] on that. [Long pause.]
000:00:45 BEAN:	There's nothing – it's nothing ...
000:00:47 GORDON:	A circuit ...
000:00:48 CONRAD:	Where are we going?
000:00:50 GORDON:	I can't see; there's something wrong.
000:00:51 CONRAD:	AC Bus 1 light, all the fuel cells ...
000:00:56 CONRAD:	I just lost the platform.

(Public Affairs Office: Altitude a mile and a half now. Velocity 1,592 feet per second.)

> **Box 3.5: (continued)**
>
> | | |
> |---|---|
> | 000:01:00 BEAN: | [Garble] Got your GDC. |
> | 000:01:02 CONRAD: | Okay, we just lost the platform, gang. I don't know what happened here; we had everything in the world drop out. |
> | 000:01:08 CARR: | Roger. |
> | (Public Affairs Office: Plus one.) | |
> | 000:01:09 GORDON: | I can't – There's nothing I can tell is wrong, Pete. |
> | 000:01:12 CONRAD: | I got three fuel cell lights, an AC bus light, a fuel cell disconnect, AC bus overload 1 and 2, Main Bus A and B out. [Long pause.] |
> | 000:01:21 BEAN: | I got AC. |
> | 000:01:22 CONRAD: | We got AC? |
> | 000:01:23 BEAN: | Yes. |
> | 000:01:24 CONRAD: | Maybe it's just the indicator. What do you got on the main bus? |
> | 000:01:26 BEAN: | Main bus is – The volt indicated is 24 volts. |
> | 000:01:29 CONRAD: | Huh? |
> | 000:01:30 BEAN: | Twenty-four volts, which is low. |
> | 000:01:33 CONRAD: | We've got a short on it of some kind. But I can't believe the volt . . . |
> | 000:01:36 CARR: | Apollo 12, Houston. Try SCE to auxiliary. Over. |
> | 000:01:39 CONRAD: | Try FCE to Auxiliary. What the hell is that? |
> | 000:01:41 CONRAD: | NCE to auxiliary . . . |
> | 000:01:42 GORDON: | Fuel cell . . . |
> | 000:01:43 CARR: | SCE, SCE to auxiliary. [Long pause.] |
> | 000:01:45 CONRAD: | Try the buses. Get the buses back on the line. |
> | 000:01:48 BEAN: | It looks – Everything looks good. |
> | 000:01:50 CONRAD: | SCE to Aux. |
> | 000:01:52 GORDON: | The GDC is good. |
>
> "SCE to auxiliary" (Signal Conditioning Equipment to auxiliary setting) came from flight controller John Aaron, credited with saving *Apollo 12*.
>
> Bean from the postflight *Technical Debriefing*:
>
> We got all the lights. I didn't have an idea in the world what happened. My first thought was that we might have aborted, but I didn't feel any g's, so I didn't think that was what had happened. My second thought was that somehow the electrical connection between the Command Module and the Service Module had separated, because all three fuel cells had plopped off and everything else had gone. I immediately started working the problem from the low end of the pole. I looked at both AC buses and they looked okay, so that was a little confusing.
>
> Officially, the NASA report on the temporary failure on *Apollo 12* blamed the incident on the need to reevaluate launch weather criteria, but certainly the political pressure of having the President of the United States sitting in the reviewing stand at a time of uncertain NASA budgets was also a factor.

Nonetheless, there would be no new moon rockets. In January 1970, NASA cancelled a lunar landing so that its Saturn V could be used to orbit space station *Skylab*. Nixon cancelled two more, announced on September 2, 1970. With these and *Apollo 13*'s failure, lunar science resources were cut in half. (J missions *Apollo 15–17* and

Figure 3.3. **Nature is boss**. (a) President Nixon and family watch *Apollo 12*'s launch in the rain, November 14, 1969. (b) Lightning striking Pad 39A at Kennedy Space Center immediately on *Apollo 12*'s launch. (c) Scene inside KSC control room immediately after *Apollo 12*'s first lightning hit. For more than a minute, the Saturn V flew without control from *Apollo 12*'s Command Module. Fortunately, control was restored (with a little known procedure: "Signal Conditioning Equipment to Auxiliary"), 56 seconds before the crucial shutdown of the rocket's first stage. (NASA photos KSC-69H-1845, S69-60068, and KSC-69P-856.)

planned Apollos 18–20 were each assigned a much larger experiment package, a lunar rover, and three moonwalks. *Apollo 11* was primarily an engineering mission.) Canceling the later two missions saved only a few hundred million dollars from NASA's $3.4 billion annual budget. NASA funding continued to decline, and the savings were demanded for other projects.

Depicting Nixon (who lost the White House a decade earlier largely as a result of the missile debate) as "anti-science" based on this decision is unfair. Four months later he announced the massive war on cancer, which he hoped would cure cancer by 1976. Signing the National Cancer Act into law on December 23, 1971, he declared, "I hope in the years ahead we will look back on this action today as the most significant action taken during my Administration." This was Nixon's moon shot, costing a quarter-trillion dollars, more than Apollo. Regrettably, its success was less complete, even if well intentioned.

As spectacular as *Apollo 8* orbiting the Moon had been, the first human lunar landing was unique in uniting much of humanity in awe, changing forever how we think of ourselves within the Universe. Unfortunately, many Americans quickly lost interest after *Apollo 11*. For instance, *Life* magazine published three articles each on *Apollo 7* and *8*, one on *9*, five on *10*, eight on *Apollo 11*, three on *12*, two on *13*, one each on *14*, *15*, and *16*, and none on *Apollo 17*. The later missions' scientific haul was glorious, but few Americans heard about it. TV coverage on the last several missions was scarce, despite many hours of transmission from the Moon. In part NASA was at fault: on *Apollo 11* TV image quality was poor and nonexistent on *Apollo 12* after the lunar surface camera was accidentally destroyed. Pictures from the final missions were beautiful and in color, but I remember frustrating hours during the summer of 1971 searching for TV coverage of *Apollo 15*'s moonwalks amid the daytime programming. Coverage dealt with much silliness: Alan Shepard hit a golf ball on the Moon on *Apollo 14*; the equipment cart he pulled was compared to a golf club trolley cart; lunar rovers were repeatedly called "golf carts." The huge science return pouring from lunar missions was rarely discussed. There was a great climax in emotion during the near tragedy of *Apollo 13*, but afterward interest trailed off quickly.

Despite its prescience, von Braun's visionary sequence of milestones was neglected: first a simple satellite (sans crew), then a reusable space shuttle, followed by a large space station, then flights to the Moon and ultimately Mars. In reality the Moon became the immediate goal after a successful satellite and the first human space flights. In von Braun's vision, huge craft bound for the Moon and Mars were assembled in Earth's orbit outside a capable space station, requiring hundreds of shuttle flights. The station might also engage in military surveillance and war fighting, an idea he later dropped in favor of robotic military satellites.[55] He designed the space station around an inflatable ring rather than solid modules, so it could grow larger faster and more cheaply than the *International Space Station*. However, his lunar craft were more conventional and might have cost a hundred times his quoted estimate.

With humans soon on the Moon, a new goal was needed. Von Braun advocated humans on Mars. A more incremental approach would expand uses for existing Saturn and Apollo hardware, the Apollo Applications Program (AAP), advanced starting in January 1965 after a White House request. The AAP ran a gamut of missions, mostly extending and expanding lunar exploration as well as developing a huge space station (which became *Skylab*), a crew in geosynchronous Earth orbit, a large robotic Mars lander, and a Venus flyby with astronauts living inside a Saturn V's third stage. The AAP required minimal booster development – a problem for humans on Mars – needing huge new rockets or nuclear rocket engines. Congress refused these Apollo-derived projects, except for approving *Skylab* in 1969.

This left NASA's post-Apollo direction unsettled. James Webb's 1964 estimate for the White House of $50 billion to send a crew to Mars was inflated arbitrarily to $200 billion before reaching the press via Congress,[56] which put Mars missions in disrepute. In 1967 another uproar ensued when NASA solicited bids from aerospace companies on the Apollo Applications Mars robotic probe, interpreted as a step toward sending a crew at great expense. (A scaled-back version became the famed *Viking* landers.) Soon after *Apollo 11*, von Braun presented his ambition to use nuclear rockets (being developed by the Atomic Energy Commission) to send humans to Mars by 1983. On August 4, 1969, he revealed this to new NASA administrator Thomas Paine and the Space Task Group, chaired by Vice President Spiro Agnew. They were so impressed that they asked von Braun to repeat his presentation to Congress. The plan would require doubling NASA's budget and completing a new Mars spacecraft and the nuclear engine. Unfortunately on July 16 (*Apollo 11*'s launch day), Agnew had declared prematurely that the United States was sending humans to Mars by the year 2000, immediately becoming a target of White House consternation and derision in the press (and even popular music[57]). The association of von Braun with Agnew was not fortuitous. In September the White House extinguished Agnew's Mars notion.

Discussion of post-Apollo space shuttles had been underway for years, and in spring 1969 requirements of NASA (primarily to build a space station someday) and the Air Force and National Reconnaissance Office (to loft large surveillance satellites) were merged into a shuttle concept larger than NASA had planned.[58] This new concept solved some of NASA's problems: carrying in its cargo bay an upper stage to send probes to the planets, housing orbital science laboratories for two-week missions, or lofting, servicing, and returning satellites. It would increase the cost and complexity of the shuttle significantly. Within the White House, cuts to NASA's budget up to 50% were being considered. The administration found the shuttle to be the project they could initiate within NASA's spending limits and proposed $110 million for this purpose in their February 1970 budget. Many Democrats, even John Kennedy's own brother,[59] refused to continue the legacy of their own recent presidents, and many Republicans wanted to cut federal spending. The fight in Congress was grueling, but ultimately the $110 million remained, at the expense of Apollo landings and other programs. Apollo would end quickly, and humans to Mars would wait for some future century. The American political system lay dormant on these goals for decades. NASA would try building cheap, quick, reliable access to space, and only when that succeeded would more ambitious goals be reachable.

After 1969 Wernher von Braun slowly departed the scene. Always willing to sell space projects to politicians, he had been warned by James Webb to stop actively lobbying Congress in violation of NASA's charter.[60] Another issue may have restrained his public persona: von Braun's Third Reich role had long been subject to public comment, focusing on the bombardment of London and Antwerp.[61] After a flattering 1966 biographical sketch of von Braun in the *Paris Match*, several readers wrote the magazine's editors accusing von Braun of cruelties against slave laborers at the V-2 plant run by the SS. The *Match* asked von Braun to respond privately and attracted no public attention to the accusations.[62] In November 1968, however, a court in West Germany demanded von Braun testify about charges filed by East German former labor camp prisoners. He exited this testimony unscathed (in New Orleans, not Germany); however, news of his appearance surfaced in U.S. wire services and West German publications. [BOX 3.6] As time passed, attentive

The New Moon

> **Box 3.6: Von Braun and German Courts**
>
> Small stories about von Braun's testimony were carried by United Press International, *New York Times* ("Von Braun Evidence in Nazi Trial Sought," January 4, 1969) and Associated Press ("Von Braun, Rocket Scientist and Man Who Helped the Nazis Develop V Weapons, Gives Evidence in War Crimes Trial," February 7, 1969). Also Neufeld (2007), p. 428.
>
> In an interview for this book Arthur Konopka, the lawyer who assisted Dembling with von Braun in late 1968 to early 1969, stated that this case, originating in East German courts, was perceived by U.S. officials (including State Department advisors) as without merit as presented and likely an effort to harass von Braun and disrupt his contribution to Apollo, then in critical phases just before *Apollo 8*'s launch. There was strong desire on the East German side to bring von Braun there, but no wish by von Braun or his U.S. advisors to comply, instead to have him give evidence in the United States if necessary. Von Braun testified at the West German consulate in New Orleans on February 8, 1969, that he was largely disconnected from actual manufacture of V-2 parts by prison labor and that the one time he might have seen these laborers was during his inspection of the assembly lines after issues arose of parts failing during V-2 launches. His U.S. advisors were confident that there was no case against him and that he was not incriminating himself by testifying, but under no conditions should he travel to East Germany. No further actions ever involved von Braun, although Arthur Rudolph, who oversaw V-2 production at Mittelbau-Dora and later important elements of the Saturn V, was investigated in 1983 for war crimes. Although never convicted, Rudolph renounced his U.S. citizenship, an action some claim was coerced.
>
> Konopka related that, during preparations for testimony, von Braun recalled being so frustrated by failing V-2 parts during the A-4 program that he decided to get closer to the rockets when they failed shortly after launch. At southern Poland's Binza A-4 launch site there was a small shack at the target where V-2 rockets under test were aimed to strike. Von Braun recalled being so compelled to see how rockets failed on reentering the atmosphere that they would actually observe from this shack, confident of so small a chance of any rocket actually hitting it.
>
> Anecdotes aside, von Braun's testimony coincided with the KGB's letters threatening sabotage of *Apollo 8* and *11*. (See note 30 in Chapter 2.) Beyond von Braun's guilt or innocence, he still seems an odd source for information: not even his harshest critics claim he had routine operational familiarity regarding criminal use of slave labor. Because U.S. participants in von Braun's case believed it a ploy originating in East Germany to interfere with von Braun and Apollo, one might detect a pattern: were other "dirty tricks" planned by the Soviets at this time, when the Soviet Union saw its best chance vanishing to beat NASA at putting humans around the Moon?

listeners could note comments broadcast about his former Nazi affiliation. After von Braun's death, pointed public accusations surfaced regarding his mistreatment of prisoners, including formal investigation of one of his formerly Nazi colleagues who moved with him all the way to at NASA.

Dr. von Braun published 370 articles for the general public about rocketry and space science in 1951–1971;[63] however, after 1972 his output of popular works declined. His concern over NASA budget cuts is obvious in writings of the time. Von Braun was told to leave his directorship of Huntsville's Marshall Space Flight

Center and move to headquarters in Washington, DC; he arrived on March 1, 1970, to become NASA's deputy associate administrator for planning. Unfortunately, his tenure there was contentious and absent many staff and daily contacts that proved effective in Huntsville. He retired from NASA on May 26, 1972, and joined a commercial aerospace firm, still giving some interviews and public lectures. In a 1975 interview he spoke of NASA's loss of impetus and his desire to benefit people on Earth with space advances.[64] Sadly, he was diagnosed with cancer at about this time and died in 1977, at age 65.

For different reasons but in a similar mood, in May 1971 Congress cancelled the supersonic transport (SST) aircraft program over objections by Nixon and two years after the first flights of the civilian Soviet Tu-144 SST and Anglo-French Concorde. Two prototype Boeing 2700–300s were partially completed, and a related design became the supersonic B-1 bomber in 1974. In contrast the smaller Concorde operated commercially for three decades, turning a modest profit, and the Tu-144 was retired in 1979 (with elements entering the Tu-160 bomber). With cancelation of Apollo and the American SST, and exit from the Vietnam War, the U.S. aerospace industry entered several years of decline, although American civilian aircraft maintained their demand worldwide. In 1975 the U.S. economy recorded its last trade surplus for decades, suffering deeply from oil price shocks in 1973 and 1979 and their pursuant recessions.

Kennedy's timing for the Moon landing goal was unintentionally prescient in other ways. Had he set 1975 as the deadline, it would have seemed much easier (and the Soviets would still not have been ready, other things unchanged, with their N1 lunar rocket program ending unsuccessfully in 1976). However, public support was drying up, and his successor in office would have been guaranteed a share of the burden in completing the project. A less aggressive Moon program would have suffered in the 1970s and would have been much more likely to fail.[65]

The Saturn rocket series, although expensive, was shut down having performed almost flawlessly. The shuttle was born a much heavier and more complicated device than originally conceived. Its operations would cost about $6 billion annually, but only for four to five flights per year – many fewer than anticipated. Although many components turned out to be non-reusable, mission turnaround required a work force of more than 6,000 as well as several times this in support roles (typically roughly 1 million hours' maintenance per mission), despite its much reduced flight schedule. In comparison, hundreds of times fewer workers per craft can maintain even critically demanding military aircraft (Figure 3.4).[66] Several shuttle designs and sizes were considered, but aspects were set (large payload, launch into polar orbit, delta wing) by mostly military applications that were never realized. The final design, seen as the only politically viable option, proved too heavy to launch desired military payloads into polar orbit from Vandenberg Air Force Base in California. (The U.S. Air Force now launches its own X37-B mini-shuttle from Cape Canaveral.) A notable specification was that the payload bay be accessible with the Shuttle ready for launch, putting the orbiter low on the shuttle component stack – a decision that may have sealed the fates of the 14 shuttle astronauts who died in flight.

Figure 3.4. **Earthbound**. (a) A Space Shuttle (Space Transportation System – STS) orbiter being ground processed for future flight inside the Orbiter Processing Facility at Kennedy Space Center. (The orbiter's nose protrudes from the right of the scaffold structure and a wing tip to the left.) (b) F18 Super Hornet fighter airplane during in-hangar maintenance. (Photos courtesy of NASA, Ace Scaffolding Services.)

Apollo's cancelation was a political strategy as well as a budgetary decision: quit while you are ahead. *Apollo 13* in April 1970 had nearly turned triumph into tragedy. However, Apollo was mature and potent, killed in its prime as exciting results were arriving, with much of the program's hardware still unused. *Apollo 17* in 1972 was the last U.S. lunar mission for a generation, and *Luna 24* in 1976 was the last for the Soviets, and the last for the Russians for two generations. Scientists gained great insight from the data and samples returned from these and earlier missions, but crucial discoveries were missed (see Chapters 5–9). Politically, a generation passed before serious effort was undertaken to restart human exploration of the Moon and Mars. That push by President Bush – not George W. in 2004, but George Herbert Walker Bush (president number 41) in 1989 – was the Space Exploration Initiative (SEI). It suffered a short and problematic life.

On January 28, 1986, Space Shuttle *Challenger* disintegrated 73 seconds into its tenth flight (the twenty-fifth shuttle launch), killing its crew of seven; again, the nation mourned. Reminiscent of Apollo 1, mistakes were revealed and blame assigned, this time severely damaging NASA's reputation. Astronauts were not launched into orbit again for 32 months. It was a time of introspection for NASA.

After serving on the Rogers Commission (Presidential Commission on the Space Shuttle *Challenger* Accident), Sally Ride, first American woman in space, headed an effort to illuminate NASA's path forward. NASA's mission should be worthy of the investment and sacrifice being made, and many felt that the Space Shuttle program was not living up to this. The August 1987 Ride report[67] recommended four prime directions: satellites to map and study Earth, robotic planetary probes especially to Mars and beyond, a permanent lunar human outpost by 2010, and a human mission to Mars.

A few months later on January 5, 1988, President Reagan approved a national space policy directive for distribution within government agencies and departments, but not to the public. Its issuance was delayed while the Economic Policy Council (a White House advisory group) finished a study on commercial space policy initiatives, and the actual policy was never made public. Instead, on February 11 a summary was released.[68] The exact content of this directive is open to interpretation regarding manned space flight, but it concerned in large part the Strategic Defense Initiative (SDI) and its relation to civilian space. It proposed building a large space station and portrayed human exploration beyond Earth's orbit as a long-range goal. It recognized the value of robotic exploration and various areas of space research and called for maintenance and improvement of the space shuttle and alternatives. The directive stressed continued preeminence of America in space, in civilian and military realms, a priority confirmed by Vice President George H.W. Bush,[69] who would soon assume leadership in the White House.

With the Cold War ending, President G.H.W. Bush issued his Space Exploration Initiative on July 20, 1989, 20 years after *Apollo 11*'s lunar landing, apparently to separate out the civilian space program and stamp it with his own vision. With elements from both the Ride report and Reagan's directive, Bush called for the construction of a large space station (Freedom), permanent return of humans to the Moon, and a long-range program to put humans on Mars. Having no national security crisis to justify the initiative, he invoked Americans' historic tradition of opening new frontiers. NASA immediately detailed study of modes and costs for the three-pronged thrust and in November 1989 briefed the National Space Council (a White House advisory committee) that they would cost about $550 billion, if spread over two or three decades. (NASA's budget was $11 billion that year.) This report was largely supported by the National Academy of Sciences, but congressional reaction was strongly negative. The White House itself found the cost estimates disappointing and abandoned the plan eventually after finding no congressional or international takers.[70]

The SEI did not die quickly, however. Following its unfortunate debut, Vice President Dan Quayle in July 1990 requested a review of NASA and civilian space exploration. This came at a bad time for NASA: the month before had seen revelation of the optical flaw in the Hubble Space Telescope and grounding of the shuttle fleet as a result of hydrogen leaks that NASA could not isolate. As a result the Advisory Committee on the Future of the United States Space Program, headed by Norman Augustine, reported in December 1990 that NASA should concentrate on space science, putting human missions beyond Earth's orbit at the bottom of its five priorities. The SEI faded away.

NASA proposed less costly lunar bases before SEI expired. The First Lunar Outpost (FLO) study in 1992 followed by the Lunar Resource Utilization concept, or LUNOX,[71] would cut lunar exploration costs by incorporating Space Shuttle hardware, involving international partners (mainly Russian), and using in situ lunar resources, especially oxygen from lunar soil. (The Congressional Budget Office in 2004 estimated that LUNOX or FLO would have cost half as much or less than Apollo or G.W. Bush's Return to The Moon.) The designs were respected by many scientists and engineers but lost politically to low Earth orbit concepts with the existing Shuttle and Russian space station cooperation.

The Clinton administration began with threats of major NASA budget cuts. A possible large space station was thrown into budget competition against the Superconducting Super Collider (SSC) particle accelerator. The space station won, with Clinton convinced that it could foster better relations with the new Russian Federation via international scientific cooperation. The SSC died in October 1993, and the *International Space Station* (*ISS*) and space shuttle became NASA's human space mission for the next decade and beyond (after 10 shuttle missions to Russia's space station *Mir*). For a decade the U.S. human space program settled into a schedule of *ISS* construction and shuttle flights.

In 2001 George W. Bush nominated Sean O'Keefe as NASA administrator. O'Keefe, a budget analyst and business professor without aerospace experience, was tapped to control the budget, especially overruns in *ISS* construction. This agenda was soon overtaken by events.

On February 1, 2003, Space Shuttle *Columbia* was reported "overdue" for landing at Kennedy Space Center after a 16-day mission conducting scientific experiments away from the *ISS*. Space shuttles, which land unpowered, cannot proceed overdue. Moments before, NASA was alerted to severe failures during *Columbia*'s reentry and usual communications blackout. A breach in its heat shield filled *Columbia*'s left wing with superheated gases, causing the shuttle to lose control and disintegrate over Texas and Louisiana. Even for a nation engaged in one war in Afghanistan and about to start another in Iraq, this was a shocking and emotional tragedy. Another seven brave astronauts had sacrificed their lives.

Later that year the *Columbia* accident board concluded not only that factors leading to the accident were mishandled, but also that NASA should ponder why it puts astronauts in harm's way. [BOX 3.7] What is the mission of NASA's human space exploration program? This hit home at NASA and influenced a January 2004 announcement of the Vision for Space Exploration. The G.W. Bush administration asked why the nation engages in a space program involving such costs and sacrifices. For more than 30 years no American, indeed no human, had ventured more than a few hundred kilometers above Earth's surface. Criticism of the goals, cost, and apparently low science return from *ISS* were common,[72] and after more than two decades and 113 space missions, the shuttles were seen as reaching the end of their useful lives. Shuttle flights began again in July 2005, after accident investigations examined flaws in NASA management decisions.

On January 14, 2004, six days preceding his State of the Union address and starting his run for reelection, President George W. Bush announced his plan for NASA sending humans back to the Moon and to Mars: the Vision for Space Exploration (VSE). Primary milestones were humans on the Moon by year 2020 in preparation for a mission to Mars at an unspecified date. For this, new technology and infrastructure would be developed, and international and commercial participation promoted. The Space Shuttle would finish the *ISS* and then retire. The *ISS* could still be used by international partners (mainly Russia) and by NASA to study space flight's effects on humans, for example, on long Mars missions. The VSE

> **Box 3.7: After *Columbia*, a Look Ahead**
>
> *Columbia Accident Investigation Board Report*, Volume I, Part 3: "A Look Ahead," 2003, 209 (http://caib.nasa.gov/):
>
> In the course of that investigation, however, two realities affecting those recommendations have become evident to the Board. One is the lack, over the past three decades, of any national mandate providing NASA a compelling mission requiring human presence in space. President John Kennedy's 1961 charge to send Americans to the moon and return them safely to Earth 'before this decade is out' linked NASA's efforts to core Cold War national interests. Since the 1970s, NASA has not been charged with carrying out a similar high priority mission that would justify the expenditure of resources on a scale equivalent to those allocated for Project Apollo. The result is the agency has found it necessary to gain the support of diverse constituencies. NASA has had to participate in the give and take of the normal political process in order to obtain the resources needed to carry out its programs. NASA has usually failed to receive budgetary support consistent with its ambitions. The result, as noted throughout Part Two of the report, is an organization straining to do too much with too little.
>
> A second reality, following from the lack of a clearly defined long-term space mission, is the lack of sustained government commitment over the past decade to improving U.S. access to space by developing a second-generation space transportation system. Without a compelling reason to do so, successive Administrations and Congresses have not been willing to commit the billions of dollars required to develop such a vehicle. In addition, the space community has proposed to the government the development of vehicles such as the National Aerospace Plane and X-33, which required 'leapfrog' advances in technology; those advances have proven to be unachievable. As Apollo 11 Astronaut Buzz Aldrin, one of the members of the recent Commission on the Future of the United States Aerospace Industry, commented in the Commission's November 2002 report, 'Attempts at developing breakthrough space transportation systems have proved illusory.' The Board believes that the country should plan for future space transportation capabilities without making them dependent on technological breakthroughs.

would also authorize robotic missions to the Moon, which became the *Lunar Reconnaissance Orbiter* (although initial VSE literature seemingly forecast a dozen missions). Robotic probes would explore Mars, and robotic telescopes would explore the planets orbiting other stars. Most costly, new spacecraft and rockets to launch them would replace the Shuttle and take people to the Moon. The prime purpose of lunar exploration would be as a Martian precursor. The new spaceship and rocket would send crews to the *ISS* by 2014, the Moon perhaps by July 2019 (*Apollo 11*'s fiftieth anniversary), and Mars after 2030.

The VSE hit snags immediately. With the *Columbia* disaster and needs to cut costs pre-VSE, in January 2004 O'Keefe announced cancelation of the Shuttle mission to repair *Hubble*. He explained that only the *ISS* had the lifeboat capacity to sustain the crew of a damaged Shuttle. Several instruments on *Hubble* would stay broken, and the entire telescope might soon fail. O'Keefe learned how dear *Hubble* was to scientists and the public. His decision was a public relations debacle, but he persisted.

NASA asked, "Why return to the Moon?" and listed six themes: extending human presence enabling eventual settlement, conducting scientific exploration (of Earth, the Solar System, and the Universe), testing techniques and technologies for Mars (and elsewhere), encouraging international collaboration, expanding the economy, and engaging the public and work force. NASA instituted a "why the Moon?" process (the Lunar Exploration Themes and Objectives Development Process, http://www.nasa.gov/exploration/home/why_moon_process.html) in which they polled scientists, engineers, educators, and an international collection of 13 space agencies, who provided them with 179 worthwhile tasks for the Moon. (There were more; it is an edited list.) This somewhat begged the question: Why had President Bush forwarded his Vision? It was incongruous that one could not answer in a simple sentence or a few why we were embarking on this journey, and as years progressed this (lack of) message affected how the public perceived (or not) The Return to The Moon and the VSE. I met many educated Americans, even in the space sciences, who knew little about the program, but polls indicated that most Americans knew of The Return to The Moon and favored its goals.[73]

The VSE seemed sorely underfunded (Figure 3.5). NASA was charged to design and build a new fleet of spacecraft and rockets without significant increases in total budget. The Congressional Budget Office estimated that the new program would cost 57% more than NASA's projection.[74] Assurances were made that science would not be

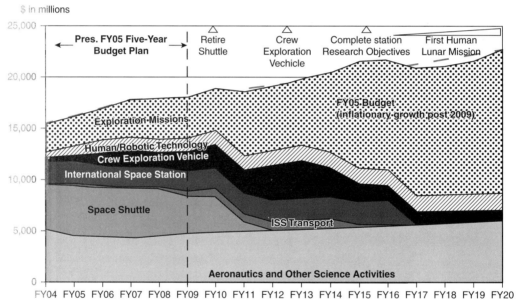

Figure 3.5. **Thin end of the wedge.** Budget forecast "sand chart" for NASA presented soon after President Bush's announcement of his Vision for Space Exploration. This budget called for building new rockets and spacecraft to explore the Moon and Mars, maintaining space science and aeronautics at near current levels, but ramping down expenditure on the Space Shuttle and *ISS*. Note major changes occurring only after President Bush left office (dashed vertical line), and promising a wholly new, large crew spacecraft designed, built, and tested for about $15 billion – a challenge. (NASA graphic.)

cut, and low cost estimates for the new systems were circulated. The VSE became an *enfant terrible* for some NASA stakeholders uninvolved in the new missions. O'Keefe initiated within NASA a new Exploration Systems Mission Directorate to institute most of the VSE and hired for this purpose, as associate administrator, Craig Steidle, retired U.S. Navy Admiral, who had overseen the Joint Strike Fighter, or JSF (at almost $400 billion, the costliest defense project in history). As with the JSF, Steidle proposed a "fly-off" between competing contractors for the crew capsule spacecraft and leaned toward VSE rockets, likely small and relatively unmodified from existing Atlases or Deltas. This foretold several rendezvous and dockings for every mission to assemble packages headed to the Moon. He also advocated *spiral development*: hardware would improve over time, incorporating new technologies and expanded mission capability.

O'Keefe hit trouble himself. Controversy arose over whether he instructed eminent NASA climate scientist James Hansen to modify his public statements about global warming's reality. O'Keefe was also found to have campaigned for a gubernatorial candidate in Alabama, home of NASA Marshall Space Flight Center. He defended his actions, but left NASA in February 2005; Michael D. Griffin took his place as Administrator. Until this time no VSE rockets or spacecraft designs had been selected. Reassigned and having differences with Griffin over basic mission architecture, Steidle resigned from NASA on June 24, 2005.

Griffin is a physicist, engineering professor, and pilot who coauthored "the book" on spacecraft.[75] He co-chaired the Planetary Society's July 2004 study calling for new rocket designs to reach the asteroids, Moon, and Mars (but landing a crew on none until after 2020).[76] At the April 18, 2005, news conference upon taking his NASA post, Griffin pointed to this study: "I view that at a few billion dollars a year, spaced out over a number of years, voyages to Mars are imminently doable, and I would urge you to download that report from the website because I don't have any better thinking to offer you than what I put into that report." The spacecraft and rocket in the Planetary Society study were essentially identical to the Crew Exploration Vehicle and Ares I soon adopted by NASA as Constellation program architecture by Autumn 2005.[77]

Constellation would consist of Ares I (a.k.a. Crew Launch Vehicle [CLV], or the "Stick") and Ares V (a.k.a., Cargo Launch Vehicle [CaLV]) (Figures 3.6 and 3.7). Ares I would carry a crew to low Earth orbit (LEO) on two stages: a solid rocket booster first stage, like a Space Shuttle SRB (or Aim for Mars first stage), except with five solid fuel segments, not four. A second stage would be liquid fueled, with modified Saturn V second/third stage main engines, the J-2X, burning liquid hydrogen (LH2) and oxygen (LOX). Ares I would lift 25.4 tonne to LEO. Ares V would be a behemoth, 109–116 meters tall, as tall as a Saturn V, with more LEO capacity (188 tonne versus the Saturn V's 119 tonne). Ares V would hold two improved Space Shuttle SRBs attached to shuttle-derived tanks feeding LH2 and LOX to five (or six) RS-68B engines from Delta rocket designs (or perhaps space shuttle main engines RS-25 SSMEs). An upper stage (the Earth departure stage or EDS) resembled a Saturn V third stage with a J-2X burning LH2 and LOX. Ares I was designed with the *Challenger* and *Columbia* accidents in mind: nothing would hit the crew capsule from above, and a Launch Abort System rocket could carry astronauts to safety in a launch failure. Ares V would be a cargo ship.

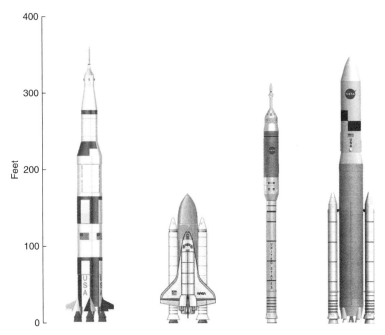

Figure 3.6. **Bigger boost**. A size comparison of the Saturn V, STS, Ares I, and Ares V launch systems. The Ares I first stage and side boosters on the STS and Ares V are solid-propellant, whereas all remaining stages are liquid-propellant, at least in part cryogenic, all using liquid oxygen. (Graphic courtesy of NASA.)

With Ares, two spaceships were planned: the Crew Exploration Vehicle (CEV) or Orion (to be tested by 2009) and Lunar Surface Access Module (LSAM) or Altair. These would be like expanded versions of Apollo spacecraft (Figures 3.8 and 3.9).[78] In addition to their larger capacity, the spacecrafts' extra power would make the Moon's whole surface accessible, including the poles, not just a narrow strip near the lunar equator. This was key in both science and practical goals of the Return to The Moon.

Tensions meeting Constellation goals pulled both from budgetary and technical sides. In half of the program's budget years, Congress funded NASA by continuing resolution or after sequestration and/or government shutdown, casting uncertainty and inflexibility into the program. During these times ramp up in project development was frozen, increasing costs and delays – effectively a budget cut. Bush administration budget requests were apparently several billion dollars short of Griffin's understanding when joining NASA.[79] Technically, shuttle RS-25 engines did not satisfy requirements to re-fire in flight, but those that could, the J-2S (recast as the J-2X), were less powerful. To compensate, a fifth Ares-I solid booster segment was added, deviating from the original Shuttle SRB. This raised fears that a longer SRB might produce thrust oscillations vibrating through the craft, worst at top where the crew rides. Unlike Apollo, an Orion capsule and its escape parachutes might encounter burning solid debris from an exploding SRB. This implied a heftier

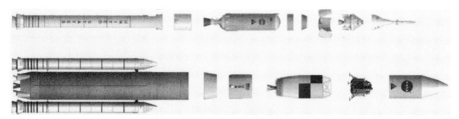

Figure 3.7. **Inside the beasts**. An exploded view of Ares I and V, to scale, showing separately each rocket's two stages and interstage adapters, spacecraft, and the Launch Escape System at the tip of Ares I. Ares I contains the Orion, and Ares V the Altair lunar lander, atop the Earth Departure Stage. (NASA graphic.)

escape system, adding mass and expense. By early 2009 Orion's first scheduled space flight slipped from 2014 to 2015 at the earliest, and Constellation's estimated cost had risen roughly $20 billion.

Orion was designed first, with flight and drop tests starting in 2007 with capsule and launch escape system mock-ups, dealing especially with simulated Earth-return capsule landings. The Launch Abort System was test fired with a capsule in May 2010. Altair and Ares V stayed highly theoretical; work began on the Ares I. A rough prototype, the Ares I-X, was test launched on October 28, 2009, to 46 kilometers altitude and splashdown in the Atlantic, not completely successfully. The four-segment SRB (not five) showed some thrust oscillations as feared.

To jump ahead in the story, Orion has now morphed into the Multi-Purpose Crew Vehicle (MPCV) as of May 24, 2011 – now scheduled for first space flight (crewless) on a Delta IV Heavy rocket in late 2014. To shed weight, the airbag for Orion's return to dry land is gone; it must splash down. The Launch Abort System is larger (Figure 3.7) but fortunately is dropped before orbit. The service module (SM) shrank commensurate with MPCV's purpose as an *ISS* lifeboat. The SM, powering the MPCV with solar panels rather than fuel cells (which make water as a by-product), will carry water for cooling and drinking and recycles water and urine.

Campaigning for president and in his first White House months in 2009, Barack Obama seemingly signaled intent to pursue the VSE. A finer reading of his policy planks indicated this was illusory. The final clause of his educational policy campaign document proposed that Constellation be delayed five years to liberate funds for pre-kindergarten education.[80] This emerged in November 2007, before most Constellation snags and budget slowdowns were apparent.[81] This would postpone the VSE to the point of irrelevance and/or add many billion dollars' cost. Counteracting this document, in January 2008 Obama endorsed Orion and Ares I.[82] In March 2008 during the campaign, after a question about his attitude toward space exploration,[83] Obama responded, "I grew up on Star Trek. I believe in the final frontier." [BOX 3.8] He continued that he disagreed with how the space program was run and thought funding should be trimmed until its mission was clear: "NASA has lost focus and is no longer associated with inspiration. I don't

Figure 3.8. **Spacecraft constellation.** (a) Artist's conception of the Altair spacecraft on the Moon. (b) The Orion spacecraft, enclosed by the Launch Escape System and atop the service module. (c) An actual test model of an Orion (now the Multi-Purpose Crew Vehicle). Compare these figures to Figures 2.6 to 2.8 showing the Apollo LM and CSM spacecraft. (Graphics and photograph courtesy of NASA.)

Figure 3.9. **More than a dream?** An artist's conception of the Space Launch System Block 1 configuration prepared for launch. This system is currently under study. (Graphic courtesy of NASA.)

Box 3.8: Obama and the Final Frontier

How evident was Barack Obama's interest in science and space exploration, or his "belief in the final frontier" (ignoring that *Star Trek* is fiction)? Although Obama values science education in the nation's economic future, less clear is that he "gets" space exploration's importance or was enthusiastic about science before reaching the White House. What is the record?

Before his political career (and election to the Illinois State Senate in 1996) Obama writes little about space exploration or natural science. The pertinent segment in his autobiography concerns his half-brother Mark, a Stanford physics student. In *Dreams from My Father: A Study of Race and Inheritance* (by Barack Obama, 1995, New York: Random House, 312, 314), Mark Okoth Obama Ndesandjo is half brother to Barack by the same father. Ndesandjo's interest in science and Western culture is a foil, if not a damper, to Barack's growing interest in his father's heritage. On page 325, 27-year-old Barack notices the Milky Way in the sky for the first time. Page 388 refers to his father's use of algebra and geometry. Although unspoken, there is discomfort with how mathematics and science distances his father's heritage from Barack; he ponders with incongruity his relatives as real individuals appreciating science.

Although neither of Obama's undergraduate transcripts from Occidental College (1979–1981) or Columbia University (1981–1983) have been released, Occidental's and Obama's own site describe his two years career there as "marked by intense courses in literature, arts, philosophy and social science." ("Occidental, Obama and the Road to

Box 3.8: (continued)

Public Service" by Jim Tranquada, September 25, 2008, http://my.barackobama.com/page/community/post/stephenfox/gG5cFK and http://www.oxy.edu/x7025.xml, quoting journalist Tom Plate speaking defensively of Obama's liberal arts training: "My friends in Asia and elsewhere in the world often offer high praise for the American system of higher education – but they mainly refer to our high-quality technical education of engineering, mathematics, physics, chemistry or medicine. They are mesmerized by the so-called 'hard sciences,' and not by the 'soft' social sciences and humanities that offer comparatively little ability of quantification. If you can't measure it or can't see it, then what is it?" – "Change, the American Way" by Tom Plate, *Khaleej Times*, August 31, 2008). Obama's last two undergraduate years focused on international relations within political science and showed little concern with science or technology. My discussions with fellow Columbia professors of the time about student Obama's natural sciences training betray nothing distinctive. His undergraduate thesis topic was the arms race. Copies of his thesis are unavailable; related writings by Obama at Columbia skirt scientific issues. ("Breaking the War Mentality" by Barack Obama, *Sundial*, March 10, 1983, contains one vaguely relevant passage: "This year, Mark Bigelow sees the checking of Pershing II and Cruise missile deployment as crucial. 'Because of their size and mobility, their deployment will make possible arms control verification more difficult, and will cut down warning time for the Soviets to less than ten minutes. That can be a destabilizing factor.' Additionally, he sees U.S. initiation of the Test Ban Treaty as a powerful first step towards a nuclear free world.")

An odd piece is "The Curvature of Constitutional Space: What Lawyers Can Learn from Modern Physics" by Lawrence H. Tribe, 1995, *Harvard Law Review*. Tribe acknowledges: "I am grateful to Rob Fisher, Michael Dorf, Kenneth Chesebro, Gene Sperling, and Barack Obama for their analytic and research assistance." It is pseudoscience, but how Obama contributed is unclear.

Clues to President Obama's interests today are his variously released reading lists. (For instance, "Obama's Reading List for Martha's Vineyard" by Mark Halperin, *Time*, July 23, 2010, from a July 19, 2010, briefing by White House Deputy Press Secretary Bill Burton.) Obama may be one of the better read U.S. presidents in recent years, but science and technology (or even science fiction) occupy little attention. To be fair, Obama is a quick study and can focus on science issues as demanded. Also, note his hosting of White House Astronomy Night in October 2009, with amateur astronomers and several astronauts in attendance. With science advisor Holdren at his side, he stressed science education and motivating young scientists but feigned ignorance (about what a pulsar is, or how to look through a telescope). More recently (February 2012) he met with science fair winners and speaks frequently of science education. There were earlier symbols: the final vehicle in Obama's January 21, 2009, Inaugural Parade was NASA's Lunar Electric Rover (LER) – the Obamas stayed to the end to watch. On February 9, 2009, the president and First Lady Michelle Obama read to second graders *The Moon Over Star* by Dianna Hutts Aston, about a young woman inspired to excel by *Apollo 11*'s Moon landing. These are but symbols and reading Washington tea leaves is an inexact science. Primarily Barack Obama is a lawyer and a politician and delegates science considerations to subordinates. On the basis of those decisions he must be judged. Unfortunately, his administration's lack of science policy consideration has hurt more than NASA.

Box 3.8: (continued)

How does Obama's administration set science-related policy, which staff is involved, and how is that accomplished? The policy victories of the Obama administration in 2009–2010 were sweeping, regardless of whether one agrees with them. (Despite my criticisms, I am uninterested in general anti-Obama diatribe.) Rewriting the nation's health care coverage and its consumer banking regulations was monumental. In science and technology policy, efforts are distracted and results unimpressive, even for vital issues of global warming and energy dependency. Obama offers support to nuclear power but terminated the Yucca Mountain radioactive waste repository with no viable alternative, abandoning federal responsibility under the Nuclear Waste Policy Act of 1982. Nuclear waste disposal took low priority, until spent fuel rod ponds in Fukushima, Japan, overheated and exploded starting in March 2011. The public is now aware of short-term as well as long-term risk in waste fuel. Japan, Germany, and Switzerland rejected nuclear power, and Italy abandoned resuming its program; America's commitment is less secure than ever.

White House technical policy failed miserably on offshore oil drilling. On March 30, 2010, the administration permitted offshore oil/gas exploration expansion off the Southeast U.S. Atlantic and Alaskan north coasts and parts of the Gulf. Three weeks later, the Deepwater Horizon oil rig exploded and collapsed, killing 11 men, spewing tens of thousands of barrels of oil daily into 1.5-mile deep sea off Louisiana for 86 days, roughly 20 times the Exxon *Valdez* in 1989. (Note: for *Valdez* up to three times the oft-quoted 260,000 barrels is proposed: "Size of Exxon Spill Remains Disputed" by Elizabeth Bluemink, *Anchorage Daily News*, June 10, 2010.) Regarding offshore drilling expansion, the administration then issued a ban rejected repeatedly in federal court as vague and overreaching despite Americans' furor over Gulf pollution. It was unimpressive administration mastery of scientific/technical policy.

To be fair, decades had passed since a major U.S. offshore oil blowout; government regulators and industry were complacent. Environmentalists focused on limiting drilling leases, not safety. With Obama's drilling expansion no blowout probability seems to have entered the calculation. With Deepwater Horizon's demise, President Obama expressed surprise that no technology of the oil companies or government, even the military, was available to suppress it. (I refer here to informed, highly placed government sources I am disinclined to identify.) Blame was deflected to the Interior Department's Mineral Management Service (MMS) by recycling old tales of sex scandals (but not in the drilling safety office) and a new report of MMS awarding drilling permits without considering endangered species. Early disaster response was largely improvised by then Energy Secretary Steven Chu, a Nobel laureate, but with no training in oil drilling and no responsibility for oil wells. Obama usually retains authority, even overriding his advisors, but not here. And even now, little has been done to prevent another Deepwater Horizon-like disaster.

Key science and technology issues have fallen in priority within the Obama administration. Comprehensive energy legislation including carbon emission cap-and-trade passed the House of Representatives (H.R. 2454) but was never pushed by the president in any major speech or public event, as it failed in the U.S. Senate. This was a prime project of John Holdren and Steven Chu. Apparently White House political advisors prevailed over science/environment ones, persuading Obama to concede to congressional pressure rather than his environmental campaign promises.

> **Box 3.8: (continued)**
>
> Obama's "reality-based" science and technology policy is often superior to his predecessor's: the Obama administration has reduced (but not eliminated) political influence over reporting federally conducted science results and stopped such silliness as discouraging the use of the terms *evolution* or even *Big Bang* in U.S. government public documents. Nonetheless, for energy and global warming issues, time is running out. His administration requires more interest and aptitude in dealing effectively with sci/tech issues, including the space program, especially in overcoming a balking opposition party. These cannot wait.

think kids are watching the space shuttle launches. It used to be a remarkable thing. It doesn't even pass for news any more." After winning nomination, the campaign told how Obama as a child was impressed by an Apollo flight returning to Hawaii. (It is unclear which mission that could possibly be; he repeated this story several times.)[84] In August 2008, in a lengthy policy paper Obama endorsed the specific goal of returning humans to the Moon by 2020 as well as shrinking the gap in U.S. human space flight post-Shuttle; one could believe he had reconsidered and now endorsed the VSE.

Entering the White House, Obama's attitude toward VSE was unclear. A heated argument between NASA Administrator Griffin and Obama transition team space advisor Lori Garver made the press in December 2008, reportedly regarding Constellation architecture, particularly concerning to Griffin.[85] He resigned in January 2009. (Charles Bolden was not sworn in as NASA administrator until six months later.) Obama's NASA budget request in February 2009 for fiscal year 2010 (even though NASA's FY 2009 budget was still unsettled) was ambiguous on the VSE. It stuck to ending the Shuttle program by 2010 and to putting Americans on the Moon by 2020, but commitment to Constellation in later years was unclear. Furthermore Constellation commitments for FY 2009 never reached original Exploration Systems Architecture System (ESAS) planned spending for the Vision and would sink even lower in FY 2010, a shortfall of over $2 billion per year.[86] How would we reach the Moon by 2020? Obama seemingly backed away from the VSE without announcing its demise. One could imagine his issue was with Constellation, not fundamental goals – or not.

On May 27, 2009, John Holdren's Office for Science and Technology Policy announced a Review of United States Human Space Flight Plans Committee (or Augustine commission), led by Norman Augustine of the 1990 commission that de-emphasized G.H.W. Bush's Space Exploration Initiative. Their report summary was delivered to Holdren on September 8, 2009, and reported to the public on October 22.[87] The committee had been instructed to assume Shuttles would end in 2010 and funds would not exceed the FY 2010 White House five-year plan. Its perspective was sobering:

The U.S. human spaceflight program appears to be on an unsustainable trajectory. It is perpetuating the perilous practice of pursuing goals that do not match allocated resources. Space operations are among the most demanding and unforgiving pursuits ever undertaken by humans. It really is rocket science. Space operations become all the more difficult when means do not match aspirations. Such is the case today.

The committee emphasized the presence of international and commercial capabilities in space, but the issue was U.S. leadership in human space, and what goal that implied. The report tendered five options (as well as variants), with three ramping up NASA's budget by 15% over five years then tracking inflation. Without more funding NASA astronauts would not leave Earth's orbit. The White House said they would review the analysis and options, then the president would decide.

Without a budget increase, all that was possible was dropping the *ISS* into the Pacific before Orion/Ares I became operational. Ares V would be unavailable for two decades. If one preserved the *ISS*, one could only reach it with commercial rockets, and a reduced version of Ares V might be available in 20 years. In either case no lunar return of NASA astronauts was possible before the 2030s. Increasing NASA's budget $3 billion over five years would allow humans on the Moon by about 2025. Finally the committee considered ramp funding leading to the flexible path (FP) of a scaled-down Ares V, a Delta-IV/Atlas-V derivative, or a new Shuttle-derived heavy rocket. The FP ranks targets according to the difficulty in accelerating to reach them: low Earth orbit, then a lunar flyby or orbital mission, then various stationary points in space in the Earth-Moon and Sun-Earth systems (their Lagrangian points: Chapter 13), then a few possible near-Earth asteroids, then a flight past Mars. Because landing and launching from the Moon requires 2.4 kilometer per second velocity changes, a lunar surface mission comes only after a Mars flyby but before entering Mars orbit, landing on one of its moons; a Venus flyby or orbit; or, hardest of all, landing on Mars. Landing on the Moon, requiring a new lander, might happen by 2030 (or not).

White House reaction was delayed. In December 2009 Congress passed the FY 2010 appropriation with a NASA budget larger than the administration request, noting that NASA was underfunded (a rare position for Capitol Hill of the day), and demanded Obama decide what he wanted for the nation's space policy. The next White House budget of January 27, 2010, contained essentially no funds for Constellation, including Ares I. Constellation was cancelled, with $9 billion already committed to abandoned hardware. An unnamed administration official was quoted as saying, "We certainly don't need to go back to the moon."[88] Along with official policy statements on the new budget, NASA released a letter from Buzz Aldrin (somewhat incongruously given his private citizen status): "The truth is, that we have already been to the Moon – some 40 years ago." NASA's new administrator Charlie Bolden, in a short statement on February 1, explained: "As we focused so much of our effort and funding on just getting to the Moon, we were neglecting investments in the key technologies that would be required to go beyond." Bolden left his deputies and Office of Science and Technology officials to answer questions. Bolden and OSTP's Holdren issued a second statement that stressed extending the

ISS until 2020, emphasizing commercial rocket development and new techniques such as orbiting fuel depots. Norman Augustine stated in the October 22nd report that the committee did not endorse specific options and was concerned primarily with underfunding compared to goals: "The President's proposed program seems to match means to ends, and should therefore be executable."

The new space policy not only abandons Constellation but also de-emphasizes United Space Alliance, a Boeing/Lockheed Martin joint venture, NASA's main booster supplier – for example, the Delta, Atlas, and until 2011's shutdown, Shuttle upkeep and processing. Instead new suppliers would be helped through the expensive process of rating rockets for human crews, particularly SpaceX (Space Exploration Technologies Corporation), with its new Falcon 1 having achieved several successful launches (1 tonne max to LEO), the Falcon 9 (10 tonnes to LEO), which successfully resupplied the *ISS*, and someday Falcon Heavy (53 tonnes LEO – INTELSAT recently bought Heavy launches for geosynchronous – GEO – satellites). For human lunar missions, Saturn V's LEO capacity was 119 tonne, so even Falcon-Heavy implies multiple launches and Earth rendezvous before heading to the Moon.[89] Orbital Sciences launched its Antares/Cygnus resupply rocket/ship in 2014.

Congress was highly skeptical of Obama's plan. House Space and Aeronautics Subcommittee Chair Rep. Gabrielle Giffords (D-AZ) summarized for many: "As I reviewed the President's budget request, I found a quite glaring omission ... My concern today is not numbers on a ledger, but rather the fate of the American dream to reach for the stars. Should we falter, should we slip, should we let our dream fade, what will we tell our children?" House Science and Technology Committee Chair Rep. Bart Gordon (D-TN) said of Obama: "My position would be to continue with the existing program until he can demonstrate there is a better alternative. The burden of proof is on the president." Reacting to Augustine's budget statement, Senate Subcommittee on Science and Space Chair (and former astronaut) Senator Bill Nelson (D-FL) offered: "It was a namby-pamby watered-down statement that was oblique at best." Reaction by congressional Republicans was even more severe. At a February 6, 2010 press conference, Administrator Bolden admitted:

> The reason Congress is angry with us, the reason that the NASA workforce was not better prepared, was because I didn't listen to people like Morrie [Goodman, NASA's Assistant Administrator for Public Affairs], who talked about how we should roll this thing out. I thought I knew better, to be quite honest ... Was it screwed up? Yes, it was.[90]

Reluctance by Obama's administration to rescue Constellation is understandable. The VSE was born abandoned. Beyond his January 14, 2004, speech announcing the Vision, George W. Bush never addressed at length the theme of the VSE and left to uncertain support from future presidents the major NASA budget manipulation needed to build several new spacecraft and launchers. During Bush's term, Constellation was insufficiently funded, even to levels promised in 2004. Bush never defended the VSE against the fester of continuing resolutions and was lackadaisical when asked about the Vision.[91] In sending humans to the Moon, Bush was no Jack Kennedy; he was no Lyndon Johnson, either. When JFK promised we would send a man to the Moon and return him safely to Earth, he could intend that this would be funded almost entirely during his presidency. JFK committed his political future to the promise. He gave two major speeches on the subject – one to Congress and one at Rice University targeted at inspiring the American people. He spent significant White House time on Apollo, despite his budgetary unease. As one of his last acts, Kennedy visited Cape Canaveral to review progress on Gemini and Apollo, six days before his murder in Dallas.

On April 15, 2010, President Obama visited Kennedy Space Center to address the changes he was making to NASA. He began his speech's substantial content with "I am 100 percent committed to the mission of NASA and its future" then called for Orion's use as an *ISS* lifeboat, commercial launchers to send astronauts to the *ISS*, $3 billion to study a new heavy-lift rocket, "breakthrough propulsion systems and other advanced technologies," astronauts to an asteroid after 2025, humans to orbit Mars by the mid-2030s, and a landing on Mars in the far future. Regarding the Moon, he concluded:

Now, I understand that some believe that we should attempt a return to the surface of the Moon first, as previously planned. But I just have to say pretty bluntly here: We've been there before. Buzz has been there. There's a lot more of space to explore, and a lot more to learn when we do. So I believe it's more important to ramp up our capabilities to reach – and operate at – a series of increasingly demanding targets, while advancing our technological capabilities with each step forward.

Obama called out Buzz Aldrin, in front of him, along with Bolden, Holdren, and elected politicians. At points throughout the speech – including near the closing – he cites the "larger purpose of space exploration in the 21st century." He never actually says what that means.

One can search public documents from NASA, the White House, and Office of Science and Technology Policy (OSTP) associated with 2010's new U.S. space policy for some larger purpose for space exploration and still be frustrated. Charlie Bolden was late to this process. Little scientific discussion occurs in the Augustine report. Much testimony on this policy has fallen to the OSTP, suggesting a search for clues from John Holden. [BOX 3.9] This yields little insight. Essentially the Obama administration promises to reevaluate the issue in 2015 – in the last half of Obama's second term – once we know more about new launch systems. Five years is a long time to await a purpose.

Congress insists on pursuing a heavy-lift booster, and its total study budget has risen to $6 billion. The bill mandating this specified a minimum LEO payload of 70 tons equals 62.5 tonnes, far beyond typical heavy-lift launchers of 20–30 tonne to LEO (Space Shuttle, Ariane 5, Delta IV Heavy, Atlas 5, Chinese Long March 5, Japanese HII-B, Russian Proton-M, or planned Angara 5). Someday Falcon Heavy might lift 53 tonnes, or Angara 7, 40 tonnes. (Ares-I was planned for 25.4 tonnes LEO, and Ares V 188 tonnes versus a Saturn V's 119 tonnes.) The MPCV fully fueled has 29 tonnes mass. The lunar Constellation package required

> **Box 3.9: Obama's Science Advice**
>
> Who sets White House space program policy? Statements from NASA head Charlie Bolden or Deputy Administrator Lori Garver bespeak little organizing principle or goals. Much falls to Dr. John Holdren, Office of Science and Technology Policy director and, hence, chief White House science advisor, since before Obama's NASA transformation. Holdren began his career as a plasma physicist and aerospace engineer but soon became associated with environmental and economic growth issues as younger coauthor of a few works with noted ecologist Paul R. Ehrlich, of *Population Bomb* fame (1968). Beyond a few works on the arms race, environmental and energy topics fill Dr. Holdren's bibliography, with scant reference to aerospace, tending to amplify Malthusian writings of Dr. Ehrlich and alarm over global warming. Paraphrasing *Ecoscience: Population, Resources, Environment* (by Paul R. Ehrlich, Anne H. Ehrlich & John P. Holdren,1977, San Francisco: W.H. Freeman, 821–823): human overpopulation cannot be overcome by expansion into the Universe. Exponential population growth would quickly overwhelm all available real estate, but Gerard K. O'Neill's vision of human habitation and manufacturing in space on the Moon and in the Earth-Moon Lagrange 5 point (Chapter 13) was then timely and their main topic of discussion.
>
> The authors were partially dismissive: "The strongest objections that will be raised against space colonization are that it cannot help humanity with the problems of the next crucial decades, that it will divert attention, funds, and expertise from needed projects on Earth, and that it is basically just one more technological circus like nuclear power or the SST." However, they consider O'Neill's argument: "There is, for instance, no sign that capital diverted from, say, a boondoggle like the B-1 bomber would necessarily be put to 'good' use. Equally, it does not follow that money for space colonies must be diverted from desirable programs." Judiciously they conclude: "Environmentalists often accuse politicians of taking too short-term a view of the human predicament. By prematurely rejecting the idea of space colonies, they could be making the same mistake." I inquired with Dr. Holdren several times regarding discussion of these and related issues, to no avail. Dr. Paul Ehrlich admitted to me, not speaking for Holdren, that if treating this today, they would likely be even more skeptical.
>
> I contacted Dr. Holdren on why he thought NASA should be following the current course but received no reply. His essential testimony includes:
>
> > A decision-support process engaging NASA and the White House was initiated to home in on and flesh out a set of options for revamping NASA's human spaceflight efforts – drawing on the Augustine committee's findings and aiming to maximize the level of activity and achievement attainable under realistic budgets – for the President's consideration in connection with the preparation of his FY2011 budget proposal. The result was the set of proposals for NASA's budgets and activities in FY2011–2015."
>
> Beyond low-Earth orbit, NASA is "pursuing a series of increasingly demanding human-exploration missions to include a mission to an asteroid by 2025 and an orbital Mars mission in the mid-2030s, demonstrating key capabilities for a later Mars landing while also achieving historical firsts in exploration and discovery; and ramping up robotic exploration of the Solar System, including missions to "scout" the human trips to follow.
>
> The more telling statement is as follows:
>
> > In addition to scrapping the $100 billion International Space Station before it had achieved more than a fraction of its scientific and technology-development potential, as well as suffering a six- or seven-

> **Box 3.9: (continued)**
>
> year gap in U.S. capability to lift its own astronauts into low Earth orbit, persisting with the pursuit of the increasingly costly Constellation program while nonetheless failing to meet its objectives would have the further liability of continuing to short-change NASA's other critical activities, including robotic missions and space telescopes, Earth observation, and aeronautics. Clearly it was time to press the reset button. (Statement on the Future of U.S. Human Space Flight, of Dr. John P. Holdren, Director, Office of Science and Technology Policy, to Committee on Commerce, Science, and Transportation, U.S. Senate, May 12, 2010).
>
> In other words, we did this because we were forced to. This is his primary explanation.
>
> Holdren can be defensive about U.S. space policy, as heard after *Curiosity*'s Mars landing on August 5, 2012: "If anybody has been harboring doubts about the status of U.S. leadership in space, well, there is a one-ton, automobile-sized piece of American ingenuity, and it's sitting on the surface of Mars right now and it should certainly put any such doubts to rest."
>
> Why do we have a space program? This plagues Obama's space policy: it lacks a mission statement. They complain of former programs' shortcomings, but a new direction is on hold. The primary response is check back in 2015. Who will decide? How? Why? Many wonder.

about 190 tonnes in LEO. When Congress passed this legislation in 2010, a NASA-associated rocket designer team had already proposed DIRECTv3 Jupiter-130 rocket, with 70 tonnes LEO. DIRECT was a Shuttle-derived series of designs, some as powerful as a Saturn V.

On September 14, 2011, NASA announced intent to design the Space Launch System (SLS) from Shuttle-derived modules into a rocket series with LEO capacity of 70–130 tonnes. These might serve most functions of Constellation, hopefully with fewer snags. Many call for the SLS study to be cancelled and funds spent on short-term missions within NASA. Because prospects for NASA's budget are significant cuts or, at best, level spending under Obama administration requests for several years, SLS's fate is tenuous. Preliminary SLS design began in mid-2012.

Congress grew impatient with Obama's plan and directed NASA to reassess its strategic direction; the National Research Council (NRC) delivered this report in late 2012, with searing criticism of U.S. space policy makers, describing NASA's 2011 Strategic Plan as so generic as to concern NASA no more than any federal research bureaucracy. It dismissed rationale for NASA's midterm goal, humans visiting an asteroid, as intellectually and strategically unsupported.[92] Scientifically motivated strategic planning is in shambles, risking NASA losing its space science edge. NASA suffers from scant direction except staying within budget, and a budget process so repressive of looking forward because of political uncertainty as to discourage any vision. Congress and the NRC criticized the retreat from the Moon, and Bolden defended it, poorly.[93] Detailed administration response is pending, while Congress introduces bills directing NASA to the Moon, for example, H.R. 1446.

NASA's new policy has effects on the ground. Visiting Kennedy Space Center coincident with *Atlantis*'s July 8, 2011, launch on the final Shuttle mission and listening to space workers discuss their future, I could hear mixtures of anger, bitterness, resignation, and acceptance. Most admit that the Shuttle must end, but reject that so little future vision replaces it and that no transition for them was established. Layoffs began a year before the last flight and peaked soon thereafter. Some 11,000 of the 15,000 Center employees (NASA and contractors) and several times more than this from surrounding areas need jobs elsewhere. The total loss may reach 100,000. Many skilled aerospace workers are seeking employment at the new aircraft plant in Charleston, South Carolina, where 4,000 employees will construct Boeing 787 Dreamliners and other aircraft. Several thousand fired Houston employees need jobs in the energy sector or elsewhere. Many skilled ex-NASA employees will be unavailable if or when new American spaceships are needed.[94]

Little was said about space policy during the 2012 U.S. presidential campaign. Obama elaborated little on previous policy, and Republican nominee Mitt Romney would appoint a commission to study the issue. Campaign statements on space policy clarified little.[95] Most prominent, and unfortunate, was Republican candidate Newt Gingrich's statement in a January 25, 2012, speech that he wanted a base on the Moon by 2020 and said later he dreamed of a lunar colony with enough population to apply for U.S. statehood. He was roundly derided, severely by rival Mitt Romney (even though the lunar base was largely in line with Bush VSE policy of a few years earlier). The reaction seems a result primarily of the U.S. press conflating the 2020 goal and the distant Moon colony dream.

The central element of Obama's human Solar System exploration plan is visiting a near-Earth asteroid after year 2025. Does this make sense? Such a human mission to an asteroid rarely appears prominently on other assessments of space exploration goals. A possible justification is understanding asteroids to better deal with them endangering Earth. There are problems to this idea: (1) plausible methods for mitigating near-Earth asteroid hazards are insensitive to their structure or composition (they involve a standoff satellite using only gravitation to deflect the asteroid)[96] and (2) asteroids that we can visit in the limited time interval accessible to astronauts are likely too small to be dangerous, so may reveal little about the hazardous ones. Other arguments for asteroid missions fail simply because the task is so difficult. Lockheed Martin Space Systems in 2009 completed the Plymouth Rock study of asteroid missions possible with Constellation architecture (with more capacity than SLS). Two Orion spacecraft joined together might suffice, with one serving as a supply ship – for a crew of only two. Although the study considers several alternatives, the only practical mission consists of finding an asteroid in an orbit very close to Earth's, chasing it for several months, visiting the asteroids for a few days, then spending several months returning to Earth. Only a handful of asteroids are suitable. Many years might pass before a target is available, and most are tiny: most under 12 meters across, the largest 60 meters. Half are actually smaller than the spacecraft. Many rotate sufficiently quickly to have effectively negative surface gravities; astronauts would be flung into space without anchoring devices. Perhaps astronomers will find more targets in time but probably smaller ones – hardly inspiring. Even many asteroid scientists consider such a mission unwise.

Many prefer investing perhaps one-third as much on robotic probes to numerous asteroids, many of them farther, for longer times.[97]

A quip heard when Obama endorsed the asteroid mission was that it might be easier to tow the asteroid back to Earth. The Keck Institute for Space Studies proposed the Asteroid Return Mission (ARM) for a robot mission to pull a small asteroid (7-meter diameter, 500 tonne) into lunar orbit for an estimated $2.6 billion.[98] Bolden testified that such a mission might fulfill Obama's asteroid goal, and $75 million was included for such a mission in the administration's FY 2014 NASA budget request.

Plymouth Rock highlights that a long mission's prime constraint is often the needs of humans, not just velocity changes of the rocket. A human mission to an asteroid is good preparation for a Mars trip, at least a flyby. Unfortunately such a mission might encounter a solar eruption unacceptably irradiating the crew, terminating the mission (if not the crew). Such an event happened in the eight months between *Apollo 16* and *17*.[99] A long-duration mission precursor for Mars or an asteroid could be accomplished at the *ISS* (shielded inside Earth's magnetosphere) or the Lagrange corotating stable points (Chapter 13) in the Earth-Moon system, 60,000 kilometers from the Moon (allowing a fast escape to a bunker below the lunar surface in the case of solar storms).

Following the 2012 election NASA detailed a lunar Flexible Path component, a Gateway Exploration Architecture, perhaps a deep space station with crew at Langrange point 2 (L2, – Chapter 13) behind the Moon's Far Side. This might send robotic probes to the Moon, stage missions to the asteroids, or inspect ARM's asteroidal haul. L2 is a jumping off point to many locations near Earth: only a few hundred meters per second velocity difference from solar orbit, other Langrange points in the Earth–Moon or Sun–Earth system, and asteroids with orbits nearest Earth's. The Sun–Earth's L2, for instance, is haven for space telescopes, for example, the James Webb Space Telescope, by about 2018. This L2 gateway would be about 0.5 second round-trip light travel time from anywhere on the Moon, so it could serve for remotely controlling lunar robots. An ion engine on the station might allow it to shift to other Lagrangian points propelled with several percent of its mass. Ideas include moving this gateway closer or farther from the Moon, depending on changing space program goals. NASA graphics consistent with the L2 gateway are shown in Figure 3.10. All such concepts are preliminary, however.

NASA's Flexible Path (FP) policy was generally supported by proponents of commercial space development. Realizations that FP would not put astronauts on Mars soon caused Mars enthusiasts to withdraw their previous FP support.[100] It is fair to say few people are certain of FP's goals or purpose except its search for a less expensive technological base. No real direction may be established until FP is revisited, planned for the penultimate year of President Obama's second term.

We leave U.S. space policy's status at early 2014; with current discord we can barely foretell the next curve in NASA's space exploration path, flexible or not. Politically acceptable options are shifting to the lunar environs: the Gateway puts astronauts six times closer to the Moon in distance and light-travel time and at least

The New Moon

Figure 3.10. **Over the Moon**. In late 2012 NASA revealed the Gateway Exploration Architecture concept of a deep space station at the Earth-Moon Langrangian 2 point fixed 60,000 km over the Far Side. From there the gateway could send robot craft to the Moon's surface (a), maybe dropping to lunar orbit if lunar missions take priority, or acting to service deep space probes or telescopes (b) such as the James Webb Space Telescope, with easy access to Sun-Earth's Lagrangian 2 point 1.5 million km beyond Earth. (Artwork credit: NASA and John Frassanito & Associates.)

20 times closer energetically. With an ARM, asteroidal exploration will occur over the Moon. The Gateway promises easier robotic lunar exploration, allowing remote-control robots with only one-half second response times, and easier access by robotic landers. For the Moon the clear issue separating the FP and VSE is the Obama team's reluctance to build a new lunar lander such as Altair, for roughly $10 billion. Looking down from the Moon from the Gateway, astronauts will see a storehouse of useful material and energy resources, including necessities

transportable from the Moon at several percent the energy required from Earth. If Obama and company permit, the Moon can become an integral and advantageous part of the FP. The possibilities require elaboration, the target of much of this book (especially Chapters 8, 12, and 13).

For half a lifetime many have waited in vain for NASA or some other agency to continue where Apollo stopped or to move on to farther shores. For this chapter's final paragraphs, let us examine postmortem why this has festered so long.

The post-Apollo story concerns mismatches between goals and means. Primarily, political support for the space program fluctuates faster than its project can possibly accommodate. A typical robotic spacecraft project takes 7–10 years from conception to data return, sometimes much longer, but rarely much less. The typical human project takes 10–30 years, much longer than 4–8 year terms of U.S. presidents[101] and certainly slower than changes of power in Congress, or business cycle swings. Canceling projects as a result of cost overruns can be healthy; killing them because of political or budgetary vagaries almost totally wastes money and careers. Most mission planners would be willing to accept major reductions in total budget if they could just plan on reliable funding 10 years into the future. The curse of Apollo is that, at least three times, goals were set to continue where *Apollo 17* ended but with insufficient will to create a program of the energy and scale of the 1960s. This would require a 20% rise in NASA's budget, but the nation seems unwilling. We often use baseline budgeting to predict and justify NASA funding, but a foundation of infrastructure and work force expenditure produces little science and exploration, yet keeps NASA centers open. Closing centers is politically painful, so the first dollars spent every year re-lays the bedrock. The last dollar is most effective at getting us into space. Small increases can have a large effect, as can minor cuts to the detriment. Of course, the current tendency to not plan budgets at all, to rely on continuing resolutions, or, worse, sequestration or government shutdown, defeats planning.

Since approximately 1988, Russia/Soviet Union has maintained a stable, relatively inexpensive space effort by incrementally improving a few rocket systems (Chapter 4) but with limited flexibility especially for beyond-LEO robot and human missions, which is a prime area of NASA accomplishment. In contrast the United States developed the Saturn series, abandoned it, built the Shuttle, abandoned it, then started and abandoned a replacement. One can imagine space station concepts using Saturns (witness *Skylab*) and smaller, reusable shuttles primarily for astronauts, avoiding issues with insulating tiles and launch hazards that killed two crews. NASA and Congress must act, consistent with the reality that a new launcher takes time and money, and a bump in funding profile is needed for the space program not to languish for years. Congress should also let the engineers design the hardware. Von Braun's 1952 space station concept suggested construction in orbit of large ships from multiple small launches rather than one massive Saturn V (to have grown even larger without LOR). On-orbit construction was discouraged for space stations Freedom/*ISS* because of politics (maintaining budgetary separation for Congress of the station from costly human Mars efforts).[102] Building a large station would also be easier with armored inflatable modules (as von Braun envisioned in 1952), but Congress banned these for NASA (more in Chapter 11).

If the United States conducted foreign policy like it decides space policy, the nation would fight more wars and lose most of them, to be rewarded with global pariahship. Instead we have treaties, an established diplomatic corps, and long-term doctrines changing rarely and gradually, for example, George F. Kennan's Soviet containment from 1947 to 1972 (if not 1991), or defensive status vis-à-vis Japan and Germany, 1945–present. This is why important security posts such as FBI director come with a interest conflict-resistant, 10-year term. Space is perceived as insufficiently important to rate such strictures on politicians, but vacillation comes at high cost in wasted dollars and inspiration. The American economy is starved for engineering talent, and one proven way to bring children into science and engineering is interest in space (see Chapter 14). Furthermore, engineers make a lifetime career investment. Aerospace engineering is not the most stable profession, but NASA employment should be seen as a source of relative stability. In reality, however, it is quickly becoming the opposite.

Early in NASA's history (1958), to coordinate and advise the White House on space policy, a National Aeronautics and Space Council (NASC) was formed, staffed by civilian and military leaders – most in posts changing with each administration, unfortunately. It provided continuity in that Lyndon Johnson became chairman in April 1961 and on assuming the presidency installed his own Vice President Humphrey as chair. Nixon ignored the NASC and abolished it in 1973. G.H.W. Bush reinstituted the National Space Council (NSC), and it helped remove NASA's administrator in 1992 but was abolished by Clinton in 1993. Presidential candidate Obama promised to reestablish it in 2008 but has not. Meanwhile the National Academy of Science/National Research Council assumes some of these advisory functions but not relating to major space hardware or military issues. NASC/NSC falls short of the needed continuity, being vulnerable and usually restaffed with each administration. NASA suffers greatly from politically motivated budget fluctuations. Funds-leveling legislation by Congress sounds appropriate, but expecting it to hold seems naïve. Congress will override its own self-strictures, maintaining space exploration as a political football.[103] Obama should fulfill his promise and establish a tradition of respect for an NASC that future presidents would hesitate to kill. Scientific/technical *gravitas* as well as political Cabinet/Pentagon strength would help.

A populace familiar with a vigorous and stable space program's economic benefits might demand more rational policy. Along with inspiration for math/science/engineering students and career stability for engineers, the space program generates technological spin-offs. The most significant spin-off from Apollo, arguably, is acceleration by several years of semiconductor integrated circuitry, especially in the development of the Apollo Guidance Computer (AGC) that kept the spacecraft on its intended trajectory. The AGC truly was essential. Although Armstrong, Shepard, and other lunar landing mission commanders would manually control the LM in the last seconds before touchdown, the economical trajectory taken by the AGC beforehand was the reason Armstrong had enough fuel to search for a satisfactory landing site on *Apollo 11*. Although integrated circuits, multiple transistors on the same silicon chip, were used in custom applications for several years in the U.S. Air Force's Minuteman missile, the AGC marked a massive, assembly-line application of identical chips. By 1963 Apollo was consuming 60% of U.S. integrated circuit production, some 110,000 units from Fairchild Semiconductor by 1964.[104] Because the rapidly expanding semiconductor industry enables several trillion dollars in annual economic output (computers, cell phones, and most electronic gadgets and applications), Apollo kick-starting this boon by several years means that expenditure on the Moon race has paid off handsomely.

Public opinion is influential but less than NASA's relations with its elected political masters. No president since 1960 has risen or fallen because of missile policy, and never because of space exploration. Public opinion usually ranks NASA in approval above nearly any other governmental organization, and space exploration seems more popular now than during the Moon Race (and certainly more than in 1970–1990).[105] Consistently, about two-thirds of people support current U.S. space program funding or think it should be increased. Approval for the space program correlates with higher educational levels. Notably, young people are more enthusiastic about space than those around 50 or more than 65 who didn't experience Apollo in their youth (but not as much as those who as youths witnessed Apollo). A common sentiment among people in their fifties is that the future did not unfold as anticipated in the 1960s. But what of today's future? We live in a phase in which leaders approximately 50 years old see their elders as passé in their Apollo-era enthusiasm, when in fact these leaders' children and younger siblings are also enthusiastic. Seeing space exploration as passé is itself becoming passé.

A useful if overly succinct summary of space history is that by 1947 von Braun envisioned incremental human progress into the Solar System – Earth satellite, space shuttle, space station, Moon base, Mars, and beyond – which NASA (and the Soviets) largely implemented except when certain U.S. presidents took the helm. Kennedy skipped steps to reach the Moon, the Bushes tried sending us back to Moon/Mars, and Obama rejected that without a coherent alternative. We struggle to find it.

One major difference between the 1950s and 1960s versus now is current absence of a major public space exploration advocate at the level of von Braun. He was less effective by 1969, suffering from personal and larger societal issues – starting largely with overexposure by his 1960 biopic – met in places with protests over his role in bombing London with V-2s. [BOX 3.10] We have strong advocates for space today, such as Senators Barbara Mikulski (D-MD) and Bill Nelson (D-FL). The Nelson case highlights difficulties of such advocacy: a scientist who might have risen to von Braun's level was James Van Allen (1914–2006), a strong advocate for space sciences especially in 1958–1994. Nelson and van Allen disagreed over the primacy of human versus robotic space exploration, conducting a noted debate on this in 1980.[106] (In Chapters 10–14 we discuss efficacy in each approach.) Divisions in the space program community require such an advocate to be consummately diplomatic, conveying a community consensus to the public and policy makers. An advocate requires unusual insight and self-confidence, because sometimes such consensus is destructively wrong. (In Chapter 8 we see such attitudes of lunar scientists toward water and lunar light elements suffering from this for decades.) Space advocates often lean on assumptions not understood by the public and must speak both to

Box 3.10: Perceptions of von Braun

Dr. von Braun contracted cancer by 1975; Apollo was essentially over, and he left NASA three years before. Until 1972 he penned inspiring public pieces about future space exploration. Of course 1966 saw *Paris Match* questions regarding his war crime involvement, and 1969, his New Orleans testimony. Seemingly he considered himself essentially innocent, although investigators after his death cast doubt on this, if not indictably so, and although close *Aggregat* associates were compelled to abandon the United States and their citizenship as a result of suspicion of war crimes (Arthur Rudolph in 1984). ("Accountability in the Aftermath of the Holocaust" by Judy Feigin, edited by Mark M. Richard, 2006, Department of Justice Office of Special Investigations, 331–341, http://documents.nytimes.com/confidential-report-provides-new-evidence-of-notorious-nazi-cases; also "Aide says von Braun wasn't able to stop slave horrors; Objection would have gotten rocket pioneer shot, Dannenberg says" by Lee Roop, October 4, 2002, *Huntsville Times*.)

Von Braun suffered public criticism of his self-promotion, but we should ask if his husbanding of space exploration's birth could have succeeded without this. Goddard could easily have been America's homegrown von Braun, perhaps delivering a long-range rocket program, but Goddard was so hounded by derision in the American press that he avoided public exposure. Goddard did his greatest work during the Great Depression, indicating his enterprising ability, but never led thousands of workers like von Braun. (From 1941 until his death in 1945, Goddard devoted much effort to jet-assisted takeoff rockets for U.S. military airplanes.)

Although von Braun became an excellent engineering manager, some biographers argue that his design creativity peaked during his early VfR years. In contrast, Goddard (or his estate) received 214 rocket technology patents. Goddard's effort raising funds and managing his research program is obvious in his private papers. Von Braun was not too far ahead of his time; the VfR immediately appreciated his breakthroughs.

Von Braun's devotion to space exploration's future landed him in jail by angering the Gestapo, but he enthusiastically revealed his incredibly ambitious dreams to America, starting with *Das Marsprojekt* and expanding to the Collier's series, Disney films, and beyond. Von Braun was not his only promoter, however. The U.S. Army drew early attention to him and his former *Aggregat* associates. Although a year passed before the press noticed, his path of celebrity already started soon after his U.S. arrival ("Outstanding German Scientists Being Brought to U.S." by Press Branch, October 1, 1945, 4:30 PM, U.S. War Department, Bureau of Public Relations press release; "Nazi Scientists Aid Army on Research" by Frederick Graham, December 4, 1946 (November 19, 1946 – delayed by War Department), *New York Times*; "Nazi Brains Help U.S.: German Scientists Are Revealed as Army Researchers," *Life*, December 9, 1946).

Certainly other personalities figure prominently in von Braun's space pioneering. U.S. Army General Bruce Medaris was crucial in launching *Explorer 1/Jupiter C* in January 1958. Medaris' vision shielded von Braun during work on the Jupiter at Redstone Arsenal despite violation, in spirit, of orders to the contrary. *Explorer 1*'s flight needed Medaris. His career path highlights the importance of von Braun's self-promotion. When NASA formed in 1958 and took over satellite development from the army, Medaris was forced into other work, whereas von Braun moved to NASA. Self-preservation skills are mandatory for a visionary compelled by events to transition between the VfR, *Wehrmacht*, U.S. Army, and NASA. Still, prominent personae in the effort do not diminish von Braun's

> **Box 3.10: (continued)**
>
> role. Samuel C. Phillips, who directed NASA's human Moon landing program, is reported to have said Americans would not have landed on the Moon without von Braun, and John Noble Wilford credits von Braun with unique ability to inspire people with his sense of purpose in humanity's destiny in outer space (*We Reach the Moon* by John Noble Wilford, 1969, New York: Bantam, 135).
>
> We may never know the precise level of von Braun's awareness and involvement in war crimes committed in building the V2. He was clearly a consummate self-promoter, but was this his motivation at the expense of morality? In February 1944 Heinrich Himmler, reichsführer second only to Hitler, demanded a meeting with von Braun in which he insisted the rocket scientist accept his help in producing the V2, to which von Braun reportedly responded by comparing the rocket project to cultivating a flower garden, and Himmler's insistence on influence to pouring too much manure over the flowers. This insanely incautious rejection of the commander of the Gestapo and architect of the Holocaust seems to indicate a man caring not only for self-promotion; three weeks later the Gestapo imprisoned von Braun. Perhaps we should give him some benefit of the doubt.

their peers and citizenry at large.[107] Perhaps Elon Musk of SpaceX could play this part if he were more vocal and inclined. These are demanding job requirements. Respected planetary scientist Carl Sagan partially filled this role. Recently, science popularizer and planetarium director Neil deGrasse Tyson has adopted this mantle. His message would be more effective if more consistent over time. His recent call to double NASA's budget is inspiring but needs careful presentation for taxpayers to pay $18 billion more annually. No evidence appears in today's White House of such awareness, far be it advocacy, for space exploration's value.

Goddard, Korolev, and von Braun shared a passion for advancing humanity into outer space. As Goddard realized, for space exploration, liquid propellants provide crucially faster rocket exhaust (made of lighter elements – hydrogen, oxygen, carbon, or nitrogen – versus solid fuels). All three men developed liquid propellant rockets over solid-fueled ones, despite the likely preference of military leaders had they appreciated liquid-propelled rockets' battlefield inadequacies (being less turn-key and less rugged than solid ones). All three men suffered due to their enthusiasm for space exploration over military applications, to differing severity.

A hard core of millions of space exploration enthusiasts permeates the United States and the world. The crowd that saved *Star Trek* (cancelled as a result of ratings but revived and maintained by fans to blossom into a multibillion-dollar industry) would be yet more enthused by an inspiring space program. Space enthusiasts look far into the future and wait decades for an opportunity for national/general interest to turn back toward space. Support is soft only in that survival is not immediately concerned (except averting asteroid impacts). Of several space advocacy organizations, none manage to rise to the level of decisive political force. NASA itself, as a federal administration, is constrained in public self-promotion. NASA accomplishments, however, must translate into public media that effectively inspires; inspiration should be NASA's prime business. Perhaps space enthusiasts' organizations should fund well-distributed public presentation that highlights exciting missions and discoveries. NASA's Internet presence could be more exciting and interactive – more video, more multimedia, more apps. If NASA cannot send more of its own people out to excite the public, they should openly encourage public spokespersons. The public cannot be at arm's length. The private partnership that runs NASA's visitors centers is effective, but the word needs to get into cities, towns, the country, and the Web.

Space exploration is one of few policy areas in the United States that needs to deal with arguments bordering on signs of mental disorder, these being various versions of Apollo hoax conspiracies. The first, fundamental form of these "theories" is that Apollo lunar landings never occurred. Various lines of evidence are presented by conspiracy theorists to argue that more than 25,000 Apollo lunar mission photographs, hundreds of video hours, and thousands of hours of audio and astronaut personal experience were all faked. In the United States these beliefs affect perhaps 5–10% of the population but form a meme prevailing in some other regions of the world and parts of the Internet. Various arguments, pro and con, are not worth presenting here in the main text [BOX 3.11], but we seem to be entering a period

---

Box 3.11: Unbelievable Feats

Sending people to the Moon was strange to accept, even for some astronauts: "Here I am, a white male, age thirty-eight, height 5 feet 11 inches, weight 165 pounds, salary $17,000 per annum, resident of a Texas suburb, with black spot[s] on my roses, state of mind unsettled, about to be shot off to the Moon. Yes, to the Moon" (*Carrying the Fire: An Astronaut's Journeys* by Michael Collins, 1974, New York: Farrar, Straus & Giroux, 361). Under these circumstances, that some people who are not involved doubting the lunar missions' reality is not surprising. Add to this an increasing tendency, starting with JFK's assassination and the *Warren Report*, of Americans to pursue conspiracy theories distrusting government, especially after the 1972–1974 Watergate scandal. Apollo was fresh in people's minds when in 1974 the first book appeared accusing NASA of perpetuating a gigantic fake lunar landing hoax. (In 1970 the idea was stated in print, however: *Of a Fire on the Moon* by Norman Mailer, 1970, Boston: Little, Brown, 130, and already an urban myth by 1969.) A small but significant part of the public – 5%–10% but 27% in younger cohorts – doubt Apollo's lunar landings ("Generation Y and Lunar Disbelief" by Anthony Young, January 22, 2007, *Space Review*, http://www.thespacereview.com/article/787/1).

A fundamental problem with these conspiracy theories is that we still await even one person with a directly operative role in Apollo to step forward with solid evidence or even a testable detailed account of how such a massive hoax occurred. (Some conspiracy theory authors were involved tangentially, e.g., in other jobs at NASA contractors.) The interest of investigative reporters or Soviet intelligence to find whistle-blowers among many thousands of hypothetical coconspirators bespeaks the improbability of such a hoax. The alternative approach has been for conspiracy theorists to sift through massive amounts of Apollo data for inconsistencies that might indicate fakery. The fact that no humans have visited the Moon since 1972 and no robots have landed at Apollo lunar sites has let this accusation live in people's minds. Recently, however, robotic lunar satellites

Box 3.11: (continued)

have returned high-resolution images of Apollo landing sites: *Lunar Reconnaissance Orbiter*'s LRO Camera (LROC) has photographed several of the sites in 50-centimeter detail (Figure 3.11), and in several years commercial spacecraft may actually visit them (Figure 4.3). Another frequently cited argument is the absence of stars photographed from the lunar surface. Figure 3.12 shows why this argument is weak. (More interesting photographs of Earth and Venus from the lunar surface are shown at "Venus over the Apollo 14 LM" by Danny Ross Lunsford & Eric M. Jones, 2007, *Apollo 14 Lunar Surface Journal*, http://www.hq.nasa.gov/alsj/a14/a14Venus.html. Other instances of planets photographed from Apollo landing sites can be found in the database.)

Despite several books expounding hoax conspiracies, attention reached its epitome in February 2001 with broadcast by Fox Television of *Conspiracy Theory: Did We Land on the Moon?* Hoax conspiracies' popularity receded with some well-produced debunking: *Mythbusters*, Episode 104: "NASA Moon Landing" (August 27, 2008, http://mythbustersresults.com/nasa-moon-landing) and *The Sky at Night*: "Twelve Men on the Moon" (April 2002). Also, potent arguments against hoax conspiracy theories are circulated by Phil Plait's *Bad Astronomy* (www.badastronomy.com). There have been no new polls on the subject, but traffic in hoax theorizing has declined. Perhaps new evidence of the landings' existence and mass media exposure of the anti-hoaxers are having their effect.

If a conspiracy theory cannot deal with why the Soviets did not expose the hoax, obviously one should bring them in as coconspirators. This is reason for aliens-on-the-

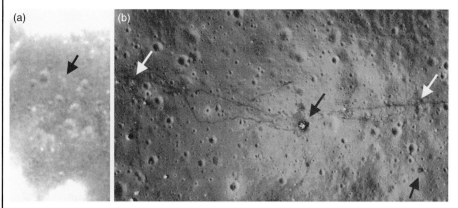

Figure 3.11. **Still there**. Most Apollo landing sites have been photographed in detail; shown is *Apollo 17*'s. (a) Taurus-Littrow Valley as imaged by JAXA *Selene* (*Kaguya*) probe's Terrain Camera. The arrow points to the bright spot from freshly agitated regolith by the Lunar Module engines; the scene is 6 km wide. (b) In September 2011 *Lunar Reconnaissance Orbiter* entered an orbit as low as 35 km above the Moon. The LRO Camera photographed *Apollo 17* in 0.5-meter detail, including many items left on the Moon by its crew. The center arrow indicates the LM descent stage, arrow at left the ALSEP site, arrow at bottom right the Lunar Rover, and center right arrow to the SEP experiment (Surface Electrical Properties). Fine traceries are foot and rover trails left by the astronauts. The scene is 0.4 km wide. *Chang'e-2* may have high-resolution images as well, but they are not released. (JAXA photo MNA_2B2_01_04493N202E0311; NASA photo M168000580RE.)

Box 3.11: (continued)

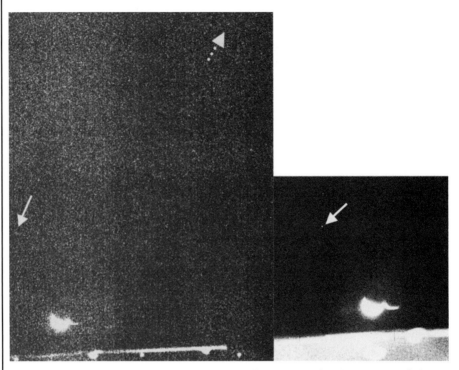

Figure 3.12. **The black sky**. On the Moon with *Apollo 14*, Shepard took two series of photos of Earth, including a set of long exposures from within the Lunar Module with lights extinguished (on February 6, 1971); two are shown. These show Earth and Venus unequivocally, but only marginally so nearby Mars, contradicting contentions of Apollo hoax believers that photos from the daylit Moon should show numerous stars. They do not; exposure times are always too short. Even in these long-exposure frames, one sees Earth overexposed (thick crescent above window frame, with small ghost image to right) and Venus (see arrows above Earth in both pictures), but Mars (near dashed arrow at top of right image) is barely detected or absent. Here Venus is 15 times brighter than the brightest star Sirius; Mars is 10 times fainter. (Photos AS14–66–9329 & AS14–66–9330, courtesy of NASA.)

Moon or American-Soviet lunar collusion theories, popularized recently by several feature films, most earnestly by *Apollo 18*. The studio cynically sells *Apollo 18* as "found footage." ("Weinstein balks at the idea that this is a work of fiction. 'We didn't shoot anything,' he says. 'We found it. Found, baby!'" – "*Apollo 18*: The Next *Paranormal Activity*?" by Tim Stack, March 4, 2011, *Entertainment Weekly*.) Audience members fooled by this did not wade through the credits (more than 300 cast and crew for "found footage") to the end disclaimer that the persons and events portrayed are fictitious. They can see such incongruities as the "secret" night launch of a Saturn V, the Sun shining straight down on Earth's and the Moon's South Poles, craters, e.g., Messier, that no such mission would overfly, anachronisms postdating 1974, and "found" footage from other missions. These "astronauts" do not moonwalk like the first twelve: bad special effects.

> **Box 3.11: (continued)**
>
> Beyond these films the biggest booster of aliens-on-the-Moon is Richard C. Hoagland, creator of the Cydonia region's "Face on Mars," which has since been debunked. Hoagland uses similar powers of apophenia (finding patterns in random data) to the Moon to find transparent lunar cities – as evidenced by what most people would see as flaws of photographic processing. Many are evident in some scans of images but not in others, which is seen by Hoagland as evidence of NASA's cover-up. A central claim is that *Apollo 17* astronauts found a rock in a 100-meter diameter Shorty Crater, which Hoagland names Data's Head – about 35 centimeters across (rather big for a human head) and the same color as surrounding regolith. (Hoagland emphasizes a red coloration on the head's "lip" that is also seen at the edge of nearly every rock in the photo – apparently another scan processing error.) This rock appears in a panoramic sequence by Eugene Cernan, so it is imaged seven times. Adding these seven together optimally to increase spatial sampling, e.g., with multidrizzle techniques, the "head" shows odd asymmetric lumps on its crown and "right cheek" that look unheadlike. The better the data on Data's Head, the more rocklike it seems – not a head. That is unsurprising: it seems Hoagland did not search a priori for a head among several thousand distinguishable boulders in Shorty Crater, like searching shovelfuls of pebbles and finding one that looks like anything familiar. LROC images (M175077349LE in particular) show an apparent boulder at the "head's" position, so someday we can retrieve it if there is reason to. Hoagland suggested Cernan had lowered Schmitt on a rope into Shorty Crater to collect Data's Head for secret NASA purposes, but apparently they did not think it worth the effort.

when the preponderance of real evidence is winning over ideas of a giant hoax instituted to fake Apollo's lunar landings. Perhaps the most compelling prima facie counterargument is that the Soviets had every reason and great capability to expose such a hoax to the world if they even suspected such a complex plot existed, involving hundreds of thousands of NASA employees and contractors. The activities described in this book of KGB attempts to delay the launches of *Apollo 8* and *11* and the attempt to extradite von Braun to East Germany underline desperate lengths to which the Soviet Union could resort to undermine Apollo. Conspiracy theorists ignore Occam's Razor, constructing ever more elaborate, improbable hypotheses to accommodate counterarguments, so following Apollo hoax conspiracy theories come American-Soviet lunar collusion theories that we discovered extraterrestrial beings on the Moon, and the two governments worked together to hide this from the public. These ideas are gifted with faux currency with recent films *Apollo 18* and *Transformers: Dark of the Moon* but have not convinced many. Younger, Internet-savvy people are not fooled and have no anti-Apollo axe to grind. Nonetheless, questions remain of why after four decades we have not returned humans to the Moon, and many people search for an answer beyond the mundane mediocrity of society.

The space exploration evokes spectra of emotional concepts: wonder, arrogance, patriotism, calamity, pride, heaven, competence, bureaucracy, waste, the unknown, and the future. We might expect that our feelings about space may change from year to year. The people studying it are teams of scientists, engineers, and managers with their own ideas of what projects to pursue but make a deal with the polity: decide what you want, and we will do it as safely and effectively as possible. If that decision vacillates, results are often correspondingly self-canceling and ill determined. To the extent that a society successfully engages in such sophisticated and cooperative tasks, they can indicate health and competence. If the decision gets bad results, unintentionally society may have asked for them.

NASA achievements are throttled by the imagination of politicians setting its goals and funding. They are risk averse, and prominent space activity has distinct downsides. Mission failures spark criticism, and Americans overestimate the budget spent on NASA. When other issues go badly, there awaits a version of "If we can land a man on the Moon, why can't we ...?" Most Americans do not oppose space exploration funding at current levels, especially when told how little that is.[108] Still, NASA maintains infrastructure ready for ambitious goals, spread over many congressional districts, with more than $10 billion per year for its upkeep. A $15 billion annual NASA budget produces a modicum of results; $20 billion, impressive ones. A $25 billion budget could produce a program to stimulate and inspire but leave politicians open to criticism by naysayers for too much space spectacular. That criterion, not results per dollar, is operative. Even if citizens want more activity and results for their tax dollar, politicians set the thermostat on space exploration.

U.S. political leaders occasionally lapse into neglect of astronautics and the space program to the peril and detriment of Americans (ca. 1930, in 1947, 1972, 2010, and perhaps 1993). In the 1930s the United States neglected Goddard's work only to be surpassed by Nazi Germany (this despite the United States buying its first military airplane only six years after the Wrights' 1903 first heavier-than-air flight). Consequently, facing the V2 and *Luftwaffe* jet and rocket-powered airplanes (as well as planned German SLBMs, submarine-launched ballistic missiles,[109] and ramjets), in 1945 the United States was intimidated into the reprehensible tactic of massive aerial bombardment of urban civilian targets. In 1947 they discouraged ICBM development only to panic a decade later facing Soviet rocketry. This radically affected U.S. policy, sharply diminished American prestige worldwide, and determined a U.S. presidential election. In 1972 the United States abandoned human deep space and non-reusable rocket development. Within two decades U.S. aerospace was displaced as the preeminent commercial satellite launcher, and Russia now leads in human space travel. We should question the 1993 decision to cooperate with Russia in orbital space stations rather than a Moon base exploiting lunar resources and Russian and Space Shuttle-derived hardware. In 2010 the United States closed its eyes to the Vision for Space Exploration, with repercussions as yet appreciated. When these outcomes are decided, it may be by other nations; the United States is losing leadership in this area. American society is capable of wasting good ideas and genius that it so proudly cherishes. Ignoring Goddard until facing the Nazis using the same technology, then leaving von Braun in the desert (50 months at Ft. Bliss), literally and figuratively, until the Soviets forced long-range rocketry's importance on them, American leaders showed symptoms of a sometimes characteristic lack of foresight, a trait that might cause history to repeat, even now. They and the public that elects them should ask how this could be avoided.

Why would we go to the Moon? There is a strange, arrogant presumption in that question now unlike ever before. No longer is it an overreach to think that we can do it. We can. Now there are finer points: who are "we"? There are several nations and teams preparing to go; "we" may not be whom we assume. Also, what is this place called the Moon? That answer has changed radically in the past few years in ways even foremost lunar scientists had difficulty anticipating. In the next few chapters we examine who is going, and what the Moon really is.

Notes

1. *The Papers of Robert H. Goddard* by Esther C. Goddard & George E. Pendray, 1970, New York: McGraw-Hill, 823. A similar thought written by Wells in 1934 closes the film *Things to Come* (1936) as the first human lunar mission is launched in year 2055, fired from a huge cannon.
2. From *True History* by Lucian, edited by Henry Morley, translated by Thomas Francklin: "About noon, the island being now out of sight, suddenly a most violent whirlwind arose, and carried the ship above three thousand stadia*, lifting it up above the water, from whence it did not let us down again into the seas but kept us suspended in mid air, in this manner we hung for seven days and nights, and on the eighth beheld a large tract of land, like an island, round, shining, and remarkably full of light; we got on shore, and found on examination that it was cultivated and full of inhabitants, though we could not then see any of them. As night came on other islands appeared, some large, others small, and of a fiery color; there was also below these another land with seas, woods, mountains, and cities in it, and this we took to be our native country: as we were advancing forwards, we were seized on a sudden by the Hippogypi, for so it seems they were called by the inhabitants; these Hippogypi are men carried upon vultures, which they ride as we do horses. These vultures have each three heads, and are immensely large; you may judge of their size when I tell you that one of their feathers is bigger than the mast of a ship. The Hippogypi have orders, it seems, to fly round the kingdom, and if they find any stranger, to bring him to the king: they took us therefore, and carried us before him. As soon as he saw us, he guessed by our garb what we were. 'You are Grecians,' said he, 'are you not?' We told him we were. 'And how,' added he, 'got ye hither through the air?' We told him everything that had happened to us; and he, in return, related to us his own history, and informed us, that he also was a man, that his name was Endymion, that he had been taken away from our earth in his sleep, and brought to this place where he reigned as sovereign. That spot, he told us, which now looked like a moon to us, was the Earth."
   *about 600 km, by the definition of Herodotus.
3. *The Sháhnáma by Firdausí*, edited by Arthur George Warner & Edmond Warner, 1912, republished by Routledge (London), 2000.
4. Kepler started *Somnium* around 1593 as a student, exploring how Earth and the rest of the Solar System would appear as seen from the Moon. His elders apparently dismissed the worth of the project. Unlike most earlier stories, his lunar inhabitants are not particularly human but are better adapted to the lunar environment. Kepler considers how humans might cope with gravitational and atmospheric effects on the Moon. Like Copernicus's *De Revolutionibus*, *Somnium* was published after its author's death.
5. *The Other World: The Comical History of the States and Empires of the Moon* by Hector Savien Cyrano de Bergerac (1657).
6. Jules Verne's works from 1865, *From the Earth to the Moon* (*De la Terre à la Lune*), and 1870, *Around the Moon* (*Autour de la Lune*), merged into *A Trip to the Moon and Around It*.
7. Goddard credited the science fiction of H.G. Wells with interesting him in rocketry:
   "In 1898, I read your *War of the Worlds*. I was sixteen years old, and the new viewpoints of scientific applications, as well as the compelling realism ... made a deep impression. The spell was complete about a year afterward, and I decided that what might conservatively be called 'high altitude research' was the most fascinating problem in existence" (*The Papers of Robert H. Goddard* by Esther C. Goddard & George E. Pendray, 1970, New York: McGraw-Hill, 23).

8. *The Exploration of Cosmic Space by Means of Reaction Devices* (Исследование мировых пространств реактиными приборами) by Konstantin Eduardovich Tsiolkovsky (1857–1935) in 1903. Tsiolkovsky also studied orientation in space via small reaction rockets in *Free Space* (*Svobodnoe Prostranstvo*, Свободное пространство) in 1883 (unpublished until 1956).

9. Rocket propelled arrows are described in *The Wu-ching Tsung-yao* (*Complete Compendium of Military Classics*) from 1045 AD, and their use is recorded in defending the town of Kai Fung Fu from the Mongols in 1232 (*A History of Firearms* by W.Y. Carmen, 2004, Mineola, NY: Courier Dover Publications), after which they quickly spread to the parts of Muslim world, then Europe. There are much earlier Chinese references to fire arrows, but whether these were rockets is less clear.

10. Oberth wrote to Goddard on May 3, 1922, asking for copies of Goddard's works before completing his own treatise in 1923 (*The Papers of Robert H. Goddard*, 485).

11. Goddard first discussed hydrogen/oxygen rockets in 1909 (ibid., 98–99 and 514).

    Modestly, Goddard's 1919 thesis discussed only rocketry's practicality, not exploring the Moon, but simply using a rocket's impact on the new Moon (his page 56) to prove via flash of impact having achieved sufficient velocity to escape Earth's gravity. He wrote scenarios for exploring the Solar System and beyond but avoided publishing it to limit controversy. His reserve was prescient. Famously, the *New York Times* (January 12, 1920) attacked Goddard's proposal for flight in a vacuum because they thought it denied a fundamental principle of physics ("and only Dr. Einstein and his chosen dozen, so few and fit, are licensed to do that"), implying that there is no reaction of the rocket to thrust in the vacuum of space. On July 17, 1969, accompanying coverage of *Apollo 11*, the *New York Times* included "A Correction" confessing: "Further investigation and experimentation have confirmed the findings of Isaac Newton in the 17th century and it is now definitely established that a rocket can function in a vacuum as well as in an atmosphere. *The Times* regrets the error." Unfortunately, Goddard died a quarter-century before.

12. *Cosmic Voyage: Fantasy novella* (*Kosmicheskiy reys: Fantasticheskaya novella* - Космический Рейс: Фантастическая Новелла, also *The Space Ship*, or *Space Journey*), directed by Vasili Zhuravlyov (1936) with Konstantin Tsiolkovsky as science advisor, was meant to inspire Soviet youth. Although filmed in black and white and without sound to save mass-production costs, it compares well in effects with *The Wizard of Oz* (1939). Unfortunately, the film went from popularity to obscurity with intercession of Soviet censors offended by the independently rebellious behavior of the central character (senior academician Pavel Ivanovich Sedikh, complete with a Christmassy Grandfather Frost beard, not unlike Tsiolkovsky's). They were equally offended by the carefree manner of cosmonauts bouncing around the Moon's surface, in violation of socialist realism ("Hijack at the Cosmodrome" by David Jeffers, July 25, 2007, http://www.siffblog.com/reviews/hijack_at_the_cosmodrome_003889.html). Fodor Krasne, who animated such irresponsible, anti-socialist frivolity, was stricken from the film's credits.

13. *Woman in the Moon* (*Frau im Mond*) directed by Fritz Lang, written by Thea von Harbou (1929). According to Nazi video, the first successful V-2 rocket bore a pinup-girl likeness of the *Woman in the Moon*. The film is said to have influenced Hitler to support developing the V-2. Hitler is also said to have suffered a nightmare about failing rockets (*Mining the Sky* by John S. Lewis, 1996, Reading, MA: Addison-Wesley, 23) and delayed full V-2 funding from 1939 until 1943, after he was shown videos of successful V-2 test flights.

    The interplay of technology, science fiction, and the film industry in the early twentieth century is significant: from Jules Verne's and H.G. Wells' novels to 1901 and George Méliès' film in 1902, with their impact on Tsiolkovsky's treatise in 1903 and inspiration for the young Robert Goddard several years after, through Oberth's and Tsiolkovsky's contribution to each of their own feature films, which had their own significant impact. Fact and fiction created each other.

14. For humans, air runs out at 19 km altitude, the Armstrong Limit (named for Harry George Armstrong, pilot and flight surgeon), above which water (and blood) boils for persons outside a pressure suit; daytime sky is noticeably darkened and Earth's curvature becomes perceptible along the horizon. Outer space starts at 100 km (60 miles in the United States; 50 miles for the U.S. military) at the Kármán line after Hungarian-American physicist Theodore von Kármán; here weak aerodynamic forces make wings impractical. (In 2009 the UN Committee on Peaceful Uses of Outer Space proposed raising this 30 km or more.) The first device exceeding this greatly was a V-2/bumper rocket reaching 127 km on May 13, 1948. The *Canadian Arrow* entry to the

Ansari X-Prize commercial spaceflight competition was a V-2-like design modified to reach 112 km. (Spacelike vacuum, where a molecule must descend before likely colliding with another molecule, starts at the base of Earth's exosphere, about 300 km or more, depending on the level of solar activity.)

German rocketry combat aspirations extended beyond the V-2. Before World War II, Eugen Sänger and Irene Sänger-Bredt, University of Vienna rocket experts, suggested the *Silbervogel* ("Silverbird" or *Raketenbomber* version of the *Amerika Bomber* or "Antipodal Bomber"), would skip repeatedly off Earth's upper atmosphere back into space and might drop a large warhead on New York. The idea was adopted by von Braun: he suggested the A-9, a winged version of the A-4/V-2, might reach a 5,000 km range with a booster stage added (*The Need for Speed: Hypersonic Aircraft and the Transformation of Long Range Airpower* by K.F. Johnson, 2005, Graduate Thesis, School for Advanced Air and Space Study, Maxwell A.F. Base). The two-stage, intercontinental A-10 rocket design (with an A-9 upper stage) intended for German attack on New York and Washington, DC, reportedly caught the eye of Joseph Stalin himself upon its discovery by the Soviets in Germany (*Red Moon Rising* by Matthew Brzezinski, 2007, New York: Times Books, 15). Von Braun said Germany started building such a rocket, weighing 100 tons with a 6-ton warhead ("Nazis Planned Rocket to Hit U.S." by United Press, November 19, 1946; *New York Times*, December 4, 1946).

15. *Red Moon Rising*, 11.
16. In public Bush expressed extreme skepticism that ICBMs would ever be sufficiently accurate as strategic bombers against targets such as cities (for example, see December 5, 1945, testimony before the Special Senate Committee on Atomic Energy, and his 1949 book, *Modern Arms and Free Men: A Discussion of the Role of Science in Preserving Democracy*, Westport, CT: Praeger, 203). Writings of the time discuss a large, high-altitude burst, perhaps loosely targeted, for example, "The rocket and the future of warfare" by Arthur C. Clarke, March 1946, *RAF Quarterly*; reprinted in *Ascent to Wonder: A Scientific Autobiography*, 1984, New York: John Wiley & Sons, 71). In modern tactics, a nuclear attack might begin with a large, high-altitude blast, disabling an adversary's electronic infrastructure with an electromagnetic pulse (EMP). Of course, advanced warheads were later targeted much more accurately. Bush later claimed to have never opposed ICBM development, despite his statements' effects (Letter to *Time*, July 10, 1964).
17. Khrushchev said ca. 1960: "What the scientists have in their briefcases is terrifying," recalls Asimov (*Isaac Asimov's Book of Science and Nature Quotations* by Isaac Asimov & Jason A. Shulman, 1990, New York: Grove Press, 283).
18. More precisely, Yeager is officially first to fly supersonically and survive. Luftewaffe pilot Lothar Seiger reportedly died breaking the sound barrier on March 1, 1945, in a Bachem Ba 349 rocket plane. Unofficially, North American Aviation test pilot George Welch likely exceeded sound speed in a XP-86 Sabre jet once or twice shortly before Yeager.
19. Include the following classics: films *Destination Moon* (1950), *The Day the Earth Stood Still* (1951), and *When Worlds Collide* (1951), and radio serial "The Martian Chronicles" on the show *Dimension X*, broadcast in 1950–1951, based on Ray Bradbury's short stories from 1947–1951.
20. For instance: "Most astronomers agree that there is primitive plant life, like lichens and algae, on Mars. The presence of this potential food supply has led a number of biologists (although not all of them) to conclude that there may be some form of animal life there, too. It is very doubtful that life of any kind exists on the other planets, however" (Willy Ley, March 22, 1952, *Collier's*, 39).

    At this time Mars was thought to have an atmosphere about 10% as dense as Earth's, water-ice polar caps, and seasonal changes in coloration often interpreted as plant life. Astronomical observations showed little Martian atmospheric water or oxygen, but *Mariner 4*'s flyby past Mars on July 14, 1965, revealed a thin atmosphere with 1% of Earth's density and heavy, lunarlike cratering of its surface. Scientists increasingly dismissed even simple life on Mars, for example, "If there is any indigenous life on Mars, it is probably very small, or even microscopic, because of the extreme scarcity of water on the Martian surface" (*Habitable Planets for Man* by Stephen Dole, 1964, Santa Monica, CA: Rand Publishing, 5) and "There is considerable feeling among some biologists (and others, as well), as expressed in Wilber's letter (9 July, p. 135), against the large expenditures being made for a quest for life on Mars – sums which might otherwise be allocated to terrestrial research projects" ("Mars and the Evolution of Life" by Joel Goodman, *Science*, August 20, 1965, 813).
21. Gallup Poll, December 1, 1949: "Do you think that men in rockets will be able to reach the moon within the next 50 years?" Yes: 15%; No: 70%; No opinion: 15%. Based on 1,500 interviews, November 27 – December 1, 1949. By

January 1955, the same question garnered Yes: 38%; No: 51%; No opinion: 11%, based on 1,446 interviews, December 31, 1954, to January 5, 1955. (On the same survey, 90% of interviewees said that they do not want to join the first men who do go to the Moon. In 1966, 80% still said they would not.)

22. An earlier book of the same title written by David Lasser in 1931 was credited by Arthur C. Clarke with convincing him at age 14 to take space travel seriously. The book is an excellent introduction to concepts of rocketry and outer space, especially for its time (*The Conquest of Space* self-published by David Lasser, 1931 versus by Willey Ley and Chesley Bonestell, 1949, New York: Viking).

23. *Das Marsprojekt* (*The Mars Project*) by Wernher von Braun, 1952, Frankfurt: Umschau Verlag. Before von Braun the Moon/Mars exploration sequence was largely a science fiction plot device, often consequence of some guidance error on a fictional rocket ship, for example, *To Mars via the Moon: An Astronomical Story* by Mark Wicks, 1911, London: Seeley & Co., or the 1950 feature film *Rocketship X-M*, directed by Kurt Neumann, starring Lloyd Bridges and Osa Massen.

24. On July 14, 1952, the executive committee of the National Advisory Committee for Aeronautics (the precursor to NASA) passed a resolution that "NACA devote *modest efforts* (my emphasis) to problems of unmanned and manned flights at altitudes from 50 miles to infinity and at speeds from Mach 10 to escape from the earth's gravity" (*NACA to High Speed Flight Research Station, Discussion of Report on Problems of High Speed, High Altitude Flight, and Consideration of Possible Changes to the X-2 Airplane to Extend Its Speed and Altitude Range*, July 10, 1953).

25. Gallup Poll (AIPO), October 15, 1957: "How long do you think it will be before men in rockets will reach the Moon?" The peak numerical answer was 10 ± 2.5 years; *Apollo 11* landed 11.8 years later. (The data – less than 2.5 years: 4%; 2.5 up to 7.5 years: 8%; 7.5 up to 12.5 years: 14%; 12.5 up to 17.5 years: 5%; 17.5 up to 22.5 years: 6%; 22.5 up to 27.5 years: 4%; 27.5 up to 72.5 years: 8%; 72.5 up to 127.5 years: 3%; 127.5 years and above: 1%; Never, silly: 14%; Couldn't say/depends/hard to say/long time: 34%. Based on 1,573 interviews, October 10–15, 1957.)

26. Cornelius Ryan, associate editor at *Collier's* and later author of *The Longest Day* and *A Bridge Too Far*, was involved in early discussions and brought the project to the magazine.

27. From *Beyond the Solar System* by Willy Ley & Chesley Bonestell, 1964, New York: Viking Press: "The first chapter predicts and describes a manned expedition to the sun's nearest neighbor, the stellar system of Alpha Centauri in the southern sky. I think that such an expedition will be made at a time when people now alive (though very young) will be able to watch the take-off on television – say, half a century from now."

To this von Braun responds in the book's foreword: "It would be necessary to devise a rocket mechanism wherein the entire mass, m, of the injected 'propellant' is converted into radiation energy.... The problem is that nobody knows how to build such a photon rocket. There is no known large-scale process that would enable us to transform matter completely into energy according to Einstein's equation. Even if we knew such a process, we would be confronted with most formidable engineering problems." In the end von Braun imagines their success someday.

28. "Conflating Space Exploration and Commercialization: Coverage of PayPal's Announcement" by John Hickman, July 2, 2013, *The Space Review*: http://www.thespacereview.com/article/2321/1.

29. Khrushchev explained in an essay to Americans: "The more planes with hydrogen weapons are flying in the air, the less the room that is left for the doves of peace and the more for the machinations of the demon of war" (*For Victory in Peaceful Competition with Capitalism: With a Special Introduction Written for the American Edition* by Nikita Sergeevich Khrushchev, 1960, Boston: Dutton).

30. "I hope you would not say that I am trying to frighten you if I remind you that the Soviet Union has rockets in a quantity and of a quality unequalled by any other country in the world. This can be confirmed by the launching of our Sputniks and cosmic rockets. Under these conditions, to settle disputable questions in the way the military revanchist quarters of West Germany want – by war – is tantamount to suicide, to the destruction of one's country": Nikita Khrushchev in a letter to Chancellor Konrad Adenauer, *Pravda*, August 26, 1959. (See *Outer Space in World Politics*, ed. Joseph M. Goldsen, 1963, New York: Praeger, 43.)

31. The first jetliner in commercial service was the de Havilland Comet in 1952. (The Avro Jetliner never reached significant production.) Unfortunately the Comet suffered several fatal crashes primarily because of design flaws, forcing it out of service in 1954 until 1958 (mainly, the wings fell off from metal fatigue). The Soviet Tupolev Tu-104 entered regular operation in late 1956. Boeing introduced the 707 and Douglas the DC-8 into commercial service in 1958–1959, and soon jetliners became common in the United States.

Moon/Mars

32. When the Soviet Union started constructing the launch site near Tyuratam, Kazahkstan, in 1955, the name chosen, Baikonur, appeared on Soviet maps as a town 300 km northeast. Baikonur workers received their mail addressed to Moscow or Leningrad and were ordered to never describe their actual location. To register Yuri Gagarin's flight internationally, Soviet officials were required to state launch (and landing) coordinates ensuring that a whole orbit had been completed. "Baikonur" was stricken from the telegram (http://www.russianspaceweb.com/baikonur_secrecy.html).
33. *Meeting the Threat of Surprise Attack: The Report to the President by the Technological Capabilities Panel of the Science Advisory Committee, Vol. II,* February 15, 1955, Washington, DC. This is also described in *Sputnik Declassified*, NOVA, November 6, 2007, WGBH Boston.
34. Khrushchev came quickly and perpetually to regret this utterance: "I once said, 'We will bury you,' and I got into trouble with it. Of course we will not bury you with a shovel. Your own working class will bury you" (from a speech in Yugoslavia, August 24, 1963).
35. *The Vantage Point: Perspectives on the Presidency, 1963–1969* by Lyndon B. Johnson, 1971, New York: Holt, Rinehart & Winston, 272. A potential military threat to homeland American territory was anathema to generations of U.S. politicians. Witness this speech excerpt by young Abraham Lincoln: "Shall we expect some transatlantic military giant, to step the Ocean, and crush us at a blow? Never! All the armies of Europe, Asia and Africa combined, with all the treasure of the earth (our own excepted) in their military chest; with a Bonaparte for a commander, could not by force take a drink from the Ohio, or make a track on the Blue Ridge, in a trial of a thousand years" ("The Perpetuation of Our Political Institutions: Address Before the Young Men's Lyceum of Springfield, Illinois," January 27, 1838). Soviet media amplified the propaganda victory, for example, with Radio Moscow announcing when *Sputnik 1* would fly over Little Rock, where the world knew President Eisenhower had sent troops to enforce racial desegregation. This was driven home by other countries, such as an Egyptian spokesman claiming that *Sputnik* "will make countries think twice before tying themselves to the imperialist policy of the United States."
36. "You cannot drive people to Communism by war. It is necessary for people to realize the need to replace the capitalist society by a Communist society. It would be madness to proceed to a new, better social system through war. We want to provide an example and I think that our example is not a bad one. Whose rocket was it that first went to the Moon? The Communist. Who first photographed the hidden side of the Moon? Communists. I shall not enumerate everything. This is quite enough. People might otherwise say that Khrushchev is making Communist propaganda" (Nikita Khrushchev, speech in Vienna, *Pravda*, July 3, 1961).
37. This is claimed in classified U.S. Army documents at least: *Project Horizon: Volume I, Summary and Supporting Considerations. A U.S. Army Study for the Establishment of a Lunar Outpost*, Redstone Arsenal, June 9, 1959. Candidly, this is the most paranoid Cold War document I have ever read. Project Horizon proposed building a 12-man lunar outpost by 1966 to beat the Soviets to the Moon, preventing their claiming it on their revolution's fiftieth anniversary. Up to six Saturn I and II (a four-stage version of a I-B) rockets would launch into Earth's orbit simultaneously, one carrying a lunar landing craft and five, the propellant to refuel the first. Other launches would send cargo directly to the Moon or cargo and astronauts to refueling stops in Earth's orbit. It proposed up to 66 Saturn launches per year and 229 total before 1968. Project Horizon cost was (under)estimated at $6 billion in 1959 dollars. Its prime justification was a distant nuclear citadel to launch a retaliatory attack during a nuclear war, unreachable for several days by enemy forces. Its mission was largely obviated by submarine-based nuclear missiles in 1960–1961.
38. Early Soviet space feats had quick and lasting effects on U.S. prestige. A few relevant polls:

    "Looking ahead ten years, which country do you think will have the leading position in the field of science?" (World Gallup Poll, February 12, 1960, percentage of respondents by country.)

	France	UK	India	Holland	Uruguay	Switzerland	Norway	W. Germany	Greece	United States
Russia?	59	48	46	43	42	40	38	36	27	16
U.S.?	18	17	8	22	27	34	22	29	29	70
Other?	14	21	7	9	16	19	9	14	27	2
No opinion	9	14	39	26	15	17	31	21	17	12

The New Moon

"Do you feel that the Russian satellite is a serious blow to U.S. prestige?" (WGP, October 27, 1957)

	India	Canada	France	Norway	Holland	U.S.	Austria	Sweden	W. Germany	Belgium
Yes	68	66	63	60	44	43	40	25	23	22
No	21	33	24	32	50	46	38	57	72	72
Don't Know	11	1	13	13	6	11	22	18	5	6

39. The number of *Life* magazine issues with astronomy or space exploration articles in 1940–1959: before 1949, 2 total; 1949, 2; 1950, 2; 1951, 0; 1952, 5; 1953, 1; 1954, 2; 1955, 3; 1956, 1; 1957, 11; 1958, 14; and 1959, 21. For the next few years these numbers held steady or increased.

40. John Glenn was on the cover of *Life* six times. This is more than some American presidents (Franklin Roosevelt with five). Excluding some presidents, a few movie stars (Liz Taylor, Marilyn Monroe, Sofia Loren), Queen Elizabeth II, and Winston Churchill, Glenn's exposure is superlative. Al Shepard and Gus Grissom are close behind at four and five covers, respectively.

41. As Kennedy explained in his May 25, 1961, speech to Congress: "Recognizing the head start obtained by the Soviets with their large rocket engines, which gives them many months of lead time, and recognizing the likelihood that they will exploit this lead for some time to come in still more impressive successes, we nevertheless are required to make new efforts on our own. For while we cannot guarantee that we shall one day be first, we can guarantee that any failure to make this effort will make us last."

42. Gallup Poll 645, May 28–June 2, 1961 (immediately after Kennedy's first Moon program speech), 3,521 Americans were asked to rank seven funding priorities. Sending a man to the Moon was tied for fifth/sixth place with "economic aid to underdeveloped countries," ahead of only "military aid for other nations." Worker retraining to offset mechanization was three times more popular.

43. These years also marked the most extreme spate of thermonuclear testing, with several 20-plus megaton explosions in late 1961 through 1962 at Novaya Zemlya in arctic Russia, including the AN602 Tsar Bomba at 60 megatons, 4,000 times the Hiroshima explosion. (The largest U.S. test, Castle Bravo at Bikini in 1954, accidentally yielded about 20 megatons.) The Saturn V was too large to usefully deliver a warhead. A reasonable extrapolation of a hydrogen bomb to 120 tonnes payload (10 times Castle Bravo's mass) gives 700 megatons yield, sufficient to blow a crater 5 km wide and 500 m deep and create a fireball taller than Earth's atmosphere. Even an airburst this large would produce a world-threatening cloud of radioactive fallout and vaporize much mass on the ground. Making an ICBM was not an ulterior motive for creating the Saturn V, even when Saturn I rockets were considered for Earth-orbital military satellites and even a military Moon base (Project Horizon).

44. For instance, arguing with NASA Administrator James E. Webb in the White House Cabinet Room on November 21, 1962, President Kennedy refers to a lunar landing as "the top priority program for the entire government" of the United States, but much of the conversation concerns his worry over rising costs of Apollo (John F. Kennedy Library Presidential Recording Log, Tape #63).

45. "Finally, in a field where the United States and the Soviet Union have a special capacity – in the field of space – there is room for new cooperation, for further joint efforts in the regulation and exploration of space. I include among these possibilities a joint expedition to the moon. Space offers no problems of sovereignty; by resolution of this Assembly, the members of the United Nations have foresworn any claim to territorial rights in outer space or on celestial bodies, and declared that international law and the United Nations Charter will apply. Why, therefore, should man's first flight to the moon be a matter of national competition? Why should the United States and the Soviet Union, in preparing for such expeditions, become involved in immense duplications of research, construction, and expenditure? Surely we should explore whether the scientists and astronauts of our two countries – indeed of all the world – cannot work together in the conquest of space, sending someday in this decade to the moon not the representatives of a single nation, but the representatives of all of our countries" (President John F. Kennedy, Address to the 18th General Assembly of the United Nations, New York, September 20, 1963).

46. "Soviets Planned to Accept JFK's Joint Lunar Mission Offer" by Frank Sietzen, *SpaceCast News Service*, October 2, 1997.

47. *Carrying the Fire: An Astronaut's Journeys* by Michael Collins, 1974, New York: Farrar, Straus & Giroux, 135.

48. NASA funding peaked in 1966 at $6 billion, 5.5% of the federal budget. Johnson discussed progress in space in his inaugural address and five of six addresses to joint sessions of Congress.

49. James E. Webb recalls discussions with Johnson over NASA's budget and increasing pressure to limit it by 1966. He cites pressure coming from elsewhere within the federal government, particularly the Department of Defense, although this was not yet primarily pressure brought on by the war in Vietnam (James E. Webb, interviewed by T.H. Baker, April 29, 1969, LBJ Library).
50. "It could cost the United States $4 billion a year for the next N years to finally put a man on the moon and to explore outer space and the other planets. All in all, do you feel the space program is worth spending that amount of money or do you feel it isn't worth it?" (Here N changes depending on when the question is asked. Errors are typically 2%–3%.)

    Harris, October 1965: Worth it: 45%; Not worth it: 42%; Don't know: 13%.
    Harris, July 1967: Worth it: 34%; Not worth it: 54%; Don't know: 12%.
    Harris, April 1970: Worth it: 39%; Not worth it: 56%; Don't know: 6%.
    "Congress has been asked to approve a program, costing 7 to 9 billion dollars during the next five years, to enable the U.S. to send a man to the moon and bring him back safely. Do you think Congress should adopt the program or reject it?"
    Gallup Poll (AIPO), June 1961: Adopt: 42%; Reject: 46%; Don't know: 12%.

    There are many other polls during the same time span indicating similar sentiments.
51. Thirteen fires were reported on the Cuyahoga in the century before the noted one in June 1969.
52. Mailer makes great hay over how poorly military test pilots-turned-astronauts serve as public spokesmen. He takes other tacks to convince the reader how soulless the program was at heart, ending his book in an unrequited love song to an Apollo lunar regolith sample, almost as if he does not realize it is made of rock. He says a few interesting things but some are not original.

    Carl Sagan wrote more succinctly: "For me, the most ironic token of that moment in history is the plaque signed by President Richard M. Nixon that *Apollo 11* took to the moon. It reads, 'We came in peace for all Mankind.' As the United States was dropping seven and a half megatons of conventional explosives on small nations in Southeast Asia, we congratulated ourselves on our humanity. We would harm no one on a lifeless rock" (*Pale Blue Dot: A Vision of the Human Future in Space* by Carl Sagan & Ann Druyan, 1997, New York: Random House, 169).
53. I recall 21 feature-length, non-documentary films based on Apollo's 12-year program (portraying Apollo missions or craft) versus 25 for the Space Shuttle's three decades; for Apollo (films in **bold** grossed over $2 million in box office): *The Reluctant Astronaut* (1967), *Countdown* (1968), ***Marooned*** (1969), ***Diamonds Are Forever*** (1971), *Stowaway to the Moon* (1975), ***Capricorn One*** (1977), *Meteor* (1979), ***Superman II*** (1980), *The Falling* (1987), *Beyond the Stars* (1989), *Moontrap* (1989), ***Apollo 13*** (1995), *Apollo 11* (1996), ***Rocketman*** (1997), ***A Walk on the Moon*** (1999), *The Dish* (2000), ***Fly Me to the Moon*** (2008), *Moonshot* (2008), ***Apollo 18*** (2011), ***Transformers: Dark of the Moon*** (2011), and ***Men in Black 3*** (2012); for the Shuttle: ***Moonraker*** (1979), *The Noah's Ark Principle* (1984), ***Space Camp*** (1986), *Moontrap* (1989), *Challenger* (1990), *The Dark Side of the Moon* (1990), ***Loaded Weapon 1 (1993)***, *Rocketman* (1997), ***Deep Impact*** (1998), *Species II* (1998), ***Armageddon*** (1998), *Max Q* (1998), ***The Astronaut's Wife*** (1999), ***Space Cowboys*** (2000), ***Nutty Professor II*** (2000), *2001: A Space Travesty* (2000), *Rocket's Red Glare* (2000), *Stranded* (2002), *Riverworld* (2003), *Threshold* (2003), ***The Core*** (2003), *Earthstorm* (2006), ***Superman Returns*** (2006), *2012: Supernova* (2009), and ***Transformers: Dark of the Moon*** (2011). We note in Chapter 4 the role *Marooned* played in Soviet-American space cooperation. *Iron Sky* (2012) portrays Constellation-era spacecraft.
54. Michael Collins makes this point about 1960s astronauts' military personæ (*In the Shadow of The Moon: Extra Material, Lieutenant Aldrin*, directed by David Sington & Christopher Riley, 2007, Vertigo Films). In Europe cynicism about space exploration took an extra decade or more to set in (see *Imagining Outer Space: European Astroculture in the Twentieth Century*, edited by Alexander C.T. Geppert, 2012, Basingstroke, UK: Palgrave Macmillan).
55. As described in Chapter 2, the Soviets took manned military spacecraft seriously, as did the U.S. Air Force with Blue Gemini and the Manned Orbital Laboratory, canceled in 1963 and 1969, respectively, before launching any crew. Some authors imagine that manned battle stations will return, for example, *The Next 100 Years: A Forecast for the 21st Century* by George Friedman, 2009, New York: Doubleday. Before von Braun, some even imagined military missile bases on the Moon, for example, "Rocket Blitz from the Moon" by Robert Richardson, October 23, 1948, *Collier's*, 23). (Richardson was a Mt. Wilson Observatory astronomer.)

Von Braun proposed Project Horizon, a 12-man lunar missile base, in a U.S. Army study in 1959, to be built in 1965. 36

56. *History of Rocketry and Astronautics*, edited by John Becklake, 1995, American Astronautical Society History Series, Vol. 17, San Diego: Univelt, 421.
57. From the lyrics of the song "It Came Out of the Sky" by Creedence Clearwater Revival (1969): "Spiro came and made a speech about raising the Mars tax."
58. From an interview with Hans Mark, NRO director from 1977 to 1979:

    HAINES: They would tell you that the redesign costs for the shuttle is what drove the cost of those satellites up.
    MARK: You know, that's certainly true, but had they designed the goddamn things for the shuttle in the first place, it was their fault that they had to redesign it. Because they said we'll never go on the shuttle, and so when I got there I said sorry fellahs, that's crazy. You compromised capability that you could have, because for reasons I don't understand you don't want to use this launch vehicle. That is how things evolve. Harold Brown was one who believed what I said, then went to persuade the President to get the shuttle out of its problems. The shuttle was in fact sized to launch HEXAGON. The size of the payload bay was determined by HEXAGON.
    HAINES: Which was a large load?
    MARK: It was a large spacecraft. HEXAGON was a compact spacecraft compared to [redacted].

    ("Hans Mark Interviewed by Gerald Haines, March 12, 1997, Chantilly, VA" released by NRO, February 9, 2012; "Between the Darkness and Light" by Dwayne A. Day, June 25, 2012, *The Space Review*.)

59. "Kennedy Puts Needs of Earth Ahead of the Space Program" by Robert Reinhold, May 20, 1969, Associated Press, regarding Senator Edward Kennedy. (Nonetheless Kennedy brought NASA centers home to Massachusetts, for example, the Electronic Research Center in Cambridge, 1964–1970.)
60. According to Paul Dembling, general counsel to NASA from 1958 to 1969, Webb first advised von Braun to use such caution in 1964 (Telephone interview, August 5, 2010).
61. This could be lighthearted, as in the famous quip by comedian Mort Sahl in rejoinder to the title of von Braun's 1960 biopic *I Aim at the Stars* ("but sometimes I hit London!" – a line taken from the script of the movie itself). I recall this joke repeated on TV by Dick Cavett, for instance. In 1960 public protests were also directed at von Braun and the V-2, and thereafter he might find himself heckled at public appearances occasionally. Tom Lehrer's satirical song *Wernher von Braun* (1965) is less lighthearted but amusing at his expense. [4.1] Even the earliest coverage of von Braun discussed his Nazi past e.g., "Nazi Scientists in the U.S." (*Life*, December 9, 1946), but for years at a time no such mention would appear, at least in the United States. Scathing articles could be read more often in Germany, e.g., "The Secret of Huntsville" by Julius Mader, 1962, *Forum*, 16, 36.
62. *Von Braun: Dreamer of Space, Engineer of War* by Michael J. Neufeld, 2007, Random House: New York, 161.
63. "A Bibliography of Wernher von Braun with Selected Biographical Supplement 1930–1969" Marshall Space Flight Center History Office, http://history.msfc.nasa.gov/vonbraun/vbbiblo.html and "Wernher von Braun, A Register of His Papers in the Library of Congress" by Allan J. Teichroew, 1978, Manuscript Division, Library of Congress.
64. *New York Times*, February 23, 1975: Von Braun said that "after the successful lunar flights, the agency lost its impetus. The challenges had apparently disappeared. Besides, there were already too many old plans to be implemented and there was too little money with which to do it. The so-called 'grand tour of the planets' was scratched and the nuclear rocket was mothballed. We were told that perhaps more funds would be forthcoming at some later date. But I had to have challenges and I suppose I couldn't wait until the grass got greener."

    A few years before, von Braun was still bullish about the Moon, e.g., "Von Braun Sees Moon Visit" *New York Times*, September 7, 1971, about his expectations for a Moon base in 10 years, and "Birth of Child on Moon Foreseen by Von Braun" *New York Times*, January 7, 1972.
65. As it was the lunar experiments, ALSEP, were turned off September 30, 1977, to save a few hundred thousand dollars a year. NASA needed ALSEP's control room to attempt reactivating *Skylab*. Many instruments were still supplying useful data (see "ALSEP Termination Report," 1977, NSSDC, NASA, http://nssdcftp.gsfc.nasa.gov/miscellaneous/documents/b32116.pdf).
66. For instance, the U.S. Air Force's 89th Airlift Wing, with an active duty work force of 1,153, maintains and flies more than 20 airplanes, notably Air Force One (a modified Boeing 747), as well as global airlift, logistics, and

communications for the president, vice president, cabinet, war commanders, and other senior officers and politicians. These must fly without compromising safety.

67. "NASA Leadership and America's Future in Space: A Report to the Administrator" by Dr. Sally K. Ride, August 1987, http://history.nasa.gov/riderep/cover.htm.
68. "Presidential Directive on National Space Policy, January 5, 1988 – Fact Sheet" (February 11, 1988). See http://www.hq.nasa.gov/office/codez/new/policy/nsdd_293.htm.
69. "National Security Policy for Space Needed for US," *U.S. Policy Information and Texts*, November 25, 1987, cited in "Dominance in Space – A New Means of Exercising Global Power?" by Stephan F. von Welck, 1988, *Space Policy*, 4, 310. Bush called for further development of SDI and continued general U.S. space dominance over the Soviet Union.
70. This history is summarized in *Mars Wars: The Rise and Fall of the Space Exploration Initiative* by Thor Hogan, 2007, NASA SP-2007–4410.
71. "Early Lunar Resource Utilization: A Key to Human Exploration" by B. Kent Joosten & Lisa Guerra, May 1993, *AIAA Space Programs and Technologies Conference*, AIAA 93–4784.
72. For instance: "Why Scientists Believe ISS is Waste of Money" by Jim Wilson, December 2002, *Popular Mechanics*, 179, 12, 44. Also, "Research planned for the *ISS* is merely an extension of the sort of science conducted on the Space Shuttle over the past 20 years. The research is not wrong, it is just not very important. No field of science has been significantly affected by research carried out on the Shuttle or on *Mir* at great cost. Much of it has never even been published in leading peer-reviewed journals." (Testimony of Robert L. Park concerning the International Space Station, October 29, 2003, before the U.S. Senate Committee on Commerce, Science and Transportation). To be fair, many American science experiments were lost on *Mir* in its June 25, 1997, collision with a Progress-M resupply ship that punctured the *Spektr* science module ("NASA's 'Can-Do' Style is Clouding Its Vision of Mir" by James Oberg, September 28, 1997, *Washington Post*).
73. In July 2004, 77% of Americans indicated awareness of the VSE and 69% approved of the plan. A year later 77% approved of the plan ("Public Opinion Regarding America's Space Program" by the Gallup Organization, July 2004 and July 2005).
74. "A Budgetary Analysis of NASA's New Vision for Space Exploration: A CBO Study" by David Arthur, Adrienne Ramsay & Robie Samanta Roy, September 2004, Washington, DC: Congressional Budget Office.
75. *Space Vehicle Design* by Michael D. Griffin & James R. French, 1991, Washington, DC: American Institute of Aeronautics and Astronauts.
76. "Extending Human Presence into the Solar System, an Independent Study for the Planetary Society on Strategy for the Proposed U.S. Space Exploration Policy" by Owen Garriott, Michael Griffin & the Planetary Society "Aim for Mars" Study Team, July 2004, Pasadena: Planetary Society; *NASA's Exploration Systems Architecture Study, Final Report* by Doug Stanley et al., November 2005, NASA TM 2005–214062.
77. "Think of it as Apollo on steroids," Michael Griffin said unveiling Constellation at a September 19, 2005, press conference. Orion would take four people to the Moon (versus three for Apollo's CSM). Altair would take all four to the lunar surface for a week versus a crew of two for three days with Apollo's LM, with 46 tonnes mass (versus 15). Alternatively, Altair could become a crewless cargo lander. Orion could carry up to six to the *ISS*.
78. "Astronomy and NASA – Rewards and Challenges" by Michael D. Griffin, January 8, 2008, Keynote address, American Astronomical Society, http://www.nasa.gov/pdf/207768main_American_Astronomical_Society_8_Jan_08.pdf.
79. One can compare early optimism ("NASA Administrator: 'We Can Go to Mars" by Associated Press, May 31, 2005) versus later gloom (testimony of NASA administrator Michael Griffin to U.S. Senate subcommittee on Space, Aeronautics and Related Sciences hearing on NASA FY 2008 budget, February 28, 2007). I have heard Dr. Griffin estimate this shortfall from the promised at several billion dollars ("The Importance of Human Spaceflight" by Michael Griffin, May 21, 2009, public lecture at Wings Club, New York).
80. "Barack Obama's Plan for Lifetime Success Through Education" by Obama'08, November 2007, BarackObama.com, https://s3.amazonaws.com/obama.3cdn.net/ a8dfc36246b3dcc3cb_iem6bxpgh.pdf.
81. http://wayback.archive.org/web/20071215000000*/http://my.barackobama.com/page/-/HQpress/112007 education plan 1.pdf.
82. "Barack Obama's Plan for American Leadership in Space" by Obama'08, January 10, 2008, BarackObama.com, http://www.spaceref.com/news/viewsr.html?pid=26647.

83. "Obama a Little Confused about Today's State" by John McCormick, March 7, 2008, http://www.swamp politics.com/news/politics/blog/2008/03/obama_a_little_confused_of_w_s.html.
84. See "President Barack Obama on Space Exploration in the 21st Century" by White House Office of the Press Secretary, April 15, 2010, 2:55 PM, John F. Kennedy Space Center, http://www.nasa.gov/news/media/trans/obama_ksc_trans.html; also "Apollo 11 Tribute at the White House" by Staff, July 20, 2011, *New York Times*.
85. "Does Obama Want to Ground NASA's Next Moon Mission?" by Jeffrey Kluger, December 11, 2008, *Time.com*.
86. "Assessing the Options of the Augustine Committee for Human Spaceflight" by Scott Pace, September 28, 2009, Space Policy Institute Symposium, http://www.gwu.edu/~spi/assets/docs/092809symposium.pdf.
87. *Seeking a Human Spaceflight Program Worthy of a Great Nation* by Review of U.S. Human Spaceflight Plans Committee, October 2009, Washington, DC; *Summary Report*, September 2009.
88. "White House Won't Fund NASA Moon Program" by Robert Block & Mark K. Matthews, January 27, 2010, *Los Angeles Times*.
89. Recall that even Constellation required Earth rendezvous, but Apollo did not. Extracting the LM on the way to the Moon simplified engineering for Apollo because Saturn's three stages need not deal with weightless propellant tanks and effects of bubbles. (The S-IVB third stage would fire gently even in orbit to settle the propellant tanks.)

    Falcon Heavy could not carry a Constellation Altair/Orion to the Moon in two flights. Earth rendezvous with three launches would require a day or more in LEO, not Apollo's one or two orbits. Apollo used a very low Earth orbit, which saved energy. Since that orbit would begin decaying in one day, Falcon Heavy would need more energy for LEO.
90. "Key Issues and Challenges Facing NASA: Views of the Agency's Watchdogs, House Committee on Science and Technology, Subcommittee on Space and Aeronautics" by Laura M. Delgado, February 3, 2010, Hearing Summaries, Space and Technology Policy Group, http://www.spacepolicyonline.com/images/stories/House_NASA_issues_Feb_3_2010_final.pdf; "Gordon Has Heavy Load in Final Year on Science" by Staff, February 7, 2010, *Murfreesboro Daily News Journal*; "An Agency in Transition" by Jeff Foust, February 8, 2010, *Space Review*, http://www.thespacereview.com/article/1560/1; "Congressmen Blast Obama Decision to Kill NASA," February 4, 2010, http://larouchepac.com.
91. Excerpt of public meeting with President George W. Bush, Cleveland, Ohio, July 10, 2007:

    QUESTION: Given all the competing demands for resources in Washington, what kind of funding do you see for NASA and its mission going forward?

    PRESIDENT BUSH: Yes. That's an awkward question to ask a Texan. (Laughter.) I think that NASA needed to become relevant in order to be – to justify the spending of your money, and therefore, I helped changed the mission from one of orbiting in a space shuttle – in a space station to one of becoming a different kind of group of explorers. And therefore, we set a new mission, which is to go to the moon and set up a launching there from which to further explore space. And the reason I did that is, I do want to make sure the American people stay involved with – or understand the relevance of this exploration. I'm a big – I support exploration, whether it be the exploration of new medicine – that would be like NIH grants – the exploration of space through NASA. I can't give you the exact level of funding.

    I would argue with you that we got a lot of money in Washington – not argue, I'll just tell you, we got a lot of money in Washington. (Laughter.) And we need to make sure we set priorities with that money. One of the problems we have in Washington is that unlike the books I saw at the hospital – of which, you're on the board – that said 'results', we're not very good about measuring results when we spend your money. A lot of time the programs sound nice; a lot of time the results don't match the intentions. So one of the things I've tried to do through the OMB is to be results-oriented, and when programs don't meet results, we try to eliminate them. And that's hard to do. Isn't it, Steve? Yes. But, no – I believe in exploration, space exploration. And we changed the mission to make it relevant. Thanks. (http://georgewbush-whitehouse.archives.gov/news/releases/2007/07/20070710-6.html)
92. The NRC urges NASA (and Congress) do one or more of four things: close infrastructure and dismiss staff inessential to a coherent strategy; share efforts with commercial, international, and interagency federal partners; abandon one of its four main areas, human space flight, Earth and space science research, technology development, or aeronautics; and/or raise NASA's budget significantly. Primarily NASA must

have coherent goals, to set priorities and plan, adopt a motivated budget, intelligently develop and incorporate new and current technologies, and get things done. Implied is that these must stem the bleeding: good people depart NASA to work in private industry, defense, and even other countries where funding is more stable and planning rational. Regarding missions: "The committee has seen little evidence that a current stated goal for NASA's human spaceflight program – namely, to visit an asteroid by 2025 – has been widely accepted as a compelling destination by NASA's own workforce, by the nation as a whole, or by the international community. On the international front, there appears to be continued enthusiasm for a mission to the Moon but not for an asteroid mission, although there is both U.S. and international interest in robotic missions to asteroids. This lack of consensus on the asteroid-first mission scenario undermines NASA's ability to establish a comprehensive, consistent strategic direction that can guide program planning and budget allocation" (*NASA's Strategic Direction and the Need for a National Consensus* by Committee on NASA's Strategic Direction, National Research Council, December 2012, Washington, DC: National Academies Press).

93. Administration response has been more authoritarian than thoughtful. In an April 5, 2013, meeting of the Space Studies Board, NRC study leader Albert Carnesale confronted NASA Administrator Bolden with the NRC findings, adding that much enthusiasm existed for lunar exploration and that mainly politicians' pride stood in the way. Bolden offered NASA potential willingness to join another country's Moon program as junior partner but not as lead. "NASA is not going to the Moon as a primary project probably in my lifetime," Bolden stated. He posited that if the next administration were to set course back to the Moon, "it means we are probably, in our lifetime, in the lifetime of everybody sitting in this room, we are probably never again going to see Americans on the Moon, on Mars, near an asteroid, or anywhere. We cannot continue to change the course of human exploration" ("NASA's Bolden: No American Led Return to the Moon 'in My Lifetime'" by Mark Whittington, April 6, 2013, *Houston Space News: Examiner*.com).

94. One KSC contract worker maintained elevators at Launch Complex 39 taking astronauts to the gangway into the Shuttle's entrance hatch. He said he might find a job maintaining elevators elsewhere, although not near Canaveral, but would never again find a job with a sense of purpose and spirit of mutual direction as at KSC. The most meaningful phase of his career is gone. Down the road in a Cocoa Beach restaurant a waitress said she had asked several NASA engineers how they were doing. One of them had responded: "Can I have your job? Really, I'm serious."

    Public relations and outreach about NASA often now appeal to science fiction's popularity. (As of 2011 the Kennedy Space Center Visitors Complex – not actually run by NASA – has three new exhibits on *Star Trek*.) When discussing space exploration's future, phrases such as "the power and innovation of American private enterprise" are used more often, but largely without detail.

    As NASA attempts to reach a sensible plan to continue human space exploration, it is demoralized by such headlines as "Is the Space Age Over?" (BBC World News promotion, 2011) or "The space-shuttle program is coming to a quiet end. Is the same true for the era of space exploration?" ("Earthbound" by Hanna Rosin, September, 2010, *Atlantic*).

95. "The Top American Science Questions: 2012; Candidates' Answers, a Side by Side Comparison" Science Debate 2012, September 4, 2012, http://www.sciencedebate.org/debate12/.

96. "Gravitational Tractor for Towing Asteroids" by Edward T. Lu & Stanley G. Love, *Nature*, 2005, 438, 177; "Using a Gravity Tractor to Help Mitigate Asteroid Collisions with Earth" by D.K. Yeomans, S. Bhaskaran, S.R. Broschart, S.R. Chesley, P.W. Chodas, M.A. Jones, T.H. Sweetser, E.T. Lu & R.L. Schweikart, 2008, *Asteroids, Comets, Meteors*, 8273. Also consider irradiation by sunlight concentrators: "Non-nuclear Strategies for Deflecting Comets and Asteroids" by H.J. Melosh, I.V. Nemchinov & Yu.I. Zetzer, 1994, *Hazards Due to Comets and Asteroids*, 1111.

97. "Plymouth Rock: An Early Human Asteroid Mission Using Orion" by Josh Hopkins & Adam Dissel, November 2009, presentation to Small Bodies Assessment Group, http://www.lpi.usra.edu/sbag/meetings/sbag2/presentations/PlymouthRockasteroidmission.pdf.

98. "Asteroid Retrieval Feasibility Study" by John Brophy, et al., April 2, 2012, Keck Institute. As new as this idea might seem, it has been receiving attention for more than 50 years: "Someday, we will be able to bring an asteroid containing billions of dollars worth of critically needed metals close to Earth to provide a vast source of mineral wealth for our factories." – Lyndon B. Johnson, in a speech at the Seattle World's Fair, May 10, 1962

99. "Sickening Solar Flares" by Tony Phillips, May 8, 2011, http://www.nasa.gov/mission_pages/stereo/news/stereo_astronauts.html.

100. Jim Bell, Planetary Society president, worries about FP: "The risk is that we'll lose all this momentum that we have in understanding the worlds around us." The Society once endorsed early versions of the plan. ("Planetary Society Statement on Obama Administration's Proposed Space Exploration Plan and Fiscal Year 2011 NASA Budget: Society Urges Congress to Endorse New Plans for Human Space Flight" by Susan Lendroth, February 1, 2010, Planetary Society, press release, versus "Scientists: Scrapping Space Programs Short-Sighted" by Paul Strand, November 6, 2011, *CBN News*). The Mars Society (Robert Zubrin, President) criticized VSE's Return to the Moon, suggesting directly targeting Mars instead: "Since a lunar-class transportation system is adequate to reach Mars using this plan, it is rational to consider a milestone mission, perhaps five years into the program, where a subset of the Mars flight hardware is exercised to send astronauts to the Moon, or more likely to a near-Earth asteroid, as a NEO mission requires no extraneous equipment that is not available in the basic Mars mission hardware set." Zubrin supported Obama's no-lunar approach, urging NASA bypass the Moon and send astronauts to Mars instead, but quickly turned against FP ("Accepting the Challenge Before Us, Testimony of Dr. Robert Zubrin to the Committee for Review of U.S. Human Space Flight," August 5, 2009, and "Obama Readies to Blast NASA" by Robert Zubrin, October 26, 2011, *Washington Times*). Buzz Aldrin, whose public rejection of lunar plans was instrumental to Obama's space policy unveiling, by September 2010 favored an International Lunar Development Corporation to promote commercial lunar development partnerships ("The Role of a Lunar Development Corporation in Facilitating Commercial Partnerships in Lunar Exploration" by Buzz Aldrin, Thomas L. Matula & Stan Rosen, September 14, 2010, *Lunar Exploration and Analysis Group*, 3075.)

101. However, not terms of leadership in Russia/Soviet Union, counting only those with power established more than about one year, hence not Rykov, Malenkov, Andropov, or Chernenko. (I am not touching the issue of Medvedev/Putin.) Effectively, terms are longer in China, as well.

102. A change in orbit would help and is plausible with Roscosmos's new site in French Guiana. The current *ISS* orbit, inclined 51.6° to the equator, requires significantly more energy to reach.

103. For a cynical but perhaps realistic view of space exploration in American two-party politics, I quote a recent opinion piece by Robert Zimmerman: "Both parties excel at feigning interest in space exploration for the purpose of justifying pork to their districts." Also: "The result is that America's incoherent space program is unable to accomplish anything except spend money the federal government doesn't have." ("No Liftoff for The Space Flights of Fancy" by Robert Zimmerman, August 13, 2013, *The Wall Street Journal*, http://online.wsj.com/news/articles/SB10001424127887324769704579008820953158510.

104. *Digital Apollo* by David Mindell, 2008, Cambridge, MA: MIT Press, 127. (DARPA made another significant contribution to microprocessor development as well.)

105. We have more polling information from the 1960s and 1970s above[49] and in Chapter 14, but consider the following:

    It is now *N* years since the United States first landed on the Moon. Do you think the space program has brought enough benefits to this country to justify its costs?

	Yes	No
1979**	41%	53%
1994*	47	47
1999*	55	40
2003	65	29
2009	51	43

    Was the Moon landing worth the effort?

	Yes	No
1979*	47%	49%
1999*	71	24
2009	70	27

    *Gallup, **NBC/AP

"Do you think spending on the U.S. space program should be increased, kept at present level, reduced or ended altogether?" (Gallup; in some cases more than once in a given year)

	1982	1986	1989	1991	1993	1998	1999	2003	2006	2009
Present or increased:	69%	76%	69%	65%	46%;53%	68%	63%;65%	74%;75%	65%	60%

("Majority of Americans Say Space Program Costs Justified, Percentage has Grown Since 1979" by Jeffrey M. Jones, July 17, 2009, Gallup, Inc.; "Thumbs Up for Apollo 11; Current Efforts, Less So" by Gary Langer, July 18, 2009, *ABC News*.) Also see "Public Opinion Polls and Perceptions of US Human Spaceflight" by Roger D. Launius, 2003, *Space Policy*, 19, 163.

106. "Space Station's Only Flight Plan Leads to Limbo" by Mike Thomas, April 11, 2002, *Orlando Sentinel*. Senator Nelson reportedly won the argument with Van Allen with the quip "No Buck Rogers, no bucks." (This inverts Tom Wolfe's 1979 *The Right Stuff* quote: "Gordon Cooper: You boys know what makes this bird go up? Funding makes this bird go up. Gus Grissom: He's right. No bucks, no Buck Rogers.") Polls show the public prefers funding human missions: "Some people say the United States should concentrate on unmanned missions like the Voyager probe. Others say it is important to maintain a manned space program as well. Which comes closer to your view?": January 29–30, 1986: unmanned: 21%, manned: 67%, no opinion: 12%; February 2, 2003: unmanned: 22%, manned: 73%, no opinion: 5%. Does the public know the relative costs of these two modes of exploration?

107. For a skeptical analysis of space advocacy assumptions, see "Reclaiming the Future: Space Advocacy and the Idea of Progress" by Taylor E. Dark III, for the Societal Impact of Space Flight Conference, NASA History Division and National Air and Space Museum Division of Space History, Hirshhorn Museum, Smithsonian Institution, Washington, DC, September 19–21, 2006.

108. When asked, Americans usually radically overestimate NASA's budget, with less than 10% answering correctly to within a factor of two, and most thinking it is more than 20% of the federal total (Robert Launius, "Public Opinion Polls and Perceptions of US Human Spaceflight"). "The question was asked without reference to the actual budget allocations, participants 'voted' for maintaining, decreasing or increasing the NASA budget in very similar percentages: 35% felt it should be retained at current levels, 30% felt it should be increased, and 35% felt it should be decreased. Later, respondents were presented with the percentage of the federal budget represented by NASA's current budget request, described as 'less than 1% – approximately seven-tenths of 1% – of the federal budget,' and asked again about funding levels. A shift occurred, with 42% now supporting an increase, 29% suggesting it remain at current levels, and 29% supporting a decrease in funding" ("Some Results from Dittmar Associates' Market Study of the Space Exploration Program" by Mary Lynne Dittmar, 2005, http://www.dittmar-associates.com/Publications/Selected Results from The Market Study.pdf).

109. Von Braun resuscitated this idea, but favored missiles on surface ships over SLBMs ("Rockets Launching from Ships" by Wernher von Braun, Herbert Axster, Hannes Luehrsen, Eduard Fischel, Helmut Schmitt & Hermann Lange, 1947, Ordinance Research and Development Division, Rocket Sub Office, Fort Bliss, Texas). Polaris and Soviet R-13 SLBMs entered service in 1961.

Chapter 4
An International Flotilla

> You too may be a big hero,
> once you learn to count backwards to zero.
> "In German or der English, I know how to count down ...
> Und I'm learning Chinese," says Wernher von Braun.
> – Tom Lehrer, 1965, "Wernher von Braun"[1]

Lehrer's song criticizes von Braun's heroic status in 1960s America despite his fickle national allegiances two decades before; the song aired when his luster in American public opinion was tarnishing (Chapter 3). The Chinese reference is biting, because, several months before, China had exploded her first atomic bomb, the fifth nation to do so, after the United States, Soviet Union, Britain, and France. (Von Braun had little problem counting but admitted before the U.S. House of Representatives having once failed physics and mathematics.[2]) Outer space was the call drawing him down his brilliant if morally ambiguous path, not patriotic inspiration. By age 17 his interest in space flight set the course that determined his career.

Lehrer did not know how ironic his reference to Chinese would become, for two years later, on June 17, 1967, China exploded her first hydrogen bomb, 150 times more powerful than the one Lehrer knew. China took 3 years to transition from fission to fusion weapons, compared to 6 years for the United States, 4 for the Soviet Union, 5 for Britain, 9 for France, and 24 years for India. In October 1966, China launched a nuclear warhead on an intermediate-range missile, detonating it at their Lop Nor test site in Xinjiang (after lofting it over populated Chinese territory).

Chinese rocketry's father was a would-be American but already knew Chinese. Tsien Hsue-shen (Qian Xuesen or 钱学森), born in Hangzhou, several months older than von Braun, moved to America at age 24,[3] helped found the Jet Propulsion Laboratory in 1943 for Caltech and the army, interrogated the captive von Braun and others late in World War II, and helped introduce German rocketry to the United States. Seeking U.S. citizenship in 1950, Tsien ran afoul of the McCarthy-era Red Scare, was denied citizenship, arrested, and stripped of security clearances required for his work. "It was the stupidest thing this country ever did. He was no more a Communist than I was, and we forced him to go," said Navy Undersecretary Dan A. Kimball.[4] Five years later he was released and deported to China, and after several years joined the Communist Party (for the first time) and convinced the

People's Republic to start a modern rocketry program. Thus the United States, having neglected modern rocketry's father (Goddard) and inherited from Germany the father of the American space program (von Braun), proceeded forcibly to repatriate Chinese rocketry's founder, more from American clumsiness and obsession than any disloyalty or treachery of Tsien's. The result several years later was the infamous Silkworm rocket, used to great effect by Iraq and Iran against each other and Kuwait, striking Persian Gulf oil installations and shipping (including U.S. flagged vessels). Tsien's work led to the *Long March*, backbone launcher of China's space program, its *Shenzhou* spacecraft. Earlier Chinese hardware was reverse engineered from Soviet designs or granted willingly, with Tsien playing a central role, especially before Mao's 1962 denunciation of Khrushchev as a capitalist sympathizer. Tsien was invited by the United States in 2002 to pay a return visit, but he refused pending a U.S. government apology for mistreating him five decades earlier. He died in 2009, aged 97.

Despite the self-defeating nature of that affair, long after the Tsien episode the U.S. government encountered sufficient real cause to suspect official Chinese motives. Someone stole for China plans for the Trident submarine's compact W88 and W70 thermonuclear warheads from the United States (presumably from Los Alamos National Laboratory where they were developed). In part this inspired prosecution of Los Alamos scientist Wen Ho Lee (born in Taiwan), later awarded a $1.6 million settlement and an apology from President Clinton for mistreatment by the United States, despite Lee's conviction for mishandling classified documents and admission of failure to report contacts with agents seeking his help in developing nuclear missiles for China.

U.S. mistrust of Chinese motives culminated with several *Long March* rocket launch failures, especially one on February 15, 1996, carrying U.S.-made telecommunications satellite *Intelsat 708*, which hit and devastated a village one kilometer downrange. Despite survival of many satellite components, never recovered was export-controlled technology from the satellite's encryption system, which many suspect was secreted away by the Chinese. The U.S. Department of State charged Boeing, Hughes, and particularly Space Systems/Loral after this and after American engineers allegedly cooperated with Chinese accident investigators, violating arms technology export control laws.[5] Since this era, International Traffic in Arms Regulations (ITAR) is strictly enforced, policed no longer by the Commerce Department but now the Department of State, covering wider ranges of technologies, practices, and knowledge. It persistently requires U.S. scientists and engineers now in international meetings and collaborations to swallow their words and ponder if they too might run afoul of ITAR. Whereas these Cold War restrictions initially concerned mainly cryptography, they now extend to many countries, restricting such varied enterprises as the British-American Virgin Galactic/Scaled Composites commercialization of suborbital space flight or launching European satellites with U.S. components. Boeing strains to cleanse ITAR technology from its airliners that are selling overseas. Non-U.S. vendors and buyers capitalize on or complain bitterly about ITAR limits. European aerospace company Thales Alenia Space builds an ITAR-free satellite technology line. Whether one agrees or not with

ITAR's implementation, it dominates American cooperation in international space missions, beyond mistrust of particular nations as a result of unfortunate history. Only recently have some of these restrictions been relaxed to what some consider more reasonable levels.[6]

Despite its rapid rise in strategic warheads and rockets, China was not next to follow America and Russia to the Moon and beyond. That was Japan. Its *Hiten* spacecraft became the first lunar probe in almost 14 years (following *Luna 24* in 1976), and although carrying only one science instrument, a dust detector (from Munich Technical University), *Hiten* released into lunar orbit a 12-kilogram microsatellite *Hagoromo* with instruments to sense lunar temperatures and electric fields. Unfortunately, *Hagoromo*'s radio transmitter failed, but telescopic observation detected it in lunar orbit. *Hiten* was an innovative engineering mission: first to use *aerobraking* – exploiting an atmosphere (Earth's) rather than rockets to change orbit (not just re-entering the atmosphere), and the first lunar mission using a low-energy "weak stability boundary" transfer trajectory from Earth to the Moon (albeit in months rather than days like Apollo), the first major innovation in lunar transfer orbit concepts in decades.[7] Many missions now use this form of transfer orbit. Although *Hagoromo* was lost, *Hiten* operated several years until it was crashed intentionally into the southeast lunar Near Side in April 1993.

Ten years later Japan's space agencies (now merged into JAXA: Japanese Aerospace Exploration Agency) flew *Hayabusa* past the Moon, snapping photographs en route to land on asteroid Itokawa and return to Earth with samples. This first ever asteroid sample-return mission succeeded in part, returning several tens of micrograms of asteroidal dust. Nonetheless, *Hayabusa* was only the second craft (after NASA's *Deep Space 1*) using an ion drive rocket beyond Earth's orbit, another major advance, as will be described later in this chapter and in Chapter 13.

The Japanese space program is simultaneously modest and ambitious. Unlike China's space program, subsumed within the military, JAXA is a civilian effort. Its major launch sites, Tanegashima and Uchinoura (Kagoshima) both near the southernmost tip of Japan's south main island Kyushu, are allowed less than 18 launches annually and until recently launched only in two periods in winter and summer to avoid disturbing the fishing catch.[8] In Earth's orbit, JAXA maintains wide varieties of Earth resource and space astronomy satellites. With 279 billion yen annually (some US$3 billion), JAXA makes impressive advances. *IKAROS* (*Interplanetary Kitecraft Accelerated by Radiation Of the Sun*), the first functional solar light sail, flew by Venus and studied the Sun, employing propulsion concepts that might allow cheaper probes of the planets and beyond. Consequently, *IKAROS* is nearly the largest craft (in area) beyond Earth's orbit, and JAXA plans a larger version later this decade to explore the outer planets. Japan maintains a major module (Kibo) on the *ISS* and automated cargo vessels (*HTV* or *H-II Transfer Vehicle*) that have supplied it. JAXA is the third largest contributor of permanent *ISS* hardware (Figure 4.1). Although Japan does not launch people into space, nine Japanese have orbited Earth, as many as any other nation except the United States, Soviet Union/Russia, and close behind China and Germany. Japan has announced plans to explore the Moon, which we discuss later.

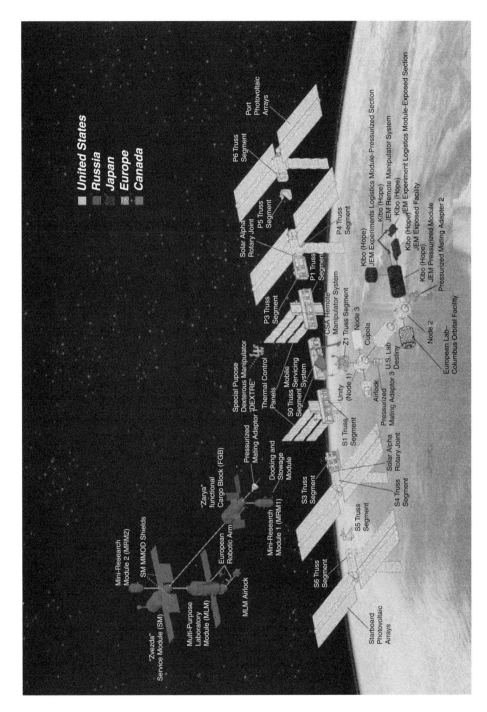

Figure 4.1. **Sum of parts.** *International Space Station* components shaded according to nation of origin. U.S. components in white stretch across most of the lower half of the picture. Russian components are dark, upper left. Japanese components are black, lower right. ESA's Columbus Orbital Facility is bottom center, its Robotic Arm is upper left. The Canadian Remote Manipulator System and DEXTRE are at center. (Graphic courtesy of NASA.)

JAXA's prime contribution to lunar exploration is the robotic mission *SELENE* (*SELenological and ENgineering Explorer*, gaining the moniker *Kaguya* – a fairy-tale princess who flew to the Moon – after a public naming competition), from September 2007 to June 2009. *Kaguya* was the then-largest lunar payload (2.9 tonnes) since Apollo. It consisted of three satellites, the main satellite and two 53-kilogram subsatellites: the *Okina* relay satellite to maintain communications with Earth even with *Kaguya* behind the Moon and the *Ouna* relative velocity satellite. *Ouna* allowed superlative detail in mapping the gravitational field of the lunar Far Side, something previously accomplished only by inferring the mass distribution by seeing how an orbit changed between disappearing and reappearing from behind the Far Side. Before the double-probe *GRAIL* mission in 2012, this was our best measurement of mass irregularities across the Moon, with implications for the lunar interior structure.

Kaguya had a powerful imaging camera and laser altimeter on board. The camera, especially, had the most diverse set of imaging filters of any high-resolution camera, useful for describing mineralogical properties and surface ages across the Moon. In particular, these data are useful for finding the age of many lunar lava flows, some only half the Moon's age. In other areas the camera discovered exposed anorthosite, thought to compose most of the underlying lunar crust. The laser altimeter made some 6.8 million height measurements across the Moon and permitted uniquely refined measurements of great use: where on the Moon (near the poles) the Sun never sets, and how large the range of elevations is on the Moon (some 10% wider than originally thought), indicating the extremes the mountain heights supported by the lunar interior. *Kaguya* carried ground-penetrating radar, uncovering surprising details of the structure of lunar crust several hundred meters underground. We examine these recent results and those of other spacecraft later.

The European Space Agency (ESA) has explored new technology en route to the Moon, as well. After NASA it was first to propel a spacecraft via ion drive (in 2001), using the technology again in September 2003 to send *SMART-1* (*Small Missions for Advanced Research in Technology*) to the Moon. This was not for the impatient; the ion rocket took 13 months to deliver *SMART-1* to the Moon and another three to settle into its working orbit. As implied, this was a relatively small satellite but carried several instruments: an optical/infrared camera and spectrometer, an X-ray spectrometer and monitor, and a dust particle detector. *SMART-1* was intentionally crashed into the lunar surface on September 3, 2006.

Although *SMART-1* started the twenty-first-century wave of missions to the Moon, there had been precursors. In addition to *Hiten*, other missions filling the long gap since *Luna 24* in 1976 included *Clementine* in 1994 and *Lunar Prospector* in 1998, both launched by the United States.

Clementine (named from the American folk lyric: "Oh, my darling Clementine, thou art lost and gone forever") is intriguing. Officially named the *Deep Space Program Science Experiment* (*DSPSE*), it was built by the Department of Defense aided by NASA at low mission cost ($80 million) and rapidly (less than two years), although it suffered a major failure. After mapping the Moon, *Clementine* left to fly past asteroid (1620) Geographos to test Strategic Defense Initiative missile interceptor technology and then to be gone forever like the fabled miner's daughter. Unfortunately it never reached the asteroid, suffering a computer failure causing a catastrophic thruster misfire. Fortunately NASA had convinced the Defense Department of the valuable lunar data that *Clementine* could contribute, and this forms its lasting legacy.

Clementine was the first mission to map the entire Moon since Lunar Orbiter in 1966–1967 and the first ever to make Moon maps at multiple ultraviolet/optical/infrared wavelengths useful for differentiating lunar minerals. This database was essential to lunar science for over a decade. *Clementine* carried other useful instruments: a laser altimeter, charged particle detector, and radar transmitter (actually, *Clementine*'s radio communications transmitter). The latter provided tantalizing results: radio waves from *Clementine* received on Earth were modified in a way consistent with ice (presumably water) on the Moon's surface close to its North and South Poles. Versions of this measurement were repeated, some agreeing and others disagreeing with *Clementine*. This first report of lunar surface water merits detailed attention in Chapter 8.

Four years later *Clementine* was followed by *Lunar Prospector* (*LP*) in January 1998 starting a 19-month exploration of the Moon. This was an inexpensive NASA mission, only $63 million. To reduce cost the craft was spin-stabilized, maintaining orientation by serving as its own gyroscope. Hence, *LP* omitted cameras (which must be pointed), instead concentrating on the lunar particle and magnetic environment. Despite damage at launch to some instruments, *LP* was a major success. Its tool kit was powerful because radioactivity produced by various elements can reveal surface composition in ways distinct from *Clementine*'s imaging. How light, nonradioactive elements scatter radiation, and heavier but radioactive gas's behavior, were also studied by *LP*. Like *Clementine* it hinted at water (or at least hydrogen) near the lunar poles. Attempting to excavate some of this water as a vapor plume that might be seen in spectra, *LP* with its last gasp of fuel plunged deep into crater Shoemaker at the lunar South Pole. *LP*'s and *Clementine*'s data together form a synergistic resource that carried lunar science through the following decade. We revisit many of these results in later chapters.

One must appreciate the long hiatus between the Apollo/Luna frenzy and revival of lunar exploration two decades later by *Clementine* and *LP*: a significant and lasting influence on lunar science and a generation of its practitioners, particularly in the United States. Data and results from Apollo were overwhelming, and the delay in follow-up so protracted as to profoundly affect what we think about the Moon. This legacy is changing with the flood of data from the new millennium's lunar missions.

The three human space-faring nations (Russia, United States, and China) are later to the game in this new lunar ascendancy. Russia has sent no spacecraft to the Moon since *Luna 24* in 1976 (but plans to by 2014 with Luna-Resurs and Luna-Grunt). The United States re-engaged in June 2009 with the *Lunar Reconnaissance Orbiter* (*LRO*) and the *Lunar CRater Observation and Sensing Satellite* (*LCROSS*) – actually three spacecraft in total. China, after three decades launching satellites, sent its first probe *Chang'e 1* beyond Earth's orbit to the Moon in November 2007 carrying an optical/

infrared imager and spectrometer, an X-ray/gamma-ray detector, a microwave radiometer measuring subsurface temperatures, and a laser altimeter. The altimeter made a detailed lunar topographic map, although *LRO/LOLA* (Lunar Orbiter Laser Altimeter) supplanted it. *Chang'e 1* ended in March 2009, crashing near the lunar equator. In October 2010 China sent to the Moon *Chang'e 2*, almost identical to *Chang'e 1*, carrying the same instrumentation (with some parameters unspecified, despite inquiries),[9] except for an improved laser altimeter and a higher resolution camera (1.3 meter versus 120 for *Chang'e 1*). *Chang'e 2* flew past asteroid 4179 Toutatis after seven months in lunar orbit. *Chang'e 3* landed on the Moon in late 2013, but failed soon thereafter.

Before discussing *LRO/LCROSS*, we should consider the other lunar newcomer: India with her *Chandrayaan-1* ("Moon traveler") lunar orbiter and hard lander. This was the first spacecraft beyond Earth's orbit of the Indian Space Research Organization (ISRO). Nonetheless *Chandrayaan-1* was complex, with the main spacecraft in low polar orbit and a 35-kilogram *Moon Impact Probe* striking the Moon near the South Pole, deep in crater Shackleton. It carried a camera, radar altimeter, and mass spectrometer analyzing neutral atoms and molecules in the lunar atmosphere. The spectrometer is claimed to have found water molecules while descending to lunar impact, at levels higher than seen by Apollo on the surface near the equator. This was claimed as evidence of water migrating in the thin atmosphere to the colder poles.

Chandrayaan-1 found good luck with bad. Although suffering catastrophic failures mainly because of design flaws prematurely ending its planned two-year mission after only six months, it first made discoveries likely to change lunar science and exploration forever. The flaw was insufficient accommodation for spacecraft heating, especially from lunar thermal radiation. To keep cool during its short life, *Chandrayaan-1* cycled through operating its scientific instruments, sometimes only one at a time, and eventually backed away from a 100-kilometer orbital height to 200 kilometers, reducing the environmental heat load. Despite this, its guidance system faltered, then power supplies overheated, ending the mission. Beforehand, *Chandrayaan-1* detected unambiguously via its radar vast amounts of water frozen into areas near the lunar poles (roughly a cubic kilometer of ice) and clues via infrared spectrometry to how some of this water is produced (by hydrogen in the solar wind impinging on the oxygen composing 40% of the lunar soil). A third instrument even measured solar wind protons on their way into the soil and some bouncing back as hydrogen atoms. These results came from international collaborations between American, European, and Indian scientists, showing how effective open, cooperative research can be. We can expect other nations to enter the lunar exploration effort. South Korea has already announced its intention for a robotic lunar orbiter and lander by about 2020, and others may join.

LRO/LCROSS was a complex mission of three spacecraft operating in tandem, especially in the last critical moments when two of them struck the Moon. The 600-kilogram *LCROSS Shepherding Satellite* followed by four minutes the 2.3-tonne *Centaur* impactor in their suicide mission plummeting 2.5 kilometers per second into crater Cabeus A near the lunar South Pole. While 3 tonnes of projectiles were smaller than the 14-tonne Saturn V S-IVB third stage hurled at the Moon in

1970–1972 to rattle Apollo's lunar seismometers, *LCROSS*'s impact was special with *LRO* observing overhead – all at a crater previously indicated as a prime candidate for water ice. *LCROSS*'s spacecraft were carefully purged before impact of propellant (oxygen and hydrogen, as well as some hydrazine: N_2H_4), minimizing artificial hydrogen and organics in the impact debris.

As television drama, *LCROSS*'s impacts were anticlimactic (with no explosion seen in real time, only later after image processing) but worth the wait scientifically. Spectra from the impact (taken by *LRO*, too) showed geysers of water, about 6% by mass – a value more typical of water-soaked sands on Earth – except on the Moon, this is retained as ice. Astounding as this result is, equally impressive are trace substances seen: hydrogen sulfide (H_2S), ammonia (NH_3), sulfur dioxide, ethylene (C_2H_4), carbon monoxide, carbon dioxide, methanol (CH_3OH), methane (CH_4), calcium, magnesium, mercury, and molecular hydrogen (H_2) (see Chapter 8). Roughly 6% of the debris mass being composed of carbon compounds is stunning: only 0.005% of typical lunar soil is carbon.

LRO is still flying and producing results, as of 2014. These are significant to lunar science (showing, for instance, that the Moon is cooling and contracting, as recently as the last 20% of its age) and in terms of lunar exploration (detailing potential future landing sites and even favorable places for human habitation). In September 2010 *LRO* entered its science-dedicated mission phase, having completed its exploration support. We discuss *LRO* results in later chapters.

NASA launched in 2011 another mission to probe deep inside the Moon. *GRAIL* (*Gravity Recovery and Interior Laboratory*) satellites are conceptual clones of their terrestrial kin, *GRACE* (*Gravity Recovery and Climate Experiment*), famous for monitoring the mass leaking from the Greenland ice sheet as a result of global warming, finding a monster 300-kilometer crater candidate under the Antarctic ice, and detecting changes in Earth's crust as a result of the December 2004 earthquake, cause of the Indian Ocean tsunami.[10] *GRACE* discovered these without detecting a single photon from its subject. It simply monitors gravity's force difference between the two satellites by watching their separation by bouncing radio waves in between (measuring the frequency shift). *GRAIL* is essentially the same system with a different satellite superstructure or "bus," in lunar orbit.[11] The two satellites, 200 kilometers apart, fall one at a time into a region of higher gravity then climb out again, always changing distance as revealed by the radio ranging system to within several microns (a tiny fraction of a human hair's width). *GRAIL* detailed the density of the deep interior, mantle, and crust of the Moon, and in its first two weeks of science operations mapped the lunar gravitational field better than all previous missions combined. We discuss in Chapter 7 what it reveals.

A new NASA mission launched in September 2013 (from Wallops Island, Virginia) is *LADEE* (*Lunar Atmosphere and Dust Environment Explorer*), designed primarily to measure one quantity: lunar atmosphere composition as a global average – actually several values because many major atomic and molecular species might exist. It also carries a camera and spectrometer to detect light scattered by dust over the lunar surface and optical/ultraviolet light emitted by gas above the Moon, for example, sodium sputtered from the surface by solar wind particles.

Although *LADEE*'s goals are interesting, it deserves criticism. Introduced in 2008 as a $70-million add-on to *GRAIL*, and not vetted as usual to avoid delaying its launch by 2011, later *LADEE* was off-loaded onto its own rocket, tripling the mission cost, but its goals were not reviewed despite a new profile.

LADEE is an internal NASA project involving several of its centers to measure the lunar atmospheric composition before its pollution by large spacecraft, for example, *Constellation*, but actually *LADEE* is mainly a technology demonstration. Its spacecraft power, mission lifetime, and payload mass are slated primarily to test the Lunar Lasercom Space Terminal (LLST),[12] beaming data at 10 kilobits per second to Earth (a downlink rate unneeded for *LADEE*'s science). Although its science justification stems from National Research Council recommendations to map sources and propagation of lunar atmospheric gases from the equator to poles,[13] *LADEE* will only measure a global equatorial average. Measured by Apollo, older results were ambiguous (see Chapter 8). *LADEE* is currently the last approved NASA lunar mission; these newer *LADEE* results might prove anticlimactic.

NASA's lunar emphasis is fading, beyond expiration of *Constellation*. NASA's Lunar Quest is a complex program of missions, instruments, research, and analysis but is slated to die in 2014. Several missions are discussed, such as spacecraft to land seismometers to sense moonquakes for the first time since 1977. NASA might provide two of these four landers in the International Lunar Network, but they are unfunded. Moon probes might win NASA competitions among universities and aerospace contractors for low-cost missions, but this is uncertain. Front-runners might include a rover to explore ultra-cold polar regions where water and other volatiles might freeze (the RESOLVE mission has preliminary funding to do this, perhaps by 2017, or might transmute into NASA's Resource Prospector mission) or a sample returned from the Far Side's South Pole-Aitken Basin. Research grant funding is healthy but falling, for universities mainly; otherwise NASA's lunar program is insecure.

Although its science justification stems from National Research Council recommendations to map sources and propagation of lunar atmospheric gases from the equator to poles, about a month into *LADEE*'s mission, the Chinese lander *(Chang'e 3)* dumped about one ton of gas into the lunar atmosphere, which *LADEE* seems not to have detected. Furthermore, *LADEE* will only measure a global equatorial average. Measured by Apollo, older results were ambiguous (see Chapter 8). *LADEE* is currently the last approved NASA lunar mission; these newer *LADEE* results might be anticlimactic.

That is the present; let us gaze to what we expect to come. Robotic lunar missions are planned, from China, India, and now Russia, as well as international collaboration and commercial competition. Remember that plans can slip months or years before fruition or fall victim to launch failure, satellite malfunction, and political/bureaucratic ennui.

With many missions from many nations, this clearly is no Moon race for two, but a diversity of players competing and cooperating. Indeed Russia, absent from lunar exploration for three and a half decades, will reenter by 2014. NASA will be there, but also at least four other national space agencies and a panoply of commercial agents. This is a new era.

Table 4.1. *Most active spacefaring nations: Population, gross domestic product, and spending on civilian space programs (2009–2011 estimates)*

Country/Agency	Population	GDP	Civilian space	% GDP	Per capita
United States/NASA	310,637,000	$14.1T	$19.0B[†]	0.13	$61[†]
Russia/RKA	141,927,000	$1.2T	$2.4B	0.20	$17
European Union/ESA*	499,723,000	$16.4T	$5.4B*	0.033*	$11*
Japan/JAXA	127,380,000	$5.1T	$2.5B	0.049	$20
India/ISRO	1,189,790,000	$1.2T	$1.3B	0.11	$1.1
China/CNSA	1,340,440,000	$5.4T	$0.5–1.3B	0.01–0.024	$0.4–1.0
Canada/CSA-ASC	34,303,000	$1.6T	$0.4B	0.025	$12
Brazil/AEB	192,273,000	$1.6T	$0.3B	0.021	$1.8

[†] per year;
* Annual budgets of individual EU member nations' space agencies total $6 billion. Adjusting for this civilian space spending is 0.07% of GDP, $23 per person averaged over EU.

The European Union (EU), Japan, China, and India do not support civilian space to levels similar in fraction of GDP as do the United States or Russia. (See Table 4.1. Unless stated otherwise -$ symbols indicate U.S. dollars.) Japan and the EU support civilian space at roughly the same per capita rate as Russia, but they are richer. The United States spends on civilian space roughly as much as all other nations combined. The distinction in China between military space projects and civilian space is not transparent, and China National Science Administration (CNSA) budget estimates depend on various accountings, especially varying definitions of labor costs.

Beyond ESA contributions, France spends an additional €1.7 billion (US$2.3 billion), Germany €1.3 billion ($1.7 billion), Italy €700 million ($900 million), the United Kingdom £230 million ($400 million), Spain €200 million ($260 million), Sweden €70 million ($100 million), and several other national agencies spend less. This may change if the European budget crises grow. In 2011, ESA announced plans to cut its budget by 25% by 2015, and the European Commission (EC) zeroed funding for the satellite program Global Monitoring for Environment and Security (GMES), estimated to cost $8 billion. Several of the most influential among the 27 Union nations (France, Germany, the United Kingdom, Italy, Spain, the Netherlands, Sweden, and Finland) protested defunding of GMES and other science programs, for example, the ITER fusion reactor. The EC wants individual nations to fund these. Some fear further ESA cuts.

ESA, like most agencies, spreads cost and increases efficiency with fortuitous opportunities to cooperate with other agencies. In the 1990s NASA and ESA scored great success with *Cassini-Huygens* to Saturn, with ESA supplying the *Huygens* Titan probe and some *Cassini* experiments. The same accolade is deserved by ESA's participation with NASA in the *Hubble Space Telescope*. Meanwhile ESA has stressed self-sufficiency, with its first interplanetary mission *Mars Express* in 2003, a wholly ESA effort except the (failed) *Beagle 2* lander from British and private funds. ESA's *SMART-1* (*Small Missions for Advanced Research in Technology*) to the Moon in 2006 carried experiments from five EU countries. Their Lunar Resources Lander has been postponed.

ESA reaches out to agencies beyond NASA, particularly the Russian Federal Space Agency (Федеральное Космическое Агенство России, a.k.a. Roscosmos, ФКА [FKA], or РКА [RKA]). Soyuz rockets now launch from ESA's Kourou facility in French Guiana, favorably closer to the equator (5.2° North Latitude) than Baikonur (46°N), Plesetsk (62°N), or the planned Vostochny site (52°N). (RKA's Baikonur lease expires in 2050, but tensions may reduce Baikonur's use in several years.) Soyuz/Kourou will loft the 30-satellite, $30 billion European Galileo system to parallel the American GPS and Russian GLONASS. ESA astronauts are also being encouraged to learn Chinese. NASA and ESA had planned to launch several Martian probes, including two rovers, in 2018 on two Atlas rockets. In 2011 NASA cancelled its rover, and ESA agreed to an RKA lander and launch to Mars for its rover in 2018. NASA/ESA collaboration persists, in part because ESA wants NASA's skycrane lander system (lowering Mars rovers suspended from a rocket platform). We discuss NASA's international collaborations later in the chapter.

As part of ESA's 2008 understanding with RKA, ESA aided the 2011 *Fobos-Grunt* probe (*Фобос-Грунт*, or "Phobos-Soil") to return samples from Mars's moon Phobos. This proved critical with *Fobos-Grunt*'s stranding in Earth's orbit, in part as a result of poor satellite communications. ESA radioed *Fobos-Grunt* from its Australian ground station and reestablished two-way communications. Unfortunately, the Mars trajectory window had closed, and the mission failed. Despite many attempts, no Russian Mars probe has succeeded in decades.

Fobos-Grunt reveals two chronic Russian space mission issues: an insufficient tracking network and frequently poor spacecraft testing.[14] NASA, ESA, and even China have worldwide space tracking networks of greater or lesser coverage, whereas Russia has only stations in its domestic territory and adjacent former Soviet republics (Ukraine and Tajikistan). The Soviet space program used up to a dozen tracking ships, but these were sold and scrapped after the Soviet Union dissolved. Space missions require exhaustive testing, roughly one-third of many NASA mission budgets, hundreds of millions of dollars in some cases. Russian missions suffer more electronic component and software failures and plagues of engine misfires. The testing problem falls to RKA, while tracking issues can be solved by international collaboration, as was Russia's inability to launch near the equator, alleviated by (limited) use of ESA's Kourou site.

Russia plans to replace the venerable Soyuz (Figure 4.2) rocket family (up to 8 tonnes to LEO) and Proton (21 tonnes LEO) with Angara rockets (up to 40 tonnes), and is designing a larger, reusable spacecraft (nicknamed "Orionski") to replace the 3-person Soyuz by 2018. Another rocket under development, the Rus-M (24 tonnes), was recently cancelled. Angara is favored for being produced with more Russian-made hardware and will compete with Falcon (10–53 tonnes) or Japanese H-2B (19 tonnes) by about 2015.

In 2007, RKA announced intentions to send cosmonauts to the Moon by 2025 with plans to establish a permanent base after several years, then human missions to Mars, but conflicting announcements from Russian space officials are common.[15]

An International Flotilla

Figure 4.2. **Workhorse**. Spacecraft Soyuz TMA-5 leaves Baikonur Cosmodrome, Kazakhstan for the *ISS* on October 14, 2004, on a Soyuz FG rocket, with Expedition 10 cosmonauts Salizhan S. Sharipov and Yuri Shargin and astronaut Leroy Chiao. The improved Soyuz, based on a 50-year-old design, has launched two dozen times, all to the *ISS* and successful so far. (NASA photo JSC2004-E-46228.)

Likely its *ISS* program will continue for many years. RKA is expanding its heritage of trips via Phobos to Mars, totaling four missions (of which only one succeeded, partially; 40% of non-Soviet/non-Russian Mars missions have also failed). RKA seems to have moved beyond extensive Venus exploration (19 missions, 1961–1984) but proposed a robotic Venus mission for 2016. These plans are in flux, but for the first time since 1976 there will again be *Luna* missions. RKA is preparing by 2015 the Luna-Grunt and Luna-Resurs/Chandrayaan-2 missions, and one or two lunar robotic missions later in the decade, for example, Luna-Glob. (Scheduling of RKA lunar missions has shifted frequently in recent years.)

Luna-Grunt and Luna-Resurs share a common lander design, with nearly the same dozen Russian-made science instruments. They will land at opposite lunar poles (Luna-Grunt in the North). Luna-Grunt's manipulator/drill is replaced on Resurs by the 15-kilogram Indian Mini-Rover to collect samples over distances depending on its lifetime – perhaps short. Also, the 1.6-tonne Luna-Grunt polar orbiter is replaced by the 1.4-tonne Chandrayaan-2, loaded primarily with copies of Indian instruments on *Chandrayaan-1*. Luna-Resurs/ Chandrayaan-2 will launch from India on their Geosynchronous Satellite Launch Vehicle (GSLV) if it is performing reliably. Luna-Grunt is slated for a Soyuz launch. After these may come a long-range rover, perhaps followed by a sample return mission by 2020.

The first Indian in space was Air Force squad commander Rakesh Sharma, who flew with two Russians in 1984 to and from space station *Salyut-7* on *Soyuz T-10* and *T-11*.[16] India is developing the three-person ISRO Orbital Vehicle to ride an upgraded GSLV into orbit by 2016, likely making India the fourth country to orbit humans. ISRO bragged it would send a crew to the Moon by 2015, but that has slipped to 2020 and beyond.[17] India also successfully launched in November 2013 a robotic probe to orbit Mars. There is much cooperation between the Indian and Russian space programs, and older Indian space scientists and leaders tend to prefer this approach, extending back decades to India's status as a nonaligned nation. More recently India has successfully collaborated with other space agencies, notably NASA.[18]

By 2050 India's population could reach 1.6 billion from its current 1.2 billion, of which 300 million are impoverished, according to the Indian government (500 million by World Bank definition). By many measures wealth in India is nearly typical for the world, but not for nations with space programs. In this context space expenditures are seen by many Indians as inappropriate, despite the relatively small cost, and are cited by foreign critics as reason for ending economic aid to India.[19] ISRO spends $1.3 billion equivalent annually, about 0.1% of India's GDP or 0.6% of the Republic's budget, three-quarters and four-fifths, respectively, of the equivalent proportions for NASA in the United States. Although the cost of a mission such as *Chandrayaan-1* is less than one-tenth dollar per capita, it feels costly in a nation where 400 million subsist on less than a dollar per day.

Few Indian public opinion polls on space exploration exist, so the following is anecdotal. Indians should proudly and enthusiastically support their space program, a relatively inexpensive, highly visible example of collective technical and scientific competence, achievement, and excellence, relatively corruption-free.[20] Salaries for space scientists should be raised; they compete in a growing Indian technology sector, and young space scientists are drawn away. Having dealt several times with Indian broadcast media about space missions, I have responses regarding space exploration's value from Indians (with access to Google searches and email). These responses fall into two categories: Indians should not waste money on space when there are many desperately poor people, or the Indian space program will likely bungle the mission. The second objection is demonstrably untrue.[21] Although *Chandrayaan-1* was not problem-free, it succeeded scientifically, especially with discoveries of lunar surface hydration. There are problems: ISRO's decade-long premier program, the cryogenic, liquid-fueled GSLV-D3, has scored fewer successes than failures, destroying several satellite payloads. Concerning the first objection, surely a billion dollars annually could immediately improve Indian lives but not change fundamentally macroeconomic reality (versus $36 billion annually for the military, or the Republic's $200 billion union budget).

Many Indians share with Americans poor awareness of how little they spend on space compared to other programs and what little good would result in diverting those funds. Preventing great actions rarely guarantees other great thing instead. A quip such as "we should divert the funds for this mission (*Chandrayaan-1*) into improving the Delhi sewage system" ignores how little $80 million affects a city of 15 million people in a nation with a government not solving these problems for decades. India's poor require the rich's attention, not hobbling Indian space exploration. Nonetheless Indian society's economic liberalization since the 1990s has left hundreds of millions behind but provided material wealth to others while costing them in ways harder to valuate. Many do not see how a space program improves this.

The challenge to Indian society and its educational system is enhancing the country's knowledge-based economy and bringing higher-salaried jobs to India without huge increases in natural resource exploitation. India has a system of capable technological institutes, excellent bulwarks against talented scientists and engineers escaping to Europe and America where they often obtain university degrees. This system keeps scientific and technological innovation within the country. Indians can succeed in new technologies where the West has not, and cases may become more common.[22] Inspiration could retain Indian scientists and engineers to make corresponding inventions in the physical sciences, energy sector, consumer electronics, or many others. As in the United States of the 1960s, excitement for space exploration might provide a hook to attract talented young people. This said, Indian space scientists have pulled back from excessive hubris: having predicted in 2009 to send Indians to the Moon by 2020, ISRO let this slide. Humility over the problematic GSLV or enormous costs of such a lunar program is sufficient cause for retreat. Perhaps China or Americans will return to the Moon before India.

The future wild card in international exploration of the Moon is China, with ambitions but only beginning to cooperate in planetary exploration. China's first interplanetary mission, the *Yinguo-1* Mars orbiter, launched on a Russian Zenit rocket on November 9, 2011, failed along with its co-manifested *Fobos-Grunt*, already delayed two years. It is unclear what conclusion Chinese leaders will draw from this first, failed launch with RKA. Sino-Russian space cooperation's complex legacy began with one-way aid before the three-decade rift ending along with the Soviet Union.[23] In 1992 Chinese leaders decided to initiate human space flight and, in 1994, bought significant space and aviation technology from Russia. In 1995 the two countries agreed to broad transfers of Soyuz technology to China.

At first it seemed China might simply launch Soyuz spacecraft on Long March rockets from their Jiuquan launch center in Inner Mongolia. The Shenzhou ("Divine Vessel," 神舟) spacecraft resembles Soyuz but is 11% longer, 22% wider, and 4% heavier than a Soyuz TM. As Chinese officials and state media emphasize,

Shenzhou is completely redesigned from Soyuz. For compatibility's sake (with Soyuz and *ISS*) the docking mechanism was largely copied from Soyuz, although since redesigned. The capsule escape system is copied directly from Soyuz, and Shenzhou space suits are identical to their Russian counterparts. The first two taikonauts were trained at Gagarin Center near Moscow. (*Taikonaut* – for *taikong*, meaning "space" – is standard in Western and English-language Chinese state media, so is used here.[24]) No taikonaut ever flew in space in Soviet or Russian craft.[25] Although Shenzhou's superstructure (service module, reentry module, orbital module) resembles Soyuz, it has two solar panels on each end allowing it to operate longer in orbit. The orbital module can operate aloft after the reentry module returns to Earth and the service module incinerates on reentry.

In the past decade China's space program transformed from a more commercial enterprise into a domestic, military, and exploration effort. International commercial Long March launches plummeted from five in 1998 to only once every year or two.[26] Long March launch fees of about $70 million are relatively expensive and no longer state-subsidized for foreign customers. Launches no longer fail at rates of 25% or more, as during the early/mid-1990s. Commercial launch sales are managed by transnational corporations, and non-commercial missions are now run by CNSA roughly equivalently to all of NASA, several defense agencies, National Oceanographic and Atmospheric Administration (NOAA), and some smaller U.S. agencies.[27]

There is essentially no collaboration between the Chinese and American space programs. CNSA would prefer working with NASA, especially on *ISS* – making statements to this effect[28] – but are long reconciled to its improbability. On the American side, many would like to consider collaboration, but others are dead set against it. The effects are disadvantageous to China, but occasionally also to the United States, such as potential damage done to *LADEE* by lack of coordination over the timing of *Chang'e 3*'s landing and contamination of the lunar atmosphere. The ITAR legacy, commingling of Chinese military and scientific space programs, and technological espionage in other arenas are incendiary for the U.S. Congress. American scientists have gone so far as to threaten boycotting government-sponsored scientific activities in the United States if Chinese scientists continue to be excluded.[29] Chinese space policy makers do not bend sufficiently on these issues and hesitate to suffer demands and interference expected from the Americans; or so it seems. ESA floated the idea of adding China to *ISS*'s 16 member nations, along with India and South Korea, but no motion resulted.[30] These wider issues are decisive in other cases; on the level of involvement of many *ISS* participating nations, one might consider Israel, for instance;[31] however, international politics override.

International tensions conflict with traditions of global cooperation between scientists even during tense times, including war. As early as the 1761 transit of Venus, proposed in 1716 by Edmond Halley as a means to measure the Sun's distance and therefore the planets', astronomers from many nations collaborated, even the British and French, despite their worldwide Seven Years War (1756–1763). Even at the Cold War's height, the International Geophysical Year (1957–1958, like international polar years 1882–1883, 1932–1933, and 2007–2009) was highly successful, engaging scientific cooperation between most major nations (except for China, protesting participation by Taiwan). These scientific forces are strong today but circumvented in some crucial cases, as we will see.

Space scientists, engineers, and administrators have created venues to discuss plans, ideas, and collaboration; these are nonbinding, as otherwise they would require involvement at ministerial levels, if not legislative. The International Space Exploration Coordination Group (ISECG) is a collaboration between 12 space agencies of most major spacefaring nations: United States (NASA), Russia (Roscosmos), Japan (JAXA), India (ISRC), Canada (CSA/ASC), South Korea (KARI), Ukraine (SSAU), Europe (ESA), and several ESA nations – France (CNES), Germany (DLR), Italy (ASI), and the United Kingdom Space Agency. Additionally, in 2007 agencies from Australia (CSIRO) and China (CNSA) attended the founding meeting along with the other 12, and they all released a preliminary document.[32] At this meeting the Chinese delegation was essentially silent and concerned primarily with collecting information; after that they dropped out. The ISECG has largely discussed lunar exploration. Recently the United States has strongly advocated human missions to an asteroid, but most agencies are unresponsive, even discouraging.

In international space cooperation, the most extensive, robust, and ambitious effort, of course, is between the Americans and Russians. *ISS* operations form the core of both human space programs and are means by which both nations have extended their expertise in constructing and maintaining structures in space (Figure 4.3). NASA astronauts could not reach the *ISS* for many years without RKA. (It is unclear how *ISS* flights by SpaceX's *Falcon 9/Dragon* or other commercial options will affect RKA's traffic to the station.) Without NASA, RKA could not afford operating the *ISS*. NASA also cooperates with other human space programs.[33] Although cooperation between NASA and the Soviet Union has ebbed and flowed with diplomatic/military bilateral relations, scientific cooperation was usually underway, and top-level discussion of grander mutual projects was common.[34] Since the G.W. Bush years yielded to the Obama administration, extensive cooperation, including humans beyond Earth's orbit, is being considered.[35] What stands in the way is uncertainty in the American program and sways in Russian foreign policy.

Russian-American cooperation penetrates commercial space. Boeing partners with RKK Energiya (formerly Korolev's OKB-1 design bureau) forming Sea Launch, which lofts several rockets per year, about 90% successfully. Sea Launch exited bankruptcy recently, recovering from a catastrophic rocket explosion in 2007 that partially destroyed its launch platform (injuring nobody). Providing one of the world's least expensive launch services, they use Zenit-3SL rockets of the Ukraine's Yuzhnoye Design Bureau (formerly OKB-586), with rocket integration in California and a Pacific launch from a repurposed oil platform owned by Aker Solutions ASA, Norway.

Not all runs smoothly between Russian and U.S. space programs. By 2006 NASA was looking for help in returning to the Moon, because it had little budget for lunar surface equipment. Particularly, it needed a partner supplying nuclear generators,[36] a potential role for Roscosmos, but did not publicly invite the Russians. Erik

Figure 4.3. **A walk outside**. (a) Stephen K. Robinson anchored to a foot restraint on *ISS*'s Canadarm2, during Space Shuttle STS-114 mission's third EVA on August 3, 2005. (b) Nicholas Patrick using tethers and hand restraints to prepare the Cupola module (also see Figures 11.7 and 14.2) during STS-130's third EVA on February 17, 2010, removing insulation blankets and restraint bolts. (Photos S114-E-6647 and ISS022-E-066880 courtesy of NASA.)

Galimov (director, Vernadskiy Institute of Geochemistry and Analytical Chemistry) strongly advocated for a Russian Moon program, arguing for profit from lunar resources, including helium-3. Not an official spokesperson, Galimov accused the United States of plotting to monopolize lunar helium-3. Parallel statements from Russian space authorities did not dull the message, for example, Roscosmos head Anatoly Perminov added: "We are ready to co-operate but for some reason the United States has announced that it will carry out the program itself."[37] NASA explained it was not excluding Russia, and several months later signed an agreement to put Russian neutron detectors on *LRO* and the *Mars Science Laboratory*.[38] The United States has issued exceptions for *ISS* cooperation that otherwise would have fouled on technological collaboration limitations due to Russia's aid to Iran. This contrasts with severe limitations of technological transfers during the Soviet era.[39] NASA and RKA will cooperate on *ISS* until the bitter end.

Beyond Luna-Grunt, Luna-Resurs, and a possible robotic lunar rover, RKA discusses with NASA and ESA possible Moon bases. Long-term Russian plans include cosmonauts on the Moon perhaps by 2025.[40] Russia and ESA discuss a common advanced crew transportation system carrying up to six people. They might also increase Russian launches from ESA's French Guiana site. In terms of public relations, ESA, NASA, CSA, RKA, and JAXA are working together to spread benefits of the *ISS*.[41] With NASA abandonment of joint missions with ESA, such as Laser Interferometer Space Antenna (LISA) in 2011 and in 2012 the Exomars 2016 orbiter and 2018 lander, RKA and ESA are collaborating, with RKA likely providing launches. NASA was to provide its Mars skycrane; now the mission may not fly. Regardless, NASA is seen as increasingly unreliable and may forfeit its position as international collaboration leader. A long line of successful NASA/ESA collaborations (*Cassini/Huygens*, *Hubble*, *ISS*, etc.) may soon end.[42]

NASA works closely with the Canadian Space Agency (CSA) on *ISS* and other projects, and CSA often adopts U.S. space policy, for example, discouraging technology transfer to China. CSA is partnered with ESA, with some collaboration with RKA, particularly experiments and flights to the *ISS*. CSA's budget can fluctuate, and at this writing apprehension regarding cutbacks is widespread. The expectation is that Canadian international space collaboration will be maintained, but not bloom.

Even though an *ISS* partner, Brazil places emphasis on developing its independent access to space. Eventually it may place people in orbit (despite numerous failures of its new VLS-1 rocket, including the deadliest launch pad explosion in the twenty-first century). Venezuela, Argentina, Peru, Chile, and Bolivia have or plan to launch satellites in connection with services in China, Russia, the United States and the United Kingdom.

One international partnership not soon to flower is the Sino-American one. U.S. congressional limitations on scientific cooperation with China make this nearly impossible. Along with NASA's FY 2011 budget came prohibition of NASA from using funds to "enter into or fund any grant or cooperative agreement of any kind to participate, collaborate, or coordinate bilaterally in any way with China or any Chinese-owned company" – not just for NASA itself, but scientists receiving NASA grants. This is no idle verbiage; in testimony before the House Commerce, Justice, Science Appropriations Subcommittee on May 16, 2011, OSTP Director John Holdren mentioned ways in which U.S.-China scientific cooperation would be useful, after which he was threatened with defunding and even prison by Rep. John Culberson (R-TX), in the congressional record:

Your office cannot participate, nor can NASA, in any way, in any type of policy, program, order or contract of any kind with China or any Chinese-owned company. If you or anyone in your office, or anyone at NASA participates, collaborates or coordinates in any way with China or a Chinese-owned company ... you're in violation of this statute, and frankly you're endangering your funding. You've got a huge problem on your hands. Huge.

Subcommittee Chairman Frank Wolf (R-VA) amplified this with an October 13, 2011, letter threatening to direct the attorney general to act against Holdren.[43] The damage has spread: since late 2011 scientists in China cannot access nasa.gov web pages, even purely scientific data that all other scientists use freely.[44]

Is this level of isolationism, ham-handed if not jingoistic, justified? It is fairly evident that China conducts national and industrial espionage, especially over the Internet. (Note that congressmen warned Holdren against using his Blackberry while in China.) China is not alone in this or even the most skilled culprit. U.S. industry and government should learn that the best place to store its most precious informational jewels might be off the web entirely. Likewise the U.S. Congress decries human rights violations in China, yet allows generous cooperation on a commercial level. What does U.S./Chinese space isolation accomplish? The United States bars China from the *ISS*, where major partner nations participate by working within their own modules (where they maintain sovereign control) and in other partners' modules with permission. If China joined the *ISS*, it would likely supply living and work space, with its modules subject to safety, environmental, and compatibility inspection and approval, but not necessarily with reciprocal inspection. It is unclear how sensitive data would flow to the Chinese. In discussing lunar outposts (Chapter 12), the Outer Space Treaty requires some exchange of information simply among neighboring personnel from different countries. Eventually the United States must retreat from its total wall against China. China might speed this with an initial gesture; clear separation between its military and civilian space programs would be hard to ignore. (Now, military personnel serve even on the Chang'e project.)

China may long delay such a gesture; its program is mostly self-sufficient. Although its long-term goal may be the Moon, its actions are building a space station. This might be a primarily civilian effort or might involve the military, like the Soviet space lab *Almaz*. Their first human launch was in 2003, second in 2005, third in 2008, fourth in 2012, and fifth in 2013. A long-duration space lab, *Tiangong 1*, has been in orbit since 2011. They visited this laboratory with three people in 2013 and plan to visit it with up to six later, then replace it with two larger modules in several years, followed by a larger space station by about 2020. A long-term Chinese space plan indicates humans on the Moon by about 2025 and on Mars by 2040. (The official CNSA plan is for a space station, not humans on the Moon.[45] The goal of Chinese on the Moon was once 2017, then 2022.) Although progress from first taikonaut to a space station with crew in nine years is rapid (the Soviets did it in 10 years, the United States in 12), early claims for Chinese space progress were unrealistic. China has launched three spacecraft beyond Earth's orbit (to the Moon) and may launch the lunar landers – Chang'e 4 in 2015, and Chang'e 5 in 2017 – followed after several years by one or two lunar robotic sample returns (which is still under review). Some Chinese leaders foresee taikonauts on the Moon, a Chinese lunar base, and even mining operations. In December 2013, *Chang'e 3*'s rover *Yutu* was supposed to explore northern Mare Imbrium for several months. So far these craft are not as sophisticated as competition from several other countries, and their scientific impact not transformational. *Yutu*'s portable ground-penetrating radar is unique, however, and will likely provide new insight into the lunar mare's regolith. Data like these could be transformational if they were acquired near the lunar poles as part of another mission. These were more engineering feats; less effort was spent on needed preflight scientific instrument calibrations or on building science teams.[46]

Chang'e and Tiangong space station expansion need the Long March 5 booster capable of 25 tonnes to LEO. For China sending humans to lunar orbit requires extra hardware: a translunar rocket stage. They could complete this in several years, making them second to send a crew to another world. To land them on the Moon requires new technology and a much larger rocket or expedition Earth-orbital assembly techniques. This requires effort comparable to that already devoted to taikonauts in space, so it might take a decade. Their using a space station for orbital construction might signal a Moon landing mission in several years but only after mastering the art and science of extravehicular activity.[47] Although the three-person *Shenzhou 7* included a spacewalk, it was a 20-minute exercise in flag waving (literally). For NASA mastering work in Earth orbit's vacuum and weightlessness required Gemini and was essential to Apollo: not only moonwalks, but scientific instrument module (Figure 2.21) data retrieval and crucial *Skylab* repairs. China has yet to develop these or robotic manipulator techniques. No book can supply these; they require long experience before work proceeds outside Tiangong. This cannot be cut short; there are few grislier prospects than losing crew during a spacewalk. This said, Apollo's Lunar Module, backed by the Saturn V, flew first in January 1968 but carried crews within 14 months to the Moon, then landing 4 months later.

China would happily surpass the United States in lunar exploration and will do so if the latter remains inactive. Ouyang Ziyuan, chief scientist of the Chinese Lunar Exploration Program, has been a strong advocate for lunar resource exploration (especially use of helium-3) and Chinese human lunar landings in the 2020s, which are officially encouraged if not specifically stated. The failed *Chang'e 3* and *Yutu* is still a source of national pride for the Chinese, and eventually some politicians in the United States will take notice. The rise of the Chinese space program, its absorption within the Chinese military, and increasing tensions between the United States and China on the high seas make the entrance of the two countries into a new space race possible. It has not happened as of 2014, but it seems a likely theme for American electoral politics, and is an effort the United States could enter easily. The Moon is the obvious venue for such competition.

China's space program should proceed carefully and safely, because a primary goal is to elevate national prestige and therefore avoid space disasters. The magnitude of the February 1996 Long March 3 disaster precipitating the ITAR clampdown was largely concealed by Chinese media (an official death toll of six despite a devastated village[48]), but taikonauts in space must be visible to excite the populace.[49] CNSA is hierarchical and focused and not the most robust organization against unexpected problems. Its program could effectively close many-decades gaps with NASA and RKA, but not adopting a best path into the frontier. We have yet to hear plans from CNSA breaking new ground for humanity. Summarizing, if von Braun miraculously revived today to choose his allegiance in advancing humanity's expansion into the Solar System, would he indeed learn Chinese and work for CNSA? Maybe not; it is unclear that China will assume the burden of expanding the human or robotic forefront in space. Given confused, counterproductive U.S. political leadership, NASA might not seem the best bet for this ghostly von Braun, either. Nonetheless there are intriguing space exploration developments in America to capture his imagination.

United States' unwillingness to collaborate with China contrasts with its initiatives to open space exploration to many players, commercial and collaborative, national and international. This is a new tactic but an old idea. The National Aeronautics and Space Act of 1958 pronounces (Sec. 20102.c): "The Congress declares that the general welfare of the United States requires that the National Aeronautics and Space Administration (as established by Title II of this Act) seek and encourage, to the maximum extent possible, the fullest commercial use of space." This began with *Telstar 1* in 1962, the first TV relay satellite and first privately sponsored space launch (by AT&T, Bell Labs, NASA, and a European consortium); now NASA supports more private initiatives, from Space-X and Orbital Sciences to deliver cargo to the *ISS* under COTS (Commercial Orbital Transportation Services) and Space-X developing the Dragon capsule as a candidate crew shuttle to the *ISS* (also the Liberty composite crew module by ATK and Lockheed Martin, and until recently Blue origin's orbital capsule). These are entirely separate from Virgin Galactic/Scaled Composites' VSS *Enterprise*, poised for commercial suborbital space flights (at $200,000 per passenger Virgin and competing companies are in a sense the anti-NASA, having achieved two human flights into space with the same vehicle in less than a week (something the Space Shuttle feigned but never achieved), but could only do so on suborbital flights, something NASA abandoned with the X-15 in 1968. (The third X-15 reached space twice in a month in 1963, but never in the same week.)

How these private efforts will impact lunar exploration is unclear. (For years RKA has offered a $100 million tourist fare around the Moon, but with no customers so far.) The Dragon capsule could orbit the Moon launched on a yet unavailable rocket, but are private companies ready to explore the Moon? Surprisingly, they are, with robots, for the purpose of winning a $20 million prize – and maybe astronauts, too.

Exploration has advanced fundamentally by prize contests. Neglecting searches for gold, glory, and mythical grails, cash prizes have inspired key developments in exploring the world. In 1714 the British government offered a prize for distant mariners to know their location East-West, as well as North-South (the latter achieved by the Vikings almost a millennium before in exploring the North Atlantic). The award of £20,000 (now equivalent to several million U.S. dollars) to find a ship's longitude to within 30 nautical miles was paid, haltingly, five decades later, to John Harrison for an accurate, ship-borne clock. In 1927 Charles Lindbergh flew solo across the Atlantic in part to win the $25,000 Orteig Prize, for which Lindbergh and his backers invested almost $20,000.[50]

The $10 million Ansari X-Prize went to the first team to launch safely a human twice in the same vehicle in two weeks into space above the Kármán line, 100 kilometers high. After eight years Scaled Composites beat 25 other groups (collectively spending over $100 million, about one-quarter of that by Scaled). This initiated Virgin Galactic's commercial space program and competing efforts approaching fruition in the next few years. NASA emulated this with its own Centennial Challenges with prize purses up to $2 million.[51] Three of these challenges present problems relevant to exploring the Moon: Regolith Excavation ($500,000) for a robot to excavate and move 150 kilograms of simulated lunar regolith in 30 minutes, Lunar Lander ($1 million) to perfect control of a robotic lunar lander, and MoonROx ($1 million) to produce breathable oxygen from chemically simulated lunar soil. This last challenge is unmet.

The big purse is the Google Lunar X Prize (GLXP), in fact a purse of purses. To win $20 million, a team must be first to land a robot on the Moon, let it rove 500 meters, and transmit video to Earth. After roving 5,000 meters, add $1 million. If the robot collects a sample containing water, add another million, as well as $1 million for photographing existing lunar hardware, for example, Apollo, another $1 million for surviving lunar night, and $1 million more for promoting personnel diversity. National space agencies cannot win, but the prize is reduced $5 million if a nation gets a new lander there before the competitors (as *Chang'e 3* did in 2013 or 2014). As of this writing 26 teams are competing, many of which are international (although almost half are based in the United States).

Twenty-six is a large number of teams (in fact reduced from 33). Likely the minority will actually build spacecraft, but several already have (see Figure 4.4). A major obstacle is a launch berth. Some teams plan hitching a ride as secondary payload on a rocket to geosynchronous orbit. Astrobotic Technology, Inc., plans on using its own Falcon 9 for about $60 million. The Barcelona Moon Team plans to send its rover to the Moon in June 2014 on a Long March 2C. GLXP teams are allowed to defray costs by offering payload to paying passengers; Astrobotic and Moon Express are doing this. Moon Express received $10 million from NASA for sharing results in developing a lunar lander from NASA's Common Spacecraft Bus, core of NASA's *LADEE*. NASA made lesser such deals with five other primarily American teams (Astrobotic, Next Giant Leap, Omega Envoy, Rocket City Space Pioneers, and FREDNET). We cannot say now who will win, but Astrobotic is often assigned an informal edge. By 2014 we should know; $5 million rides on victory before CNSA.

GLXP encourages a commercial path to the Moon, and this may aid scientific exploration with more flexibility. Now scientific groups must justify typically several hundred million dollars' expenditure on a new mission, and NASA supports one of these every few years. Commercial missions might cost much less and could be assembled among several science teams. It will still be expensive; NASA probably must bankroll these efforts in the United States. NASA and others encourage small satellites (CubeSats, SmallSat) to co-manifest to LEO or GEO, even continuing to lunar orbit. There will still be chances for collaboration mixing payloads and spacecraft from different nations, and these should be encouraged.

Figure 4.4. **Private space**. (a) the Odyssey Moon Google X Prize team is adapting for their landers the NASA Ames Research Center's Common Spacecraft Bus, here seen being prepared for landing simulation tests. (b) Astrobotic team's rover undergoing field tests. (c) Launch of a SpaceX Falcon 9 rocket testing the first Dragon capsule for COTS on December 8, 2010, from Cape Canaveral.
(d) Dragon capsule approaching the *ISS* with 520 kg of cargo on May 25, 2012. (Photographs from NASA, Astrobotic GLXP team press package.)

Commercial mission collaborations might be less fraught than some international ones; we will see. This is no excuse for xenophobia over working with scientists and space agencies from other countries.

NASA has adopted a new strategy for exploring the Moon, being frustrated from doing so on its own, of cooperating with private industry to encourage lunar mining. NASA will encourage private development of prospecting robots through its Cargo Transportation and Landing by Soft Touchdown program (Catalyst). NASA will not provide direct funding, but provide logistical support for a privately developed rover/spacecraft.

Several nongovernmental groups have seriously considered sending people to the Moon. Shackleton Energy Company planned spending $30–50 billion extracting water for propellant and eventually supplying Earth with solar electrical power from the Moon, starting with a lunar crew of a dozen by 2019 (or seven years after securing funds). To gauge interest, Shackleton tried "crowd funding" to collect $1.2 million during November–December 2011 to seed $40 million needed for setup and design studies. Unfortunately, it raised only $5,517. Bigelow Aerospace proposes lunar habitats based on inflatable structures, and recently Golden Spike announced plans for privately funded trips for astronauts to the Moon's surface. These may reach fruition in the 2020s. Further private space plans sound more audacious: Planetary Resources, Inc., and Deep Space Industries would mine asteroids, and SpaceX and Dennis Tito's Inspiration Mars Foundation propose private human expeditions to Mars. Although Shackleton Energy was not propitious, these ideas are worth long-term discussion (see Chapter 12).

The history of private space ventures for pure science and exploration is not encouraging; many fail: *Beagle-2* on Mars (contact lost at Mars), the Planetary Society's 2005 solar sail mission (launch failed), and Living Interplanetary Flight Experiment on the (failed) *Fobos-Grunt* mission in 2011 produced no useful results, for reasons sometimes under the private group's control and in others, not. Many efforts, mostly smaller, on the Shuttle and *ISS* have succeeded. Disappointing results can often be blamed on low-cost options. Space exploration is an expensive and hazardous business, so far. Improving spacecraft safety to human-rated standards is exorbitant, and we have no statistics on private astronauts' safety. Can commercial space afford this or handle catastrophic failure? For now, exploring other worlds is expensive and scientific discoveries will not produce quick economic return. The possibility of profitable activities on the Moon distinguish it from other worlds; we consider this after learning more about our natural satellite in Chapters 5–11.

For now we must ask which organization will spend hundreds of millions of dollars to robotically explore the Moon, or many billions to send people there. Russia, India, China, the EU, Japan, and even Iran have proposed a human presence there by 2025 or 2030.[52] Americans, who rejected such commitments, arguably are the most active lunar explorers, via NASA, private consortia, and mixes of the two. Any of these parties (except Iran[53]) could lead if they made lunar exploration a national (or commercial) priority. Officially lunar exploration and their space station is a key Chinese technological priority for 2006–2020, but how much human presence that implies is unclear.

Space exploration is sufficiently expensive to offer obvious venue for international cooperation. Regrettably the United States, uniquely positioned to lead a new effort beyond low Earth orbit, has abandoned the Moon as a core target, the one destination on which other spacefaring nations agree. The United States should drop its paranoid avoidance of space cooperation with China, which in turn should unambiguously demilitarize its space exploration. The Moon is worthless as a military base; rocket weapons are now largely divorced from space exploration. Spacefaring nations should declare an International Lunar Decade to build the next step beyond the *ISS*. As unacceptable as some may find it, the United States has nurtured a *pax Americana* in international relations for seven decades, and a peaceful influence over low Earth orbital space. The Moon is sufficiently interesting that nations will behave themselves in order to join the lunar club. American politicians should support this objective; it is cheaper than war.

International exploration rivalry has transformed now that competitors know what others are doing within milliseconds, or at least at the next press conference. The Greeks' rivals in Persia or Rome could not know that Pytheas was exploring the Arctic, nor did Portugal's Prince Henry the Navigator realize that pressing his explorers harder to round Africa to India might have shown Zheng He's Chinese fleets coming the other way. In 1804 Spanish forces tried intercepting the Lewis and Clark expedition only after being warned by a U.S. Army spy.[54] By 1910, however, Scott and Amundsen had prompt information (within days) about what the other was doing to reach the South Pole. The U.S. government was well aware of the Soviet Moon program despite its secrecy. Indeed, a major payback of a national space program is the fame and prestige it earns by appearing in the news. With few exceptions today, one nation could know what the others' space program is planning from their hardware, probably with a couple of years warning for major turns. Unlike America, Antarctica, or Australia's distance from Europe centuries ago, the Moon is a few seconds away.

When NASA was intent on returning to the Moon, it provided direction for other space agencies, in cooperation or competition. As in the 1960s when the space program was a less expensive technology demonstration and a less catastrophic proxy for war, the *ISS* is a means for cementing international collaboration and aspiration that can continue if new direction and investment arises. The United States has now adopted a plan that few, if any, nations care to follow; it is confusing. There is a natural role for the United States to lead space exploration, but it is shrinking from it. It is becoming an unreliable partner; it is overly xenophobic; it has shut down its human launch capability; its goals are too long term and diffuse. If this goes on too long, might it become irrelevant?

Could China adopt this leadership role? CNSA can grow impressively in the future but is not overwhelming now. Although claiming it is headed to the Moon, China's place in space exploration might be overstated. In fourteen years, the Chinese space program orbited five crews and a space lab, achievements roughly equivalent to Soviet feats over 1961–1971, excluding the planetary probes, robotic lunar landers, and (secret) human lunar program. CNSA has built no heavy-lift booster to take taikonauts beyond low Earth orbit. They have little spacewalk experience. Even the People's Republic of China (PRC) admits its future as a force of scientific innovation is unproven (see Chapter 14). It is unwise to dismiss China, but its space dominance is supposition. It is becoming a less sympathetic force in international relations and maintains significant tensions with spacefaring nations Japan and India – at least as severe as with the United States. These relationships may not improve. Nonetheless in 1992, China, Pakistan, and Thailand founded the Asia-Pacific Multilateral Cooperation in Space Technology and Applications, which in 2003 expanded to the Asia-Pacific Multilateral Cooperation in Space (APSCO) including Bangladesh, Indonesia, and Iran. Deng Xiaoping's famous maxim was "China should hide its strengths and bide its time," but it is doing so much less these days, and too soon. The United States could cooperate with all major players if it could deal more intelligently with its sino-paranoia. Its role in space is potentially more tolerable internationally than a nation justifying after centuries a mythos of resurgent, historically dominant, and justified world power.

In lunar exploration, Americans still exercise their strength in robotics. How will this impact their concept and mythos of the frontier? Does exploration have power without human presence? In most countries *frontier* refers simply to the border with foreigners of neighboring societies. In Russia the Wild East was frequently a forlorn place of despair and death as much as fortune (yet space exploration was a significant idea in Russian thinking as early as the 1800s).[55] The Australian outback was a trackless wilderness to absorb you. In other lands the jungle is a vast ecosystem to digest you. In traditional China, the frontier marks hinterland, hardly worth bother. In America you headed out to be transformed, to make your fortune, to dance with wolves.[56] It has been the mythos since Lewis and Clark and the clipper ships two centuries ago, that another people might interpret as hubris. Outer space is the logical, spiritual continuation.[57] It was this American attitude that in part inspired Jules Verne's important *De la Terre à la Lune* and *Autour de la Lune* of 1865 and 1870 (with main protagonists in them Baltimore Gun Club). Space fantasies were vicarious means to express expansionist longings in an America with imperialism more confined to adventurism in the Americas and Pacific, whereas other colonialist powers subdued whole continents. In a few chapters we will consider attitudes to the new space frontier with robots.

The frontier of the West is now sepia-toned, departing with Buffalo Bill a century ago, but Americans struggle to preserve it, in their minds and in John Muiresque conservation and environmental spiritualism. The twentieth century maintained the mythos in histories, fictions, and Hollywood and in balancing development and preservation of wild spaces. In much of rural and wilderness United States, it all happened just yesterday. The frontier had unique, fundamental, and largely beneficial effect on the American mind. U.S. society offers a rare mixture of technological forefront, commonsense technical intuition, and appreciation of nature that space exploration needs and contact with wilderness roots nurtures. What of this spirit might continue in the new millennium of globalization and ubiquitous communication? Outer space, particularly the Moon, is an obvious venue for continued frontier, limitless, uncrowded, and truly wild, not the homeland of other peoples with the misfortune of standing before a human tidal wave. It is not the West; it is unlike any land we have known, but it has promise.

This is not a book primarily about space politics, although they rule, but about lunar science and especially the reality of the Moon: what it is like, the effects it will have on us, and we on it. What is this place that we call the Moon? It seems so familiar that we specify no name for it. (Witness Eudora Welty. [See note 5 in Chapter 1.]) It is not "Moon" as in "Venus" or "Saturn." It is *the* Moon as in "the Sun" or "the moons of Mars" – categorial, not particular. Having no name for it as a separate entity in modern speech, for example, Luna, threatens hidden subjectivity when we think about this world. To pin it down we appeal to ancient connotations of goddesses and harvests, not the huge magma and rock spheroid orbiting through the vacuum. We are so accustomed to this alien planet hanging close overhead, and would be so horrified if we thought daily of its true magnitude, that we do not easily focus on it. It is too big and too reliable. Our sense of familiarity is an illusion. Humans make no difference to the Moon, save a few miniscule spacecraft. For all six Apollo missions in total, both pairs of astronaut's boots tread the lunar surface for barely 72 hours. What is the Moon, separate from what we want it to be? What do we know about this proximate but strange world?

Notes

1. There are recordings online of Lehrer singing "Wernher von Braun." His characterization was popular but harsh: "Wernher von Braun: a man whose allegiance is ruled by expedience. Call him a Nazi, he won't even frown." The song continues, "'Once the rockets are up, who cares where they come down. That's not my department,' says Wernher von Braun." Lehrer does not charge him with war crimes but mocks "some think our attitude should be one of gratitude, like the widows and cripples in old London town, who owe their large pensions to Wernher von Braun."
2. In testimony before two House education subcommittees, von Braun urged "more attention to bright students. He said a two-level educational system with stiffer classes for bright pupils might be the answer. He said one problem is that some children develop very late. 'I flunked classes in math and physics when I was 12 or 13,' he admitted" ("Eggheads, Take Heart – Von Braun Likes You" *Washington News*, March 15, 1958). At Senate Committee on Labor and Public Welfare hearings on federal education assistance, January 23, 1958, von Braun confided: "I also do not remember having considered my school years as always the most enjoyable, or the purpose of my schooling always to have fun. At times I found it pretty rough to go to school. I think there were years when I outright hated school." And there is no evidence that he learned Chinese.
3. Tsien engaged in the active Sino-American educational exchange (between the 1911 Revolution and Japanese invasion before World War II) leading to the establishment of Tsinghua University and education of foremost scholars such as C.N. Yang, Nobel physics prize recipient.
4. "Sea Change" by Bradley Perrett, 2008, *Aviation Week and Space Technology*, 168, 57.
5. Congress's *Intelsat 708* investigation produced the Cox Report ("U.S. National Security and Military/Commercial Concerns with the People's Republic of China, Declassified" by Rep. Christopher Cox & Select Committee of U.S. House of Representatives, May 25, 1999, U.S. GPO). Chinese reaction appears in "Facts Speak Louder Than Words and Lies Will Collapse by Themselves – Further Refutation of the 'Cox Report'" by Information Office of the State Council, July 15, 1999, *Xinhua*.
6. William J. Broad, "Communication Satellites Made Legal for Export," *New York Times*, January 3, 2013.
7. *Hiten*'s mission history and novel transfer orbit are described by the orbit's inventor in *Fly Me to the Moon: An Insider's Guide to the New Science of Space Travel* by Edward Belbruno, 2007, Princeton: Princeton University Press; *Capture Dynamics and Chaotic Motions in Celestial Mechanics: With Applications to the Construction of Low Energy Transfers* by Edward Belbruno, 2004, Princeton: Princeton University Press. We discuss weak stability boundary transfers in Chapter 13. Previously, simpler figure-8 transfers and Hohmann orbits were used in lunar missions (*Die Erreichbarkeit der Himmelskorper* by Walter Hohmann, 1925, Munich: R. Oldenbourg Verlag; also, NASA Technical Translation F-44, 1960: *The Attainability of Celestial Bodies*).
8. "Tanegashima, Uchinoura Centers to Open for Year-Round Rocket Launches," *Kondo News*, July 29, 2010.
9. Official news on *Chang'e 2* is positive, if vague: "China's Second Moon Orbiter Outperforms Design" March 13, 2012, *Xinhua*. *Chang'e 2*'s lunar map at 7 m resolution was released: "China Publishes High-Resolution Moon Map" by Zhang Dan, February 6, 2012, *Chinese National Television*. (The map is browsable at lower resolution, and data retrievable by registered users.)
10. "Spread of Ice Mass Loss into Northwest Greenland Observed by GRACE and GPS" by Shfaqat Abbas Khan, John Wahr, Michael Bevis, Isabella Velicogna & Eric Kendrick, 2010, *Geophysical Research Letters*, 37, L06501; "GRACE Gravity Evidence for an Impact Basin in Wilkes Land, Antarctica" by Ralph R. B. von Frese, et al., 2009, *Geochemistry Geophysics Geosystems*, 10, Q02014; "Crustal Dilatation Observed by GRACE After the 2004

Sumatra–Andaman Earthquake" by Shin-Chan Han, C. K. Shum, Michael Bevis, Chen Ji & Chung-Yen Kuo, 2006, *Science*, 313, 658.

11. GRAIL carried no other instruments except MoonKAM: four cameras used primarily for education and student research. ESA also proposes an educational lunar imaging satellite: ESMO.
12. The LLST has a mass of about 30 kg and power consumption of 50–140 W. The science payload is less than 20 kg, power with 60–100 W and will function only about 150 days in science mode (http://nssdc.gsfc.nasa.gov/nmc/spacecraftDisplay.do?id=LADEE).
13. The NRC finds "Processes involved with the atmosphere and dust environment of the moon are accessible for scientific study while the environment remains in a pristine state" and recommends that lunar exploration should

   "**8a**. Determine the global density, composition, and time variability of the fragile lunar atmosphere before it is perturbed by further human activity.
   8b. Determine the size, charge, and spatial distribution of electrostatically transported dust grains and assess their likely effects on lunar exploration and lunar-based astronomy.
   8c. Use the time-variable release rate of atmospheric species such as ^{40}Ar and radon to learn more about the inner workings of the lunar interior.
   8d. Learn how water vapor and other volatiles are released from the lunar surface and migrate to the poles where they are adsorbed in polar cold traps."

   At the equator, *LADEE* ignores 8c or 8d, but addresses 8a and 8b. Where gases ("volatiles") originate and how they reach the poles inspire more recommendations, which *LADEE* largely skips: "The lunar poles are special environments that may bear witness to the volatile flux over the latter part of solar system history." The NRC urges that exploration:

   "**4a**. Determine the compositional state (elemental, isotopic, mineralogic) and compositional distribution (lateral and depth) of the volatile component in lunar polar regions.
   4b. Determine the source(s) for lunar polar volatiles.
   4c. Understand the transport, retention, alteration, and loss processes that operate on volatile materials at permanently shaded lunar regions.
   4d. Understand the physical properties of the extremely cold (and possibly volatile rich) polar regolith.
   4e. Determine what the cold polar regolith reveals about the ancient solar environment."

   ("The Scientific Context for Exploration of the Moon: Final Report" by Committee on the Scientific Context for Exploration of the Moon, National Research Council, 2007, Washington DC: National Academies Press, 31–36 and 43–46.)
14. Some analyses cover this: "Red Planet Blues" by Dwayne A. Day, November 28, 2011, *The Space Review*, http://www.thespacereview.com/article/1980/1; "Time for Russia to Rethink its Mars Exploration Plans" by Lou Friedman, December 5, 2011, *The Space Review*, http://www.thespacereview.com/article/1984/1; "Phobos-Grunt Mission" by Anatoly Zak, 2011, *RussianSpaceWeb*, http://www.russianspaceweb.com/phobos_grunt_scenario.html; "Russian Space Industry Shakeup of 2011," http://russianspaceweb.com/centers_industry_2010s.html. Russian politicians have weighed in. President Medvedev spoke to reporters on nationally broadcast TV: "The recent (space) failures are a big blow to our competitiveness. This does not mean that anything fatal happened. It just means that we have to find those responsible and punish them. I am not suggesting that we line people up against the wall like we did when Joseph Vissarionovich (Stalin) was in power. Nevertheless, we have to punish them seriously." ("Medvedev Suggests Prosecution for Russia Space Failure" by Maria Kiselyova, November 26, 2011, *Reuters*). Fear is no motivator for careful work but ensures that employees keep their heads down (which may factor into the culture of Russian spacecraft testing). Medvedev was more likely concerned with how the spate of space disasters would affect the December 4, 2011, Russian Duma elections.
15. For instance, in March 2012, RKA sent the government a draft plan for robotic probes to Venus, Jupiter, and Mars (including a permanent Mars facility), humans in lunar orbit, and an asteroid mining facility, all by 2030. Later that week, RKA General Director Vladimir Popovkin announced that RKA's main priority was Earth resources and global positioning satellites, de-emphasizing human missions, but that no actual 2030 plan would soon be released ("Roscosmos Takes on NASA" by Alexei Timoshenko, March 20, 2012, *Moskovskiye Novosti*; "Russia to Focus on Its Orbital Cluster – Popovkin" by Staff Writers, March 27, 2012, *RIA Novosti*). RKA stated in 2010 that Russia has no plans for human lunar flights or settlements, despite numerous

upcoming robotic lunar missions. ("Russia: No Plans for Moon" by Andrew E. Kramer, February 3, 2010, *New York Times*).

16. Kalpana Chawla was born in Haryana state, northern India. As an American citizen she flew on *Columbia* on STS-87 in 1997 and STS-107 in 2003, on which she died with six other astronauts.
17. "India's Giant Leap Over China in Moon Race" by Bhargavi Kerur, October 8, 2008, *Daily News and Analysis India*; "India's Manned Moon Mission by 2020: ISRO" January 12, 2009, *Hindu*.
18. Some are sensitive about this, for example: "In *Chandrayaan-1* many Indian scientists regretted that their achievements were sidelined especially with regard to the discovery of water and NASA took away the credit." ("We're Launching Chandrayaan-2 for Total Coverage of the Moon" by Srinivas Laxman, *Times of India*, September 10, 2010). On the contrary, ISRO is usually credited before NASA in science talks I hear on M^3 hydration results, and *Chandrayaan-1* and ISRO scientists appear at the head of the author list for the 2009 M^3 hydration discovery paper, not NASA.
19. "UK to Halt India Aid and Focus on Trade" by Hélène Mulholland & Jason Burke, November 9, 2012, *Guardian*.
20. I dislike discussing corruption, but it matters. Many Indians confront governmental corruption as bribes or influence peddling (*India Corruption Study 2005: To Improve Governance: Volume I – Key Highlights* by Transparency International India, June 30, 2005, New Delhi: Centre for Media Studies). Tens of billions of dollars in bribes are paid yearly. Transparency International's Corruption Perceptions Index rates India as the 87th least corrupt nation of 178, tied with Albania, Jamaica, and Liberia. Along with corruption, Indians complain often of technical incompetence.
21. This success challenges opinions of senior Indian journalists, for example: "Beleaguered Indian Space Research Gathers Star Dust" by Pinaki Bhattacharya, September 29, 2009, *Huffington Post*, and many skeptical web comments from writers with seemingly Indian names, for example, responding to "After Water, Now Indian Scientists Find Cave on Moon" *SiliconIndia*, February 8, 2010.
22. For instance, there is unfulfilled demand for a "male pill" – male contraceptives less permanent than a vasectomy but more so than condoms. Soon entering the market is a long-term but reversible male contraceptive, developed by Sujoy K. Guha, biochemical engineering professor (Indian Institute of Technology, Kharagpur, West Bengal). Another example: IBM India leads in Internet navigation by voice control. ("IBM: Telecom and Mobile Research: Spoken Web," http://www-07.ibm.com/in/research/emergingsols_research.html) India also benefits from "reverse innovation" – less expensive approaches for the Indian market succeeding in the West.
23. Sino-Russian space cooperation mixes with issues of Chinese espionage, but not to Sino-American levels. Academic and chief executive of TsNIIMASh-Export, Igor Reshetin, was arrested for transferring "sensitive dual-use technology" to China while working with RKA and sentenced to 11.5 years in prison, along with three other men. His offense involved increasing nuclear missile and space vehicle accuracy by improving their reentry vehicles. Human rights activists accuse the Russian officials of bogus prosecution. Reshetin was paroled after four years. A similar case against Russian physicist Valentin Danilov resulted in longer imprisonment.
24. *Taikonaut* is a more westernized term. Chinese use *yǔhángyuán*, 宇航员 or *hángtiānyuán*, 航天员 for "crewmember sailing the universe/sky." *Taikonaut* is amusing, because in Japanese *taiko* refers to a big drum of similar shape and not much smaller than a Shenzhou reentry capsule.
25. The same is true of Chinese on American spacecraft; however, the first Chinese-born astronaut is Taylor Gun-Jin Wang, a naturalized American who flew on *Challenger* (STS-51B) in 1985. (William Anders of *Apollo 8* was born in Hong Kong.)
26. "Long March International Commercial Launch Record," China Great Wall Industrial Corporation, 2011, http://www.cgwic.com/LaunchServices/LaunchRecord/Commercial.html.
27. The CNSA is overseen by the State Administration for Science, Technology and Industry for National Defense, a civilian bureaucracy setting policy for defense procurement and technology. See http://www.gov.cn/english/official/2009–01/20/content_1210227_13.htm.
28. "There's Room for China in Space" by Jeffrey Manber, *Los Angeles Times*, January 16, 2008.
29. "Chinese No Longer Banned from NASA Astronomy Meet" by Staff Writers, October 21, 2013, Space Travel.com, http://www.space-travel.com/reports/Chinese_no_longer_banned_from_NASA_astronomy_meet_999.html.
30. *ISS* member nations are Brazil, Canada, Japan, Russia, the United States, and ESA member countries Belgium, Denmark, France, Germany, Italy, the Netherlands, Norway, Spain, Sweden, Switzerland, and the United

Kingdom. In 2010 ESA Director-General Jean-Jacques Dordain proposed inviting China, India, and South Korea to join the *ISS*. He said, "These three nations have been active in the multilateral discussion of future space exploration architecture. It seems that these three would be a good place to start widening the partnership. But this needs to be a collective decision by all the current partners" ("ESA Chief Lauds Renewed U.S. Commitment to Space Station, Earth Science" by Peter B. de Selding, *Space News*, February 3, 2010).

31. Israel has launched several scientific research satellites and dozens of communications and reconnaissance satellites, many on its own rockets, and collaborated with international partners France, India, the United States, corporations in the Netherlands, and users in Africa. Ilan Ramon was the first Israeli to fly in space and first non-American/Soviet to die there, aboard *Columbia* in 2003.
32. *The Global Exploration Strategy: The Framework for Coordination*, 2007. See http://www1.nasa.gov/pdf/178109main_ges_framework.pdf. More recently the ISECG released a draft of *The Global Exploration Roadmap*. Lunar discussions are summarized in *Advancing the Global Exploration Strategy: Human Exploration of the Moon* (2008) and *The ISECG Reference Architecture for Human Lunar Exploration – Summary Report* (2009).
33. NASA has flown many other space agencies' astronauts on the Space Shuttle. At least half who have flown in space have done so on the Shuttle, from each of the following *ISS* partner nations: Belgium, Canada, France, Germany, Italy, Japan, the Netherlands, Spain, Sweden, Switzerland, and the United Kingdom, as well as the non-*ISS* countries Israel, Mexico, Saudi Arabia, and (post-Soviet) Ukraine.
34. Roald Sagdeev, for decades key to space exploration as director of the Space Research Institute of the USSR (and science advisor to President Mikhail Gorbachev) recounts the history of Soviet-American relations in an article for NASA: Since NASA was created as a civilian effort, it was easier for them to collaborate internationally than Soviet space scientists who worked under the Ministry of General Machine Building, performing largely classified work for the Soviet military as its principal client and owner/operator of most military launch sites and ground control centers.

    Kennedy at several times considered space cooperation with the Soviets, as in his inaugural address: "Let both sides seek to invoke the wonders of science instead of its terrors. Together let us explore the stars." Khrushchev warmed slowly to the idea, leading not to joint lunar missions but exchange of weather satellite data and their coordinated launches, joint mapping of the geomagnetic field, and experiments in communications relays. With Brezhnev's rise a more aggressive, less cooperative phase ensued for the rest of the 1960s.

    Sagdeev recounts, in a 1970 meeting with Soviet science academy President Mstislav Keldysh, his U.S. counterpart Philip Handler described how in the film *Marooned* the fictional Soviet crew rescues stranded American astronauts from their Apollo capsule. Ultimately this led to Apollo 18 becoming the Apollo-Soyuz Test Project with its first Soviet Union-American joint mission in July 1975 and development of common docking hardware. The 1970s cooperative efforts led to several bilateral space science and engineering working groups, especially in the life sciences, permitting many U.S. experiments onto Soviet satellites.

    During the Reagan administration the U.S.-Soviet space cooperation agreement lapsed, and the Soviet Union viewed U.S. space programs with suspicion because of the Strategic Defense Initiative. Still, the nations collaborated on using satellites to locate missing ships and airplanes, saving over 400 lives. Some mutual space biomedical research continued. Not flying a dedicated mission to Comet Halley in 1986, NASA did aid communications with Soviet comet probes.

    In April 1987 the United States and Soviet Union signed a five-year pact cleaving civilian space from strategic defense. Gorbachev proposed joint human Mars missions but was dissuaded by Reagan ("United States-Soviet Space Cooperation During the Cold War" by Roald Sagdeev & Susan Eisenhower, 2008, *NASA 50th Magazine*, http://www.nasa.gov/50th/50th_magazine/coldWarCoOp.html). Sagdeev and Eisenhower, granddaughter of the 34th president, wed in their own collaboration.
35. "Russia Seeks Cooperation with U.S. in Space Effort" by Andy Pasztor, May 19, 2010, *Wall Street Journal*.
36. "NASA Seeks Partners as Budgets Tighten" by Maggie McKee, February 8, 2006, *New Scientist*.
37. "Russia Sees Moon Plot in NASA Plans" by Adrian Blomfield, May 1, 2007, *Daily Telegraph*; "NASA 'Rejects Russia Moon Help,'" April 30, 2007, *BBC News*.
38. "NASA Denies Snubbing Russia's Moon Offer" by David Shinga, April 30, 2007, *New Scientist*; "U.S., Russia Sign Pact to Hunt for Water on Mars, Moon" by Michael Stott, October 3, 2007, *Reuters*.
39. For instance, Norman Friedman's Cold War history (*The Fifty Years War*, 2000, Annapolis: Naval Institute Press) describes the lengths to which the United States limited Soviet access to computer technology beyond simple pocket calculators and occasionally fed them sabotaged chips.

40. "Roscosmos Revives Permanent Moon Base Plans," January 19, 2012, *RIA Novosti*; see also "Marsianskie Khroniki po Russkomu Scenariy" by N. Yachmennikova, January 21, 2009, *Rossiyskaya Gazeta*.
41. "*International Space Station*: Benefits for Humanity," http://www.esa.int/SPECIALS/ISSBenefits/; *International Space Station Benefits for Humanity* by Nicole Buckley, et al., 2012, NASA Pub. NP-2012–02–033-JSC.
42. NASA/ESA collaboration will continue on *Mars Express*, *Rosetta* (comet mission), and *BepiColombo* (Mercury), albeit at levels below what was once expected for LISA or Mars 2016/2018.
43. http://wolf.house.gov/uploads/Holdren_10_13_2011.pdf.
44. I verified this with six sources within the People's Republic of China. Nasa.gov sites including planetary data sites are blocked from the PRC inside the Internet "Great Wall." The block occurs in Southern California because of the Jet Propulsion Laboratory, says NASA unofficially.
45. "Chinese Manned Space Engineering – Future Missions," CNSA Web site, 2012, http://en.cmse.gov.cn/list.php?catid=64. In 2002 the lead Chinese scientist on their Moon program claimed they would establish a lunar base soon, although the anticipated date was unclear ("China Sets Timetable to Launch Manned Space Flight to Moon," May 25, 2002, *Xinhuanet*, http://news.xinhuanet.com/english/2002-05/20/content_400263.htm).

    China announced a space plan through 2016 including staged development of a larger station, the Long March 5 (25 tonne to LEO), a "fast response" Long March 6 to quickly launch satellites for the military, a regional positioning satellite system, lunar rover and sample return missions, a "study" of human lunar landings, and an interplanetary robotic mission (target unspecified) ("China's Space Activities in 2011," *China Daily*, December 30, 2011).
46. Recently CNSA made public some *Chang'e 1* data, so some non-Chinese scientists may use them. Other missions' data largely superceded them. Their topographic lunar map might have had influence, but delays in release caused it to fall behind Japanese and U.S. results.
47. I quip that in his science fiction novel *2010: Odyssey Two*, Arthur C. Clarke portrays Chinese space station *Tsien* suddenly launching out of Earth's orbit, flying to Jupiter, and landing on Europa to claim the moon for China. The mission ends badly, however, as the *Tsien* refuels on Europan water when simple Europan life forms under its ice find the *Tsien* to be a tasty energy source.
48. The Xinhua news agency reported the launch disaster killed 6 and wounded 57, but an eyewitness Western reporter photographed soldiers removing many bodies that he says amounted to at least 100: "China Space Chiefs Ready to Expend Human Lives" by Nigel Hawkes, June 15, 1996, *Times (of London)*. An unofficial video of the village hit by the rocket shows many destroyed buildings (http://www.youtube.com/watch?v=OOtSwQkybVw).
49. Indeed, Chinese media can engage in gross propagandistic manipulation of space program news, such as reporting before launch the success of *Shenzhou 7*. ("In Cyberspace, No One Can Hear You Lie: How Taikonauts' Success Blasted Off Early" by Jane MacArtney, September 26, 2008, *Times (of London)*.
50. *The Spirit of Saint Louis* by Charles A. Lindbergh, 1950, New York: Scribner, 25, 31, and 73.
51. In addition to those listed, current challenges include Sample Return Robot ($1.5 million) to retrieve geological samples autonomously in rugged terrain, Nano-Satellite Launch ($2 million) to loft two 1 kg satellites in a week, Strong Tether ($2 million) for a high tensile-strength tether, and Night Rover ($1.5 million) for a rover to power by stored solar energy. Past challenges include Power Beaming ($900,000) to climb a tether powered by a light beam, Green Flight ($1.35 million) to fly 200 miles on less than a gallon of fuel, and Astronaut Glove ($250,000) for a flexible vacuum glove (see "NASA Office of the Chief Technologist: Centennial Challenges," http://www.nasa.gov/offices/oct/early_stage_innovation/centennial_challenges/index.html). Some challenges are partnerships with other organizations, for example, NASA and the Spaceward Foundation for the $250,000 Telerobotic Challenge, which in small part compensates NASA for the closing of its telerobotics research group ("NASA Announces Telerobotic Construction Competition," December 5, 2005, *Space Ref*, http://www.spaceref.com/news/viewpr.html?pid=18433).
52. *The National Medium- and Long-Term Plan for the Development of Science & Technology*. Anatoly N. Perminov, director of RKA from 2004 to 2011, stated in 2010 that Russia has no plans for human lunar flights or settlements, despite their numerous robotic future lunar missions. ("Russia: No Plans for Moon" by Andrew E. Kramer, February 3, 2010, *New York Times*)
53. Iran claims having sent a monkey into space on a suborbital flight, far from an orbital mission. The United States and Soviet Union achieved similar prowess ca. 1950 with World War II–era rockets.

54. *Undaunted Courage: Meriwether Lewis, Thomas Jefferson and the Opening of the American West* by Stephen Ambrose, 1996, New York: Simon & Schuster, p. 334: Gen. Wilkinson sabotaged U.S. exploration, aiding the Spanish for many years (see *An Artist in Treason: The Extraordinary Double Life of General James Wilkinson* by Andro Linklater, 2009, New York: Walker).
55. Russian society developed an old, influential cosmic mythos. Russian Orthodox philosopher Nikolai Fyodorovich Fyodorov (1827–1903) promoted *cosmism*: expansion into space as means for humanity to reach perfect, immortal existence, with space exploration even offering recapture of the atoms allowing dead humans to be reconstituted. He mentored pioneering space scientist Konstantin Tsiolkovsky (1857–1935) who peppered his physical and mathematical space science concepts with visions of humanity's cosmic expansion. The Soviet Union promoted Tsiolkovsky's technical works but not his cosmism. Other cosmists, especially with occult leanings, escaped the Soviet Union or were exiled to the gulag. The Soviet Union killed cosmism and replaced it largely by the 1930s with socialist heroism (personified by Heroes of the Soviet Union, first largely aviators and including most cosmonauts). Earlier idealism echoes in modern writing, but these now can be dystopian, such as the novel *Omon Ra* by Victor Pelevin,1992, Moscow: Tekst, in which the protagonist looks to space to escape his Soviet existence but finds himself caught in cruel exploitation of cosmonauts in the cause of "heroic achievements."
56. After Tom Sawyer's spat with "fiancée" Becky Thatcher: "What if he went away – ever so far away, into unknown countries beyond the seas – and never came back any more! How would she feel then! The idea of being a clown recurred to him now, only to fill him with disgust. For frivolity and jokes and spotted tights were an offense, when they intruded themselves upon a spirit that was exalted into the vague august realm of the romantic. No, he would be a soldier, and return after long years, all war-worn and illustrious. No – better still, he would join the Indians, and hunt buffaloes and go on the warpath in the mountain ranges and the trackless great plains of the Far West, and away in the future come back a great chief, bristling with feathers, hideous with paint, and prance into Sunday school, some drowsy summer morning, with a blood-curdling war-whoop, and sear the eyeballs of all his companions with unappeasable envy. But no, there was something gaudier even than this. He would be a pirate! That was it! NOW his future lay plain before him, and glowing with unimaginable splendor. How his name would fill the world, and make people shudder!" (*The Adventures of Tom Sawyer* by Mark Twain, 1876, chap. 7)
57. "Our youths leave their long-cherished homes, and cities spring up before them almost by magic, and the towering forests give place to beautiful and cultivated fields. They annihilate space, navigate nearly every stream, ascend the rugged brows of the Allegany, and the craggy peaks of the Rocky Mountains, and sail unmolested amid far ice-bergs and burning tropical waves. No project is too hazardous to be attempted, and few obstacles too great to be removed. Yankee enterprise has already navigated the aerial vault of heaven, and we playfully anticipate a time when they will attempt to plant a colony on the Moon's dark side, and send flying ships and winged cars along the Milky Way!" ("American Character, An Oration Delivered at the Exhibition of the Clinton Liberal Institute, August 30, 1837" by J.T. Goodrich, 1837, *Evangelical Magazine and Gospel Advocate*, 8, 388). Goodrich, a Universalist pastor in New York and Pennsylvania, was lost in the Great Chicago Fire of 1871 – that ironic nemesis of American expansion. Goodrich was far from the only advocate of American exploration of his time: 1838 saw the first U.S. Exploring Expedition, a federally funded mission to scientifically investigate the Pacific and Antarctic.

Chapter 5
The Moon Rises from the Ashes

From his mind the Moon was born,
And from his eye the Sun.
– "The Sacrifice of Primal Man" in the Rig Veda[1]

"The moon is backing away from us
an inch and a half each year. That means
if you're like me and were born
around fifty years ago the moon
was a full six feet closer to the earth.
What's a person supposed to do?
I feel the gray cloud of consternation
travel across my face. I begin thinking
about the moon-lit past, how if you go back
far enough you can imagine the breathtaking
hugeness of the moon, prehistoric
solar eclipses when the moon covered the sun
so completely there was no corona, only
a darkness we had no word for"
– Dorianne Laux, 2004, "Facts About the Moon"[2]

Mythology in most traditional cultures tells of the Moon's birth. It seems appropriate that, of all creation's elements, the Moon was born from the Hindu primal entity's mind. Of all scientific accounts of creation of major extraterrestrial Solar System bodies, the Moon's has most generated debate, and still does. Foremost physicists attacked the problem, and it has orbited basic scientific questions such as Earth's age and how the Sun and stars generate energy. The effect of the Moon's birth on Earth was so profound and unbelievably violent that it took time to accept it. Before Apollo, understanding the Moon's birth was largely a mental exercise, starved for data, while stellar or galactic astrophysics progressed via remote telescopic and spectroscopic observations. Now we ask how much our understanding of the Moon's origin is complete and correct.

The scientific Moon is so unaffected by humans that we are irrelevant. Little on the Moon, except 70 small sites since 1959, feels our influence. This world is more than indifferent; it is alien and can surprise or fool us, and as Laux implies, it can generate great and changing earthly influence.

René Descartes (1596–1650), inventor of rationalism (in part), the mathematical coordinates connecting geometry with algebra, and *cogito, ergo sum*, receives credit for the first scientific theory of lunar creation. Like Copernicus before him and cautioned by Galileo Galilei's persecution in the 1630s, Descartes delayed publishing his Solar System treatise until he died.[3] His theory neglected gravitation (or saw it as vortices acting between proximate bodies). In his picture Earth and the Moon arose equal distances from the Sun but orbited at different speeds, until the Moon became trapped in Earth's vortex. Obviously Descartes' work could have benefited from the laws of motion and gravitation in Newton's *Principia Mathematica* (1687). His treatise violated Kepler's laws of planetary motion (1609, 1619) but was written in approximately 1633.

The issue of the Moon's evolution to its current state became a central issue in interpreting gravitational theory. Sir Edmond Halley suggested in 1695 and Richard Dunthorne measured in 1749 (by looking at ancient lunar eclipse timings) that the Moon's orbital period relative to Earth's day was decreasing (which changed most, the day, was still unknown). The theoretical physicist who formalized the mathematics of Newton's theory, Pierre-Simon, Marquis de Laplace (1749–1827), explained this quantitatively as the Sun's effects on the Moon's orbit, given Earth's changing orbital shape about the Sun. This weighed into a central debate in Newtonian mechanics, whether quantities such as the Moon's period undergo permanent changes or only cyclical ones. Newton thought the equations implied permanent changes, whereas Laplace's derivation argued changes were cyclic. Newton thought that to avoid permanent Solar System instabilities, God must be at work. Famously, when asked by Napoleon of Newton's conjecture, Laplace, buoyed by his calculation, is said to have replied, "I have no need of that hypothesis."[4]

Regrettably for Laplace another force was at work, but not divine. In 1754 philosopher Immanuel Kant theorized that the Moon slows Earth's rotation via tidal forces (Figure 1.3). Physicist Hermann von Helmholtz publicized this effect in 1854, clarifying that the increase in energy of the Moon's growing orbit is fed by deceleration of Earth's rotation, which lengthens the day. In 1859 astronomer John Couch Adams noticed that Laplace had neglected a mathematical term, so the good match between the tide-free calculation and the Moon's observed orbital acceleration was spurious. Later measurements of this acceleration showed its true value (about 0.01 degree per century) to be several times larger than eighteenth-century determinations, further ruining the correspondence.

Appropriately, a founding father of modern science, Charles Robert Darwin (1809–1882) begot the central figure in a half-century's thought on the Moon's origin, George Howard Darwin (1845–1912). George Darwin argued that tides raised by the Moon imply an Earth much closer to the Moon far in the past.[5] Modeling Earth (and Moon) as a viscous liquid rather then a system dominated by oceans and basins (Chapter 1), Darwin calculated that the two bodies merged about 50 million years ago, with the system rotating every five hours. Darwin used the analysis of Sir William Thomson (1824–1907, known as Lord Kelvin) to show that the Earth-Moon merger would tend to oscillate with a 1.7-hour period, resonating with the tidal cycle driven by rotation with respect to the Sun. Darwin proposed that recently the Moon fissioned from the rapidly rotating Earth, like a droplet falling from a spout. This fueled disputes between proponents of Charles Darwin's evolutionary theory (published in 1859) and Kelvin (who actually upheld evolution) over whether Earth was billions of years old as indicated by evolution and the fossil record. George Darwin was caught in the middle.

Geologist Osmond Fisher elaborated on George Darwin's theory by hypothesizing that a crater survived from the Moon emerging from Earth: the Pacific Ocean.[6] The continents represent remnants of original crust, with the Americas, Antarctica, Australia, and Malay Archipelago pulled away from the central Eurasian/African landmass in the flow as the Moon left Earth. By 1912 the theory of continental drift was proposed and over several decades displaced the Pacific-born Moon idea.

Laplace appears here in another guise, with his nebular hypothesis invoked to create the Solar System from a rotating gas/dust cloud. A miniature version of this model was the major rival to Darwin's, with the Moon forming from a gaseous ring circling Earth. Edouard Roche (who showed how planets' tidal forces disrupt moons, Chapter 1) advanced this in 1873. An elaboration of this theory by Grove Karl Gilbert in 1893 proposed that the girdle around Earth condensed first into solid fragments, then accumulated into the Moon. Alternatively, Earth and the Moon might arise simultaneously from the nebula as twin planets.[7]

Darwin's idea largely dominated until the Apollo landings, when lunar samples were obtained. To summarize, during this time science offered several alternatives for the Moon's origin:[8] (1) the Moon was captured as a single, composed body from an independent orbit,[9] (2) Earth and the Moon were formed out of the primordial nebula as a double planet, (3) the initial rotation of the Earth was so fast that the Moon was ejected from it via fission, or (4) perhaps the Moon was even formed by the capture, disintegration, and reaggregation of incoming planetesimals (subplanetary fragments destined to gravitate into planets). Finally, we arrived at the currently favored model of (5) Earth impacted by a Mars-sized planetesimal,[10] producing a disk of debris that formed the Moon, the alternative seemingly best explaining much of the evidence derived from Apollo-era data and before. These models are portrayed in Figure 5.1.

This list is not comprehensive; planetary scientists are creative in developing alternative lunar origin hypotheses. For instance, one model supposes that Earth's radioactive elements (uranium, thorium, even plutonium) collect at the mantle base, near the core, incorporated in calcium silicate perovskite ($CaSiO_3$), a minor mantle component that sinks.[11] The hypothesis is that locally these radioactive elements produce enough neutrons to generate plutonium by becoming a natural breeder reactor. If Earth is already spinning at almost the speed sufficient to cause it to break up into two objects orbiting each other (2.7 hour period), the heat from this natural reactor might rise so rapidly that it explodes, not just convecting or conducting the heat away, splitting Earth.[12] This defeats an objection to the third option that the spinning Earth has insufficient energy to eject the Moon from the planet. Extra energy comes from the nuclear explosion, heating the lunar material to thousands of degrees.

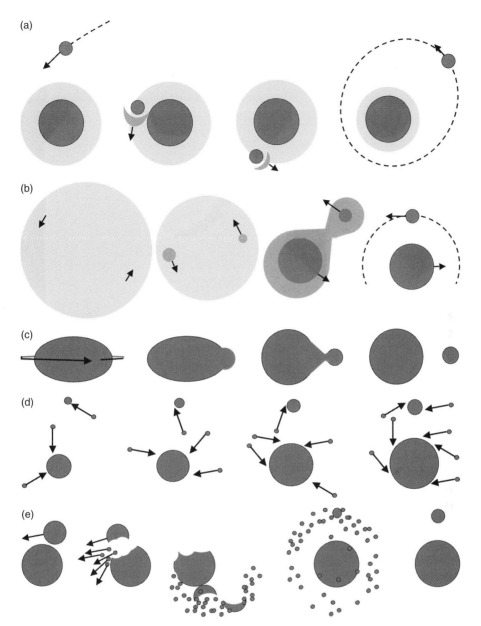

Figure 5.1. **Moon maker**. Five alternative lunar origin hypotheses: (a) capturing the entire Moon into Earth's orbit, requiring at least half of the proto-Moon's kinetic energy be lost, in this case "aerobraking" by hypersonic shock in Earth's thick primordial atmosphere, stripping much of the atmosphere (alternatively, one could imagine the Moon ejecting a preexisting moon, begging the question of that moon's origin); (b) accretion of two large mass concentrations into Earth and the Moon within the primordial nebula; (c) fission of proto-Earth spinning at breakup speed, ejecting the Moon and with it much angular momentum; (d) planetesimal co-accretion of the Moon and Earth; (e) the Giant Impact model in which a Mars-sized planetesimal (about 10% Earth's mass) glancingly strikes proto-Earth, with most debris falling to Earth; about 10% of the debris stays in orbit and accretes into the Moon.

Figure 5.2. **The Big Whack**. Artist's conception of the Moon-forming Giant Impact between proto-Earth and Theia several minutes after first impact. In reality the splash of debris would not yet extend to the distances shown, and the actual impact might be more obscured by debris and radiation. (Graphic courtesy of NASA.)

Most lunar scientists now subscribe to the fifth option, the Giant Impact hypothesis (Figure 5.2). This picture has been challenged recently, as we examine later. The Moon is so familiar we can almost imagine it always accompanying Earth; however, we describe it as a relative latecomer to an established proto-Earth, with lunar-destined matter ripped from it in collision with a second body. How was this dramatic theory formulated? What is the evidence for it? Is it true?

Looking to other planets for truth on the ground, literally, we see that Earth's immediate neighbors all suffered tortured infancies. Mars, 7,792 kilometers in diameter, has a hole punched in its surface 2,100 kilometers wide (Hellas) and likely one 8,000 kilometers across covering the entire northern polar region. Mercury (4,879 kilometers in diameter) suffered a hit, digging Caloris Basin (1,400 kilometers). The Moon, however, bears some of the largest scars: not just Imbrium (1,100 kilometers), but also the South Pole–Aitken basin (SPA; 2,500 kilometers), extending over most of the Moon's 3,576 kilometer diameter (as well as maybe a larger basin covering the Near Side, Chapter 7). Most of these impacts occurred hundreds of millions of years after birth of their host planets. Even the planetesimal excavating the SPA, however, was relatively small, 200–300 kilometers in diameter assuming it hit at velocities typical of inner Solar System bodies, about 20 kilometers per second.[13] Theia, the Earth Killer, was nearly 6,000 kilometers wide and struck proto-Earth maybe 50 million years after the Solar system's formation, after most large, unstable bodies had already met their fate. Earlier, when planets were still assembling, huge impacts were common. In contrast the biggest impact crater (astrobleme) on Earth is South Africa's 300-kilometer-wide Verdefort crater.[14] Volcanism and erosion on Venus along with plate tectonics on Earth efficiently erase craters. Venus' largest astrobleme, named after anthropologist Margaret Mead, is 270 kilometers across.

Despite tortured early lives of all inner planets, Earth differs from its neighbors, Mercury and Venus. These two planets barely turn (58.6 and 243 day periods, respectively – in fact Venus rotates in reverse). They rotate so slowly that one could keep a fixed constellation overhead just by walking quickly over either planet's surface. In contrast Earth has 300 times more angular momentum; something jolted Earth to spin.[15]

The Earth-Moon system is unique among major planets because Earth has not only a fast spin and the largest fraction of mass in moons[16] but also the largest fraction of angular momentum invested in its satellite(s), four times more for the Moon's motion than in Earth's rotation alone.[17] This suggests that the system's angular momentum is tied to the Moon's origin. Earth's rotational direction versus the Moon's orbit is another clue. They are tilted versus each other by 18.3°–28.6° (around the Earth's obliquity of 23.44°, see Figure 5.3), whereas the Moon's orbit around Earth tilts by only 5° versus Earth's orbit around the Sun.[18] This seems odd if Earth once spun so fast to eject the Moon; those two motions should be aligned. These put pressure on the second hypothesis (Earth and Moon formed together from primordial nebula) and the third (Moon ejected from Earth

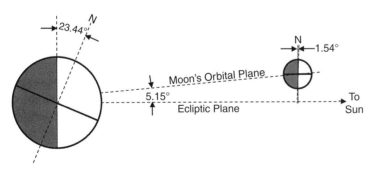

Figure 5.3. **Tilt of the world**. Earth has seasons, but the Moon almost none, because of the negligible tilt (1.54°) of the lunar equator versus direction to the Sun. Because lunar orbit around Earth is inclined versus Earth's orbit around the Sun, the Moon appears in Earth's sky up to 23.44° + 5.15° = 28.59° from Earth's celestial equator (sometimes north, sometimes south, since the Moon's orbital plane precesses – keeping its 5.15° tilt versus the ecliptic but oscillating in direction in the third dimension, not shown here). The Moon's rotational axis and orbit precess together, staying 6.69° apart. The Moon's position observed from particular places on Earth is offset by parallax versus Earth's center, so it can reach 29.6° from Earth's celestial equator. The distance between the Moon and Earth are not shown to scale here; the Moon is about 60 Earth radii away.

by rotational fission). This leaves possibilities involving the Earth capturing the Moon, particularly via a glancing collision (the fifth hypothesis).

Further aspects of Earth and the Moon beg explanation and bear on lunar origin. The two bodies' compositions are radically different, with the Moon 40% less dense than Earth. Paradoxically, the Moon has much lower abundances of low atomic-numbered elements, which almost always have lower density. In fact the Moon is the second densest satellite in the Solar System after Io. This is true despite the Moon's small inner core – a radius about 350 kilometers or less – and Io being one of the most volcanically reprocessed bodies in the Solar System. The Moon's interior must lack the heaviest substances, such as nickel-iron. How?

To deepen matters, we can investigate compositional ratios not of the elements but of isotopes of the same element. A good case is oxygen – the sole abundant light element on the Moon. Most oxygen (99.76% on Earth) is ^{16}O, with eight protons and eight neutrons, but small amounts (0.04%) are ^{17}O with nine protons, and 0.2% ^{18}O, with ten. From meteorites and Apollo/Luna samples, we have pieces of many bodies throughout the inner Solar System: Earth, Moon, Mars, and primordial and evolved asteroids.[19] Varying environments, especially different temperatures, can alter isotopic fractions of the same element, for example, ^{17}O versus ^{16}O. The effect for ^{18}O versus ^{17}O is nearly identical to ^{17}O versus ^{16}O because of their same mass differences, so materials starting with set ratios of ^{16}O to ^{17}O to ^{18}O will only change $^{17}O/^{16}O$ in concert with the $^{18}O/^{17}O$ value. Thus, even differently processed materials starting from the same origin will disperse along a well-constrained track graphed in $^{17}O/^{16}O$ versus $^{18}O/^{17}O$. Carbonaceous chondrite meteorites form such a track, whereas non-carbonaceous chondrites form a different track. Eucrites, mainly from the asteroid Vesta, select a particular region along the latter track. Martian meteorites fall just off this track, and nickel-iron meteorites fall farther still. Farthest of all are terrestrial and lunar samples, but Earth's and the Moon's samples cluster closely together. One can play the same game with other elements, for example, chromium or manganese, and find similar results.[20] Exceptions, with their own interesting story, will be discussed later.

To understand the Moon's origin, we must explain how it acquired a non-terrestrial chemical composition but the same isotopic ratios as Earth. Does this rule out the Giant Impact theory? If a planet from another part of the Solar System struck Earth, would it have a different isotopic composition? Not necessarily. Presumably these ratios are set primarily by temperature, which is determined mainly by distance from the Sun. One interesting variant of the Giant Impact theory proposes a natural mechanism wherein Theia was formed in Earth's orbit as a sister planet.

When two bodies orbit each other (in ellipses), specific Lagrange points indicate where a third, smaller body can co-revolve in a fixed configuration with the other two. Two of these five points, L4 and L5, lead and follow the smaller of the two main masses in its orbit, 60° ahead and behind, and matter can accumulate there. Normally a mass's position at L4 or L5 is stable if the two larger masses have a mass ratio larger than 24.96. Belbruno and Gott argue that once the third mass reaches about 10% of the mass of the second, the force between them can cause them to collide, especially if "harassed" by other bodies. Theia could form at L4 or L5 in the proto-Earth/Sun system, to be perturbed by other planetesimals into hitting proto-Earth.[21] In this scenario Theia and Earth orbit equally distant from the Sun, so they might share similar isotopic compositions.

Although not universally accepted, this version of the Giant Impact puts Theia and the proto-Earth in the same thermal environments where they might develop similar isotopic ratios. This is no guarantee, because inner, terrestrial planets (Mercury, Venus, Earth, or Mars) are theorized to assemble from distinct, individual planetesimals not necessarily from the same part of the Solar System.[22] A planet can be constructed of isotopic components different from neighbors equally distant from the Sun. We will see that the Big Whack may not succeed in mixing Theia and proto-Earth into homogeneously identical Earth and Moon, so the Belbruno-Gott model takes what was perceived as a strength but is really a weakness, the same isotopic ratios on Earth and Moon, and at least makes it neutral evidence for/against the Giant Impact. To some, the importance of the Belbruno-Gott model is unappreciated; without it the significance of isotopic uniformity is problematic (more on this later in the chapter).

Nonetheless other theories are challenged even more. If the Moon's angular momentum were deposited within the Earth-Moon system, Earth would rotates every five hours, still two times slower than what is needed for breakup by fission. The remaining angular momentum is missing from the system. Although this model would explain the matching isotopic ratios, the light-element depletion is problematic. Furthermore, external capture of a separately formed Moon does not explain depletion of heavy elements such as iron. Whereas the Earth is about 30% iron, the Moon is probably less than 10% (about one-fifth of this in a small core[23]); much is likely rust, FeO. There is little evidence that the Moon's early orbit was elliptical – the shape expected if it were captured gravitationally as an intact body.[24]

Initially co-formation was often considered the most plausible model, in which the Moon grew in Earth's orbit from swept-up, smaller bits from the disk orbiting the Sun, as thought to have built outer planets' satellites. Although such models successfully produce Moons, they struggle with both the lunar iron deficiency and Earth-Moon angular momentum, because growth via small-body accretion typically produces slow planetary rotation.[25] (This last condition can be circumvented if material originates preferentially from outer edges of the zone feeding into Earth, where the material with the highest angular momentum resides.[26])

Shortly after Apollo ended, two groups proposed models in which proto-Earth was struck by a large external object, thereby creating the Moon. In 1975 William Hartmann and Donald Davis resuscitated the idea that a Moon-sized object impacting early Earth ejected iron-depleted mantle debris into Earth's orbit, from which the Moon formed, heated to the temperatures driving volatile elements from the system as vapor.[27] Alistair Cameron and William Ward in 1976 suggested a larger impactor could account for Earth's rapid initial rotation and that vaporization of even heavier elements might provide a mechanism for ejected material to avoid impacting Earth but eventually settle into bound orbit.[28] This Giant Impact theory of lunar origin was largely ignored until it emerged from underdog status to surprise favorite, mainly because of the other theories' problems, at a 1984 conference on the Moon's origin in Kona, Hawaii.

Crucial since 1984 are improvements in detailed numerical modeling to simulate giant impacts and the use of increasingly powerful computers.[29] Good news from these models is that specific quantitative results are produced and compared to increasingly detailed data. Less encouraging is the number of relevant variables: impact speed, angle (or distance off-center) of the impact, and the spin of impactor and proto-Earth. Some are constrained a priori, but others a posteriori, limiting the models' power of prediction. We need to assume an initial composition, mass, and other properties for proto-Earth and the impactor. Although these models' spatial detail improves with larger, finer calculations, some important questions depend on small-scale properties estimated separately.

Broadly, the Moon is severely depleted in materials vaporizing below 1,300 K (about 1000°C). We should expect the impact speed to reach at least proto-Earth's escape velocity, about 11 kilometers per second, as expected from a body falling from great distance onto its surface. For hydrogen atoms, this corresponds to a temperature of 5,000K, and hotter for heavier atoms. The typical temperature reached in these simulations is 7,000K for material slamming to rest on Earth's surface and 3,000K for material in orbit. For material to enter orbit around Earth, it must decelerate by about 30% over Earth's diameter, hence feeling about 1/3 g (where g is the standard acceleration of Earth's surface gravity). This is sufficiently nonviolent that large chunks of rock, especially from the impactor, might survive the deceleration intact. This effect is not simulated; the minimum simulation particle size is about one-millionth Earth's mass, or about a 100 km cube. Of course, such huge particles are demolished by the collision's forces, but much smaller particles are not.

In simulations the debris disk around Earth forms within 24 hours of the collision and starts to cool significantly. Although ices have a difficulty surviving temperatures seen in the collision, volatile substances trapped in large rocks cannot escape unless temperatures approach those melting the rock, not just ice. Below these temperatures, even at 1,000K, volatiles stay locked in the rock lattice unless rock is pulverized to centimeter size or smaller. Simulations now ignore these effects and cannot state reliably whether volatiles survive in orbit trapped in rock (see Chapter 8).

Although every collision simulation is different, let us discuss a typical example. Continuing from Chapter 1 (see note 29 in that chapter), we now concentrate primarily on Theia, the Mars-sized impactor. In the first 20 minutes after impact, Theia is eviscerated, with its core plunging into Earth, and decapitated, with about half of its mantle flying back into orbit, a relatively cool 2,000K–3,000K, whereas Earth heats to triple this temperature. During the next few hours Theia stretches into a thin peninsula of demolished rock, one end raining onto Earth, the rest stretching 50,000 kilometers into space. As the disk forms in the first day, concentrations start accumulating. By the first week's end, one might identify the mass seeding the Moon; in 30 days, most fragments still in vacuum have formed into the Moon, sitting just beyond the Roche tidal breakup radius (about 3 Earth radii). Meanwhile the disk circling Earth starts with much of its mass, maybe 30%, as rock vapor. Some rains onto Earth for hundreds of years, some onto the Moon. The Moon forms so quickly that matter accretes with little chance to cool. Matter falling onto the Moon melts all surface rock with liberated gravitational energy, and a global magma ocean forms. Chapter 6 discusses how the Moon then changes.

Giant Impact models contain within themselves fuel for controversy, if not necessarily their own demise. Model collisions produce Moon-like objects containing primarily matter from the impactor, not Earth, at 70%–90% levels, whereas Earth contains 90% or more proto-Earth material.[30] Is this consistent with the isotopic similarities of Earth and the Moon, even if their elemental compositions are different? The Belbruno-Gott model helps, but planetesimals in similar orbits need not share identical composition. Instead researchers try various impacts to increase proto-Earth's share in the Moon. For instance, a faster (30% more than escape velocity), more head-on collision can produce Moons more of proto-Earth than impactor (with higher impact temperatures, not surprisingly).[31] This is paralleled by active debate about how much Earth-like material composes the Moon. The argument swung to either side just in 2012; for example, one study showed that titanium isotopic ratios, as for oxygen, are much more uniform on the Moon than in meteorites and agreed with terrestrial values, implying that most lunar material derived from proto-Earth. Several months later another paper showed that zinc, which should evaporate in the Big Whack and thereby produce heavier residual isotopic mixes, lacks light isotopes on the Moon compared to Earth (and Mars and chondritic meteorite varieties).[32] This implies lunar materials were subject to higher temperatures and that this occurred throughout, as expected for material accreting from the Giant Impact.[33]

Four decades after Apollo and almost 30 years since the Giant Impact model's birth, debate continues about how the Moon was formed. Most lunar scientists accept the Big Whack's reality, but the issue is surprisingly dynamic. Increasingly at lunar science meetings, this model, once elevated nearly to the level of dogma, is openly challenged. Perhaps the Giant Impact will prevail; time will tell. More surprising is how even basic questions about the Moon, even ones that would occur to a child, are so unsettled, although we have clues and ideas. For instance, what made the Man in the Moon and why does it face Earth? We confront this in Chapter 7. Chapters 8 and 9 explore issues even more unsettled. There are real mysteries about the Moon.

Do we care about the Big Whack for practical reasons? Chapter 1 describes how the Moon's birth affects Earth profoundly, likely altering plate tectonics, Earth's atmosphere makeup, and ultimate regulation of Earth's thermostat by the carbon cycle (upset by anthropogenic carbon dioxide, methane, and other greenhouse gases). Whereas crust composes more of the Moon's volume (about 7%) compared to crust on Earth (about 1%), no study yet shows this resulting directly from the Giant Impact. The standard model places proportionally more light material on the Moon and nickel-iron in Earth and alters Earth's mantle's elemental mix by several tenths of a percent. How this translates to different structure of Earth's crust is not established. Earth's early crust was likely made more of basalt (not the current granite) and reached its current state after differentiation and plate tectonics. However, Earth has less material to make continents because much of the original resides on the Moon.

Even if plate tectonics' Big Whack connection is weak, the Moon's creation supplied Earth's fast rotation, necessary for strong magnetic fields protecting our atmosphere from direct stripping by solar wind. Earth's hydrogen, hence our oceans, depends on the Moon's birth, radically differentiating Earth from Venus, where hydrogen's lack in the atmosphere, crust, and mantle has severe consequence. Earth differs from Venus because of the Moon.

Details of the Big Whack's influence on surface structure are erased by plate tectonics' very action in remaking ocean basins every hundred million years or so, even the Pacific. Is the Moon's impact, literally, only skin deep? Probably not, and we need to understand this someday. In several centuries, hopefully after humans solve global warming problems, the next geoscientific urgency may be rapid evolution of Earth's magnetic field. We have not caused this, but it may profoundly affect life on Earth. The Moon may tell us about this.

Frequently in Earth's existence, magnetic North and South Poles trade places. This can happen in less than 1,000 years, not by the poles drifting 180° but by the global structure of Earth's magnetic field collapsing as field strength plummets, being replaced by new global magnetic structure where polarity swaps North versus South. Meanwhile Earth's magnetic field will be weak, patchy, and variable. Temporarily it no longer reliably shields Earth from energetic solar particle radiation. We see this global magnetic collapse's advent now. Earth's magnetic field strength drops 10% every 70 years, changing significantly during your lifetime. How will this affect our great-great-grandchildren? Magnetic North is moving 10° per century and accelerating, no longer in Canada but zooming across the Arctic Ocean. Few people notice this. (Given GPS, who uses magnetic compasses anymore?) There are worrisome implications, however.

Earth's magnetic field traps two roughly concentric bands of radiation. The innermost of these Van Allen radiation belts is actually centered offset 450 kilometers from Earth's center, with radiation stretching down within 200 kilometers of Earth, brushing its upper atmosphere. This South Atlantic Anomaly (SAA) is familiar to satellite operators who must safeguard their craft flying through its heightened radiation field. The *International Space Station* requires extra shielding for this reason. The SAA affects the ground, too, raising radiation levels over the South Atlantic near the Argentine/Brazilian coast.[34] Perhaps because of a slight difference in rotation rates for Earth's core versus its lithosphere (its crust and upper mantle), the SAA drifts westward at 0.3° (33 kilometers) annually, directing it toward Rio de Janeiro, Sao Paulo, Buenos Aires, and Montevideo. Soon it will rest over most of South America.

Perhaps Earth's magnetic field weakening will reverse, and effects such as the SAA will disappear. Typically the field strength collapses every few 100,000 years, but with no clear polarity swap in the past 800,000. Several times during this interval the field strength approached zero but climbed again; maybe our luck will hold. Or perhaps collapse will continue, and the SAA will grow. Other magnetic anomalies as a result of iron mineral deposits and so forth, now less than 1% of the geomagnetic field strength, will become important in relative importance as the global field disappears. Scientists speculate that a patchy, locally shielding magnetic field structure will emerge, similar to a scaled-up version of the Moon's today. This has never happened in recorded history, but life on Earth, even pre-homo-sapiens hominids, has survived such events.

Understanding this evolution might benefit from details of the effects that the Moon's birth had on Earth. We suspect concentrations in Earth's mantle of siderophilic elements ("iron-loving," e.g., cobalt and nickel) rose from injection of Theia's core. How this separated between Earth's mantle and core affects how the dynamo generating Earth's magnetic field undergoes polarity reversals. The Moon might retain enough record of this to help. We cannot affect the reversal but may better anticipate how to build infrastructure and locate population centers in response to the changing environment if we better understand a central event that configured Earth's magnetic dynamo: the creation of the Moon.

The Giant Impact model arose during a time when lunar science persisted with little new data for a generation. The theory of the Moon's origin may seem matured, but as a result in part to this hiatus. We may yet be missing salient facts about the Moon's formation. New data abound, and the theory is under stress, not because evidence has weakened but because some evidence (isotopic similarity of Earth and the Moon) is too good. Although isotopic similarity may doom simple lunar capture (hypothesis 1), it is so fine that it threatens the Giant Impact as well. This was uncomfortable concerning oxygen isotopes, but as more elements are analyzed the situation grows more critical (including silicon, chromium, and tungsten).[35] For stable isotopes titanium-47 and titanium-50, their abundance ratio for the Moon equals that for Earth to within 1/150 the range of the ratio for asteroids, representing materials assembled into Theia and proto-Earth. This is either a huge coincidence (at the 1% level) or implies that material is homogenized between Earth and the Moon nearly perfectly.[36] The Giant Impact model must accommodate.[37]

Elaborations of the Giant Impact reduce needs for fortuitous matching of proto-Earth's and Theia's initial isotopic composition, mixing them after the impact, but the quandary is unresolved. Work since Pahlevan and Stevenson suggests how the hot atmosphere and accretion disk after impact mix materials further,[38] but huge amounts must be lifted from Earth to the Moon to equilibrate isotopes. The Giant Impact model is also incomplete. How much of the Moon's core (below the magma ocean formed by infalling rock, vapor, and liquid) is mixed in the impact? Is the deep interior mixed like the mantle, or is it composed of bodies maintaining integrity against accretion of the remaining Moon? Might earlier satellites of proto-Earth be incorporated into the Moon without ever melting? Such a preexisting moon would more efficiently collect Big Whack debris. The Giant Impact might not have been so gigantic in this case; alternatively the collision could have been more head on, with more terrestrial matter ejected into space, solving the isotope equality problem, as originally hoped for the Big Whack. These processes are not specified by the Giant Impact theory.

Radically new Giant Impact models increase Earth-Moon mixing by altering the proto-Earth and impactor themselves. In one model, two roughly half-Earth masses strike nearly head on making a symmetrically spiraling mass reminiscent of merging galaxies, and an orbiting disk that forms a Moon (and accretes onto Earth).[39] Another model employs a rapidly spinning Earth (flattened 2-to-1 by centrifugal force) struck by a small mass that helps fling Earth's outer parts into orbit.[40] These help homogenize Earth-Moon abundances, and the liquid/gaseous disk atmosphere of Pahlevan and Stevenson mix things further.

These models can produce too much angular momentum. A new factor, the early Moon orbiting in an ellipse precessing once per year, allows roughly half of Earth-Moon's angular momentum to transfer to the Sun-Earth system in as little as 100,000 years (while heating the Moon internally with tides). Dumping angular momentum weakens the objections to the old fission lunar origin hypothesis. George Darwin's old model would not have been discarded so rapidly if it cold have appealed to this convenient mechanism. Ćuk and Stewart's model approaches a proto-Earth spinning at near breakup speed harassed into forming a Moon by a smaller impact. Of course a previous, huge impact (not a succession of smaller ones) would give proto-Earth so much spin. We have not yet returned to Darwin's fission model, but near-homogenous isotopes have led to compromise of the Giant Impact idea.

New data may address these issues. The superlative view of the lunar interior from mapping its gravitational field with *GRAIL* will constrain how material is distributed, especially when combined with new analyses of deep moonquakes recorded by Apollo (Chapter 7). A proposed mission by NASA Ames Research Center would seek residual Big Whack debris in or near Earth's orbit – would-be pieces of Earth/Moon relatively unaltered for 4.5 billion years.

The Giant Impact provides a natural heat source for the melting of the lunar mantle: the magma ocean that we probe at length in Chapter 6. Questions remain, however. A straightforward analysis of the thermal contraction of a thoroughly melted Moon may contradict limits from contraction-induced geological faults.[41] New data from recent missions are revising our view of this, as we will discuss in Chapter 7.[42] We will see in Chapter 8 that the depletion of light elements in surface samples appears to fail at great depth. Is this consistent with the Big Whack? What does this apparent inconsistency imply?

Notes

1. This selection from the Veddas of the second millennium BCE tells the Hindu story of the creation of existence from the universal Primal Man, with his possessions and parts of his body becoming deities, human social castes, and components of nature: "A thousand heads had Primal Man, a thousand eyes, a thousand feet: encompassing the Earth on every side. He exceeded it."

2. Continuing, more darkly:

   > "And future eclipses will look like this: the moon
   > a small black pupil in the eye of the sun.
   > But these are bald facts.
   > What bothers me most is that someday
   > the moon will spiral right out of orbit
   > and all land-based life will die.
   > The moon keeps the oceans from swallowing
   > the shores, keeps the electromagnetic fields
   > in check at the polar ends of the earth.
   > And please don't tell me
   > what I already know, that it won't happen
   > for a long time. I don't care. I'm afraid
   > of what will happen to the moon.
   > Forget us. We don't deserve the moon.
   > Maybe we once did but not now
   > after all we've done. These nights
   > I harbor a secret pity for the moon, rolling
   > around alone in space without
   > her milky planet, her only child, a mother
   > who's lost a child, a bad child,
   > a greedy child or maybe a grown boy
   > who's murdered and raped, a mother
   > can't help it, she loves that boy
   > anyway, and in spite of herself
   > she misses him, and if you sit beside her
   > on the padded hospital bench
   > outside the door to his room you can't not
   > take her hand, listen to her while she
   > weeps, telling you how sweet he was,
   > how blue his eyes, and you know she's only
   > romanticizing, that she's conveniently
   > forgotten the bruises and booze,
   > the stolen car, the day he ripped
   > the phones from the walls, and you want
   > to slap her back to sanity, remind her
   > of the truth: he was a leech, a fuckup,
   > a little shit, and you almost do
   > until she lifts her pale puffy face, her eyes
   > two craters and then you can't help it
   > either, you know love when you see it,
   > you can feel its lunar strength, its brutal pull."

   (Reprinted with permission from W.W. Norton & Company.)

   Actually if Earth and the Moon were to remain otherwise unchanged, Earth would donate angular momentum to the Moon in its orbit until the day and month both lengthen to equal each other at roughly 50 days. The Moon would hang over one side of Earth and no longer change its orbit via Earth-Moon tidal forces. However, this requires billions of years, while Earth will heat under the expanding Sun, its oceans will boil, and tidal interactions will be thereby largely suppressed.

   The roughly 50-day period is easily understood. Earth's rotational angular momentum is $J_E = 7.1 \times 10^{33}$ kg m² s⁻¹ and varies inversely with Earth's rotational period. The Moon's orbital angular momentum is 2.9×10^{34} kg m² s⁻¹ (4.06 times J_E), varying as the cube root of its orbital period. To transfer all angular momentum in the Moon requires adding 24.6% to its current value. Cubing 1.246 gives 1.935; multiplying this by the Moon's 27.3-day orbital period gives 52.8 days. Giving 1/53 of this angular momentum to Earth's rotation and 52/53 to the Moon means Earth would rotate and the Moon would orbit every 52 days.

3. *Le Monde, ou Traité de la Lumière* by René Descartes, 1664, Paris.
4. Hugh Doherty, *Organic Philosophy, Volume 1: Man's True Place in Nature: Epicosmology*. London: Tubner, 1864, 379.
5. "On the Procession of a Viscous Spheroid and on the Remote History of the Earth" by George Darwin, 1879, *Philosophical Transactions of the Royal Society of London*, 170, 447.
6. "On the Physical Cause of the Ocean Basins" by Osmond Fisher, 1882, *Nature*, 25, 243.
7. *A Theory of the Origin of the Earth* by O. Y. Schmidt, 1959, London: Lawrence & Wishart.
8. See "Moon over Mauna Loa – A Review of Hypotheses of Formation of Earth's Moon" by J. H. Wood, in *Origin of the Moon*, eds. W. K. Hartmann, R. J. Phillips & G. J. Taylor, 1986, Houston: Lunar and Planetary Institute, 17.
9. "Uber Gezeitenreibung beim Zweikorper-problem" by H. Gerstenkorn, 1955, *Zeitschrift für Astrophysik*, 36, 245.
10. Theia is the name commonly used for the Mars-sized planet hypothesized to have struck Earth and formed the Moon. In ancient Greek mythology Theia, the goddess of the sky, and Hyperion, god of light, were parents of the titan goddess of the Moon, Selene. (In some versions Helios, god of the Sun, was father to both Selene and Hyperion. There are a few alternative genealogies.)
11. "Phase Transformations and Differentiation in Subducted Lithosphere: Implications for Mantle Dynamics Basalt Petrogenesis and Crustal Evolution" by A. E. Ringwood, 1982, *Journal of Geology*, 90, 611.
12. "An Alternative Hypothesis for the Formation of the Moon" by R. J. de Meijer & W. van Westrenen, 2009, *Lunar and Planetary Science Conference*, 40, 1847; "Forming the Moon from Terrestrial Silicate-Rich Material – 2012 Edition" by W. van Westrenen, R. J. de Meijer, V. F. Anisichkin & D. V. Voronin, 2012, *Lunar and Planetary Science Conference*, 43, 1738.

13. "Lunar Impact Basins – Numerical Modeling" by B. A. Ivanov, 2007, *Lunar and Planetary Science Conference*, 38, 2003.
14. Some geophysicists think the 500 km diameter Shiva crater on the ocean floor west of India might have resulted from an impact.
15. Mars is equally anomalous, rotating 97% as fast as Earth, with a 25.2° rotational axis tilt. Earth's rotational axis tilt is 23.44° (from its orbital plane perpendicular), at least nine times more than Venus's or Mercury's.
16. The Moon is 1.2% of its system's mass, versus about 0.02% for the moons of Jupiter, Saturn, or Neptune. Dwarf planet Pluto's moon Charon has 10% of its system's mass.
17. In contrast, the angular momentum of Venus's rotation is only 1/2000 that of Earth-Moon's. Jupiter's orbital motion around the Sun carries most of the Solar system's angular momentum (2×10^{43} kg m^2 s^{-1} per second); the angular momentum within Jupiter and its satellites is much smaller (about 5×10^{36} kg m^2 s^{-1} for its satellites and a huge 7×10^{38} kg m^2 s^{-1} in Jupiter's rotation) but still enormous compared to Earth and the Moon: 3.6×10^{34} kg m^2 s^{-1}. The binary dwarf planet Pluto-Charon has an angular momentum of about 6×10^{30} kg m^2 s^{-1}, about 1/6000 of Earth-Moon's, but about 1/500 of its mass. It beats the Earth-Moon system, however, in having even a larger fraction of angular momentum invested in the satellite, about 99%.
18. Because of perturbations known as the Kozai Mechanism ("Eccentricity Evolution of Extrasolar Planets by Kozai Oscillation" by Genya Takeda & Frederic A. Rasio, 2005, *Astrophysical Journal*, 627, 1001), lunar orbits tilted much more than this would be unstable, and the Moon would not exist. Indeed, if one placed the Moon in an orbit the size of its current one but tilted 90° versus Earth's equator, the Moon would follow an ever-elongating path until crashing into Earth, within about 10 years. This would contradict the Moon's existence if orbital tilt exceeded about 60°.
19. Yes, rocks do fall from the Moon to Earth. Currently there are some 84 independent samples of lunar meteorites. This amounts to about one sample per 800 km diameter patch on the Moon.
20. "On the ^{53}Mn Heterogeneity in the Early Solar System" by Alexander Shukolyukov & Günter W. Lugmair, 2000, *Space Science Reviews*, 92, 225.
21. "Where Did the Moon Come From?" by Edward Belbruno & J. Richard Gott, III, 2005, *Astronomical Journal*, 129, 1724.
22. "High-Resolution Simulations of the Final Assembly of Earth-like Planets, I. Terrestrial Accretion and Dynamics" by Sean N. Raymond, Thomas Quinn & Jonathan I. Lunine, 2006, *Icarus*, 183, 265, and "II. Water Delivery and Planetary Habitability" 2007, *Astrobiology*, 7, 66; "Asteroidal Sources of Earth's Water Based on Dynamical Simulations" by J. Lunine, A. Graps, D. P. O'Brien, A. Morbidelli, L. Leshin & A. Coradini, 2007, *Lunar and Planetary Science Conference*, 38, 1616.
23. "Recent Refinements in Geophysical Constraints on Lunar Origin and Evolution" by L. L. Hood & M. T. Zuber, in *Origin of the Earth and Moon*, eds. Robin M. Canup & K. Righter, 2000, Tucson: University of Arizona Press, 397.
24. "Quantifying the Oldest Tidal Record: The 3.2 Ga Moodies Group, Barberton Greenstone Belt, South Africa" by Kenneth A. Eriksson & Edward L. Simpson, 2000, *Geology*, 28, 831.
25. "On the Origin of Planetary Spins" by Luke Dones & Scott Tremaine, 1993. *Icarus*, 103, 67; also "The Origin of the Systematic Component of Planetary Rotation. I – Planet on a Circular Orbit" by Jack J. Lissauer, David M. Kary, 1991, *Icarus*, 94, 126.
26. "Sources of Planetary Rotation: Mapping Planetesimals' Contributions to Angular Momentum" by R. Greenberg, M. Fischer, B. Valsecchi & A. Carusi, 1997, *Icarus*, 129, 384; also "Planetary Rotation by Accretion of Planetesimals with Nonuniform Spatial Distribution Formed by the Planet's Gravitational Perturbation" by Keiji Ohtsuki & Shigeru Ida, 1998, *Icarus*, 131, 393.
27. "Satellite-sized Planetesimals and Lunar Origin" by William K. Hartmann & Donald R. Davis, 1975, *Icarus*, 24, 504; American astronomer Henry Norris Russell (1877–1957, of Hertzsprung-Russell diagram fame) expressed the idea, quoted in "Origin of the Moon and Its Topography" by Reginald A. Daly, 1946, *Proceedings of American Philosophical Society*, 90, 104.
28. "The Origin of the Moon" by A. G. W. Cameron & W. R. Ward, 1976. *Lunar and Planetary Science Conference*, 7, 120.
29. An effective summary of results is found in "Dynamics of Lunar Formation," by Robin M. Canup, 2004, *Annual Reviews of Astronomy and Astrophysics*, 42, 411.
30. "Origin of the Moon in a Giant Impact Near the End of the Earth's Formation" by Robin M. Canup & Erik Asphaug, 2001, *Nature*, 412, 708.

31. "A Hit-and-Run Giant Impact Scenario" by Andreas Reufery, Matthias M.M. Meier, Willy Benz & Rainer Wieler, 2012, *Icarus*, 221, 296.
32. "The Proto-Earth as a Significant Source of Lunar Material" by Junjun Zhang, Nicolas Dauphas, Andrew M. Davis, Ingo Leya & Alexei Fedkin, 2012, *Nature Geoscience*, 5, 251.
33. "Zinc Isotopic Evidence for the Origin of the Moon" by Randal C. Paniello, James M. D. Day & Frédéric Moynier, 2012, *Nature*, 490, 376.
34. "Effects of the South Atlantic Anomaly on the Muon Flux at Sea Level" by C. R. A. Augusto, J. B. Dolival, C. E. Navia & K. H. Tsui, 2008, arXiv:0805.3166, http://arxiv.org/abs/0805.3166; "The South Atlantic Field Anomaly and Its Effect on the Calculated Production of Atmospheric Neutrinos" by J. Poirier, 1999, *26th International Cosmic Ray Conference*, 2, 253.
35. "Oxygen Isotopes and the Moon-Forming Giant Impact" by U. Wiechert, A. N. Halliday, D. C. Lee, G. A. Snyder. L. A. Taylor & D. Rumble, 2001, *Science* 294, 345; "Silicon in the Earth's Core" by R. Bastian Georg, Alex N. Halliday, Edwin A. Schauble & Ben C. Reynolds, 2007, *Nature*, 447, 1102; "On the ^{53}Mn Heterogeneity in the Early Solar System" by Alexander Shukolyukov & Günter W. Lugmair, 2000, *Space Science Reviews*, 92, 225; "Late Formation and Prolonged Differentiation of the Moon Inferred from W Isotopes in Lunar Metals" by M. Touboul, T. Kleine, B. Bourdon, H. Palme & R. Wieler, 2007, *Nature* 450, 1206.
36. "The Proto-Earth as a Significant Source of Lunar Material," by Junjun Zhang, Nicolas Dauphas, Andrew M. Davis, Ingo Leya & Alexei Fedkin, 2012, *Nature Geoscience*, 5, 251.
37. "Simulations of a Late Lunar-Forming Impact" by Robin M. Canup, 2004, *Icarus* 168, 433; "Obtaining Higher Target Material Proportions in the Giant Impact by Changing Impact Parameters and Impactor Compositions" by A. Reufer, M. M. Meier, W. Benz, & R. Weiler, 2011, *Lunar and Planetary Science Conference*, 42, 1136; c.f. "Identification of the Giant Impactor Theia in Lunar Rocks" by Daniel Herwartz, Andreas Pack, Bjarne Friedrichs and Addi Bischoff, 2014, Science, 344, 1146."
38. "Equilibration in the Aftermath of the Lunar-Forming Giant Impact" by Kaveh Pahlevan & David J. Stevenson, 2007, *Earth and Planetary Science Letters*, 262, 438; "Lunar Accretion from a Roche Interior Fluid Disk" by Julien Salmon & Robin M. Canup, 2012, *Astrophysical Journal*, 760, 83.
39. "Forming a Moon with an Earth-like Composition via a Giant Impact" by Robin M. Canup, 2012, *Science*, 338, 1052.
40. "Making the Moon from a Fast-Spinning Earth: A Giant Impact Followed by Resonant Despinning" by Matija Ćuk & Sarah T. Stewart, 2012, *Science*, 338, 1047.
41. "On the Early Thermal State of the Moon" by S. C. Solomon, in *Origin of the Moon*, eds. W. K. Hartmann, R. J. Phillips & G. J. Taylor, 1986, Houston: Lunar and Planetary Institute, 435.
42. "Thermal Aspects of a Lunar Origin by Giant Impact" by M. E. Pritchard & D. J. Stevenson, in *Origin of the Earth and Moon*, eds. Robin M. Canup & K. Righter, 2000, Tucson: University of Arizona Press, 179.

Chapter 6
Moons Past

> The Moon is in plain sight. It is the first object in the Universe for us to explain.
> – Dr. Harold C. Urey, as quoted by Dr. Stuart Ross Taylor[1]

HARTSFIELD: And, Ken, the guys are back inside. I don't know whether you heard me a while ago or not, but EVA-1 was a total success. They had a seven hour and 11 minute EVA.
MATTINGLY: Outstanding. Did they have anything particularly significant to say or . . .
HARTSFIELD: I didn't catch all of it . . . Let me ask . . .
MATTINGLY: Did they have any surprises in the things they saw or that they didn't expect?
HARTSFIELD: I guess the big thing, Ken, was they found all breccia. They found only one rock that possibly might be igneous.
MATTINGLY: Is that right? (Laughs)
HARTSFIELD: Yeah. I guess the guys are a little bit surprised by that.
MATTINGLY: Well, that ought to call for a session with the – yeah (laughs). Well, it's back to the drawing boards or wherever geologists go.

– *Apollo 16* Command Module Pilot Thomas Kenneth "Ken" Mattingly (in lunar orbit) and Capsule Communicator Henry Hartsfield (Houston) discussing samples from the mission's first Extra-Vehicular Activity (moonwalk), April 21, 1972[2]

A basic prediction of the Moon's violent birth is crystalline minerals solidified from the giant magma reservoir of molten rock from its creation. Such things happen on Earth even now when magma bodies cool in place into *plutons* or bodies of igneous rock (hence the definition of *igneous*: cooled from liquid rock magma) over sufficient time to separate into different materials that sink or float depending on their density. Earth's crust, like the Moon, is made mostly of silicates (both crusts having densities of about 3 grams per cubic centimeter). When plutons form on Earth, what commonly floats to the top is a feldspar like anorthosite,[3] made primarily of anorthite with some other *plagioclase feldspars* – silicate minerals containing

Moons Past

Figure 6.1. **Genesis rock**. This is the nickname of the first large anorthositic rock found on the Moon, by *Apollo 15*. At 269.4 grams, #15415 is the seventh heaviest anorthositic sample recovered (99% pure and the 174th heaviest rock overall, the rest mostly breccias and basalts). The rock's white color is typical of its highlands origin, brighter than the iron-rich maria. (NASA photo AP-S71–42951.)

aluminum and sodium and/or calcium (with a density of about 2.7 grams per cubic centimeter). Lunar composition is enough like Earth's crust that anorthosite should collect on the Moon's surface as well. A primary distinction between terrestrial plutons and what formed lunar rocks is size. Plutons on Earth tend to be a few kilometers across, rarely more than 100 kilometers,[4] whereas the lunar magma ocean presumably spanned the entire 11,000-kilometer lunar circumference, like the one that probably covered early Earth.

Lunar primordial anorthosite rocks became apparent only after the fourth human lunar landing. Although some small anorthosite fragments were discovered by *Apollo 11*, the Genesis Rock (Figure 6.1) from *Apollo 15* was the first large anorthositic sample (8 centimeters across),[5] later shown to be more than 4 billion years old – older than most large impact basins that scarred the Moon.

The hunt's intensity puts the Genesis Rock at par with the Holy Grail worthy of any crusading knight or avaricious conquistador. *Apollos 11* and *12* had yielded few plagioclase fragments, and *Apollo 14* was directed to a site seemingly more indicative of deep lunar layers and their extrusion onto the surface, hence not anorthositic. Nonetheless small bits of anorthosite from *Apollo 11* (and new rock types, as we detail shortly) led to the idea of rocks separating out of a plutonic-like melt as on Earth and set anorthosite as the goal. In contrast *Apollo 15*, 4 million mission kilometers later, landed at Mare Imbrium's edge, presumably which ejected large clumps of crust to be later fragmented and strewn onto the landing site by smaller, subsequent impacts. Genesis Rock was beautiful: 97% plagioclase, with crystalline grains up to 3 centimeters across, and at 269 grams, one of the largest pieces of lunar anorthosite recovered. Yet many anorthositic rocks should reside in the *Apollo 15* samples; finding only one Genesis Rock was surprisingly scant.

Where is old lunar crust? The oldest surfaces, showing high cratering density from past impacts (randomly distributed), are the highlands. Landing on such chaotic surfaces is challenging (witness *Apollo 11*'s difficulty avoiding the boulder field near its smoother mare landing site), but one might accomplish the same thing by alighting in less cratered plains just below the highlands, again expecting fragments to be thrown down from above. This strategy was employed to select *Apollo 16*'s site.[6]

Four years before *Apollo 16*, *Surveyor 7* found highland material high in aluminum and calcium, presaging anorthosite. *Apollo 16* encountered little of the rock: only 10% of its sample mass. Surprisingly, it found primarily *breccia* (hence the earlier quote): many smaller rock fragments fused together by heat, crudely resembling concrete, and rock perhaps nearly completely remelted. Breccia and remelted samples composed 40% of the mass of rock returned to Earth by all missions. In contrast the soil from *Apollo 16* was 69% anorthosite. Highlands material might be there but mainly in small fragments.

Luna 20 had found 52% anorthositic soil in the Near Side highlands four months before. *Apollo 15* and *17* returned only 0.5% anorthositic rocks, but 33% breccias. Apollo in general (mostly *Apollo 16*) found only 4% of its rocks anorthositic, with breccias and impact melt 58% and basalts 33%. Of course breccias' composition varies depending on local rocks.

Science's splendor is finding the unexpected, and Apollo's early landings turned up some, too. In addition to primitive rocks like those on Earth (anorthosite, minerals pyroxene and olivine[7]) came a relatively unknown class high in potassium (K), rare earth elements, phosphorus (P), thorium (Th), and uranium (U). This is awarded the moniker KREEP (for K, Rare Earth Elements, and Phosphorus, not for being creepy), with Th and U neglected more to keep the acronym pronounceable than because of their rarity. KREEP rocks appear mainly in igneous form or mixed in breccias and are much more commonly associated with plagioclase and pyroxene than olivine. Where exactly KREEP rocks originate in the Moon's bedrock is being isolated, although they obviously reside on the surface primarily in the western part of the Near Side, more in the maria than highlands, for reasons we explore in Chapter 7.

Pyroxene and olivine are igneous silicates found on Earth and in some meteorites as primitive, unevolved minerals. They are incorporated into many lunar rock types but are often found in relatively pure samples. With KREEP's discovery and the finding of anorthosite, a consistent picture unfolded for the origin of the lunar surface and remains the dominant idea of the early structure and evolution of the Moon.

If the Moon was molten to great depth, it might take tens of millions of years to solidify, which is adequate time for various rock types to form and separate via gravitational settling and buoyancy (*differentiation*, in other words). There is a logical progression to this. All of these silicate rocks melt somewhere between 600°C and 1,200°C at low pressure (although this varies at lunar interior pressures), and in the mantle this order develops in reverse. First, olivine (usually an olive green mineral) solidifies, and with density about 3.3 grams per cubic centimeter, sinks and fills in the deep mantle. The Moon is not made of green cheese, but most of its mantle, the largest part of its structure, is probably olivine.[8] Iron-rich and magnesium-rich pyroxenes then solidify, falling out of solution. With about 80% of the magma now rock, more aluminum-rich minerals, particularly plagioclase feldspar, leave solution, but float upward.[9] This seems to imply a primarily anorthositic crust, but not exactly: in rising, anorthosite can carry much pyroxene and even olivine.

The remaining magma ocean burgeons with substances not bound to olivine or pyroxene, particularly potassium, hence the KREEPy elements (including uranium and thorium). These incompatibles should remain just below the anorthositic crust. Much of the Moon's water, as much as may be, is concentrated in this layer and forced to the base of the crust.[10]

This basic model predicts a crust rich in anorthosite, which composes roughly half to three-quarters of the highlands (with some pyroxenes and less olivine and oxides). On the maria, huge vents spewed forth KREEP-heavy lava over roughly half of lowlands and one-twelfth of the lunar surface, especially the western Near Side mare. Olivine, which may compose most of the Moon's interior, makes up much less than 10% of the surface rocks. The magma ocean model succeeds in placing each kind of rock in roughly its proper place relative to the Moon's surface.

One surprise about the Moon was less that its rocks were different but that they were not (nothing to match kryptonite, dilithium crystals, or unobtainium[11]). Only three new lunar minerals – armalcolite, tranquillityite, and pyroxferroite – were found by *Apollo 11* (and at other Apollo sites), and all three were later discovered on Earth.[12] Although all lunar minerals are also found on Earth, testifying to the worlds' common origins, roughly 30 times as many Earth minerals are known as lunar ones. Earth is unique among known planets with its plate tectonics and liquid surface water. The Moon has neither of these.

A telling example is silicon dioxide (SiO_2: quartz, silica, sand, glass, e.g., obsidian) in rocks such as granite, sandstone, quartzite, chert, and many others, and some of the most common minerals on Earth (12% of the crust). Pure SiO_2 is rare on the Moon but mixes microscopically with other substances, for example, in olivine $(Mg,Fe)_2SiO_4$. Granite is rare on the Moon but is the most common quartz-bearing rock in Earth's crust; however, this quartz component does not form easily. Instead, quartz stays in solution until the last crystallization, greatly influenced by the amount of water in the melt. Pure quartz is readily formed by water in veins and other deposits where it crystallizes. This is less likely on the Moon, so lunar silicates mix with other material in magma melts, as do many Earth minerals. Even obsidian, sometimes a pure silicate glass, is often laced with other earthly substances by 10–30%. SiO_2 feels forces on Earth unknown on the Moon: erosion and sedimentation by air and water to form sandstone and chert/flint/chalcedony. Tectonics cause subduction, and sandstone is cooked to quartzite underground, but not on the Moon.

As on Earth, not all lunar magmas are identical: some igneous silicates are higher in silicon (and other light elements such as aluminum, sodium, or potassium, like feldspars), therefore, they are felsic, whereas other silicates are high in iron and magnesium, therefore, they are mafic (magnesic and ferric). As time progressed, simple lunar magma ocean layering broke down as heat continued to escape and magmas mixed.

Differentiation of the magma ocean into floating and sinking layers (*cumulates*) would be stable if nothing entered or left the Moon, but heat radiates into space from the surface causing the interior to change. The balance of buoyancies is upset by further cooling, contraction, and crystallization, and layers rearrange. Ilmenite ($FeTiO_3$) is a prime culprit: the most common silicon-free mineral on the lunar surface (several percent of Apollo samples) and the biggest source of titanium. At these pressures, solid ilmenite is 3.7 times denser than water but layers into the magma ocean below anorthosite (density of 2.7) but above olivine/orthopyroxene (density about 3.3). Once ilmenite crystallizes, it overlies a lighter layer and is unstable, with a tendency to sink.

Scientists have differing ideas about mantle settling. Perhaps the ilmenite-rich layer, tens of kilometers thick, close below the anorthositic crust, simply plunges hundreds of kilometers to the mantle base over several tens of millions of years. Mathematical models indicate layer overturn proceeds in roiling motions on scales of several tens of kilometers.[13] This is a complicated process: not just movement of the ilmenite, but consequences of gargantuan friction driven by rearrangement within the Moon's gravity. This heat will melt other rock. Ilmenite itself might mix with lower olivine and pyroxene or run in finger-like magma flows through the underlying rock.[14] This complexity is seen in collections of samples from the same site showing titanium abundances varying by factors of 50 or more – larger than expected from a simple, large-scale mixing roil. Ilmenite sinking inside the Moon releases huge amounts of energy, generating lava. In the process, lunar basalts can entrain varying amounts of titanium ("high titanium," "low titanium," or "very low titanium") depending on the contribution of ilmenite, which varies significantly in surface basalts between 3 billion and 4 billion years old.

Reheated mantle materials generate melts from minerals liquifying easiest at mantle pressures. These partial melts differ in composition from the original mantle material around them and are usually basalts, usually with low SiO_2 versus FeO content, unlike higher-SiO_2 lavas produced by melting in Earth's deep crust.

Although the magma ocean cooled in millions of years, lava eruptions continued for billions. The first lava flows were likely the magnesium-suite mix: high in magnesium versus iron. The Moon formed 4.52 billion years ago, the magma ocean hardened 4.4 billion years ago (iron-rich anorthosites' age), and magnesium-suite flows died 4.1 billion years ago. The magnesium-suite flows are distinct among eruptions from inside, being distributed throughout the crust, particularly the highlands, not just low points (although magnesium concentration is highest there). The magnesium suite arose as plutons into the anorthositic crust, not always at the surface, and produced some common rock types found by Apollo.[15]

Generations of violent eruptions continued for a billion years. Giant lava flows extruded 10 million cubic kilometers of basalt[16] – more than the largest flow known on Earth. What powered these? There was enough heat: magma-ocean overturn generated heat sufficient to melt roughly 100 million cubic kilometers of basalt, while radioactive decay generated even more.[17] Magma would liquify simply as pressure fell in rising through the mantle. The Moon was still wracked by titanic impacts, melting and vaporizing huge expanses of material and, more importantly, sculpting basins into which lavas would pour, forming the setting for the next phase of lunar evolution.

A clever tool betrays which layer of magma ocean produced a mineral despite the complex motions and melting that followed. The rare-earth elements (REE or *lanthanides*, atomic numbers 57–71) are trivalent (can donate three electrons to other atoms in a molecule or mineral lattice), with one exception: europium (number 63, Eu), which can be divalent in reducing (versus oxidizing) environments, as the Moon. Thus europium concentrates differently into magma ocean minerals than other REEs: this "europium anomaly" traces this point of origin. For KREEP minerals, REEs are enriched by large factors, often 100 times or more, whereas europium is down relative to the other 14 REEs by factors of up to ten. In anorthositic crust, REEs are at roughly their lunar average abundance, but europium can be enhanced versus the other 14 by a factor of two (Figure 6.2). In mare basalts, REEs are greatly curtailed, and europium can be cut slightly more (or less, or the same). This can distinguish if REEs are depleted because the rock (or source magma) is formed somewhere other than the KREEP layer, versus situations where some material is KREEPy but diluted by other rock.

Because the KREEP layer was nearly the last to solidify, its material also contributed significantly to early volcanic eruptions. Original KREEP material (urKREEP: *ur-* means primordial) mixed with anorthositic crust above to make new, KREEPy minerals and to remelt into new magma, even before mantle overturn. Deluges of this material poured onto the Moon into giant impact basins as well as immense surroundings, millions of square kilometers, evident in the distribution of potassium, REEs, and associated radioactive element thorium seen in gamma rays (Figure 6.3).[18] The concentration by several times of radioisotopes such as potassium-40, thorium-232, and uranium-238 in this region hints at the role of radioactive heating in melting this material. The high KREEP abundance region (Procellarum KREEP Terrane: PKT in the western maria) is also site of the most recent volcanism, so this activity persisted for billions of years.

In contrast were other early, episodic eruptions – some more extensive than KREEP extrusion. High-aluminum basalts erupted about 4.2 billion years ago and may represent mixing of highlands rock with deeper magma. High-titanium basalts were erupting 3.9 billion years ago from deep in the Moon into mare basins even

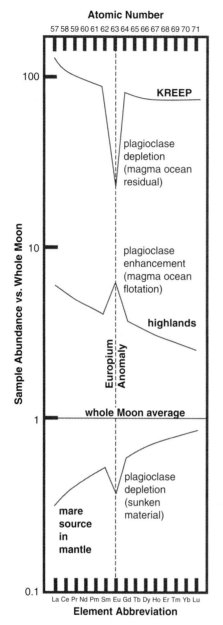

Figure 6.2. **Odd element out**. Because of its different chemical bonding, europium concentration varies versus other rare earth elements depending on where the mineral originates in the lunar magma ocean: much less in KREEP minerals, more in the predominant highlands crust, and less in the mantle source of mare basalts. The Whole Moon values are based on magma ocean models.

beyond the PKT, spewing forth almost 500 million years, trickling another billion years more, producing the most mare mass. In some areas, especially in eastern maria, low-titanium basalt began erupting as high-titanium flows ebbed, from about 3.5 to 2.5 billion years ago or even later in Mare Imbrium. Up to 0.9–1 billion years ago, low-flow eruptions continued around crater Aristarchus and westward in the PKT, near craters Flamsteed and Lichtenberg. In the Moon's history's last

Moons Past

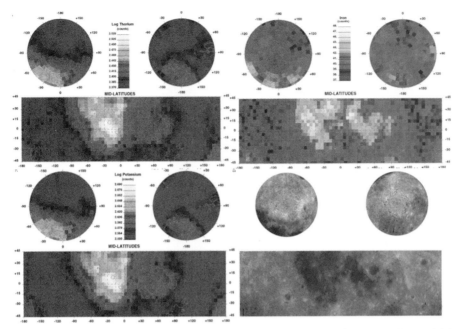

Figure 6.3. **Hot stuff**. *Lunar Prospector* gamma ray maps trace (*upper left*) thorium, (*upper right*) iron, and (*lower left*) potassium, as well as (*lower right*) an albedo map showing maria and highlands (all split into polar ant equatorial regions 45 degrees from the equator). Note that KREEP element potassium and associated thorium trace each other and the western maria closely, whereas iron also extends over eastern maria. (Maps courtesy of NASA and Los Alamos National Laboratory; *Clementine* images courtesy of NASA/DOD.)

quarter, lava flow largely ceased, certainly on scales above several thousand square kilometers, a faint echo of the 7 million square kilometers once covered. As massive as lunar eruptions were, they stretched over billions of years.

Flowing basaltic lava surface betrays secrets of the lunar interior, because temperatures beyond 1,000°C (at depths of about 200 kilometers) are required. The occurrence of these eruptions indicates the mantle's thermal state and thus the Moon's thermal evolution. This implies that most mantle locations cooled below temperatures sustaining partial melting by about 3 billion years ago. These being partial melts, chemically complementary to unmelted mantle materials left behind, we can use thermodynamic principles involved in the equilibrium of partial melting; the residue can be characterized, and the overall mantle composition calculated.

The ancient, massive mare eruptions and subsequently quiet lunar environment leave us with massive volcanic structures unlike any on Earth. Schröter's Valley (*Vallis Schröteri*) is comparable in size to the Grand Canyon but gouged by up to 1 million cubic kilometers of flowing lava, not running water.[19] The Moon has large volcanoes, for example, the Marius Hills with 5,000 cubic kilometers of volume, but only 6% the volume of the Hawaiian shield (Mauna Loa) and 0.2% of Mars' Olympus Mons (although the Marius Hills might partially hide under basalt they produced). Lunar eruptions filled huge mare volumes over long æons with

Figure 6.4. **Solid sea**. A typical view of mare landscape, looking across southern Mare Imbrium toward Copernicus (large crater near horizon), with crater Pytheas in foreground (20 km in diameter). The mountains Montes Carpatus run along the margin of the mare. The light streaks and elongated craters on the mare surface are caused by secondary impacts and ejecta rays from Copernicus. (NASA photograph AS17-M-2444.)

low viscosity lava, not ideal for volcano building. Additionally, huge areas filled with pyroclasts, ejected eruptive particles, including cinders, ash, and glass beads, with more than 1,000 cubic kilometers on the Aristarchus Plateau alone.[20] The Moon has many volcanic landforms preserved over billions of years. We examine several.

The prime relic of eruptive volcanism is the maria themselves, almost 7 million square kilometers, 16% of the lunar surface (but only 1% of the lunar crust's volume). From Earth they appear smooth and lightly cratered but have structures betraying past activity (Figure 6.4). In addition to craters, maria are also crossed by wrinkles, faults, ridges, scarps, and valleys, best revealed at oblique illumination angles (Figure 6.5). Wrinkle ridges betray forces at work in the mare lava because of contraction, settling, and buckling. Scarps (cliffs) could indicate where more viscous lava plains end, but more often reveal more recent shrinkage of the entire Moon (Figure 7.4). Different eruptive episodes fill mare areas, indicated by differing, adjacent lava compositional units. Some Apollo landing sites show thin strata, indicating successive eruptions of runny lava (Figure 6.6). These and other areas

Moons Past

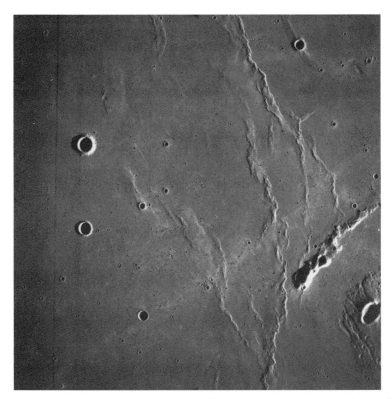

Figure 6.5. **Frozen wrinkles**. At oblique illumination many expanses of maria betray uneven surfaces; here wrinkle ridges are seen running roughly vertically. Many are about 250 m tall. (The elevated corner at lower right is the edge of the Aristarchus Plateau, and the diagonal ridge below right of center is Montes Agricola.) This view is about 130 km wide. (NASA photo AS15-M-2487.)

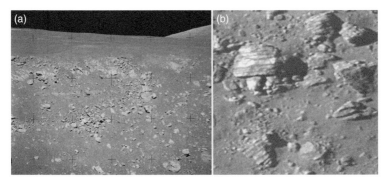

Figure 6.6. **Strange strata**. (a) A view of the wall of Hadley Rille from the other side, 1,350 m away. The width of wall shown is about 100 m, revealing fine-grained soil, talus blocks, and rock strata outcrops. (b) Blocks ejected by Aristarchus' impact, in a scene about 100 m across. One explanation for the striped structure is layers from successive pyroclastic eruptions on the Aristarchus Plateau.[21] (NASA photos AS15–89–12045 and NAC M120161915L.)

183

The New Moon

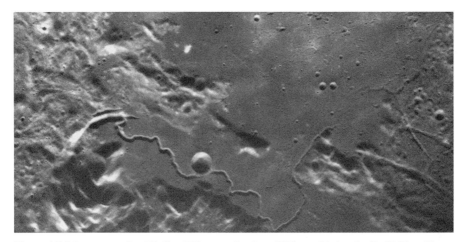

Figure 6.7. **Moon meander**. Hadley Rille, seen in view 150 km wide by *Apollo 15*. The rille extends from the deep double-arc gash at left to the faint, shallow depression at upper right. Figure 2.16 shows the view up the rille where it hits the bottom of this image right of center. Figure 6.6 was taken looking across the rille at the small bend 3 km up and to the right from there. *Apollo 15* landed on the plain at bottom right. Other rilles are seen in Figures 8.2, 9.3, and 9.6. (NASA photo AS15-M-1137.)

capture past eruptions as sinuous rilles through which lava flowed, cutting river-like paths in channels or collapsed lava tubes (Figures 6.7 and 6.8). Smaller rilles are found on Earth (Figure 6.9); lunar rilles can be huge, up to 340 kilometers long and 5 kilometers wide.

As on Earth, volcanic eruptions involve not only lava but also gas and pyroclasts (cinders, glass spray, ash). Maria's iron-rich basalts are dark, but pyroclasts are more, called *dark mantle deposits*, with dozens of sites clustering maria edges, prominently the Aristarchus Plateau, Alphonsus, Mare Humorum, Sinus Aestum, Rima Bode, Mare Vaporum, Sulpicius Gallus, and Taurus-Littrow, all on the Near Side. *Apollo 17* visited the last site, finding ubiquitous glass beads, formed when gas bubbling from lava into the vacuum explodes into droplets quickly hardening to glass beads before hitting the ground (Figures 6.10 and 11.3). Dark mantles are a promising source of material for future lunar habitats, as discussed later.

Titanic forces were also loosed on the Moon by impacts of other Solar System objects. Although hundreds of craters formed volcanically, uncountable numbers on all scales resulted from impacts, up to the largest in the entire Solar System.[22] Many were flooded by lava (Figure 6.11), but most sit in the highlands far from these flows, where cratering densities are so high in most places that new impacts destroy as many old craters as they create the new (Figure 6.12). Maria provide a clean slate, containing all-sized craters (except the largest) with little interference from other craters. Crater morphology predictably progresses with size, closely tracking the impactor mass (typically ejecting about 1,000 times its own mass; see Figure 6.13). Size also depends on impact velocity, as small as lunar escape velocity, 2.38 kilometers per second (twice the speed of a high-powered rifle bullet), up to head-on

Moons Past

Figure 6.8. **Lava land**. View from Mare Imbrium into Oceanus Procellarum, with flooded crater Prinz (46 km wide) and fresh crater Aristarchus (42 km across, 2 km deep). Lava covering Prinz probably flowed from the sinuous rilles (rimae) seen flowing off the Aristarchus Plateau to the right and sourced from lava ponds in the foreground, seen with Montes Harbinger. Note ray patterns in the background ejected from the Aristarchus impact. See Figure 9.8 for an adjacent view. (NASA photo AS15-M-2606.)

lunar-comet collisions at 73 kilometers per second. Identical masses hitting at these two speeds would make two craters about a factor of three different in diameter. At typical velocities of 18 kilometers per second, the crater is 25 times wider than the impactor (for stony meteorites).[23] Rebound in the crater as a result of compressed bedrock produces a central peak for craters larger than about 20 kilometers, growing more complex for larger craters (Figure 6.14).

Most meteoroids striking the lunar surface burrow inside, heat, largely vaporize, and explode. (For iron, 20 kilometers per second corresponds to 900,000°C.) Exploding a bomb underground makes a circular crater regardless of entry angle; most craters form fairly round, with exceptions for impactors entering almost parallel to the ground (Figure 6.15).

Figure 6.9. **On Earth, too**. Aerial view of a volcanic crater and lava channel near Craters of the Moon National Monument, Idaho. Note the similarity to rilles in Figure 6.8. This crater is about 100 m across and the channel more than 5 km long. (NASA photo S-69–42867.)

Along with craters come mountains, including impressive ones: Mons Huygens, the highest at 5.5 kilometers tall, and Mons Hadley and Bradley at about 4.5 kilometers (compared to Earth's Mount Everest, 8.85 kilometers). One might expect lunar mountains to equal Earth's given the Moon's lower gravity thick crust, typically 40 kilometers deep versus 7 kilometers under the ocean and 40 for Earth's continents (and 55 in the Himalayas). The Moon's lower gravity is less crucial but makes mountains even taller. By the principle of *isostasy* (in equilibrium the portion of a mountain sticking above the surface is balanced by a dense root protruding below the crust to support it as a result of buoyant force, like an iceberg floating on the ocean), the Moon's mountains could protrude to about 15% of the crust's thickness. Mountains on the Moon, in contrast, are born not by gradual tectonics but catastrophic impacts. Although not as tall, they are impressive: the Aristarchus Plateau, for instance, 2 kilometers high above the maria and covering 40,000 square kilometers, which is apparently a block of lunar crust ejected by the titanic explosion of 1,145-kilometer-diameter Mare Imbrium, 300 kilometers away.[24]

The interplay of volcanic eruptions, impacts, and rock chemistry is crucial for understanding the Moon. Geologists find the relative ages of different structures or

Moons Past

Figure 6.10. **Rain of fire**. A fire fountain over 300 m high (*background*) feeding lava flows (foreground, pouring over ledge) into 'Alo'i Crater, Kilauea east rift zone, Hawaii. This eruption lasted for five years, 1969–1974. Extensive areas of the Moon were affected by fire fountains as evidenced by widespread volcanic glass beads, spread either as a result of a long-distance ballistic throw or extensive eruption sites. (USGS photo by Don Swanson.)

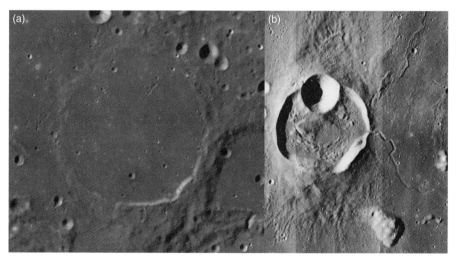

Figure 6.11. **Over the top**. Lava can rise to overflow onto maria, evident for flooded crater Wargentin (a), more elevated than mare to its left. Several such overflows are found, e.g., crater Spörer. Lava can also breach crater rims as rilles, e.g., crater Kreiger (b) or Posidonius (Figure 8.2). (NASA photo mosaics from *LRO*/LROC-WAC and AS15-P-10320, 10322, 10324.)

Figure 6.12. **Not swiss cheese**. A view near crater Chavenet amid Far Side highlands, with overlapping craters. Note crater chains as a result of secondary impacts (*foreground*). (NASA photo AS17–155–23702.)

"units" by comparing which overlays what, because few processes invert time ordering of strata (although they may disrupt them). Volcanic flows are important in this scheme, but more important is relative crater age, measured not so much by which crater obliterates the other, but more gently, which sprays the other with ejecta. Between eruptions and impacts, a sequence of events has been constructed, more activity when the Moon was younger, but better preserved more recently, describing a roughly, evenly distributed period sequence in lunar evolution (see Table 6.1). With a relative series of events, radioactive sample dating pegs much of the lunar timetable extending back 4.5 billion years.

Basin impacts, mare flooding, and overlying impact ejecta patterns form the basis whereby we establish which layers overlie older layers and thereby extend across the Moon knowledge of ages determined from lunar samples. Finer degradations are made by counting craters on surfaces and assuming that global impact rates apply, giving the surface's age. Figures 6.4 and 6.8 show the pattern of bright ejecta rays from craters Copernicus (800 million years old) and Aristarchus (175 million years old).[25] Figure 6.4 shows several examples where large fragments entrained in the ejecta excavated secondary craters within the rays.

Rays are crucial to setting ages in lunar science. Rays and jets of secondary ejecta can fly thousands of kilometers: Tycho's rays (108 million years old) extend more than 1,500 kilometers in radius. Several more craters have ray systems beyond

Moons Past

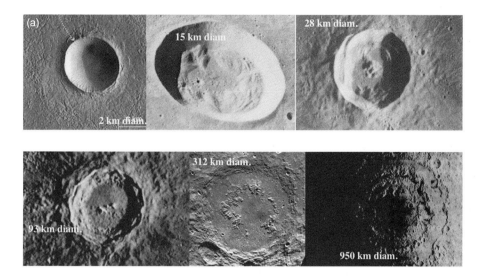

Figure 6.13. (a) **Family portrait**. Sequence of typical, largely unflooded, relatively isolated craters of increasing size, starting with a small bowl crater, followed by Bessel, Euler, Copernicus, Schrödinger, and Mare Orientale, of diameters 2, 15, 28, 93, 312, and 950 km, respectively. Note progression from smooth spherical bowl to flatter floor, to terraced walls and simple central peak, to complex central peak, to no peak but several concentric rings. (NASA photos.) (b) **Lunar terraces**. A 25 km wide view of wall of 42 km diameter crater Aristarchus (also in Figures 6.8 and 9.8) with multiple layers of terraces sloping down to the crater floor, as well as large boulders and rockfalls. (NASA photo LROC-NAC M1755659775.)

300 kilometers, together forming a loose global web of rays.[26] When a ray visibly overlaps a feature, that feature is older. If a ray overlies a sampling site, one expects the sites' samples to contain some from that distant crater. This was employed in selecting Apollo sites, for example, *Apollo 12* and *17*.

Seeing fresh impacts from afar depends on ejecta's optical properties (partially because crater shapes remain robust over billions of years (e.g., Eratosthenes, Figure 6.16, versus Aristarchus, Figure 6.8). Changes occur mainly in surface properties of *regolith* (debris from previous impacts) aging as a result of accumulated effects of even smaller (micrometeorite) impacts, as well as solar wind and

The New Moon

Figure 6.14. **On the rebound**. Side view of crater Tycho's central peak. (a) 16 km wide view of the central peak. (b) 1.5 km close-up of same view, featuring a 120 m wide boulder resting on the peak's summit. (c) In analogy, splashes from water drops into a deep tank produce centrally focused rebounds lofting droplets, like an impact forming a lunar crater several tens of kilometers across. (NASA photo LROC-NAC M162350671LE and flash photo courtesy of Andrew Davidhazy, Rochester Institute of Technology.)

cosmic rays. Depending on regolithic iron content, fresh surfaces often contain loosely bound iron ions, Fe^{+2}, vibrating at several characteristic frequencies, principally 315 trillion vibrations per second (hence absorbing 0.95-micron wavelength light). Particles hitting the regolith cause tiny explosions that melt its surface to glass (Figure 6.17). Additionally, solar wind ions and cosmic rays evaporate dust material, and this solidifies on adjacent grains as glass. This patina, or *rim*, on dust changes regolith's optical properties, and surface iron consolidates into small metal clusters (loosing the 0.95-micron feature). Grains grow darker (in part because of iron nanoparticles) but brighten at 0.95-microns as a result of Fe^{+2} destruction; this is called *optical maturity* (OMAT),[27] part of the space weathering process.

Ground truth comes from Apollo and Luna samples, using radiometric age determination. In radioactive decay, an isotopic form of one element transforms into another isotope (usually of another element), and the ratio of that which decays

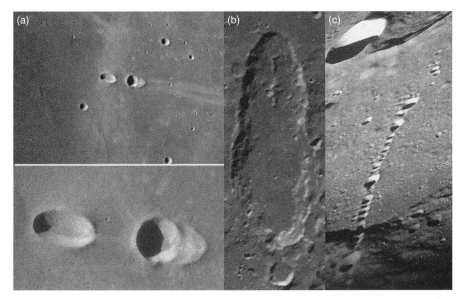

Figure 6.15. **Slantwise**. Three rare, highly noncircular impact features. (a) Craters Messier and Messier A (20 km apart) result from the same impactor ricocheting off the Moon at a glancing angle, entering from the left. Note the vertical splash from the first impact and the horizontal rays from the second. (b) Crater Schiller (180 km long) is arguably the most elongated major crater, once thought to be volcanic but now explained by an oblique impact. Schiller seems partially filled by lava, but with anomalous composition compared to surrounding maria. Artificial craters resembling Messier and Schiller are seen in laboratory hypervelocity oblique impact experiments.[28] (c) Impactors themselves can be highly elongated, as seen in the case of Catena Davy, 50 km long, likely an impactor (not a volcano) disrupted by tidal forces before hitting, like Comet Shoemaker-Levy 9 hitting Jupiter in 1994. Could a slightly tidally disrupted impactor have caused Schiller? (NASA photos AS11–42–6233, *LRO*/LRO-WAC mosaic, and AS12-H-51–7485.)

compared to that which remains tells the sample's age. Three things are required: a radioisotope with decay time not short compared to the Moon's age, a resulting isotope (daughter product) from this decay that was absent from the original mineral, or whose initial abundance is estimated independently, and, finally, the inability of either daughter or parent isotopes to enter or leave the system. A basic reaction is uranium decaying to lead, and there are two options: uranium-238 goes to lead-206 (with a decay half-life of 4.47 billion years, the time for half of the initial uranium-238 atoms to decay), and uranium-235 goes to lead-207 (704 million year half-life). A mineral containing uranium but not incorporating lead is zircon ($ZrSiO_4$; also baddeleyite, ZrO_2, common on the Moon). (Other interesting systems include potassium-40 decaying to inert gas argon-40 with a 1.3 billion years half-life – and also in part to calcium-40, rubidium-87 to strontium-87: 49 billion years, and samarium-147 to neodymium-147: 106 billion years.) Zircons found to contain lead must have acquired it from radioactive decay. If the zircon is heated, lead leaks out, so measuring lead-uranium abundance ratio ($^{206}Pb/^{238}U$) across the zircon grain is a

Table 6.1. *Lunar geological timeline.*

Period	Past time interval	Initial defining event	Major processes
Pre-Nectarian	4.533–3.92 Gigayear	Formation of the Moon	Highlands and some basins form
Nectarian	3.92–3.85 Gy	Nectaris basin impact	Many basins form
Early Imbrian	3.85–3.80 Gy	Imbrium basin impact	Most remaining basins form
Late Imbrian	3.8–3.2 Gy	End of Late Heavy Bombardment	Most maria fill
Eratosthenian	3.2–1.1 Gy	Eratosthenes impact	Decline in mare filling
Copernican	1.1 Gy–present	Copernicus impact	Easily visible crater rays form

(Copernicus is now estimated at 800 million years old.) *Early* and *late* actually split geological periods into epochs, hence, Imbrian is a period but Early Imbrian an epoch. On Earth a period is longer than an epoch, which is longer than an age. Periods can be grouped into eras and eras into eons then into supereons (informally), and ages can be subdivided into chrons, informally. On Earth we live in the Holocene epoch (12000–0 years ago) in the Quaternary period (2.6–0 million years ago) of the Cenozoic era (65.5 million–0 years ago) in the Phanerozoic eon (542 million–0 years ago). If the Holocene divides into ages in terms of human development, not geology, a new Anthropocene epoch is proposed.

Figure 6.16. **Aging archetype**. Crater Eratosthenes (58 km diameter) shows much structure similar to Aristarchus, 3 billion years younger (Figure 6.8) but is absent bright rays. Chains of secondary craters from the impact are visible. (NASA photo AS17–145–22285.)

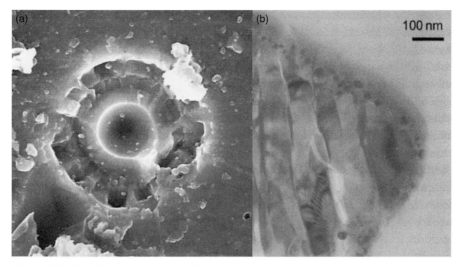

Figure 6.17. **Magnifying glass**. (a) A 10-micron-diameter crater on a lunar glass surface includes a smooth melt pit and a larger zone of spalled glass, surrounded by a zone of ejected material, some melted. (b) 100 nm thick glassy surface rim deposited by vapor and sputtering from local impacts and particle bombardment, with dark, round iron nanoparticles, 10–20 nm across. (NASA photos, including by Sarah Noble, NASA/Marshall Space Flight Center.)

useful cross-check. A half-millimeter zircon from an *Apollo 17* breccia has a ^{206}Pb/^{238}U age of 4.417 ± 0.006 billion years.[29] This indicates the lunar crust formed only 100 million years after the Giant Impact, although this zircon's original host rock was destroyed long ago. Lunar rocks competing for oldest include anorthosites at 4.29 ± 0.06 to 4.562 ± 0.068 billion years old (samarium/neodymium dates), and others at 4.44 ± 0.02 and 4.53 ± 0.12 billion years. Earth's oldest known mineral is a zircon – 4.404 ± 0.008 billion years old – from Western Australia.[30]

As amazing as 4.4-billion-year-old rock fragments are, as old as any seen on Earth, in only 382 kilograms, also amazing is the fraction of samples dating to a narrow but ancient interval 3.85–3.92 billion years ago. The later date, 3.85 billion years, is seen as when a massive impact formed Imbrium basin, second or third largest on the Moon, and centered amid all of the Apollo landing sites (but not the *Luna 16, 20,* and *24* samplings). Radiometric dates in these ranges are accurate to a fraction of a percent, several million years. This bespeaks intense activity in a narrow time range, starting roughly when another impact formed the Serenitatis basin, next door to Imbrium and almost as big. Ejecta from Serenitatis at one time would have overwhelmed *Apollo 15* and *17*'s sites and partially *Apollo 11* and Luna sample sites, and Serenitatis seems about 50 million years older than Imbrium.[31]

The conclusion that *Apollo 17* samples rocks formed in Serenitatis' impact is crucial to the concept that many large impacts wracked the Moon about 3.9 billion years ago, the Late Heavy Bombardment (LHB), altering not just the Moon and Earth but the entire inner Solar System. Earth's oldest crustal rocks, in Greenland and the Canadian Shield, are 3.8–4.0 billion years old,[32] and models predict a rain of asteroids from disruption of planetary stability at Jupiter and beyond.[33] Crucially, if

The New Moon

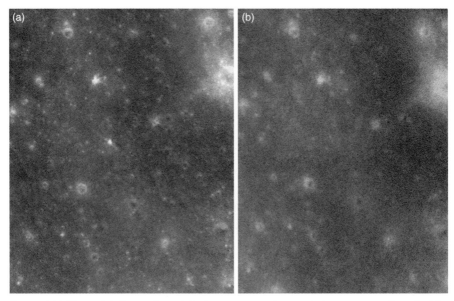

Figure 6.18. **Not so big bang**. One of the largest craters known to have appeared on the Moon's surface since Apollo, sometime between August 1971 and September 2009. (a) The LROC-NAC image of the region (on the rim of crater Franz) showing the new crater, the white splash just left of center, about 10 m in diameter. (The scene is 1 km across.) (b) Same scene viewed at a similar angle of illumination from *Apollo 15*'s orbital panoramic camera, with no sign of the crater. If the crater formed via an impactor of typical velocity and composition, it had a mass of about 50 kg. Several smaller new craters have also been found. (NASA photos LROC-NAC M108971316 and AS15-P-9527.)

the *Apollo 17* site is dominated by the effects of Imbrium and not Serenitatis, then perhaps Apollo samples are primarily sensitive to the Imbrium impact about 3.9 billion years old. In this case Serenitatis could have formed much earlier, and a tight sequence of maria formation is less evident, and an LHB not indicated. Recent work arguing for this chronology ambiguity for the Moon (therefore, the planets) is unsettled.[34]

Changes have occurred more regularly on the Moon over the last 3 billion years, primarily resulting from impacts. Unlike the highlands, which suffered through early epochs of frequent cratering, mare surfaces formed as a blank slate late in lunar history so are not heavily cratered, certainly not to the level of crater saturation, with impacts destroying as many craters as they create. To the extent that impacts occur at random at roughly uniform rate on average, they are useable as a clock: more craters over more time. Based on crater counts on maria of known ages, one finds an impact rate perhaps surprisingly small. Few impact craters have been seen to form while humans have photographed the lunar surface in detail, over five decades (Figure 6.18).

Cratering rates have plummeted since mare basins formed, but objects constantly hit the Moon. Interplanetary spacecraft show a typical square meter is hit hourly by one micrometeorite of only 10 quadrillionths of a gram, smaller than many viruses.

Figure 6.19. **Battered existence**. Even mare surfaces are heavily cratered, as seen in this field in Oceanus Procellarum (near 26°N, 58°W) with the Sun shining only about 1° above the horizon. This is not far from one of the youngest mare surfaces (31°N, 66°W).[35] Compare this with seemingly smooth mare with the Sun at higher angles: Figures 6.8, 6.11, 6.15, and 6.16. This scene is about 25 km across. (NASA photo AS14–98–13349.)

Annually a typical square meter is struck by a few microgram mass, barely enough for a human to notice.[36] On larger scales, this is more dramatic: every year a typical square kilometer is struck by an object about 1 gram in mass, making a crater about 20 centimeters across. Every billion years a typical square kilometer suffers a crater about 0.1 kilometer in diameter, from an impactor of several hundred tonnes.[37] Looking at pictures of maria, one might conclude that much of the lunar surface has not been hit, but this is illusory. Even young maria surfaces illuminated at glancing solar illumination (Figure 6.19) reveal craters nearly everywhere, most in low relief. This must be recalled when counting craters to measure age. Regolith is everywhere on the Moon.

There are exceptions and modifications in using cratering to measure age. As seen in Figure 6.4, secondary craters following large impacts add to counts of small craters. Many secondary craters land in soft impact melt. This means they formed within minutes of the major impact (not before or else they would be destroyed, not long after or else the melt would have hardened). Furthermore, crater counts vary significantly with the surface material in which they form. Also, the western limb of the Moon faces forward into its orbit so is hit 20% more often than average, versus a deficit of about 10% at the poles.[38]

Why does cratering continue? Why are objects in unstable orbits not cleared from the Solar System? Impactors arise from rare collisions between asteroids, but these do not suffice. Instead, strangely, asteroids are equipped with weak, solar-powered rockets driving them slowly from their stable orbits. The Yarkovsky effect results from the Sun heating an asteroid on one side, which then rotates into the direction it is orbiting. If the asteroid rotates in the same sense that it revolves around the Sun, heat radiating from its recently solar-irradiated side thrusts the asteroid along its orbit, shifting it farther from the Sun. (If the asteroid spins retrograde to its revolution, it falls toward the Sun.) Most asteroids drift slowly from their stable orbits. Even nonrotating asteroids start spinning as a result of a related effect, the YORP effect (Yarkovsky-O'Keefe-Radzievskii-Paddack), because the rocket reaction usually changes the rotation rate.

This does not imply that the cratering process has not changed. In fact studies that divide lunar craters into Nectarian, Imbrian, and post-Imbrian impacts find that the crater size distribution is not identical for these different periods.[39] One hypothesis is that before 3.9 billion years ago main belt asteroids were being ejected by interactions with Mars that sent them toward Earth and the Moon, but 3.9 billion years ago a major main belt collision occurred sending a pulse of impactors, or a disruption in the outer planets' orbits sent ripples through the asteroids, wreaking inner Solar System havoc (the LHB), either way sending different impactor populations to the Moon.

A popular LHB model holds that many comets orbited in the outer Solar System (roughly where Uranus and Neptune are now), and outer planets slowly consumed these, changing the planets' orbits. Eventually this led to catastrophic changes in which Uranus and Neptune switched places, and Saturn and particularly Jupiter rapidly changed orbital radius in response.[40] A prediction of such models, along with a pervasive pelting of objects throughout the Solar System, is two different populations over time: comets at first, followed by asteroids as Jupiter perturbed the main belt. Increasingly, lunar scientists discuss two impactor populations excavating large lunar basins,[41] although not always as asteroids versus comets. They interrogate the record of ancient events preserved on the Moon, relating to changes billions of kilometers away, involving planets that would otherwise swallow the evidence. We can compare the lunar impact record to asteroid populations remaining, or impacts on Mercury, Mars, and other moons, but our Moon offers the most detailed evidence, which can be calibrated with ages from lunar samples (from Apollo or in the future). LHB model variants discuss planets being destroyed or ejected from the Solar System.[42] Similar processes are suspected in other solar systems, but we can only study ours closely. On the Moon we might probe such distant but world-shattering events.

These impacts generate heat to melt hundreds of times the impactor's mass and, maria were once conjectured to fill with such melt. Now we know that most maria filled millions of years after their basin formation. One of the largest, youngest basins, Orientale (Figure 6.13), 950 kilometers across and 3.8 billion years old, is

Moons Past

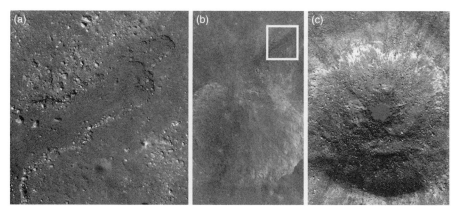

Figure 6.20. **Minor meltdown**. Even small craters can produce enough melt to make temporary flows. (a) 500 m wide view of an anonymous Far Side crater's outer rim, south of Ingalls G, showing dark impact melt flowing from lower left to upper right, pushing aside impact debris. (b) The flow in context of the 2 km crater: box shows place of panel at left. (c) Anonymous 80 m diameter crater near Reiner Gamma, with a prominent impact melt pool solidified in the crater's central 10 m. (NASA photos LROC-NAC M153863408LE and M111972680RE.)

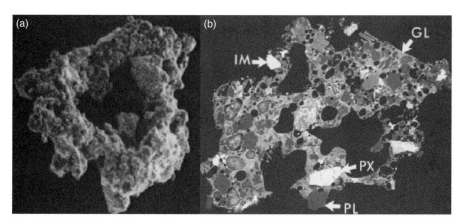

Figure 6.21. **Mixed grit**. Two photomicrographs of agglutinate particles (both about 0.5 mm tall). (a) Scanning electron image of surface of agglutinate from *Apollo 11* soil 10084 (NASA photo S87–38812). (b) Cross-sectional backscatter electron image of agglutinate in *Apollo 17* soil sample 78221,8 showing smaller phases of glass (GL), ilmenite (IM), pyroxene (PX), and plagioclase (PL) (NASA photo). Both particles are cemented by thin melt glass films.

only 7% mare covered.[43] Impact melts constitute no more than several percent of usual mare volumes. Conversely, impact melt features appear even in small craters (Figure 6.20), and melts are nearly ubiquitously associated with young, many kilometer craters. Even on microscales these melts are significant, fusing dust grains into agglutinates and rock fragments into breccias (Figures 6.21 and 6.22). Lunokhod's rover wheels could crush many breccias rolling over them (weight equivalent of 16 kilograms).

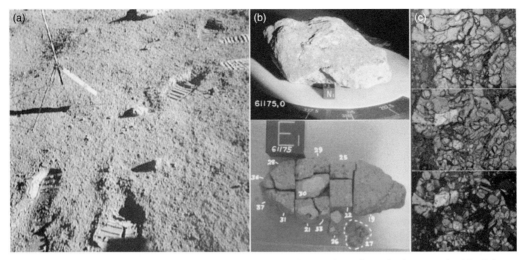

Figure 6.22. **Career of a breccia.** (a) Sample 61175 about to be collected, photographed by John Young at Station 1 near Plum Crater, on EVA 1 of *Apollo 16* on April 21, 1972. (b) 61175 was documented in the sealed environment of Houston's Lunar Receiving Laboratory in 1972. The 543-gram breccia consists of many ancient components: anorthosite, clinopyroxene, orthopyroxene, olivine, ilmenite, spinel, and glass beads, all in a melt glass matrix. In 1973 it was split into several slabs, one being further sectioned. Estimates for how long the rock lay exposed on the surface vary from 10 million years to less than 1.5 million, coincident with nearby South Ray Crater's age. (c) Thin slices of 61175 are shown, from top to bottom, in reflected light, transmitted normal light, and transmitted polarized light, showing many different component clasts up to 3 mm across. Although not special, 61175 was featured in several small publications by 1977. (NASA photos AS16-114-18401, S72–40966, S73–25606, JSC02518, JSC02517, JSC02516; http://www.lpi.usra.edu/lunar/samples/atlas/detail/?mission=Apollo%2016&sample=61175 shows some of these in color.)

Brecciation's importance inspired the quote at this chapter's opening, and breccias constitute 48% of the rock sample mass returned by Apollo astronauts, despite their tendency to visit (less brecciated) maria and desire to return non-breccia samples. A similar fraction of dust is agglutinate, with the top several meters of lunar regolith dominated by agglutinates. Relatively pure fragments of individual rock types tend to be smaller. Even when rock fragments are not fused together, they are immersed in this new entity, the regolith. Even though covering in a relatively thin layer, it is the material lunar explorers first encounter, and have never penetrated in substance.[44]

Regolith is sometimes compared to beach sand, but inaccurately. A closer analog is Portland cement, one reason being its particle size distribution. Billions of years of impacts grind regolith particles into finer sizes, whereas small particles can fuse together or melt into agglutinates. The number of regolith particles of given size follows a function that is a power law in size over a wide range. That is, if one reduces the particle size by a factor a, the number of particles of that smaller size increases by a factor a^{-n}, where n is constant (not necessarily an integer), acting nearly as a fractal geometry. This geometry gives regolith particles an enormous total area, about 0.6 square meter per gram of regolith, equivalent to only 1.7-micron radius particles.[45]

Packing many equal-sized (radius a) spheres in the tightest configuration (hexagonal close-packed), smaller spheres as large as $0.22a$ fill gaps between without requiring extra space.[46] Two such smaller spaces accompany every large particle, so if n exceeds 0.51, enough small particles exist to fill the spaces between large ones (continuing to even smaller particles packed in smaller spaces). For regolith, n is 1.6, so there are sufficient small particles.[47] For Portland cement, n is also 1.6,[48] and the mean particle size in regolith and cement is about 20 microns. When wet, regolith is more reactive than simple silica sand and even bleeds calcium into spaces between particles like cement's gypsum (more about this in Chapter 8). In principle, regolith particles could pack nearly all space. Does this happen in reality?

Knowing directly how well lunar regolith is packed in situ on the Moon is difficult. Although we have soil samples, they have been agitated and more loosely packed than in nature. Estimating how much space exists between particles by driving a metal shaft into the soil to increasing depth indicates that regolith is about two-thirds solid material (except within 0.2 meters of the surface, where it is half solid/half space).[49] In contrast a set of hexagonally close-packed spheres is 74% solid.

In brief, regolith is tightly packed, mostly microscopic particles, unweathered, therefore, often sharp-edged (Figure 6.23). Their composition varies with lunar location, but everywhere, especially in the highlands, many consist of composite agglutinates of angular particles fused in a glass matrix, which is also abrasive. We have drilled many regolith core samples less than 1 meter down, but only a few

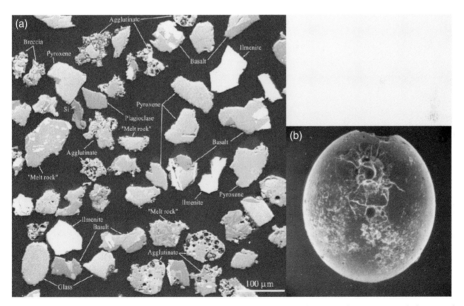

Figure 6.23. **Rough edges or smooth as glass**. (a) A variety of relatively large regolith particles, roughly 50–100 microns across, including agglutinates, basalts, pyroxenes, ilmenites, glasses, breccias, and melt particles (photo courtesy of NASA). (b) Glass spherule, 600 microns in diameter, from meteoritic impact in the regolith, showing splashed glass, microcraters, and mineral fragments welded onto the surface. (Electron micrograph by D.S. McKay, NASA Photo S71–48109.)

2–3 meters deep (from *Luna 24* and *Apollos 15–17*). Probing deeper regolith, or even how deep regolith extends, is harder. One test is how craters in various areas transition from bowl-shaped to flat-bottomed inconsistently (Figure 6.13), indicating loose regolith's depth over more solid rock. Also, some craters eject many boulders in their debris, whereas other like-sized craters do not, indicating varying bedrock depths.[50] We probe the regolith with long wavelength radar, or sense its thickness with long wavelength thermal emission.[51] These methods indicate regolith typically 10–15 meters thick, but several times thinner in the maria, with parts of Oceanus Procellarum only one to a few meters thick, and Far Side regolith north of the giant South Pole-Aitken impact maybe 40 meters thick.

We know even less about what underlies the regolith. Presumably pulverized regolith merges into fractured, larger fragments – the *megaregolith* zone – probably extending several kilometers deep before merging into crustal bedrock. We picture the megaregolith from long wavelength radar units on *Kaguya* and *Apollo 17*,[52] for example, Figure 6.24. Regolith is finely ground and impermeable versus megaregolith pervaded by networks of cracks, many interconnected. Many times and at many places volcanic eruptions have penetrated the megaregolith to extrude lava and vent gas during mare eruptions or later (Chapters 8 and 9). In these and other places, regolith is born as pyroclasts: glass beads and volcanic ash. As discussed earlier, these dark mantle deposits (DMDs) have different composition and mechanical properties from pulverized crustal bedrock.

One DMD encompasses Taurus-Littrow visited by *Apollo 17*, including ilmenite-rich ($FeTiO_3$) deposits that can be dissociated to iron and titanium to build structures by high-temperature application (up to 1,000°C) of hydrogen, converted to

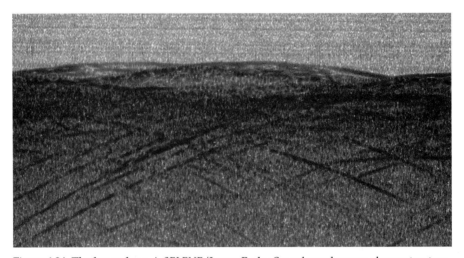

Figure 6.24. **The lunar deep**. A *SELENE*/Lunar Radar Sounder radargram shows structure beneath a typical highlands site. The uniform light gray area at top is vacuum, and the dark gray/black region denotes subsurface radar echoes. This view is 38 km wide and 6 km from bottom to the surface. Strong radar echoes create parabolae extending below each feature. Discontinuities in the megaregolith are common in the upper 1–1.5 km, but the lower 4 km is smooth. (Image from author's research, data courtesy of JAXA.)

water. Because hydrogen is light (1% of the reactant mass), it could be shipped economically from other lunar locations such as the poles, or even Earth. (Few DMDs appear close to the poles.[53]) Furthermore, with water electrolysis, hydrogen is reusable effectively as a catalyst, with net reaction: $2FeTiO_3 \rightarrow 2Fe + 2TiO_2 + O_2$, and TiO_2 reduced to Ti with more hydrogen, producing more water.[54] In large DMD such as the Aristarchus Plateau, pyroclasts are fine-grained and freer of large rocks, therefore, it might be easy to surface mine. At *Apollo 17*'s site glass beads are relatively uniform in size (and easier to process).

Ilmenite is the prime TiO_2 source, and more efficient than typical regolith in absorbing hydrogen and helium, by 10 times or more,[55] so lunar industrial mining might provide its own renewable hydrogen source for producing iron, titanium, and oxygen. In addition, it could produce helium, particularly the helium-3 isotope, which might eventually provide optimal fuel for nuclear fusion reactors (see Chapter 12). While mining ilmenite, hydrogen, and helium, one might also harvest some carbon and nitrogen. A large enough DMD mine might provide many materials required to maintain lunar human habitation. Some DMD glasses are coated by thin layers of industrially useful zinc, copper, lead, gallium, and so forth. DMD sites might be some of the best to start lunar industry.

New lunar probes detailed in Chapter 4 transform lunar surface compositional data into a book to be read from afar. Rocks have no spectral lines betraying their composition like gases (as in a neon sign) but do have spectral bands, like lines but spread by changes in atomic states in the crowded mineral lattice environment. Examples include 0.95-micron (and 1.9-micron) wavelength bands by Fe^{+2}, the basis of OMAT (see note 27). There are others, and new electronic detectors invented since Apollo probe these in detail (despite excellent spatial resolution in Lunar Orbiter and Apollo images). Near 8 microns, the Christiansen feature shifts to higher or lower frequency depending on a mineral's silica content, for example, 7.85 microns for anorthite, 8.11 microns for orthopyroxene, 8.13 for clinopyroxene, and up to 9.1 microns for olivines. These wavelengths are studied with the DIVINER instrument on *LRO* and corrected for soil maturation effects using OMAT (from *Clementine* or others). Although these do not uniquely identify mineral content, there are other measures. Anorthite's spectrum has a valley at 1.25 microns, whereas pyroxene rises between 0.95 and 1.9 microns. Magnesium-rich spinel rises at 1.25 microns but has no corresponding 0.95–1.9 micron dip.

These spectral measures (and more) combine to remotely map the Moon, allowing more discoveries about lunar minerals without touching the Moon again. Anorthosite is common in the highlands, especially the Far Side. It also resides in large craters' central peaks, likely excavated by massive impact rebounds: the long-sought anorthositic lunar crust. In addition, several new lunar minerals have been discovered (although nothing exotic), even recently. These data combined with maps from X-rays, gamma rays, neutrons, and other probes further constrain composition. In this regard regolith's role is key, because it mixes signals from many depths. Basin impacts pull material from mantle and melt it with crust, and smaller craters mix the megaregolith, regolith, and bedrock. It is hard to hide from our remote sensing; however, confusion still occurs. Several teams construct competing spectral measures of iron versus magnesium concentration, known as *Mg#*, and they do not agree. Rather, they agree sometimes, but disagree qualitatively even more. Further work is needed; more samples and even rovers across the Moon (transects) are proposed[56] to fill gaps in "ground truth."

Remote-sensing data are tied to ground truth via not only Apollo and Luna samples but also lunar meteorites and laboratory studies. We can examine samples returned from Apollo then compare them to inferences of a site's composition from remote sensing. Or minerals' laboratory spectroscopic signatures are calibrated for terrestrial samples, or minerals are synthesized and measured. Surprisingly, we also have 77 meteorites ejected from the Moon (broken into 170 stones) as of 2012. These were not recognized as lunar until 1979, with wisdom gained from Apollo (but anticipated since 1963).[57] Ten are marelike basalts, and 16 are breccias (mare or highland), but 51 are anorthositic highlands rocks. With highlands covering 84% of the Moon, these proportions are unsurprising.

With remote sensing data, future lunar inhabitants will know where to find minerals they need, but even generic regolith properties make it useful regardless of compositional variations. Earth's most common economic commodity (other than water), by mass, is construction aggregate – that is, concrete without the cement and water.[58] Lunar habitats will need graded roads, regolith-covered structures, and berms to deflect flying dust,[59] requiring unmodified regolith to be moved. Lunar concrete is also possible. What is the cement? Those iron nanoparticles in regolith (Figure 6.17) make excellent microwave absorbers. Regolith can form into bricks with enough heat to fuse regolith particles together (without melting them entirely).[60] Microwaving could also pave roads or seal walls against air leakage, especially for underground spaces, or fuse dust to walls and floors.

Like cement, lunar regolith is calcium rich, especially in the highlands (10–15% by mass). Portland cement is about 50% calcium but mixes to 15% with aggregate, making concrete with final calcium concentrations like highlands regolith. Calcium leaches from cement (gypsum) to harden in the gaps between grains. Regolith has properties of a good concrete: a particle size distribution in which smaller particles tend to fill gaps between large ones, as well as good amounts of calcium binder. Whereas low energy particles sputter and coat regolith grains with thin glassy rims (Figure 6.17), high-energy heavy particles leave narrow radiation damage tracks up to a millimeter through the mineral lattice, at high densities of about 10^{11} tracks per square centimeter. Water etches these tracks, tripling the effective grain area by expanding the tracks, which channel dissolved calcium to the inter-grain interstitial gaps from deep in the grain, leaving the grain size unaltered.[61] Simulated anorthositic lunar regolith has been steam treated to make cemented material, and structures have been made using regolith aggregate.[62] Regolith and water (and energy) might make good lunar concrete. In Chapter 8 we discuss lunar water sources, so precious that a method to recover excess water from setting concrete is wise. Once set, the concrete is stable even in vacuo.[63] These justify more tests of regolith concretes.

We discuss regolith's additional practical uses in Chapter 11, as shielding, insulation, and building material, but consider regolith now as a scientific tool. Its great utility is that it is buried and, therefore, has recorded past processes: lunar, solar, asteroidal, cometary, and beyond – even material from nearby Earth. Indeed several Apollo lunar samples likely contain cometary or asteroidal material. Although regolith depths are typically 10 meters, present impact rates have uncovered little material deeper than a meter since the LHB.[64] Hence, below the first meter is a repository of relics of impacts and other processes from the first 600 million years of Solar System history for which we may find no other record (except perhaps Mercury's regolith). Other planets have been too geologically active. Such evidence on asteroids is destroyed or blown into space by the very impacts we want to study. Several meters below the lunar surface, therefore, rests a unique record of objects that no longer exist: early Solar System asteroids and comets, trapped solar wind from the colder, early Sun (we might also look in regolith caught beneath mare basalts, trapping light elements from the Sun – *cryptovolatiles*[65]), or fantastically, even material from completely beyond the Solar System.

The burial by regolith is not necessarily just random mixing ("gardening"). Imagine a hole in the lunar surface in the shape of a straight vertical well. Debris from nearby impacts will accumulate in this well in a steadily thickening layer. In such depressions, strata will develop into a time-ordered record of what hit the Moon, extending backwards perhaps billions of years. Such wells actually exist. Subsurface lava flows can evacuate lava tubes, and these can collapse or be punctured by impacts to create "skylights" (Figure 6.25). These might also provide ready shelter for future inhabitants from temperature fluctuations, impacts, and radiation (Chapter 11). In these places, regolith accumulates in organized, sequential strata to be used by scientists to betray an explicit time sequence of lunar deposition. We might find this behavior in other areas: On inclined surfaces such as crater walls, debris from impacts tends to shift downhill, whereas no net motion occurs on flat surfaces. At the base of the incline, material will accumulate. One can calculate that this deposition is most extreme for places closest to the hill's base.[66] For craters 10 or 20 meters in diameter, the crater bottom collects debris raining from all sides and accumulates such strata.

Occasionally over the Solar system's history, Earth (and the Moon) will orbit in the interstellar medium permeating our Galaxy. Usually we live interior to the *heliopause* where solar wind outflow arrests the oncoming interstellar medium. This boundary occurs where *ram pressure* (force per unit area from oncoming material) from the Sun, like a fire hose, balances pressure from material flowing the opposite way from interstellar space. The *Voyager* probes that visited Jupiter, Saturn, Uranus, and Neptune in 1979–1989 now approach the heliopause, about 80 times Earth's distance from the Sun. Ram pressure is proportional to density times velocity squared, and the density in the local interstellar environment is about 0.1 atom per cubic centimeter. Portions of our Galaxy, called giant molecular clouds, have typical densities of about 1000 atoms per cubic centimeter, and in some places up to a million. Because the solar wind's density drops as the inverse of the square of distance from the Sun, if the interstellar density shoots up by 10,000 times (= 1,000 cm^{-3}/0.1 cm^{-3}), the heliopause will collapse to 1% its current radius, hence, within Earth's orbit. The Solar System enters such clouds about

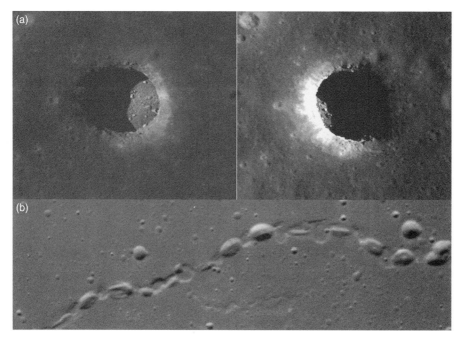

Figure 6.25. **Look out below**. (a) Two views with different sun angles of one of several known lunar skylight pits, here 100 m in diameter, in Mare Tranquillitatis. Note that the depression's walls are not visible; its floor sits 70 m below the top entrance. (b) 15 km long, partially collapsed lava tube Rima Gruithuisen between Mare Imbrium and Oceanus Procellarum. (LROC/NAC images M126710873R, M137332905R, and M181459858C, courtesy of NASA, Arizona State University.)

every billion years; that is when interstellar material can be buried on the Moon. Mostly gases will strike the Moon from giant molecular clouds, and places on the Moon as cold as 20 degrees above absolute zero (20 K = −424°F) can cause these gases to stick and freeze. We discuss this in Chapter 8.

We find few rocks on Earth older than 3.9 billion years. In the LHB, huge impacts on Earth blew material into space, including rocks, with some landing on the Moon. Estimates for Earth material deposited in lunar regolith amount to 0.2% by mass – an immensity.[67] We have no record of Earth's organic chemistry before 3.9 billion years ago, yet terrestrial meteorites on the Moon should preserve this chemical record. We may uncover the origin of life on Earth by exploring the lunar regolith.

There may be stranger stuff in the lunar regolith, repository of any process with lasting surface effects, over billions of years. Some authors ponder finding alien artifacts: Lunar SETA (Search for Extra-Terrestrial Artifacts).[68] The difficulty finding an artifact depends on its size and special signatures and whether it is buried or penetrates the surface. We have found no obvious, unexpected artifacts at current levels of imaging detail (whole surface at 20 meters resolution, much at 1 meter).[69] An artifact more than 20 meters in size would probably stay exposed until impacts fragment it. On smaller scales, it joins the regolith, perhaps becoming apparent in some future mining operation but nearly invisible now except in several special sites. In reality 70 lunar sites contain artifacts from hard or soft landings of spacecraft from Earth. We have found 29 and suspect four more,[70] mostly because we knew their position from the start. (None exceed 20 meters in size, although some of their impact craters do.) We have only weak limits on numbers of small artifacts on the Moon and probably cannot improve this quickly without moving large regolith volumes or developing special tools (see Chapter 13).

Notes

1. "The Origin of the Moon Revisited" by S. R. Taylor & C. Koeberl, 2013, *Lunar and Planetary Science Conference*, 44, 1165; also *Destiny or Chance: Our Solar System and its Place in the Cosmos* by Stuart Ross Taylor, 2000, Cambridge: Cambridge University Press, 149. Chapter 8 discusses Harold Urey further.
2. *Apollo 16 Flight Journal* edited by David Woods & Tim Brandt, 2008, Chapter 20, Day 6, Part 2 (126 hours 10 hours 38 seconds MET), http://history.nasa.gov/ap16fj/20_Day6_Pt2.htm.
3. Anorthite has the chemical formula $CaAl_2Si_2O_8$. If one replaces the calcium in the crystal structure with potassium, potassium feldspar, $KAlSi_3O_8$, is produced. On Earth these often mix with the sodium version of the crystal albite, $NaAlSi_3O_8$. Plagioclase feldspar is any mixture of anorthite and albite, whereas alkali feldspar is a mixture of albite and K-feldspar. Lunar anorthosite is white, which helps give the highlands its bright hue compared to the maria.
4. "The Relationship Between Length and Width of Plutons within the Crustal-Scale Cobequid Shear Zone, Northern Appalachians, Canada" by Ioannis K. Koukouvelas, Georgia Pe-Piper & David J. W. Piper, 2006, *International Journal of Earth Science*, 95, 963.
5. *Apollo 15* sample 15415 was found to consist of 97% plagioclase and 3% pyroxene ("Mineralogy of Apollo 15415 'Genesis Rock': Source of Anorthosite on Moon" by Ian M. Steele & J. V. Smith, 1971, *Nature*, 234, 138). We describe pyroxene shortly.
6. We mentioned the highlands before (Figure 2.4) but delve into this later.
7. Unlike in the Moon, olivine $((Mg,Fe)SiO_4)$ deep in Earth's mantle compresses into a succession of minerals until decomposing to silicate perovskite $((Mg,Fe)SiO_3)$ and ferropericlase $((Mg,Fe)O)$.
8. Olivine consists of magnesium or iron in layers of silicate crystal, denoted $(Mg,Fe)_2SiO_4$. Pyroxene is a complex mixture that can be denoted $XY(Si,Al)_2O_6$ (where X represents calcium, sodium, magnesium, and/or iron in the +2 oxidation charge state, and Y smaller ions: chromium, aluminum, iron^{+3}, magnesium, manganese, titanium, etc.) and has a more interlocked crystal structure. Some atoms will not fit, such as potassium. Pyroxene encompasses many minerals.
9. This solidification sequence is described by "Water in the Lunar Mantle: Results from Magma Ocean Modeling" by L. T. Elkins-Tanton, 2010, *Lunar and Planetary Science Conference*, 41, 1451.
10. Ibid.
11. For you fans of Superman, Star Trek, and *Avatar* (or *The Core*, or *Skyrates*).
12. The minerals' formulæ are $(Mg,Fe^{2+})Ti_2O_5$, $(Fe^{2+},Ca)_8(Zr,Y)_2Ti_3(SiO_4)_3O_4$, and $(Fe^{2+},Ca)SiO_3$, respectively. Armalcolite honors *Apollo 11*'s Armstrong, Aldrin, and Collins, and tranquillityite, Tranquility Base. ("Armalcolite-Ti-Phlogopite-Diopside-Analcite-Bearing Lamproites from Smoky Butte, Garfield County, Montana" by Danielle Velde, 1975, *American Mineralogist*, 60, 566; "Tranquillityite: The Last Lunar Mineral Comes Down to Earth" by Birger Rasmussen, Ian Fletcher, Courtney J. Gregory, Janet R. Muhling & Alexandra A. Suvorova, 2012, *Geology*, 40, 83; "Lunar Mineral Found at Franklin, Sussex Co., New Jersey" by Tony Nikischer & Joe Orosz, 2007, *Mineral News*, 23, 8.)
13. "Overturn of Magma Ocean Ilmenite Cumulate Layer: Implications for Lunar Magmatic Evolution and Formation of a Lunar Core" by P. C. Hess & E. M. Parmentier, 1993, *Lunar and Planetary Science Conference*, 24, 651; "On the Scale of Lunar Mantle Overturn Following Magma Ocean Fractional Solidification: The Role for Multiple Scales of Convective Motion" by E. M. Parmentier, 2009, *Lunar and Planetary Science Conference*, 40, 1781.

14. "Re-examination of the Lunar Magma Ocean Cumulate Overturn Hypothesis: Melting or Mixing Is Required" by Linda T. Elkins Tanton, James A. Van Orman, Bradford H. Hager & Timothy L. Grove, 2002, *Earth and Planetary Science Letters*, 196, 239.
15. Examples include norite, made of labradorite, $(Ca,Na)(Al,Si)_4O_8$, and enstatite, $(Mg,Fe)SiO_3$, with olivine. Norite can come with gabbro, mixing pyroxene, plagioclase, amphibole, and olivine. With less pyroxene, one finds troctolite: olivine, calcium-rich plagioclase, and a bit of pyroxene.
16. "Lunar Mare Deposits: Areas, Volumes, Sequence, and Implication for Melting in Source Areas" by James W. Head, in *Origins of Mare Basalts and their Implications for Lunar Evolution*, 1975, Houston: LPI, 66. The largest volcanic flow volume known on Earth is the Siberian traps feature, about 2 million km^3, likely leading to the extinction of most life 251 million years ago, ending the Permian period. Of course, the ocean bottoms are larger, about 2 billion km^3.
17. The subcrustal ilmenite-rich layer was ~50 km thick, with excess density ~0.4 g cm^{-3} and mass ~8×10^{17} tonne. Falling 400 km in the mantle, this generates ~4×10^{23} J. Basalt's heat of fusion ~500 J g^{-1} (and accounting for temperature rise) would melt ~300,000 km^3. For ^{232}Th and ^{238}U abundance of a few ppm (half-lives 14 and 4.5 Gy – gigyears = billion years – and ignoring ^{26}Al, ^{235}U, etc.), decay makes 4×10^{19} J y^{-1}, outstripping mantle overturn in ~10,000 y. Basin-forming impacts generated ~10^{24} J. ("Measurement of Heat of Fusion of Model Basalt in the System Diopside-Forsterite-Anorthite" by Hiroshi Kojitani & Masaki Akaogi, 1995, *Geophysical Research Letters*, 22, 2329.) As well as melting as a result of heat, magma forms from hot rock as pressure is released, mainly because of change in depth during mantle overturn, or because of overlying material blown away by basin-forming impacts.
18. Elements Al, Ti, Mg, Si, Ca, and O are also mapped by gamma rays. Ti follows Fe closely, whereas Ca and Al have distributions complementary to Fe. Mg covers even more maria area than Fe. ("Elemental Composition of the Lunar Surface: Analysis of Gamma Ray Spectroscopy Data from Lunar Prospector" by T. H. Prettyman, J. J. Hagerty, R. C. Elphic, W. C. Feldman, D. J. Lawrence, G. W. McKinney & D. T. Vaniman, 2006, *Journal of Geophysical Research*, 111, E12007.
19. "Emplacement of Long Volcanic Features on the Moon: A Review of the Mare Imbrium Lava Flows and Vallis Schröteri" by W. B. Garry, N. H. Warner & J. R. Zimbelman, 2007 Fall, *American Geophysical Union*, abs. P13A-1039; "Emplacement Scenarios for Vallis Schröteri, Aristarchus Plateau, the Moon" by W. Brent Garry & Jacob E. Bleacher, 2011, *Geological Society of America*, Special Paper 477.
20. "Clementine Observations of the Aristarchus Region of the Moon" by Alfred S. McEwen, Mark S. Robinson, Eric M. Eliason, Paul G. Lucey, Tom C. Duxbury & Paul D. Spudis, 1994, *Science*, 266, 1858.
21. *The International Atlas of Lunar Exploration* by Philip J. Stooke, 2007, Cambridge: Cambridge University Press, 464, and personal email, March 30, 2012.
22. Most lunar scientists held that craters were volcanic rather than of impact origin until Ralph Baldwin's (1912–2010) *The Face of the Moon* (1949, Chicago: University of Chicago Press) and *The Measure of the Moon* (1963, Chicago: University of Chicago Press). An impact origin was confirmed by Eugene Shoemaker (1928–1997) studying the structure of Meteor (Barringer) Crater, Arizona, in 1952–1960, nuclear explosion craters in Yucca Flats, Nevada, and lunar craters.
23. The reader can experiment with this using the impact simulator Web site tool "Lunar Cratering" by Keith A. Holsapple, 2007, http://www.lpi.usra.edu/lunar/tools/lunarcratercalc/.
24. "The Aristarchus Plateau on the Moon: Nature and Stratigraphy of the Substratum" by S. D. Chevrel, P. C. Pinet, Y. Daydou, S. Le Mouélic, Y. Langevin, F. Costard & S. Erard, 2009, *Lunar and Planetary Science Conference*, 40, 1234.
25. "Aristarchus Crater: Mapping of Impact Melt and Absolute Age Determination" by M. Zanetti, H. Hiesinger, C. H. van der Bogert, D. Reiss & B. L. Jolliff, 2011, *Lunar and Planetary Science Conference*, 42, 2330.
26. Other ray systems include Copernicus (600 km radius), Proclus (600 km), Jackson (600 km), Byrgius A (400 km), Aristarchus (400 km), Ohm (400 km), Kepler (300 km), McKellar (300 km), and Giordano Bruno (200 km). Giordano Bruno bridges the Far and Near Side; Jackson, Ohm, and McKellar are Far Side. The lunar circumference is 10,920 km.
27. OMAT is measured by graphing reflectance (ratio of reflected light to incident light intensity) at 0.75-micron wavelength versus the ratio at 0.95 microns divided by that at 0.75 microns. OMAT is the distance on this plot for a sample to the point for an infinite-age sample. Different mineral FeO contents describe different slopes on this plot because the 0.95-micron feature depends on Fe abundance ("Imaging of Lunar Surface Maturity" by Paul G. Lucey, David T. Blewett, G. Jeffrey Taylor & B. Ray Hawke, 2000, *Journal of Geophysical Research*, 105, 20377).

Iron composes typically about 10%–15% of lunar regolith, and its surface state Fe^{2+} is responsible for absorption bands at 0.95 and 1.8 microns. This state is damaged when glass rim containing Fe^0 rather than Fe^{2+} replaces surface states, and the absorption disappears. Formally this is measured using the comprehensive data set of lunar surface photometry from *Clementine* in 1994, including a band centered on 0.95 micron, and using the nearby 0.75-micron band as an overall reflectivity reference. As regolith ages, it darkens at most wavelengths, 0.75 microns in particular, but the ratio of 0.95 micron to 0.75 micron reflectivity increases. At infinite age, this ratio approaches 1.26 as 0.75 micron reflectivity approaches a few percent. Depending on regolith composition, the approach to this infinite-age point differs, but a sample that is far from this point is reliably young. This distance allows one to rank regolith samples in age. OMAT changes rapidly over the first few million years of exposure. One can test this looking at OMAT aging of rays ejected from fresh crater impacts; OMAT is nearly halved going from 2-million to 50-million-year-old deposits. After a billion years OMAT no longer changes appreciably ("The Optical Maturity of the Ejecta of Small Bright Rayed Lunar Craters" by J. A. Grier, A. S. McEwen, M. Milazzo, J. A. Hester & P. G. Lucey, 2000, *Lunar and Planetary Science Conference*, 31, 1950). Some craters older than 1.1 Gy still show rays, e.g., Lichtenberg, roughly 1.8 Gy old; this depends on composition ("The Origin of Lunar Crater Rays" by B. Ray Hawke, D. T. Blewett, P. G. Lucey, G. A. Smith, J. F. Bell III, B. A. Campbell & M. S. Robinson, 2004, *Icarus*, 170, 1).

28. Donald E. Gault and John A. Wedekind, "Experimental Studies of Oblique Impact," *Lunar & Planetary Science Conference*, 9, Vol. 3 (1978), 3843.
29. "Timing of Crystallization of the Lunar Magma Ocean Constrained by the Oldest Zircon" by A. Nemchin, N. Timms, R. Pidgeon, T. Geisler, S. Reddy & C. Meyer, 2009, *Nature Geoscience*, 2, 133.
30. "Isotopic Studies of Ferroan Anorthosite 62236: a Young Lunar Crustal Rock from a light Rare-Earth-Element-Depleted Source" by Lars Borg, Marc Norman, Larry Nyquist, Don Bogard, Greg Snyder, Larry Taylor & Marilyn Lindstrom, 1999, *Geochimica et Cosmochimica Acta*, 63, 2679; "An Ancient Sm–Nd Age for a Ferroan Noritic Anorthosite Clast from Lunar Breccia 67016" by Chantal Alibert, Marc D. Norman & Malcolm T. McCulloch, 1994, *Geochimica et Cosmochimica Acta*, 58, 2921; "The Age of Ferroan Anorthosite 60025: Oldest Crust on a Young Moon?" by R. W. Carlson & G. W. Lugmair, 1988, *Earth and Planetary Science Letters*, 90, 119; "Evidence from Detrital Zircons for the Existence of Continental Crust and Oceans on the Earth 4.4 Gyr Ago" by Simon A. Wilde, John W. Valley, William H. Peck & Colin M. Graham, 2001, *Nature*, 409, 175.
31. "Argon-40/Argon-39 Age Spectra of *Apollo 17* Highlands Breccia Samples by Laser Step Heating and the Age of the Serenitatis Basin" by G. B. Dalrymple & G. Ryder, 1996, *Journal of Geophysical Research*, 101, 26069.
32. "Priscoan (4.00–4.03 Ga) Orthogneisses from Northwestern Canada" by Samuel A. Bowring & Ian S. Williams, 1999, *Contributions to Mineralogy and Petrology*, 134, 3; "The Origin of Decoupled Hf–Nd Isotope Compositions in Eoarchean Rocks from Southern West Greenland" by J. Elis Hoffmann, Carsten Münker, Ali Polat, Minik T. Rosing & Toni Schulz, 2011, *Geochimica et Cosmochimica Acta*, 75, 810. There is some debate about a rock 4.2 Gy old ("Neodymium-142 Evidence for Hadean Mafic Crust" by Jonathan O'Neil, Richard W. Carlson, Don Francis & Ross K. Stevenson, 2008, *Science*, 321, 1828; "Comment on 'Neodymium-142 Evidence for Hadean Mafic Crust'" by Rasmus Andreasen & Mukul Sharma, 2008, *Science*, 325, 267; "Response to Comment on 'Neodymium-142 Evidence for Hadean Mafic Crust'" by Jonathan O'Neil, Richard W. Carlson, Don Francis & Ross K. Stevenson, 2008, *Science*, 325, 267).
33. "Origin of the Cataclysmic Late Heavy Bombardment Period of the Terrestrial Planets" by R. Gomes, H. F. Levison, K. Tsiganis & A. Morbidelli, 2005, *Nature*, 435, 466.
34. "The Sculptured Hills of the Taurus Highlands: Implications for the Relative Age of Serenitatis, Basin Chronologies and the Cratering History of the Moon" by Paul D. Spudis, Don E. Wilhelms & Mark S. Robinson, 2011, *Journal of Geophysical Research*, 116, E00H03.
35. Schultz and Spudis, *Nature* (1983), 302, 233.
36. *LSB*, 83.
37. "Cratering in the Earth–Moon System: Consequences for Age Determination by Crater Counting" by G. Neukum, B. König, H. Fechtig & D. Storzer, 1975, *Proceedings of Lunar Science Conference*, 6, 2597; "The Comparison of Size–Frequency Distributions of Impact Craters and Asteroids and the Planetary Cratering Rate" by B. A. Ivanov, G. Neukum, W. F. Bottke Jr. & W. K. Hartmann, in *Asteroids III*, eds. W. F. Bottke Jr., A. Cellino, P. Paolicchi & R. P. Binzel, 2002, Tucson: University of Arizona Press, 89; "Lunar Mare Basalt Flow Units: Thicknesses Determined from Crater Size-Frequency Distributions" by H. Hiesinger, J. W. Head III, U. Wolf, R. Jaumann & G. Neukum, 2002, *Journal of Geophysical Research*, 29, 89.

38. "New Observational Evidence of Nonuniform Cratering of the Moon" by M. A. Kreslavsky, S. C. Werner, J. W. Head & C. I. Fassett, 2012, *Lunar and Planetary Science Conference*, 43, 1193; "Nonuniform Cratering of the Moon and a Revised Crater Chronology of the Inner Solar System" by Mathieu Le Feuvre & Mark A. Wieczorek, 2011, *Icarus*, 214, 1.
39. "Chronology and Sources of Lunar Impact Bombardment" by Matija Ćuk, 2012, *Icarus*, 218, 69.
40. R. Gomes, H. F. Levison, K. Tsigania, and A. Morbidelli, "Origin of the Cataclysmic Late Heavy Bombardment Period of the Terrestrial Planets," *Nature* (2005), 435, 466.
41. For instance, "Lunar Impact Basins: Stratigraphy, Sequence and Ages from superposed Impact Crater Populations Measured from Lunar Orbiter Laser Altimeter (LOLA) Data" by C. I. Fassett, J. W. Head, S. J. Kadish, E. Mazarico, G. A. Neumann, D. E. Smith & M. T. Zuber, 2012, *Journal of Geophysical Research*, 117, E00H006.
42. "The Formation of Ice Giants in a Packed Oligarchy: Instability and Aftermath" by Eric B. Ford & Eustace Chiang, 2007, *Astrophysical Journal*, 661, 602.
43. "Characteristics, Affinities and Ages of Volcanic Deposits Associated with the Orientale Basin from *Chandrayaan-1* Moon Mineralogy Mapper (M3) Data: Mare Stratigraphy" by J. Whitten, J. Head, M. Staid, C. Pieters, J. Mustard, L. Taylor, T. McCord, P. Isaacson, R. Klima, J. Nettles & the M3 Team, *Lunar and Planetary Science Conference*, 2009, 1841.
44. Al Bean, *Apollo 12* Lunar Module pilot, describes what he found: "The entire lunar surface was covered with this mantle of broken-up material, fine dust of varying depth. As a result, everything looked pretty much the same – sides of the craters, tops of the craters, flat lands, and ejecta blanket. If you're going to do any geology, you're going to have to dig through this mantle of brown or black and to look beneath the surface a little bit. We had a shovel that we used for trenching, but because of the length of the extension handle and the inability to lean over and what have you, we never could trench more than about 8 inches. That was about the best we could do, and that was a pretty big effort." (*Apollo 12 Technical Debrief*, December 1, 1969, Houston: NASA Manned Space Center, 10–42.) Also, see Figure 2.20.
45. "Specific Surface Area as a Maturity Index of Lunar Fines" by R. B. Gammage & H. F. Holmes, 1975, *Earth and Planetary Science Letters*, 27, 424.
46. "Density and Packing in an Aggregate of Mixed Spheres" by Douglas Rennie Hudson, 1949, *Journal of Applied Physics*, 20, 154.
47. See *LSB*, 306, 477.
48. "Measurement of Particle Size Distribution in Portland Cement Powder: Analysis of ASTM Round Robin Studies" by Chiara F. Ferraris, Vincent A. Hackley & Ana Ivelisse Avilés, 2004, *Cement, Concrete and Aggregates*, 26, 11920.
49. *LSB*, 483–503. Regolith particles have density = 3 g cm^{-3}, but regolith itself has 1.8 g cm^{-3}.
50. "Thickness Determinations of the Lunar Surface Layer from Lunar Impact Craters" by W. L. Quaide & V. R. Oberbeck, 1968, *Journal of Geophysical Research*, 73, 5247; "Constraints on the Depth and Variability of the Lunar Regolith" by B. B. Wilcox, M. S. Robinson, P. C. Thomas & B. R. Hawke, 2005, *Meteoritics and Planetary Science*, 40, 695.
51. "Regolith Thickness over the Lunar Nearside: Results from Earth-based 70-cm Arecibo Radar Observations" by Wenzhe Fa & Mark A. Wieczorek, 2012, *Icarus*, 218, 771; "A primary Analysis of Microwave Brightness Temperature of Lunar Surface from *Chang'e* 1 Multi-channel Radiometer Observation and Inversion of Regolith Layer Thickness" by W. Fa & Y.-Q. Jin, 2010, *Icarus*, 207, 605.
52. "Local Lunar Topography from the *Apollo 17* ALSE Radar Imagery and Altimetry" by C. Elachi, M. Kobrick, L. Roth, M. Tiernan & W. E. Brown, Jr., 1976, *Moon*, 15, 119; "The Lunar Radar Sounder (LRS) Onboard the *KAGUYA (SELENE)* Spacecraft" by T. Ono, et al., 2010, *Space Science Reviews*, 154, 145.
53. "Lunar Regional Dark Mantle Deposits: Geologic, Multispectral and Modeling Studies" by Catherine M. Weitz, James W. Head III & Carle M. Pieters, 1998, *Journal for Geophysical Research*, 103, 22725.
54. "Production of Oxygen from Lunar Ilmenite" by Y. Zhao & F. Shadman, 1991, *Industrial and Engineering Chemical Research*, 30, 2080.
55. "Trapped Solar Wind Gases in Lunar Fines and A Breccia" by P. Eberhardt, J. Geiss, H. Graf, N. Grögler, M. D. Mendia, M. Morgeli, H. Schwaller & A. Stettler, 1972, *Lunar Science Conference*, 2, 1821; "Gas Ion Probe Analysis of Helium Profiles in Individual Lunar Soil Particles" by H. W. Mueller, J. L. Jordan, S. Kalbitzer, J. Kiko & T. Kirsten, 1976, *Proceedings of 7th Lunar Science Conference*, 2, 937; "Mapping Pyroclastic Deposits and

Other Lunar Features for Solar Wind Implanted Helium" by J. L. Jordon, in *Workshop on Lunar Volcanic Glasses: Scientific and Resource Potential*, eds. J. W. Delano & G. H. Heiken, 1989, Houston: Lunar and Planetary Inst., 43; "Hydrogen, Helium and other Solar-Wind Components in Lunar Soil: Abundances and Predictions" by L. A. Taylor, in *Engineering, Construction and Operations in Space II: Proceedings of Space '90*, 1990, New York: American Society of Civil Engineers, 68.

56. Promising robot mission proposals include a Far Side sample return from the South Pole-Aitken basin. Rover sampling transects from ranges of adjacent sites include Teleprospector ("A Teleoperated Robotic Field Geologist" by G. Jeffrey Taylor & Paul D. Spudis, in *Engineering, Construction and Operations in Space II: Proceedings of Space'90*, 1990, New York: American Society of Civil Engineers, 246) and "Intrepid: Lunar Roving Prospector Providing Ground Truth and Enabling Future Exploration" by M. S. Robinson, S. J. Lawrence, E. J. Speyerer, & J. Stopar, in *Lunar Exploration Analysis Group Meeting*, 2011, Houston: Lunar and Planetary Institute, 2042.

57. "At least a small amount of material from the lunar surface and perhaps as much or more than the impacting mass is probably ejected at speeds exceeding the escape velocity by impacting objects moving in asteroidal orbits. Some small part of this material may follow direct trajectories to the earth, some will go into orbit around the earth, and the rest will go into independent orbit around the sun. Much of it is probably ultimately swept up by earth" ("Interplanetary Correlation of Geologic Time" by E. M. Shoemaker, R. J. Hackman & R. E. Eggleton, 1963, *Advances in Astronautical Sciences*, 8, 70).

58. Annual global concrete output is 9 billion tonnes versus 4 billion for petroleum, about 3 billion each for coal and natural gas, and about 2 billion each for wood and food grains. Also, aggregate fill is used in asphalt, roads, and drainage.

59. An example of rocket damage via exhaust-accelerated dust is *Apollo 12*'s effect on *Surveyor 3*. Analysis indicates that dust grains hitting *Surveyor* from LM *Challenger*'s exhaust were flying at 2 km s^{-1}, almost escape velocity. These abraded roughly half of *Surveyor*'s 200-micron paint thickness and dug many small pit craters on one side, this despite NASA's intent to avoid damaging *Surveyor 3*. Most particles flew at such small angles that a berm could stop them. See "Apollo 12 Lunar Module Exhaust Plume Impingement on Lunar Surveyor III" by Christopher Immer, Philip Metzger, Paul E. Hintze, Andrew Nick & Ryan Horan, 2011, *Icarus*, 211, 1089.

60. "Sintering Bricks on the Moon" by Carleton C. Allen, John C. Graf & David S. McKay, in *Engineering, Construction and Operations in Space IV*, 1994, New York: American Society of Civil Engineers, 1220.

61. "Blocking of the Water-Lunar Fines Reaction by Air and Water Concentration Effects" by R. B. Gammage & H. F. Holmes, 1975, *6th Lunar Science Conference*, 3, 3305, and references therein.

62. "Lunar and Martian Resource Utilization – Cement and Concrete" by T. D. Lin, S. Bhattacharja, L. Powers-Couche, S. B. Skaar, T. Horiguchi, N. Saeki, D. Munaf, Y. N. Peng & I. Casanova, in *Workshop on Using In Situ Resources for Construction of Planetary Outposts*, 1998, Houston: Lunar and Planetary Institute, 12; "Physical Properties of Concrete Made with Apollo 16 Lunar Soil Sample" by T. D. Lin, H. Love & D. Stark, in *Space Manufacturing 6*, 1987, American Institute of Aeronautics and Astronautics, Houston: Lunar and Planetary Institute, 361.

63. "Lunar Concrete for Construction" by Hatice S. Cullingford & M. Dean Keller, in *2nd Conference on Lunar Bases and Space Activities of the 21st Century*, ed. W. W. Mendell, 1992, Houston: NASA Conference Publication 3166, 497.

64. "Mixing of the Lunar Regolith" by D. E. Gault, F. Hörz, D. E. Brownlee & J. B. Hartung, 1974, *5th Proceedings of Lunar Science Conference*, 3, 2365, especially Figure 9.

65. "Preservation Potential of Implanted Solar Wind Volatiles in Lunar Paleoregolith Deposits Buried by Lava Flows" by Sarah A. Fagents, M. Elise Rumpf, Ian A. Crawford & Katherine H. Joy, 2010, *Icarus*, 207, 595.

66. The deposition rate varies as $r^{-0.7}$, where r is the distance from the foot of the hill along the horizontal flat below the hill: "Exploring Volatile Deposition in Lunar Regolith" by J. A. Alford, A. R. Hodges, E. Heggy & A. Crotts, 2012, *Lunar and Planetary Science Conference*, 43, 2938.

67. "On the Survivability and Detectability of Terrestrial Meteorites on the Moon" by I. Crawford, E. Baldwin, E. Taylor, J. Bailey & K. Tsembelis, 2008, *Astrobiology*, 8, 242; "The Moon: A Repository for Ancient Planetary Samples" by John C. Armstrong, Llyd E. Wells, Monika Kress & Guillermo Gonzalez, 2002, *The Moon Beyond 2002: Next Steps in Lunar Science and Exploration*, 2.

68. "Searching for Alien Artifacts on the Moon" by P. C. W. Davies & R. V. Wagner, 2011, *Acta Astronautica*, doi:10.1016; "Earth–Moon System as a Collector of Alien Artifacts" by A. V. Arkhipov, 1998, *Journal of British Interplanetary Society*, 51, 181.
69. There is one exceptional claim, of a rectilinear complex near Far Side crater Lovelace (ibid.). Unfortunately, this region is marked by two roughly perpendicular ejecta scour patterns, so much of the landscape appears rectilinear.
70. "The International Atlas of Lunar Exploration" by Philip J. Stooke, 2007, Cambridge University Press & personal email, 2012 March 30.

Chapter 7
The Pull of the Far Side

> With a holy host of others standing around me
> Still I'm on the dark side of the moon
> And it seems like it goes on like this forever
> You must forgive me
> If I'm up and gone to Carolina in my mind
> — James Taylor, 1968, "Carolina in My Mind"

People seem forever confused about the Far Side versus dark side of the Moon.[1] Let us be clear – here is an easy mnemonic: the dark side is darker, and the Far Side is farther! They are not the same except once per month, at full Moon. The Moon, like Earth (and other planets), has at any time a side pointed away from the Sun. That is the dark side.

The Moon has another effect in play; the same lunar Near Side always turns to face Earth, with the opposite side – the Far Side – turned away. This matter of physics is common to many worlds. A consequence is that almost half of the Moon's surface[2] remained hidden to humans until 1959, two years to the day after *Sputnik 1*'s launch, when the third Soviet lunar probe attempt, *Luna 3*, photographed the Far Side for the first time (Figure 2.1). Beforehand, that side had been an abyss to human knowledge, more unknown to us than distant reaches of the Universe, and a place where one might peer into space and never see Earth. This sense of isolation James Taylor invokes with the "dark side."

Nature may not abhor a vacuum, but the human mind does. Humans people the Far Side, at least in their minds. In Disney's 1955 film with von Braun *Man and the Moon* (see Chapter 3), the Far Side holds huge, artificial-looking structures, ominously radioactive, seen by the first astronauts to orbit behind the Moon. Humans constructed ideas about what resides on the Far Side long before we knew enough to justify them. Consider the sad case of Peter Andreas Hansen's lunar hypothesis.

Hansen, respected Danish astronomer/mathematician, argued in 1856 that the Moon bulged towards Earth.[3] Based on discrepancies between observed and predicted lunar positions, Hansen found the Moon's center of gravity offset from its geometric center by 59 kilometers, away from Earth, with the Near Side higher in the lunar gravitational field. He posited a Near Side devoid of atmosphere,

The New Moon

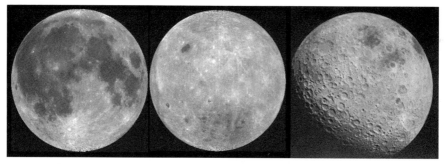

Figure 7.1. **The Moon seen and unseen**. The (*left*) Near Side and (*center*) Far Side Moon and an arbitrary view (Right) over the eastern limb of the Moon as seen from Earth (actually appearing to Earth's west because of the orientation flip going from Earth's surface to the lunar surface). Whereas the Near Side is one-third dark maria, especially on portions facing Earth, the Far Side is 99% highlands. The view at right, seen by *Apollo 16* leaving the Moon, shows Mare Crisium at upper right (also seen at upper right on the Near Side image, but rotated about 180°). Mare Smythii is below left of Crisium and Mare Marginis above left of that. (Images courtesy NASA: *Clementine* composites and AS16-M-3021.)

whereas the Far Side, deeper down, pooled air and water, supporting life. Hansen's hypothesis was popular until Simon Newcomb revealed it was based on erroneous interpretation of the data,[4] killing the idea in 1868.

Hansen's idea was not totally wrong, even if his evidence was. A huge displacement exists between the Moon's center of mass and the midway point between its Far and Near Sides ("center of figure" or geometric center), but it is only 1/30th of Hansen's value and in the opposite direction. This was discovered in 1964 when the Ranger probes hit the Moon one second later than expected.[5] The Near Side has mare concentrations shifting the center of mass toward Earth, and the Far Side has thicker but lighter crust, producing a 2-kilometer offset.[6]

What is the Near/Far asymmetry (Figure 7.1)? It has been obvious since the beginning in 1959: the Far Side is relatively devoid of maria, but the Near Side is dominated by "seas" – the Man in the Moon.[7] The Near Side has four huge maria basins of 700–1,100 kilometers diameter (and many smaller ones):[8] Mare Imbrium, Tranquilitatis, Serenitatis, and Nubium. [BOX 7.1] The largest Far Side basins have little or no maria, including Sharanov, Korolev, Hertzsprung, and Apollo, 430–590 kilometers across. Most exceptionally, South Pole–Aitken basin, a 2,500-kilometer Far Side monster, 12 kilometers deep, is nearly devoid of mare. Also Orientale, straddling Near and Far, is 350 kilometers across but rests in a 950-kilometer basin, largely unfilled. The largest Far Side maria are tiny: Maria Moscoviense and Ingenii, both 280 kilometers across, and the flooded crater Tsiolkovskiy, only 150 kilometers. Excluding area within 5° of the Moon's edge seen from Earth (the limb), about 2% of the Far Side but 31% of the Near Side are maria–covered. There are diverse ways the Near/Far asymmetry emerges, therefore: crustal thickness, number of basins, or mare-flooded area. (Figure 7.2).

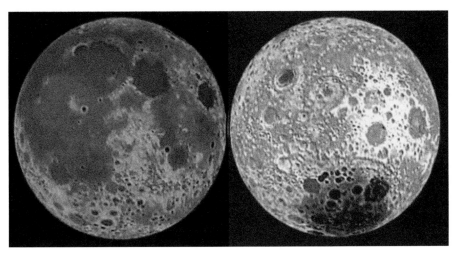

Figure 7.2. **Smooth side, rough side**. The asymmetry between Near Side (*left*) and Far Side (*right*) extends beyond simple appearance to topography, shown here with the darkest shade 8 km below average and the lightest 8 km above. Note that the deepest areas are on the southern Far Side (in the South Pole-Aitken basin), whereas the highest areas are just north of that. In contrast the Near Side is much smoother. (NASA graphic from *LRO*/LOLA data.)

BOX 7.1: What's in a Name?

Unlike most features on the Moon, maria are not named after people but primarily after the weather (except Mare Smythii after nineteenth-century astronomer William Henry Smyth). An old adage offers that a waxing Moon (when the eastern side is visible) brings pleasant weather, whereas the waning Moon (west side visible) brings storms. Thus stormy-sounding maria – Imbrium (rains), Nubium (clouds), Procellarum (storms) – are on the western half, and Tranquillitatis and Serenitatis are eastern. Confusingly, Crisium (crises) is eastern. These were named by Giovanni Baptista Riccioli and Francesco Maria Grimaldi in their 1651 lunar map *Almagestum novum*. (Both were Jesuits yet named the Moon's largest Near Side craters after key astronomers in the heliocentric revolution: Copernicus, Tycho, Kepler, and also Lansberg.) Maria named later, around the limb or on the Far Side, ignore this convention.

According to Rükl the adage states "when the Moon is waxing the weather will be fine and when it is waning the weather will be cloudy, rainy, stormy and unpleasant" (*Atlas of the Moon* by Antonín Rükl, 2007, Cambridge, MA: Sky Publishing, 22). This stems from the fancy that a Moon concave upwards in the Northern hemisphere's night sky (waxing) holds water, and concave down (waning) lets water pour out (*Ghosts and Goosebumps: Ghost Stories, Tall Tales and Superstitions from Alabama* by Jack Solomon, Olivia Solomon & Mark Brewton, 1994, Athens: University of Georgia Press, 116; "Lunar Influences" by A.W. Buckland, in *Hardwicke's Science-Gossip*, Vol. 25, 1889, London: William Clowes & Sons, 50). Curiously, there is evidence for (another) correlation between river volume and lunar phase ("Lunar Tidal Influence on Inland River Streamflow across the Conterminous United States" by Randall S. Cerveny, Bohumil M. Svoma & Russell S. Vose, 2010, *Geophysical Research Letters*, 37, L22406).

> **BOX 7.1: (continued)**
>
> There are about 8,800 named lunar features, primarily 7,000 satellite craters echoing nearby major feature's names (followed by a letter, e.g., "Ross A"). There are 1,546 primary craters named, mostly after men: primarily scientists and explorers, but only 28 women (as well as another 40 feminine names; note that features on Venus are named after women). Beyond craters, named features include one Albedo Feature (Reiner Gamma), 20 crater chains (Catena) 40 wrinkle ridges (Dorsum), 20 lakes (Lacus), 22 seas (Mare), 49 mountains (Mons), one oceans (Oceanus), three marshes (Palus), one plain (Planatia), nine promontories (Promontorium), 50 fissures (Rima), eight scarps (Rupes), 11 bays (Sinus), and 14 valleys (Vallis). Features are named officially by the International Astronomical Union, with most named on the Near Side by 1935 and Far Side by 1976, but more named every year or a few. Many names have been withdrawn, and an entire category is defunct (Fossa: long, narrow depression).

Do aspects of the Earth-facing side cause this asymmetry? For instance, can Earth cause more objects to strike the Moon on its Near Side? At first intuition, the opposite might apply: Earth will block some objects from hitting the Moon simply by Earth being hit. Indeed if the impacting objects traveled at infinite speed, this would occur, like Earth's shadow on the Moon during a lunar eclipse. On second thought, the case seems even worse: slower moving objects will be pulled from their straight-line path into an Earth-collision course.[9] There is an opposite effect, however. Consider this: an object passing by the Earth to the north will be pulled by Earth's gravitation into a hyperbolic path that cuts south, and a path to the south will be pulled north. If the Moon happens to sit where the paths cross, it will be hit by more objects. At the velocities of passing asteroids and comets, this surplus overwhelms the shadowing effect's deficit. For objects striking at 20 kilometers per second and at the current distance of the Moon from Earth, the Near Side surplus is about 10%.[10] At early times when the Moon and Earth were closer, this would be slightly larger.

Another simple effect of geometry and orbits is even stronger. The Moon orbits the Earth itself at 1 kilometer per second so is moving through the collection of smaller objects. There is a tendency for impacts to occur on the leading side, the western face of the Moon straddling the Near and Far Sides. Recent estimates converge on roughly 25% asymmetry between the leading and trailing points.[11] The difficulty here is the concentration of impacts is halfway between Near and Far, not on the Near Side. The Moon, however, occasionally suffers large impacts sufficient to push a new side into facing Earth.[12]

Why is the dense part pointed toward Earth? Although the lunar center of mass orbits in an ellipse around the Earth, points on the Moon closer to (or farther from) Earth revolve neither precisely along an ellipse nor at the period that would correspond to their orbital distance. This means they are not weightless but feel a small force because they are displaced from their ellipse (where they would be weightlessly in orbit). This tidal force will pull most strongly on the largest masses that are displaced furthest from their preferred orbit, so tidal forces will orient the Moon so that parts protruding farthest align along the Earth-Moon axis. A precise way to describe this invokes the *moment of inertia*, the measure of how much torque is required to spin up an object. If the object is not a perfect sphere, its moment of inertia will depend on the choice of axis about which to spin the object. Tidal forces will select the axis with the smallest moment of inertia. The Moon's actual spin axis, perpendicular to this (and pointing roughly northward on the sky), will become the axis with the largest moment of inertia. If one kicks the Moon to force it into another orientation, it will eventually oscillate down in this preferred direction to equilibrium, over several dozen millennia.[13] Which side of the axis the Moon chooses, however, is essentially symmetrical. The Far Side could have once been the Near Side, and vice versa, given a big kick, an impact sufficient to dig a mare basin.[14]

If mass on a body like the Moon is unevenly distributed, because of tidal forces the side where the mass protrudes (the Far Side) has the lowest energy configuration in pointing toward Earth. Intuitively one would guess that Earth and the Moon should fall into this lowest-energy state, but the situation is more complicated. The current state of the Moon, Near Side toward Earth, is a *local energy minimum*, not global. Rotating the Moon 180° from its current orientation would release energy, but only by passing through even higher energy states than where the Moon is now. The Moon stays in its current state, Near Side facing Earth.

How does the Moon select which side to present toward Earth? There is a protruding side (highlands) and a depressed side (maria). If the Moon spins too quickly, its rotational angular momentum will carry it past either, in a condition of free rotation, impossible to trap.[15] The unrotating Moon's energy as a function of orientation is lowest in the state with highlands toward Earth but also has a local minimum in the state of maria towards Earth. Between these two states are two energy barriers separating these minima, roughly 90 degrees on either side of the Moon's current orientation. The Moon's rotational energy dissipates as a result of tides from Earth and eventually drops below the highest energy barrier's peak. The Moon then oscillates like a pendulum through an almost 360° swing, never again crossing this barrier, but eventually its rotational energy will dissipate below the second barrier's peak. With this event the Moon has chosen: one side of the second barrier is the highlands side, the other is the maria. Which prevails is simply a matter of luck, depending on whether the Moon is swinging to the left or right when its energy drops below the second peak.[16]

We tackle the Man in the Moon soon in this chapter. There is no obvious way to pummel the Near Side sufficiently just because it faces Earth, and impacts favoring the western Moon seem incapable of thinning its crust enough or making enough maria. How does one measure the crust's thickness, anyway? There are three common methods: mapping gravity, using seismology (quakes), and measuring the crust's electromagnetic structure using radar and related probes. Seismology uses the speed of soundlike vibrations to measure a combination of the density of materials and their mechanical stiffness. Radar measures the speed of light (radio waves, actually), and associated probes measure electrical conductivity and related quantities (but not density). The third method, directly measuring density using gravity, takes several steps. First, one measures the topography (heights of mountains and valleys) and computes the implied gravitational field (called the *Bouguer correction*). Then one observes the free-air gravitational field of the planet, perhaps

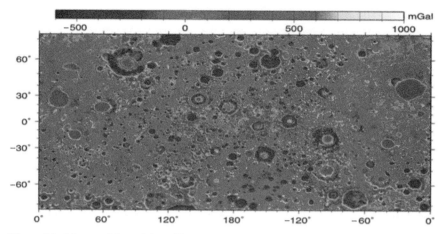

Figure 7.3. **A lumpy Moon**. Map of lunar gravitational field strength, based on the first month of science data from GRAIL. This projection is centered on the Far Side, showing the unprecedented detail in these data. Particularly, large impact basins vary radically in structure. Some are centers of overgravity, such as Imbrium (35° lat, −15° long, near right edge above equator) versus undergravity, such as Schwarzschild (70° lat, 122° long, top edge, left center) versus overgravity ringed by undergravity, such as Humboldtianum (57° lat, 81° long, near top edge, left). The South Pole-Aitken basin corresponds to the cluster of undergravity regions near bottom center.

by watching the force on a spacecraft orbiting overhead. The free-air gravitational field (the gravitational acceleration felt by a mass near the Moon's surface) is shown in Figure 7.3 from *GRAIL* data, accurate to about 1 part per million (averaged over an area of about 1,000 km^2). This is the fundamental result from *GRAIL*, almost 100 times more accurate than previous results on the Near Side and 1,000 times more accurate on the Far Side. The difference between this map (corrected for distance between the satellite and ground) and the Bouguer correction is the *Bouguer anomaly*, which maps real density anomalies in the Moon, which can be interpreted to dissect crust from mantle. A strongly negative Bouguer anomaly usually indicates a mountain, for instance, resting on a deep root of lower-density crust, expected if the mountain is buoyant gravitationally rather than pushed up by compressed rock exerting overpressure. For *GRAIL* this map is exceedingly smooth, indicating that most lunar topography is actually uncompensated.

If a feature is isostatically balanced (Chapter 6), a protrusion above the lunar surface balances a low-density "root" extending below the crust to support it, like buoyancy acting on a floating iceberg. One can imagine mapping the implied depth of the base of the crust, which portrays a negative ghost of the surface topography: a depression into the mantle below each mountain. Accounting for this implies 98.5% of all gravitational structure sensed by *GRAIL* is explained by topography. The extra 1.5% of the signal is intriguing, and we discuss that momentarily.

Many missions have mapped global topography (with laser altimetry or binocular imaging) and/or gravity (from radio tracking spacecrafts' orbits): *Clementine* (1994), *Lunar Prospector* (1998), *Kaguya* (2007), *Chang'e-1* (2007), and

Lunar Reconnaissance Orbiter (2009). The grandparents of these were *Lunar Orbiter I* through *V*, first discovering the lunar gravitational field's lumpiness in 1966–1967. This caused consternation, especially in landing *Apollo 11*, until it was soon detailed during Apollo. A new consensus has formed around the *GRAIL* gravity map and *LRO/LOLA* topography. Crustal thickness peaks about 45 kilometers on the Far Side and is thinnest at about 30 kilometers on the Near Side, with deep holes in some basins – smaller than what scientists thought before, but consistent with Apollo seismography. Sharpest variations extend from the South Pole-Aitken basin thinning on the southern Far Side next to the largest crustal bulge: the entire central Far Side. This is a huge elevation change, too: a 1% grade stretching 1,500 kilometers, steep enough to roll a bowling ball if the slope were planar. The *GRAIL* data also indicate that the Moon's upper crust is less dense than thought: 2.516 grams per cubic centimeter, not 2.9 as assumed.

The extra 1.5% in gravity lumps arise in large impact basins and are called *mascons* (mass concentrations). Standing at a mascon's center an astronaut weighs some 0.5% more than standing at its edge, roughly 60 times the largest variations on Earth. Mascons' origins are unresolved but differ from other large-scale topographic effects. They mostly arise in large maria, for example, mare-filled Imbrium, Nectaris, Crisium, Humorum, and Serenitatis, but also "drier" Orientale. For flooded maria, theories have long posited repeated lava floodings build a mascon: lavas rising to a certain "sea" level via hydrostatic pressure of overlying rock, followed by contraction of the lava sea by solidification, then more flooding to sea level, now higher because of the greater overburden of basalt, and so on.[17] For Orientale and mascons generally, basin impact is followed by rebound nearly to pre-impacts elevations by denser mantle material, involving complex hydrodynamic/lithospheric evolution; computer models are still being developed.[18] Either way, mare or mantle materials are denser than highlands anorthosite, therefore, producing mascons.

Might Earth's tides have induced lunar crustal asymmetry, especially when the Moon was 10 times closer and tidal forces 1,000 times larger? In the Moon's magma ocean in early millennia after formation, its crust would flex and generate heat, unless the lunar orbit were perfectly circular. Lunar crust should grow thickest near the axis connecting Earth and the Moon, including the Far Side (but unfortunately also the Near Side).[19] A later process must explain the thinner Near Side.

Internal processes can alter crustal thickness. We discussed mantle overturn in Chapter 6: as minerals solidify in the magma ocean, their relative densities invert and stratigraphic instability may occur – the heavier layer now on top. Because the mantle is plastic to long-range deformation, heavier rock on top drifts downward and lighter upward. Ilmenite ($FeTiO_3$), armalcolite (($Mg,Fe)Ti_2O_5$)[20], and related minerals are crucial, being common and growing denser while solidifying, driving mantle overturn.[21] Can this colossal roiling of the mantle alter the lunar crust's structure, akin to Earth's plate tectonics? Can this imprint determine the crust's distribution on scales of many thousands of kilometers? This overturn layer is only about 30 kilometers thick, so influencing placement of crust on 10,000-kilometer scales (the Moon's circumference) would require unreasonable values for its viscosity. The argument is made, however, that the first overturn does not eliminate all overdensity in the topmost layer, and mixing progresses, expanding to a 100-kilometers thickness or more. Instability can then drive a single roil across the whole Moon, with dense material sinking at one point and sucking in lighter material from great distances.[22] Riding this flow, crust will thicken where the mantle descends and thin on the opposite side. For many lunar scientists this is the preferred explanation for Near/Far asymmetry. There are issues with this: (1) Ilmenite and related minerals are dominant forms of lunar titanium, and titanium concentration varies radically (up to a factor of 100) over short distances and even among rocks at the same site. Does the smearing of ilmenite concentration just beneath the lunar surface because of this model's sideways flow contradict this? (2) Overturn generates heat from release of gravitational energy, which in this model will concentrate below the thicker crustal region. Is this consistent with the maria seen melting where crust is thin?

Maria tend to flood lunar low points, inconsistently. Mare Crisium and Smythii are about 5 kilometers below average elevation (versus the Moon's center of mass), whereas Procellarum is only 3 kilometers below and only 2 kilometers below the surrounding highlands. Typically maria are 4–5 kilometers below the highlands. Most lower points are mare flooded. In contrast, the difference between mid-Near and Far Sides is 7 kilometers. However, the South Pole-Aitken basin (SPA) is sparsely flooded but 6 kilometers below average. The obvious Far Side maria, Ingenii and Moscoviense, are about 3 kilometers below average, and Tsiolkovskiy 1 kilometer below, but large Far Side basins that are not maria (Hertzsprung, Sharanov, and Korolev) are 0–3 kilometers above. The Apollo basin, in the SPA and only 8% maria, is fully 5 kilometers below. Most regions lower than about 3 kilometers below average are flooded, except in the SPA, which does not start flooding until about 5 kilometers below. Amazingly, the SPA's center shows little flooding even 6–7 kilometers below.

With the odd exception of SPA, maria correlate with depth below the average Moon, in which Near and Far Sides differ radically. Is this elevation difference of several kilometers crucial to mare flooding? A few kilometers might seem small compared to depths where mare magmas form. Unlike on Earth, magma probably does not accumulate much in chambers a few kilometers below the surface. Calculating the equilibrium, chemical composition of basalt magma as a function of pressure yields the equivalent of 5,000 to 25,000 atmospheres (at sea level). With lunar density and gravity, this implies a source depth of 100–500 kilometers.[23] Driving these magmas to the surface seems problematic (because they are not buoyant in the surrounding mantle) and is still debated.[24] As discussed, some magma holds dissolved gas that expands and drives it to the surface after rising through most of the crust.

Given maria occurring at different elevations in different places on the Moon, and elevation determining hydrostatic pressure on lava and its ability to erupt, how do we understand eruptions being driven by hydrostatic pressure? Simultaneously, mantle-overturn theory might predict composition (hence density) varying significantly over the Moon. Does higher density offset shallower depth? Charles

Shearer[25] explains how magma buoyancy (combining depth and density) is the operative driving force. Magma of low density (relative to the crust) will seek weak zones, for example, impact fractures, and flood the surface as maria. The reasonable hypothesis is that mare magmas were less dense than the anorthositic upper crust. Another factor is heat; hotter magma is less dense. This would explain how Oceanus Procellarum, composed in part of high-density basalts, could flood over extensive areas of anorthositic upper crust, despite no deep basins there. As maps of radioactive elements (Figure 6.3) show, the Procellarum region (Procellarum KREEP Terrain: PKT) is the Moon's most radioactive area. Perhaps the extra heat generation by the PKT superheated its magma, reducing its density enough for it to rise through the anorthositic crust. Several eruptive hot spots contribute to this: the huge Aristarchus Plateau eruptive feature, the Marius Hills, Mons Rümker, and domes near Copernicus. The Far Side has only one small, analogous, feature: the Compton–Belkovich Thorium Anomaly.[26] Pete Schultz and collaborators suggest a mechanism for the hot spots together on the Near Side. These occur antipodally opposite the monstrous impact that excavated SPA. This explosion produced enormous seismic waves that would refocus on the Moon's opposite side. (This explains the "chaotic terrain" opposite Mercury's Caloris Basin.) A result would be enormous cracks extending from the surface deep into the Moon, which might channel lava to the maria.[27]

Despite our imperfect understanding of mare basaltic magma's rise from the mantle, the last few kilometers ascent appears crucial in setting where maria flood. Why is the Far Side higher? Might this arise from another giant collision, accreting on the Far Side? Yes, if it was slow enough. An object can strike the Moon as slowly as lunar escape velocity, 2.38 kilometers per second, its debris covering the impacted hemisphere in an hour, at roughly the speed of sound. In such cases the collision is less an explosion and more a splat, like a giant dirt cake. An object with 4% of the Moon's mass covers an entire lunar hemisphere typically 80 kilometers thick. If the mantle below is still plastic, it will deform, flowing toward the opposite hemisphere.[28] Such a collision could occur between two moons in Earth's orbit, but the second, smaller moon can survive at Langrange point L4 or L5 (see Chapter 5) after forming there from the Moon-forming giant collision. Such a "Trojan" can persevere, but suffers intervals of instability if the Moon drifts away from Earth.[29] This can happen so slowly that the lunar magma ocean might largely solidify, and the bulge left by this cosmic splat will persist and interact with Earth's tidal force to become the Far Side.

A primary issue with splat models is their likely implication of strong compositional variations across the lunar surface. Because the second moon is smaller, it would cool more quickly than the Moon and likely differentiate into crust and mantle (if not a core of nickel-iron, which might be absent). Even if undifferentiated, it certainly would not consist of anorthositic rock as does most lunar crust. Either way, there should be huge compositional variations across the Far Side, visible to remote sensing from spacecraft, and would already be easily detected.

Along similar themes, giant impacts might blast material from the future Near Side rather than just trowel it onto the Far Side. This might involve a large, high-velocity impact and explain puzzling features of the Near Side. In contrast to prominent basins (Imbrium, Serenitatis, etc.), large mare regions on the Near Side periphery are elongated, noncircular expanses: Oceanus Procellarum and maria Frigoris and Fecunditatis (as well as a long depression separating the central and southern Near Side highlands). Many basalt plains appear relatively shallow and not easily explained as superpositions of smaller basins. Could a whole hemisphere with thin crust result from a single, gargantuan impact onto what is now the Near Side?[30] This "Near Side megabasin" would start with a roughly 400-kilometer impactor hitting at perhaps 20 kilometers per second and dig a crater actually larger than the Moon's diameter (wrapping halfway around its circumference). This is unprecedented, the Solar system's largest known impact depression. A slightly smaller proposal, an impact feature underlying Procellarum, Frigoris, and most of the PKT, is sometimes referred to as the Procellarum Impact Feature.

The Near Side megabasin hypothesis was judged a dark horse candidate by many lunar scientists, but evidence from studies support it. Recent compositional maps from *Kaguya* favor a picture in which differences between Near Side and Far Side crust might arise from recrystallization of anorthosite after the enormous heat from a collision with enough energy to blow material across half of the Moon's surface.[31] One unsettled trouble is the Central Highlands midst the Near Side in this huge basin's center. Smaller basins suffer extensive rebound producing a central complex of peaks (Figures 6.13 and 6.14), but models of impacts gigantic enough for the Near Side megabasin produce no central uplift. The SPA, the largest extant basin in the Solar System otherwise, has no such uplift.

The Near Side/Far Side asymmetry debate is amazingly active for such a fundamental lunar question. Children will gaze at our natural satellite and wonder why there is a Man in the Moon. The opening lunar mystery unleashed by space exploration, as *Luna 3* returned those first, grainy Far Side pictures (Figure 2.2), was why the two hemispheres differ so much. We are still uncertain. Apollo gave us a new picture of how the Moon (and Earth) formed, but even this is debated. There are many things we still do not understand about the Moon.

Is this asymmetry only "skin deep"? Unlike crustal moonquakes, most deep moonquakes occur 700–1,200 kilometers underground (versus the 1,737 kilometer lunar radius), tending to cluster spatially.[32] These deep quakes are detected overwhelmingly on the Near Side, as expected from seismometer placement by Apollo. Whereas these data overlook asymmetry, we will learn about internal structure from *GRAIL*, with results forthcoming. For instance, as the shape of the Moon varied with varying Earth tides over the course of *GRAIL*'s mission, the variation in *GRAIL*'s orbit detected this and provided the mechanical strength of the material in the lunar interior in resisting such forces. Deep moonquakes can map core structure, particularly its liquid content, which cannot transmit shear waves (just pressure waves). These show a tiny but complex core: partially melted below 480 kilometers, liquid at 240–330 kilometers, and solid below that.[33] We might expect real asymmetry in deep quakes, because they correlate closely with mare edges (Chapter 9), nearly exclusively Near Side. These bear on with lunar water, because deep quakes occur at high temperatures and pressures where brittle fracture, cause of shallow earthquakes, is impossible. Water and other volatiles widening cracks in plastic rock (dehydration embrittlement) could work instead.

A revolutionary transformation in the past several years is the degree to which lighter elements, volatiles, are known on the Moon. Water, hydroxyl, hydrogen in other forms, forms of sulfur, nitrogen, carbon compounds, and other volatiles are found on the Moon, much to the surprise and difficult acceptance of many lunar scientists. Much of these arise from the Moon, within the Moon, and are not just transported or produced by comets, meteorites, or the solar wind. These are the results, and mysteries, considered in Chapters 8 and 9.

Indeed that story depends partially on the basic structure and Near/Far nature of the Moon. Distinct Near/Far Side separation leads to material freezing at the lunar poles: the Moon has no seasons (with an axial tilt of only 1.54°, see Figure 5.3). If a depression near a pole passes a whole month without solar illumination, it will remain in darkness for billions of years. Consequentially much of both polar regions is in permanent darkness, never sunlit, cooling to nearly interstellar temperatures only a few dozen degrees above absolute zero. Hence, at the poles where the Near and Far Side, and the lunar dark and bright side come together, is a special place – one that the human race will soon explore and may exploit. The Far Side's major feature, the SPA superbasin, raises a huge ridge though the South Pole (known unofficially as the Leibnitz Mountains[34]). This makes contours of elevation more extreme near the South Pole – not only the deep regions deeper (and more shadowed) but also peaks that are illuminated nearly 100% of the time by the Sun, which skirts the horizon of this special place. Here humans can benefit from nearly uninterrupted solar power. These fortunate and fundamental confluences of cosmic circumstance make the polar regions (and, hence, the Moon as a whole) an interesting and welcoming place for humans and their robotic envoys.

SPA is mysterious, and many lunar scientists consider it a compelling target for future sampling missions. The transient crater that formed this 2,500-kilometer depression stove deep into the mantle, likely excavating material difficult to reach otherwise. However, do not expect pure mantle rock awaiting us in SPA. Its impact melted about 70 million cubic kilometers of rock, and its surface is likely a soup of primarily pyroxene, olivine, and anorthite, about 45 kilometers thick, punctured by only the largest subsequent impacts.[35] SPA is clearly anomalous in composition, and we probably need ground truth samples to understand our SPA remote sensing data. SPA sample return can determine the basin's age, SPA being so huge that it affects nearly the whole lunar surface. Was SPA part of the Late Heavy Bombardment, or an earlier feature?

SPA is far from the only enigmatic feature on the Moon. There are several *swirls*, the most striking being Reiner Gamma (Figure 7.4). These seem to be artifacts of regolith aging, via OMAT (Chapter 6). Recent studies indicate their connection to magnetic anomalies, with the solar wind being deflected by local magnetic fields and so producing a structured staining of the surface regolith according to the direction of magnetic field lines. How these local magnetic field lines were thus organized is open to debate but seems associated with the placement of each swirl

Figure 7.4. **Magnetic mystery**. Reiner Gamma, on the northwestern Near Side, appears to result from one of several strong lunar magnetic field regions, apparently deflecting charged solar wind particles away from the lighter areas, preventing them from darkening. This view is about 100 km across. (NASA *Clementine* image.)

directly antipodally to large basins on the opposite side of the Moon. A plasma cloud will sweep the lunar surface from each of these massive explosions, carrying with it magnetic field, and concentrate this field at its antipode. How strong these fields were and how fixed they remain on the Moon potentially probes the lunar fluid nature at the time of the impact billions of years ago.

Not all breaking news about the Moon relates to its volatiles. For instance high resolution imaging from *LRO* reveals that many "shrinkage marks" or *lobate scarps* across the Moon are young. Detailed imaging shows many craters being consumed by this consolidation of lunar crustal surface, and some of these craters appear young (Figure. 7.5). The Moon appears to have contracted appreciably, probably a result of cooling, in the past billion years. Is this cooling a result of settling of unstable strata, loss of heat via volcanism, or changes in tidal or radiogenic heating?

The Moon's shape is in disequilibrium with its rotation, with a fossil bulge consistent with rotation every 3.5 days, presumably supported by a lithosphere responding by accumulating stress.[36] This is important in several ways. First, this might supply a force behind deep moonquakes. Secondly, this component of the Moon's shape will determine how it interacts tidally with Earth over billions of years, which establishes how its orbit has changed over geological time and how its rotational axis might have drifted, determining if the polar permanently shadowed

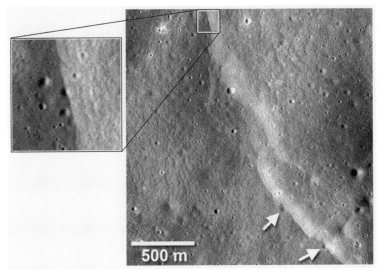

Figure 7.5. **The shrinking Moon**. Lunar lobate scarps, tectonic features in which one side rises over the other as a result of shrinkage, have destroyed or damaged craters, some relatively young, indicating that shrinkage continued even in the last billion years, decreasing the Moon's radius by about 100 meters in this time. (LROC/NAC image courtesy of NASA.)

regions really were shadowed billions of years ago. The issue of moonquakes also relates to the role of volatiles in the lunar interior, and we consider this in the next two chapters.

Notes

1. Many connect Dark Side of the Moon with the rock group Pink Floyd's 1973 album, but bad usage predates this (including Taylor's), i.e., a 1947 history by this name by Helena Sikorska & T. S. Eliot. This mix-up was rarer in the nineteenth century but dates at least to 1810: "Men may be found possessing great professional knowledge, much integrity, and yet be as utterly unnoticed as though they tenanted the dark side of the moon" (Clovis – a pseudonym? – October 1, 1810, *Rural Visitor*, Burlington, NJ: David Allison & Co., 2). "Everyone is a moon, and has a dark side which he never shows to anybody" (*Following the Equator: A Journey Around the World* by Mark Twain, 1897, Hartford: American, 654). See also note 55 to chapter 4 from an 1837 work. Scientists and teachers try correcting the error (possibly reversed by popular films such as *Transformers: Dark of the Moon* willfully reintroducing the confusion). In fairness Pink Floyd muses on lunacy, not lunar observation, but explains in their lyrics: "There is no dark side of the moon. As a matter of fact, it's all dark. The only thing that makes it look alight is the Sun"– all true, especially given the low lunar reflectivity (albedo): 9–12% for highlands, 5–7% for maria, more typical of asphalt. The Moon looks bright in the night sky but only in relation to outer space's blackness.
2. The mean lunar rotational and orbital periods are exactly equal, hence we always see the Near Side, but an extra 9% of the Moon's surface is seen at one time or another as a result of libration and parallax. Libration allows us to see sometimes 6° around the limb (edge) of the Moon and occurs because of the Moon's variable speed in its elliptical orbit and because its orbital plane is not in Earth's equatorial plane. In the first case, rotation cannot always match revolution, so we can peek around the eastern or western edge of the Moon. In the second, we sometimes look up or down on the Moon, seeing past its North or South poles. Parallax also occurs because our vantage point on Earth changes (because of its rotation), so we can peer an extra 1° around the limb.
3. One can calculate the Moon's distance gravitationally based on its period: $r = P^{2/3} (GM/4\pi^2)^{1/3}$, where r is orbital radius (semimajor axis, actually), P is the period, M is Earth's mass, and the rest constants. Distance can also be

estimated geometrically, e.g., from the Moon's parallax seen from places on Earth. Errors comparing these led to Hansen's hypothesis. See "Life on the Moon? A Short History of the Hansen Hypothesis" by Daniel A. Bech, 1984, *Annals of Science*, 41, 463.

4. "On Hansen's Theory of the Physical Constitution of the Moon" by Simon Newcomb, 1868, *Proceeding of American Association for Advancement of Science*, 17, 167. Hansen's hypothesis fed the popular Plurality of Worlds debate on extraterrestrial intelligence and Christianity, e.g., *On the Plurality of Worlds: An Essay* by William Whewell, 1859, London: Parker & Sons.

5. "Physical Constants as Determined from Radio Tracking of the *Ranger* Lunar Probes" by W. L. Sjogren, D. W. Trask, C. J. Vegos & W. R. Wollenhaupf, 1966, NASA Technical Report 32–1057. Apollo 15's altimeter showed that the Far Side was correspondingly higher: "Apollo Laser Altimetry and Inferences as to Lunar Structure" by W. M. S. Kaula, G. Schubert, R. E. Lingenfelter, W. L. Sjogren & W. R. Wollenhaupt, 1974, *Proceedings of Lunar Science Conference*, 5, 3049.

6. This is the preferred interpretation, although others have been offered about internal structure, e.g., "Lunar Asymmetry and Paleomagnetism" by D. J. Stevenson, 1980, *Nature*, 287, 520.

7. By 1903, the paucity of maria even on the lunar limb led to speculation that the Far Side had few maria: "It therefore appears probable that if such maria exist on the unseen portion they are less extensive than on the part of the orb which we see" (*A Comparison of the Features of the Earth and the Moon* by Nathaniel S. Shaler, 1903, Washington: Smithsonian, p. 15).

8. Some investigators consider Mare Tranquillitatis, 870 km across, to be a complex of smaller impact basins. Studies disagree on even large basins, and their diameters are debatable. (See "Crustal Thickness Evidence for More Previously Unrecognized Large Lunar Basins" by H. V. Frey, 2009, *Lunar and Planetary Science Conference*, 40, 1687.) Several impact basin lists exist, e.g., "A Global Catalog of Large Lunar Craters (≥ 20 km) from the Lunar Orbiter Laser Altimeter" by S. J. Kadish, C. I. Fassett, J. W. Head, D. E. Smith, M. T. Zuber, G. A. Neumann & E. Mazarico, 2011, *Lunar and Planetary Science Conference*, 42, 1106; *The Geology of Multi-Ring Impact Basins* by Paul D. Spudis, 1993, Cambridge: Cambridge University Press.

9. We can estimate this effect's size using angular momentum. If a small, stationary object at infinite distance from Earth falls to Earth's surface (rather, the top of its atmosphere), it will hit at Earth escape velocity v_{esc}. If instead the object starts out at speed v_0 and hits Earth, it strikes at $v_E = \sqrt{v_0^2 + v_{esc}^2}$. In the minimal case of an object barely hitting Earth, it skims Earth's surface at a tangent at Earth's radius r_E. Because the angular momentum is (mass) × (distance from Earth, perpendicular to velocity direction) × (velocity): $L = m\, r_\perp\, v$, and because L and mass do not change, the ratio of the perpendicular distances (distance from the object to a straight line passing through Earth's center) is the inverse of the velocity ratio: $r_{\perp 0} v_0 = r_E v_E$, so $r_{\perp 0}/r_E = v_E/v_0$. Because $v_{esc} = 11.2$ km s^{-1} and objects travel versus Earth at ~20 km s^{-1}, $v_E/v_0 = \sqrt{1 + v_{esc}^2/v_0^2} \approx 1.15$, so $r_{\perp 0} \approx 1.15\, r_E$ ($r_{\perp 0}$ is the "impact parameter"). Any object with $v_0 \leq 20$ km s^{-1} and $r_{\perp 0} \leq 1.15\, r_E$ will hit Earth.

10. "The Asymmetric Cratering History of the Moon" by M. Le Feuvre & M. A. Wieczorek, 2005, *Lunar and Planetary Science Conference*, 36, 2043.

11. "Current Bombardment of the Earth–Moon System: Emphasis on Cratering Asymmetries" by J. Gallant, B. Gladman & M. Ćuk, 2009, *Icarus*, 202, 371; "Nonuniform Cratering of the Moon and a Revised Crater Chronology of the Inner Solar System" by Mathieu Le Feuvre & Mark A. Wieczorek, 2010, *Icarus*, 214, 1.

12. "Did a Large Impact Reorient the Moon?" by Mark A. Wieczorek & Mathieu Le Feuvre, 2009, *Icarus*, 200, 358.

13. "Large Impact Craters and the Moon's Orientation" by H. J. Melosh, 1975, *Earth and Planetary Science Letters*, 26, 353.

14. Wieczorek & Le Feuvre, "Did a Large Impact Reorient the Moon?" 358.

15. The minimum incremental energy to send the Moon into free rotation is about 2×10^{20} J. This is small compared to the current lunar rotational energy, about 3×10^{23} J, so the Moon would only need to rotate once every 225 years versus its monthly period to present a regular progression of sides to Earth. Putting this in perspective, this minimal extra energy is roughly that consumed by humanity annually, so spinning the Moon would be a monumental task. (Readers might recall this fictional event from Arthur C. Clarke's science fiction novel *Childhood's End*.)

16. "Why Do We See the Man in the Moon?" by Oded Aharonson, Peter Goldreich & Re'em Sari, 2012, *Icarus*, 219, 241. Aharonson et al. estimate that the probability of the Moon falling into its current configuration instead of the highlands facing Earth is 63%. This depends critically on how efficiently the Earth/Moon system dissipates the Moon's rotational energy, because the energy inequality between barriers and the monthly energy dissipation are roughly equal.

17. "On the Origin of Mascons and Moonquakes" by S. K. Runcorn, 1974, *Proceedings of Lunar Science Conference*, 5, 3115; "Mascons and the History of the Moon" by Harold C. Urey, 1968, *Science*, 162, 1408; "The Deep Structure of Lunar Basins: Implications for Basin Formation and Modification" by Steven R. Bratt, Sean C. Solmon, James W. Head & Clifford Thurber, 1985, *Journal of Geophysical Research*, 90, 3049.
18. "Modeling the Origin of the Orientale Basin Mascon" by David M. Blair, Brandon C. Johnson, Andrew M. Freed, H. Jay Melosh, Gregory A. Neumann, Sean C. Solomon & Maria T. Zuber, 2013, *Lunar and Planetary Science Conference*, 44, 2821; "The Origin of Lunar Mascon Basins, Part I. Impact and Crater Collapse" by Brandon C. Johnson, David M. Blair, H. Jay Melosh, Jeffrey C. Andrews-Hanna, Gregory A. Neumann, Roger J. Phillips, David E. Smith, Sean C. Solomon, Mark A. Wieczorek & Maria T. Zuber, 2013, *Lunar and Planetary Science Conference*, 44, 2043; "The Origin of Lunar Mascon Basins, Part II: Cooling and Isostatic Adjustment" by Andrew M. Freed, David M. Blair, H. Jay Melosh, Jeffrey C. Andrews-Hanna, Gregory A. Neumann, Roger J. Phillips, David E. Smith, Sean C. Solomon, Mark A. Wieczorek & Maria T. Zuber, 2013, *Lunar and Planetary Science Conference*, 44, 2037.
19. "Structure and Formation of the Lunar Farside Highlands" by Ian Garrick-Bethell, Francis Nimmo & Mark A. Wieczorek, 2010, *Science*, 330, 949.
20. See note 12 to Chapter 6.
21. "A Chemical Model for Generating the Sources of Mare Basalts: Combined Equilibrium and Fractional Crystallization of the Lunar Magmasphere" by Gregory A. Snyder, Lawrence A. Taylor & Clive R. Neal, 1992, *Geochimica et Cosmochimica Acta*, 56, 3809; "Numerical Convection Modelling of a Compositionally Stratified Lunar Mantle" by J. de Vries, A. P. van den Berg & W. van Westrenen, 2011, *Lunar and Planetary Science Conference*, 42, 1745.
22. "On the Scale of Lunar Mantle Overturn Following Magma Ocean Fractional Solidification: The Role for Multiple Scales of Convective Motion" by E. M. Parmentier, 2009, *Lunar and Planetary Science Conference*, 40, 1781; "Gravitational Differentiation Due to Initial Chemical Stratification: Origin of Lunar Asymmetry by the Creep of Dense KREEP?" by E. M. Parmentier, S. Zhong & M. T. Zuber, 2002, *Earth and Planetary Science Letters*, 201, 473. This occurs because of Rayleigh-Taylor instability. Because of dilution of density differences in this model, the huge spatial scale, and weakness of lunar gravity, the mantle will flow very slowly and may conflict with lunar evolution time scales. For more about R-T instability, see "An Overview of Rayleigh-Taylor Instability" by D. H. Sharp, 1984, *Physica D: Nonlinear Phenomena*, 12, 3.
23. Given that the Moon's surface gravity is 1/6 Earth's, its density is about three times water's, and one atmosphere's pressure on Earth is equivalent to a 10 m water column, you can see that these depths produce such pressures, e.g., (5,000 atmospheres) × (10 meters/atmosphere) × (6) / (3) = 100,000 m = 100 km ("Origin of Green Glass Magmas by Polybaric Fractional Fusion" by J. Longhi, 1992, *Lunar and Planetary Science Conference*, 22, 343; "Experimental Petrology and Petrogenesis of Mare Volcanics" by John Longhi, *Geochimica et Cosmochimica Acta*, 56, 2235).
24. This is summarized in "Thermal and Magmatic Evolution of the Moon" by Charles K. Shearer, 2006, *Reviews in Mineralogy and Geochemistry*, 60, 325, especially 463–464.
25. "The Role of Magma Buoyancy on the Eruption of Lunar Basalts" by Mark A. Wieczorek, Maria T. Zuber & Roger J. Phillips, 2001, *Earth and Planetary Science Letters*, 185, 71.
26. "Small-area Thorium Features on the Lunar Surface" by D. J. Lawrence, R. C. Elphic, W. C. Feldman, T. H. Prettyman, O. Gasnault & S. Maurice, 2003, *Journal of Geophysical Research*, 108, 5102; "Compton-Belkovich: Nonmare, Silicic Volcanism on the Moon's Far Side" by B. L. Jolliff, et al., 2011, *Lunar and Planetary Science Conference*, 42, 2224.
27. "Lunar Activity from Recent Gas Release" by Peter H. Schultz, Matthew I. Staid & Carlé M. Peters, 2006, *Nature*, 444, 184.
28. "Forming the Lunar Farside Highlands by Accretion of a Companion Moon" by M. Jutzi & E. Asphaug, 2011, *Nature*, 476, 69.
29. "The Fate of Primordial Lunar Trojans" by Matija Ćuk & Brett J. Gladman, 2009, *Icarus*, 199, 237.
30. "A Large Basin on the Near Side of the Moon" by Charles Joseph Byrne, 2007, *Earth, Moon and Planets*, 101, 153; "The Near Side Megabasin: Topography and Crustal Thickness" by Charles J. Byrne, 2008, *Lunar and Planetary Science Conference*, 39, 1302.
31. "Compositional Evidence for an Impact Origin of the Moon's Procellarum Basin" by Ryosuke Nakamura, et al., 2012, *Nature Geoscience*, DOI: 10.1038.

32. "Farside Deep Moonquakes and Deep Interior of the Moon" by Yosio Nakamura, 2005, *Journal of Geophysical Research*, 110, E01001.
33. "Seismic Detection of the Lunar Core" by R. C. Weber, P. Lin, E. J. Garnero, Q. Williams & P. Lognonné, 2011, *Science*, 331, 309.
34. Wilhelms tells of how the Leibnitz Mountains played a role in William Hartmann's anticipation of the existence of the SPA, and how this was confirmed by Soviet spacecraft. (See *To a Rocky Moon: A Geologist's View of Lunar Exploration* by Don E. Wilhelms, 1993, Tucson: University of Arizona Press, 244.)
35. "Modelling the South Pole-Aitken Basin Subsurface" by W. M. Vaughan & J. W. Head, 2013, *Lunar and Planetary Science Conference*, 44, 2012. A prime candidate for SPA sample return is Moonrise, perhaps by 2016–2017.
36. "The Physical Mechanisms of Deep Moonquakes and Intermediate-Depth Earthquakes: How Similar and How Different?" by C. Frohlich & Y. Nakamura, 2009, *Physics of the Earth and Planetary Interiors*, 173, 365; "Evidence for a Past High-Eccentricity Lunar Orbit" by I. Garrick-Bethell, J. Wisdom & M. T. Zuber, 2006, *Science*, 313, 652.

Chapter 8
Water in a Land of False Seas

"Beside the Mare Crisium, that sea
Where water never was, sit down with me
And let us talk of Earth, where long ago
We drank the air and saw the rivers flow
Like comets through the green estates of man,
And fruit the color of Aldebaran
Weighted the curving boughs."

– Adrienne Rich, 1952, *The Explorers*[1]

A century ago water on the Moon began as an idea with the worst possible intellectual pedigree. In 1894 Hans Hörbiger, a successful engineer (who invented a valve to control blast furnace airflow), had a curious vision of the Universe. His *Glacial-Kosmogonie*, published in 1912, propounded the *Welteislehre* ("World Ice") theory, with the Moon, our Galaxy, and even space itself dominated by water ice, apparently inspired by the icelike appearance of the Moon in the night sky.[2]

Hörbiger's book was championed by respected German amateur selenographer Philipp Fauth, aided by Hörbiger and family. Public extravaganzas promoted the theory to common knowledge. Its cold, northerly tenor in opposition to Einstein's relativity (and even Newtonian physics) attracted Nazi leaders.[3] Welteislehre became party doctrine, and Fauth was promoted by S.S. Reichsführer Heinrich Himmler to university professor (having never taught at that level or conducted sufficient research). Fauth named a lunar crater Hörbiger (following his death in 1931). During and after the Third Reich, Hans Schindler wrote several books expanding the World Ice theory, soon discredited. In 1948 Hörbiger's name was stripped from the crater (although a crater Fauth remains).

Fauth should have known better. The Moon's surface can be hot, and by his time we knew it had little atmosphere. The Moon, as close to the Sun as is Earth, actually absorbs 40% more solar flux because of its lower reflectivity. Under these conditions water ice on its surface would more than melt; it would sublime directly into vapor and likely escape violently into vacuum.[4] Fauth knew this did not happen. In 1907 after thousands of hours sketching lunar maps (even after photography made this increasingly obsolete), he proclaimed, "as a student of the Moon for the last twenty years and as probably one of the few living investigators who have kept in practical touch with the results of selenography, he (Fauth) is bound to express his conviction that no eye has ever seen a physical change in the plastic features of the Moon's surface."[5]

Several years after Welteislehre's demise, prospects for water on the Moon were adopted by a more competent and respected scientist, Harold Urey, 1934 Nobel Chemistry Prize recipient for discovering deuterium (heavy isotope of hydrogen), major contributor to techniques in uranium isotopic enrichment, and major originator of the fields of meteoritics, cosmochemistry, and isotopic analysis and the idea that life sprang from Earth's primordial atmospheric chemistry. In space exploration's early years, Urey was decisive in persuading NASA to explore the Moon scientifically.[6] He began writing about the Moon in 1952 in his book *The Planets*,[7] arguing that to understand planetary origins, we should study the Moon – which Urey called "Rosetta Stone of the Solar System" – a primitive world that might hold samples indicative of early Earth, based in part on the Moon and planets being formed perhaps by accretion of cold material like dust. In this scenario he imagined water playing important roles.

Urey imagined many processes influencing the lunar surface. By 1956 he was entertaining the idea of aqueous effects on the Moon's surface features.[8] He was later influenced by new data, starting with *Ranger 7* pictures of the Moon in 1964.[9] *Ranger 9* in 1965 returned close-ups of flooded crater Alphonsus revealing a complex of features atypical of the Moon: rivers and channels, some apparently cut by fluids (Figures 8.1 and 8.2). He jumped to their interpretation:

Various lines of evidence indicate that the material of the maria floors and especially of the Alphonsus floor consist of fragmented material to a very considerable depth, with substantial crevasses below the surface. It is not possible to decide whether such crevasses are the result of lava flows or the evaporation of massive amounts of water from beneath the surface. It is, however, the author's opinion that the water interpretation is the more likely of the two.

He summarized: "They [lunar maria] may have been subjected to water at some time in their history, but the evidence of pictures alone is not sufficient to make a firm decision in regard to these conclusions."[10]

Urey was not the sole proponent of a watery Moon. A topic common to several authors was the similar look of lunar sinuous rilles and terrestrial meandering rivers, with their oxbow loops (Figure 8.2). These need not form by water but by any sufficiently nonviscous, turbulent fluid. A river rounding a bend forces water outward centrifugally so water level rises in outer radii of the bend versus inner ones. On the river bottom, water flows slowly because of riverbed drag. That water feels high pressure on the outside of the bend, so it flows toward inner radii, taking silt with it. Thus the riverbank erodes at outer radii, and the oxbow grows. This also occurs for lava if the ratio of viscous force acting on the lava divided by the lava's inertia of motion (called the *Reynolds number*) is sufficiently similar to that ratio for terrestrial rivers.[11] Lunar magma is basaltic so is less viscous than many on Earth, therefore, lunar lava flows and terrestrial meandering rivers share common turbulence conditions. Their resemblance arises from similar physical parameters, not the same fluid. Evidence for lunar water does not depend on landform shape.

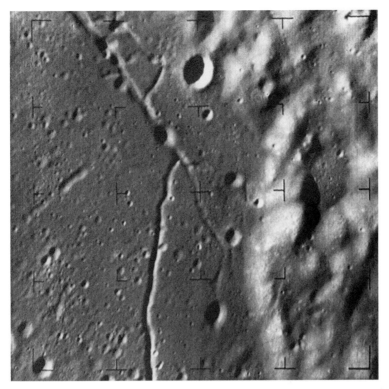

Figure 8.1. **Lunar close-up**. *Ranger 9* image showing a 7.3 km wide portion of western Ptolemaeus crater, just north of where *Ranger 9* impacted 72 seconds later in crater Alphonsus. Note the straight and arcuate rilles. Some craters along these rille paths, e.g., upper left of center and near bottom center, are really volcanic vents, not impacts, highlighting the rilles' volcanic nature. Such vents tend to lie near boundaries between maria and highlands (see Chapter 9). (Photograph B-074, courtesy of NASA. For more details, see *Ranger IX Photographs of the Moon* by Gerard P. Kuiper, R.L. Heacock, E.M. Shoemaker, H.C. Urey & E.A. Whitaker, December 15, 1965, NASA Publication SP-112.)

Urey suffered sharp criticism for his openness to ideas of water on the Moon. It wore on him, so much that he confessed (in a letter to the prestigious journal *Nature*) that some thought him under the influence of more intoxicating liquids.[12] The most substantial attack came from NASA planetary scientist John O'Keefe (who suffered his own controversy regarding the lunar origin of glassy globules called *tektites*, now thought ejected from large terrestrial impacts). O'Keefe showed how an ice layer under the lunar surface would distort and flow like a glacier if it exceeded a kilometer thickness, in contradiction to craters 2 kilometers deep absent signs of their walls flowing onto their floors.[13] The criticism is not definitive: a thinner layer would not deform but might conceivably produce effects Urey claimed in Ranger photos. Urey grew haggard defending himself and authors of similar ideas.[14]

This changed with the first samples returned from the Moon, showing few hydrated minerals. The only sedimentary rocks resulted from layered lava (Figure 6.6). Hadley Rille, visited by *Apollo 15*, was obviously made by flowing lava,

The New Moon

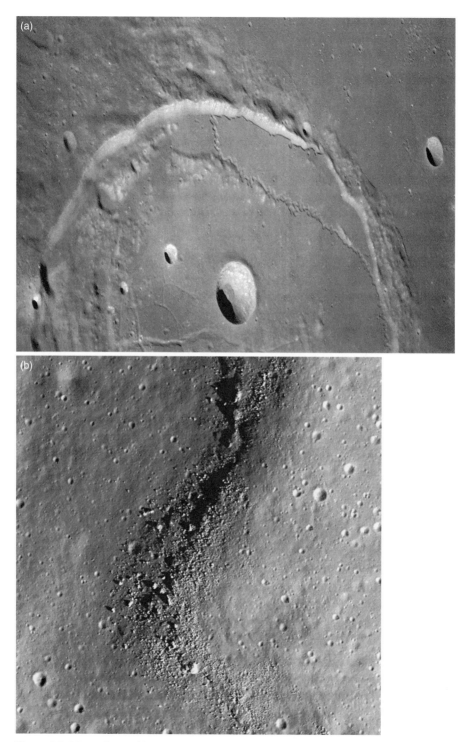

Figure 8.2. **Moon river, of lava**. Two views of the sinuous rille in flooded crater Posidonius:
(a) An 80 km wide view of southern Posidonius, showing rille Rimae Posidonius arising in the
lower right corner, deepening and meandering to the upper left and reversing to flow along the

Figure 8.3. **Gas in the stone**. Vesicular mare basalts, in the laboratory and in situ. (a) *Apollo 15* sample 15556, with a 1.5 kg mass, collected near Hadley Rille's rim and 3.4 billion years old. Hadley Rille is sinuous and displays layered basalt exposed in the rille wall. (Courtesy of NASA.) (b) A vesicular basalt boulder at EVA Station 1 in the Taurus-Littrow Valley visited by *Apollo 17*. Astronaut Gene Cernan found this rock curious because "the vesicularity changes from a hummocky vesicularity to a very fine vesicular" (*Apollo 17* flight transcript, 122:72:22) along a boundary seen here running near the rock center roughly vertically (*upper left to lower right*) indicating an inhomogeneous change of phase. Cernan used sample tongs at left to set the camera focus distance. (Courtesy of NASA; photograph AS17-134–20403.)

not water. Urey relented, abandoning any idea of lunar water.[15] Eventually, Urey and O'Keefe published scientific papers together – but not about lunar hydration. Urey died in 1981, long before minds changed once again regarding water on the Moon.

The evidence from Apollo against water on the Moon was varied and manifest, and little was in favor. Although astronauts left instruments on the Moon that detected a substantial lunar atmosphere, most of its mass was the inert gas argon, formed by the decay of radioactive potassium.[16] Water amounted to less than 1% of this, barely detectable at a density of 600 molecules per cubic centimeter.[17] An unknown gas dissolved in the lava flowing onto the lunar maria expanded into foamy *vesicular basalt* as the lava neared the surface, where the gas expanded under low pressure (Figure 8.3). If these vesicles were steam-filled, the mineral that should result is *amphibole*, essentially like the mineral pyroxene, common on the Moon, but

---

Figure 8.2. (cont.)
upper crater wall and spill through a gap onto Lacus Somniorum's mare. (Photo AS15-91–12366, courtesy of NASA.); (b) A 500 m wide portion of Rima Posidonius (zooming into Figure 8.2a above and right of center where the rille encounters the rightmost part of a small, horizontal ridge) showing the rille's marginal wall, indicated by mass wasting talus. Note that surface texture and cratering density is similar inside and outside the rille, supporting the idea that some rilles are lava erosional features, not collapsed lava tubes. (Courtesy of NASA/GSFC and ASU, LROC photograph M113771795RE.)

hydrated. Amphibole's absence from vesicular basalts indicated the gas in the lava consisted almost entirely of something other than water.[18]

Arguments developed that lunar rocks contained essentially no water at all; lunar water content could be limited to levels in the range of parts per billion or less. Exceptions existed; in fact, many samples contained water at 250 to 500 parts per million (by weight).[19] H_2O in samples was identical to "Pasadena water vapor" in ways noted by Caltech authors of one analysis: ratios of different types, or isotopes, of oxygen and hydrogen are the same as on Earth.[20] Terrestrially, most hydrogen nuclei consist of just a proton, but 0.015% of hydrogen nuclei have a neutron attached to this proton (hence form deuterium: ^2H), and 0.2% of oxygen has an extra two neutrons (^{18}O, versus the usual isotope ^{16}O). On the Moon, the number of atoms of oxygen isotopes have essentially the same ratio as on Earth, so the argument becomes one mainly of deuterium.[21] (Remember, 45% of the lunar soil is composed of oxygen.) The deuterium ratio in the solar wind striking the Moon is not well known but is much smaller than on Earth. Many lunar samples have a tiny deuterium fraction, so their hydrogen is assumed to derive from solar wind. If they show high hydrogen abundance, they also tend to have a larger deuterium fraction, close to the terrestrial value, which is interpreted as contamination. Even minerals that seem to evidence aqueous processing, for example, rust, show the same isotope ratios, so they were suspected to result from terrestrial water.[22]

This interpretation is missing any way to ever conclude water was lunar if it has the same oxygen and hydrogen isotopic ratios as on Earth (and the Moon), as is true of so many chemical elements, and lesser fractions of deuterium can be explained away by solar wind origin. The dry Moon hypothesis lacks sufficient scientific falsifiability: wide ranges of hydrous isotopic ratios were explainable in terms of solar wind or terrestrial contamination regardless of whether the water's origin was lunar.

Was terrestrial contamination a problem? Definitely, but one that was exaggerated. Dust covers essentially all lunar surfaces (Chapter 6) and quickly spreads to equipment (and people) from Earth unless carefully prevented (Chapter 10). This dust led to terrestrial contamination in a subtle way: dust fouled the air seals on some sample boxes intended to return lunar dust and rocks to Earth isolated from air (and moisture).[23] Apollo lunar sample return containers, used to carry soil and rock home on all six landing missions, were aluminum boxes with triple seals protected by a cloth and Teflon cover removed just prior to closing. Even so, some seals were compromised by the ubiquitous, hard-grained dust.[24] Samples were held in passive quarantine for six weeks along with returning astronauts to prevent lunar biological infection, and over the years samples have been contaminated in processing and analysis. This issue grew to a common myth that all samples were corrupted with terrestrial atmosphere,[25] which is untrue. Today some Apollo samples remain almost certainly uncontaminated since containment.[26] One can imagine, unfortunately, mild contamination from the Lunar Module's exhaust, or astronauts' backpacks, which vent water vapor to cool.[27]

Beyond isotopic ratios, there was further reason to think the Moon contained little water. For instance, lunar rocks are depleted of low melting-point elements such as lead, thallium, bismuth, and indium (melting at 327°C, 304°C, 272°C, and 157°C, respectively), to levels about 100 times less than terrestrial basalts'. Because these would accrete last onto the Moon as it cooled, one might use these as a guide to water abundance (assuming that they all accrete from carbonaceous chondritic meteorites, for instance). This implies water of only a few tenths part per million, equivalent to an ocean only 3.7 meters deep covering the entire lunar surface.[28] Results might be much different, and wetter, if water arrived via comets. With few exceptions, however, Apollo samples seemed to contain much less water than terrestrial rocks, and the bulk of that was suspected to be terrestrial contamination.[29] Three decades passed before the issue was revisited, and changed dramatically.

Of course, the Giant Impact lunar formation hypothesis, gaining favor for decades, explained why so little volatile elements (lead, etc.) or compounds like water remained. They boiled into space. Typical temperatures in proto-lunar material reached a few thousand Kelvins at which any mineral would melt and tend to release dissolved gas. Thus, with hydrogen evaporated and the Moon's core largely lost inside Earth, only intermediate mass elements remained abundant, starting with oxygen bound tightly to some heavier elements such as silicon. We know impact ejecta are desiccated: tektites, from impacts on Earth, are among the driest rocks, with 0.005% average water content. Together the Big Whack and dry Moon were copacetic.

Exceptions to the dry Moon were seen even during the Apollo era. Rusty Rock 66095, discussed earlier and later (see notes 22 and 87), is the most volatile-laden Apollo rock, an impact melt breccia typical of highland rocks. It has rust (often goethite: FeO(OH)), hence hydration, and schreibersite ($(Fe,Ni)_3P$), common in meteorites but rare on Earth. Because 66095 seems to result from meteoritic impact, it was dismissed as hydrated by the meteorite. Most highland rocks are breccias, however, and contain both rust and schreibersite. Volatiles in highland rocks might be alien in origin but are not hard to find.[30]

Minerals found occasionally in lunar rocks usually indicate water's presence: hematite (Fe_2O_3), magnetite (Fe_3O_4), and goethite are more consistent with a low-temperature mix with hydrogen (a few hundred degrees Celsius) than cometary impact.[31] Relics of comet impacts are also found; soil sample 61221 from *Apollo 16* contains H_2O, CO_2, H_2, CH_4, CN, and perhaps CO, totaling 300 parts per million.[32]

Not all lunar hydration evidence is bound in rock and soil samples; sometimes it flies through the vacuum. *Apollo 14* astronauts Shepard and Mitchell installed SIDE (Suprathermal Ion Detector Experiment) to detect accelerated, charged atoms in the lunar atmosphere and measure their mass and energy. Other versions of SIDE were deployed on *Apollo 12* and *15*. On March 7, 1971, 29 days after *Apollo 14*'s LM *Antares* blasted off the Moon, another blast hit SIDE: millions of ions, all with energies (about 49 electron volts) several times that sufficient to blow a typical atom apart, with masses of 17 atomic mass units, the same as hydroxyl.[33] *Apollo 12*'s SIDE, 183 kilometers west, also detected an ion blast (but unfortunately was set to measure the total number of ions, not their mass). The event was also seen by another *Apollo 14* experiment.[34] In their 1973 paper, SIDE scientists considered several artificial and extra-lunar sources, eliminating all but natural lunar outgassing. Two of the authors revisited the question 18 years later and found the likely ion source was *Apollo 14*'s engine exhaust.[35] *Antares* landed in the morning of the previous lunar day, and the March 7 event started 30 hours after the Sun rose again over the site. Freeman and Hills argued that water vapor from *Apollo 14* sequestered in the regolith over the previous lunation escaped via heating after sunrise. This is also speculative but cancelled for most people earlier claims of natural hydroxyl outgassing.

By the early 1970s the dry Moon was settled science, especially among lunar geologists in the United States. Had that gone differently, the last mission of the Moon Race, by the Soviets, might have changed everything. It discovered water within the Moon. Instead it was ignored.

On August 9, 1976, *Luna 24* launched toward the Moon on a Proton rocket and nine days later landed safely in unexplored southern Mare Crisium.[36] Within 24 hours, it deployed a drilling rig, extracted a core sample from 2 meters into the Moon, stowed it in its return capsule, and blasted off again with 170 grams of lunar soil. Four days later it successfully reentered Earth's atmosphere over Siberia, with the core sample taken to Moscow intact and uncontaminated (as far as we know). It was the Soviet Union's last lunar mission, and last from Earth to soft-land on the Moon in the twentieth century.

It returned special payload. The core sample was found by scientists M. Akhmanova, B. Dement'ev and M. Markov of the Vernadsky Institute of Geochemistry and Analytic Chemistry to contain about 0.1% water by mass, seen in infrared absorption spectroscopy (at about 3 microns wavelength), a detection about 10 times above the threshold. The water signal tended to increase deeper below the lunar surface. Their paper's original title in the February 1978 Russian-language journal *Geokhimiia* translates to "Water in the Regolith of Mare Crisium (Luna 24)?"[37] and in the English language journal, "Possible Water in Luna 24 Regolith from the Sea of Crises"– but the abstract claims water detected fairly definitively. They indicate the sample shows no tendency to absorb water from the air but are unwilling to stake their reputations on absolute statements on terrestrial contamination being absent. Nonetheless, they claim every possible precaution and stress that this result must be followed up.[38] Three Soviet lunar sample return missions (*Luna 16, 20,* and *24*) from 1970–1976 brought back 327 grams total in lunar soil. The six Apollo lunar landing missions in 1969–1972 returned 381,700 grams of rock and soil. Apollo won the samples race. No other author has ever cited the *Luna 24* work, as of this writing.[39]

A generation passed without further lunar hydration research. After Apollo the next lunar science mission was *Clementine* in January 1994.[40] Along with distinctions described in Chapter 4, *Clementine* was the first lunar probe to perform radar since Apollo. Radar was not designed into the craft but innovated after launch by rigging a *bistatic* system, with the transmitter on *Clementine* (its communications channel) and the receiver elsewhere (Deep Space Network antennae back on Earth).[41] *Clementine*'s radio signals bouncing off the Moon might reveal ice.

How can radar detect ice? Radio waves cannot reflect from uniform media, so radar essentially probes material nonuniformities (such as storm cells among the clouds, or airplanes in air). Electromagnetic radiation, such as radio, can be imbued with *circular polarization* in which the direction of the electric part of the wave disturbance rotates either clockwise or counterclockwise (known as right-handed circular polarization [RCP], or left-handed [LCP], respectively). *Clementine* transmitted only RCP radio waves. If RCP radio reflects once from a surface, polarization is reversed as in a mirror and the radio signal returns LCP. If the radio signal bounces twice, it tends to return RCP. Bouncing multiple times, it returns mixed in RCP and LCP. Pure ice scatters few waves within its volume, but ice containing imbedded rocks scatters radar more effectively. Before the wave heads outwards to the receiver, it undergoes more bounces than a wave reflecting from the topmost regolith at the Moon's surface. An RCP wave reflecting from surface regolith returns more LCP than one reflecting from an ice/rock mix, a combination of RCP and LCP. Pure regolith volume scatters radar less strongly and tends to return LCP. When the angle between the radar transmitter, target, and receiver is allowed to change, the ratio of RCP to LCP will change much differently for ice and regolith than for regolith alone. This ratio of RCP to LCP is known as the *circular polarization ratio*, or CPR.

CPR can be ambiguous in detecting ice, because regolith surfaces replete with boulders scatter more times than those without, partially mimicking rocky ice's effects. This provoked controversy: *Clementine*'s bistatic radar in 1994 showed strong evidence for water, supported by more data in 2001.[42] Other radar observations at different angles to the target, this time with both transmitter and receiver on Earth, produced contradictory results. This investigation found no correlation between CPR and lunar terrain's ability to retain ice in deep freeze. The authors say this implies no slabs or "lakes" of ice, but at best isolated crystals.[43]

Controversy between bistatic and Earth-based radar went unresolved for years until Mini-SAR radar orbited on ISRO's *Chandrayaan-1*. This instrument (and its companion on *LRO*) produced high-resolution CPR maps of both poles, and this detail is crucial. Many fresh craters show high CPR because of rough surfaces covered by ejected boulders on both the crater's flanks and interior. Near the poles, however, many craters show high CPR only in their interiors. This seems clear evidence for ice. In some craters this signal is patchy, perhaps like the ice itself, or just its form more as a frost in places than as solid thickness.

Why expect water ice in craters near the poles? The idea dates to Urey's 1952 book: some areas near the poles never see sunshine, so volatiles might condense there.[44] The Moon has essentially no seasons (with its rotation axis tilted only 1.54° from perpendicular to Earth's orbital plane around the Sun), so one lunar day (a month, actually) is nearly identical to any other. A day without sunlight implies all days are dark. In 1961 Watson, Murray, and Brown[45] showed water vapor might stick in such cold traps for geologically long times. (Amazingly, this also happens near the poles of otherwise hellish Mercury: see discussion in Chapter 13 [see note 25].) Water, more even than much heavier sulfur dioxide, carbon dioxide, or hydrogen chloride (although not mercury), sticks longer in these cold traps.[46] Water forms ice easily, 78°C warmer than ammonia with almost the same mass (usually 18 atomic mass units for H_2O vs. 17 for NH_3), and is so abundant (with hydrogen, most common element in the Universe, and oxygen, 45% of lunar soil's mass) that water ice should form the foundation of volatiles frozen near the poles.

Hydrogen, being the lightest element, offers another detection scheme. When cosmic rays (here, atomic nuclei) from distant space strike lunar soil, they often eject or *spall* a neutron, which has the same mass as a hydrogen nucleus (to within 0.1%). These neutrons bounce off nuclei in the regolith. If a neutron collides with a heavier nucleus (like oxygen), it recoils with most of its momentum, like a ball bouncing off a wall. If it strikes a hydrogen atom, however, it loses much momentum, like a cue ball on a pool table hitting another ball and stopping instantly. These spallation neutrons start with kinetic energies of millions of electron volts (eV), whereas only 5 eV will knock a hydrogen atom from a water molecule. The kinetic energy of a neutron at lunar regolith temperatures is about 0.02 eV. In between, these neutrons are *epithermal*, with tens, hundreds, or even thousands of eV. Epithermal neutrons usually bounce out of the regolith after several collisions, unless they hit hydrogen.

The first epithermal neutron detector seeking hydrogen on the Moon (indeed, any extraterrestrial body) was *Lunar Prospector*'s neutron spectrometer in 1998. It mapped epithermal neutrons over the entire Moon from polar orbit, spending much of its 18-month mission over the poles. It saw a deficit up to 4.6% in epithermal neutrons concentrated near the North Pole and more diffusely at the South Pole. The investigators estimated that it detected hundreds of millions of tonnes of water, or up to several billion tonnes depending on the model.[47]

Details matter in determining the amount of water detected by epithermal neutrons. Some questioned whether *Lunar Prospector* detected lunar hydrogen at all. For instance, fewer epithermal neutrons should accompany increased thermal neutrons, not reported by the 1998 paper.[48] It is complicated: if neutrons penetrate too deeply, they never exit the soil. Most neutrons sample less than 1 meter of regolith, whereas some sensitivity extends 2 meters deep. This depends on the energy of cosmic rays and their neutrons; higher energy neutrons are less affected by hydrogen. Indeed at more than 1,000,000 eV, neutrons are more sensitive to oxygen than hydrogen. Iron affects neutron flux, and elements such as rare-Earths gadolinium or samarium correlate with epithermal effects. Hodges (see note 48) argued high SiO_2 content could mimic hydrogen. However, one can detect hydrogen and use different energies to decide how deeply it extends underground. *Lunar Prospector* indicated that potential water on the Moon likely rests a large fraction of a meter underground, covered by dry soil. Later probes imply similar results; we will consider those soon.

With epithermal neutrons and radar pursuing lunar water, additional methods developed. One invokes techniques used on *Luna 24* samples by Akhmanova and colleagues (see note 37): absorption of light by hydration of dust grain mineral lattices. This absorption band occurs around 2.9 microns wavelength in the near infrared, caused when increasing vibrational (and rotational) energy of water molecules. Because these vibrations correspond in part to stretching and shrinking the bond between oxygen and a hydrogen atom, similar wavelengths are also absorbed by the free radical hydroxyl (OH). Water (and hydroxyl) also appears in Earth's atmosphere, so a similar atmospheric band absorbs at many of these wavelengths (not an identical list, because water molecules are distorted within the mineral matrix compared to Earth's atmosphere).

Faith Vilas and collaborators investigated several absorption bands from hydration in phyllosilicates – minerals common on Earth but rare on the Moon – consisting of silicate sheets containing hydroxyl and metals (magnesium, iron, aluminum, manganese, potassium, calcium, lithium), among these mica, serpentine, chlorite, and clays, for example, talc. These show 2.9-micron bands, plus overtones at half and two-thirds of this wavelength, and another at 0.7 micron from ionized iron transitions (Fe^{2+} into Fe^{3+}) as in hydrated minerals.[49] This 0.7-micron feature is used to classify phyllosilicates in the laboratory and identify them in hydrated asteroids.[50] Vilas and coauthors published several papers showing this absorption near the lunar poles.[51] Regolith hydration signals seem strongest on Sun-facing surfaces, suggesting hydrogen implanted as energetic solar-wind protons. They published the results in a refereed journal after a decade's struggle, and only then in one less known, available online only for paid subscribers. Most lunar scientists remained unaware of this discovery, but not because of the authors' obscurity: Dr. Vilas directed MMT Observatory, with one of the world's largest optical telescopes. Some readers dismissed the result as explained by other mineral features faking phyllosilicates; considering hydrated regolith was too extreme.

Vilas and collaborators used images from *Galileo*'s lunar flyby (headed to Jupiter), but confirmation of regolith hydration came from another spacecraft, to much surprise among lunar scientists. India's first lunar probe, *Chandrayaan-1*, carried U.S.-made M^3 (Moon Mineralogy Mapper). Not specified to find hydration, that is indeed what M^3 saw. Confirming this, investigators employed data from two other lunar flybys: *Deep Impact* heading to comet Tempel 1 and *Cassini* to Saturn. The M^3 spectrometer covered visual/infrared wavelengths, 0.4–3.0 microns, splitting light into 260 wavelength bins or "colors." At any instant M^3 viewed a 40 x 0.07 kilometer rectangle of Moon, dispersing this into 600 individual spectra, 0.07 kilometers wide. Milliseconds later, M^3 viewed the next 0.07-kilometer-wide strip passing beneath *Chandrayaan-1*, slowly compositing the entire lunar surface (in 300 orbital swaths) in each of 260 colors. Although the M^3 team knew of hydration issues,[52] wavelength coverage terminated clumsily amid the 2.9-micron band. Amazingly, hydration seems ubiquitous over the Moon, stronger near the poles. (The 3-micron cutoff made defining hydration's nature, H_2O versus OH or even water ice, difficult.) Hydration was so unexpected that confirmation was needed from earlier *Deep Impact* and *Cassini* missions: hydration everywhere across the Moon, varying in longitude depending on solar illumination such that solar wind might produce it. However, M^3 found variation toward the poles (where the Sun barely shines) in the opposite sense: some places near the poles show OH concentrations (or H_2O) up to 0.07%, versus typically 0.002% equatorially. *Deep Impact* and *Cassini* supported this (without distinguishing H_2O versus OH).[53]

The *Chandrayaan-1*/M^3 hydration result in 2009 created great excitement in the United States and India. Lunar hydration's detection by NASA instrumentation received public endorsement by the agency and also demonstrated scientific success for India's first extraterrestrial probe (despite its demise from overheating and power failure), marking first awareness by many of lunar H_2O (or at least hydroxyl). Indeed *Chandrayaan-1* deployed its *Moon Impact Probe* ten months before the M^3 announcement, and it too found water (not announced until the M^3 publication).

MIP carried a mass spectrometer (CHACE; Chandra Altitudinal Composition Explorer) to detect gas and found water from 45° latitude and 98 kilometers altitude down to impact in the South Pole's Shackleton crater – the first time any instrument found water vapor at the Moon since marginal detections by Apollo. Unfortunately it detected gases consistent with rocket exhaust contamination.[54]

Let us be clear: before *Chandrayaan-1* and separate from radar, neutron, or infrared spectroscopy, we knew lunar water existed, coming from inside the Moon – in the rocks. Despite skepticism about hydrated Apollo samples, we knew by 2008 of surprisingly large water/hydroxyl amounts locked in some lunar minerals. This derives from better instrumentation and techniques to dissect contents internal to rock samples, clarifying that some minerals have prodigious water content, separate from conceivable terrestrial contaminants, in Apollo samples.

Previous technique vaporized a sample for analysis, ionized it with an electron beam, and then ran these charged particles through electromagnetic fields in a mass analyzer to measure the charge/mass ratio of these atoms, molecules, and molecular fragments. A newer method, SIMS (secondary ion mass spectrometry), uses an ion gun to blast tiny subsamples from material being studied, sending the resulting shrapnel of ions into the mass analyzer. The ion gun focuses to a small spot scanned across the sample, analyzing each particle blast, thereby mapping each constituent across the sample face as the beam scans the sample. Rocks often contain a jumble of tiny mineral inclusions or grains; SIMS offers potent study of each mineral separately. (Electron beams carry more charge per unit kinetic energy so may problematically distort charge distribution over the sample.) SIMS maps compositional variation on micron scales, smaller than most grains.

Erik Hauri of the Carnegie Institution improved SIMS for volatiles and with Alberto Saal (Brown University) and collaborators persuaded NASA in 2007 (after years of trying) to fund analyzing lunar samples. They studied *picritic* glasses – picrite composed of olivine and pyroxene deep in the Moon – and glasses quenched rapidly at the lunar surface – tiny spherules from *fire fountains*, lava droplets spewing into space, solidifying, then hitting the ground as beads. Expanding gas propelled these eruptions; what could that gas be, originating deep in the Moon?

Surprisingly, gas erupting from the Moon was high in water and sulfur, even if low in carbon dioxide or monoxide. Fire fountain beads, typically submillimeter in diameter, came from *Apollo 15*'s green glass (of olivine) and *Apollo 17*'s famous orange soil (ilmenite/olivine). Slicing the beads in half and pecking an ion microprobe every 15 microns across their diameter, Saal and collaborators found volatiles concentrated on the beads' interior: volatiles had leaked out (presumably as the droplets flew through the vacuum), not leaked in (from contamination). The amounts were large: 115–576 parts per million of sulfur (presumably associated with SO_2) and 4–46 ppm water. There was scant chlorine (0.06–2 ppm) but significant fluorine (4–40 ppm), and surprisingly negligible carbon dioxide,[55] important in many theoretical models.[56] Modeling how these gases diffused based on their concentration profile across the beads, Saal and associates extrapolate 260–745 ppm of water originally in the beads and up to 700 ppm sulfur. These huge volatile amounts approach those found in basalt extruded from Earth's mid-ocean ridges.[57]

Some investigators suspected picritic glass beads might be anomalous, but hydration is also high in some crystalline lunar minerals. Francis McCubbin (Carnegie/University of New Mexico) and collaborators have studied apatite, a lattice of calcium and phosphate, with locations filled by negative ions (hydroxide, fluoride, chloride, or bromide). Apatite is familiar in defining level five of the "scratch" or Mohs mineral hardness scale. Using samples from *Apollo 14* and *15*, the researchers found hydroxyl at levels of 220–2,700 parts per million and up to 7,000 ppm (or 0.7%) for lunar meteorites recovered in northwest African deserts.[58]

Apatite forms at lower pressures, usually within 10 kilometers of the lunar surface (accounting for lower rock overburden pressures at $^1/_6$ g). Magma melts yielding apatite begin deep and differentiate; accounting for this the whole Moon's water content is probably lower. McCubbin and collaborators estimate 0.05–17 ppm water for the magma's interior source, higher than Apollo-era interior water estimates by 100 times or more. Saal and associates results imply 2–20 ppm. Hauri, Saal, and collaborators later published analyses based not on glass beads but on mineral crystals within them, not degassed and likely preserving undiluted magmatic water, reaching higher volatile concentrations: 615–1,410 ppm water, 612–887 ppm sulfur, and 50–68 ppm fluorine (as well as 1.5–3 ppm chlorine). Correcting to mantle melt concentrations, these correspond to 79–409 ppm water, 193–352 ppm sulfur, 7–26 ppm fluorine, and 0.14–0.83 ppm chlorine.[59]

Apatite and lava bead hydration is unexceptional. Anorthosite, dominating the crust, produced by the magma ocean itself, contains typically 2–6 ppm water, implying a magma ocean roughly 100 ppm water and a concentrated layer near 1% (urKREEP, see note 75). While apatite results are still being interpreted, even olivine inclusions in mare basalts typically hold about 100 ppm water.[60]

Results since 2007 of McCubbin and associates, Hauri, Saal, and collaborators revolutionize our understanding of the lunar interior's volatile content (especially water). As recently as 2006 the consensus value for bulk lunar water content was below 1 part per billion.[61] Most estimates now far exceed 1 part per million, a radical change (see note 75). Before these results sank in by 2010, most lunar scientists' attitude regarding water in the Moon was "extraordinary claims demand extraordinary evidence." Was water inside the Moon so extraordinary? No, beyond the Sun every major Solar System body displays water's presence, as do many lesser ones, including all comets and many asteroids. Mercury, with 430°C daytime temperatures but much like the Moon, hordes many billion tonnes of water ice at its poles.[62] Earth, ostensibly baked by the Giant Impact, is the "water planet." From whence came this water? For many years scientists imagined terrestrial water was delivered by comets, because magma was presumed to degas. Now we appreciate that loss of water vapor, entrained in convecting magma, is discouraged by the difficulty of forming bubbles that escape into space. The Big Whack seemingly desiccated the Moon, but how much? Are there loopholes? Besides, this is a theory, and a relatively young one (compared to, for instance, five-decades' delay accepting continental drift). Evidence before 2007 did not exclude lunar water to the extent commonly held and allowed for gas, even water vapor, leaking recently from the Moon, especially now that we know water helped drive lunar fire fountains. Many lunar mysteries remain. This has long fascinated me, been a major research effort of mine since 2006, and will be a focus of Chapter 9.

Recent lunar water results have been challenged; chlorine is crucial. Hydrogen and chlorine combine strongly, and hydrochloric acid (HCl) reacts strongly in magma melts. Chlorine consists of two stable isotopes, ^{35}Cl and ^{37}Cl (with 18 and 20 neutrons, respectively), which everywhere on Earth reside in a 76.8:24.2 abundance ratio, nearly without exception. Being lighter, ^{35}Cl vaporizes faster, but ^{37}Cl reacts more easily with H. Together these factors cancel in terms of chlorine isotopes remaining in melt or vapor, if hydrogen is present in greater abundance. Absent H, this balance breaks, and ^{37}Cl/^{35}Cl ratios can vary. This occurs on the Moon: some minerals have Earth's ^{37}Cl/^{35}Cl ratio, but others up to 2.4% more ^{37}Cl.[63] This suggests that hydrogen is rarer than chlorine, despite hydration results. Alternatively, the Moon may be nonuniform, either in ^{37}Cl/^{35}Cl ratio or hydrogen concentration. In this work nonuniformity is dismissed because of magma ocean mixing. (We discussed this assumption in Chapter 6.) There is no discussion of some magmas being hydrated.

Results indicate at least two and likely three lunar hydration sources. Hydroxyl resides in some minerals (Saal, Hauri; McCubbin and collaborators). Surface hydration, seen in infrared/optical absorption, may arise when solar wind protons join regolithic oxygen (Vilas; Pieters and collaborators). Hydrated asteroids and comets certainly hit the Moon, presumably adding atmospheric water vapor, temporarily. If those water molecules (or others) enter ultra-cold craters permanently shadowed near the poles, they stick. Water can also come from other sources.[64]

In 2006 any presence of water on the Moon was debated. (I overheard at the time that some would not believe lunar water existed until they drank it from a glass.) In April 2006 NASA proposed probing lunar water nearly as directly as pouring a glass. *LCROSS* (*Lunar CRater Observation and Sensing Satellite*) would slam into a permanently shadowed lunar crater and eject tons of regolith, and perhaps water. Relatively inexpensive, riding with *LRO* on the same rocket, *LCROSS* in part was the rocket, or at least its Centaur top stage. The *Centaur* was prepared to minimize contamination, purged of fuel, and guided to its doom by the *Shepherding Spacecraft* (*S-S/C* or *SSc*), a modified, detachable ring joining the *Centaur* to the payload (*LRO*). Hitting the Moon with 2.3 tonnes at 2.5 kilometers per second, the *Centaur* blasted a 25-meter crater, 3.5 meters deep, on October 9, 2009. *LRO* flew overhead, examining the debris. *S-S/C*, with instruments observing the impact, followed the *Centaur* to a nearby impact four minutes later.

LCROSS's results contained several surprises. Target site choice optimized epithermal neutron absorption signal and visibility from Earth (consistent with a permanently shadowed depression), but observations from Earth largely missed the event.[65] The densest debris stayed hidden behind a 2-kilometer-tall massif towering on South Pole-Aitken's edge. Signals from *S-S/C* were faint but informative. It carried five cameras covering the optical to mid-infrared, an optical photometer, and three ultraviolet to near-infrared spectrometers. Several of these detected a faint flash within 0.3 seconds of impact, then more brightening a couple of seconds later as an ejecta sheet rose into the sunlight. Thousands of tonnes were heated, typically to 1,000 K, and roughly 1,000 tonnes were sunlit. Strong emission lines indicated about 5.6% of the mass was water.[66]

LCROSS fascinates not just because of water on the Moon, increasingly accepted by 2009, but other substances: 5.7% carbon monoxide, 1.4% molecular hydrogen, 1.6% calcium, 1.2% mercury, and 0.4% magnesium. Sulfur forms hydrogen sulfide (H_2S) and SO_2, at levels 1/6 and 1/30 of water, respectively. Nitrogen resides in ammonia (NH_3), at 1/16 water's abundance. Trace amounts (less than 1/30 of water) are detected for ethane (C_2H_4), CO_2, methanol (CH_3OH), methane (CH_4), and OH.[67] Volatiles compose at least one-eighth of the soil mass. The poles differ radically from any part of the Moon we have visited or sampled.

LCROSS liberated abundant water, not by the glass but by multiple barrels full. Still, you would not want to drink it unfiltered, being tainted with chemical and even isotopic poisons (more on isotopes soon) – a carbonated soup stocked with metals and organics. These contaminants are scientifically interesting as well as potentially useful, but we are unready to deploy a backhoe and start mining. Complex processes must first be understood (more on this later). NASA and other agencies plan sending rovers and probes into these permanently shadowed regions (PSRs) soon. As compelling as *LCROSS* results are, we cannot presume comprehensive command of important processes; hints of unanticipated and fundamental factors beyond the simple story abound. Having denied lunar water's existence for decades, we presume suddenly that we can narrow our vision to confirming restricted predictions – not so fast.

LCROSS hints of greater volatile concentration at increasing depths, up to 3 meters underground.[68] Indeed epithermal neutron results indicate that the first meter's regolith depth contains a desiccated layer in the top 0.4 meters. *LRO*'s LEND (Lunar Exploration Neutron Detector) found 4% water-equivalent hydrogen in Cabeus, and *Lunar Prospector* Neutron Spectrometer (LPNS) even less,[69] versus 5.6 ± 2.9% dug out by *LCROSS*. These may indicate more water at greater depth at the impact.

Few realize how little correlation LEND and LPNS found between lunar soil hydrogen and PSRs. First, many PSRs (about 70%) betray little hydrogen, and secondly some strongest neutron-absorbing areas are not PSRs but highlands outside deep craters in polar regions but hundreds of kilometers from the poles.[70] Disagreement and controversy exist between LEND and LPNS, but spots exist where they agree in both of these cases.[71] There is no evidence that the non-PSR hydrogen sites mark impacts of volatile-laden asteroids or comets. Plausibly, hydrogen here leaked from inside the Moon rather than froze from the thin lunar atmosphere.

We posit several origins for lunar hydration: delivery by comets and asteroids, solar wind conversion, lunar interior (fire fountains, seepage, minerals), Solar System dust, even interstellar clouds. Surface hydroxyl totals more than 10 million tonnes instantaneously, and much more over lunar history.[72] OH/H_2O inside the Moon is huge, perhaps 40 trillion to quadrillions of tonnes,[73] relatively little reaching PSRs (holding at least a billion tonnes). Reasonable estimates of water in comets and asteroids hitting the Moon are 0.1–6 trillion tonnes, perhaps 10–100 billion tonnes surviving impact.[74] Interstellar giant molecular clouds might contribute significantly but not overwhelmingly.[75] Four or more basic sources might each provide at least 10% of known PSR water. Let us consider each and how to distinguish them.

If trillions or quadrillions of tonnes of water form inside the Moon, can a billion tonnes outgas and reach the poles? Water in the magma ocean behaves as an incompatible component, accumulating in the topmost urKREEP layer (pristine KREEP after 99% magma ocean crystallization) 50 kilometers subsurface, under the anorthositic crust.[76] This analysis implies average water content below 10 ppm; otherwise concentrations in hydrogen-attracting minerals exceed those seen. (This is enormous compared to 1 ppb favored before 2008; reconciliation with 100+ ppm results still awaits.) The oxygen concentration available to react (its *fugacity*[77]) determines which minerals form and may indicate hydrogen-dominating water in the mantle (although water can form later).

Could water collected at the crust's base reach the surface? Water-bearing magma could dehydrate by foaming while ascending to low pressure, degassing from the urKREEP. Also, rock can be reheated via later volcanism, as in mare, with volatiles baking out. We must consider more than water; sulfur dioxide contributed as much fire fountain gas, which might have also included carbon monoxide (and some carbon dioxide).[78] *LCROSS* released as much CO as H_2O, versus volatiles from comets (or solar wind interaction) from Cabeus crater. Comets consist more of water, outweighing CO by 5–100 times and CO_2 by 10–50 times.[79] Most carbon compounds are detected at roughly similar abundances in Cabeus as comets, CO being exceptional. High CO might betray gas from the interior; its chemistry is complex, needing further investigation.

Atomic argon still escapes the Moon, from radioactive decay, outgassing from the deep interior, possibly sweeping molecular gases with it (see Chapter 9). Gas transits the vacuum to PSRs from distant lunar surfaces (see note 90); gas inside the Moon might suffuse the regolith from below. Regolith's impermeability forces gas molecules into microscopically tortuous paths as they percolate through it. A single molecule requires roughly a trillion collisions with regolith grains before penetrating its entire depth. If a molecule sticks for even microscopic times to each grain it hits, the molecule's transit time is macroscopic. If the molecule sticks for a macroscopic time, its transit time can last geological intervals. Some gases can react with the regolith, not always a passive medium (Chapter 6).

In this way water is special. Unlike most substances, its freezing temperature is found in the regolith.[80] The equatorial temperature just underground is 250 K = –23.16°C, below freezing even at low pressures (more than 0.006 atmosphere[4]), and temperatures rise above freezing to about 5°C at the regolith base as a result of internal heating.[81] Water is "sticky" – with molecules attached to regolith long durations between each flight segment in their percolation, with highly temperature-dependent sticking times.[82] At 0°C this lasts about 1 microsecond; water transits the regolith in several days. At temperatures typical of lunar polar regions, about –140°C, sticking lasts about a day, so regolith transit requires billions of years. Suffused with water vapor, regolith grains grow a thin water film. If enough water molecules stall in transit and accumulate, ice will form. PSR temperatures reach −245°C, so ice can form even near the surface.

Ice forming underground depends strongly on latitude and how quickly water vapor enters the regolith. Primarily, vapor must reach minimal pressure, 0.006 atmosphere at 0°C, 0.0001 atmosphere at −50°C, and so forth. The highest seepage rate consistent with vapor density measured in the lunar atmosphere is about 0.1 gram per second. Given regolith's impermeability, water vapor entering the regolith's base flowing at 0.1 g s^{-1} rates will form ice (if the flow extends over a patch smaller in diameter than the regolith's thickness). Incoming flow can maintain this pressure over a certain area, limiting ice growth. An ice patch will add mass at its edges until so much vapor sublimes from its surface that this balances the incoming flow. At the equator, this area is small (as big as a football field for 0.1 g s^{-1}), but in polar regions it increases rapidly, for example, 30 square kilometers at 80° latitude (Figure 8.4).[83] The depth where ice forms may vary with latitude, 5–10 meters on the equator and more shallow near the poles (depending on details of regolith properties versus depth).

Water ice in regolith can alter it, however. Few experiments exist on lunar regolith's aqueous chemistry, because water was presumed absent. With air's oxygen and nitrogen gone, water etches into regolith grains along damage tracks from solar wind and cosmic rays, pulling calcium from the grains to fill interstitial spaces between them, maintaining grain sizes (Chapter 6) and frustrating water vapor from reaching the vacuum. Unsurprisingly, liquid water etches faster than vapor, but ice establishes a pseudo-liquid layer on grain surfaces.[84] Over time regolith hydration leading to interstitial space-filling may slow water vapor from escaping.

Over geological times water can leach silicates, which can migrate.[85] It may make calcium hydroxide, magnesium hydroxide, or iron (II) hydroxide ($Fe(OH)_2$), all alkaline. $Fe(OH)_2$ will not oxidize to insoluble rust, $FeO(OH)$. Anorthosite might make clay. Sulfur dioxide may outgas with water as sulfuric acid, making gypsum ($CaSO_4 \cdot 2H_2O$) from leached calcium. Theory suggests carbon dioxide, hence carbonic acid (H_2CO_3), implying olivine (($Mg,Fe)_2SiO_4$) or pyroxene (typically ($Ca,Mg,Fe)SiO_3$) can make talc ($Mg_3Si_4O_{10}(OH)_2$) or serpentine ($Mg_3Si_2O_5(OH)_4$ plus methane) or may slowly dissolve.

Hydration-affected minerals are seen, albeit rarely, mostly by *Apollo 16*, more highland-like. Two meters underground in core sample 60002 various oxyhydrates of iron were found.[86] Rusty Rock 66095, from a large boulder 1 billion years old but excavated only about 1 million years ago, contains various hydration-affected minerals, particularly goethite ($FeO(OH)$, also found by *Apollo 14*). Terrestrial contamination was suspected, then cometary impact; evidence disfavors both. Rusty Rock 66095's chemistry is mysterious, but fumarolic origin seems plausible.[87]

This complex chemistry may remain mysterious without samples or simulations, but water seems to plug regolith's interstices and hydrate its minerals. More complexity enters over geological time: the Moon cools, and the Sun grows hotter.[88] These make ancient temperatures colder just subsurface but temperature gradients steeper, with ice growth closer to the surface. Over time, ice might move deeper into the regolith, leaving a water-modified rock cap above it. Meanwhile regolith

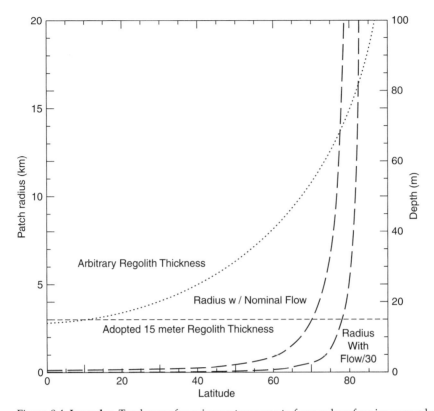

Figure 8.4. **Icy poles**. Tendency of seeping water vapor to form subsurface ice versus lunar latitude.[83] For point-source seepage, the ice patch radius formed is shown by the solid curve (use left axis) and the preferred depth by short dashed curve (use right axis). The arbitrary 15 m limit is adopted assuming that the low diffusivity regolith overlays a higher diffusivity megaregolith discouraging ice growth. If regolith is actually deeper, or if an ice cap might actually encourage ice growth at greater depth, the ice layer might extend to the dotted curve (reading right axis); this would likely encourage more ice growth at a given flow rate. (If regolith were surprisingly deep, the ice patch area might grow larger by a factor roughly the ratio of the dotted curve to the dashed curve.) The long-dashed curve is similar to the solid curve, showing the ice patch size if the flow rate is reduced by a factor of 30 (to 0.0033 g s^{-1} of water). This curve does not account for time required to reach equilibrium radius. Smaller than this flow, ice patches might not grow near the equator.

thickness grows, and impacts slowly fracture this rock cap.[89] If water vapor continues to flow, does it leak out or tend to fill the cracks?

This model predicted possibly hydrated materials toward the lunar poles, before they were dramatically revealed by *Chandrayaan-1*/M^3 in 2009, and predicts another effect: near the poles, one might expect patches of elevated hydrogen concentration even far from PSRs. (Neglecting that terrain slopes create small scales PSRs up to 32° from the poles.) These were found via neutron absorption (see note 69). Although other models may explain the first pattern, the second may require outgassing from the interior.

How do gas molecules propagate over the surface to the poles? They jump. The Moon has a *ballistic atmosphere* – uncharged atoms and molecules leap via thermal vibrations off the surface into elliptical orbits (nearly parabolic) usually until landing elsewhere on the Moon (see Chapter 9). Its atmosphere is a prime example of an unbounded exosphere (solid planet below, near perfect vacuum above). Except for hydrogen and helium, atoms/molecules bounce many times (or stick, or ionize), never reaching speeds sufficient to escape the Moon forever (more than 2.38 kilometers per second). Heavy gases bounce in small areas (for radon, only about 100 kilometers across, making it decay radioactively before reaching the poles).

How much water propagates to the poles? Ballistic transport models of water molecules find only small fractions sticking in PSRs,[90] with different studies varying by a factor of 10. Crider and Vondrak study water created by solar wind protons reacting with regolithic oxygen and find an intermediate value of 4.2% reaching a pole. Of water delivered by comets, roughly 95% escapes immediately before even joining the atmosphere. For other molecules, we expect to find PSR deposits of atoms/molecules of 10 to 100 atomic mass units (AMU): less than this, and molecules achieve escape velocity during lunar day; more than this, and they rarely reach polar regions.

With surficial regolith hydroxyl results in 2009, scientists realized that we should look at production, transport, and sticking of OH as much as H_2O. A roughly 100-nanometer rim of glassy, amorphous material covers regolith grains, often studded with refined-iron inclusions and covered by vapor-deposited volatiles. Solar wind protons typically penetrate roughly 10 nanometers so react in this rim, which is also the barrier overcome by escaping products. OH or H_2O production has been hypothesized for decades,[91] but recent analyses detail more how solar wind hydroxyl is produced (see note 72). Nonetheless, the solar wind's surface flux is smallest at the poles, but surficial hydroxyl concentrations are greatest there, despite daytime surface temperatures more than 0°C (at latitudes exceeding 15° from the poles). How solar wind hydroxyl produce observed polar concentrations has yet to be shown.

Hydroxyl on the Moon's surface seen by *Deep Impact*'s and *Cassini*'s infrared spectrometers hinted that hydroxyl concentrations vary during a lunar day and that surface temperatures matter (see note 53). Also, at any time of day different minerals generate or maintain OH by varying amounts: highlands more than mare but less than some fresh craters (but not all). Near Mare Orientale appears water and hydroxyl rich; South Pole-Aiken appears water/hydroxyl poor.[92] These effects require more study.

As intriguing as hydration by cometary/asteroidal or solar wind origin may be, the water reservoir inside the Moon was unanticipated by lunar scientists until several years ago. Water as an incompatible magma ocean component concentrating near crust's base (see note 76) may explain hydroxyl in apatite or anorthosite at shallow depths. For picritic fire fountain glasses and their inclusions within them, an origin at least 400 kilometers deep is indicated.[93] However, europium, which can act like calcium in anorthosite (floating to the surface early in the magma ocean), is relatively depleted in fire fountain glasses,[94] which shows that magma producing these glasses arose in the magma ocean below the crust (but above the earliest, deepest, olivine-rich layer).

Many seek caveats to explain water in the Moon without destroying the Giant Impact hypothesis. Models of water's magma ocean role support ideas that the Moon is drier, perhaps 10 ppm compared to Earth's several hundred ppm. This level might agree with the Giant Impact. As material fell back to Earth or congealed into the Moon, it formed an accretion disk primarily of material from the extraterrestrial impactor. Although this material was hot and lost most of its volatiles, some remained in the environment and with some entrained to support a magma ocean bulk composition of 10 ppm water.[95]

Some matter need not pass through the Giant Impact's violent heating. If proto-Earth or the Mars-sized impactor (Theia) had a satellite, it likely merged into the Moon with too little energy to dehydrate its rock (about 1,200 K). Such a satellite of proto-Earth would likely stay in Earth's orbit, and satellites of Theia (entering at proto-Earth's escape velocity) have slightly more than 50% probability of remaining in Earth's orbit (depending on whether their velocity points with or counters Theia's). Previous satellites might provide the core around which the Moon accreted and would heat insufficiently to catastrophically liberate volatiles.

Water deep in the Moon is intriguing, because it might explain the preponderance of deep lunar quakes – on Earth thought to originate with phase transitions caused by water.[96] Deep moonquakes arise in clusters 700–1,200 kilometers the Moon's surface, far below likely depths of the magma ocean.[97] There may be two interior water reservoirs: one consolidated from the drier magma ocean, and another from the deep Moon below.

Much knowledge of water's and hydrogen's behavior in the Moon is summarized by the oxygen fugacity f_{O_2}, the effective pressure of oxygen as if it were an ideal gas, measuring the availability of oxygen to react.[98] High oxygen fugacity implies highly oxidized systems. Oxygen fugacity can be measured by the equilibrium of oxides of iron, silicon, and/or titanium, for example, ilmenite dissociation ($FeTiO_3$). Lunar samples tend to imply mantle f_{O_2} values several orders of magnitude below those for Earth's mantle. For Earth this implies mantle hydrogen largely oxidized to water, but in the Moon one expects much more H_2. Because hydrogen diffuses much faster through mineral lattices than water, it is lost rapidly. With hydrogen loss, water vapor tends to dissociate (primarily by oxidizing iron). If magma is quenched rapidly, this might not occur. This might explain hydrated glass beads, but lunar fire fountain glasses show the highest f_{O_2} values of any lunar samples (see note 56). Elkins-Tanton and Grove[76] argue that hydrogen must be important, perhaps dominant, in partition with water, but still allows for significant water.[99] We might not appreciate how hydrogen is lost or processed, however.

Even tiny amounts of water inside the Moon have profound effects on some systems, particularly chlorine isotopes, mentioned earlier. Sharp and colleagues[100] argue water's importance in maintaining constant $^{35}Cl/^{37}Cl$ ratios on Earth, and its lunar heterogeneity implies water's absence from the lunar interior. Low lunar oxygen fugacity encourages decoupling of hydrogen and chlorine degassing, however, demonstrated experimentally,[101] or this may indicate high inhomogeneity, as a function of depth or selenographic location. Perhaps the Moon is more complex than we imagine.

How did lunar science stray on the issue of the Moon's water content for 40 years? Most still think the Moon is relatively dry versus Earth, perhaps by a factor of 30. How did this reach the discrepancy between bulk water content favored today, maybe 10 parts per million, and only a few years ago, a part per billion or less. Oxygen fugacity data indicate low hydration, but not lower by a factor of 10,000. As described, this rests primarily with the dismissal of any water as contamination. In general, persuasion employed a preponderance of evidence, not any clinching argument. Everything indicated that the Moon was dry: minerals (little amphibole or hematite), landforms (no glacial/aqueous alteration), atmospheric (only traces of water vapor), and compositional (only marginal water detections, still sometimes tens of parts per million). However, the case was overstated, by factors of thousands.

Returning to water seen in PSRs, its origin is amenable to isotopic analysis. Solar wind hydrogen is low in deuterium. Endogenous (that is, native) lunar oxygen has isotopic abundance ratios like Earth's. This offers a signature for solar proton-generated hydroxyl. Terrestrial water ranges in fractional deuterium enhancement (δD) versus the global ocean average by –50% to +10%, with most values at least twice as close to zero.[102] Cometary δD values cluster around +100% or larger, whereas chondritic meteorites span a wide range. Interstellar δD varies with molecule and state of matter in complex ways.[103] Meteoritic oxygen isotope ratios cluster close and symmetrically around lunar/Earth values, however.[104] In contrast, lunar apatite δD varies from +39.1% to +101%, with highlands apatite +24% to +34%, and apatites in lunar meteorites as low as +10%.[105] Some mare basalts fall to –17.2% to –21.5%. Some argue that lunar water comes from comets, but many processes affect hydrogen isotopes. δD agrees with comets, but they can change later; this is not the only interpretation.[106] A difficulty is poor understanding of where Earth's water originates (with δD like chondritic meteorites more than comets). It is hard to know why the Moon is different.

How will we make scientific progress? We imagine sampling volatiles in PSRs robotically, but contributions from candidate sources (meteoritic, cometary, solar wind interactions, endogenous, or interstellar) are mixed by regolith impact gardening, destroying the isolation of these different contributions. This might be ameliorated. In Chapter 6 we discussed topographic features sitting dramatically lower than adjacent terrain, like lava tube skylights (Figure 6.25). Unfortunately lava tubes are more products of mare; in polar highlands they would be old and damaged. Might other structures accumulate regolithic strata over time (called *secular deposition*) versus random mixing ("gardening")? We found that, at the base of slopes like crater walls, secular versus gardening improves approaching the wall, becoming greatest within 10–20 meters.[107] A 10–20 meter crater bottom is hit by secular deposition from all sides, perhaps optimal for strata deposited in a time-ordered way. We need a tool to estimate the relative importance of secular deposition, given topographic data (digital elevation models). Asteroidal/cometary impact, incursion of dense molecular interstellar clouds in the inner Solar System, episodes of anomalous solar wind flux, or endogenous outgassing could be recorded within these vertical strata. Such a tool might tell us where best to drill. Volatiles from various sources will be layered with different telltale substances (SO_2 and CO/CO_2 from outgassing, organics from comets/asteroids versus interstellar clouds). Together these time-ordered compositional and isotopic clues might suffice to describe lunar atmospheric sources, when, and how much.

Robotic drilling and sampling in PSRs is not the only way to settle these issues. Some more examples include the following:

1. An old important drill sample needs work, the unconfirmed hydration signal from *Luna 24*'s 2-meter core (see note 36). This should be probed with infrared spectroscopy, pyrolysis, and/or SIMS.
2. Samplers are not the sole mission architecture to elucidate lunar volatiles, especially water. Water's lunar surface interactions inherently involve regolith, and current probes extend a meter or less into lunar soil. Ground penetrating (or sounding) radar at 600 Megahertz could map regolith's full depth in meter detail and unearth signatures of subsurface ice or volatile-altered soil.
3. Ground-clutter problems (Figure 6.24) limit sounding radar to smoother areas. Other electromagnetic probes avoid this, yet probe the regolith's depth. Time-domain electromagnetic sounding, used on Earth to map soil in slices down to several meters, could fly on a satellite (but with surface resolution depending on altitude).
4. *Kaguya* carried an alpha particle spectrometer, sensitive to outgassing radon-222 and its daughter polonium-210, that delivered few useful data as a result of a power failure. A new, global alpha particle map is needed.
5. The *Lunar and Dust Environment Explorer* will study lunar atmospheric constituents in only a narrow equatorial strip. To understand hydroxyl and other species migrating to the poles, or to localize lunar outgassing sources, a mass spectrometer in low polar or nearly polar orbit is required.
6. A follow-up instrument to M^3 could finish mapping hydroxyl, study how it changes over the month, and extend wavelength range around 3 microns to distinguish hydroxyl versus water in mineral lattices versus water ice.
7. A more sensitive and resolved epithermal neutron absorption map is needed. These last five or six instruments could share a lunar polar orbiter.

Excitement over lunar volatiles arises not just from their science, but their potential utility. We see high concentrations in regolith's upper meters in PSRs and suspect up to 0.1% in the upper millimeter in larger polar regions. There may be more; if outgassing from the interior into regolith near the poles has or does occur, larger concentrated resources may rest several meters down. Potentially these sites may maintain temperatures of 150 K or 200 K, more like Antarctic temperatures than the cryogenic PSRs (20–50 K). We may find solid ice bodies, unlike surficial OH evident at 3 microns. Scenarios are cast for exploiting PSR or surficial hydration; subsurface reservoirs might be easier.

Why be excited about lunar hydration resources? There are several reasons – obvious and not so obvious. Lunar transport costs some $30,000 per kilogram; so necessary for life, earth water's price is prohibitive. In addition, water is propellant. With abundant electricity from solar energy, water provides oxygen and hydrogen, nearly optimal rocket bipropellants. Even for water not mined from PSRs, cold storage of oxygen and hydrogen liquid is easier (with critical temperatures 154.6 K and 33.2 K, respectively, warmer than many PSRs). In addition, oxygen and hydrogen make an easy "battery" for energy storage and electricity production in fuel cells. Water is useful together with regolith, with chemical and mechanical properties (for example, particle size distribution) usefully similar to cement, potentially making fully indigenous and adaptable building materials. Water also provides excellent shielding against high-energy protons, especially solar wind, and spallation products, for example, neutrons.

Efforts proceed to make lunar water an exploration resource, even commercially. Issues abound: the optimal hydration source is uncertain. Surficial hydroxyl is handy but poorly concentrated, several kilograms per hectare. Dense sources reside in PSRs (hundreds or thousands of tonnes per hectare), available only to machinery operating below 50 K. Water frozen into the regolith from outgassing needs exploration. Space law on lunar water use is unclear, maybe awaiting test cases before the United Nations or elsewhere to determine legal uses. Should water in PSRs serve as humanity's common legacy, not subject to exploitation by any one country or company? Beyond these hurdles, water from PSRs is tainted isotopically (deuterium fraction) and chemically (mercury, other heavy metals, perhaps organic poisons). Some PSR organic compounds promise value in their own right, with carbon and nitrogen rare on the Moon. Beyond material exploitation, PSRs' conditions themselves can be assets, among the coldest and most stable in the Solar System, where objects might remain unaltered for gigayears. Humanity's cultural or genetic legacies, or bodily remains of the super wealthy seeking personal pharoanic eternity, can settle the lunar poles with unprecedented security.

Abundant liquid oxygen and hydrogen offer many a boon. With this propellant, easiest transportation on the Moon beyond tens of kilometers is via a ballistic "hopper," accelerating and decelerating itself, and payload to/from ballistic arcs in an hour or less. A North and/or South Pole moon base could access the surface with robotic or human expeditions. Access to free space from the Moon is some 30 times easier than from Earth; the Moon offers a supply depot for missions preparing to accelerate toward distant Solar System destinations, with their parting impulse provided by lunar chemical fuel. An industry of spacecraft using lunar propellant transiting cislunar space could service satellites in geosynchronous orbit, which otherwise die after running empty and drifting from orbits useful for communications or remote sensing. "Space tugs" using lunar oxygen and hydrogen could restore them to operational orbits and extend their lifetimes manifold, then retreat to lunar vicinity for propellant recharge and a new cislunar cycle.[108]

Impressive progress made in understanding volatiles such as water comes primarily from new sample analysis techniques expanding on work largely from the 1970s. Further phenomena associated with volatiles in the lunar environment sit close to new discovery promising new views into the Moon's interior and evolution; we treat these next. The recent revolution in understanding of the Moon's origin and volatile content derives from several major influences: the new Apollo sample reanalyses, data from recent lunar missions, and gradual shifts in what lunar scientists would accept as evidence and hypotheses that they once dismissed. We delve into these science sociology issues in Chapter 9.

Notes

1. Aldebaran, in Arabic "the follower" of the Pleiades star cluster, is a binary star dominated by a red giant, the brightest star in the constellation Taurus. *The Explorers* continues:

   > "Weighted the curving boughs. The route of stars
   > Was our diversion, and the fate of Mars
   > Our grave concern; we stared throughout the night
   > On those uncolonized demesnes of light.
   > We read of stars escaping Newton's chain
   > Till only autographs of fire remain.
   > We aimed our mortal searchlights into space
   > As if in hopes to find a mortal face.
   > O little Earth, our village, where the day
   > Seemed all too brief, and starlight would not stay,
   > We were provincials on the grand express
   > That whirled us into dark and loneliness.
   > We thought to bring you wonder with a tale
   > Huger than those that turned our fathers pale.
   > Here in this lunar night, we watch alone
   > Longer than ever men have watched for dawn.
   > Beyond this meteor-bitten plain we see
   > More starry systems than you dream to be,
   > And while their clockwork blazes overhead
   > We speak the names we learned as we were bred;
   > We tell of places seen each day from birth –
   > Obscure and local, patios of the Earth!
   > O race of farmers, plowing year by year
   > The same few fields, I sometimes seem to hear
   > The far-off echo of a cattle bell
   > Against the cratered cliff of Arzachel,
   > And weep to think no sound can ever come
   > Across that outer desert from my home!"

   (*The Diamond Cutters and Other Poems* by Adrienne Cecile Rich, 1955, New York: Harper, 45.)

2. The notion is common, e.g., this anonymous, untitled poem from Edo period Japan (1603–1868):

The purity of the moonlight,	(*Oo zora no*
Falling out of the immense sky,	*Tsuki no hikari shi*
Is so great that it freezes	*Kiyokereba*
The water touched by its rays.	*Kage mishi mizu zo*
	Mazu koori keru)

   Additionally, Gustave Flaubert's magnum opus, *The Temptation of St. Anthony* (1874), describing the sky:
   "ANTHONY. 'Yet from below the vault seemed solid as a wall! – on the contrary I penetrate it, I lose myself in it!'
   (*And he beholds the moon, – like a rounded fragment of ice filled with motionless light.*)
   THE DEVIL. 'Formerly it was the sojourn of souls! Even the good Pythagoras adorned it with magnificent flowers, populated it with birds!'
   ANTHONY. 'I can see only desolate plains there, with extinct craters yawning under a black sky!'"
   Fortunately, this idea rarely becomes cosmological theory.

3. Nazism traces its origin to the German Workers' Party, sponsored by the Thule Society, a romantic, occultist group established near the end of World War I to promote racial supremacist and anti-Communist ideals and believing Aryan peoples arose in a northerly, icy land (Thule).

4. Water like most substances is characterized by a triple point in pressure and temperature, marking the minimum conditions at which a liquid can form (0.01°C, or just above freezing at sea level, and 0.006 atmospheres, the pressure about 40 km above sea level, but still about 200 trillion times denser than the Moon's surface atmosphere but about the surface pressure of Mars). Lunar daytime temperatures typically exceed 0.01°C, so pressure must be even higher for surface water ice to form or remain stable.

    Lunar water was dubious even in the nineteenth century: "The absence of any bodies of water on the moon is placed beyond doubt, both by actual telescopic examination and by inference from absence of clouds. There are no streams, lakes or seas. An eminent astronomer has remarked that the heat of the surface exposed to the sun would occasion a transfer of any water the moon might contain to its dark side, and that there may be frosts in this part, and perhaps running water near the margin of the illuminated portion. But in such a case, would not clouds appear about the margin at times in telescopic views?" ("On the Volcanoes of the Moon" by James D. Dana, 1846, *American Journal of Science and Arts*, 11, 335).

5. *The Moon in Modern Astronomy; A Summary of Twenty Years Selenographic Work, and a Study of Recent Problems* by Phillipp Fauth, 1907, London: A. Owen, 156.

6. "Harold Urey and the Moon" by Homer E. Newell, 1972, *Earth, Moon and Planets*, 7, 1; "Nickel for Your Thoughts: Urey and the Origin of the Moon" by Stephen G. Brush, 1982, *Science*, 217, 891.

7. *The Planets, Their Origin and Development* by Harold Clayton Urey, 1952, New Haven: Yale University Press.

8. "The Origin and Significance of the Moon's Surface" by Harold C. Urey, 1956, *Vistas in Astronomy*, 2, 1667. See also "The Moon's Surface Features," 1956, *Observatory*, 76, 232, which among other ideas suggests exploding an atomic bomb on the Moon's surface and collecting the resulting lunar meteorites reaching Earth.

9. Urey, one of the few Ranger project co-experimenters, had ready access to the data: "But if water were present on the moon, one may ask how much and for how long. Since river valleys or stream structures of any kind are not present on the moon, it seems certain that the amount was small and the time was short. Small effects of this kind could have been destroyed by the erosion processes shown to be present by the *Ranger 7* pictures. Could it be that the comparatively smooth floors of the maria are the beds of ancient temporary lakes? Their smooth structure has led most students of the subject to assume that the maria are lava flows, and anyone not subscribing to this view is compelled to try to devise other explanations for this smoothness. The *Ranger 7* pictures have made many people, including me, think seriously that Mare Cognitum consists of fragmented material rather than lava flow material. We must account for the crater Wargentin, which is full of smooth material to the brim. Could it be water or ice covered with some layer of dust and could it have become filled with water by temporary rains, and are its walls impervious to water while those of other craters are not? It has always seemed odd to me that the moon could produce hot lavas to fill Wargentin and at the same time be sufficiently rigid to support differences of 10 km in elevation of the lunar surface. Kopal and Gold have proposed that water has diffused from the lunar interior to fill the maria basins, and they compare this to water coming from the interior of the earth. However, water probably comes from the earth's interior through its numerous volcanoes and lava flows which have covered the original surface of the earth to a mean depth of some 15 km. Craig finds that terrestrial hot springs consist mostly or entirely of meteoric and not juvenile water. It is difficult to believe that diffusion of water through rocky material either of the earth or moon would supply more than very limited quantities of water to their surfaces. Also the estimates of the amounts of water that have come to the earth's surface during geological time rest on very uncertain evidence. But these suggestions of Kopal and Gold have stimulated me to consider the possibility that contamination of the moon with water from the earth was larger than I intended to suggest previously. Only the Surveyor and Apollo missions to the moon can answer the questions raised in this way." ("Meteorites and the Moon" by Harold C. Urey, 1965, *Science*, 147, 1262.) We show crater Wargentin in Figure 6.11.

10. "Study of the Ranger Pictures of the Moon" by Harold C. Urey, 1967, *Proceedings of the Royal Society of London, Series A*, 296, 418.

11. If L is the size of the system being considered, the rate at which momentum density ρv (where ρ is mass density and v the velocity) flows through area L^2 at velocity v is $\rho v^2 L^2$. Viscous force on an object at low velocity is proportional to v, size of the object L, with proportionality constant given by the viscosity μ, so $F_{drag} = \mu v L$. Reynolds number R is the ratio of momentum density to drag force: $R = \rho v^2 L^2 / \mu v L = \rho v L / \mu$, which is dimensionless. If $R \leq 500$, the flow is turbulent, usually the case for water. Basaltic magma is at least some 10,000 times more viscous than water. Magma that is *rhyolitic* (alternatively *siliceous* – high in silica) can reach μ values 10 million times water's. R exceeds 500 for basaltic lava only if the flow width L and mass flow rate

(which determines v) are also large, but almost never does so for siliceous magma. ("Lunar Sinuous Rille Formation by Thermal Erosion: Conditions, Rates and Duration" by J. W. Head & L. Wilson, 1981, *Abstracts of Lunar and Planetary Science Conference*, 12, 427; "The Formation of Eroded Depressions Around the Sources of Lunar Sinuous Rilles: Theory" by J. W. Head & L. Wilson, 1980, *Abstracts of Lunar and Planetary Science Conference*, 11, 1260.)

12. "The possibility that water has existed on the Moon for varying lengths of time, both in liquid and in solid form, and both beneath the surface and on the surface, has been widely discussed during the past 10 years. The subject has been discussed repeatedly at scientific meetings and has been received mostly with great skepticism. Evidence supporting this view has recently become quite overwhelming and, in fact, no communication seems necessary to point out the evidence from the *Orbiter* [4] and [5] pictures. Because many people are not aware of this evidence and suggest that the effects are caused by other liquids, that is, lava, dust-gas or possibly even vodka, a brief discussion of the evidence may be in order" ("Water on the Moon" by Harold C. Urey, 1967, *Nature*, 216, 1094).

13. "Water on the Moon and a New Nondimensional Number" by John A. O'Keefe, 1969, *Science*, 163, 669.

14. Urey retorted to O'Keefe: "All right, attack if you wish to. This is, so far as I recall, my suggestion, not that of my good friend, T. Gold. Possibly Lingenfelter et al. considered some modification of this idea. I am not at all convinced that Gold's mechanism may not contribute to the problem to some extent" ("Water on the Moon" by Harold C. Urey, 1969, *Science*, 164, 1088). Lingenfelter and colleagues postulated that a 1 km thick ice layer on the Moon, shielded from sublimation into the vacuum by a 100 m overburden of regolith, could melt when impacted by a meteorite and flow underground to form rilles as in Figure 8.2 ("Lunar Rivers" by Richard E. Lingenfelter, Stanton J. Peale & Gerald Schubert, 1968, *Science*, 161, 266). Gold detailed similar ice sheet/overburden geometry to explain odd-looking lunar craters that seemed to contain features similar to *pingos*, ice mounds caused by water forced out of permafrost. ("The Moon's Surface" by T. Gold, in *The Nature of the Lunar Surface*, edited by Wilmot N. Hess, Donald H. Menzel & John O'Keefe, 1966, Baltimore: Johns Hopkins Press, 107.)

15. From discussion after an address by Harold Urey in Philadelphia on April 24, 1970:

    QUESTIONER: "What is the origin of the crooked rilles? Possibly water?"

    UREY: "The crooked rilles on the Moon looked to me, for a long time, as though they were due to liquid water. Since water cannot be found in the rocks, these rilles must have been produced by melted silicates of some kind flowing across the surface of the Moon. I am immensely surprised that such materials as this would flow for 250 kilometers across the surface of the Moon without freezing, and at the end, just disappear somewhere. On the other hand, other people suggest that the rilles are due to lava flows from the interior of the Moon. In fact, I think we must be rather skeptical in regard to all sources of water on the Moon, unless we get other evidence pointing in a different direction; water on the Moon seems not to have been present in any important amounts at any time in the past. We may have to revise that opinion again in the future, but that certainly is my opinion at the present time" ("A Review of the Structure of the Moon" by Harold C. Urey, 1971, *Proceedings of the American Philosophical Society*, 115, 67). Urey wrote more papers about the Moon but never again about water. Even Tommy Gold at this April 1970 symposium ceased mention of water but persisted in his theory of the lunar surface dominated by dust transported by electrostatic force ("The Nature of the Lunar Surface: Recent Evidence" by Thomas Gold, 1971, *Proceedings of the American Philosophical Society*, 115, 74).

16. ^{40}K, potassium with 21 neutrons (and 19 protons) decays with a half-life of 1.277 billion years, either becoming ^{40}Ca (with 20 protons) by emitting an electron (β particle) in 89.28% of decays or by capturing an electron out of its atomic orbital to become ^{40}Ar (with 18 protons) in 10.72% of decays (unless the ^{40}K is ionized). Thus after 1.277 billion years, from a sample of ^{40}K atoms, 50% will still be ^{40}K, 44.64% will be ^{40}Ca, and 5.36% will be ^{40}Ar, an inert gas at temperatures above 87.3 K at 1 atmosphere pressure, and above 150.87 K at under 50 atmospheres.

17. "Molecular Gas Species in the Lunar Atmosphere" by J.H. Hoffman & R.R. Hodges, Jr., 1975, *The Moon*, 14, 159. The molecular water number density at sunrise (the peak signal) is 600 ± 300 cm^{-3} (1σ) versus ^{40}Ar with 30,000 cm^{-3} at sunrise, which drops to under 1,000 cm^{-3} at night.

18. "H$_2$O in Lunar Processes: The Stability of Hydrous Phases in Lunar Samples 10058 and 12013" by R. W. Charles, D. A. Hewitt & D. R. Wones, 1971, *Proceedings of 2nd Lunar Science Conference*, 1, 645.

19. "Deuterium, Hydrogen and Water Content of Lunar Material" by L. Merlivat, M. Lelu, G. Nief & E. Roth, 1974, *Geochimica et Cosmochimica Acta*, 2, 1885.

20. "$^{18}O/^{16}O$, $^{30}Si/^{28}Si$, $^{13}C/^{12}C$ and D/H Studies of *Apollo 14* and *15* Samples" by S. Epstein & H. P. Taylor, Jr., 1972, *Proceedings of 3rd Lunar Science Conference*, 2, 1429. We discuss more recent measurements of D/H shortly, which tell a slightly different story.
21. If one strips away increasingly deeper layers of regolith grains (by reacting them with fluorine gas), the $^{18}O/^{16}O$ ratio is at first very high, then decreases to the terrestrial value as center of the grains is reached ("$^{18}O/^{16}O$, $^{30}Si/^{28}Si$, $^{13}C/^{12}C$ and D/H Studies of *Apollo 16* Lunar Samples" by S. Epstein & H. P. Taylor, Jr., 1973, *Abstracts of Lunar and Planetary Science Conference*, 4, 228).
22. "D/H and $^{18}O/^{16}O$ Ratios of H_2O in the 'Rusty' Breccia 66095 and the Origin of 'Lunar Water'" by S. Epstein & H. P. Taylor, Jr., 1974, *Proceedings of 5th Lunar Science Conference*, 2, 1839.
23. From Alan Bean after *Apollo 12*: "Closing of the sample return containers was not difficult and was similar to that experienced during one-sixth g simulations in an airplane. The seal for the sample return container lid became coated with considerable dust when the documented samples were being loaded into the container. Although the surface was then cleaned with a brush, the container did not maintain a good vacuum during the return to earth" (*Apollo 12 Mission Report*, March 1970, Manned Space Center, Houston, NASA MSC-01855, p 9/19).
24. From Lunar Curator of samples at NASA Johnson Space Center, Gary Lofgren: "Most of the Apollo Lunar Sample Return Containers (ALSRCs) sealed." He lists Earth-received status of all 12 boxes; one of two ALSRCs on each of *Apollo 12, 14, 15*, and *16* leaked; eight maintained less than 0.00025 atmosphere ("Overview and Status of the Apollo Lunar Collection" by Gary E. Lofgren, 2009, *Lunar Exploration and Analysis Group* meeting, #2075; also *Catalog of Apollo Experiment Operations*, Thomas A. Sullivan, January 1994, NASA Reference Publication, 1317, 60).
25. For instance, "Many also believe a lunar sample return will be necessary. True, the Apollo astronauts brought back some 800 pounds of lunar rocks from six landing sites. But the dust played a dirty trick: The gritty particles deteriorated the knife-edge indium seals of the bottles that were intended to isolate the rocks in a lunar-like vacuum. Air has slowly leaked in over the past 35 years. 'Every sample brought back from the moon has been contaminated by Earth's air and humidity,' Olhoeft says. The dust has acquired a patina of rust, and, as a result of bonding with terrestrial water and oxygen molecules, its chemical reactivity is long gone" ("Stronger than Dirt: Lunar Explorers Will Have to Battle an Insidious Enemy – Dust" by Trudy E. Bell, September 1, 2006, *Air and Space Magazine*).
26. Special Environment Sample Containers (SESCs) excluded atmospheric contamination (with a stainless steel edge pressed into an indium seal mentioned earlier[24]). About half of the samples in the six SESCs are unused, with one container completely unopened. Four other containers for regolith cores and other samples are also pristine. ("Special, Unopened Lunar Samples: Another Way to Study Lunar Volatiles" by Gary E. Lofgren, 2011, *A Wet vs. Dry Moon: Exploring Volatile Reservoirs and Implications for the Evolution of the Moon and Future Exploration*, #6041).
27. "That's terrestrial water, which has a much, much larger amount of deuterium. We kept on trying to get down to the least contaminated samples, but we couldn't find any lunar soil sample that didn't have a little tiny bit of deuterium in it, and we finally concluded that it was just contamination from the astronauts' backpacks, because the samples were carefully preserved on the way back. And after they got to Houston, we don't think it was added, although you can never be sure, because it doesn't take much contamination to put a little bit of deuterium in there." (Hugh P. Taylor, 1932, interviewed by Shirley K. Cohen, June – July 2002, Archives of California Institute of Technology, Pasadena, p. 83)
28. "Water on the Moon?" by Edward Anders, 1970, *Science*, 169, 1309.
29. Some context: Earth's igneous minerals range from tens of ppm to more than 5% water. Fresh basalts in mid-ocean ridges range more than 0.1%–0.5% water, but from hot mantle plumes can be several times wetter ("Recycled Dehydrated Lithosphere Observed in Plume-Influenced Mid-Ocean-Ridge Basalts" by Jacqueline Eaby Dixon, Loretta Leist, Charles Langmuir & Jean-Guy Schilling, 2002, *Nature*, 420, 385). Earth's upper mantle is likely 20–200 ppm water ("Water in Anhydrous Minerals of the Upper Mantle: A Review of Data of Natural Samples and Their Significance" by Anne H. Peslier, 2007, *Workshop on Planetary Basalts*, #2003). Martian meteorites indicate an origin in Mars' mantle with 70–300 ppm. ("Hydrous Melting of the Martian Mantle Produced both Depleted and Enriched Shergottites" by F. M. McCubbin, E. H. Hauri, S. M. Elardo, K. E. Vander Kaaden, J. Wang & C. K. Shearer, 2012, *Geology*, 40, 683–686). "Dry as a bone" is not so dry: animal bone matrix is some 0.55% water ("Water Content Measured by Proton-Deuteron Exchange NMR Predicts Bone Mineral

Density and Mechanical Properties" by Maria A. Fernández-Seara, Suzanne L. Wehrli, Masaya Takahashi & Felix W. Wehrli, 2004, *Journal of Bone and Mineral Research*, 19, 289).
30. "Rusty Rock 66095 – A Paradigm for Volatile-Element Mobility in Highland Rocks" by Robert H. Hunter & Lawrence A. Taylor, 1982, *Lunar and Planetary Science Conference*, 12, 261, and "Rust and Schreibersite in Apollo 16 Highland Rocks – Manifestations of Volatile-Element Mobility" by Robert H. Hunter & Lawrence A. Taylor, 1982, *Lunar and Planetary Science Conference*, 12, 253.
31. "The Origin and Stability of Lunar Goethite, Hematite and Magnetite" by Richard J. Williams & Everett K. Gibson, 1972, *Earth and Planetary Science Letters*, 17, 84. See also "The Occurrence of Geothite in a Microbreccia from the Fra Mauro Formation" by S. O. Agrell, J. H. Scoon, J. V. P. Long & J. N. Coles, 1972, *Abstracts of the Lunar and Planetary Science Conference*, 3, 7.
32. "Volatile-Rich Lunar Soil: Evidence of Possible Cometary Impact" by Everett K. Gibson, Jr. & Gary W. Moore, 1973, *Science*, 179, 69.
33. "Observations of Water Vapor Ions at the Lunar Surface" by J. W. Freeman, Jr., H. K. Hills, R. A. Lundeman & R. R. Vondrak, 1973, *Earth, Moon and Planets*, 8, 115.

    Another event on February 22 1973, by LACE (Lunar Atmospheric Composition Experiment) at *Apollo 17*'s site showed un-ionized CH_4, C_2H_6, and N_2 (or CO), apparently releasing 10–500 kg of gas, 100–300 km from *Apollo 17*, a crowded neighborhood: *Luna 21* is in this annulus, with *Apollo 11*, *Surveyor 5*, and *Ranger*s 6 and 8 within some 600 km. Many contamination sources are relevant ("Summary of Conference: Interactions of Interplanetary Plasma with the Modern and Ancient Moon" by David R. Criswell & John W. Freeman, 1975, *The Moon*, 14, 3; see p. 12).
34. The Charged Particle Lunar Environment Experiment is sensitive to ions and electrons of 60–50,000 eV in energy, coming from two directions (zenith and 60° west of zenith). The particles were detected in the low energy channel (60–300 eV per electronic charge).
35. "The Apollo Lunar Surface Water Vapor Event Revisited" by J. W. Freeman, Jr. & H. K. Hills, 1991, *Geophysical Research Letters*, 18, 2109. In 2009 with new evidence of lunar surface hydration, Freeman said the new findings were consistent with their earlier results, seemingly contradicting Freeman and Hills, 1991. ("Water on the Moon? Pfft. We Saw that 40 Years Ago" by Eric Berger, September 24, 2009, SciGuy blog, *Houston Chronicle*, http://www.chron.com). Note that the astronauts "backpack" (PLSS) emitted up to 4 kg of water vapor.
36. *Luna 15* and *Luna 23* had both landed in Mare Crisium but failed.
37. "Water in the Regolith of Mare Crisium (Luna 24)?" by M. V. Akhmanova, B. V. Dement'ev & M. N. Markov, February 1978, *Geokhimiia*, 285, and "Possible Water in Luna 24 Regolith from the Sea of Crises" by M. V. Akhmanova, B. V. Dement'ev & M. N. Markov, 1978, *Geochemistry International*, 15, 166. I asked Boris Dement'ev in March 2010 about *Luna 24*'s core sample integrity; he said the sample might have absorbed water vapor in the laboratory, but their tests showed that it had little such tendency. It was transferred in a lab coat/glove environment, not anhydrous glove boxes like for Apollo samples. *Luna 24* would be an excellent case for pyrolytic mass spectrometry to distinguish terrestrial hydration from that strongly bound in the regolith.

    Luna 24 was propelled by UDMH (unsymmetrical dimethylhydrazine: $(CH_3)_2NNH_2$) and dinitrogen tetroxide (N_2O_4) plus nitric acid (HNO_3), burning to N_2, H_2O, and CO_2, suggesting the site might be contaminated by water vapor from rocket exhaust. To hydrate to 0.1% the volume of regolith to 2 meters depth requires at least 15 kg of water, versus about 300 kg of propellant used to land (hence about 100 kg of water). The great majority of water vapor exhaust escaped directly into the vacuum, so the core sample's hydration by exhaust seems unlikely.
38. Recently, we learned where *Luna 24* actually landed on the flank of a 64 m diameter crater. LROC shows *Luna 24* only 15 m beyond this crater's lip, which probably excavated material at this point from about 12 m below the surface. Scientists conjecture this is a secondary crater from the Giordano Bruno impact several million years ago and 1,300 km away.
39. *Explorer 49* entered lunar orbit six months after *Apollo 17*. Its primary mission was radio astronomy, not lunar science. The only other mission to visit the Moon between Apollo and *Clementine* was *Hiten* in October 1991, primarily a technology demonstration, carrying only one scientific instrument, the Munich Dust Counter.
40. Bistatic lunar radar was performed for other reasons on *Lunar Orbiter 1*, *Luna 11* and *12* in 1966, *Explorer 35* in 1967–1969, *Luna 14* in 1968, and *Apollo 14–16* in 1971–1972. See "Spacecraft Studies of Planetary Surfaces Using Bistatic Radar" by Richard A. Simpson, 1993, *IEEE Transactions on Geoscience and Remote Sensing*, 31, 465, for history and theory.

41. "The Clementine Bistatic Radar Experiment" by S. Nozette, C. L. Lichtenberg, P. Spudis, R. Bonner, W. Ort, E. Malaret, M. Robinson & E. M. Shoemaker, 1996, *Science*, 274, 1495; and "Integration of Lunar Polar Remote-Sensing Data Sets: Evidence for Ice at the Lunar South Pole" by Stewart Nozette, Paul D. Spudis, Mark S. Robinson, D. B. J. Bussey, Chris Lichtenberg & Robert Bonner, 2001, *Journal of Geophysical Research*, 106, 23253.
42. "No Evidence for Thick Deposits of Ice at the Lunar South Pole" by Donald B. Campbell, Bruce A. Campbell, Lynn M. Carter, Jean-Luc Margot & Nicholas J. S. Stacy, 2006, *Nature*, 433, 835.
43. "Results of the Mini-SAR Imaging Radar, *Chandrayaan-1* Mission to the Moon" by P. D. Spudis, D. B. J. Bussey, B. Butler, L. Carter, M. Chakraborty, J. Gillis-Davis, J. Goswami, E. Heggy, R. Kirk, C. Neish, S. Nozette, W. Patterson, M. Robinson, R. K. Raney, T. Thompson, B. J. Thomson & E. Ustinov, 2010, *Lunar and Planetary Science Conference*, 41, 1224.
44. *The Planets: Their Origin and Development* by Harold C. Urey, 1952, New Haven: Yale University Press, 17.
45. Scientists use knowledge of the Moon to teach themselves about Mercury, not just by analogy but by using lunar techniques to analyze Mercurian data. Mercury has lunar-like highlands, maria and perhaps a similar interior (with a larger core), a similarly thin atmosphere, and permanently shadowed craters near the poles containing some trillion tonnes of water ice. Mercury's axial tilt is even closer to perpendicular to its orbital plane, only 0.01 degree off. These similarities allow investigators to use similar computer models for how sunlight, topography, regolith, and heat allow ice formation, replacing lunar data with that from *MESSENGER* orbiting Mercury. Among other discoveries this indicates abundant organic molecules, e.g., "Thermal Stability of Frozen Volatiles in the North Polar Region of Mercury" by D. A. Paige, J. K. Harmon, D. E. Smith, M. T. Zuber, G. A. Neumann & S. C. Solomon, 2012, *Lunar and Planetary Science Conference*, 43, 2875.
46. "On the Possible Presence of Ice on the Moon" by Kenneth Watson, Bruce Murray & Harrison Brown, 1961, *Journal of Geophysical Research*, 66, 1598; "The Behavior of Volatiles on the Lunar Surface" by Kenneth Watson, Bruce Murray & Harrison Brown, 1961, *Journal of Geophysical Research*, 66, 3003.
47. "Fluxes of Fast and Epithermal Neutrons from Lunar Prospector: Evidence for Water Ice at the Lunar Poles" by W. C. Feldman, S. Maurice, A. B. Binder, B. L. Barraclough, R. C. Elphic & D. J. Lawrence 1998, *Science*, 281, 1496.
48. "Reanalysis of Lunar Prospector Neutron Spectrometer Observations over the Lunar Poles" by R. Richard Hodges, Jr., 2002, *Journal of Geophysical Research*, 107, 5125; "Improved Modeling of Lunar Prospector Neutron Spectrometer Data: Implications for hydrogen Deposits at the Lunar Poles" by David J. Lawrence, W. C. Feldman, R. C. Elphic, J. J. Hagerty, S. Maurice, G. W. McKinney & T. H. Prettyman, 2006, *Journal of Geophysical Research*, 111, E08001; "Correlation of Lunar South Polar Epithermal Neutron Maps: Lunar Exploration Neutron Detector and Lunar Prospector Neutron Spectrometer" by T.P. McClanahan, 2010, *Lunar and Planetary Science Conference*, 41, 1395.
49. Liquid water's absorption features shift about 0.06 micron to longer wavelengths versus corresponding water vapor features, like features for water in mineral lattices. Ice features shift even further. Water features also appear at 0.73, 0.82, 0.93, 0.97, 1.45, and 1.95 micron.
50. "Classification of Iron Bearing Phyllosilicates Based on Ferric and Ferrous Iron Absorption Bands in the 400–1300 nm Region" by L. Stewart, E. Cloutis, J. Bishop, M. Craig, L. Kaletzke & K. McCormack, 2006, *Lunar and Planetary Science Conference*, 37, 2185; "Phyllosilicate Absorption Features in Main-Belt and Outer-Belt Asteroid Reflectance Spectra" by Faith Vilas & Michael J. Gaffney, 2006, *Science*, 246, 790.
51. "A Search for Phyllosilicates Near the Lunar South Pole" by E. A. Jensen, F. Vilas, D. L. Domingue, K. R. Stockstill, C. R. Coombs & L. A. McFadden, 1996, *Bulletin of American Astronomical Society: Division of Planetary Sciences*, 28, 1123; "Evidence for Phyllo-silicates near the Lunar South Pole" by Faith Vilas, E. Jensen, Deborah Domingue, L. McFadden, Cassandra Coombs & Wendell Mendell, 1998, *Workshop on New Views of the Moon*, Houston: Lunar and Planetary Institute, 73; "Aqueous Alteration on the Moon" by F. Vilas, D. L. Domingue, E. A. Jensen, L. A. McFadden, C. R. Coombs & W. W. Mendell, 1999, *Lunar and Planetary Science Conference*, 30, 1343; "A Newly-Identified Spectral Reflectance Signature near the Lunar South Pole and the South Pole-Aitken Basin" by F. Vilas, E. A. Jensen, D. L. Domingue, L. A. McFadden, C. J. Runyon & W. W. Mendell, 2008, *Earth, Planets and Space*, 60, 67.
52. See http://m3.jpl.nasa.gov/Volatiles/ (retrieved before the *Chandrayaan-1* mission).
53. "Character and Spatial Distribution of OH/H_2O on the Surface of the Moon Seen by M^3 on *Chandrayaan-1*" by C. M. Pieters et al., 2009, *Science*, 326, 568; "Detection of Adsorbed Water and Hydroxyl on the Moon" by Roger N. Clark, 2009, *Science*, 326, 562; "Temporal and Spatial Variability of Lunar Hydration as Observed by the Deep

Impact Spacecraft" by Jessica M. Sunshine, Tony L. Farnham, Lori M. Feaga, Olivier Groussin, Frédéric Merlin, Ralph E. Milliken & Michael F. A'Hearn, 2009, *Science*, 326, 565.

54. *Chandrayaan-1*'s main engine and thrusters used monomethyl hydrazine and mixed oxides of nitrogen (MON-3; 97% dinitrogen tetroxide, http://www.isro.org/chandrayaan1/page19.aspx), making nitrogen, carbon dioxide, and water: $4CH_3NHNH_2 + 5N_2O_4 \rightarrow 9N_2 + 4CO_2 + 12H_2O$, with masses 18, 28, and 44 AMU. CHACE saw peaks at 17–18, 28, and 44 AMU, also at 1 AMU ("The Sunlit Lunar Atmosphere: A Comprehensive Study by CHACE on the Moon Impact Probe of *Chandrayaan-1*" by R. Sridharan, S. M. Ahmed, Tirtha Pratim Das, P. Sreelatha, P. Pradeepkumar, Neha Naik & Gogulapati Supriya, 2010, *Planetary and Space Science*, 58, 1567). *Apollo 17*'s LACE found scarce H_2O, N_2, or CO_2 but in this mass range mainly ^{40}Ar and ^{20}Ne, which should also dominate at CHACE's altitude ("Molecular Gas Species in the Lunar Atmosphere" by J. H. Hoffman & R. R. Hodges, Jr., 1975, *Moon*, 14, 159). The CHACE signal appears contaminated.

55. "The Volatile Contents (CO_2, H_2O, F, S, Cl) of the Lunar Picritic Glasses" by A. E. Saal, E. H. Hauri, M. J. Rutherford & R. F. Cooper, 2007, *Lunar and Planetary Science Conference*, 38, 2148; "Volatile Content of Lunar Volcanic Glasses and the Presence of Water in the Moon's Interior" by Alberto E. Saal, Erik H. Hauri, Mauro L. Cascio, James A. Van Orman, Malcolm C. Rutherford & Reid F. Cooper, 2008, *Nature*, 454, 192.

56. "The Driving Mechanism of Lunar Pyroclastic Eruptions Inferred from the Oxygen Fugacity Behavior of *Apollo 17* Orange Glass" by Motoaki Sato, 1979, *Lunar and Planetary Science Conference*, 10, 311 (page 321 especially).

57. "Recycled Dehydrated Lithosphere Observed in Plume-Influenced Mid-Ocean-Ridge Basalts" by Jacqueline Eaby Dixon, Loretta Leist, Charles Langmuir & Jean-Guy Schilling, 2002, *Nature*, 420, 385.

58. "Nominally Hydrous Magmatism on the Moon" by Francis M. McCubbin, Andrew Steele, Erik H. Hauri, Hanna Nekvasil, Shigeru Yamashita & Russell J. Hemley, 2010, *Proceedings of the National Academy of Sciences*, 107, 11223.

59. "High Pre-Eruptive Water Contents Preserved in Lunar Melt Inclusions" by Erik H. Hauri, Thomas Weinreich, Alberto E. Saal, Malcolm C. Rutherford & James A. Van Orman, 2011, *Science Express*, 10.1126/science.1204626.

60. "Equilibrium-Exchange Apatite Hygrometry and a Solution to the Lunar Apatite Paradox" by J. W. Boyce, S. M. Tomlinson, F. M. McCubbin, J. P. Greenwood & A. H. Treiman, 2014, *Lunar and Planetary Science Conference*, 45, 2096; "Water in Lunar Anorthosites and Evidence for a Wet Early Moon" by Hejiu Hui, Anne H. Peslier, Youxue Zhang & Clive R. Neal, 2013, *Nature Geoscience*, 6, 177; "Initial Water Concentration and Degassing of Lunar Basalts Inferred from Melt Inclusions in Olivine" by Y. Chen & Y. Zhang, 2012, *Lunar and Planetary Science Conference*, 43, 1361. Note that Hui et al. used Genesis Rock.

61. "Earth–Moon System, Planetary Science and Lessons Learned" by S. Ross Taylor, Carle M. Pieters & Glenn J. MacPherson, in *New Views of the Moon*, edited by Bradley L. Jolliff, Mark A. Wieczorek, Charles K. Shearer & Clive R. Neal, 2006, Chantilly, VA: Mineralogical Society of America, 657–704, summarizing on page 663: "The Moon is dry, with less than one ppb water, except for some possible amounts trapped in permanently shadowed craters."

62. "External Sources of Water for Mercury's Putative Ice Deposits" by Julianne I. Moses, Katherine Rawlins, Kevin Zahnle & Luke Dones, 1998, *Icarus*, 137, 197.

63. "The Chlorine Isotope Composition of the Moon and Implications for an Anhydrous Mantle" by Z. D. Sharp, C. K. Shearer, K. D. McKeegan, J. D. Barnes & Y. Q. Wang, 2010, *Science*, 329, 1050.

64. Suggestions include water escaping Earth's atmosphere, reaching the Moon via the magnetotail ("The Source of Water Molecules in the Vicinity of the Moon" by T. Földi & Sz. Bèrczi, 2001, *Lunar and Planetary Science Conference*, 32, 1148), and water occasionally entering the Solar system from interstellar giant molecular clouds[74]. Interplanetary dust can also be hydrated and strike the Moon ("The Poles of the Moon" by Paul G. Lucey, 2009, *Elements*, 5, 41).

65. Resonance line emission from 2 gm of sodium in the ejecta was observed from Earth ("Observations of the Lunar Impact Plume from the LCROSS Event" by R. M. Killen, A. E. Potter, D. M. Hurley, C. Plymate & S. Naidu, 2010, *Geophysical Research Letters*, 37, L23201).

66. "The *LCROSS* Cratering Experiment" by Peter H. Schultz, Brendan Hermalyn, Anthony Colaprete, Kimberly Ennico, Mark Shirley & William S. Marshall, 2010, *Science*, 330, 468; "Detection of Water in the *LCROSS* Ejecta Plume" by Anthony Colaprete et al., *Science*, 330, 463.

67. "LRO-LAMP Observations of the *LCROSS* Impact Plume" by G. Randall Gladstone et al., 2010, *Science*, 330, 472; "Detection of Water in the *LCROSS* Ejecta Plume" by Anthony Colaprete et al., *Science*, 330, 463. Members of this collaboration expect some implied abundances to decline.

68. "A Model for the Distribution of Volatiles at the LCROSS Impact Site" by A. Colaprete, J. Heldmann, D. H. Wooden, K. Mjaseth, M. Shirley, W. Marshall, R. Elphic, B. Hermalyn & P. Schultz, 2011, in *A Wet vs. Dry Moon* Lunar and Planetary Institute, June 13–16, 2011, #6011.

69. "Hydrogen Mapping of the Lunar South Pole Using the LRO Neutron Detector Experiment LEND" by I. G. Mitrofanov et al., 2010, *Science*, 330, 483; "Not All that Is Cold Glitters: Cold Traps Are Necessary But Not Sufficient for Near-Surface Polar Ice Sequestration" by Richard Elphic, 2013, *NLSI Workshop Without Walls: Lunar Volatiles*, #1.15.

70. "Which Spot on the Moon has the Highest Content of Hydrogen?" by Anton Sanin for the LEND Team, 2010, *Lunar Exploration and Analysis Group*, http://www.lpi.usra.edu/meetings/leag2010/presentations/WedAM/saninEtAl.pdf; "Relationship between Hydrogen-enriched Areas and Permanently Shadowed Regions near the Lunar South Pole" by William Boynton, Jerry Droege, Igor Mitrofanov & LEND team, 2010, *Lunar Exploration and Analysis Group*, http://www.lpi.usra.edu/meetings/leag2010/presentations/WedAM/boyntonEtAl.pdf: despite its weakness, the PSR-hydrogen correlation (25% of area) is stronger than random (6%), but this is tied almost entirely to craters Shoemaker and Cabeus.

71. The debate between LPNS and LEND neutron spectrometer teams involve data from LEND that are *collimated* (narrowed in acceptance angle) by shielding around the detectors. LP scientists are fairly persuasive in showing that differences between LPNS and LEND epithermal neutron maps involve high-energy neutrons leaking into LEND.

72. One estimate is typical OH surface concentration is about 100 ppm (Pieters et al. 2009, Clark 2009) for the upper 0.5 mm of regolith particles penetrated by 3-micron light. This translates to 150 mg m^{-2}, or 5 million tonnes over the Moon's surface. Estimating the reaction of solar wind protons and regolithic oxygen yields 10–50 million tonnes ("Sources and Physical Processes Responsible for OH/H$_2$O in the Lunar Soil as Revealed by the Moon Mineralogy Mapper, M^3" by T. B. McCord, L. A. Taylor, J.-P. Combe, G. Kramer, C. M. Pieters, J. M. Sunshine & R. N. Clark, 2011, *Journal of Geophysical Research*, 116, E00G05; see page 20). If this were replenished monthly, in several billion years it amounts to almost a quintillion tonnes, 1% of the Moon's mass.

73. "Nominally hydrous magmatism on the Moon" by Francis M. McCubbin, Andrew Steele, Erik H. Hauri, Hanna Nekvasil, Shigeru Yamashita & Russell J. Hemley, 2010, *Proceedings of the National Academy of Sciences*, 107, 11223; "Volatile Content of Lunar Volcanic Glasses and the Presence of Water in the Moon's Interior" by Alberto E. Saal, Erik H. Hauri, Mauro L. Cascio, James A. Van Orman, Malcolm C. Rutherford & Reid F. Cooper, 2008, *Nature*, 454, 192; "High Pre-Eruptive Water Contents Preserved in Lunar Melt Inclusions" by Erik H. Hauri, Thomas Weinreich, Alberto E. Saal, Malcolm C. Rutherford & James A. Van Orman, 2011, *Science Express*, 10.1126/science.1204626.

74. "Volatile Retention from Cometary Impacts on the Moon" by Lissa Ong, Erik I. Asphaug, Donald Korycansky & Robert F. Coker, 2010, *Icarus*, 207, 578.

75. Volatiles from giant molecular clouds (GMCs) might collect as the Moon sweeps up 10^{23} cm^3 s^{-1} of interstellar volume across a GMC in a few million years. At GMC density (~10^3 atoms cm^{-3}), it encounters ~10^{16} moles of gas per GMC. Oxygen composes 1/2000 of all atoms; if all oxygen swept up becomes water ice in lunar cold traps, 100 million tonnes result. Earth resides in GMCs ~0.1% of the time, so the Moon may have transited several GMCs. GMCs deposit material only if they compress the Sun's heliopause, where the solar wind arrests oncoming interstellar medium, to within 1 AU (the average radius of Earth's orbit around the Sun). *Voyager 1* is finding the heliopause sits at ~100 AU. In a GMC, the heliopause will shrink in radius as the inverse square root of interstellar density compared to the density now, hence to ~0.01 times its size now. (Our Local Interstellar Cloud is 0.1 atom cm^{-3}, versus GMCs at roughly 10^3 cm^{-3}.) Dense molecular cores, although small, can exceed 10^6 cm^{-3}. (See "The Interaction of the Solar Wind with a High Density Cloud" by G. P. Zank & P. C. Frisch, in *Solar Wind 9*, eds. S. R. Habbal, R. Esser, J. V. Hollweg, & P. A. Isenberg, 1999, AIP Conference Proceedings, 471, 831.)

76. "Water (Hydrogen) in the Lunar Mantle: Results from Petrology and Magma Ocean Modeling" by L. T. Elkins-Tanton & T. L. Grove, 2011, *Earth and Planetary Science Letters*, 307, 173; "Water in the Lunar Mantle: Results from Magma Ocean Modeling" by L. T. Elkins-Tanton, 2010, *Lunar and Planetary Science Conference*, 41, 1451.

77. See notes 56, 98 and 99 to this chapter.

78. "The Driving Mechanism of Lunar Pyroclastic Eruptions Inferred from the Oxygen Fugacity Behavior of *Apollo 17* Orange Glass" by Motoaki Sato, 1979, *Lunar and Planetary Science Conference*, 10, 311.

79. "The Molecular Composition of Comets and Its Interrelation with Other Small Bodies of the Solar System" by Jacques Crovisier, in *Asteroids, Comets, Meteors 2005: Proceedings of IAU Symposium 229*, eds. D. Lazzaro, S. Ferraz-Mello & J. A. Fernández, 2006, Cambridge University Press, 133: comets are mostly water, with CO composing 1%–20% as much mass as water; CO_2, 2%–10%; CH_3OH, 1%–7%; and roughly 1% each for CH_4, H_2CO, NH_3, and H_2S.

80. Organics nonane (C_9H_{20}) and benzene (C_6H_6) freeze at –13.5°C and 5.4°C, respectively, at low pressure, but are rare in the Moon. Sulfuric acid freezes at 16.8°C but reacts easily and changes freezing point rapidly with impurities, especially water. (Water's special temperature is unsurprising: Earth is poised near 14°C, at the same distance from the Sun as the Moon.)

81. From *Apollo 15* and *17*'s ALSEP heat flow data ("In-Situ Measurements of Lunar Heat Flow" by M. G. Langseth & S. J. Keihm, in *Soviet-American Conference on Geochemistry of the Moon and Planets*, 1977, NASA SP-370, 283), we know that just underground regolith temperatures remain 247–253 K (depending on latitude), with gradients (below 1 m) of 1.2–1.8 deg m^{-1}, extrapolating to 5°C at 13–16 m underground, roughly the regolith's depth. More recent analyses put the 0°C level twice as deep, meaning ice could live longer ("Lost Apollo Heat Flow Data Suggest a Different Lunar Bulk Composition" by Y. Saito, S. Tanaka, J. Takita, K. Horai & A. Hagerman, 2007, *Lunar and Planetary Science Conference*, 38, 2197).

82. "Subsurface Migration of H_2O at Lunar Cold Traps" by Norbert Schorghofer & G. Jeffrey Taylor, 2006, *Journal of Geophysical Research*, 112, E02010.

83. "Lunar Outgassing, Transient Phenomena and the Return to the Moon. II. Predictions and Tests for Outgassing/Regolith Interactions" by Arlin P. S. Crotts & Cameron Hummels, 2009, *Astrophysical Journal*, 707, 1506.

84. Water forms a pseudo-liquid monolayer in several seconds even at 10^{-9} atmosphere pressure. This might also be considered in terms of etching mildly contaminated Apollo samples in general.

85. See "Reaction Kinetics of Primary Rock-forming Minerals under Ambient Conditions" by S. L. Brantly, 2004, *Treatise on Geochemistry*, 5, 73.

86. "*Apollo 16* Deep Drill – A Review of the Morphological Characteristics of Oxyhydrates on Rusty Particle 60002, 108, Determined by SEM" by Stephen E. Haggerty, 1978, *Lunar and Planetary Science Conference*, 9, 1861.

87. "*Apollo 16* 'Rusty Rock' 66095" by L. A. Taylor, H. K. Mao & P. M. Bell, 1973, *Abstracts of Lunar and Planetary Science Conference*, 4, 715; "Origin of Inert Gases in 'Rusty Rock' 66095" by D. Heymann & W. Hübner, 1974, *Earth and Planetary Science Letters*, 22, 423; "Chlorine Isotope Composition of 'Rusty Rock' 66095 and *Apollo 16* Soil. Implications for Volatile Element Behavior on the Moon" by C. K. Shearer & Z. D. Sharp, 2011, *A Wet Vs. Dry Moon*: LPI, 6006.

88. "Solar Interior Structure and Luminosity Variations" by D. O. Gough, 1981, *Solar Physics*, 74, 21.

89. This process likely does not dominate, because overturn to even 1 m depths takes over 1 Gy ("Mixing of the Lunar Regolith" by D. E. Gault, F. Hoerz, D. E. Brownlee & J. B. Hartung, 1974, *Abstracts of Lunar and Planetary Science Conference*, 5, 260; "Development of the Mare Regolith – Some Model Considerations" by William Quaide & Verne Oberbeck, 1975, *The Moon*, 13, 27). Craters 75 m in diameter permanently excavate to 15 m depths (e.g., "On The Scaling of Crater Dimensions. II – Impact Processes" by K. A. Holsapple and R. M. Schmidt, 1982, *Journal of Geophysical Research*, 87, 1849, and ignoring effects of crack and breccia formation), forming at rates of ~1 Gy^{-1} km^{-2} (Shoemaker number/size power-law index = 2.9, extending Neukum et al.: "Cratering Records in the Inner Solar System in Relation to the Lunar Reference System" by G. Neukum, B. A. Ivanov & W. K. Hartmann, 2001, *Space Science Reviews*, 96, 55).

90. "Volatile Retention from Cometary Impacts on the Moon" by Lissa Ong, Erik I. Asphaug, Donald Korycansky & Robert F. Coker, 2010, *Icarus*, 207, 578; "The Migration of Volatiles on the Surfaces of Mercury and the Moon" by Bryan J. Butler, 1997, *Journal of Geophysical Research*, 102, 19283; "The Solar Wind as a Possible Source of Lunar Polar Hydrogen Deposits" by Dana Hurley Crider & Richard R. Vondrak, 2000, *Journal of Geophysical Research*, 105, 26773; "Simulations of a Comet Impact on the Moon and Associated Ice Deposition in Polar Cold Traps" by Bénédicte D. Stewart, Elisabetta Pierazzo, David B. Goldstein, Philip L. Varghese & Laurence M. Trafton, 2011, *Icarus*, 215, 1. Early models are found in "Orbital Search Lunar Volcanism" by R. R. Hodges, J. H. Hoffman, T. T. J. Yeh & G. K. Chang, 1972, *Journal of Geophysical Research*, 77, 4079.

91. "Proton-Induced Hydroxyl Formation on the Lunar Surface" by E. J. Zeller, L. B. Ronca & P. W. Levy, 1966, *Journal of Geophysical Research*, 71, 4855; "Water Detection on Atmosphereless Celestial Bodies: Alternative Explanations of the Observations" by L. Starukhina, 2001, *Journal of Geophysical Research*, 106, 14701.

92. "Water and Hydroxyl on the Moon as Seen by the Moon Mineralogy Mapper (M^3)" by R. Clark, et al., 2010, *Lunar and Planetary Science Conference*, 41, 2302; also 2011, *A Wet vs. Dry Moon: Exploring Volatile Reservoirs and Implications for Evolution of Moon and Future Exploration*, 6047.
93. "Magmatic Processes that Produced Lunar Fire Fountains" by Linda T. Elkins-Tanton, Nilanjan Chatterjee & Timothy L. Grove, 2003, *Geophysical Research Letters*, 30, 1513.
94. "Basaltic Magmatism on the Moon: A Perspective from Volcanic Picritic Glass Beads" by C. K. Shearer & J. J. Papike, 1993, *Geochimica et Cosmochimica Acta*, 57, 4785.
95. "A Model of the Moon's Volatile Depletion" by Steven J. Desch & G. Jeff Taylor, *A Wet vs. Dry Moon: Exploring Volatile Reservoirs and Implications for Evolution of Moon and Future Exploration*, 6046.
96. "The Physical Mechanisms of Deep Moonquakes and Intermediate-Depth Earthquakes: How Similar and How Different?" by C. Frohlich & Y. Nakamura, 2009, *Physics of Earth and Planetary Interiors*, 173, 365.
97. "Seismic Detection of the Lunar Core" by Renee C. Weber, Pei-Ying Lin, Edward J. Garnero, Quentin Williams & Philippe Lognonné, 2011, *Science*, 331, 309.
98. Oxygen fugacity is related (exponentially) to the chemical potential. Chemical potential μ (of a chemical reactant) is a measure of the energy that might be liberated in a reaction. In chemical equilibrium, the chemical potentials of the reactants on one side of a reaction balance those on the other side. For an ideal gas one can easily show that $\mu = \mu_0 + kT \ln(P/P_0)$, where μ_0 is a constant (energy), k is Boltzmann's constant, T is temperature, P is the pressure of the particular substance, and P_0 is a constant reference pressure, so that for a hypothetical reaction in equilibrium among ideal gases $X + Y \leftrightarrow Z + W$ implies $\mu_X + \mu_Y = \mu_{XY} = -kT \ln(P_X/P_{X0}) - kT \ln(P_Y/P_{Y0}) = -kT \ln(P_Z/P_{Z0}) - kT \ln(P_W/P_{W0})$, so $e^{-kT}(P_X/P_{X0})(P_Y/P_{Y0}) = e^{-kT}(P_Z/P_{Z0})(P_W/P_{W0})$, hence $[(P_Z/P_{Z0})(P_W/P_{W0})] / [(P_X/P_{X0})(P_Y/P_{Y0})] = K$, a constant regardless of how reactants or products vary in strength.

 What if some reactants and products are not ideal gases? Then the expression for K no longer holds. Other gas models such as a Van der Waals gas (accounting for the finite size of molecules and the forces between them) result in pressures often a few percent lower than for an ideal gas, so the fugacity for those gases in a reaction is a bit lower than the pressure. In general, fugacity can replace pressure and the validity of the constant K still applies. For a reaction even more complicated, gases can be buffered to almost constant levels, and fugacities can be tiny compared to the corresponding pressure. Sometimes fugacity is computed thermodynamically but is often measured in the laboratory. In more complicated systems such as mineral melts, different buffer compositions even at the same temperature and total pressure can yield fugacities for the same substance, e.g., O_2, that differ by factors of 10^{25}. Fugacities can vary by a similar amount over temperature changes of 100°C. Fugacities, like the pressure in the definition of K, are defined versus a standard pressure, chosen as 1 atmosphere.
99. The water/hydrogen system is complicated in silicate melts by the loss of hydrogen as hydroxyl, significantly reducing concentration of hydrogen relative to water. The rapid diffusion of hydrogen suggests that it outgasses quickly. (See "Formation of Carbon and Hydrogen Species in Magmas at Low Oxygen Fugacity" by A. Kadik, F. Pineau, Y. Litvin, N. Jendrzejewski, I. Martinez & M. Javoy, 2004, *Journal of Petrology*, 45, 1297; "Influence of Oxygen Fugacity on the Solubility of Carbon and Hydrogen in $FeO-Na_2O-SiO_2-Al_2O_3$ Melts in Equilibrium with Liquid Iron at 1.5 GPa & 1400°C" by A. A. Kadik, N. A. Kurovskaya, Y. A., Ignat'ev, N. N. Kononkova, V. V. Koltashev & V. G. Plotnichenko, 2010, *Geochemistry International*, 48, 953; "Nitrogen and Hydrogen Isotope Compositions and Solubility in Silicate Melts in Equilibrium with Reduced (N+H)-Bearing Fluids at High Pressure and Temperature: Effects of Melt Structure" by B. O. Mysen & M. L. Fogel, 2010, *American Mineralogist*, 95, 987; "'Water' in Lunar Basalts: The Role of Molecular Hydrogen (H_2), Especially in the Diffusion of the H Component" by Y. Zhang, 2011, *Lunar and Planetary Science Conference*, 42, 1957; "Diffusion of H, C and O Components in Silicate Melts" by Y. Zhang & H. Ni, in *Diffusion in Minerals and Melts*, eds. Y. Zhang & D. J. Cherniak, 2010, Chantilly, VA: Mineralogical Society of America, 171–225.)
100. "Differential Degassing of H_2O, Cl, F and S: Potential Effects on Lunar Apatite" by Gokce Ustunisik, Hanna Nekvasil & Donald Lindsley, October 2011, *American Mineralogist*, 96, 1650.
101. Ibid.
102. Here we use percentages. Isotopic concentrations appear as parts per thousand, e.g., deuterium enhancement $\delta D/1000 = [(D/H)_{sample}/(D/H)_{ocean}] - 1$, where "ocean" is the Vienna standard mean ocean. Terrestrial water samples, therefore, range in δD from −500‰ to +100‰.
103. "Interstellar Ices as Witnesses of Star Formation: Selective Deuteration of Water and Organic Molecules Unveiled" by S. Cazaux, P. Caselli & M. Spaans, 2011, *Astrophysical Journal Letters*, 741, L34.

104. "Oxygen Isotopes in Meteorites" by R. N. Clayton, 2003, *Treatise on Geochemistry*, 1, 129. This glosses over interrelationships of stable oxygen isotopes: ^{16}O, ^{17}O, and ^{18}O, evenly spaced in mass. If a process enhances ^{17}O over ^{16}O, the same process enhances ^{18}O over ^{17}O by equal amounts. A reservoir of oxygen evolves along a well-defined line in a plot of $^{17}O/^{16}O$ ratio versus $^{18}O/^{16}O$. For the same $^{18}O/^{16}O$, for instance, Mars has ~400 ppm more ^{17}O than Earth or the Moon. Meteorites from asteroid Vesta have ~250 ppm less. Other meteorites (CAIs and chondrules), subject to different processes, trace different ratio-ratio lines, even with different slopes. The lunar $^{17}O/^{16}O$ and $^{18}O/^{16}O$ ranges overlap various meteoritic and cometary sources, so different reservoir mixes can merge to produce an oxygen isotopic ratio on the Moon the same as a completely different mixture. Oxygen isotopic ratios alone might not uniquely determine origin.

105. "Abundance, Distribution and Isotopic Composition of Water as Revealed by Basaltic Lunar Meteorites" by M. Anand, R. Tartèse, J. J. Barnes, N. A. Starkey, I. A. Franchi & S. S. Russell, 2013, *Lunar and Planetary Science Conference*, 44, 1957; "Distinct Petrogenesis of Low- and High-Ti Mare Basalts Revealed by OH Content and H Isotope Composition of Apatite" by R. Tartèse, M. Anand, J. J. Barnes, N. A. Starkey & I. A. Franchi, 2013, *Lunar and Planetary Science Conference*, 44, 2222. Hydroxyl content varies from about 0.1% to 1.5%, versus +10% to +120‰ in δD. Drier samples have heavier hydrogen, as can occur in evaporation (by Rayleigh fractionation). Because apatite forms near the surface (within a few kilometers), this might be a consideration. Others might say that the largest OH concentrations, in meteorites, might be a sign of terrestrial contamination.

106. "Hydrogen Isotope Ratios in Lunar Rocks Indicate Delivery of Cometary Water to the Moon" by James P. Greenwood, Shoichi Itoh, Naoya Sakamoto, Paul Warren, Lawrence Taylor & Hisayoshi Yurimoto, 2011, *Nature Geoscience*, 4, 79.

107. "Exploring Volatile Deposition in Lunar Regolith" by J. Alford, A. Hodges, A. Crotts & E. Heggy, 2012, *Lunar and Planetary Science Conference*, 43, 2938; "Regolith Gardening vs. Secular Deposition: A Numerical Model" by Andrew Hodges & Arlin P.S. Crotts, 2014, in preparation.

108. We outline technical details of such a system in Chapter 12 and pending patent documents.

Chapter 9
Inconstant Moon

> In observational fields fortune favors the prepared mind. (*Dans les champs de l'observation le hasard ne favorise que les spirits préparés.*)
> – Louis Pasteur, December 7, 1854, lecture, Université de Lille[1]

Now we are mapping the Moon. Unlike past times when mapping meant sailing or walking to geographic features to mark them on a map, or peering through a telescope from Earth, our robots orbit tens of kilometers overhead and remotely characterize each lunar surface parcel (pixel) in optical light, infrared, particle flux, radar, elevation, and a dozen other techniques. It may seem we will soon describe the Moon completely. It is secure science: if a scientist discovers something, it will still be there on the next orbit in two hours, or when the spacecraft passes overhead two weeks or a month later, or next year, or when another lunar craft orbits with a similar instrument. What we discover depends critically on not only how we look with our choice of instrumentation (like Oersted's magnetic needle) but also how observers are trained and what they acknowledge.

We will soon enter a new phase. Since Soviet robotic lunar rover *Lunokhod 2* crashed and overheated in May 1973, and the last lunar sample reached Earth on *Luna 24* in August 1976, no spacecraft and certainly no human has operated significantly on the Moon's surface. (Recently the Chinese *Chang'e 3* and Yutu rover worked on the Moon, only briefly; several lunar craft have crashed landed, too.) NASA shut down Apollo's lunar surface ALSEP instruments on September 30, 1977. Soon robotic spacecraft will once again be roving and sampling the Moon, and eventually humans will leave new trails of nearly eternal footprints. Gene Cernan and Jack Schmitt left the lunar surface more than 40 years ago, and are in some sense the most recent human space explorers, certainly of an alien world. The Apollo lunar astronauts' experience is key to understanding how exploration is possible, especially by humans – one of the key questions in the next few chapters.

Twelve humans walked on the Moon and were no accidental tourists. Not just jet pilots; they trained in field geology (Figure 2.22). The cognizant lunar scientists rate the astronaut's training equivalent to a master's degree in geology (training in simulated lunar traverses to refine exploration procedures, orientation in settings of comparable geology, sampling, sample suite selection, documentation, data communication). *Apollo 15* Commander Dave Scott (not previously a geologist) notes that he and Jim Irwin spotted famed Genesis Rock 10 meters away and a unique "green glass clod" (sample #15426) at distance. Dramatically, he spotted dark-black vesicular basalt #15016 while driving the LRV and feigned a loose seat belt to allow time (103 seconds) to collect it. The one professional geologist on the Moon, Jack Schmitt, found important samples: to everyone's surprise, orange soil from 3.5-billion-year-old volcanic fire fountains, and discovery of the contact between two types of impact-generated debris units, one intrusive into the other. We will evaluate these and other, less successful examples from Apollo.

Chapters 10 and 11 deal with humans (not necessarily scientists) working and living on the Moon. Should new lunar environments be primarily domains of scientists? What can we do with telerobots monitored by scientists? Must scientists be the explorers to warrant the expense and risk of humans in space? Do we trust only trained scientists to accurately report their findings? Some disagree with Pasteur. Thomas Henry Huxley offered: "Trust a witness in all matters in which neither his self-interest, his passions, his prejudices, nor the love of the marvelous is strongly concerned. When they are involved, require corroborative evidence in exact proportion to the contravention of probability by the thing testified."[2]

One problem is that some science changes too quickly to be mapped. Scientists and their instruments are not always available when the phenomena occur. When considering where to put our scientists and astronauts, let us first consider examples of how we dealt with new but transient scientific phenomena.

Restrictive attitudes have drawbacks: for decades airplane pilots and other non-scientists reported lightning-like flashes traveling from thundercloud tops to the upper atmosphere. Only when observed by atmospheric scientists in 1989 were these "sprites" accepted as a real phenomenon. Even now claims by nonscientists, for example, "ball lightning" with more than 200 reports, fall in this phenomenological nether realm. Infrequent, short-lived phenomena rarely perform for scientists, whereas nonscientists' reports are discounted. Thus meteorite falls were ignored for centuries[3] as were sightings of rare animals. Being non-repeatable, such rare event reports are viewed as invalid scientific evidence.

Lunar geologists are trained to study relics of processes that occurred eons ago and/or over geological time and are unaccustomed to rapid changes in the time domain. The idea that the Moon changes quickly (excepting external causes, e.g., meteoroids) is strange and one most lunar scientists do not consider. However, interesting future discoveries about the Moon likely reside in this domain.

The Apollo astronauts trained to identify interesting geology fulfilled their promise. For instance, Jack Schmitt on *Apollo 17* identified the famous orange soil, seen to be volcanic glass. Less appreciated was that he saw a similar effect earlier that was an illusion, but seeing the orange soil was able to make subtle assessments that the second effect was real. [Box 9.1] The astronauts were able to make rapid and accurate assessments of potential discoveries. Would this be the case if observations were made robotically?

Astronauts were trained to evaluate observations even from orbit in terms of geology. Farouk El-Baz, who supervised Apollo lunar landing site selection at ATT Bellcom for NASA, describes how even command module pilots, never landing on the Moon, were trained in lunar geology and made meaningful contributions. For

> **Box 9.1: The Color Orange**
>
> About 144 hours into *Apollo 17*'s mission, near the start of the second of three lunar EVAs, Schmitt is collecting samples while Cernan is taking a photo panorama:
>
> SCHMITT: A chunk of yellow-brown rock that apparently has several spots behind it, probably indicating direction from which it came ... Oh, no ... What is that? That's a reflection. (Laughs) That really fooled me. A reflection off the Mylar. (Laughs) Crazy. Well, what the heck, I'll sample it anyway.
>
> (A gold-lined Mylar sheet reflecting spots of yellow-brown light onto the rock was the thermal blanket on the Lunar Communications Relay Unit on the Lunar Rover.)
>
> An hour and a half later, Schmitt encounters what first appears to be a similar situation, but it proves to be one of the most significant discoveries of Apollo.
>
> SCHMITT: Oh, hey! Wait a minute ...
> CERNAN: What?
> SCHMITT: Where are the reflections? I've been fooled once. There is orange soil!!
> CERNAN: Well, don't move it until I see it.
> SCHMITT: (Excited) It's all over!! Orange!!!
> CERNAN: Don't move it until I see it.
> SCHMITT: I stirred it up with my feet.
> CERNAN: (Excited) Hey, it is!! I can see it from here!
> SCHMITT: It's orange!
> CERNAN: Wait a minute, let me put my visor up. It's still orange!
> SCHMITT: Sure it is! Crazy!
> CERNAN: Orange!
> SCHMITT: I've got to dig a trench, Houston.
>
> (The astronauts quickly performed a number of intuitive tests to validate the discovery of orange soil. Within several seconds, Schmitt checked the geometry of the situation, ascertained that the orange patch was too large to be a reflection from the Rover and that the rover was too far away. Furthermore, despite Cernan's instructions, Schmitt established that the orange color extended below the soil's surface so was not an optical effect. Furthermore Cernan double-checked what he suspected might be subjectivity on Schmitt's part and also established the soil's color independent of the protective colored visor on the astronauts' helmets.)

instance, Ken Mattingly, *Apollo 16* CMP, who trained extensively in geological survey from aircraft, is credited by El-Baz with recognizing from orbit that the *Apollo 16* site's nature was different than originally suspected, more felsic, which allowed El-Baz and geologist Don Wilhelms to realize the true nature of the site,[4] as mentioned in the opening of Chapter 6. We will see shortly that orbital observations were not so successful for phenomena for which the astronauts were not trained.

Prompt, on-site evaluation of potential discoveries is crucial for several reasons. First, rovers or astronauts do not linger forever but continue traversing the Moon. If one later questions a discovery's meaning, the opportunity may be gone to return and collect more evidence. Prompt and accurate evaluation is premium; the results may influence science for years. Second, transient phenomena might change and disappear and should be recorded and documented rapidly and accurately. Situations can change because explorers enter the environs, and might even be hazardous. Lunar lava tube caverns or landscapes affected by volatile buildup and evaporation might collapse or change rapidly.

In this book we ask what might occur by mid-century on the Moon, a challenge because the relevant science is still unfolding. Transient events on the Moon, being understudied, are a good place to start. By this chapter's end we will survey a wider range of potential breakthroughs.

Apollo astronauts reported several transient events but with results less memorable than the orange soil's. Unlike static situations in geological sampling, these transients suffered from poor record keeping and flawed on-site interpretation. This occurred for several reasons: primarily lack of astronaut training, likely originating in and exacerbated by low priority given to such phenomena by scientists at the time, and lack of equipment or missing equipment in the short time of observation. Sometimes observing conditions preclude a well-documented report.

A singular example of a reported transient occurred during *Apollo 11* on July 19, 1969, some 25 hours before lunar landing. Astronomers on Earth reported a sudden brightening in crater Aristarchus, and the astronauts were asked to observe it. *Apollo 11* reported an unusually bright appearance of Aristarchus, but Aldrin interpreted it as "anomalous backscatter" – a brightening of material when viewed at an angle 180° from the Sun. (You can observe this as the halo around your airplane's shadow on your next flight, more than about a half kilometer from the ground.) As you will see, [Box 9.2] this interpretation was incorrect. The astronauts could re-observe Aristarchus for changes on the next orbit but did not, leaving the matter unresolved.

On later Apollo flights astronauts in orbit reported three brief flashes of light from the lunar surface. Response to these developed informally; the quality and utility of data recorded were mixed. [Box 9.3] *Apollo 16* Command Module Pilot Ken Mattingly saw a flash on the lunar Far Side, in lunar shadow, below the horizon (hence, from the surface). This flash's location on the Moon is highly uncertain. On *Apollo 17* two such flashes were reported, both from the sunlit lunar surface. Lunar Module Pilot Jack Schmitt saw a flash near crater Grimaldi on the western Near Side and verbally described its rough location versus lunar landmarks, eventually marking this on a lunar map. Command Module Pilot Ron Evans saw a similar flash near Mare Orientale, also near the westernmost Near Side, 25 hours later and immediately marked the flash's map location. Accurate flash locations are crucial in their interpretation. (Unfortunately, after searching for this map at the National Air and Space Museum and National Space Science Data Center, I think it is missing.) Even after *Apollo 16* no observation regimen was established to record these. A flash this bright might indicate a several kilogram meteoroid impact, digging a crater 1 meter in diameter and fresh ejecta spread several times wider, detectable by LROC-NAC or Apollo metric cameras. These events could be closely localizable in space and

Box 9.2: The Cislunar Transient Report

A unique naked-eye and telescopic incident, reported simultaneously by independent observers 385,500 kilometers apart at Earth and the Moon, occurred with *Apollo 11* on its second lunar orbit. It was reported by two astronomers in West Germany (Pruss, Witte) and two of three *Apollo 11* astronauts. (The mission transcript does not indicate Neil Armstrong as witness.) Buzz Aldrin and Mike Collins reported a strange dark side surface appearance near Aristarchus during the 1–2 minute period in which ground-based observers saw a similar phenomenon at likely the same location. Discussion at this time was recorded verbatim, so one can understand some of the human factors involved with the report, unlike most cases. The crew had been alerted to other activity near Aristarchus, so this was not fully independent. Aldrin dismissed the phenomenon as explained by anomalous backscatter, but the geometry was in correct for this.

Communications transcript between *Apollo 11* and Mission Control, July 19, 1969; list below with 6-digit timestamp code: hours:minutes:seconds after launch (Julian Date 2440419.0639 geocentric). Speakers are Capsule Communicator (CAP COMM) Bruce McCandless, Commander Neil A. Armstrong, Command Module Pilot Michael Collins, and Lunar Module Pilot Edwin E. Aldrin, Jr.

076:57:07 **MCCANDLESS**: Roger. And we've got an observation you can make if you have some time up there. There's been some lunar transient events reported in the vicinity of Aristarchus. Over.

076:57:28 **ALDRIN**: Roger. We just went into spacecraft darkness. Until then, why, we couldn't see a thing down below us. But now, with earthshine, the visibility is pretty fair. Looking back behind me, now, I can see the corona from where the Sun has just set. And we'll get out the map and see what we can find around Aristarchus.

076:57:54 **ARMSTRONG**: We're coming upon Aristarchus right now –

076:57:55 **MCCANDLESS**: – Okay. Aristarchus is at angle Echo 9 on your ATO chart. It's about 394 miles north of track. However, at your present altitude, which is about 167 nautical miles, it ought to be over – that is within view of your horizon: 23 degrees north, 47 west. Take a look and see if you see anything worth noting up there. Over.

076:58:34 **ARMSTRONG**: Both looking.

076:58:36 **MCCANDLESS**: Roger. Out.

(*Note*: Here we skip 14 minutes and roughly 33 lines of dialog between Aldrin, Armstrong, and McCandless regarding navigation and communications channels. Aristarchus is mentioned during this time only in terms of when *Apollo 11* can view it from orbit.)

077:12:51 **ARMSTRONG**: Hey, Houston. I'm looking north up toward Aristarchus now, and I can't really tell at that distance whether I am really looking at Aristarchus, but there's an area that is considerably more illuminated than the surrounding area. It just has – seems to have a slight amount of fluorescence to it. A crater can be seen, and the area around the crater is quite bright.

077:13:30 **MCCANDLESS**: Roger, 11. We copy.

> **Box 9.2: (continued)**
>
> *077:14* (*Note*: This is the time of the Pruss and Witte Aristarchus TLP report from Germany.)
>
> | | |
> |---|---|
> | *077:14:23* **ALDRIN**: | Houston, Apollo 11. Looking up at the same area now and it does seem to be reflecting some of the earthshine. I'm not sure whether it was worked out to be about zero phase to – Well, at least there is one wall of the crater that seems to be more illuminated than the others, and that one – if we are lining up with the Earth correctly, does seem to put it about at zero phase. That area is definitely lighter than anything else that I could see out this window. I am not sure that I am really identifying any phosphorescence, but that definitely is lighter than anything else in the neighborhood. |
> | *077:15:15* **MCCANDLESS**: | 11, this is Houston. Can you discern any difference in color of the illumination, and is that an inner or an outer wall from the crater? Over. |
> | *077:15:34* **COLLINS**: | Roger. That's an inner wall of the crater. |
> | *077:15:43* **ALDRIN**: | No, there doesn't appear to be any color involved in it, Bruce. |
> | *077:15:47* **MCCANDLESS**: | Roger. You said inner wall. Would that be the inner edge of the northern surface? |
> | *077:16:00* **COLLINS**: | I guess it would be the inner edge of the westnorthwest part, the part that would be more nearly normal if you were looking at it from the Earth. |
> | *077:16:20* **MCCANDLESS**: | 11, Houston. Have you used the monocular on this? Over. |
> | *077:16:28* **ALDRIN**: | Stand by one. |
> | *077:17:59* **ALDRIN**: | Roger. Like you to know this quest for science has caused me to lose my E-memory program, it's in here somewhere, but I can't find it. |
>
> (*Note*: E-memory was the spacecraft's erasable memory that held temporarily programs for control of the spacecraft, e.g., for guidance.)
>
> | | |
> |---|---|
> | *077:18:08* **MCCANDLESS**: | 11, this is Houston. We're – we're hearing only a partial COMM. Say again please. |
> | *077:18:20* **ARMSTRONG**: | I think … |
> | *077:18:41* **ARMSTRONG**: | Houston, we will give it a try if we have the opportunity on next – when we are not in the middle of lunch, and trying to find the monocular. |
> | *077:18:51* **MCCANDLESS**: | Roger. Copied you that time. Expect in the next REV you will probably be getting ready for LOI 2. |
> | *077:19:09* **MCCANDLESS**: | So, let's wind this up, and since we've got some other things to talk to you about in a few minutes. Over. |
>
> Aldrin guesses that Aristarchus looks bright in lunar night by strongly scattering Earthlight, with *Apollo 11* in line between Aristarchus and Earth, causing anomalous backscattering. Unfortunately this was untrue: at observation time the phase angle was about 63°, whereas enhanced backscattering is significant only for angles of a few degrees. Anomalous backscatter does not explain these observations. During the observation, the Moon's phase was 5.2 days past new, with Aristarchus in darkness 26° from the

> **Box 9.2: (continued)**
>
> antisolar point and 57° from the sub-Earth point on the Moon. The spacecraft was about 245 kilometers above the mean lunar equatorial surface and some 750 kilometers from Aristarchus's center, which appeared inclined only 5° from edge-on. Only the north-northwestern part of the inner rim would be seen easily.
>
> At mission elapsed time 077:14 (to the minute), Gail Pruss and Manfred Witte (at the Institute for Space Research, Bochum, West Germany) reported seeing a 5–7 s brightening in Aristarchus in a 150 millimeter diameter telescope. (The Pruss and Witte report did not lead to the alert by NASA, contrary to media reports, for example, "The Moon, A Giant Leap for Mankind," *Time*, July 25, 1969.) It is unclear which earlier event the capsule communicator indicates to *Apollo 11*. He probably refers to a pulsing glow in Aristarchus reported by Whelan from New Zealand 12.3 hours earlier. That night we know of nine transients reported, mainly from LION (see Appendix B), seven involving Aristarchus over a 14-hour interval, including independent reports (some with photographs) over 077:58–078:58, of Aristarchus being brighter than normal. There was no apparent attempt on *Apollo 11* to observe Aristarchus on the next revolution, 2.15 hours later in its initially wide, uncircularized lunar orbit. In 2008 I asked the astronauts, and neither recall a second attempt to observe it. Collins has no clear recollection of the entire Aristarchus event, and Aldrin recalled no attempt to re-observe it. (The LOI 2 burn – Lunar Orbit Insertion – at mission time 080:11:36 accomplished the circularization.)
>
> Some issues include observers can dismiss real temporal anomalies because they have a mental model for normal appearance, in this case, the direct, 180° backscatter – really caused by less understood effects. To what extent can simultaneous, independent reports differ in description and still constitute a confirmation? Many selenographers made careful transient observations with written, redundant records: Do more incidental observers provide useful reports? In the end this singular case is not a strong transient report even though simultaneous to another report on the same feature, because it is unclear that the Apollo observers saw truly transient activity. Issues regarding astronauts as scientific witnesses include: did they not re-observe Aristarchus on the next orbit two hours later because they were too busy, or because Aldrin generated a (false) explanation for the phenomenon that obviated the need to re-observe? Does it matter that the astronauts, although trained in field geology, had not studied how to record transient events? This case illustrates some important issues at play in transient reports, particularly in the Apollo era. This is a unique example not only because of the setting but also because of how well the information flow is documented. Is it a transient report if observers are told to pay special attention to a specific area? Did observers know to distinguish the exceptional crater Aristarchus as a spatial anomaly rather than a temporal one in comparison to other craters?

time, so the lunar interior could be probed with ALSEP's seismometers. *Apollo 17*'s flashes correspond to no known moonquakes, but *Apollo 16*'s matches a weak seismic signal.

We discussed poorly understood sites where explorers might find things changing "under their feet," literally or figuratively. Most were revealed by recent lunar orbital observation, for example, lava tube interiors, or volatile-rich polar regions. A long history of other lunar transients, even predating the space age, offers more clues about what might be changing on the Moon.

Box 9.3: The Lunar Flash Cases

Three instances of rapid, bright flashes apparently from the lunar surface were observed on *Apollo 16* (by Ken Mattingly) and *Apollo 17* (by Jack Schmitt and Ron Evans – not Cernan as per some sources). They are documented in the mission transcripts, debriefings, preliminary science reports, and catalogs of transient phenomena. The two *Apollo 17* reports were seen at flooded crater Grimaldi and Mare Orientale, respectively. The first locus, and even the second, although indistinct, are sites of some of the few outgassing events detected by nonoptical transient means (both during *Apollo 15*). Grimaldi is a persistent optical transient site; Orientale is too close to the limb to be seen well from Earth. This is interesting because few, if any, flashes were seen not coming from the direction of the lunar surface, so their cause being cosmic ray interactions with the retina or vitreous humor of the eye is problematic. *Apollo 16*'s event is hard to localize being on the dark side and purely visual. Being on the Far Side, it cannot be tied explicitly to other Earth-observed transients. However, Mattingly saw it as coming from below the horizon, therefore, ostensibly the lunar surface. As I reconstruct from his description, Mattingly was looking near Far Side crater Korolev. This is highly uncertain. Note Mattingly's description to himself from the mission transcript (*Apollo 16* Command Module Onboard Voice Transcription):

MATTINGLY: That was a flash. I know it was. 123:07, that had to be below the horizon. At 123:07, I was looking in an area – let me see where it started at now. Well, I don't want this to get out, but I think you might – you might make a note that at 123:07, I was looking out the window, and I was looking at the horizon and there was a horizon, and there was a bright flash that I saw. It was below the horizon. Now whether that was . . . Maybe I saw one of these light flashes that everyone else has been seeing all day and that I have not seen yet, or maybe I saw a flash. I don't know. It was a . . . was a bright flash. It could have been one of these light flashes that everyone else sees. But I'm going to look at this same area again. Now, back to the zodiacal light.

I discussed these events with Schmitt and Mattingly. (Evans died in 1990.) Both were well aware of cosmic ray-induced flashes within the human eye. Schmitt is nearly certain this is not what he saw; Mattingly is somewhat less sure because he never saw a cosmic ray flash otherwise but describes his event as fully pointlike and instantaneous, which varies from cosmic ray event descriptions. Schmitt also described his event as pointlike and instantaneous, although the *Apollo 17* transcript is slightly contrary (*Apollo 17* Technical Air-to-Ground Voice Transcription, MET 087:38:09):

SCHMITT: Hey, I just saw a flash on the lunar surface! It was just out there north of Grimaldi. You might see if you got anything on your seismometers, although a small impact probably would give a fair amount of visible light. It was a bright little flash near the crater at the north edge of Grimaldi; the fairly sharp one to the north is where there was just a thin streak of light.

CAPCOM FULLERTON: How about putting an X on the map where you saw it.

SCHMITT: I keep looking occasionally for – Yeah, we will. I was planning on looking for those kind of things.

> **Box 9.3: (continued)**
>
> Evans also described a flash (*Apollo 17* Command Module Onboard Voice Transcription, MET approximately 112:54).
>
> **EVANS:** Hey! You know, you will never believe it. I am right over the edge of Orientale. I just looked down and saw a light flash myself. Right at the end of the rille that is on the east of Orientale. Yes, you know, you don't suppose that could be a Vostok? I'll be darned. I got to mark that spot on the map.
>
> Rather than dwell on memory of four-decade-old events, we can ask if they are easily explained by meteorite impacts on the lunar surface. Quantitatively, this seems likely for two of three cases.
>
> Based on the verbal descriptions, one can localize Mattingly's event to within roughly 10° on the lunar surface (about 300 kilometers) versus roughly 0.25° for Schmitt's and Evans' events. (The map they annotated is scaled about 1:7,500,000, and I assume they marked it accurately to 1 mm.) The later two reports occur during the relatively intense Geminid meteor shower (peaking December 14 versus the two reports on UTC December 10, 1972, 21:11:09, and December 11, 1972, 22:28:27), and occur only 20° and 18° in longitude, respectively, from the point on the Moon's surface directly under the Geminid radiant, close to the point most likely to be struck by Geminids. Mattingly's event (at UTC April 21, 1972, 19:01) occurs 142° in longitude from the leading point of lunar motion, far from the most likely point for meteoritic impacts, but not conclusively so. (Note that the two *Apollo 17* events were 104° and 99° from the leading point but could easily be Geminids.) The two *Apollo 17* reports might be explained by Geminids. The *Apollo 16* report has no such obvious explanation but might also be a result of impact.

Nearly as long as some scientists have argued that lunar activity exists (or water, or life), others have thought the Moon a dead, inactive world.[5] The question arises in several forms: Is there a detectable atmosphere? How does the lunar surface change? Where is the seismic activity?

Astronomers have agreed for centuries that there is little evidence of lunar weather seen from Earth, nor evidence of a thick atmosphere.[6] How would one detect the Moon's atmosphere from Earth? An obvious means is atmospheric refraction. As you watch the Sun set, its image is deflected by the bending of light by Earth's atmospheric refraction from the position it would have without Earth's atmosphere, by 0.6 degree at the horizon (slightly more than the Sun's angular diameter). On the Moon, any atmosphere would analogously deflect a star passing behind the Moon during occultation (multiplied by two because light travels both into and out of the lunar atmosphere). Furthermore, lunar gravity is one-sixth Earth's, so its atmosphere is six times more extended. However, the Moon's radius is one-quarter Earth's, so the atmospheric path shrinks by four times. Timing an occulted star's disappearance, nineteenth-century astronomers found it never deviated more than 1.5 arcsecond (about 0.0004 degree) from its predicted (vacuum) position, so atmospheric pressure at the Moon's surface must be less than $(0.0004°/0.6°) \times 1/2 \times 1/6 \times 4$ or about 1/4000 of Earth's atmospheric pressure (assuming both atmospheres have similar composition and temperature).[7] Modern occultation tests using radio wave refraction from spacecraft find densities 10 billion times smaller (for charged particles, especially electrons).[8]

Lunar instruments answer whether the Moon has an atmosphere. To determine atmospheric composition, mass spectrometry can be amazingly sensitive. The Lunar Atmospheric Composition Experiment (LACE) on *Apollo 17* could sense several hundred molecules per cubic centimeter, about a 10-trillionth of Earth's atmospheric density at sea level.[9] An amazing result from LACE was the "wave atmosphere" of un-ionized gas on the Moon, with a cyclic density profile moving around the globe, sweeping past each site each month, highest at sunrise, lowest at night. As per Chapter 8, the cold nighttime lunar surface is sticky, and few molecules fly freely in ballistic bounces once they hit this cold ground. In contrast, the night's atmosphere is much thinner than the day's. The CCGE (Cold Cathode Gauge Experiment, or CCIG, Cold Cathode Ion Gauge) was placed on the Moon with ALSEP on *Apollo 12*, *14*, and *15* and measured the total number of particles hitting the detector: a density of 10 million per cubic centimeter during the day, 200 thousand at night.[10] At night these molecules were *ad*sorbed (stuck on the soil's surface, not *ab*sorbed inside).[11] At sunrise, however, lunar soil heats, and stuck gas molecules vibrate free into the atmosphere again, bouncing over the Moon. The atmosphere is densest then for most atomic/molecular species, 10 times the daytime average.

Millions of molecules per cubic centimeter sound huge, and monthly giant waves of gas sweeping westward around the Moon are dramatic but produce nothing seen from Earth. However, sunrise has other consequences. The Sun irradiates the lunar surface intensely with ultraviolet photons, many having sufficient individual energy to knock electrons from regolith grains photoelectrically. Losing electrons produces positive charge in sunlit areas, and charged dust grains are repelled from charged soil to actually levitate and fly to less positive, dark areas. (Also, solar wind electrons can impart negative charges in dark regions.) Electric fields are strongest at the boundary between light and dark, the local terminator, and mainly at sunrise or sunset (but elsewhere near rocks, craters, and local topographic relief). Thin clouds of dust hug the terminator, seen when illuminated at low sun angles looking toward the rising or setting sun (Figure 9.1). Seen throughout the 1966–1968 Surveyor program (by *Surveyor 1*, *5*, *6*, and *7*), this horizon phenomenon went unexplained[12,13] and unpublished in science journals, whereas similar observations of the solar corona (the Sun's extended atmosphere) were published.[14] Horizon glows observed in 1973 by *Lunokhod 2* were largely overlooked outside the Soviet Union.[15] In 1972 David Criswell (Lunar Science Institute) theorized photoelectric dust levitation, and later that year Gene Cernan, *Apollo 17*'s commander, was asked to draw from lunar orbit the setting Sun, to recreate the Surveyor observations. His results were shocking: one could not only see the corona but also huge light plumes stretching far beyond the Moon, even beyond the *zodiacal light* – dust in solar orbit in the inner Solar System. This implied dust elevated tens of kilometers, scattering sunlight toward Apollo, and was published (simultaneous to the Surveyor horizon glow paper)[16] but garnered little prompt attention. Huge plumes were only reported visually, not seen on all missions (seen on *Apollo 8*, *10*, *15*, and *17*, but

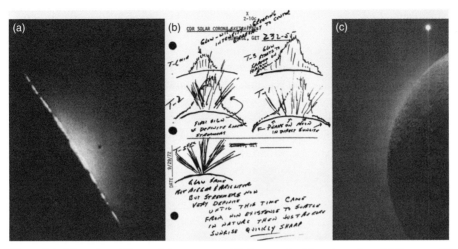

Figure 9.1. **Glow dust**. Views of dust-scattered light over the Moon's surface. (a) *Surveyor 6* took this TV image of the lunar horizon on November 24, 1967, about 40 minutes after sunset. Note the line of bright patches on the horizon, now interpreted as scattering by dust levitated by photoelectric charging of soil by solar ultraviolet light. (b) Sketches by Gene Cernan on *Apollo 17* as the Sun rose over the Moon as seen from orbit on December 16, 1972, around 22:28 Universal Time Coordinated (UTC), at 6, 5, 2, 1, and $1/_{12}$ minutes before sunrise. Note the extension of light along the horizon several minutes before sunrise, followed by bright streamers immediately before sunrise. The rapidly changing structure, even over several seconds, indicated effects of local lunar dust. (c) The Moon occulting the Sun (with Venus near the top of the image creating a vertical signal saturation streak), taken by the *Clementine* star tracker camera on April 18, 1994, at 01:12:52 UTC, about 4,500 kilometers from the Moon. One can see diffuse light possibly from dust scattering near the Moon's surface, but far above it mixes with zodiacal light from dust in solar orbit and seems to be weaker than the glow seen by Cernan. (Photos 67-H-1642, S-83–15138, LBA5881Z, courtesy of NASA.)

not *Apollo 16*; no observations were attempted on 8, *11*, *12*, and *14*). Results indicate small dust particles, about 0.1 micron = 0.0001 millimeter across, spread large distances from the Moon, highly variable but not artificial contamination. It remained a mystery.

In 1994, *Clementine* ended a generation's hiatus in lunar exploration, carrying imaging optical CCD detectors (charge-coupled devices) found in today's digital cameras but unavailable during Apollo. They could easily detect faint surface brightness signals such as Cernan's lunar glow and found a weaker signal, causing confusion but reinvigorating attention to the Moon's dust atmosphere.[17] New missions about 2009 promised progress (e.g., *Clementine*'s star tracker CCD had no filters, limiting analysis), and interest increased again, especially when NASA chose a mission devoted to lunar dust and gas, *LADEE*, in 2013–4 (see Chapters 4 and 8). Inspired by this, NASA developed a model to study how dust particles are lofted many kilometers off the Moon. Whether *LADEE* saw dust plumes above the Moon is still being debated.

A new model of lofted lunar dust starts with the same photoelectric effect of solar ultraviolet light but follows the path of charged dust.[18] Simply, dust motes do not rise until electrostatic forces overcome the sum of gravity and mechanical "sticking" forces. Once unstuck by a sufficiently strong force, the mote accelerates from the charged lunar surface with constant force. The velocity it attains depends on a quantity called the *shielding* or *Debye length*: the plasma that fills all space in the Solar System is composed of separate positive and negative charges, and, over the positively charged lunar surface, negative charges cluster closer to the ground until the surface's positive charge appears cancelled or shielded from the point of view of a distant charge. At the terminator this distance is about 10 meters, and escaping charged dust motes (as small as 0.01 micron) can reach 1 kilometer per second, almost half lunar escape velocity, enough to loft them 100 kilometers. Strong electric fields reside in the solar wind; perhaps these sweep lofted dust from the Moon altogether.

These little dust storms likely have long-term consequences in moving dust mass. Usually shadowed areas in bright plains (or vice versa) will collect (or lose) dust. Criswell's 1972 analysis (see note 14) implies typical dust loss of 1 milligram per square centimeter per year (1 tonne per square centimeter per billion years), more than all lunar regolith and many times faster than micrometeorite erosion. Most grains are too large to be lofted, and only surface dust feels the electrostatic force, so the same dust may move repeatedly. Long ago Thomas Gold was derided for hypothesizing that electrostatic forces in moving large amounts of dust might re-sculpt the lunar surface[19]; perhaps he was half correct.

The interaction of the plasma around the Moon with its surface is complex and leads to surprises. Solar wind plasma can flow along the surface, over a crater rim, then into the crater, leaving a wake of low plasma density just below the crater lip. Surprisingly, this plasma void under the wake becomes negatively charged, because electrons flow into the void easier than heavier ions.[20] Charged particles appear around the Moon at dawn and dusk as a result of material flowing along Earth's magnetic field.[21] Most generally, the Moon is always influenced by the solar wind, tending to hit the noonday surface at an angle perpendicular to the wind's entrained electric field. This field direction points from the sunrise side of the Moon to the sunset side, so electrons far above the Moon fly off toward sunset, and positive ions toward sunrise. Once an atom or molecule picks up a charge, it leaves the Moon quickly.[22] This is not surprising, considering that whole dust grains are flying from the Moon.

The Moon has two coextensive but separate atmospheres: its ionized plasma and electrically neutral gas, rarely interacting between the two except when an atom/molecule is ionized or deionized (so switching membership among the two). Plasma is quickly swept away (*advected*) from the Moon, whereas most neutral atoms/molecules engage in the monthly wave: ballistically bouncing over the surface (until escaping in too fast a bounce, or ionizing), often sticking and unsticking. This is our local example of a *solid-bounded exosphere*, an atmosphere so thin that atoms/molecules rarely interact with each other but do interact with the solid surface and surrounding electric and magnetic fields. Such atmospheres typify nearly all large Solar System bodies (except Earth, Mars, Venus, Saturn's moon Titan, gas giant planets, and Neptune's moon Triton). Mercury, Pluto, the asteroids, and hundreds of other moons have solid-bounded exospheres. We have the Moon nearby to tell us how these work.

The neutral atmosphere is steadier, being less affected by solar wind, so let us discuss it. Data about neutral gas come from *Apollo 17*'s LACE, from 10 months in 1973, revealing the wave atmosphere, measuring neutral atmospheric composition of molecular/atomic gas and of 1–110 atomic mass units (1 AMU being about a proton or neutron's mass),[23] and finding several codominant atmospheric components. Atomic hydrogen is surprisingly rare; the dominant form is molecular H_2 at roughly 60,000 per cubic centimeter.[24] Like numbers of neon atoms are found, also from the solar wind, mostly neon-20 (of 20 neutrons and protons) but 7% neon-22. Comparable numbers of helium-4 atoms are seen, from solar wind as well as some radioactive decay inside the Moon. In terms of mass, a large component is argon-40 from radioactive decay of lunar potassium-40, and some argon-36 common in primordial Solar System composition (solar argon is 86% argon-36 versus 0.3% on Earth, the rest being argon-40). Day/night variation is evident: 10,000 by day versus 200 at night for argon; 40,000 by day versus 2,000 at night for helium (per cubic centimeter).

Strongest atmospheric signals occur near sunrise – the best time to find most species. Solar wind species do not necessarily behave this way, and not all species stick well at night, for example, N_2, so atmospheric composition ratios change over lunar phase. Nonetheless, LACE looked for all of these, detecting species (sunrise number density, excluding hydrogen) ^{40}Ar, 20,000; ^{4}He, 15,000; ^{36}Ar, 1600 ± 400; CO_2, 1,400 ± 1,000; CH_4, 1,200 ± 400; O_2, 1,000 ± 1,000; N_2, 800 ± 800; CO, 800 ± 800; NH_3, 600 ± 600; and H_2O, 600 ± 600 (number per cubic centimeter, errors after ± symbol of about 2 standard deviations, usually 95% of probability range). In summary, argon-40 and helium-4 are strongly detected at similar count levels, but argon has 10 times the mass and dominates mass density. Argon-36 is detected at levels expected for solar wind, but at only 8% that seen for argon-40. If both argon isotopes originated in the solar wind, one would see 80 times less argon-40, so it must originate from lunar radioactive decay. Methane (CH_4) is detected and might originate from solar wind/regolith reactions. Other molecules (carbon dioxide, oxygen, nitrogen, carbon monoxide, ammonia, and water) are detected marginally.

Three instruments provide varied (conflicting?) measures of lunar atmospheric gas: SIDE (spectra of ions only, Chapter 8), LACE (spectra, neutrals only), and CCGE/CCIG (total particle count).[25] The *LADEE* mass spectrometer orbiting the Moon in 2013–4 might break this "tie." (It has detected considerable argon-40.) Meanwhile *LRO*'s LAMP (Lyman Alpha Mapping Project) has seen helium at one-third the concentration from LACE, detected by light emitted because of solar ultraviolet light excitation.[26] SIDE and CCGE/CCIG results are even higher than LACE; the new LAMP result better supports LACE. Suffice now saying that roughly equal numbers of hydrogen molecules helium-4 and neon-20 and somewhat fewer argon-40 exist, still large in terms of mass. Summing all species, day or night, yields a total lunar atmospheric mass around 20 tonnes (the mass of air in a 50,000-square-meter building, a soccer/football field's area, one story high).

Argon-40 density varies wildly over a month but also varies by a factor of three between consecutive months. Because typical argon-40 atoms reside in the lunar atmosphere about this long, argon-40's source must vary roughly 100% between months, leaking from the Moon highly episodically.[27] Helium is also produced radioactively, but more arrives via solar wind. Argon-40 is the purer tracer, with surprising implications. On average several tonnes of argon reside in the lunar atmosphere, so 2×10^{21} atoms per second must be produced. This is huge considering potassium-40 composes only 1/100,000,000 of the Moon's mass and decays with a long half-life (1.25 billion year). About 8% of argon-40 produced in the Moon leaks out, so some originates deep inside.[28]

There are more ways to detect gas from radioactive decay, and these also indicate gas leaking episodically from the Moon. Uranium-238 and thorium-232 decay by emitting helium-4 nuclei (alpha particles), which mix confusingly with solar wind helium.[29] In several steps uranium-238 decays to radon-222 gas, also radioactive and decaying eventually to polonium-210, then finally to lead-206, a stable isotope. Radon-222 has a half-life of 3.8 days, usefully long, whereas thorium-232 produces radon-220, with a 56-second half-life decaying before it can do much.

In 3.8 days, can radon-222 leak to the lunar surface? Apparently it does. Several times in lunar orbit, spacecraft observed radon-222 gas pooled on the ground. Radon is so heavy that it disperses only about 100 kilometers before decaying. (Radon atoms' ballistic bounces are only a few hundred meters.) At least two such events were seen by *Apollo 15* and at least two more by *Lunar Prospector*. Similar preliminary results are reported by *Kaguya*.[30] These were detected with alpha particle spectrometers, measuring the energy of those helium-4 nuclei decay products. These alphas have insufficient energy to penetrate even a sheet of paper but can launch from the Moon through space to hit an orbiting detector. Measuring their energy, the detector identifies which atomic isotope produced them. Half of radon-222 decays will shoot an alpha into space, then recoil into the ground, depositing polonium (then bismuth and finally lead). Half of alphas fire into regolith and rocket the radon (now polonium) beyond lunar escape velocity into the Solar System.[31] Alphas from the first sort of decay geometry can be detected.

Because radon-222 vanishes in several weeks, these gas concentrations must be fresh. Since the distribution of gas fluctuates over time, this does not indicate constant leakage. Still, these signals require only a gram per event of radon-222, assuming all gas reaches the surface. Laboratory experiments in vacua with regolith simulants show radon moving slowly through the regolith, implying either much more radon released into the regolith than reaching the surface or other gases released to carry the radon through the regolith into space.[32] Lunar radon outgassing and argon (a much larger contribution) both appear episodic. The Moon seems to vent much of its gas in short-lived events.

Sufficient gas leaking through the soil can create a situation in which the whole regolith layer is unstable to gas explosively expanding into the vacuum. Lunar regolith is tremendously impermeable to gas, so even a small flow (less than 1 gram per second) will accumulate. Even a tonne of typical lunar gas (e.g., mass 20 AMU, between 4 for helium and 40 for argon) creates a situation in which an explosion is energetically favored, blowing a plug of regolith into space as a dust cloud. It will expand to several kilometers in radius in several minutes. If gas reaches the soil by delivery at moderate or high flow rates to a small area at the regolith base, several such explosions might take place per month, considering that roughly 10 tonnes leak from the Moon monthly.[33]

An important coincidence is that phenomena resembling this are reported. Although the term *transient lunar phenomena* (TLPs or LTPs [see note 11 in Appendix B]) is a catchall category for any temporary change in lunar appearance (or even its gas release events), the reported appearance range is more limited. More than 1,000 such reports, collected by the exhaustive works of Winifred Sawtell Cameron and of Barbara M. Middlehurst,[34] are primarily brightenings (white or color-neutral increases in surface brightness), reddish (red, orange, or brown color changes with or without brightening), bluish (blue, blue-green, or violet color changes with or without brightening), or gaseous (obscurations, misty, or darkening changes in surface appearance). Nearly all TLPs are highly localized, usually to a radius much less than 100 kilometers, often unresolved points (roughly 1 kilometers or less). Aristarchus covers a full quarter of all reports; other sites on the Aristarchus Plateau (Cobra's Head, Schröter's Valley, and Herodotus) add another 8%. Hence one-third of reports inhabit the plateau, only 0.1% of the lunar surface. Roughly another 20 sites each contribute 1% or more of the reports, and some 80 other sites only 1–3 reports each. TLP report site frequency is shown in a Near Side map (Figure 9.2) and listed in Table 9.1.

Now some readers have started groaning, because TLPs have a bad reputation in select audiences. By nature TLP reports are anecdotal, an issue of difficulty with the scientific method because they are not testable by repeat experiment. The subjectivity of many TLP reports does not engender trust for their reliability. Establishing their consistent behavior is required to assess their applicability in understanding physical processes on the Moon. What is the consistent behavior in these reports? This requires lengthy historical discussion and statistical analysis that some readers may prefer to skip. These are located in Appendix B, which the reader should scrutinize. (Other Appendix B topics include lunar meteoroid impact observations, naked-eye TLPs, TLP photography, and emission of light by atoms in the lunar atmosphere.)

Few TLPs were reported before 1610 and the invention of astronomical telescopes, and TLPs emerged primarily via visual lunar mapping using telescopes, as selenographers described surface features, then used shadow lengths to map topography (heights of mountains and crater walls), and exploited lunar librations to peer around edges of the Moon at the extra 9% in marginal area. Summarizing a major point in Appendix B, these early observers with few exceptions reported their TLPs such that few other observers were aware of them, until 1957, when several notable reports and incipient space age enthusiasm radically increased awareness of TLPs. The unfortunate result of this interest is a change in TLP reporting, making it less reliable after 1957.

The analysis in Appendix B discovers consistent behavior among observers before 1957. Regardless of how one subdivides these observers by their observational vantage point or when they observed, they report activity consistently at the same TLP sites. Amazingly, we discovered later that these same sites are nearly identical to sites showing a transient physical behavior local to the Moon: outgassing of radon.

The New Moon

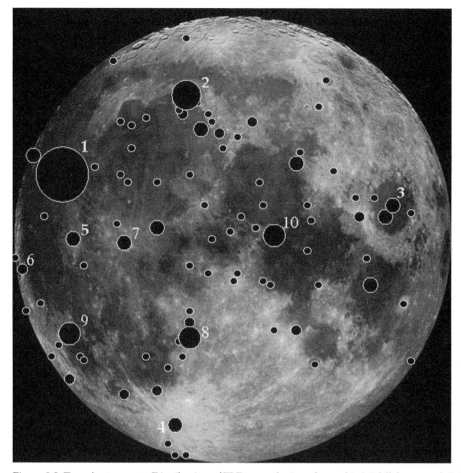

Figure 9.2. **Transient reports**. Distribution of TLP report loci catalogued in Middlehurst et al. (1968) (see note 36), with exception of a minority of cases rejected for reasons detailed in the text. The symbol sizes encode the number of reports per features, listed in Table 9.1. Marked features include (1) Aristarchus, including Schröter's Valley, Cobra's Head, and Herotus, (2) Plato, (3) Mare Crisium, (4) Tycho, (5) Kepler, (6) Grimaldi, (7) Copernicus, (8) Alphonsus, (9) Gassendi, and (10) Ross D. (*Galileo* photo PIA00405 courtesy of NASA.)

By *consistent* we mean TLPs behave independently of factors characterizing various properties of the observers, separate from physics on the Moon. By requiring that TLP sites maintain consistent behavior throughout the report catalog regardless of when historically or where on Earth observations were made, the following features survive (Table B.9, with statistical confidence of 99% or better): Aristarchus Plateau 46.7% ± 3.3%, Plato 15.6% ± 1.9%, Mare Crisium 4.1% ± 1.0%, Tycho 2.8% ± 0.8%, Kepler 2.1% ± 0.7%, Grimaldi 1.6% ± 0.6%, and Copernicus 1.4% ± 0.6% (errors are ±1 standard deviation). Portraits of these are shown in Figure 9.3, and we discuss each. Aristarchus and vicinity are responsible for roughly half of reports and worth more discussion (see Figure 9.8). Four of these form the youngest, largest Near Side impacts: Aristarchus, Tycho, Kepler,

Table 9.1. *Number of transient lunar phenomenon reports, by feature: Raw report count from Middlehurst et al. 1968 (see note 36).*

Number of TLP reports	Feature (latitude, longitude)
122	Aristarchus (24N 48W)
40	Plato (51N 9W)
20	Schroter's Valley (26N 52W)
18	Alphonsus (13S 3W)
16	Gassendi (18S 40W)
13	Ross D (12N 22E)
12	Mare Crisium (18N 58E)
6 each	Cobra Head (24N 48W); Copernicus (10N 20W); Kepler (8N 38W); Posidonius (32N 30E); Tycho (43S 11W)
5 each	Eratosthenes (15N 11W); Messier (2N 48E)
4 each	Grimaldi (6S 68W); Lichtenberg (32N 68W); Mons Piton (41N 1W); Picard (15N 55E)
3 each	Capuanus (34S 26W); Cassini (40N 5E); Eudoxus (44N 16E); Mons Pico B (46N 9W); Pitatus (30S 13W); Proclus (16N 47E); Ptolemaeus (9S 2W); Riccioli (3S 74W); Schickard (44S 26E); Theophilus (12S 26E)
2 each	1.3' S.E. of Plato (47N 3W); Alpetragius (16S 5W); Atlas (47N 44E); Bessel (22N 18E); Calippus (39N 11E); Helicon (40N 23W); Herodotus (23N 50W); Littrow (21N 31E); Macrobius (21N 46E); Mare Humorum (24S 39W); Mare Tranquilitaties (8N 28E); Mons La Hire (28N 26W); Montes Alps, S. of (46N 2E); Montes Teneriffe (47N 13W); Pallas (5N 2W); Promontorium Agarum (18N 58E); Promontorium Heraclides (14N 66E); South Pole (90S 0E); Theaetetus (37N 6E); Timocharis (27N 13W)
1 each	Agrippa (4N 11E); Anaximander (67N 51W); Archimedes (30N 4W); Arzachel (18S 2W); Birt (22S 9W); Carlini (34N 24W); Cavendish (24S 54W); Censorinus (0N 32E); Clavius (58S 14W); Conon (22N 2E); Daniell (35N 31E); Darwin (20S 69W); Dawes (17N 26E); Dionysius (3N 17E); Endymion (54N 56E); Fracastorius (21S 33E); Godin (2N 10E); Hansteen (11S 52W); Hercules (47N 39E); Herschel (6S 2W); Humboldt (27S 80E); Hyginus N (8N 6E); Kant (11S 20E); Kunowsky (3N 32W); Lambert (26N 21W); Langrenus (9S 61E); Leibnitz Mts. (unofficial: 83S 39W); Manilius (15N 9E); Mare Nubium (10S 15W); Mare Serenitatis (28N 18E); Mare Vaporum (13N 3E); Marius (12N 51W); Menelaus (16N 16E); Mersenius (22S 49W); Mont Blanc (45N 0E); Montes Carpatus (15N 25W); Montes Taurus (26N 36E); Peirce A (18N 53E); Philolaus (72N 32W); Plinius (15N 24E); Sabine (1N 20E); S. of Sinus Iridum (45N 32W); Sulpicius Gallus (20N 12E); Taruntius (6N 46E); Thales (62N 50E); Triesnecker (4N 4E); Vitruvius (18N 31E); Walter (33S 0E)

Not counted: 4 (global lunar changes), 14 (cusp events), and 43 (events w/ unknown coordinates).

The New Moon

Figure 9.3. **Diverse activity**. Various lunar missions' views of robust TLP report activity sites: (1) southeast Aristarchus Plateau: Herodotus in lower (distant) left, Aristarchus crater in lower right, Cobra Head above Herodotus extending into Schröter's Valley at left. Notice many other rilles running off the plateau. Image is foreshortened, roughly 160 km on a side; (2) Plato. Note the rilles at right. Image is 230 km wide; (3) Mare Crisium. Foreshortened image is 760 km across; (4) Tycho (image 130 km wide); (5) Kepler (42 km); (6) Grimaldi (280 km); (7) Copernicus (95 km). Aristarchus, Copernicus, Kepler, and Tycho are similar in appearance; this figure shows the effects of illumination angle.

and Copernicus. Aristarchus appears a class in itself; the other three craters total 6.2% ± 1.2% of reports. Plato and Grimaldi are both old, flooded craters at edges of large mare areas (Imbrium and Procellarum, respectively). Mare Crisium is a unique, confusing case.

This list's utility is its potential reliability; at least it is consistent and varies less with observer parameters than the full list. It can be compared to known physical phenomena. A localizable physical transient relevant to lunar outgassing is radon alpha-particle events. Their correlation with the TLP list is stunning. All four prominent radon-222 outgassing events land at this list's features: Aristarchus (twice, on *Apollo 15* and *Lunar Prospector*), Grimaldi, and Kepler. Indeed, these four radon events are entirely consistent with a random draw from the frequencies listed above for robust TLPs, especially because *Apollo 15* never flew over Plato or Tycho. Chances of these four events chosen totally at random and matching the robust TLP list are minuscule. A generous estimate for the area associated with each TLP site implies the probability of such a random selection is about 0.0001; a more reasonable estimate is 30 times less. Almost certainly some reason explains outgassing events' association with robust TLP sites; the most plausible explanation being some physical association between TLPs and outgassing. *SELENE/Kaguya*'s Alpha Ray Detector also detected Aristarchus radon-222 in 2009 (see note 32).

Another way of measuring radon-222 outgassing is counting alpha particles from its daughter product polonium-210, produced in no other significant ways. Even if radon-222 leaks out slowly, accumulated activity over the past century will appear as polonium-210 alpha particles at that location. A polonium-210 alpha flux map shows good correlation with the boundary between mare and highlands.[35] The polonium-210 distribution was also measured on *Lunar Prospector* and also correlates well with mare edges, with only 0.0065% probability of being random.[36]

TLPs obviously correlate with mare edges, as even casual inspection reveals (Figure 9.2). Also the robust sample (Figure 9.4) correlates with mare edges at 99.9% to 99.99% confidence; all events, robust or not, correlate with essential certainty (see note 38). However, as robust TLP sources correlate with sites of episodic radon-222 outgassing, even low-level TLP activity correlates with the mare edges like long-term radon outgassing traced by polonium-210. Other features correlate with this boundary: (1) surviving volcanic vents (elongated depressions with V-shaped depth profiles, often associated with rilles; Figure 8.1) and dark deposits from pyroclastic eruptions (driven by escaping volcanic gas) and (2) pyroclastic deposits themselves.[37] Given mare structure (basalt caps overlying volcanic activity), perhaps gas escaped at their edges.

Correlation with outgassing in two different isotopes supports the idea that TLPs are physically tied to the Moon, not an observer/physiological or otherwise Earthbound effect. What is the nature of this association, causal or merely coincidental? What outgassing effects are expected, and could they cause TLPs? Remember that argon-40 appears to outgas episodically as well as the tracer gas radon-222, so plausibly most lunar atmospheric mass comes from the interior episodically.

Another physical association with mare edges is moonquakes (see Chapter 8). Although shallow moonquakes correlate somewhat with mare edges, correspondence of deep (700–1,200 kilometers underground) quakes to mare edges is obvious (Figure 9.5). Few TLPs are tied to quakes temporally, not convincingly beyond randomness. This is unsurprising given the depth of these quakes versus TLPs with

The New Moon

Figure 9.4. **Consistent activity**. Distribution of robust TLP sites, in Table A.9 and as explained in the text. Symbol sizes encode the fraction per feature of total reports. Symbols correspond to (1) Aristarchus, including Schröter's Valley, Cobra's Head, and Herotus, (2) Plato, (3) Mare Crisium, (4) Tycho, (5) Kepler, (6) Grimaldi, and (7) Copernicus. The dashed curved contour is the adopted boundary between mare and highlands, as explained in the text. (Photo *Galileo* PIA00405 courtesy of NASA.)

which they correlate geographically. TLPs correlate weakly with times of maximal tidal stress. Separately, moonquakes have been tied to episodic ^{40}Ar release.[38]

We do not know the true TLP rate but discuss constraints on this in Appendix B. There are redundant reports of the same TLP, indicating a large fraction of events being reported, when the report rate approaches one to several per month, consistent with explosive TLPs providing the mass of gas leaking into the atmosphere, at several tonnes per month.

Assuming no chemical reactions or phase changes of gas in the regolith (Chapter 8), behavior divides between uneventfully slow gas seepage at one extreme and outgassing explosions or jets on the other. Between, gas can "bubble" through the regolith, fluidizing it, which agitates and sculpts it without moving it long distances. The

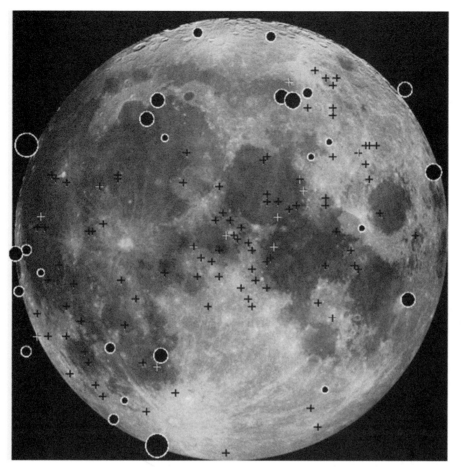

Figure 9.5. **Moon shaker**. The distribution of moonquakes, primarily on the Near Side (where detections concentrate as a result of the Near Side placement of seismometers). Shallow moonquakes (less than 700 km deep) are marked by filled circles, some off the lunar surface by an amount equal to the distance they wrap onto the Far Side. Shallow quakes' marker diameters scale with their magnitudes. Deep moonquakes are shown as small crosses. Their correlation with mare edges is obvious. Although the chance that shallow moonquakes' correlation with mare edges is random has a probability of 0.0002, the deep moonquakes almost certainly correlate with mare edges, to a certainty of one part in 10^{14}.

complex onset and behavior of fluidization has been studied.[39] Can we distinguish between these alternatives by how they mix and move regolith?

A tool to study exposed regolith, measuring regolith grain freshness, the Optical MATurity (OMAT) parameter (see Chapter 6), measures the age of surfaces utilizing changes in light at 0.95 microns wavelength because of surface iron deactivation. Shocks from outgassing discussed earlier are insufficient to fracture regolith grains; they only agitate them. OMAT might show when volatiles mix fresh, unweathered grains to the surface. Outgassing agitation could be so rare that young OMAT either fades away or exists as single, isolated surface features, over 100 million years. This depends in part on how deeply regolith is agitated. Explosive outgassing will mix

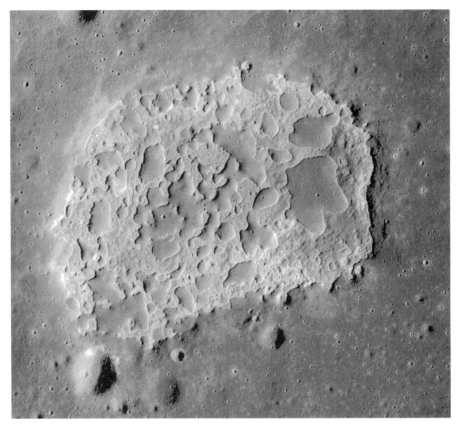

Figure 9.6. **Scrubland**. The Ina feature is a depression not excavated by impact. It is 3 km wide and 60 m deep and has an age measured by OMAT of several million years or less. The lighter areas are deeper than the grey areas they surround, and some of the grey areas rise about 40 m, approaching the level of the surrounding regolith surface. (NASA/ASU/GSFC LROC-NAC images M113921307LE and M113921307RE.)

the entire regolith thickness, at least several meters thick, so meteorite impact mixing (gardening) is overwhelmed by gas action. If fluidization occurs only at shallow depths for gas approaching vacuum, gardening could overwhelm volatile mixing in this thin layer before OMAT ages significantly (10 centimeters per 100 million years). More outgassing could homogenize the entire regolith depth to uniform OMAT age, and further mixing would hardly change surface OMAT, not unusually young, for the entire several meters (in several billion years since maria flooding). If observed TLPs arise from outgassing, their observed rate (many per year) likely corresponds to frequent mixing.

OMAT and outgassing may join in a dramatic circumstance: zones where outgassing from the interior erodes regolith down to the bedrock (or at least the megaregolith underlying the regolith). An example may be the Ina feature in Mare Imbrium (called a crater, caldera, or the D feature; Figure 9.6). Its strangeness was noted as early as *Apollo 15*, but OMAT showed its surface to be several million years old or less. (Now we know of several dozen similarly eroded features.) Subsequent high-resolution images show the rough, low-lying surface to be almost crater free, young and rocky, whereas the softer, elevated landscape has a cratering density closer to the surrounding maria's.[40] The hypothesis of regolith eroded by outgassing from low-lying areas is still the favored hypothesis, but not by all investigators.

There are methods sensitive to changes from outgassing, some being used. Unfortunately LROC has no 0.95- or 1.8-micron filter (nor did *Lunar Orbiter* or *Apollo*). *Kaguya/SELENE*'s Multiband Imager had a 0.95-micron filter, but at 20-meter resolution. A way to look for small-scale changes, given the tight constraints on filters, employs broadband reflectivity (as in the 0.75-micron OMAT band). Comparing features at different epochs has little meaning unless illumination and viewing geometries are similar, because reflectivity varies with these passive conditions. We have found one new feature at a TLP site that changed too radically for passive effects alone;[41] we hope for LROC-NAC to recreate the conditions for this site as first imaged by *Lunar Orbiter V*.

Further work in progress might catch outgassing in action. We have models of explosive outgassing producing TLP-like events and TLP historical records indicating optical transients' magnitude and time scale. Automated telescopes and CCDs enable continuous monitoring of the Near Side from Earth, and on TLP time scales several images should be taken per minute. Changes on the Moon in an hour from normal diurnal shadow motion are negligible except at the terminator, and software can analyze changes via image difference algorithms (Figure 9.7). To study TLPs

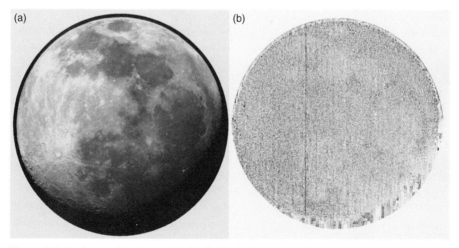

Figure 9.7. **Staring at the Moon**. (a) Flat-fielded, dark-corrected but otherwise raw image of a typical lunar Near Side image obtained by our robotic imaging monitor. (The image is trimmed to a standard circular region.) (b) The difference in signal between the image at left and similar one obtained 5 minutes later. The noise in the residual signal is essentially at the photon shot-noise limit. Because of a slight error in the photometric calibration between the two images, there is a very slight ghost of high-contrast global features, especially Imbrium, Humorum, and the eastern maria. Note that even bright smaller features, e.g., Tycho left and below center, are subtracted nearly identically. Even subtle features not apparent in the image above, e.g., a slightly nonlinear image column, just left of center, are readily apparent. There are errors along the lunar limb because of the rapid gradient in signal level.

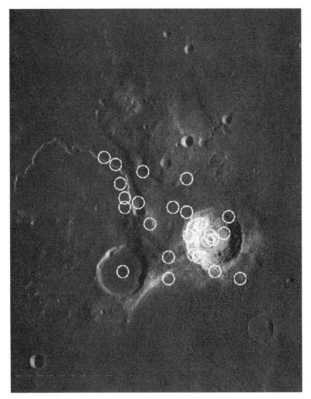

Figure 9.8. **Realm of activity**. A *Lunar Orbiter* image of Aristarchus (bright crater, right of center), Herodotus (below left of center), Vallis Schröteri (extending from center to left edge), and the southeastern quadrant of the Aristarchus Plateau (its southern edge running horizontally below Aristarchus and eastern edge vertically to the right). White circles indicate TLP reports from historical records before 1957 localized to within 10 km. (The view is about 220 km across.) Note report concentrations within the crater Aristarchus and on the plateau adjacent and running down Vallis Schröteri.

and potentially use them to probe outgassing, we need a sample that can be evaluated digitally, objectively, and with known observer selection biases. Robotic monitoring may replace the historical database and put the field on new footing. Our group has collected a quarter million such Near Side monitor images and will present data on TLPs later. Monitoring lunar optical transients can also have other purposes.[42]

Obviously the Aristarchus region is special, with roughly half of TLPs as well as half the radon-222 events. Figure 9.8 maps TLPs localized within 10 kilometers (others mentioning the whole crater, for instance). Evidently these populate the plateau's southeasternmost 10,000 square kilometers. Aristarchus is among the youngest large lunar craters, 175 million years old, and the plateau adjoins south, east, and west some of the youngest lunar surfaces, evidenced by crater counts, 1 billion years old or less.[43] The plateau is mantled by pyroclastic deposits from huge volcanic effusions evident in many volcanic rilles, on the plateau, in the flow

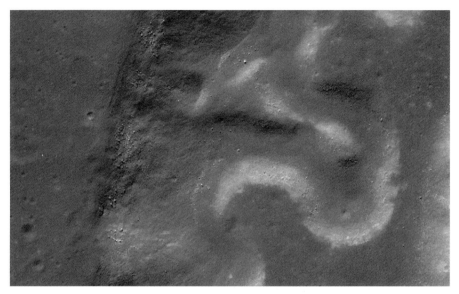

Figure 9.9. **Down in the valley**. View of Vallis Schröteri seen from LROC-NAC (image M104848322L), showing, from the left, the surface typical of the Aristarchus Plateau, with high density of cratering and dark mantling. About one-quarter of the way from the left one encounters crest of the outer rille showing mass wasting and lower crater density, sloping down to the level of the outer rille just left of center. The outer rille's left half is occupied by the inner rille, with its oxbow, meandering shape. The outer rille's floor continues to the right edge, meeting the base of upslope back to the plateau at bottom right. The floor of the inner rille, at the oxbow bottoms, are lightly cratered, whereas the cratering density on the outer rille's floor is intermediate between the inner rille's and plateau's. The field of view is 4.8 km across.

off it, and in surrounding landscape.[44] Vallis Schröteri, by some definitions the largest lunar rille, slices through the plateau and is associated with many reported optical transients (Figure 9.8).

Several units compose Vallis Schröteri: the large rille up to 10 kilometers wide and a longer, more sinuous inner rille about 1 kilometer wide, tending to flow from wall to wall of the large rille, having a flat bed several hundred meters below the large rille's floor. The large rille's walls display great mass wasting (avalanching), with several meters to 30-meter boulders rolling onto both rilles. Figure 9.9 shows examples of these units. Their crater counts vary greatly. The plateau itself has the highest crater count, the outer rille a density about 50% this, but the inner rille about 10% of the plateau's, one of the smoothest lunar surfaces. Although the rilles are recessed and therefore subject to smaller angles of incoming impactors, this is only a 10–20% effect. On the plateau, many features are secondary impacts from the Aristarchus impact itself. However, both rilles are often aligned radially to Aristarchus crater (including in Figure 9.9), so they received a full flux of secondary impacts. It seems material within the rilles, especially the inner one, either erases or cannot sustain cratering. We speculate that this area shows effects of volatiles, either disturbances from outgassing, modification of the inner rille's bed, or response to

Aristarchus's impact, 175 million years old – 20 times younger than the plateau mantle. There is much unknown about lunar volatiles' role, and we leave the reader with that mystery.

Many mysteries remain about the Moon, and the Moon can settle many mysteries about the larger Universe. What are the contents and structure of those lava tubes (Figure 6.25)? They are amazing in preserving a moment from several billion years ago, likely free of the slow grinding of regolith from impacts. Some may even be sealed against the vacuum: What atmosphere do they contain? (Helium and argon are prime candidates, but water vapor and carbon dioxide/monoxide are possible.) These seem prime structures to explore but also challenges to robots or humans. Clever designs for lava tube robots are underway. The first step should be ground-penetrating radar to map their entire three-dimensional structure (Figure 6.24). Radar on upcoming Chandrayaan-2 seems promising. Until we establish lava tubes' safety, we should not send astronauts (although lunar caverns are likely more stable than Earth's). Lava tube rovers would likely not roll on wheels because of debris on the tube floor: boulders and rocks, not dusty regolith. A walking rover is better, a long-legged quadruped (Figure 9.10) with sets of spare feet. If a foot is caught in lava tube talus, the rover could release it and attach a spare to continue. Unbreached lava tubes would be intriguingly pristine, likely containing an atmosphere. Scientists would desire drilling into this space, carefully. Depending on the tube's atmospheric density, we might imagine inflating a small blimplike drone to explore this inner lunar world.[45]

Lava tubes are not the sole pristine, ancient environs. In some places scientists have identified ancient regolith buried by later lava flows (*paleoregolith* versus buried mare under thick regolith *cryptomaria*). Paleoregolith was sandwiched between lava flows anywhere from about 4 billion to 1 billion years ago, and may preserve particles deposited by ancient solar wind, galactic cosmic rays, ancient lunar atmosphere, and maybe even ancient terrestrial material from Earth impacts, otherwise unobtainable samples from definite times in the distant past.

Near the poles even ephemeral atmospheric gases are preserved for billions of years, in the coldest places on the planets, colder than distant Pluto. The deep freeze entombs even molecules from the vacuum, which should collect ultimately in the upper 50–100 centimeters of soil. Some were excavated by the *LCROSS* impacts, which may have released even deeper volatiles. We must avoid tendencies to explore no farther than predisposition suggests. Send a robot to drill down only a meter, we may miss some of the most interesting material. If gas leaks up from the lunar interior, it will stick in the regolith even deeper. Our drills should explore this volatile component as well (how ironic to drill down a limit of 1 meter, only to find the interesting stuff just at the edge of our reach). As discussed, we might explore gases from the interior, from comets/asteroids, the solar wind, probably even interstellar medium, and even in a time sequence. Again lunar lava tubes might play a part, accumulating (not expelling) material falling in: a "clock" betraying what was deposited in layers over millions or billions of years.

Recently we learned that water (and sulfur and probably carbon monoxide/dioxide) spewed from the Moon prodigiously several billion years ago in fire

Figure 9.10. **Exploring inner space**. Artist's conception of an ambulatory rover exploring a sealed lunar lava tube. A sealed cavern might accumulate an atmosphere of several percent of Earth's sea level density; one can imagine launching a simple reconnaissance drone (*foreground*), a balloon with several grams devoted to a control unit and camera. The quadruped rover's chassis has a flood lamp, multi-wavelength camera head, and array of sampling and sensing instruments. Each of its four feet can be jettisoned if it befouls on obstacles; the chassis holds several replacement feet. The lava tube's thermal, radiation, and meteoritic environment is benign, so the rover carries little insulation or shielding once injected into the tube. Communications to the surface are maintained via radio relay. (Composite of NASA and National Park Service photos.)

fountains, some of them gargantuan. (Indirect measurement of cooling fire fountain glasses indicate some flew 100 kilometers before hitting the surface, erupting at hundreds of meters per second.) What happens in quieter cases? If magma rises to within several kilometers of the surface, it can release its water vapor and other gases, and these will interact with the crust and regolith. We do not understand how much water is locked into the regolith or mineral lattices of rocks themselves. We are only starting to ask these questions.

Of course hydration (hydroxyl) arises from solar wind interacting with the regolith, leaving a chemical "dew" in and on regolith grains. We know this from infrared spectra from Moon Mineralogy Mapper (M^3) and other investigators in the previous decade. Uncertainties remain: the conversion from how much hydration forms to how strong an infrared signal is produced (a subtle calculation of light propagating through regolith grains). How hydroxyl propagates to the poles is unsettled. Some wait for a new M^3 version orbiting the Moon pole to pole at perhaps the end of this decade years (because *Chandrayan-1* died prematurely), but maybe we will learn sooner. The interpretation of M^3 data depends crucially on correcting for spectral contributions by thermal emission. *LRO*/Diviner thermal infrared data will make this easier.

Mysteries about volcanism exist beyond fire fountain lava. The comparatively tiny Far Side Compton-Belkovich region of several thousand square kilometers mimics massive eruptions from Near Side Procellarum-KREEP Terrane covering 10% of the Moon. Different in scale, both seemingly expose magma from the mantle at the same depth. Compton-Belkovich penetrates some of the thickest lunar crust. Is this a lunar analog of hot spots under Hawaii, Yellowstone, or the Galápagos?

A panoply of geological mysteries remains about why compositions and landforms appear in particular places – for example, what causes iron-rich pockets in primitive crust – but lunar geologists feel secure in the minerals predicted by magma ocean models. We do not know how deeply this magma ocean reached. We have poor understanding of how deep moonquakes occur. The deep Moon is mysterious. *GRAIL*, new seismometers perhaps such as those of the International Lunar Network (ILN), and better understanding of magma rising from the deep, from where and when, might tell what occurs down deep. We are recently converging on consistent understanding of the Moon's core: liquid or solid, how big, and what composes it. Seismometers spread further around the Moon than Apollo's will tell. (See note 32 in Chapter 7.)

The Moon has done some of our work for us: the SPA basin is the 12-kilometer deep hole that briefly dug hundreds of kilometers below the surface. What did this dredge up? Remote sensing helps; samples returned from SPA will provide ground truth and a check.

Basic questions remain: Why is there a Near Side/Far Side asymmetry, or more commonly, why a man in the Moon? Why does the side with maria face us, and why does the Moon have one face with maria? One theory says this occurred because of large-scale unevenness in the internal overturn of magma ocean. Another theory says a huge but slow collision layered crustal material onto the now Far Side. Another says a huge, fast collision blew material from the lunar side now containing maria. The jury is undecided.

Evidenced by lobate scarps seen by LROC, the Moon is apparently shrinking. Why? We will learn much about the lunar interior, its mass distribution, internal stresses, limits on its strength and therefore its makeup, from new data from *GRAIL*. From the core to the mantle and Man in the Moon, we might solve some of these.

How was the Moon born? Was it a Giant Impact, or is the match between isotopes from Earth and the Moon too good? Improving our insight is challenging, because we need data from Earth and the Moon's earliest times, from which the least remains. The straightforward way to constrain this is better information on the lunar interior's structure and chemistry to decide which of many numerical models for the Big Whack fits best. One desires radically new data that make a definitive statement. A mission is proposed to sample potential debris from the Whack remaining around stable points in Earth's solar orbit, some from parts of proto-Earth and most from the impactor Theia. Such data could pick the winning theory.

A major victory of Apollo and lunar exploration for planetary science was establishing impact chronology over Solar System history, especially the Late

Heavy Bombardment. The Moon itself is a unique test particle to study gravitation. With completion of *LRO* topography and *GRAIL*, we have superlative knowledge of lunar distributions of mass and surface locations. Reflectors on *Apollo 11*, *14*, and *15* and *Lunokhod 1* and *2* originally located by laser travel time from Earth within 20 centimeters, now approach 1-millimeter precision with improved lasers. Better, cleaner reflectors (without decades of dust), might reach 0.01 millimeter, astonishing precision to dissect Earth-Moon gravity so intricately to test amazing ideas. Recall Galileo and his apocryphal Tower of Pisa experiment: Do all objects fall identically in a gravitational field? Does space have extra dimensions? Does gravitation strengthen or weaken over time? Do extra forces act as pseudo-gravity? Mirrors and the Moon are and could still be best to test these with superlative sensitivity.

Not only is the Moon a test particle, but it is a particle detector – for the highest energy particles known, a single subatomic particle having the kinetic energy of a speeding motorcycle. No particle can travel faster than the speed of light in vacuo, but these particles travel faster in the Moon than lightspeed in lunar material. This produces Čerenkov radiation, seen as bluish light from energetic particles streaming through nuclear fuel-rod storage tank water. Čerenkov light is analogous to sonic booms from airplanes exceeding the speed of sound. For an ultra-high energy particle this spreading photon cone is so energetic that it forms non-photonic particles such as electrons, called Askaryan radiation (for Russian-Armenian physicist Gurgen Askaryan). These charged particles interact with the Moon to emit low frequency radio waves detectable from Earth.[46] This excellent way to detect the Universe's highest energy particles could use radio antennae on the lunar surface to add information on the extent of this Askaryan spray, further informing us of these particles' nature.

An optical telescope on the Moon has been a powerful dream for decades. It could stare intensively at the sky for the faintest, most distant objects in the Universe. The Moon excels for this in many ways: it has too little atmosphere to affect observation, is extremely dark all month long in places, and rotates so slowly that easy effort keeps a telescope on the same patch of sky. The telescope must deal effectively with dust, but this can be simple. The telescope would have competition: NASA/ESA plan the 6.5-meter diameter James Webb Space Telescope (JWST) at Earth-Sun's Lagrangian point L2, 1.5 million kilometers from Earth, by 2018. A lunar telescope could be larger, primarily via liquid telescope mirrors. This technology uses gravity to define the telescope's main mirror shape rather than finely polished glass. Instead one pours liquid into a dish and rotates it, and it rises up the dish's sides to form a paraboloid (primary element in a Newtonian telescope and other designs). Such telescopes are inexpensive: the Magellan telescopes in Chile with 6.5-meter glass mirrors cost $50 million apiece. The 6-meter liquid-mirror Large Zenith Telescope (LZT) near Vancouver costs about $900,000. (JWST will cost initially about $8 billion.)

Lunar liquid mirror telescopes (LLMTs) have issues. What is the liquid? The LZT uses mercury, which on the Moon would freeze (reflecting poorly) or sublime into the vacuum if liquid. Ionic liquids are like salts liquid at low temperatures,[47] have near-zero evaporation rates, and can hold reflective metal coatings. Possible dust contamination is mitigated by LLMT's potential to resurface itself quickly, keeping peak reflectivity. LLMTs only point nearly straight up, because gravity defines their mirrors. An LLMT at the North/South Pole would trace a celestial circle 1.54° in radius every 18.6 years (the lunar precession period), only 0.52 degrees annually or 5 arcseconds daily, versus Earth rotating 15 arcseconds per second. LLMTs could stare at one patch of sky for months or years at a time, tens or hundreds of times fainter than JWST, depending on mirror size. Scientists propose LLMT diameters of tens or hundreds of meters, with lightweight, modular, or even inflatable dishes. A forefront LLMT might be built robotically for $1 billion.

The Moon is portal to the Universe at other electromagnetic frequencies. A new radio probe of the distant universe is unfolding, at meter wavelengths and longer. This is the transition within hydrogen atoms of the electron and nucleus changing spin orientation relative to each other, at a 21.1-centimeter wavelength, but in the early Universe at red shifts 5–100, hence wavelengths about 1–20 meters observed. So long ago, most atomic matter was confined in hydrogen gas (not stars or galaxies), but eventual collapse into galaxies had begun, hundreds of million years after the Big Bang. Subtle temperature imbalances in these clouds between gas, photons, and 21-centimeter spin states create signal differences against otherwise uniform emission background, at 0.01% background levels. Such observations' utility is not only studying collapse itself of structures leading to galaxies but also tracing large-scale structure (larger than galaxies) imprinting the matter distribution. Cosmology now studies this structure in the low-redshift galaxy distribution and in cosmic microwave background radiation from redshift 1100. The redshifted 21-centimeter radiation corresponds to huge emission volumes, so it potentially can better describe how large-scale structure clusters allow better description of major components of matter and energy in the Universe.[48] Redshifted 21-centimeter photons come from several sources: purely cosmological hydrogen at redshifts near 100 (in the 20-million-year-old Universe), radiation from sources re-ionizing the Universe's hydrogen at redshifts near 10, and familiar 21-centimeter sources from lower-redshift galaxies.[49]

As compelling as science might be for long-wave radio telescopes, they are technically challenging. These wavelengths overlap common artificial transmission bands (about 3 meters for FM, 0.3–6 meters for television, etc.), so radio interference is distracting. Luckily, these telescopes are interferometric arrays, meaning that unless a signal comes from a small range of directions, it cannot affect the final result. Still, too much noise overwhelms this, so these telescopes are often built in remote areas. The lunar Far Side is often suggested. More basically, 21-centimeter radiation redshifted to more than 10 meters has difficulty even penetrating Earth's ionosphere. To study these early times in the Universe, one must enter space, such as the lunar surface. The Far Side accomplishes both this and maximal radio noise shielding. Such a telescope could be relatively inexpensive, such as metal foil on thin, flexible substrates, deployed robotically. One could even scatter simple dipole antennae connected to transponders and collect their signals at a monitoring station over the Far Side, for example, at L2. Results from such telescopes could sense the Big Bang's earliest moments, the first billionth of a billionth of a second when exponential expansion (inflation) ruled. One could determine subtle but important properties of dark matter: neutrino masses or the presence of more exotic particles. Such telescopes could explore solar physics[50] or planets orbiting other stars.

The Moon's unique characteristics and location make it valuable in observing many kinds of objects. For instance, when Venus transited in front of the Sun as seen from Earth in June 2012, the *Hubble Space Telescope* was used to sense the event by observing the Moon. How does that make sense? Venus's transit is a valuable analog for how we discover many planets around other stars, by observing an extrasolar planet transit in front of its host star, for example, with the *Kepler* space telescope. But why observe the Moon? When an extrasolar planet transits its star, it not only blocks some of the star's light but also absorbs some in the planet's atmosphere. This absorption is selective, producing absorption lines in the star's spectrum. Likewise, Venus produces absorption lines in the Sun's spectrum. The Sun is too bright for *Hubble*, but *Hubble* can observe the Moon. Reflected sunlight from the Moon will faithfully convey these absorption lines, introducing nearly no lunar absorption lines. A patch of the Moon has reflective properties smeared by billions of years of regolith dispersal. A spectrum of this patch before Venus's transit and a spectrum taken during the transit will differ in Venusian atmospheric absorption lines, whereas lunar absorption will be essentially constant (even if *Hubble*'s pointing is not perfect – accurate to tens of meters at the Moon). Making this observation for familiar Venus tells us how to interpret similar measurement for distant planets.

A similar trick measures a closer light source: Earth itself. Whereas half of the Moon is always in shadow from the Sun, the Near Side always bathes in Earthshine. At new Moon, someone on the Near Side would observe a full Earth (versus a new Earth when at full Moon). By observing the Moon's dark side visible from Earth, scientists can determine the Earth's brightness.[51] This is vital in studying global climate change, because Earth's brightness relative to the Moon's sets Earth's surface reflectivity, setting how much solar radiation it absorbs. This is fiendishly difficult to measure accurately, because satellites looking down on Earth do not last forever and drift in sensitivity and calibration. An image of the Moon, for example, at crescent phase, is highly repeatable in appearance if Earth does not change and if one's camera does not scatter much light from the Moon's sunlit side to its dark side image illuminated only by Earth. Monitor changes in earthly cloud cover, a vital but frustrating variable in constraining global warming, can be done accurately. As with Venus's transit, a spectrum of Earthshine would tell even more about changes on Earth's surface. One can track spectral features because of various atmospheric constituents, compounds such as chlorophyll, or other surface constituents. This tells us about Earth's changes on global scales and provides an exoplanetary analog.

The two prior examples show how observing the Moon offers unique benefits, but we can also benefit from observing Earth from the Moon with a highly stable Earth-monitoring platform. A geosynchronous satellite might work, but satellites break or drift from orbit.[52] Global climate monitoring requires stable Earth observation over decades, not just years. With a lunar base (human or robotic) and an adjacent Earth-observing facility, scientific instruments could run consistently for decades (maybe two or three monitors running in parallel to maintain consistency even while one monitor is repaired). This includes not just optical telescopes and spectrometers but also monitors at other wavelengths. For instance, NASA proposes a radar satellite (OASIS: Orbiting Arid Subsurface and Ice Sheet Sounder) to measure water under Earth's surface as ground water or ice.[53] This satellite's designers suggest that lunar radar, a simple mast antennae array, is simpler to build and maintain than an orbital version, mapping Earth at several kilometers resolution.

Clearly many scientific discoveries can result from lunar exploration. Some investigations can occur robotically in lunar orbit and on surface with rovers and other probes. Many could benefit from human presence. Distinguish this from telepresence: remote human operation from Earth (or elsewhere). There are diverse research topics: lunar mineralogy and evolution and exploration of unique environs (such as lava tubes and volatile-laced regolith). Time-variable phenomena can be characterized, with changes seen visually or otherwise. Also, physics experiments (some just described) might detail the distant Solar System and Universe via data accessible to humans after intense computer processing. What role will humans play here, versus humans interacting telerobotically, versus robots operating with little human intervention?

The science done on the Moon (or elsewhere) depends on the sociology of scientists: who is included, and which science is not. From Appendix B we see examples of science passing in and out of favor for reasons not entirely evidentiary or logical. A strength of telerobotics is a data feed potentially available, promptly, to many scientists, widening the discovery range. Which scientists will contribute, and, recalling Pasteur's quote, which equipment and sensors will be available for the variety of discoveries, the eyes attached to "prepared minds"?

Clearly discoveries were made because of exquisite field geology training of Apollo astronauts, but other field events were not handled in usefully scientific ways in which the astronauts were not trained exhaustively, such as optical transients. The *Apollo 11* event went unresolved when more careful or repeated observation (two hours later) might have settled the issue and was dismissed with a mistaken, untested hypothesis. The positions of later flash transients on *Apollo 16* and *17* were recorded on a map later lost, even though it might have proved useful in later analysis and constitutes basic scientific record keeping procedure. Human explorers seem useful to the extent they are trained observers and field scientists. Scientists often argue that humans provide increased opportunities for serendipitous discoveries, being uniquely able to recognize new situations or phenomena, even if unanticipated. Need they be there to enable this serendipity, looking through a space suit helmet faceplate, or will telerobotic virtual presence suffice? To understand this and more, we must appreciate the lunar working environment for humans and robots. We explore this in Chapter 10.

Notes

1. Louis Pasteur (1822–1895) spoke on being appointed professor and dean: "Do you know how the electric telegraph was born, one of the most beautiful applications of modern science? In that memorable year 1822, Oersted, the Danish physicist, held in his hands a copper wire with its ends connected to the two poles of a Voltaic battery. On his desk was a magnetic needle, placed on a pivot, and he suddenly saw (by chance, you say maybe, but remember that in observational fields fortune favors the prepared mind), he suddenly saw the needle move and take a very different position from that picked by terrestrial magnetism. A wire through which electric current flows deflects the position of a magnetic needle. Voilà, gentlemen, the birth of today's telegraph." *La vie de Pasteur* by Renée Vallery-Radot, 1911, Paris: Hachette, 88.
2. "Agnosticism" by Thomas H. Huxley, 1889, *Popular Science Monthly*, 34, 750. Thomas Henry Huxley (1825–1895), grandfather of Aldous and Julian Huxley, was self-taught but became a foremost comparative anatomist. He argued first based on Archaeopteryx skeletons that birds evolved from dinosaurs, which is still thought today. Aggressively defending evolutionary theory, he was nicknamed Darwin's Bulldog. He contributed instrumentally to develop science education and defended against religiosity, originating the term *agnosticism*.
3. An interesting essay on these issues is "Collecting Data on Meteors and Fireballs" by Raymond E. Crilley, 1937, *Popular Science*, 45, 190. One recalls a quote attributed to Thomas Jefferson in 1807 on hearing from Benjamin Silliman and James Kingsley of a meteorite fall near Weston, Connecticut, that he found it easier to believe that "two Yankee professors would lie than that stones would fall from Heaven." This attribution appears spurious ("Who is the Liar Now?" by Anna Berkes, November 14, 2008, *Thomas Jefferson's Monticello*, http://www.monticello.org).
4. "Training Apollo Astronauts in Lunar Orbital Observation and Photography" by Farouk El-Baz, 2011, *Geological Society of America Special Paper*, 483, 49. Also, "Impressions of the Lunar Highlands from the Apollo Command Module" by T. K. Mattingly & F. El-Baz, 1973, LPSC, 4, 513.
5. E.g., "Physical Changes upon the Surface of the Moon" by Edmund Neison, 1877, *Quarterly Journal of Science*, 7, 1: "The present condition of the surface of the moon is one of the most interesting and important questions within the whole range of Astronomy . . . The question of the present condition of the surface of the moon has engaged, therefore, the attention of some of the most eminent astronomers. . . . It is a remarkable circumstance, in relation to this question, that whereas those astronomers who have devoted much time and labour to the study of the moon's surface, and to whom astronomers in general are mainly indebted for our present knowledge of the surface of our satellite, hold in general one view as to the present condition of the lunar surface, astronomers as a body hold a very different opinion. To take a striking instance, scarcely any astronomer known to have devoted time to the study of selenography doubts that many processes of actual lunar change are in progress, and it is doubtful if there is one who could not promptly instance one or more such cases. Yet the general opinion of astronomers appears to be against any such physical changes having occurred. And another instance, almost as striking, exists in connection with the subject of the lunar atmosphere: whilst all selenographers appear to have detected instances where the existence of this atmosphere is revealed, astronomers in general appear to question almost the possibility of its existence, and this in face of the absence of any evidence whatever that there is no atmosphere of the nature supposed.

    It would be an interesting inquiry to ascertain in what manner arose this direct conflict of opinion on this subject, between those who have systematically studied the appearances presented by the moon, and those who have not in the same systematic and assiduous manner examined the lunar surface. It appears to have originated in a short summary by Mädler of the appearances presented by the moon, wherein the differences between the condition of the moon and earth were forcibly stated, and where he pointed out the impossibility of the view that was held in the time of the earlier astronomers, that the moon might be a mere copy of the earth, containing a dense atmosphere, large oceans, abundant vegetation, and animal life, or even human inhabitants. In a condensed and more unqualified form these remarks crept into all astronomical textbooks, and were the main basis of the views commonly held by astronomers."

    Neison quotes textbooks of the day calling the Moon an "airless, waterless, lifeless, volcanic desert." Johann Heinrich von Mädler did his lunar research in the 1830s.
6. For instance, "It has been a question which has been long debated among astronomers, whether the Moon has an atmosphere or not, and, as far as I have been able to learn from reading and verbal inquiry, the question is yet

undecided. The best astronomers I have talked with about this matter have told me they could never discover any atmosphere about the Moon" ("Certain Reasons for a Lunar Atmosphere" by Samuel Dunn, 1761, *Philosophical Transactions*, 52, 578).

7. "The Lunar Atmosphere and the Recent Occultation of Jupiter" by William H. Pickering, 1892, *Astronomy and Astrophysics, Carleton College Goodsell Observatory*, 11, 778.

8. This is contentious. Charged particle density may vary greatly with time; nonetheless various measurements may conflict. Early results ("Measurement of the Lunar Ionosphere by Occultation of the *Pioneer 7* Spacecraft" by J. C. Pomalaza-Díaz, 1967, in *Scientific Report SU-SEL-67–095*, Stanford Electronics Lab) set an upper limit of 40 electrons per cubic centimeter (vs. 1.45×10^{16} cm^{-3} for gas at Earth's surface). *Luna 19* and *22* occultation experiments detect several hundred electrons per cm^3 ("Preliminary Results of Circumlunar Plasma Investigations" by M. B. Vasylyev et al. 1973, *Doklady Akademii Nauk SSSR*, 212, 67; "Preliminary Results of Circumlunar Plasma Research by the *Luna 22* Spacecraft" by A. S. Vyshlov, 1975, *Space Research*, 16, 945). New results are closer to *Pioneer 7*'s ("Studying the Lunar Ionosphere with *SELENE* Radio Science Experiment" by Takeshi Imamura, 2010, *Space Science Reviews*, 154, 305).

9. "Molecular Gas Species in the Lunar Atmosphere" by J. H. Hoffman & R. R. Hodges, 1975, *Moon*, 14, 159.

10. "Lunar Atmosphere Measurements" by Francis S. Johnson, James M. Carroll & Dallas E. Evans, 1972, *Proceedings of 3rd Lunar Science Conference*, 3, 2231.

11. One can show that atmospheric density n varies with temperature T such that $n \propto T^{-5/2}$ ignoring sticking effects, which reduces density when T is small (when $T^{-5/2}$ would be large, otherwise).

12. "Lunar Theory and Processes: Post-sunset Horizon 'Afterglow'" by D. E. Gault, J. B. Adams, R. J. Collins, G. P. Kuiper, J. A. O'Keefe, R. A. Phinney, & E. M. Shoemaker, 1970, *Icarus*, 12, 230.

13. "*Surveyor I* Observations of the Solar Corona" by Robert H. Norton, James E. Gunn, W. C. Livingston, G. A. Newkirk & H. Zirin, 1967, *Journal of Geophysical Research*, 72, 815. Surveyor horizon dust observations were reported in internal publications: "Astronomy" by R. H. Norton, J. E. Gunn, W. C. Livingstone, G. A. Newkirk & H. Zirin, 1967, in *Surveyor V Mission Report, Part II: Science Results*, JPL Technical Report 32–1246, 115; "Post-Sunset Horizon 'Afterglow'" by D. E. Gault, J. B. Adams, R. K. Collins, G. P. Kuiper, H. Masursky, J. A. O'Keefe, R. A. Phinney & E. M. Shoemaker, 1968, in *Surveyor VII Mission Report, Part II: Science Results*, JPL Technical Report 32–1264, 171; "Post-Sunset Horizon Glow" by D. E. Gault, J. B. Adams, R. K. Collins, G. P. Kuiper, H. Masursky, J. A. O'Keefe, R. A. Phinney & E. M. Shoemaker, 1968, in *Surveyor Project Final Report, Part II: Science Results*, JPL Technical Report 32–1265, 401.

14. "The Measurements of Sky Brightness on Lunokhod 2" by B. Severny, E. I. Terez & A. M. Zvereva, 1975, *Earth, Moon and Planets*, 14, 123.

15. "Lunar Dust Motion" by David R. Criswell, 1972, *Proceedings of 3rd Lunar Science Conference*, 3, 2671.

16. "Surveyor Observations of Lunar Horizon-Glow" by J. J. Rennilson & D. R. Criswell, 1974, *Moon*, 10, 121; "Evidence for a Lunar Dust Atmosphere from Apollo Orbital Observations" by J. E. McCoy & D. R. Criswell, 1974, *Abstracts of Lunar and Planetary Science Conference*, 5, 475. A later paper showed ALSEP detector LEAM (Lunar Ejecta And Meteorites) found dust clouds as the local terminator passed *Apollo 17*'s site ("Lunar Soil Movement Registered by the *Apollo 17* Cosmic Dust Experiment" by Otto E. Berg, Henry Wolf & John Rhee, 1976, in *Interplanetary Dust and Zodiacal Light, IAU Colloquium 31*, Berlin: Springer-Verlag, 233).

17. "The Lunar Dust Exosphere and Clementine Lunar Horizon Glow" by H. A. Zook & A. E. Potter, 1995, *Lunar and Planetary Science Conference*, 26, 1577; "Clementine Observations of the Zodiacal Light and the Dust Content of the Inner Solar System" by Joseph M. Hahn, Herbert A. Zook, Bonnie Cooper & Bhaskar Sunkara, 2002, *Icarus*, 158, 360.

18. "A Dynamic Fountain Model for Lunar Dust" by Timothy J. Stubbs, Richard R. Vondrak & William M. Farrell, 2006, *Advances in Space Research*, 37, 59.

19. "The Lunar Surface" by T. Gold, 1955, *Monthly Notices of Royal Astronomy Society*, 115, 585.

20. "The Plasma Wake Downstream of Lunar Topographic Obstacles: Preliminary Results from 2D Particle Simulations" by M. I. Zimmerma, W. M. Farrell, T. J. Stubbs & J. S. Halekas, 2011, *Lunar and Planetary Science Conference*, 42, 1836.

21. "Bow Shock Protons in the Lunar Environment" by J. Benson, J. W. Freeman & H. K. Hills, 1975, *Moon*, 14, 19; "Lunar Dust Grain Charging by Electron Impact: Dependence of the Surface Potential on the Grain Size" by Z. Němeček, J. Pavlů, J. Šafránková, M. Beránek, I. Richterová, J. Vaverka & I. Mann, 2011, *Astrophysical Journal*, 714, 14.

22. "Measurements of Lunar Atmospheric Loss Rate" by Richard R. Vondrak, John W. Freeman & Robert A. Lindeman, 1974, *5th Lunar Science Conference*, 3, 2945.
23. LACE scanned three simultaneous mass ranges: 1–4, 12–48, and 28–110 AMU, including all species (except atomic lithium, beryllium, and boron) up to and including argon, sulfur dioxide, chlorine, hydrocarbons (up to octane), and krypton, but not xenon or radon. Accidental tungsten contamination in the instrument was used to maintain calibration.
24. "Helium and Hydrogen in the Lunar Atmosphere" by R. R. Hodges, Jr., 1973, *Journal of Geophysical Research*, 78, 8055.
25. *Apollo 12* and *15* ALSEP (Figure 2.15) included the Solar Wind Spectrometer (SWS), collecting (in seven directional horns) high-energy charged particles (6–1330 eV electrons, 18–9780 eV ions), originating mainly in the solar wind. *Apollo 14* 's Charged Particle Lunar Environment Experiment (CPLEE) studied both ions and electrons of 50–50,000 eV. *Apollo 15* and *16* launched lunar subsatellites carrying charged particle detectors and deployed SIM bay boom Orbital Mass Spectrometer Experiments (OSME) for neutral species, detailed later (Figure 2.22).
26. "Lunar Atmospheric Helium Detections by the LAMP UV Spectrograph on the Lunar Reconnaissance Orbiter" by S. A. Stern, K. D. Retherford, C. C. C. Tsang, P. D. Feldman, W. Pryor & G. R. Gladstone, 2012, *Geophysical Research Letters*, 39, L12202. Stern et al. hope to detect argon the same way, stressing it is much harder. Helium was detected in a 50-orbit observation.
27. "Formation of the Lunar Atmosphere" by R. R. Hodges, Jr., 1975, *Moon*, 14, 139.
28. Of course, half-life is the time required for 50% of the substance to decay. ^{40}K atoms decay into ^{40}Ar 10.72% of the time and 89.28% into calcium (^{40}Ca). About 0.01% of potassium is ^{40}K. See "Implications of Atmospheric ^{40}Ar Escape on the Interior Structure of the Moon" by R. R. Hodges, Jr. & J. H. Hoffman, 1975, *Lunar and Planetary Science Conference*, 6, 3039.
29. Apollo LACE results indicate a gas leakage rate from the Moon of 10^{23} s^{-1} in ^{4}He and 2×10^{21} s^{-1} in ^{40}Ar (see note 28).
30. "Detection of Radon Emanation from the Crater Aristarchus by the *Apollo 15* Alpha Particle Spectrometer" by Paul Gorenstein & Paul Bjorkholm, 1973, *Science*, 179, 792; "Radon Emanation from the Moon: Spatial and Temporal Variability" by Paul Gorenstein, Leon Golub & Paul Bjorkholm, 1974, *Moon*, 9, 129; "Recent Outgassing from the Lunar Surface: The *Lunar Prospector* Alpha Particle Spectrometer" by Stefanie L. Lawson, William C. Feldman, David J. Lawrence, Kurt R. Moore, Richard C. Elphic, Richard D. Belian & Sylvestre Maurice, 2005, *Journal of Geophysical Research*, 110, E9009. The Alpha Particle Spectrometer (APS) also flew on *Apollo 16* but in a nearly equatorial orbit covering a tiny fraction of the lunar surface. Radon outgassing events were detected on *Kaguya*. All three missions (*Apollo 15*, *Lunar Prospector*, and *Kaguya*) detected radon outgassing at the Aristarchus Plateau ("In-orbit Performance of Alpha-Ray Detector (ARD) Onboard SELENE and the Early Results" by Katsuyuki Kinoshita, Yusuke Haruki, Masayuki Itoh, Takeshi Takashima, Takefumi Mitani, Kunishiro Mori, Jun Nishimura, Toshisuke Kashiwagi, Syouji Okuno & Kenji Yoshida, 2011, Asia Oceania Geophysics Society, Abstract PS10-A013).
31. The decay chain from uranium-238 to stable lead-206 is circuitous: ^{238}U (half-life = 4.5×10^9 years) $\to_\alpha$ ^{245}Th (25 days) $\to_\beta$ ^{234}Pa (1.1 minute) $\to_\beta$ ^{234}U (2.3×10^5 years) $\to_\alpha$ ^{230}Th (83,000 years) $\to_\alpha$ ^{226}Ra (1,590 years) $\to_\alpha$ ^{222}Rn (3.8 days) $\to_\alpha$ ^{218}Po (3.1 minutes) $\to_\alpha$ ^{214}Pb (27 minutes) $\to_\beta$ ^{214}Bi (20 minutes) $\to_\beta$ ^{214}Po (0.00015 second) $\to_\alpha$ ^{210}Pb (22 years) $\to_\beta$ ^{210}Bi (5 days) $\to_\beta$ ^{210}Po (140 days) $\to_\alpha$ ^{206}Pb, where U stands for uranium, Th for thorium, Pa for protactinium, Ra for radium, Rn for radon, Po for polonium, Pb for lead, α for alpha emission (helium-4 nucleus), β for beta emission (an electron or anti-electron). Of these only radon is gaseous at lunar temperatures. Betas alter the element but not mass (not significantly); alphas alter both mass and element.
32. "Low Pressure Radon Diffusion: A Laboratory Study and Its Implications for Lunar Venting" by Larry Jay Friesen & John A. S. Adams, 1976, *Geochemica et Cosmochemica Acta*, 40, 375; "Radon Diffusion and Migration at Low Pressures, in the Laboratory and on the Moon" by Larry Jay Friesen, 1974, PhD thesis, Rice University; "Model for Radon Diffusion through the Lunar Regolith" by L. J. Friesen & D. Heymann, 1976, *Moon*, 3, 461.
33. "Lunar Outgassing, Transient Phenomena and the Return to the Moon. II. Predictions and Tests for Outgassing/Regolith Interaction" by Arlin P. S. Crotts & Cameron Hummels, 2009, *Astrophysical Journal*, 707, 1506. We discuss this model further on pages 281–283.
34. "Chronological Catalog of Reported Lunar Events" by Barbara M. Middlehurst, Jaylee M. Burley, Patrick Moore & Barbara L. Welther, July 1968, *NASA Technical Report TR R-277*; "Lunar Transient Phenomena

Catalog" by Winifred Sawtell Cameron, July 1978, *NASA-TM-79399*, Greenbelt, MD: National Space Science Data Center/World Data Center A for Rockets & Satellites, Pub. 78–03; "Lunar Transient Phenomena Catalog Extension" by Winifred Sawtell Cameron, July 2006, unpublished, with 579, 1,463, and 475 reports, respectively. Middlehurst's list is largely covered by Cameron 1978 but little of the 1978 list is in the 2006 list.

35. "Detection of Radon Emission at the Edges of Lunar Maria with the Apollo Alpha-Particle Spectrometer" by Paul Gorenstein, Leon Golub & Paul Bjorkholm, 1974, *Science*, 183, 411.

36. "Recent Outgassing from the Lunar Surface: The *Lunar Prospector* Alpha Particle Spectrometer" by Stephanie L. Lawson, William C. Feldman, David J. Lawrence, Kurt R. Moore, Richard C. Elphic, Richard D. Belian & Sylvestre Maurice, 2005, *Journal of Geophysical Research*, 110, E09009 shows the ^{210}Po data, and "Lunar Outgassing, Transient Phenomena and the Return to the Moon. I. Existing Data" by Arlin P. S. Crotts, 2008, *Astrophysical Journal*, 687, 692 shows the ^{210}Po/mare edge correlation.

37. "'New' Volcanic Features in Lunar, Floor-fractured Oppenheimer Crater" by Lisa R. Gaddis, Lynn Weller, Janet Barrett, Randy Kirk, Moses Milazzo, Jay Laura, B. R. Hawke, T. Giguere, Briony Horgan & Kristen Bennett, 2013, *Lunar and Planetary Science Conference*, 44, 2262; "V is for Vents: Cloud-Sourcing the Discovery, Description and Distribution of Lunar Vents" by C. A. Wood, P. Leon, D. González, M. Zambelli, R. Hentzel & M. Collins, 2013, *Lunar and Planetary Science Conference*, 44, 1710; "Characterization of previously Unidentified Lunar Pyroclastic Deposits Using Lunar Reconnaissance Orbiter Camera Data" by J. O. Gustafson, J. F. Bell III, L. R. Gaddis, B. R. Hawke & T. A. Giguere, 2012, *Journal of Geophysical Research*, 117, E00H25. Pyroclasts have high H_2O: "Evidence for Increased Water Content in Pyroclastic Deposits from M^3 Data" by Shuai Li, July 17, 2013, *Lunar Science Institute Virtual Forum*.

38. "Transient Lunar Phenomena, Deep Moonquakes and High-Frequency Teleseismic Events: Possible Connections" by Barbara M. Middlehurst, 1977, *Philosophical Transactions of Royal Society of London – A*, 285, 1327. "Shallow Moonquakes – Argon Release Mechanism" by A. B. Binder, 1980, *Geophysical Research Letters*, 7, 1011; "Release of Radiogenic Gases from the Moon" by R. R. Hodges, Jr., 1977, *Physics of Earth and Planetary Interiors*, 14, 282.

39. "Fluidization Phenomena and Possible Implications for the Origin of Lunar Craters" by A. A. Mills, 1969, *Nature*, 224, 863; "Experimental Studies of the Formation of Lunar Surface Features by Fluidization" by S. A. Schumm, 1970, *Bulletin of Geological Society of America*, 81, 2539; "Parameters Critical to the Morphology of Fluidization Craters" by Barry S. Siegal & David P. Gold, 1973, *Moon*, 6, 304.

40. "Lunar Activity from Recent Gas Release" by Peter H. Schultz, Matthew I. Staid & Carlé M. Peters, 2006, *Nature*, 444, 184; "High Resolution Imaging of Ina: Morphology, Relative Ages, Formation" by M. S. Robinson, P. C. Thomas, S. E. Braden, S. J. Lawrence, W. B. Garry & the LROC Team, 2010, *Lunar and Planetary Science Conference*, 41, 2592; also "The Geology and Morphology of Ina" by P. L. Strain & F. El-Baz, 1980, *Lunar and Planetary Science Conference*, 11, 2437, and the *Apollo 15 Preliminary Science Report*.

41. "Search for Short-Term Changes in the Lunar Surface: Permanent Alterations over Four Decades" by A. P. S. Crotts, 2011, *Lunar and Planetary Science Conference*, 42, 2600.

42. Outgassing or impact events might start signals detectable by lunar seismometers. ALSEP's four seismometers were deactivated in 1977, but two or three seismometers could map the interior if seismic events' place and time are known. Imaging monitors could supply this event localization.

43. "Aristarchus Crater: Mapping of Impact Melt and Absolute Age Determination" by M. Zanetti, H. Hiesinger, C. H. van der Bogert, D. Reiss & B. L. Jolliff, 2011, *Lunar and Planetary Science Conference*, 42, 2330; "Ages and Stratigraphy of Mare Basalts in Oceanus Procellarum, Mare Nubium, Mare Cognitum and Mare Insularum" by H. Hiesinger & J. W. Head III, 2003, *Journal of Geophysical Research*, 108, 5065, and citations therein.

44. See "Compositional Diversity and Geologic Insights of the Aristarchus Crater from Moon Mineralogy Mapper Data" by John F. Mustard, et al., 2011, *Journal of Geophysical Research*, 116, E00G12; "Clementine Observations of the Aristarchus Region of the Moon" by Alfred S. McEwen, Mark S. Robinson, Eric M. Eliason, Paul G. Lucey, Tom C. Duxbury & Paul D. Spudis, 1994, *Science*, 266, 1858.

45. Studies of potential lunar lava tube exploration include "Reference Mission Architecture for Lunar Lava Tube Reconnaissance Missions" by Samuel W. Ximenes, J. O. Elliott, O. Bannova & R. Y. Nakagawa 2011, *1st International Planetary Caves Workshop*, LPI Contrib. No. 1640, 46, http://www.lpi.usra.edu/meetings/caves2011/pdf/8013.pdf; "Lunar and Martian Lava Tube Exploration as Part of an Overall Scientific Survey" by Andrew W. Daga, M. M. Battler, J. D. Burke, I. A. Crawford, R. J. Léveillé, S. B. Simon & L. T. Tan, 2009, *Lunar Exploration and Analysis Group*, 2065, http://www.lpi.usra.edu/meetings/leag2009/pdf/2065.pdf. Lunar lava

tubes' utility is treated by "Lunar Lava Tubes as Prepared Emergency Shelters" by Austin Mardon and Kenneth Nichol, 2008, 37th COSPAR Scientific Assembly, p. 16, and "Lava Tubes" by Peter Kokh, May 1989, *Moon Miners' Manifesto*, #25: http://www.asi.org/adb/06/09/03/02/025/lavatubes.html and other issues (also see Chapter 12).

46. Several radio telescopes do/will observe this lunar radiation: GLUE (Goldstone Lunar Ultra-high energy neutrino Experiment, http://www.physics.ucla.edu/~moonemp/public/; RESUN, http://resun.physics.uiowa.edu/; NuMoon, https://www.kvi.nl/~scholten/numoon/numoon.html; and LUNASKA (Lunar UHE Neutrino Astrophysics using the Square Kilometer Array), http://www.physics.adelaide.edu.au/astrophysics/lunaska/index.html. See "Radio Cherenkov Signals from the Moon: Neutrinos and Cosmic Rays" by Yu Seon Jeong, Mary Hall Reno & Ina Sarcevic, 2012, *Astroparticle Physics*, 35, 38.

47. An ionic liquid example is 1-ethyl-3-methylimidazolium ethyl sulphate ($C_6H_{11}N_2$ and $C_2H_5SO_4$), sold by BASF as ECOENG 212, remaining liquid above –98°C. See http://www.sigmaaldrich.com/catalog/product/ALDRICH/51682.

48. "How Accurately Can 21 cm Tomography Constrain Cosmology?" by Yi Mao, Max Tegmark, Matthew McQuinn, Matias Zaldarriaga & Oliver Zahn, 2008, *Physical Review D*, 78, 023529.

49. "Background Reionization History from Omniscopes" by Sébastien Clesse, Laura Lopez-Honorez, Christophe Ringeval, Hiroyuki Tashiro & Michel Tytgat, 2012, astro-ph/1208.4277, http://arxiv.org/abs/1208.4277.

50. "A Radio Observatory on the Lunar Surface for Solar studies (ROLSS)" by R.J. MacDowall, T.J. Lazio, S.D. Bale, J. Burns, W.M. Farrell, N. Gopalswamy, D.L. Jones & K.W. Weiler, 2011, astro-ph/1105.0666, http://arxiv.org/abs/1105.0666.

51. "The Earthshine Project: Update on Photometric and Spectroscopic Measurements" by E. Pallé, P. Montañés Rodriguez, P. R. Goode, J. Qiu, V. Yurchyshyn, J. Hickey, M.-C. Chu, E. Kolbed, C. T. Browne & S. E. Koonin, 2004, *Advances in Space Research*, 34, 288.

52. We describe a way to maintain geosynchronous satellites by using lunar resources (Chapter 12).

53. "NASA Announces Groundwater Discovery in the Arabian Peninsula" by Rana Khalifa, September 28, 2011, *Yalla Finance*.com.

Chapter 10
Moonlighting

> CERNAN: We're getting a repeat.
> PARKER: I said, 'Close the covers, please.'
> CERNAN: That's right. I heard what you said, but you're turning our voice around.
> SCHMITT: (singing) I was strolling on the Moon one day – in the merry, merry month of December
> – *Apollo 17* Commander Eugene Cernan and Lunar Module Pilot Harrison ("Jack") Schmitt (at lunar crater Camelot) with Capsule Communicator Robert Parker (in Houston)[1]

Work on the Moon is serious business. In a strange, potentially hostile environment, many simple tasks become challenges: communicating with Earth, manipulating objects, or just moving around. However, that does not mean it cannot be enjoyable. It is an interesting work environment, despite differences with Earth at first seeming bleak: "Magnificent desolation," as Aldrin said, first stepping out as second man on the Moon. Apollo astronauts spent 160 person-hours outside on the Moon with no significant injury or major accident (but several minor ones). They accomplished much and even had fun. Unfortunately, it remains one of the most exclusive human experiences. (How many of you have heard some variation on the saying, "More people have walked on the Moon than have ... *fill in the blank with some rare activity*"?) Fortunately, it is one of the better documented experiences. What is involved in successfully working on the Moon?

Another glance at Appendix A is worthwhile, particularly items 7–17. Living and working on the Moon involves many challenges, and it is testament to von Braun and collaborators that in 1952 they anticipated these, except lunar dust's unusual nature. How humans adapt to this new environment to perform familiar tasks is an interesting story.

Consider walking. Muscles are trained to alternate between stability and imbalance, propelling one forward while fighting gravity's downward force. Leg bones forestall falling just long enough for the other leg to take the load. On the Moon that force is 16% of Earth's, so expending the same energy lifts one six times higher off the ground. The time to fall from a height, however, varies inversely as the square root of this 16%, so only 2.5 times longer.[2] In a moonwalk, swinging one's legs on the same path as on Earth would feel underpowered and slow.

People who have walked on the Moon compare it to striding on a trampoline. Too much power in low gravity will send the walker into a step flying free of the ground. This is difficult to control; one can tilt and rotate beyond where landing on one's feet can succeed. Striking one's head (or helmet visor) or vital moon suit part on a rock might prove punishing, especially because impact hits about as fast as one pushed off the ground. A compromise strategy is a gait longer than minimum but power below maximum.

To use your legs effective in one-sixth gravity, change your gait. Walking requires less energy in low gravity. How can one use earthly strength? Energy expended depends not only on muscle force provided but also the stroke length of muscular contraction (multiply the two together.) Use shorter strokes.

Rather than derive theoretically how best to moonwalk, human physiology is talented in finding good solutions. Early science fiction films portrayed astronauts leaping around the landscape – too dangerous in reality. When NASA trained astronauts for first lunar surface activity, they encouraged a double-legged hop using reduced muscle stroke, sending one into successive arcs above ground, 1–2 meters in length. The hop is useful locomotion in some lunar circumstances (Figure 10.1a).

Moonwalkers say 10 minutes suffice to become accustomed to walking in lunar gravity. They favored another mode: a long, loping gait – the stroll (Figure 10.1b). Put one leg in front of the other, then alternate, as on Earth but using longer strides (like cross-country skiing). For short strides, alternating which leg leads is optional. Legs need bend less on the Moon: when the foot hits the ground, it bounces. Using only one leg to push off the ground, muscular power spans more earthly ranges, with safety margins. Crucially, strolling requires little time with feet off the ground, making orientation control easier. Aldrin described this trade-off soon after stepping onto the Moon.[3] For slow strolls, the motion devolves into nearly a normal Earth walk. Neil Armstrong, first to use it, called it the "lope." For small steps, keep one foot on the ground and the other nearby, moving tens of centimeters per step in a "shuffle." Shuffles cover 1–2 kilometers per hour, running strolls 4–5.

Aldrin's checklist included evaluating walking modes, recalling his development of extravehicular techniques during Gemini in 1965–1966. *Gemini IX–XI* suffered snags in crew work outside the spacecraft, after America's first spacewalk by *Gemini IV*'s Edward White. His initiative led to his successful *Gemini XII* spacewalks, practicing them with neutral buoyancy underwater simulations and new hardware.

Reduced gravity's effects are profound. One can lift and move objects much larger than on Earth (Figure 10.2). Normal exertion can send one flying, intentionally or not (Figures 10.2 and 10.3). Once flying through space, touching the ground again (hopefully with one's feet) is how to control one's trajectory (traced by center of mass, a point usually located behind the navel if one is standing or prone). The difficulty on the Moon is that friction, the controlling force, depends on one's weight on the ground and the slipperiness of surfaces on which one stands, for example, boot soles on regolith. Frictional force is computed by multiplying these two factors.

The New Moon

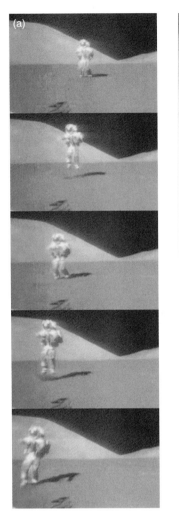 

Figure 10.1. **Doing the Moonwalk**. Two iconic sequences showing how to walk (or run) on the Moon. (a) Hippity-hoppity sequence of Gene Cernan epitomizes the two-legged hop, including leaning back in the last frame to brake and change direction. (b) "While strolling on the Moon one day," Jack Schmitt sang the praises of the one-leg-at-a-time stroll like that shown here. (Shown is not the actual "strolling on the Moon" sequence.)

When weight shrinks to one-sixth of its Earthly value, but surfaces remain unchanged, friction falls to one-sixth of its usual, as if ground surfaces were radically more slippery versus Earth. In contrast the mass of a person or other object is not reduced on the Moon; stopping, starting or changing motion requires as much force as on Earth. On the Moon control over motion is reduced to by five-sixth. Video sequences (in Figure 10.3) show consequences of this.

In Figure 10.3 *Apollo 15*'s Dave Scott throws an empty experiment pallet from ALSEP. The pallet weighs little, but its unchanged mass means as much force as on Earth is needed to accelerate it. Unfortunately, this force produces an equal but opposite force on Scott, counteracted by reduced friction on his feet. Consequentially

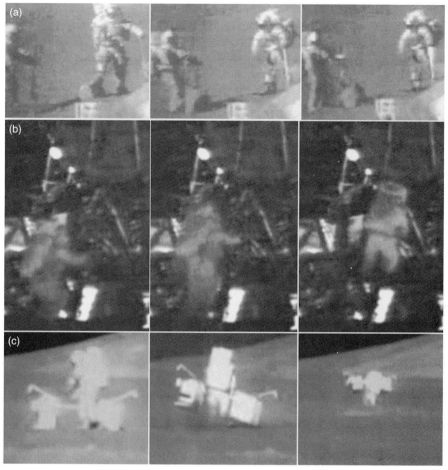

Figure 10.2. **The hard made easy**. Low lunar surface gravity allows seemingly superhuman feats. (a) Cernan and Schmitt easily kick-pass a rock that on Earth would weigh about 100 kg (220 lb.). (b) Jim Irwin (with a 96 kg suit – 212 lb.) jumps 3 ft. (1 m) in a single bound onto the LM ladder. (Neil Armstrong on *Apollo 11* made a giant leap of 5 ft. back onto the ladder.) (c) Jack Schmitt picks up and carries ALSEP instruments, about 160 kg (350 lb.) on Earth.

Scott's feet slip from underneath and he falls. Likewise, below that in Figure 10.3, Jack Schmitt applies force to a drilling rig, counteracting this with friction on his feet. When too much force overcomes that friction, his feet slip and he falls.

These factors influence walking. One cannot brake instantly if moving quickly, with limited control at each step. Running can cause control loss over one's orientation (Figure 10.4, left). An advantage of the hop is both feet touching the ground simultaneously; stopping and turning are easier, especially by digging in heels. Cernan does this ending the "hippity-hop" sequence (Figure 10.1).[4] Key is that stopping force must act through the body's center of mass near the navel, but forces often act on the feet. Braking rapidly will flip one's heels over head. Astronauts slow the stroll by separating their feet front to back, touching ground simultaneously,

Figure 10.3. **The easy made hard**. (a) Dave Scott throws an empty experiment pallet. Although the pallet weighs much less on the Moon, its mass is unchanged, causing a large opposing force to push Scott off balance. Fortunately, he immediately pushed off the ground with his right hand to stand again (not seen here). (b) Jack Schmitt and Gene Cernan use a jack and treadle to raise a deep core tube; Schmitt leans on the jack handle but cannot resist the opposing sideways force on himself because of the very low frictional force on his feet in lunar gravity. Schmitt disappears momentarily behind Cernan in a cloud of dust and hardware.

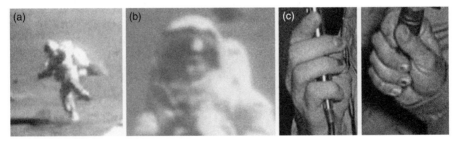

Figure 10.4. **More problems on the Moon**. (a) Gene Cernan rotates in midflight after a bad hop while moving rapidly down a slope. (b) One sees Jack Schmitt through his helmet several seconds before Capsule Communicator Robert Parker in Houston warns him to lower the gold-tinted visor that shields his face from solar ultraviolet light. (c) Close-ups show Dave Scott's and Jack Schmitt's right hands photographed the day each returned to Earth. Because of forces of their moon suit gloves resisting their manipulations, not only were their forearms fatigued after working on the Moon, but bruises formed under their fingernails because of the gloves' resisting pressure.

Moonlighting

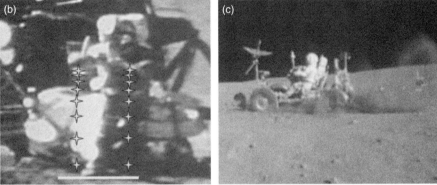

Figure 10.5. **Newton rules**. Free objects' motions are simple in near-total lunar vacuum. (a) Video of Cernan tossing a flimsy heat flow probe wrapper. It spins end over end, captured in multiple frames when it is vertical (every 0.13 second). Its image blurs because of motion. It follows a parabola, which on Earth distorts by air drag. (b) Scott drops a rock hammer and falcon feather together, recalling Galilei's famous experiment. Both follow the same accelerated fall to the ground, their positions every 0.13 second indicated by star symbols. (c) As Young guns the LRV's motor, each dust particle grain follows its individual parabolic path, creating dust pillars and loops falling immediately to the ground rather than persisting suspended in a cloud as in Earth's air.[40]

keeping a particular foot in front at least at slow speeds: the shuffle (mentioned earlier). It is easier to keep one's balance with feet spread apart.

The lunar surface sits in nearly total vacuum; even tiny objects feel no air resistance, usually dominating their motion on Earth. Figure 10.5 shows relatively insubstantial objects following trajectories through space as if they were heavy masses. Feathers fall as rapidly as hammers, and flimsy wrappers arc through space as if steel rods. This applies to minuscule dust particles (Figures 10.5 and 10.7), although dust reacts to electrostatic forces, as well. We discuss dust later in this chapter and Chapter 11.

Survival on the lunar surface is as good as one's moon suit. It protects from vacuum and micrometeorites and is water-cooled against extreme daytime temperatures. It does not fully protect one from radiation, not without weighing too much for motion. It is a wearable spacecraft, providing oxygen and removing carbon dioxide, and has (or should have) emergency oxygen, connections for your buddy if he loses oxygen, as well as plumbing and electronics for control and communications. Apollo's lunar extravehicular suit weighed 212 Earth pounds (96 kilograms), but on the Moon fortunately weighed only 34 pounds (16 kilograms). Nonetheless, Apollo's EVA suit (A7L and A7LB) and NASA's current Extravehicular Mobility Unit (EMU) contain so many layers that three-quarters of the energy in weightless movement involves just flexing the suit. (This is not extreme: the Soviet moon suit Kretchet was so inflexible and back heavy that concern arose that the cosmonaut might become stuck on his back like a tortoise. On the LK lunar lander, the usual Soyuz hatch was replaced with a large, custom one to accommodate the suit. Although built, Kretchet was never used on a mission.) EVA suits differ from flight suits or escape suits (used near Earth launch and landing) in being bulky but capable of extensive motion, hence, usually with articulated joints, often rotating bearings or collars, requiring less energy bending limbs.

To understand what one tugs against, consider the EMU's layers: next to one's skin is tricot lining, then a nylon/spandex layer holding plastic tubes for cooling. Outside that is the polyurethane-coated nylon pressure bladder, then the woven Dacron garment for primary mechanical support. Next comes a neoprene-coated, anti-rip nylon liner, then multiple, aluminized Mylar/Dacron thermal insulation layers. Finally, enveloping all, the Gore-tex/Nomex fiber weave on Kevlar backing stops abrasion, punctures, flame, and micrometeorites. Counting multiple thicknesses, this suit contains 15 layers. Along with plumbing, air supply, water supply, helmet, and communications and control system, it truly is a spacecraft that one wears, even if it cannot stop some harmful radiation.

New Constellation suits were planned differently, addressing two issues: the EMU totals 136 kilograms, too heavy for the Moon, and the new suit should be more flexible than Apollo's EVA suit. Constellation's EVA suit was still being designed but had a back hatch for entering the suit quicker and easier than Apollo A7L's zippers around the abdomen. Its hatch's outer wall had a "backpack" containing air and water supplies, radiators, and pumps (officially, the Portable Life Support System [PLSS]; call it the backpack). In contrast Constellation escape suits closely resembled the shuttle's. Plans call for suits more interchangeable among astronauts, almost one size fits all versus Apollo's, custom-made for each astronaut. Normally, increasing air pressure in suits stiffens them like over-inflated balloons. At 8 pounds per square inch (more than half Earth's atmospheric pressure, 14.7 psi), the new joint design would allow greater flexibility despite being inflated to twice the pressure for the EMU. This alleviates concern over "the bends" (decompression sickness) well known from deep-sea diving, when nitrogen dissolved in blood and tissues bubbles in the body as pressure drops. Constellation lunar suit concepts even had a dust-free provision allowing astronauts to enter suits from a spacecraft without exposure to dust, with the suit completing the seal, keeping dust out. A new seal closes behind the astronaut before the suit detaches from its seal and the spacecraft.

ZPS Mark III was a pre-Constellation design (late 1980s) with hard shells in the upper and lower torso, hard limb joints, and soft layers elsewhere. Its joints' range of motion approaches an unsuited human's (97° bend at the hip, 116° at the elbow, 89° at the knee; compare to your joints). In contrast Apollo's A7L prevented astronauts from reaching more than 6 feet or less than 2 feet off the ground without kneeling. An astronaut wearing Mark III gloves could pick up a pencil or push buttons a centimeter apart. The suit leaked more than *ISS* EMUs, a problem being addressed. Another Constellation candidate was ILC-Dover's Rear Entry I-Suit (REI), with more soft components, therefore lighter than Mark III, but with similar flexibility and rear-entry features. No NASA suits flexes for an astronaut to look straight up; Roscosmos's Orlan suit on *ISS* has a helmet top window for astronauts/cosmonauts to see overhead. New NASA suits have internal gas connections (no snagging external hoses), standard on Russian/Soviet suits for decades.

Current moon suits cannot eliminate all problems. Consider the most serious moonwalk hardware accident (Figure 10.6). Starting the first of three *Apollo 16* moonwalks, John Young was deploying ALSEP's heat flow experiment to measure heat leaking from the lunar interior. These flows are surprising; the Moon leaks interior heat nearly as intensely as the much more molten Earth. The same experiment succeeded on *Apollo 15* and *17*, and *Apollo 16* sits uniquely on older, highland terrane, evading later mare volcanic eruptions. Placing heat probes in holes drilled for previous sample cores, Young caught his toe on the main cable of the experiment, destroying it. Four seconds elapsed between Young snagging the cable and when it snapped, meanwhile not feeling the cable until almost tripping. The experiment's loss was unlucky, but under different conditions we could have lost the astronaut's suit integrity and maybe the astronaut himself. Similar accidents were narrowly avoided, for example, Neil Armstrong snagging a TV camera cable with his toe but aided by Edwin Aldrin. On *Apollo 15*, Dave Scott also tripped on a heat flow probe wire, shifting the electrical box but not disconnecting it. It was realigned on the next EVA.

Astronauts should feel contact through their suits. Consider implementing Smart Suit, consisting of a lightweight pressure sensors mesh on the suit's outside, and a network of small pressure transducers touching the astronaut's skin. Signals of exterior contact travel across the suit, through a computer, then through wires near the suit's inside surface, but to the astronaut pressure seems to propagate immediately through the suit's layers to his skin at the same point the suit makes contact. This technology is similar to that giving tactile sensation to virtual realizations of computer-generated environments.

Second Skin is another approach, pursued by Professor Dava Newman and her group at MIT. Instead of suits like air-filled balloons, Second Skin (or BioSuit) fits tightly around limbs and the lower abdomen like a scuba suit, with suit fabric tension fighting pressure differences because of the vacuum outside. Suit fibers are aligned to provide mechanical strength although not requiring the wearer to stretch them when bending the limbs or waist. One still needs a "balloon" around one's lungs to breath comfortably, and a helmet. The suit is so thin that one can feel contact through it. If punctured, its air does not leak out, but the less extreme problem is one's flesh being exposed to the vacuum, not fatal except in the worst

Figure 10.6. **Unsuitable situations**. Apollo moon suits hinder astronauts' motion. (a) Young accidentally destroys a heat flow experiment by tripping on its main cable. At top, he steps over the cable (arrow) with his right foot but catches it with his left. By the fourth frame its chassis yanks toward him, but he reacts only in the fifth, verbally acknowledging it in the sixth: "I didn't even know it." The cable was tugged 4 seconds before breaking at the chassis connector. Second Skin or Smart Suit could prevent this. (b) Schmitt shows how easy dropping but how difficult picking up an object can be. Note the dust he kicks up scrambling for balance. (c) Duke shows how hard rising from a fall is.

cases. (In everyday life, one can plug small vacuum leaks temporarily with a finger and little injury occurs for some time.)

Second Skin allows more flexibility. This solves a serious problem with Apollo moon suits (Figure 10.6). Bending over in the suits was hard, as was standing again, even in one-sixth gravity. On Apollo, this required design of special tools to pick up and manipulate objects. In new, demanding work environments, the Apollo approach makes some tasks impossible.

Both Smart Suit and Second Skin solve the minor but irritating problem of an itch you cannot scratch. To appreciate how irritating this can be, stand for two minutes not allowing yourself to touch any area that starts to itch. Distracted by other activity, you might not mind, otherwise it can be maddening (for many people). Most itches, if not a skin ailment or other medical condition, are temporary phenomena of the nervous system relieved by local stimulation.[5] Smart Suit or Second Skin would allow this stimulation. Pre-Apollo suit concepts with enough internal space to allow reaching the itch inside the suit proved too bulky to be practical.

Another new technology potentially useful to astronauts is three-dimensional sound or acoustic holography, which with only two loudspeakers allows the listener to locate a noise source's direction.[6] This can work with simple headphones but would require different design considerations for a space helmet. Sound does not travel in a vacuum, of course, and hearing is useful in listening to communications and sensing vibrations and sound through solid objects. Radio receivers configured for directionality (relative to the astronaut's head) and fed into the helmet's 3-D speakers could permit the astronaut to locate his colleagues, and even objects tagged with 3-D position radio transmitters. A low hum might tell the astronaut where a vital object might be, such as a critical component or dropped tool. This is an artificial sense: real, tagged objects and hazards are sonically invisible and might present problems without warning. Alternatively, 3-D sound could convey output from radar location.

Gloves are special challenges: they must protect hands but still allow sense of touch and flexibility of movement. Figure 10.4 (right) shows that Apollo's solution was imperfect. J mission astronauts (*Apollo 15–17*) with three moonwalks apiece left the lunar surface with bruises under their fingernails from gloves compressing the fingers as they worked. Apollo A7L gloves reduced mobility to 30% of bare hand response. Astronauts complained of exhausted hand muscles after excursions.

A new approach to space gloves was developed by NASA's 2007 Astronaut Glove Challenge winner, Peter K. Homer (see note 51 to Chapter 4). The glove's balloon portion is thin but supported and constrained by ribbing crossing at the finger joints, making them flexible where they must bend. These gloves still need layers for thermal insulation and puncture/abrasion protection but are much more flexible than Apollo-era gloves.

Apollo's experience distorts our view of a prime challenge to lunar surface activity: extreme temperatures. All six Apollo landings occurred with the Sun just above the horizon (5°–20°), with the Moon's surface neither too hot nor cold. On the longer J missions, the Sun reached 48°. Otherwise, conditions grow lethal. Mean equatorial surface temperature varies from 110°C (boiling water) at midday to –150° C (nearly liquefying oxygen) at midnight. In permanently shadowed craters near the poles, −250°C is possible, while crater floors at equatorial noon can reach 123°C.[7] No existing space suit handles all of these temperatures, but A7L performs well: −175°C to 150°C. In contrast, natural Earth surface temperatures range from −89°C to 58°C. (Mars averages about −60°C, typically reaching −80°C at night and −140°C at the poles.) In 24 hours near lunar sunrise, surface temperature jumps 200 degrees, then 15–20 degrees per day for several days, reversing at sunset, less violently. Apollo targeted more temperate periods. (Low Sun angles also made shadows of obstructions far easier to see and spacecraft landings safer.)

We will see in Chapter 11 that lunar surface temperatures are ameliorated by digging a little, which also shields from radiation and meteorites. However, some tasks require the surface, and working there much of the month requires more robust suits. Temperatures 60°C below coldest Antarctic nights require suits with active heating or more insulation. The same applies to Martian nighttime conditions. Lunar PSRs, 80°C colder still, necessitate new technologies, more like articulated hard-shell spacecraft, because few flexible materials exist at these temperatures. We may leave these extreme cryogenic environments to machines (Figure 10.8).

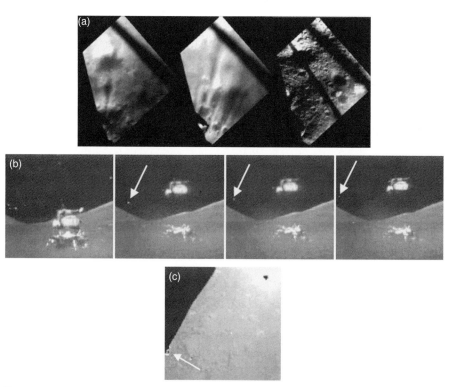

Figure 10.7. **Dust and vacuum**. Strange behavior of dust and vacuum. (a) Sequence illustrates dust streams flying toward the horizon away from *Apollo 12*'s LM descent engine, then clearing within one second. Compare to Figure 2.10. (b) Sequence shows the *Apollo 17* LM ascent engine throwing flakes of metal foil in nearly ballistic arcs. (Note one particle indicated by arrow.) (c) View from *Apollo 17* LM looking down on LM descent stage (*arrow, lower left*), and shadow of LM ascent stage (black dot, upper right, surrounded by bright halo of heiligenschein).

Moonlighting

Figure 10.8. **Red Rover, Moon Rover**. Rovers, post-Apollo's LRV. (a) Lunokhod with solar panels, 8-spoked wheels, Earth-pointing antenna. (b) *Mars Pathfinder* Sojourner and a Mars Exploration Rover, both preparing to analyze a rock. (c) Scarab, Carnegie Mellon's prototype for exploring ultra-cold lunar polar craters. (d) Design for LUNOX drilling rover and crew. (e) Constellation-era ATHLETE rover carrying mobile human habitat (*left*) and a heavy equipment and personnel carrier.

Near lunar noon, hefty radiators are needed. Even at comfortable temperatures an active human dumps some 200 watts of heat, growing intolerable if not vented. In vacuum, air cooling fails, and infrared-emitting radiators are too bulky or fragile for space suits. Apollo's A7L sustained humans in all but hottest lunar environs, but by depleting a prime consumable: water. With water wicking to the backpack surface and evaporating, rapid cooling is possible, using 8 liters in 6 EVA hours. One can reduce this with infrared radiators, but not eliminate it. Because body heat generation varies by a factor of 10, active astronauts risk heat exhaustion without reserve water cooling. Between extra insulation against lunar night and radiators at noon, the push for lighter suits à la Second Skin or Constellation is prescient, allowing humans to do more, more of the time. This might prove critical on Mars, nearly as cold as lunar night, but with suits weighing 2.3 times more.

Space suits protect one against extreme temperatures by minimizing heat transfer via radiation. Pressing one's body or limbs in a suit against a hot or cold surface forces aluminized Mylar sheets inside to contact each other and conduct heat through the suit. Similarly, one must guard against injury to feet or hands by heat flow through boots and gloves; this is a common complaint among astronauts.

As in outer space, the Moon's surface is bathed in solar radiation, particularly higher-energy ultraviolet (UV) light otherwise filtered by Earth's atmospheric ozone layer. Although this UV composes only 3% of sunlight's energy, it is most destructive. Human skin will sunburn badly in several minutes' direct exposure and suffer markedly increased carcinoma risk in under an hour. To prevent this, space helmets all sport sun visors, coated with a gold layer 0.0002 millimeters thick (about 0.2 grams over the visor) or other, newer metal alloys. For this reason Jack Schmitt (Figure 10.4) was admonished by Houston to lower his visor after raising to see better.[8] Nobody was injured by solar UV on the Moon (although TV viewers of *Apollo 12* were disappointed when Al Bean accidentally destroyed the TV camera by pointing it at the Sun). Materials, especially space suits' outer layer, must be UV resistant. Astronauts must be able to raise their sun visors when looking into shadows, and visors can be hard to see through and flip when dusty (see note 8 of this chapter and note 9 in Chapter 11). They also play tricks with the apparent colors of objects (see note 8 and Box 9.1).

The most dangerous radiation has invisible consequences. Heading to the Moon, one transits the Van Allen radiation belts around Earth, which inflict a dosage equivalent to about 10 days in deep space or 20 on the lunar surface (50% shielded by the Moon), mainly a result of cosmic rays from our galaxy. For long lunar stays, one needs effective shielding. Occasionally, the Sun spews forth coronal mass ejections, solar wind storms sufficient to sicken or kill unshielded humans with radiation within hours. We will discuss mitigating these in Chapter 11. Space suits cannot offer effective shielding against either mass ejections or galactic cosmic rays; space-suited astronauts must enter a radiation shelter.

Moonlight has unique characteristics, making lunar surface work a special experience. In the *heiligenschein* or "opposition effect,"[9] the surface looks exceedingly bright for light reflecting 180° from the source, usually the Sun. Viewing your head's shadow on the lunar surface, you see it surrounded by an impressive halo (Figure 10.7, bottom). Full Moon seen from Earth shows this behavior, appearing five times brighter than without this effect. The upper dust layer is rife with gaps and cavities between outermost grains; looking directly into these from the Sun's direction, all area you see, even cavity bottoms, is illuminated. Viewing even slightly to one side implies seeing much of each cavity filled with its own shadow. This occurs on Earth and other planets, but moon dust is particularly effective.

Moonlight is dim despite one's impression seeing the Moon at night. Accounting for heiligenschein, the Moon reflects about 10% of visible-wavelength light it receives, varying within a factor of two between locations[10] – fairly dark compared to other planets but brighter than many asteroids. Between low Sun angles in Apollo landing photographs and heiligenschein in the opposite direction, one can develop false impressions of lunar landscapes bright with glare. In reality the surface resembles powdery Portland cement, dark and colorless (Figure 11.6).

Without atmosphere the Moon's surface is about 5% brighter than similar surfaces on Earth. Bothering astronauts more, near the horizon no obscuration occurs with distance, so far objects appear equally distinct as close ones, excepting angular size differences. Mountains look like hills, and big craters look small,[11] giving the Moon an otherworldly feel, disorienting without good maps and distance indicators. Seemingly short rises can become major ascents. Similar effects punished Alan Shepard and Ed Mitchell hiking 700 meters up the 10% incline to Cone crater's lip, 1.5 kilometers from the LM. They thought they had failed (in fact coming close), in part from disorientation on the crater's slope.[12] Sense of scale is usually provided by nearby objects, so confusion is a risk. (Jack Schmitt exploited his own shadow's length by knowing the Sun angle.) This justifies redundant information layers (good maps, lunar GPS satellites, computer displays showing terrain's appearance along the local horizon, reference beacons, dead-reckoning/gyroscopic positioners, and star finders). Magnetic compasses cannot work. On the Near Side one can always tell direction toward Earth. Few named lunar features exist, so computer maps and navigators depend on coordinates.[13] Novices cannot be allowed roam freely to become disoriented if they cannot use these tools.

Without moonlight, surprising things happen. As discussed, nights can reach –153° C, but shadows can grow frigid even during the day. Unlike Earth's, the Moon's sky produces negligible illumination (away from the Sun). In the recesses of a shadowed cave or overhang (even under a spacecraft), heat provided by reflection from small illuminated areas may not suffice to keep the recess's temperature from plummeting 100°C or more. John Young reportedly discovered this accidentally when he rested in the LM's shadow a 1-kilogram rock he collected, then returned later to load aboard. Inside the LM he found the rock to be extremely cold. This is a problem, or an opportunity; such drastic temperature differences can easily drive a heat engine to generate power. Several other opportune lunar energy sources exist: photovoltaic solar cells, fuel cells, and nuclear fission reactors. All four have advantages and disadvantages, worthy of more discussion (see Chapter 12); let us summarize them briefly.

Solar power can be cheap and abundant – during the day. Power is needed during the 14-day night, too. Carnot-cycle or thermocouple heat engines can work night or day, in principle, but need big temperature differences. Unfortunately, at night this requires digging deep below the surface for warmth. Natural engines are effectively daytime-use only. Fuel cells have disadvantages because they use, well, fuel. Hydrogen and oxygen combine to yield electricity and water. Of course, water is useful elsewhere. Hydrogen and oxygen can be liquefied and stored, and water can be electrolyzed back into oxygen and hydrogen during the day. Nuclear power can operate day and night but is expensive to adapt for lunar application. An exception is the Apollo ALSEP stations, operating day or night using a combination nuclear/heat engine: thermocouples heated by radioactive decay of low-grade plutonium. Rechargeable batteries are key – regardless of how we charge them – to get around the lunar surface, either in vehicles or space suits.

Power and water supply are EVA's primary limitations, with Apollo's A7L restricted to 8 hours in vacuo. This limited EVA distance from home base to how far one could move in remaining space suit time using one's backup transportation. If one has two rovers, one must be close enough to return in case the other breaks. On Apollo J missions, despite the rover, astronauts could only venture as far from the LM as they could walk straight back in the remaining space suit time. This kept distance from the LM to about 5 kilometers, compared to total distance traveled via rover on each outing up to 12 kilometers (except *Apollo 17* where both limits were relaxed). Apollo's EVA suit was once intended to rescue one's companion if his oxygen failed – a buddy system – but this was dropped because of design difficulties. Fortunately need for such contingency was never realized.

For EVA consumables, Man does not live by oxygen alone, and water is not just for cooling, but also drinking. This is no joking matter; perhaps the sole fatality attributable to Apollo in space resulted from dehydration on *Apollo 15*, seemingly a likely contributor to Jim Irwin's ensuing heart attacks and death 20 years later.[14] Irwin complained of dehydration, particularly after his first lunar EVA, with no water for seven hours. During the flight, fears rose of Irwin approaching incapacitation, even a heart attack.[15] Heart irregularities developed for him and Scott, and after the mission he was diagnosed with heart disease from which he never fully recovered. He suffered a heart attack three months later.[16] Regardless of this case's causality of events, dehydration is deadly serious and must be avoided via sufficient water and monitoring of electrolytes. Opinions vary as to the causality between Irwin's dehydration and heart attacks. In an interview for this book Dave Scott reports that he thought during the mission that the problem was not so serious, but does not claim top authority.

Even less deadly consequences can threaten a mission: on his highly stressful space walk on *Gemini IX*, once humidity fogged Gene Cernan's helmet, he could see only by wiping a small patch clear with the tip of his nose.[17] (Subsequent missions used anti-fogging coatings on helmet faceplates.) Suits must maintain proper humidity and a drinking supply.

In Apollo's A7L and the Shuttle's EMU, food and water are available. In the A7L helmet a fixture allowed a nutritional fluid mix and/or food slurry to be sucked into the mouth, later augmented by a food stick. The EMU has a similar, solid nutrition stick and fluid valve in the helmet. Connoisseurs differ regarding the cuisine, but it beats sucking on nothing. Even this system can malfunction, for instance, when *ISS* Expedition 36 astronaut Luca Parmitano found his helmet soaking with water, with his EVA aborted, and him being helped inside by fellow astronaut Chris Cassidy.

You have been wondering about space suits and bodily waste. Apollo astronauts were male, so they used a condom-like attachment to the Urine Collection and Transfer Assembly storage bags, in addition to the Fecal Containment System, essentially a big diaper. Space Shuttle EVA suits use a single method, a Maximum Absorbency Garment (MAG), another diaper (made famous because of a certain criminal indictment[18]). In reality on the Space Shuttle and *ISS*, astronauts preparing for EVA usually pace their meals and drinking to avoid using the MAG.

It is obviously time to change the subject. A reality easily overlooked is that astronauts working on the Moon are the tip of a very large spear, with many people pondering the moon walkers' actions, both at present and in preparation. Their medical condition is monitored, as is their life support hardware and supplies. Strategies and procedures are considered and implemented to maximize their progress and minimize risk. We have all seen pictures of Mission Control in Houston – the Mission Operations Control Room (MOCR) of 19 divisions with one or several people – but supporting them were seven Staff Support Rooms: Flight Dynamics, Vehicle Systems, Life Systems, Flight Crew, Networks, Operations and Procedures, ALSEP/Lunar Science. Representing science in MOCR was the experiments/lunar surface operations officer. The Science Room held some dozen geologists and similar numbers of other specialists, for example, photogrammetrists (perhaps quickly assembling lunar science station panoramas when the Lunar Rover TV remotely scanned each stop), as well as representatives of each science instrument team, usually the principal investigator. Altogether hundreds were involved. Lower levels of such activity occur today for robotic explorers: several thousand people built the MER *Opportunity* (compared to 100 times this for Apollo), some 300 interact with the rover(s) on first landing, reduced to about 40 more routinely.

The Moon is the only "planet" where immediate backup is even possible. At the speed of light, a trip to the Moon and back is 2.6 seconds. (Even communicating on Earth via satellite suffers typical half-second delays.) Asking a question via radio of someone 384,000 kilometers away, a delay of several words' length is the cost of awaiting an answer. On Apollo one hears frequent instances of colliding dialog (this chapter's opening), but this is minor. Except several live televised slots per mission, Apollo crews talked with Earth only via Capcom in Mission Control, Houston.[19] For Mars, delays range more than 6–45 minutes, too slow for many purposes. Furthermore, a point on Mars lacks direct contact with Earth half of the time, requiring a relay satellite to improve this. Without a Martian relay network, this takes many hours. In contrast, a lunar machine with electronic eyes and mechanical legs/wheels operated from Earth might almost make it seem like we are there. This is *telerobotics*, and the 2.6-second *latency* (round-trip light travel time) between Earth and the Moon allows Earth-based telerobotics far beyond any other planet's capacity.

Telerobotics is a form of *telepresence*: communicating human control to a distant place and transmitting information back. Thus Apollo's astronauts, their rover's remote-controlled TV, and people back home interacting with the astronauts and their TV together formed a telepresent system. This pertains to the Surveyor landers and their team on Earth, or Lunokhod.[20] Likewise the Mars rovers or even the Voyagers at the Solar system's edge and their operators on Earth have some telepresence. We are becoming familiar with robotics' utility in many roles in our economy and society: manufacturing, mining, farming, construction, defense, animatronics, toys, energy production, automobiles, medical, patient care, logistics, nuclear power and waste, and sewer maintenance. Less familiar are more hybrid telerobotic approaches in which humans are in ultimate control, but robotic manipulators and senses have some autonomy in instant contact with subjects and environs of the effort. This mix is useful in cases in which the job is too dangerous, remote, or expensive for humans but requires more adaptation or dexterity than autonomous robots now master. In some cases now and more in the future, simply saving labor costs by turning humans into superhuman laborers is sufficient cause to consider telerobotics. Now telerobotics engage in drone warfare and reconnaissance, oceanography and underwater archeology, underground and underwater mining and oil industry (such as closing the Deepwater Horizon oil leak), surgery and medical examination, nuclear reactors and waste, and, in cases, space exploration including the Moon.

The case of oceanography is instructive in which exploration done by divers was extended to humans in submarines, then largely supplanted by telerobotic surveys and manipulation. Telerobotic and autonomous robotic surveys via sonar, magnetometers, and optical cameras are now central and prerequisite to ocean exploration but have not eliminated humans in submersibles. Deep-ocean, human-capable submersible *Alvin* (*DSV*-2) has been in nearly continuous human service since 1964 (except for 10 months on the ocean floor after a 1968 accident, as well as some upgrades). Although many more robotic craft exist than ones carrying crews, several human deep submersibles operate, and an *Alvin* replacement is planned. Humans at the ocean floor are partially displaced but not obsolete.

This begs questions of why we need humans on the Moon, because remotely controlled machines are so capable. This persists over coming chapters, but let us anticipate momentarily. At this point some authors would wax on human presence endowing exploration with near magical perception both irreplaceable and justifying vast expenditure. We will see; industry makes excellent perceptual machines, and virtual reality technology improves human-machine perceptual interfaces every year. Space suits remain numbing shrouds around explorers until we upgrade technology: limited mobility, sensing only through helmet visors, with little hearing, and without smell, or taste (an important field geology's tool) and little touch. Human vision needs 10–20 megabits per second (Mbps) versus *LADEE*'s lasercom at 600 Mbps. Simultaneous observers are able to monitor telerobots, receiving various data streams such as different electromagnetic wavelengths, other particles (subatomic, molecular spectroscopic, etc.), radar response, lidar, and so on, and even sophisticated real-time data interpretation, such as ratios, spectral index, composition, temperature, radioactivity, 3-D relief, and so forth. Astronauts could also use these via heads-up displays or eavesdroppers.[21]

Telerobotics from Earth can conduct scientific research and remote sensing, find and exploit lunar resources, unload/load service spacecraft, build structures, excavate regolith, aid astronauts, run and service production facilities, grow foodstuff, pave roads, and transport material between stations. Telerobots will park at recharging stations particularly during lunar night for repeated, varied missions – not be discarded when consumables or batteries exhaust. Facilities should exist to upgrade and repair them. They are more flexible if modularized, for example, appliance units riding on a separable transport chassis. Futuristically, they could be designed to self-replicate.

This said, situations arose during Apollo that no robot, even telerobot, could negotiate successfully. Certainly *Apollo 13*'s rescue was a human tour de force. A scientist in situ on *Apollo 17* enhanced scientific return. A telerobot might have noticed the famous orange soil [Box 9.1], but Jack Schmitt brought other samples to the LM, reevaluated them, then corrected methodology on succeeding EVAs. We quote an example.[22]

This example is compromised, with reexamination in the cramped LM, not a laboratory. Field geologists in Houston (not Capcom) should have collaborated, and transmitted data would help. With the nearest laboratory 384,000 kilometers away, this on-site evaluation trumps post-mission correction, too late and expensive. However, as in Chapter 9, for astronauts untrained in observation in other science disciplines their value on site is unclear. This could improve with experts eavesdropping from Earth, either on telerobots or astronauts. Is on-site human control and insight worth associated risks and costs, with one human lunar mission as expensive as dozens of (tele)robotic ones? Tele-operators on Earth wearing virtual reality goggles and Smart Suit gloves could sense multidimensionally the circumstances on the Moon. With a lunar telerobotic laboratory, operators might hear (surface microphones), smell (mass spectrometers or new "electronic noses"), and sense panoplies of scientific measurements. Earth operators could also sense visor TV and Smart Suit glove outputs on astronauts, not overriding, but helping interpret scientific or technical activity. Is human exploration more efficient? Does it introduce too much expense and complications, for example, planetary protection/ contamination? Should we instead justify human space exploration on its own merits, not because we make better science probes? Much space telepresence experience relates to Mars and distant targets, and short-latency (nearby) legacy on the Moon is now four decades old. However, work on Earth's ocean bottom – where humans need rarely go – is done telerobotically. We must consider not just the better worker, humans or robots, but which better encounters unknown situations, never before experienced, perhaps never imagined.

In the opinion of Jack Schmitt, astronauts are more easily trained in the necessary operational skills than in scientific ones: "[D]uring Apollo, the scientists acquired ~75% or more of the aircraft and spacecraft operations skills (other than professional test flight and combat skills) of the pilots in the program, while the latter, after a 15–20 month training cycle, attained ~25% of the field geology skills typical of active field geologists."[23] A telerobotic alternative a trained astronaut backed up by "eavesdropping" scientists via sensors with the astronaut, as would be easily implemented for lunar astronauts and scientists on Earth on time scales that might suffer some motor coordination loss if the scientists were steering a telerobot, but with sufficiently short latency for interpreting information from the astronaut and sensors, and for picking the best strategy for scientific sampling and measurement.

Telerobotics for space might be considered in their infancy, although used extensively in industry (see Chapter 12). On *ISS* since March 2008, Canada's telerobotic Special Purpose Dexterous Manipulator (SPDM), or Dextre, is essentially a torso and foot with two long, seven-joint arms on either side, "hands" consisting of a gripper and toolkit, and a 7-meter total span, controlled from Earth (Figure 10.9). It can self-propel around *ISS* or ride Canadarm2 (Figures 4.1 and 4.2). Dextre was less active until late 2010 but now has been used in several operations.[24] It was joined in late 2011 (*ISS* Expedition 30) by NASA's telerobotic/quasi-autonomous R2 (Robonaut2), now largely experimental or used for routine tasks such as simple monitoring. Upgrades and experiments are planned in 2013–4 and after. (This is

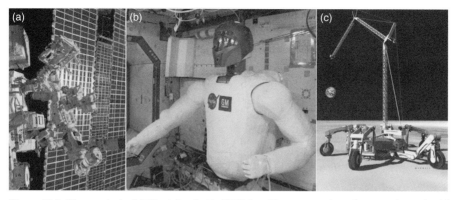

Figure 10.9. **They, robots**. (a) The telerobotic SPDM, or Dextre, consists of a torso (*center*) with two arms with 7 m total span and can move itself around the *ISS* or by Canadarm2 placement (see Figures 4.1 and 4.2). (b) Telerobotic/quasi-autonomous R2 (Robonaut2) on *ISS* (Expedition 30) where experimental operation began in late 2011. Further upgrades and experiments are planned. (c) Artist conception of a Lunar Surface Manipulator System (LSMS) on an ATHLETE rover. LSMS has done field tests and can lift loads up to 3 tonnes on Earth and reach heights up to 9 m. (NASA images.)

separate from JAXA's quasi-humanoid robot Kirobo which is promoted as a conversational companion.) Several telerobotic systems were built with the Moon in mind. ATHLETE (All-Terrain Hex-Legged Extra-Terrestrial Explorer) was built by JPL as a lunar telerobotic test bed for most lunar terrain, with six legs (usually wheeled) each with six degrees of motion freedom, 450 kilogram load capacity, and 10 kilometers per hour top speed. NASA Langley's telerobotic Lunar Surface Manipulator System can lift up to 18 tonnes on the Moon and reach 7.5 meters horizontally and 9 meters high. Of course the Lunokhod rovers in 1970 and 1973 were telerobotic; we discuss these later. The short history of space telerobotics is good news and bad: it is not yet standard practice, but industrial operations indicate potential, especially on the Moon.

How distant is the *tele* in telerobotics? The farthest the Near Side can be from anywhere on Earth is 415,000 kilometers, and signals would likely relay via geosynchronous satellite, adding some 40,000 kilometers. Light speed over this distance is 3.0 seconds round trip. On the Far Side, one must communicate via another relay. Although no selenosynchronous orbit exists, a special point (Lagrange point 2, or L2) sits behind the Moon such that an object at L2 orbits Earth once per month. L2 is blocked from Earth by the Moon, but objects placed beside L2 will follow halo orbits some 60,000 kilometers from the Moon. Relaying a signal via the Far Side, halo orbit, geosynchronous, and then Earth takes 3.8 seconds round trip. Inefficient routes sometimes take 6 seconds. Near Side to Far Side signals would likely proceed via halo orbit and L1 (the point between Earth and Moon opposite the Moon from L2). Far to Near Side signals need about 1.6 seconds round trip versus about 0.04 seconds for a lightspeed signal through a (costly) link along the lunar surface.

People encounter everyday reaction times of 3.0–3.8 seconds, such as driving a car. At 40 kilometers per hour (24.9 mph), an earthly driver suffers similar, maddening delays between seeing a hazard threatening ahead and the car finally screeching still. Lunar teledriving will demand attention like driving at legal speed through residential neighborhoods, except for rocks and pothole-like craters (some big). Shadows are excellent in distinguishing craters and rocks, except driving into the Sun (glare) or away (heiligenschein contrast loss). Differences with direction are so severe one might consider zigzag "tacking" to maintain good visibility. Teledriving displays can be augmented with radar imaging to spot obstructions and craters, aided by rapid pattern-recognition software to map hazards on the landscape, for example, painting obstacles on the display screen in flashing red if there is danger of running into them.

Three seconds corresponds to something dropping 7 meters in lunar gravity (see Figure 10.2). Catching a dropped object telerobotically will be hard, moving 2.5 meters per second before the teleoperator sees it falling. Safety nets, safety lines, and procedures aware of this will be necessary.

Lunar latency is sufficiently short for many applications. Virtual-reality lunar regolith bucket wheel excavator tests proceed almost unhindered with 2.6-second latencies.[25] Some argue that operators at L1 or L2 with 0.4-second latency could better perform some tasks, following shorter times required now for surgery, drone navigation and attack, or online combat games.[26] In human surgery, latency times less than 0.4 second can double rates on dexterity tests but is largely mitigated by intelligent programming.[27] To mitigate latency effects (at 1-second levels) the Rio Tinto mining corporation augments direct telerobotics with intelligent control systems: collisions avoidance and other motion constraints.[28] Without this, some telerobotic mine operators claim accidents increase with only 0.035-second delays, typically short for global relay times. Human operators stations at L1 or the Moon will cost extravagantly compared to simple office environments on Earth; this requirement requires rigorous study and justification. An effort to improve automated reflexes that telerobots need on second- and sub-second timescales, like muscular reflexes in humans faster than conscious reaction, would have growing commercial application. Developing lunar telerobots can advance this new economic requirement.

Telerobotic rovers require new approaches to space exploration. For surgery, we already have advanced haptic devices sending feedback to the user informing them of conditions at the robotic end: applying forces, vibrations, or motions signaling the operator through gloves or visually. This requires significant operator training, yet more such technology is demanded, for the Moon and space, but also Earth. This has begun: *Curiosity* has "doppleganger" rover *Scarecrow* on Earth for testing maneuvers and human/robot system training, but the process is much slower than needed for the Moon (or Earth).

A shortcoming is painfully slow robotic control operation in space because of special circuitry robust against radiation and heat. Aiding human teleroboticists with computers on the Moon or operating modern robots on the lunar surface are hobbled by limited processing power with hardened units like the RAD750 single board computer. These are necessarily slower, expensive because of low demand, and generations behind current CPUs. The RAD750 – more than 10 years old but the fastest hardened processor commercially available – costs about 1,000 times more than typical one-processor CPUs and runs ten times slower.

The Moon provides other options. A computer beneath 1 meter of regolith is shielded from temperature swings and most harmful radiation. Even this can be lightened if shielding includes a hydrogen-rich layer, for example, plastic or water, stopping neutrons created by energetic particles striking atomic nuclei in the shield. Imagine a plastic enclosure with outer ribs retaining a soil layer on top, perhaps a half meter deep. Inside, modern, unhardened electronics could function, or the enclosure might even garage a rover waiting through lunar night. Electronics in the enclosure could play home base, controlling robotics in surrounding areas. (Such an enclosure is portrayed as the pyramid in Figure 12.3.) Water or organic liquid could serve as both coolant and neutron shielding.

By midcentury, robots will clean our houses, build our widgets, mend our injuries, fight our wars (in fact they already do these), and do things hard to predict. Improving robotics will speed their infusion into the general economy, like Apollo-inspired integrated circuits accelerating the microprocessor revolution. Telerobotics divide robotics into mechanical (robot) and control (human) issues, providing natural ways to attack hardware/software problems in autonomous robotics. This could produce major earthly benefits of lunar exploration. This is evolutionary and welcome. The online virtual world Second Life and games, for example, World of Warcraft, have been popular; our youth have been raising themselves on them. Attaching the interaction's other end to a telerobot seems natural.

We have raised many issues about (tele)robots and humans working on the Moon and will consider more. Along with quantitative issues like latency times and expense are subtler, qualitative ones. Apollo only scratched the surface of human planetary exploration, with an impressive record of geological exploration and uneven records in some other cases. Astronauts make great explorers, in areas for which they are trained and equipped. It is undemonstrated that one or several astronauts on the frontier outperform a panel of talented investigators looking over the shoulder of in situ telerobotic probes. In addition, there are issues of planetary protection/alien contamination. There are several criteria on which humans and (tele)robots might be compared.[29]

Beyond planetary protection/contamination, we may be lucky to encounter alien life and need to compare how (tele)robots and humans respond. We expect humans to exercise more sophisticated judgment than autonomous robots; humans versus telerobots with experts monitoring their sensing data is hard to compare. We will not appreciate this fully without realistic trials, for example, on Earth in real situations. Despite telerobotic use in mining, warfare, and undersea, telerobotic exploration is still in its infancy but it has been used on the Moon.

Apollo J missions' (15–17) color TV camera on the LRV played the role of almost a third explorer, a prototype of one telerobot variety. In future applications a remote camera could present data in different electromagnetic bands (infrared, X-ray, etc.) or preparations (polarized, image processed). Along with Capcom, the LRV TV operator could pan and zoom the camera independently, although the astronauts were stars of the show. The astronauts, Capcom, and the TV operator often worked in tandem. The camera provided valuable records of astronaut activity and occasionally followed other objects of interest, in brief, parallel studies. (Figure 10.5 [top panel] shows an LRV TV sequence merged in a multiple exposure. The TV operator tried following the projectile in its entire trajectory but was frustrated by the 3-second delay.) It produced high quality TV coverage (with which the public grew bored, unfortunately). With the lunar surface mission completed, it could still communicate with Earth, for example, filming the LM ascent stage lift-off as the astronauts started homeward.

The LRV hit top speed at 13 kilometers per hour (ably benchmarked on the Moon by John Young, Figure 10.5, lower right), about 2.5 times faster than astronauts could safely run. The LRV's heaviest use (4.4 hours) was on *Apollo 17*, covering 36 kilometers. It was fairly simple: a lightweight aluminum chassis (3 meters by 2.3 meters) on four 0.8-meter diameter steel-mesh/titanium-tread wheels, each driven by a separate motor with two more motors steering the front and rear wheels in pairs, with a tight 3-meter steering radius. Two non-rechargeable batteries ran the whole LRV, including the Earth radio link, TV camera, and numerous sensors. Its mass was 210 kilograms but could carry a 490-kilogram load. (See LRV close-up in Figure 11.2.)

Apollo's experience with dust and machinery is unrepresentative. Its hardware was lightweight more than robust or easily serviced, and not over designed against dust's effects (more on dust in Chapter 11). We must consider systems operating for long times on the Moon, not just several days. Modularity in design is one important approach, with vulnerable components replaceable. A telling case is Lunokhod, operating for 16 months.

The first (and still only) truly telerobotic lunar vehicles were *Lunokhod 1* and *2* (Figures 10.8 and 10.10), carried to the Moon on *Luna 17* and *21*. The roughly 800--kilogram rovers together traveled 47 kilometers on the Moon (in Mare Imbrium in 1970–1971 and crater Le Monnier on Mare Serenitatis's edge in 1973, operating altogether 16 months). Each of eight steel-mesh wheels was self-powered; the Lunokhods were so maneuverable that the entire rover could rotate in place. Powered by batteries recharged by solar cells, they hibernated through lunar night with warmth from radioisotopic decay heaters. They carried several scientific experiments to measure soil consistency, solar X-rays, cosmic rays, magnetic fields, optical/ ultraviolet illumination, as well as laser retroreflectors to accurately measure the distance between Lunokhod and Earth, in collaboration with French scientists.

The Lunokhods were driven by crews of five (pilot, commander, navigator, radio operator, and vehicle engineer) conferring, two hours per shift, on each next path segment, perhaps 30 meters and several minutes apiece. They drove "slow" or "fast," at 0.8 or 1.9 kilometers per hour (0.2 or 0.5 meters per second), with four cameras taking images (mostly panoramic, black and white) every few seconds, every meter or several meters. Despite this, there were unfortunate blind spots. In a good 24 hours, a Lunokhod moved 0.5 kilometer. A bad maneuver into a crater while driving away from the Sun (into heiligenschein) by *Lunokhod 2* led to its overheating from dust contaminating its radiator after accidentally scraping the crater wall, after four months on the Moon (Figure 10.10). *Lunokhod 2*'s demise is key in telerobotic exploration; we should understand it.[30] A telerobotic rover operated by experienced humans can run afoul in an ill-considered move, despite what we mentioned earlier. *Lunokhod 1* ran 11 months, well beyond its 3-month design life.[31]

The New Moon

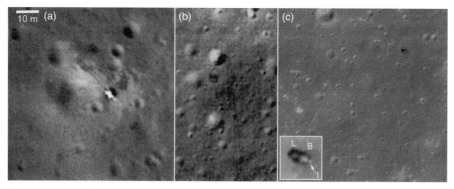

Figure 10.10. **Birth and death of *Lunokhod 2*.** (a) *Luna 21*, here shown as its rover tracks and lander, landed on January 15, 1973, in crater Le Monnier in Mare Serenitatis. Two hours later *Lunokhod 2* drove onto the regolith. (b) (45 m wide) Exploring the environs of a rille on May 9, after 35 km of roving, *Lunokhod 2* slipped into the 7 m crater at top. Before ascending it was covered with dust. Note that narrow, dark rover tracks seem to breach the crater from east (*right*) and tracks to the north (*top*). The rover took a complex journey around the fatal crater (see text) to exit south. About 880 m later it closed its lid for lunar night, pouring dust onto its radiators. (c) (170 m wide) The rover overheated, unable to proceed at sunrise on June 3. It rests where it failed (upper right), detailed in inset (6 m wide): *L* being its photocell lid, *B* the rover chassis, and *I* its instrument boom. (LROC NAC images M122007650LE and M175070494LR.)

Except for the brief travel of *Yutu*, no rover have visited the Moon since *Lunokhod 2*, and rovers have landed beyond Earth only on Mars: *Mars Pathfinder* in 1997, two Mars Exploration Rovers (MER) *Spirit* and *Opportunity* in January 2004 (Figure 10.8), and *Curiosity*, August 2012. The seemingly unstoppable MERs each explored independently over 6 years, in utter contempt of their 90-day design life,[32] each with 185 kilograms' mass but requiring a 1,063-kilogram spacecraft to transit Mars' atmosphere to a soft landing. In total they have traversed more than 45 kilometers, about 7 meters per day. They are semi-autonomous, executing a daily program unaided once uploaded from Earth and shutting down during Martian night (being 3% longer than Earth's). As such they are more robots than telerobots.[33] Covering much less ground than their speedier lunar rover kin, they nonetheless produced prodigious quantities of science results with the suite of about 10 instruments onboard each.

A limitation to both people and machines working on the Moon is its dusty surface characteristics. Once, cavities between grains, explaining the heiligenschein, were imagined to threaten human activity. Theories anticipated that, in dry vacuum, grains would form loose structures ("fairy castles") unable to support weight. However, other theories held that grain surfaces, absent coatings of humidity or other contaminants, would eventually fuse solid like rock. The truth is intermediate; craft do not sink below the lunar surface, but they can have problems with traction.

On *Apollo 15* (and two other J missions), the LRV stayed largely horizontal, with 90% of travel on slopes under $2.5°$[34]. At its nominal 20° limit the LRV would slip, but it traveled only 1% of its distance (of 28 kilometers total) at more than a 10° slope. The power consumed fighting dust was equivalent to the LRV running up a $0°.4$ slope, not overly significant. *Lunokhod 2* had other problems at crater flanks. On level ground, Lunokhod's wheels sank a few centimeters into the dust. At some crater flanks, however, it would sink 20 centimeters in dust and stall.[35] *Apollo 15*'s LRV lost traction in one loose dust patch near the LM.[34] Machines traversing the Moon need reserve traction to free themselves from hazards or "buddies" to tow them out. Alternately, like the LRV, a vehicle light enough can be lifted and carried by astronauts.[36]

As civil engineers know (or anyone reading of construction mishaps), the angle that one can dig a hole or trench in soil is crucial in determining safety and stability in that landscape. This maximum stable angle or "angle of repose" depends on the depth one digs and crucially on whether soil is excavated, compacted, or dumped. (The angle of repose, technically, is limited to the latter.) Apollo astronauts had neither tools nor time to experiment with this, but earthly experiments and models lend a clue. An excavated trench in lunar regolith can maintain a vertical wall (90° slope) up to 3 meters (Figure 11.6), amazingly, and 60° slopes up to 10 meters.[37] These angles are much greater than for rover operations or stable angles for dumped soil (60° at 3 meters), meaning that there are conditions where rovers might fall down a slope or initiate an avalanche. This is crucial for telerobotics to study early in lunar exploration to be programmed into rules about driving and excavating the surface.

Dust in a vacuum moves unlike expectation from earthly experience. Landing *Apollo 11*, Neil Armstrong noticed two aspects of lunar dust's behavior blasted from the surface by his rockets in those crucial seconds maneuvering the LM away from boulders as his fuel ran low. ("Picking up some dust" was all that he said.) First, escaping dust made the Moon difficult to see (Figure 10.7). Secondly, when he gunned the engine, a puff of dust would dissipate simply by following freely ballistic trajectories over the horizon, as if metal fragments. Dust "clouds" influence little the motions of their individual grains. When *Apollo 12* landed within 200 meters of *Surveyor 3*, despite carefully avoiding the robot craft, LM exhaust accelerated dust to impact and pit surfaces with tiny craters on one side of *Surveyor*.

On smaller scales moving dust is also surprising. Astronauts describe lunar dust between their fingers as being surprisingly slippery, because of the extremely fine grain size. In contrast, boots on Apollo EVA suits were significantly abraded by relatively brief moonwalks. Lunar dust grains are hard, with edges not eroded by wind or water. For billions of years, dust has been shattered into sharp-edged grains by meteorite impacts and fused into irregular, jagged clusters by heat of impact. The effects of dust will be discussed in Chapter 11.

Even with humans, work on the Moon is aided by machines; we consider prime examples and lessons learned. Before the LRV came MET (Modularized Equipment Transporter) on *Apollo 14*, not a rover but more an unpowered two-wheel rickshaw or golf pull-cart, 75 kilograms mass fully loaded. MET allowed Shepard and Mitchell to carry more equipment with greater ease – until they reached Cone crater's flank. Taking turns pulling MET up the flank's 10% grade, their heart rate soared and progress slowed until Houston advised them to turn around, short of their nominal goal. They nearly reached the crater lip, but disorientation, and their heart and time stress prevented them from then knowing this. Along with the LRV and MET, NASA built a lunar motorcycle on two pneumatic tires for *Apollo 15* but abandoned it for the LRV.

Some lunar features are craggy, and for these wheeled vehicles, they are inappropriate. Pre-space age writers imagined astronauts carrying large loads in one-sixth gravity, for example, Arthur C. Clarke's two-man "stretcher" equipment carrier from 1954.[38] This is possible over short distances à la Jack Schmitt in Figure 10.2 (bottom), but the MET/Cone crater experience argues ambitious applications. Sometimes rovers must be crawlers, more spiderlike. For most lunar terrain, however, wheels are the way, even if each needs its own drive motor and independent steering.

As discussed, machines have limited lunar power options: batteries and/or photovoltaic solar cells, radioisotopic decay/thermoelectric generators, or fuel cells (given a fuel supply or fuel regenerator). They must rely on radiative cooling unless they consume precious water and cope with 300°C or greater temperature swings, most likely with heaters for surviving lunar night. Alternatively, if machines can simply sink themself even a few centimeters below ground (or be covered even temporarily), they remain at Earth-like temperatures, night or day. Machines' surfaces must be protected against ravages of solar UV radiation, and their interiors from effects of cosmic rays and solar wind. This is crucial in their microelectronics in which space-rated semiconductor chips are now used to avoid "soft failures" as a result of digital data corrupted by charged particles deposited by cosmic ray hits.

For machines more than humans, another hazard is anticipated. Blue and UV light consists of photons sufficiently energetic to kick electrons from surfaces of materials, depending on their composition. This famous photoelectric effect proceeds based on whether the photon energy is greater than the material's work function electron voltage. In the lunar surface's vacuum, an illuminated surface can develop a positive charge leading to electrical discharges between it and shadowed material. When the Sun rises or sets, or machines pass into/out of shadows, sparks can fly and coronal discharges glow and unprotected circuitry can be destroyed. Careful electrical grounding and material choice is essential.

In the vacuum, space-gloved astronauts have several options to fasten materials. Duct tape holds the world together and also works on the Moon, even in a vacuum. Nonetheless duct tape will not stick to dusty surfaces. As Gene Cernan described (Chapter 11), he used a piece of duct tape to pull off the dust, then a second piece to stick to the newly cleaned surface. Velcro was developed in 1941–1951 by Swiss engineer George de Mestral, but the space program rescued it from relative obscurity. Although it works well in vacuo, Velcro can also become choked with dust. On Apollo lunar EVAs, Velcro on radiator surfaces could become so dust-filled that heat buildup could create problems. For easier operation by gloved astronauts, designers use fasteners, for example, captive screws and quick release pins with easy grips. These also degrade from dust, slowly. In weightlessness, astronauts even attached objects such as documents using Silly Putty, but this stopped with *Apollo 8* and was perhaps most useful advertising the toy putty. Dust affects other surfaces: laser ranging retroreflector mirrors (Figure 2.18) have degraded in reflected signal by 10 times in 40 years since left by Apollo and Lunokhod.

Astronauts put these techniques into practice. *Apollo 16*'s LRV suffered a damaged fender (while stationary when John Young walked into it near the end of the second of three EVAs), spraying dust over the LRV when it moved. On the first of

Apollo 17's three EVAs, Gene Cernan accidentally pulled off a fender panel with the rock hammer looped in his suit leg. To reduce the spray of dust on the rover he duct-taped an unused geology map as a replacement fender panel. It worked somewhat, but the astronauts were still sprayed with dust.

PARKER: Okay, copy that. We're ready to press on with ALSEP INTERCONNECT. And Geno, how are you doing on that fender?
CERNAN: Bob, I am done. If that fender stays on, I'm going to take a picture of it because I'd like some sort of mending award. It's not too neat, but tape and lunar dust just don't hang in there together.
PARKER: Okay. Copy that.
CERNAN: Well, let's hope. Keep your fingers crossed, and I'll be more careful around the fenders. (*Apollo 17* Onboard Voice Transcription – Lunar Module)

There were rare instances during Apollo of frayed tempers. Dealing with dust tried patience. Set procedures were practiced many times, eliminating most surprises. Houston and sometimes the entire world listening must influence behavior. Rarely the routine itself of completing another checklist produces a little snap, but nothing significant.[39] These crews worked together for months; such problems were largely smoothed out. In the future, this may not always be possible; it is not a military hierarchy. For people who work together all day, Moon bases will need sufficient room for an opportunity to spend time away from each other.

There was no room to escape one another inside the LM – only about 2 meters across. There was no privacy and even sleeping at first was difficult. On *Apollo 11* Armstrong and Aldrin had no sleep equipment and did not sleep. By *Apollo 15*, astronauts could strip out of their exterior suits to relax between their three 7-hour EVAs and were able to get adequate rest (more in Chapter 11).

Each of the 12 people who worked on the Moon did so no more than a few days, returning to Houston to their real job and their real life. The vision advanced for a new generation of explorers is to remain on the Moon six months or more, long enough to appreciate living there. This is unprecedented, challenging, and even revolutionary. For the first time humans will begin to call another world home.

Notes

1. Cernan and Schmitt were loading the rover to drive beyond the LM, with Cernan stowing a TV camera. Schmitt is walking to the LM with drilling equipment. Here is a continuation of this interchange (repeating the quote's last line):

    SCHMITT: (Singing) I was strolling on the Moon one day –
    SCHMITT AND CERNAN: (Singing in unison) – in the merry, merry month of –
    CERNAN: (Singing) – May
    SCHMITT: (Singing) – December
    CERNAN: No, May.
    SCHMITT: May.
    CERNAN: May's the month this year.
    SCHMITT: May – that's right.
    CERNAN: May is the year, the month.

SCHMITT:	(Singing) When much to my surprise, a pair of bonnie eyes – be-doop-doo-doo...
PARKER:	Sorry about that, guys, but today may be December.
CERNAN:	Okay, the battery cover –
SCHMITT:	(Humming) Da-da-da-da-da-da-da-dee-da-dee...
CERNAN:	Roger. We can confirm that.
CERNAN:	— and you want me to leave those two blankets open 100 percent, right?
PARKER:	Roger. That's affirm.
CERNAN:	Okay, now I got – now I got to mount my camera and tether my tongs. Boy, Jack, I can bare – I can't see you at all. Looking into the east is terrible. All I can tell you is that there's a LM there.
CERNAN:	Okay, mount camera, tether tongs. See if my camera's going to work. Bob, I'm on – on Bravo – mag Bravo and frame count 19.
PARKER:	Okay; copy that, Gene.
CERNAN:	And for EMU status, I can give you about 36 percent, no flags, 3.85, and I'm on – I'm still INTERMEDIATE cooling.
PARKER:	Okay. Copy that.
CERNAN:	Okay, inventory. Camera, tongs, gnomon. Okay, I'm ready to get on. Ready to get on.
SCHMITT:	Okay, you want us to take the – Ouch! That rock by your front porch is really a major nuisance.
CERNAN:	Oh, doggone it.
SCHMITT:	What's the problem?
CERNAN:	Oh! Every time I get on, I get dust around. I still haven't learned how to get on yet. You'd think after three times, I'd know better. I know better, but it's –
SCHMITT:	Okay, I've got the transmitter. I'm heading west, or east. (Laughs) Heading east. Heading west – or east. (Laughs) Heading east. Sorry about that. (Hums)

It is heartening that the crew worked well together and enjoyed themselves despite small problems. Note references to difficulties looking east into the Sun, talking on a single radio channel, and to environmental hazards. To clarify, the strolling sequence in Figure 10.1 does not correspond to this dialog. (From *Apollo 17: Technical Air-to-Ground Voice Transcription*, Test Division, Apollo Spacecraft Program Office, December 1972, NASA publication MSC-07629.)

2. Energy E needed to lift a mass m to a height h against gravitational acceleration g is $E = m\,g\,h$.

    The height fallen during time t because of gravity (starting motionless) is $h = \frac{1}{2} g\,t^2$, so fall time is $t = \sqrt{2h/g}$. Lunar g is six times smaller, so t is 2.5 times longer.

3. After 30 minutes on the surface, Aldrin describes walking on the Moon (*Apollo 11* Technical Air-to-Ground Voice Transcription):

ALDRIN:	(To Houston) I'd like to evaluate the various paces that a person can (garbled) traveling on the lunar surface. I believe I'm out of your field-of-view. Is that right, now, Houston?
MCCANDLESS:	That's affirmative, Buzz. (Pause) You're in our field-of-view now.
ALDRIN:	Okay. You do have to be rather careful to keep track of where your center of mass is. Sometimes, it takes about two or three paces to make sure you've got your feet underneath you. (Pause) About two to three or maybe four easy paces can bring you to a fairly smooth stop. (Pause as he turns at the TV camera then heads toward the LM) (Garbled) ... change directions, like a football player, you just have to put a foot out to the side and cut a little bit. (Pause as he turns toward the TV camera and starts to do a two-footed hop.) The so-called kangaroo hop does work, but it seems as though your forward mobility is not quite as good as it is in the more conventional one foot after another. (Pause as he turns at the TV camera and heads back to the LM) As far as saying what a sustained pace might be, I think that one that I'm using now (as he turns at the LM and runs toward the TV again) would get rather tiring after several hundred (garbled). But this may be a function of this suit, as well as the lack of gravity forces.

4. In the *Apollo Lunar Surface Journal* (http://history.nasa.gov/alsj) for *Apollo 17*:

   The video sequence on the left in Figure 10.1 is accompanied by the corresponding audio:

   PARKER: We'd like you to leave immediately, if not sooner.
   CERNAN: Hippity-hoppity, hippity-hoppity, hippity-hopping over hill and dale. (Singing) Hippity-hopping along. Okay, my golly, this time goes fast."
   PARKER: That's affirm. Okay, and when you leave here —
   CERNAN: Where did it go?
   PARKER: — 17, remember that we want to pick up —
   CERNAN: I'm giving you readings.
   PARKER: — EP number 8. Roger. We're ready.

   Things can sound silly, but the astronauts returned immediately to work. Incidentally, this song is Gene Cernan's version of "Mule Train" from the western film *Singing Guns* (1950), according to Andrew Chaiken and Brian Lawrence, from the *Apollo Lunar Surface Journal*.

5. "Relief of Itch by Scratching: State Dependent Inhibition of Primate Spinothalamic Tract Neurons" by S. Davidson, X. Zhang, S. G. Khasabov, D. A. Simone & Glenn J. Giesler, Jr. 2009, *Nature Neuroscience*, 12, 544.
6. "The Binaural Performance of a Cross-Talk Cancellation System with Matched or Mismatched Setup and Playback Acoustics" by Michael A. Akeroyd, John Chambers, David Bullock, Alan R. Palmer, A. Quentin Summerfield, Philip A. Nelson & Stuart Gatehouse, 2007, *Journal of the Acoustical Society of America*, 121, 1056; also "Optimal Crosstalk Cancellation for Binaural Audio with Two Loudspeakers" by Edgar Y. Choueiri, 2008, submitted to *Journal of the Audio Engineering Society* (see http://www.princeton.edu/3D3A/Publications/BACCHPaperV4d.pdf).
7. *LSB*, 28, 34.
8. *Apollo 17* astronauts Gene Cernan and Jack Schmitt discuss UV exposure hazards (and less probable lunar hazards) with Capsule Communicator Robert Parker in Houston (*Apollo 17* Onboard Voice Transcription – Lunar Module):

   PARKER: Hey, Jack. And we see your gold visor up? You may want to put it down out here in the Sun.
   SCHMITT: Well, I think I might – I can't see with it down; it's scratched. Bob, I'll use it.
   CERNAN: I think I can monitor that one.
   SCHMITT: Hey, I'm standing on a boulder track. How does that make you feel?
   CERNAN: That makes me feel like I'm coming over to do some sampling. Think how it would have been if you were standing there before that boulder came by.
   SCHMITT: I'd rather not think about it.

   Astronauts do not necessarily need to be admonished:

   CERNAN: Okay. Man, I forgot I had my visor up. Zowie! See if I can get back in the shade.

   Caution is crucial on deep-space EVAs such as Ron Evans retrieving a film canister halfway to Earth, with no moonlight filling shadows (*Apollo 17* Onboard Voice Transcription – Command Module):

   EVANS: Hey, there's the Earth, right up ahead.
   CERNAN: Okay, Ron. You've got a —
   EVANS: The crescent Earth.
   CERNAN: You've got a GO for egress.
   EVANS: Beautiful.
   CERNAN: And just take it slow.
   EVANS: Okay. First of all, I got to get back in, and get the old TV camera. Oop. Yes. That's right. I don't even know where the Sun is. Which way is the Sun? Okay, yes. Sun's on the right. Okay.
   CERNAN: Okay, Houston. Ron's putting the camera out there on the pole now – pole out there, rather.
   PARKER: Roger. We see the EVA light out there.
   EVANS: ... stay lower?
   CERNAN: Okay.

EVANS: Man, that Sun is bright. Whoooo!
CERNAN: Pull down that visor, Ron. You're going to need it.

Sometimes esthetics prevail, temporarily (*Apollo 17* Onboard Voice Transcription – Lunar Module):

SCHMITT: Okay, get the visor down, Geno.
CERNAN: Get the visor down – Holy Smoly. Think it'd be better to leave it up. Beautiful out here today, Bob. We can look to the east for a change – a little bit, anyway.

In the LM's shadow, Cernan deploys the *Apollo 17* lunar rover:

CERNAN: Oh, boy. There it goes. The Rover looks in good shape. ETB is down there. Okay. I've got all my visors down. Jack, I wouldn't lower your gold visor until after you get on the porch, because it's plenty dark out here.

Even a simple mechanism such as a visor can succumb to lunar dust:

PARKER: Okay. Copy that. And you guys may still have your visors up. We can't tell, but you might be better off with them down, if you've forgotten they're up.
SCHMITT: Well, boy, I can't see. My hands work just as well as my visor, as a matter of fact.
CERNAN: No, I can't believe – mine could be up.
SCHMITT: You've got a crater right in front of you.
CERNAN: Yes. I got it.
SCHMITT: Okay. Well, you can't tell much about the countryside going into the Sun, can you?
CERNAN: Put your upper visor down. That's what – that'll give you a whole different perspective.
SCHMITT: It doesn't vise very well. It's stuck.
SCHMITT: Soil sample. Gene, if you hit them off in there, it's going to be awful hard to find them. That's the problem.
CERNAN: Did you pick a spot – a good spot while you were over here?
SCHMITT: No, I didn't. I just was looking at it. I think we need to get in the light, though.
CERNAN: I – I can see with my gold visor.

9. "Wavelength Dependence of Polarization. III. The Lunar Surface" by T. Gehrels, T. Coffeen & D. Owings, 1964, *Astronomical Journal*, 69, 826; "Detailed Photoelectric Photometry of the Moon" by R. L. Wildey & H. A. Pohn, 1969, *Astronomical Journal*, 69, 619.
10. "Polarimetric Properties of the Lunar Surface and Its Interpretation, Part I. Telescopic observations" by Dollfus, A. & Bowell, E. 1971, *Astronomy and Astrophysics*, 10, 29.
11. Heading away from the LM on the lunar rover beginning the second EVA on *Apollo 17*, the effects of distance disorientation become apparent (*Apollo 17* Onboard Voice Transcription – Lunar Module):

CERNAN: Boy, is that getting big.
SCHMITT: Whoo-ee
CERNAN: Hold on.
SCHMITT: Whoooee. Oh, boy, that really gives me a strange feeling (laughter).
CERNAN: Gives me a strange feeling too. Those are not intentional.
SCHMITT: I understand.
CERNAN: I'm not sure I've got enough guts to make them intentional. Man everything's getting to look big the closer you get. Hole-in-the-Wall looks more promising, though, Bob.
SCHMITT: Yes, I don't think that's going to be any problem until we get up and look back. Oh, man, what a trip this is going to be. Golly.

(There are other effects at play here that we leave to the reader to discuss.)

12. *Apollo 14* **CAPCOM (HAISE)**: Very good, Ed. Next question. The difficulty you had at the last there, climbing up to Cone rim, was that due primarily to the terrain slope or did the soil conditions change again that caused you to have some greater problems?

MITCHELL: I think probably both. I think we just entirely underestimated the difficulty in going that far and getting that high in such a short period of time. It's a darn hard climb to try to do rapidly, and the soil is a little bit thin and mushy. And the suits are bulky; it's all those problems rolled in, Fred; we just… It was too ambitious, I guess.

SHEPARD: Let me say that I don't really think that the composition of the soil changed very much. Matter of fact that was one thing that struck me about the whole area: the consistency of the texture of the regolith – outside of soft areas, of course, in crater rims. I think as far as the progress up there is concerned, it was due to the grade and the boulders and the rocks that we had to go around. But, really, as far as the surface texture is concerned, as far as the bearing strength of the surface, I thought that, outside of the crater rims, that it was unusually consistent all the way through. And the thing that surprised me was the raindrop pattern with these very small sort of pebbles, which is decidedly different than we had down here in this area where we landed.

(From *Apollo 17* Onboard Voice Transcription – Lunar Module)

Similar experiences were encountered by *Apollo 15–17*'s crews. With the Rover for hill climbing, J mission crews could reach sites on slopes steeper than Cone's 10°. At such sites they needed caution moving downhill from the Rover because of effort in returning. In places, the slope and soil softness made work more difficult. The best plan of action was moving cross-slope as much as possible, and then, if possible, standing in a place to lean into the hillside, chipping a rock or bagging a sample. In hindsight, allocating more time to the Cone climb could have allowed cross-slope traverses. They could have made better time, walking farther with fewer rest stops.

13. There are about 8,800 named lunar features, or one every 4,500 square kilometers = $(67 \text{ km})^2$.
14. *Disaster and Accidents in Manned Spacelight* by D. Shayler, 2000, New York: Springer, 209.
15. *Flight: My Life in Mission Control* by Chris Kraft, 2001, New York: Penguin, 343.
16. *On the Moon: The Apollo Journals* by G. Heiken & E. Jones, 2007, New York: Springer, 473.

    Shortly before dying from his third heart attack, Jim Irwin mused about his heart and *Apollo 15*:

    "Even knowing the heart problems, I'd gladly do it again. In fact, I wish I had a chance to do it again, because I think I'm a little cooler than I was then and I think I'm really in better shape now than I was when I made the flight. (Chuckling) I was surprised that it looked as if my resting heart rate was up in the 60s and 70s and, now, my resting heart rate is in the 40s. I'm 30 pounds lighter than I was when I made the flight. And back then, I thought I was in good shape. Hopefully, we learn more as we progress through life. but I realize, man, I could have brought back 30 pounds more of lunar material. If only I'd known that, I'd have been smaller. I would have had a little more room to move around, (chuckling) and I probably could have gotten out of the hatch a little easier, too. I would like to go back again, to complete the exploration that we had planned, get up to the Northern Complex to bring back some of the things that I forgot and left there, and then to see if the experience has as much meaning today as it did 18 years ago."

17. *The Last Man on the Moon: Astronaut Eugene Cernan and America's Race in Space* by E. Cernan & D. A. Davis, 1999, New York: St. Martin's, 139. A brief account of Gemini program EVAs is noted earlier. More is found in *On the Shoulders of Titans: A History of Project Gemini* by B. C. Hacker & J. M. Grimwood, 1977, NASA SP-4203.
18. "NASA Astronaut Lisa Nowak Charged with Attempted Murder in Bizarre Love Triangle" *Fox News*, February 7, 2007; "Lisa Nowak: Astronaut's Lawyer Calls Diaper Story 'A Lie'" *People*, June 29, 2007. Fortunately, there are no more stories like this one.
19. Apollo crews even received their daily world and family news report via the Capsule Communicator, for example, for *Apollo 17* on December 17, 1972 (MET 249h 57m 17s): "CAPCOM: Okay. We start today's newscast out with this historical fact. Today marks the 69th anniversary of man's first flight in a heavier-than-air powered machine. Back on December 17, 1903, Wilbur and Orville Wright, of Dayton, Ohio, took three historic flights on the sand beaches of Kitty Hawk, North Carolina. The brothers will be honored today at a ceremony at a visitor's center near the flying site. Now, a look at the news. There's apparently been a serious hitch in the peace talks between the U.S. and North Vietnam. Dr. Henry Kissinger, in a curt news conference at the White House, has said that the North Vietnamese have reneged on earlier agreements and have brought our – brought the peace talks to a halt. Dr. Kissinger said the unresolved problems which center around the number of peace supervisors and their placement, is not acceptable to the President, and Kissinger feels, and we quote, 'We have not yet reached an agreement that the President considers just and fair.' Final unofficial returns from the Federally

supervised election name Arnold Miller the new president of the United Mine Workers. His victory over Tony Boyle appears to be only his first step in his promise to clean up the union. Former President Truman remains in serious condition at a Kansas City hospital. Doctors say that the 88-year old Truman is not responding to treatment." The newscast continues with national politics, weather, and sports.

20. Works concerning telepresence, including specific missions: *Intelligent Motion and Interaction within Virtual Environments*, edited by Stephen R. Ellis, Mel Slater, Thomas Alexander, 2007, NASA/CP–2007–213468; *Digital Apollo* by David A. Mindell, 2008, Cambridge: MIT Press; *Voyager: Seeking Newer Worlds in the Third Great Age of Discovery* by Stephen Pyne, 2010, London: Penguin; *Working on Mars, Voyages of Scientific Discovery with Mars Exploration* by William Clancey, 2012, Cambridge: MIT Press.

21. A "partisan" comparison of human versus robotic exploration (neglecting telerobotics) can be found in various works. Here are some relevant metrics – distance traveled: the record breakers have a near tie as of mid-2013 – human: *Apollo 17* LRV traveled 35.7 km in three days (at 30% duty cycle); telerobotic: *Lunokhod 2* traveled 37 km in 139 days (at roughly 50% duty cycle because of lunar nights); robotic: *MER-B Opportunity* has traveled 38.5 kilometers in 3,570 days (at 15% duty cycle due to solar cell illumination). In comparison, these missions cost approximately $7 billion, $500 million, and $750 million, respectively, in 2013 dollars. In terms of distance per dollar, Lunokhod was a bargain; in terms of traveling a distance rapidly on a set dollar amount, Lunokhod and Apollo were comparable. *Luna 16*, *20*, and *24* returned a total lunar sample of 0.3 kilogram for about $2 billion, whereas *Apollo 11–12* and *14–17* returned 382 kilograms for about $150 billion. Thus Apollo was somewhat more economical per kilogram, but Luna was radically less costly per site. Apollo and Luna could both drill samples at 2–3 meters depth. Apollo carried radically more and varied scientific instruments, but not when scaled for cost. I think these numbers argue for some advantage for telerobotic operations, especially on the Moon. Some readings to support the point: "Dispelling the Myth of Robotic Efficiency" by Ian Crawford, 2012, *Astronomy and Geophysics*, 53, 2.22; "Human Spaceflight: Science or Spectacle? by David Clements, *Physics World*, 22, 11, 16; "Is Human Spaceflight Obsolete?" by James A. Van Allen, 2004, *Issues in Science and Technology*, 20, 4. We discuss this issue more in Chapter 13.

22. In the LM after the first EVA on *Apollo 17* at MET 128:08:04:

    SCHMITT: Joe, I just took a – quick look with the hand lens at that large rock I brought in, and I'll – I don't think there's much more than 30 percent plagioclase. I'll go back – could be more of a standard basalt or gabbro. It has a fair proportion of ilmenite in it, I believe. There's a bright platelets – in the vugs or vesicles – of ilmenite. Now it could be that the glass – if the soil is very glassy, that it's developed the darker color from the contribution of the – of the basic minerals through the glass, particularly the iron and the titanium.
    ALLEN: Roger, Jack. Copy that. Sounds interesting.
    SCHMITT: All it means is that we don't yet know the origin of the dark mantle.
    SCHMITT: That rock – looks I may have, by accident, sampled the front side of one of the parting planes that I mentioned. Very, very sharply bounded on one side by a planar surface.

    Further dialog is found at *Apollo 17 Voice Transcript Pertaining to the Geology of the Landing Site* by N. G. Bailey & G. E. Ulrich, 1975, Flagstaff: U.S.G.S. Branch of Astrogeology.

23. "Motives, Methods and Essential Preparation for Planetary Field Geology on the Moon and Mars" by H. H. Schmitt, A. W. Snoke, M. A. Helper, J.M Hurtado, K. V. Hodges and J. W. Rice, Jr., 2011, Geological Society of America *Special Paper*, 483, 1.

24. "Canada's 'Dextre' Robotic Hand Performs First Ever Operational Tasks" by Pete Harding, February 4, 2011, *NASA Spaceflight.com*, http://www.nasaspaceflight.com/2011/02/canadas-dextre-performs-first-operational-tasks/.

    NEO telerobotics, for example, ESA's METERON and NASA's HET via *ISS* and TDRSS (Tracking and Data Relay Satellite System) suffer 2 to 5 second latency times.

25. "Digital Spaces Simulation-Based Lunar Telerobotics Design, Acquisition and Training Platform for Virtual Exploration" by Bruce Damer, July 18, 2005, *Final Technical Progress Report for SBIR I: NNA05AC13C*, http://www.digitalspace.com/reports/sbir04-phase1-finalreport/.

26. "Human Space Exploration and Human Spaceflight: Latency and the Cognitive Scale of the Universe" by Dan Lester & Harley Thronson, 2011, *Space Policy*, 27, 89; "On-Orbit Control of Lunar Surface Telerobots from

Earth-Moon Lagrange Points" by D. F. Lester, K. V. Hodges & M. L. Raftery, 2011, *LEAG Meeting*, 2012, http://www.lpi.usra.edu/meetings/leag2011/pdf/2012.pdf.
27. "Telerobotic Surgery: An Intelligent Systems Approach to Mitigate the Adverse Effects of Communication Delay" by Frank M. Cardullo, Harold W. Lewis III & Peter B. Panfilov, in *Intelligent Motion and Interaction Within Virtual Environments*, eds. S. R. Ellis et al., 2007, NASA/CP-2007–213468, 45.
28. "Innovate to Succeed: Interview with John McGagh" by Matthew Brace, 2000, *Earthmatters: CSIRO Exploration and Mining Magazine*, 19, 10, http://www.csiro.au/files/files/pp19.pdf.
29. Comparing exploration capabilities of humans and (tele)robots we must consider how well (1) they collect scientific data, judged both by instrumentation available and decisions made on their basis; (2) these data are stored and conveyed for later use; (3) human or (tele)robotic explorers correlate data with location where they are taken; (4) the explorer follows mission and operational procedures and guidelines in routine and exceptional circumstances; (5) the explorer copes with situations where they work autonomously; (6) do limitations of human versus (tele)robotic explorers feed back into planning of missions by scientists and mission designers ab initio; and (7) how well and how much can human explorers *in situ* and (tele)robots work together and with distant humans, once all of this is established.
30. *Lunokhod 2* encountered its fatal crater on May 9, 1973, with sunrise in the east. Evident in Figure 10.10 (center), it never passed west of the crater. In accident accounts, illumination effects were blamed in part, heiligenschein obscuring detail looking west. Reviewing the figure carefully, it appears *Lunokhod 2* was driving north when contacting the crater's east rim, with the crater bottom lost in shadow and its top in heiligenschein. It drove a complex track around the crater, to the north and east. One can even see small circles in the tracks left by rover wheels when they turn. This path seems to exit the crater on its north rim. Others claim the small loop south of the crater is part of this track ("Game Developer's Lost Electric Buggy Found on Moon" by Lewis Page, March 17, 2010, *TheRegister.co.uk*). If true, *Lunokhod 2* entered and exited the crater twice.

Author Anthony Young quotes Vyacheslav Dovgan, one of two Lunokhod drivers: "A situation arose in which it was contemplated as to whether or not to proceed down into what looked to be a fairly steep crater. General Dovgan recommended that the lid be closed in order to protect the solar cells before proceeding. This was not normal written procedure, and, with a recent personnel shift change having occurred at Lunokhod Mission Control, the lid closing was not approved. As Lunokhod-2 started down into that crater, it began to slide, and it was decided to immediately reverse its direction as soon as this could be accomplished. As the vehicle was being remotely driven out of the crater, it hit the side of the crater and a large amount of lunar soil deposited on the solar cells of the doomed Lunokhod-2" (*Lunar and Planetary Rovers: The Wheels of Apollo and the Quest for Mars* by Anthony Young, 2007, New York: Springer-Praxis, 293).

There were two Lunokhod crews of five and one crew member in reserve. Sergei Gerasimenko of Kharkov University and Alexander Basilevsky of the Vernadsky Institute are collecting original records and interviewing crew members; perhaps we will have a telerobotic first-person account.
31. "The Other Moon Landings" by A. Chaikin, 2004, *Air and Space*, 18, #6, 30.
32. *Spirit*, with two of its six wheels misbehaving, bogged down in Martian soil after 5 years and died about a year later.
33. For several years NASA ran a telerobotics research group until its 1997 closure. In part to compensate, it sponsored with the Spaceward Foundation a public challenge for US$250,000 (http://www.spaceref.com/news/viewpr.html?pid=18433) and a workshop in telerobotics in May 2012. NASA sponsors challenges to encourage areas it cannot fully support, described further in Chapter 4 and in the Astronaut Glove Challenge.
34. *Mobility Performance of the Lunar Roving Vehicle: Terrestrial Studies Apollo 15 Results* by N.C. Costes, J.E. Farmer & E.B. George, 1972, NASA Pub. TR-R-401, 87 pp. 35.
35. "The Floor of Crater Le Monier: A Study of *Lunokhod 2* Data" by C. P. Florensky, et al. 1978, *Proceedings Lunar and Planetary Science Conference*, 9, 1449.
36. *LSB*, 529.
37. *LSB*, 521.
38. *The Exploration of the Moon* by R. A. Smith & A. C. Clarke, 1954, New York: Harper, 76.
39. Gene Cernan and Jack Schmitt preparing for the first EVA on *Apollo 17:*

CERNAN: Okay. You're thermally – Let me double check that. The helmet is locked. Your visor is locked. It's one thing you don't want to lose among some others. Okay. Okay. You want to give me a hand?

SCHMITT: Not particularly. (Laughter)
CERNAN: Oh, man. Where did that come from?
SCHMITT: Watch your nose, drink bag, candy bars, popcorn. Click, click, click.

This is not an issue of concern; moments earlier in the checklist, the two exchanged the following and more meaningful comments, attaching Schmitt's helmet and food supply:

CERNAN: Okay. Okay. All your candy bars, and lemonade, and all that jazz are all clear. Water, I should say.
SCHMITT: That sounded good.
CERNAN: Okay. Try it. Okay. It looks good here, Jack. Okay. And what's your LEVA? (LEVA = helmet visor assembly)
SCHMITT: Okay.
CERNAN: Enjoy it in there; you're going to be in there for a few hours.
SCHMITT: Can't think of any place I'd rather be right now.
CERNAN: Sounds like you're in there, too. Darn, too far back. Okay, that's better. I'm freezing my you know what off.
SCHMITT: Me, too. (Laughter)

(From *Apollo 17* Onboard Voice Transcription – Lunar Module)

Chapter 11
Lunar Living Room

lunarian n.: An inhabitant of the Moon, as distinguished from Lunatic, one whom the Moon inhabits. The Lunarians have been described by Lucian, Locke and other observers, but without much agreement. For example, Bragellos avers their anatomical identity with Man, but Professor Newcomb says they are more like the hill tribes of Vermont.
– Ambrose Bierce, 1911, *The Devil's Dictionary: A Cynical View of the World*

Despite Ambrose Bierce's sardonic treatment, the real Professor Simon Newcomb did not think much of possible lunarians. Newcomb was Bierce's elder contemporary, a foremost American astronomer (founding president of the American Astronomical Society) and noted author on lunar motion. In 1870 he visited Europe and found a theory running rampant allowing for possibly habitable conditions on the Far Side of the Moon. He effectively crushed it.[1]

In 1824 Franz von Paula Gruithuisen published observations of what he saw as buildings, streets, and other structures of a massive city in jumbled terrain north of crater Schröter (Figure 11.1).[2] Despite his campaign to show other scientists these structures, few were convinced. Gruithuisen wrote more about the Moon; he is credited with first realizing that lunar craters formed largely via impacts with other objects orbiting the Sun, more than a century before Eugene Shoemaker showed it.[3]

Lunar cities lost any credibility after New York's *The Sun* in 1835 published six articles describing faked observations from a nonexistent, 24-foot diameter optical telescope revealing animals, forests, intelligent beavers, winged humans, temples, and monuments.[4] *Sun* writer Richard Adams Locke was the author, to whom Bierce refers (as he does Lucian of Samosata; Chapter 3). As late as 1880 Flammarion spoke of advanced lunar life, and in approximately 1920 W.H. Pickering considered lunar plants and insects.[5] Now optical telescopes, wider than 24 feet, even with modern instruments could hardly detect a large building on the Moon, much less human-sized beings, their wings, anatomical genders, or expressions on their faces.[6] In 1835 few thoughtful astronomers were fooled, but the farce accelerated the *Sun* into the world's top circulation newspaper.

From the *Sun* hoax until the space age, Moon people were topic of intentional, guileless science fiction. Since 1645, authors referred to lunar inhabitants as

The New Moon

Figure 11.1. **Images and imaginings**. (a) Gruithuisen's Wallwerk sketch, June 12, 1822. (b) Image of same area, 60 km across, by *Lunar Orbiter IV* (image 114-H1). Wallwerk appears more a group of gradual hills than walls, less organized than in Gruithuisen's mind. *Lunar Orbiter*'s image is sunlit from 20° above the horizon, whereas in the 1822 sketch the Sun is nearly setting. This feature is near the center Near Side, up and left (uninverted) at 4.5° north, 7.5° west of center. Get a telescope and look yourself. (c) Drawing from New York *Sun* based on its 1835 Moon hoax, with trees, waterfalls, and winged humanoids. (You will never see this scene on the Moon no matter how hard you look.)

"selenites"[7] (selenite also being a calcium sulfate crystal, a gypsum). H.G. Wells used this term in *First Men in the Moon* (1901), and it appeared in Ian McDonald's 1999 retro *Breakfast on the Moon, with Georges*.[8] Webster's 1913 dictionary lists *lunarian* so the choice is confusing. Some call them "lunans," sensible with "terrans."

Lunar Living Room

This chapter's question is how humans can survive on the Moon biologically (not economically; see Chapter 12). This is new, not a space station. Nothing surrounds the *ISS*, an entirely artificial environment. A lunar outpost will have regolith below and vacuum above (maybe far above). Regolith has advantages and drawbacks; how are these advantages best accommodated? NASA's latest lunar habitat concepts seemed primarily transporting *ISS*-like modules to the Moon's surface, neglecting local resources. Better results are possible by "living off the land" in outpost construction and operation, recalling the difficulty of travel to the Moon versus into low Earth's orbit (to the *ISS*). How will we live on the Moon?

In reality, lunar life is less poetic than imagined by Locke and others. When people live on the Moon more than several days, they might start calling themselves "moon dusters." That is the feeling one gets listening to astronauts spending time there. Witness *Apollo 16* astronauts Charlie Duke and John Young, and *Apollo 17*'s Gene Cernan and Jack Schmidt reentering the LM from the Moon's surface, and other outtakes from their lunar stay.[9] (Joe Allen and Robert Parker, Capcoms in Houston, both flew later on Space Shuttles):

DUKE: Houston, the lunar dust smells like gunpowder...
DUKE: But it is a really strong smell.
YOUNG: Boy, I'll tell you.
CERNAN: Pick up your right foot. Jack, you're just going to have to get up on that ladder somewhere so I don't – don't get the dust all over this thing.
SCHMITT: Well, I've got to dust you, too.
CERNAN: Well, okay. Go ahead and get me.
SCHMITT: Yeah, where's your brush?
CERNAN: Right on the hook.
CERNAN: Right; see what I can do. (I'll) kick most of it off, I hope.
SCHMITT: You have to go anywhere else, now?
CERNAN: Just right around here; no place but right around here. Man! That's a – that (dusting)'s like a super-endless task.
CERNAN: And the reason tape won't stick is that everything's got a fine coating of dust, and the only way I could finally get it to stick was to put tape on it (and) rip the tape off – or take the tape off – which took some of the dust off and then (another piece of) tape would tend to hold it.
CERNAN: Wait until that dust hits the sweat of your hands.
SCHMITT: (Smells) like gunpowder, just like the boys said.
CERNAN: Oh, it does, doesn't it?
CERNAN: Smells like someone's been firing a carbine in here.
ALLEN: Sounds like you've got hay fever sensors, as far as that dust goes.
SCHMITT: It's come on pretty fast just since I came back. I think as soon as the cabin filters most of this out that is in the air, I'll be all right. But I didn't know I had lunar dust hay fever.
ALLEN: It's funny they don't check for that. Maybe that's the trouble with the cheap noses, Jack.
SCHMITT: Could be. I don't know why we couldn't have gone and smelled some dust in the LRL just to find out. (The Lunar Receiving Lab was the ground facility where rocks and dust from Apollo missions were stored initially.)

CERNAN: No. Boy, we're sure giving this suspension system a workout. Whew! I can't even see it. Well, everything's getting awful dusty. Boy, everything is stiff. Everything is just full of dust. There's got to be a point where the dust just overtakes you, and everything mechanical quits moving.

CERNAN: I have never seen so much dirt and dust in my whole life. Ever. (Pause) Ron's not going to be able to see through either one of these helmet visors. (Laughs) Yes, he will. ("Ron" refers to Command Module Pilot Ronald Evans.) [10]

SCHMITT: Bob, sample 15 Echo is a bunch of dust that gradually accumulated in my pocket.

PARKER: No fair, Jack, you can't go collecting samples after the EVA's over.

Astronauts on the lunar surface busied themselves with dust, cleaning it from themselves and all exposed surfaces before reentering their spacecraft. On *Apollo 17* in particular, they accidentally lost a LRV fender panel, and dust from its wheel pelted many surfaces (Figure 11.2). Even partially repairing this, they spent an hour dusting themselves and equipment, consuming space suit oxygen, before returning to the LM and other duties. Astronauts on most landings, especially later ones with more surface time, suffered from dust.

Cernan and Schmitt summarized their thoughts about lunar dust:

"I think dust is probably one of our greatest inhibitors to a nominal operation on the Moon. I think we can overcome other physiological or physical or mechanical problems except dust." – Gene Cernan (1972)

"A common sense, layered, engineering design defense can solve any apparent problem with dust during long-term human activity and habitation in the lunar environment." – Jack Schmitt (2004)[10]

Dust was subject to experimentation on Apollo.[11] Apparently *Apollo 12*'s Command Module pilot told the LM's crew that they were so dirty that they could transfer into his spaceship only after stripping off their space suits, which they did, wearing little but their medical sensor wiring![12] Indeed *Apollo 12*'s LM raised so much dust on landing that Pete Conrad set it down flying blind the last 30 meters. (Unfortunately, dust was so bad it affected landing radar as well as vision.[13]) *Apollo 15*'s landing was dust obscured, but *Apollo 14* and even dust-plagued *Apollo 17* landed almost dust free. (*Apollo 11* and *16* were intermediate.) Dust's severity varies with location but can be ameliorated (or worsened).

One thing is sure, if you smell something like gunpowder on the Moon, you have dust. Dust irritated astronauts' eyes, nostrils, sinuses, and mouths. Dust ruined seals of boxes designed to maintain the lunar rock and dust samples in their pristine state so scientists could examine them unaffected by air, moisture, and organic material. Dust repeatedly clouded and scratched camera lenses, helmet visors, and display windows. Dust contaminated and reduced scientific instruments' effectiveness. Dust clogged tool components so they would not fit together, Velcro so that it would not stick, and the LRV's radiator surfaces so it cooled less efficiently. Dust jammed zippers and made mating surfaces of astronauts' glove and helmet couplings grind against their space suits so dressing was harder for lunar surface activity. Dust eroded space suits' outer surfaces and increased suit leakage into the

Figure 11.2. **Dust buggy**. (a) Gene Cernan drives *Apollo 17*'s LRV, showing how dust it kicks up is directed down toward the ground. (Otherwise, dust would simply fly in parabolae until hitting something.) (b) One of the LRV fender panels was accidentally knocked off, but necessity required a solution: duct tape and a geological chart. (c) Tired and dusty, Cernan relaxes after a hard (Earth) day's work on the Moon, and the astronaut's dusty external suits (d) are stored to the side, under the docking hatch at the top of the lander.

vacuum. Had astronauts stayed longer, likely other problems would have arisen. Both with broken LRV fenders, *Apollo 16* and *17* suffered more dust problems, mostly with suits and equipment; *Apollo 15*'s LM floor was perhaps the dirtiest – all of this after three days on the Moon.

Lunar dust comes in many sizes, shapes, and compositions (Figures 6.21 and 6.23); some containing glass beads from ancient volcanic eruptions and highly abrasive. Moon dust is so tiny that it insinuates itself nearly anywhere and remains suspended in mid-air seemingly forever. By mass, half of lunar soil consists of particles smaller across than a human hair (about 50 microns),[14] and, depending on the sample, 5–10% of particles by mass are so small (under 5 microns)[15] that human upper respiratory tracts cannot filter them. This most unhealthy size, especially

0.5–2 microns,[16] can stay suspended for days mid-air; on Earth drifting for hundreds of kilometers on the wind.[17]

This property was turned to useful advantage on *Apollo 12*; depressurizing the cabin for the next EVA would evacuate floating dust until it was agitated in the air again. When the LM ascent stage launched into weightlessness, its cabin would fill again with floating dust. The astronauts had dust brushes, but Conrad noted that eventually these primarily drove dust into space suit fabric. Shepard said that stomping feet and slapping suit exteriors removed as much dust as brushing. By then, further actions were taken (from *Apollo 14* debriefing):

SHEPARD: We did find that we had to take the boots off because there's so much dust in your overshoes that we did take those off before we went to bed.
MITCHELL: In training, we thought that maybe was an unnecessary time-consuming step and we'd probably sleep with the boots on, but they were so covered with crud that I didn't want it sifting down in my face during sleep. We took them off.

Being meticulous, *Apollo 14* astronauts avoided major dust problems, but the longer J missions suffered. *Apollo 15* was very dusty, but Scott and Irwin wrapped their legs in plastic bags to control it. With the rover fender accidents, *Apollo 16* and *17* were so dusty that large amounts sifted between EVAs into suits then onto the astronauts. Young and Duke sprayed the LM floor with water and scraped up mud. By *Apollo 17* NASA added equipment to EVA suits to reject dust from seals (initially with quiet objections from astronauts, but eventually they approved).

Apollo taught how to control dust. Beyond boots, bags, brushes, stomping, and slapping, we need airflow control and floor-level dust traps. Air filtration is paramount; we must evaluate if electrostatic dust eliminators add more contaminants – for example, ozone – than they are worth. Obviously, habitats should incorporate entrance "dust locks," ejecting dust from inhabited areas. Jack Schmidt suggested that inrushing air-refilling airlocks' vacuum be exploited for jets eliminating dust. Suit outer garments could be cleaned or recycled to remove dust.

Is lunar dust dangerous? Cigarette smoke particles this size enter and irritate lung air sack aveolae, causing cancer and other disease; however, smoke particles are organic structures, for example, fullerenes (after Buckminster Fuller, inventor of geodesic domes, which fullerenes resemble on molecular scales, hence, terms like *buckyballs* and *buckytubes*). Fullerenes and other polycyclic aromatic hydrocarbons (PAHs) are sheets of benzene molecules linked and often curved into cylinders or spheres (actually polygons like dodecahedra and soccer balls, truncated icosahedra – impress your nerdier soccer friends with that).

Benzene is poisonous and carcinogenic, but Moon dust is neither. Silica dominates Earth's crust, whereas on the Moon anorthite, pyroxene, and olivine are common. Fortunately, although related to it, none include asbestos, causing cancer and other chronic lung diseases. Life on Earth evolved to ignore chemically most silicates, but fine dust is rare and only became significantly problematic on Earth with the introduction of mining and manufacturing, raising much freshly created dust, leading to silicosis and even cancer (from quartz dust). Natural dust and sand are rounded by years of weathering, and unlike on the Moon or Mars, settled by water into mud and sediment. (Natural silicosis from fresh volcanic dust is called *pneumonoultramicroscopicsilicovolcanoconiosis*, English's longest word, which I think is simply supercalifragilisticexpialidocious, because this indicates it must be rare.) Silicosis is no laughing matter, however, because small silicate particles stuck in the lungs lead to immune reactions along with pain, fatigue, fever, shortness of breath, expansion of heart chambers, and susceptibility to tuberculosis. Physiotherapy (like chest massage) and bronchodilators can help, but the only sure therapy is lung transplantation. Silicosis is serious. Furthermore, surface rim iron nanoparticles caused by space weathering may worsen matters.[18]

Inhalation tests of lunar dust on animals show no ill health effects, but these were limited, especially for particles less than 5 microns.[19] Lunar materials are so rare that realistic dust tests are difficult, furthermore, even those lunar regolith samples which do exist have changed in surface properties having spent four decades in a lab rather than a lunar environment, so future tests in a lunar environment may be required.[20] (Of Apollo's 382 kilograms of samples, about 2 kilograms were particles less than 5 microns.) Having never undergone weathering, lunar dust's geometry tends to be jagged, likely making it more dangerous.

Not all lunar dust minerals are silicates, and composition varies with location. Ilmenite ($FeTiO_3$) is common and more reactive, although is not shown to produce more negative effects than silicates. On Earth, quartz particles react more if their surfaces hold hydrogen,[21] which is rare on the Moon; likewise hydroxide (rare on the Moon but not so on Mars) forms peroxides, immediately damaging. Little in lunar dust is explicitly poisonous, but the dust covering Mars looks potentially reactive and dangerous.

More tests are necessary with real lunar material, probably specific to individual lunar sites. On Earth, hazards are so specific that two substances with the same name, diatomaceous earth, differ radically in effect. Freshwater diatomaceous earth, mainly silica, is sprinkled in homes to discourage insect pests (scratching their exoskeletons to cause dehydration) with no human health effects and used to grow hydroponic foods and even added to some feed. Saltwater diatomaceous earth, with sharp-edged, crystalline silica, causes silicosis and likely cancer.

Small differences produce large results, and regolith varies with location. However, safety seems a matter of dust geometry and chemistry. We should be able to assay this with small robotic landers (whereas remote sensing is unlikely to suffice). We need not test every human landing site, only those for longer stays. Exposures of a week seem unlikely to sicken people; longer ones are worrisome.

Humans evolved not only biological means but also diverse social and tactical memes and technologies to counter infection; analogous and easier approaches will ameliorate lunar dust, which unlike bacteria cannot grow exponentially or adapt. Lunar dust, unlike smoke, asbestos, or perhaps Martian dust, is inert (except perhaps for a micron-thin for surface [see note 18]). Martian dust might contain not only peroxides, causing foaming and burning in lung tissue, but also more grains less than 5 microns. The thin Martian atmosphere excludes micrometeorites that fuse dust into agglutinates and grinds particles in the wind without water to bind them. We may develop a good long-term lunar solution, and lunar dust is no immediate danger. That solution may aid later human explorers on Mars.

Clever and practical lunar dust abatements exist, and many involve surface nanophase iron metal. Lunar dust can be moved with magnets and removed with magnetic brushes. Lunar vehicles and habitats will have air locks also serving as dust locks. Magnetic air filters can effectively remove lunar dust. More simply, a most effective method is for lunar astronauts to shower. Hopefully, (recycled) water will be more plentiful on future missions than on Apollo, which had no such luxury.

Iron in lunar dust absorbs microwaves better even than water in foodstuff in microwave ovens. With enough power (likely available in abundance from solar cells), one imagines irradiating smoothed areas of regolith with a portable microwave source, transforming it into dust-free glass via the process of sintering.[22] One could microwave sinter structures' interior walls to seal dust out or high traffic surfaces in the vacuum outside to suppress agitated dust (as plagued *Apollo 17*).[23]

One must alter sintered floors, because of their low effective friction in one-sixth Earth gravity. From Chapter 10, frictional stopping capacity is denoted by the "coefficient of kinetic friction" – how much force is generated to slow forward motion (parallel to the floor) as a fraction of the object's weight pushing down on the floor. Because lunar acceleration of gravity is 0.1654 times that on Earth, downward force for a given mass is 0.1654 times that on Earth, as is the friction one feels. If you try to stop using friction on some lunar floor surface, it feels six times more slippery than that surface on Earth. If sintered surfaces remain as glass (with a kinetic friction coefficient for metal, e.g., nickel, sliding on glass of 0.56), on the Moon this feels like a coefficient of 0.09, as slippery as greased glass sliding on glass. (The kinetic frictional coefficient for dry glass on glass is 0.4.) Literally, walking surfaces feel like they have been greased.

To fight this, as with the hop, "dig in your heels." (Imagine leaning back on your heels, like the old Road Runner cartoon character Wile E. Coyote trying to stop short of a cliff.) In doing so, one converts forces usually perpendicular to the direction of motion into parallel forces, so they work directly against one's momentum of motion, rather than via the coefficient of friction (much less than one). If sintered glass is smooth, one cannot do this; one's feet simply slide out from underneath one's body. To help this, mold treads or ruts into freshly sintered surfaces, like treads on a tire, so one has something to stop against.

This is generic to the Moon, even beyond sintered surfaces. One needs high-friction "booties" for more effective stopping, even in one's own living space. (Imagine waking to answer a telephone on the other side of the room and forgetting that the floor is effectively as slippery as ice.) One can grow accustomed to this (or be born into it), although it presents challenges. Consider that floor surfaces (and walls) with treads or ridges allowing one to stop quickly will also have greater tendency to collect dust. Some clever person will invent a way to ease dust removal from ridge-textured surfaces. (Nanotechnologists take heed.)

Fundamental challenges to living on the Moon are meteoritic impacts, vacuum, temperature fluctuations, and solar and cosmic radiation. Despite regolithic dust's worrisome issues, building with regolith can solve these problems, as we discuss shortly. First, what is being built?

Architectures proposed for Constellation lunar missions were either sorties, typically seven days on the Moon, or eventual habitation of a lunar outpost in six-month shifts. An issue was the longer lunar explorers' surface stay, the more efficiently they use their transportation to the Moon. For example, when Apollo LMs returned from the Moon to the Command Module in orbit, it carried 300 kilograms' payload (mostly astronauts). Instead, if astronauts and LM stayed on the Moon, aided by local resources to sustain themselves, each new LM could deliver 5 tonnes of supplies, out of 15 tonnes for the total LM. (For Constellation's Altair, these masses would have been three times greater.)

Consider what early long-term human inhabitants might expect on the Moon. In a South Pole analogy, Apollo echoed the Amundsen/Scott's generation, when small groups of men, supported by many, succeeded (or failed) in reaching their goal, stayed briefly, returned with data or samples, only once individually. This occurred a century ago (in 1912 for Amundsen and Scott), followed only in 1956 when George Dufek led a team in a DC-3 aircraft (after the Richard Byrd team's overflight in 1929), and the Amundsen-Scott South Pole Station was established over the next year. ASSPS started with an 18-man U.S. Navy crew, then roughly doubled in 1975, and more than doubled again (Figure 11.3). McMurdo Station on Antarctica's Ross Island was founded simultaneously, with support stations extending to New Zealand, and the United States. South of 75°S latitude exist six permanent stations, operated by Russia, New Zealand, the United Kingdom, Japan, and France/United States. New Zealand, the United States, the United Kingdom, and Russian stations started in 1956–1957 (the International Geophysical Year), and others much more recently. Returning to the Moon would do well to recreate the Antarctic success, crucial scientifically (justifying itself with insights into global warming, as well as astronomy, astroparticle physics, biology, and geophysics). Also, administration over Antarctica (read: international/U.S. sway) suppresses sovereign territorial claims on the ice-free 2% sliver of Antarctica, primarily the Antarctic Peninsula. (Or is it the Palmer Peninsula, Tierra de O'Higgins, Tierra de San Martín, or British Antarctic Territory? Thank the status quo for absence of war over this.) The Moon/Antarctic analogy is a useful one for Moon base planners. (Note: 1957−1912 = 45 years; if we build a lunar base by 2020, it would follow a similar schedule given *Apollo 17*'s departure in 1972.)

Extending the lunar/Antarctic analogy, recall that one expedition to first visit the South Pole succeeded whereas the other perished, in large part because of Amundsen's better use of resources appropriate and indigenous to the poles, rather than depending solely on engineered solutions from foreign materials. As on the Moon (South Pole or otherwise), Antarctic inhabitants deal with unnatural sunlight cycles, difficult transportation, isolated small communities, and extreme environments that can prove readily fatal if plans go wrong.

Vacuum, radiation, temperature, meteoroids: four primary, potentially apocalyptic lunar surface threats all mediated by wise use of regolith in burying construction under a thin layer. Recall that this limits mobility: tonnes of regolith

The New Moon

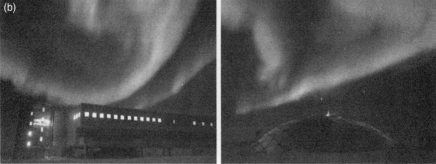

Figure 11.3. **South Pole, Earth**. South Pole Amundsen-Scott Station. (a) Aerial view of the new station (blocky building), 1975–2007 main station dome (with entrance trench above right of center, partially buried). South Pole is indicated by flags and marker above left of center. Not shown are the landing strip and astrophysics research station. (b) New station building during 6-month winter night (during impressive aurorae) and (*bottom right*) old station main structure, which one can see is a geodesic dome.

removed and reapplied to change location. For each phase of exploration, how mobile must everything be?

These four dangers plague other inner Solar System worlds, including Mars (except Venus with atmosphere density 90 times Earth's, making the vacuum inviting in contrast). On the Moon the average temperature is acceptable, −20°C at the equator (potentially offsetting heat generated by humans and their machines). This is unsurprising; the Moon is equal distance from the Sun as Earth but slightly cooler because of absence of atmospheric greenhouse warming.

As in Chapter 8, dig a few centimeters into the soil and mad monthly surface temperature swings (roughly ±130°C) damp to several degrees, with the temperature rising slowly with depth, starting at −20°C (depending on latitude). Hollow a room 20 meters subsurface to find temperatures like an earthly home. In addition, regolith is so impermeable that atmosphere in this space needs hours to leak out, longer if sealed. Because regolith is a good thermal insulator, one might excavate less deep because waste heat from one's appliances and body heat

340

will accumulate, making preferred outside temperatures slightly cooler. One might even need radiators if humans live closer together than every 5 meters.[24] Sending light from the surface via mirrors or light pipes, one could grow plants and produce oxygen. This would raise the temperature also, implying that simply excavating a space several meters down and sealing the dust out and air in with sintering, one could populate it with humans and plants such that they maintain themselves with just sunlight (at least during lunar day) without additional power or technology – a lunar living room. Machines have even been designed that would dig and seal such underground spaces.[25] The Moon is a fairly hospitable place once one digs down a few meters, more even than Mars or elsewhere in the Solar System (except Earth itself).

Recall those pesky meteorite impacts. At rates of one per square meter per year, they blast 100-micron diameter pits, about 50 microns deep.[26] At one per square kilometer annually, expect craters about 1 centimeter deep – big enough to challenge structures lightweight enough for payload from Earth. For structures with regolith depths several meters or more, penetrating impactors are 1% as common; even structures 100 meters across are penetrated only once per million years.

Regolith itself has little mechanical strength, a factor in seismic design. Moonquakes are usually far underground, only reaching equivalents of magnitude 4 or 5 on the Earth-based Magnitude Moment scale[27] – barely noticeable – or just able to crack plaster or shift heavy objects. Many moonquakes last longer than earthquakes, so lunar building materials and joints should be flexible. During earthquakes some ground itself can "fluidize," losing ability to support loads, causing buildings to sink. Regolith is much more cohesive than terrestrial waterfront or landfill.

More worrisome are galactic cosmic rays, solar wind, and coronal mass ejections, because the lunar surface has little natural shielding like Earth's magnetic field.[28] A flare often produces enough radiation to kill a person not hiding behind several meters water-equivalent of shielding, too heavy to carry to the Moon (unless everyone hides in spaces smaller than phone booths whenever the Sun is active). This plagues outer space generally beyond low Earth's orbit: heading to the Moon, on its surface, or traveling to more distant goals like Mars. Usually one day's warning precedes radiation actually reaching Earth/Moon – enough time to take cover. (This is unclear for an asteroid- or Mars-bound astronaut.)

Radiation levels permitted for astronauts are already higher (50 rem in one year, 25 rem in one month) than for radiation workers on Earth (5 rem per year). Radiation dissociates molecules (particularly water) into free radicals (like OH) that can damage DNA strands. If slow enough, only half the DNA helix is damaged, and natural repair uses the other half to reconstruct the damaged site. Truly dangerous are heavy ions (mainly atomic nuclei in cosmic rays) creating extensive damage tracks through the body that break both helix strands, much harder to repair, and likely making affected cells nonviable or mutagenic/carcinogenic.

NASA monitors radiation doses for each astronaut in space. This will grow more crucial and difficult on the Moon: crucial because exposures above Earth's magnetosphere accumulate to potentially dangerous levels absent precautions and difficult because lunar magnetic fields, although weak, can focus radiation locally, changing each astronaut's dosage depending on where they move. Solar wind pulses reach Earth within several days of eruption, but violent solar coronal mass ejections (large, super fast pulses of charged particles) can arrive in much less than a day. Maxima of solar activity will occur in 2021–2025 and 2032–2037, during which ejections will occur some 15 times more often than during solar minimum. The most dangerous of these can supply lethal radiation doses in less than 10 hours. Most such events can be anticipated by observing the Sun optically, but this is not foolproof.

A ready remedy is shielding – mass between the person and radiation source. Solar wind stops in several centimeters of solid/liquid, but stopping half of high-energy heavy nuclei requires 0.6 tonnes per square meter of regolith or light metal such as aluminum.[29] (Earth's atmosphere is 10 tonnes per square meter.) While these particles stop, they generate secondary neutrons. Lighter material (containing hydrogen) is needed to stop these, at least 50 kilograms per square meter of water or equivalent. Prolonged missions such as the first U.S. space station *Skylab* carried radiation shield vaults (1.5 tonnes). On the Moon we have prodigious shielding, the soil itself, and need not pay to launch it. Instead, buildings must sustain extra loads of soil on top (and means to put it there). In one-sixth gravity this is not hard; buildings are best covered with regolith.

NASA envisioned early lunar habitats as mobile and wanted to avoid difficulties and uncertainties from covering structures with regolith. The agency studied active shields deflecting radiation by electrostatic fields.[30] These shields have problems: they attract dust, cannot stop high-energy cosmic rays, and probably need regolith barriers anyway to stop particles hitting the structure from angles near the horizon. Maintaining mobility, most studies involve shields brought from Earth not using regolith (Figure 11.4).

Extending the lunar/Antarctic analogy, as on the Moon, Antarctic inhabitants deal with unnatural sunlight cycles, difficult transportation, isolated small communities, and extreme environments that can prove readily fatal if plans go wrong. These can be ameliorated in base construction.

Attitudes and philosophy are implicit in base design. Constellation base concepts were criticized as "Motel 6 on the Moon." This is unfair, although Constellation base ideas looked much like the *International Space Station* – on the Moon. Funding priorities for the Vision for Space Exploration left little for habitats and funding in situ resource development suffered. Instead of launching many modules to fit together or separately on the Moon, a sensible alternative would be rapid transition to maximal local material use once a foothold is established. Engineering justification for this is manifest. This permits longer stays, lowering transportation costs, which is the major expense. Operationally and psychologically, this base's crew would be highly motivated to maintain and improve their environment. With a viable base established, give them tools to build and improve their environs and make their stays a year or two, so they see fruits of their effort and innovation. This crew would hold the epitome of motivation and competence in technically oriented humanity and would do inspiring work given the opportunity. Emphasis should go

Figure 11.4. **Setting up house**. Different approaches to lunar habitats. (a) NASA LUNOX pressurized modular habitat concept ca. 1993, with robotically landed mobile units, including logistics and living/working and space suit maintenance modules. (b) Terrestrial field test of mobile habitat with two ATHLETE rovers and pressure modules, ca. 2007. (c) Concept of stationary integrated habitat landed as unit, from NASA Global Exploration Lunar Architecture, December 2006. (d) Planetary surface habitat and air lock developed by ILC Dover for NASA in 2007, the main unit 3.7 m in diameter composed of a multilayer fabric. (e) Bigelow Aerospace model of inflatable modules and landers. (Graphics by John Frassanito and Associates for NASA, photos by NASA and Bigelow Aerospace.)

to a precursor telerobotic base to demonstrate and perfect these techniques, at several percent of the price.

NASA is exploring moving regolith. Two technology challenges focus (at low cost) on excavating and transporting regolith small distances.[31] Another approach uses gas pressure to move surprisingly large regolith masses with small amounts of gas, with mass ratios over 100.[32] Such gas could come from inevitable excess margin in lunar landers, or local lunar volatiles or oxygen production. We must evaluate if these could pollute the lunar atmosphere and raise fine dust, befouling equipment and instruments. Most NASA Challenge approaches use a mechanical scoop and bucket wheels or ladders using less gas but still agitating dust. Large dust

The New Moon

Figure 11.5. **Digging in**. Sketch based on description of lunar base concept in 1969 Soviet study by Design Bureau of General Machine Building (KBOM, and reported by Anatoly Zak) showing covered structure built into crater. Draped over the base is a sheet holding 1–2 m regolith layer for thermal, radiation, and micrometeorite protection. The base includes a greenhouse for food and oxygen production and a water and waste recycling facility. It is powered by both nuclear and solar power. Rovers are garaged at the base, as well as supplies and scientific equipment including an astronomical observatory. It would support 4 to 12 people. (Drawn by author from Zak's description.)

particles act ballistically, but small particles appear suspended longer, possibly via electrostatic effects.

Soviet engineers designed a lunar base that makes generous use of regolith; it is built into a crater then draped with membranes supporting a 1–2 meter regolith shield (Figure 11.5). Its 4–12 cosmonaut crew would serve at least a year and be provided with life support, power, thermal control, a kitchen, gym, dorm, shop, science lab, power plant, control center, communications, data center, rovers, greenhouse garden, and water and waste recycling. Power would be both nuclear and solar, and the base would include an astronomical observatory. Construction materials shipped from Earth totaled 52 tonnes, added to 300 tonnes of regolith.[33]

A useful concept for lunar bases is inflatable spaces. With regolith shielding and support frameworks, an inflatable offers huge living, storage, and operations space at low cost. NASA once actively pursued inflatable structures (TransHab), but for political reasons related to *ISS* funding, Congress outlawed further work on inflatable space structures, despite NASA and White House objections.[34] NASA sold relevant patents to Bigelow Aerospace, now preparing inflatables for space tourism in low Earth's orbit and someday the Moon. Bigelow designed a base housing

18 astronauts in three connected 12-meter by 7-meter diameter BA-330 inflatable modules, based on an Earth orbital station design.[35] Exposed on the lunar surface, these require multilayered walls against meteoritic impact. Burying them allows more space per mass, using thinner-walled modules, but this is not immediately planned.

We can imagine complex use of regolith in construction. Cement is activated by sintering or *calcination* – converting calcium so it will bind the cement – and indeed experiments with lunar regolith simulant show that sintering does this.[36] Solar energy aplenty on the Moon could provide heat for sintering; this may be unneeded. Gammage and collaborators (see note 58 in Chapter 6) showed that calcium leaches from regolith in the manner and quantity crucial to Portland cement. We should be optimistic about the efficacy of such concrete; even if calcium binding falls short (and concentrations in highlands regolith are close to useful levels), we can supplement water with calcium-rich leachate from other regolith. One approach is a sandwich of regolith and water/leachate concrete slab in the middle surrounded by microwave-sintered dry regolith skin (dry because nanophase iron rusts in water) and then covered with loose regolith insulation against violent temperature swings (and resulting contraction/expansion cycles). Amazing properties of such construction include enough water to shield against neutrons. This should be done in vacuo if possible while conserving water; perfecting the process requires research and more lunar regolith than available now.

On simpler levels using regolith can be easy. With low lunar gravity and regolith's cohesiveness, its angle of repose (steepest stable slope) is greater than for Earth's soil (Chapter 10). Excavated natural regolith maintains vertical walls 3 meters tall and 60° slopes 10 meters tall. Apollo astronauts discovered these properties digging trenches and walking in lunar soil (Figure 11.6). Construction

Figure 11.6. **Boots on the ground**. (a) Two successively zoomed views of footprint treads left by EVA boots, illustrating structural integrity of impressions in lunar regolith. (b) Vertical-walled footprints made deep into regolith. Lunar dust can maintain a natural vertical wall 3 m high without collapsing.

trench cave-ins, the bane of earthly building sites (60 U.S. deaths annually on average), are less a lunar peril (in part because of terrestrial trench collapses caused by humidity changes). Cleverly, Soviet base designs (Figure 11.4) used natural crater slopes, but even intricate regolith use can maintain safety margins.

Moon base designers will appreciate regolith; gritty and irritating, it is a construction boon. Build it into your habitat; it surrounds you anyway. Nonetheless, other building materials can be made on the Moon. Studies exist for producing lunar fiberglass,[37] ceramics,[38] glass, and metals including iron, aluminum, and titanium. Chapter 12 covers some of these, as well as other manufacturing and energy. We see how membranes are useful. Woven fiberglass utilizes the substance's strengths (fiberglass being weak against shear but strong in tension). It also needs no binder or resin, usually organic (although silicone uses only half as much carbon). Fiberglass can be composited with metal foils (perhaps also lunar products) and is flexible but extremely strong.

By digging the regolith we can make habitats on the Moon protecting us from temperature extremes, radiation, meteoroids, and the vacuum itself, with modest power requirements easily supplied. Beyond Earth itself, this offers the easiest sustained place for human life in the Solar System, contrary to the impression from Apollo, summarized by Aldrin after scanning through the window of the LM, descending the nine rungs of its ladder and dropping to the surface: "Magnificent desolation."[39] On Earth many places are desolate above, but life thrives underground.

Subtle, unexpected things might go wrong; things that do not work the same in lunar environments. Plumbing does not behave the same in one-sixth gravity as it does on Earth, where pipes half-filled with water flow with air on top. On the Moon, as in weightlessness, small pipes gurgle in a bubbly flow, with air and water mixed. Likewise lunar flames will differ. On Earth stronger gravity draws gas from burning material, but weightless combustion products are not removed by convection, but collect to slow burning, proceeding more by radiative heating than on Earth. On the Moon, conditions are intermediate, meaning some materials and objects will be flammable that are not on Earth.[40] We need to learn many behaviors by trial and error; much engineering and design depends on experience.

Based on *ISS* data, humans use mostly water: 23.4 of the 24.8 daily kilograms per person in consumables versus 0.6 kilogram per day of food and 0.8 of oxygen (3.6 kilograms water in food/drink, 6.8 in hygiene, 12.5 washing clothes, 0.5 flushing). In a sustainable habitat, most must be recycled. NASA developed a water recycler for six people, costing $154 million. This Water Recovery System failed after *ISS* installation in 2009 and is not yet perfected. When working properly, it recycles about 85% of its water input. *ISS* waste water is now often fed to another system producing oxygen; wash water is recycled.

Lunar agriculture will likely be hydroponic: grown in water, or at least in water-dominated environments with solids serving as inert carrier more than a nutrient reservoir. There are reasons to favor hydroponics: more organic mass in plants rather than soil, plants more compactly grown and easily handled mechanically, nutrients precisely applied as needed, and less perch for disease and pests. Hydroponics require less water and turn out to need less light as well. Emphasis on nutrient conservation is crucial because carbon and nitrogen will be dear, coming either from Earth or the lunar poles or extracted in small quantity from regolith. This strength becomes a drawback when things unbalance because no soil nutrient reservoir buffers rapid changes. Pump failures can be fatal. Experiments with lunar regolith simulant show that it is too finely packed for root growth; regolith sieving seems necessary. Lunar regolith's iron content (and perhaps leached calcium), far from fatal, could add a distinctive *terroir* to food – maybe a "hard" taste. NASA is taking the idea of lunar gardens seriously with an experiment by Lockheed and Ames Research Center to grow turnips, basil, sunflowers and the plant Arabidopsis (the importance of which is explained later in the chapter) on the Moon on a Google Lunar X Prize mission in 2015.[41]

Lunar and space hydroponics have long histories. Past decades of food and oxygen regeneration studies focused on major crops, for example, wheat, rice, potato, peanut, and so forth, but recently focus has turned to fruits and leafy vegetables more for supplemental "salad machines" than comprehensive nutrition, using radish, small lettuce, onion, and so on; *Mir*'s cosmonauts ate mustard greens. Animal husbandry is not considered. Experiments included recycling nutrient solution, not human waste. Hydroponics yield is several times conventional agriculture's versus space and energy use, so it will remain the method of choice, expandable when space/lunar demand rises. High efficiency lamps are used, with broad-spectrum LEDs thought preferable to line spectra like high-pressure sodium. These will be mandatory during 14-day lunar nights. Recent experiments show highest yields at low atmospheric pressure (1/5 atmosphere) but more oxygen (30%), and high carbon dioxide but not too high. Sometimes yields decline apparently from accumulated ethylene or other volatiles from plastics.[42]

Other plants are useful for other reasons. Small, flowering Eurasian *Arabidopsis thaliana*, not particularly edible, might be "the canary in the coal mine" because its genome is decoded and subject of an international consortium. If it changes, we will know why. Its life cycle of only six weeks will promptly trace genetic changes. Classes of subtle pollutants might arise in lunar environments. Radioactive radon leaks from some minerals; others contain heavy metals. With pesticides and herbicides unneeded in lunar food production, lunar habitants will still be replete with plastics, electronics and other synthetics releasing pollutants, some insidious, for example, perhaps Bisphenol A, a mock hormone. NASA and ESA maintain extensive constraints on outgassing of various substances, greatly limiting possible raw materials in space environments and raising hardware costs in ways perhaps impossible for lunar manufacture.

Humans need not suffer illness or contamination before reducing trace pollutants. Many plants are dynamic accumulators pulling minerals from the soil.[43] Rarer are hyperaccumulators (such as *Arthrocnemum macrostachyum*, Maltese glaucous glasswort, or South African *Berkheya coddii*), concentrating on metals such as nickel, zinc, selenium, lead, arsenic, cadmium, or cobalt.[44] For bodies of water, some microbes such as yeasts can remove heavy metal toxins.[45]

A useful idea for lunar habitats is large repositories of biochemically interacting material to absorb pollutants of long residence time, allowing them to decay from the system, for example, by soft ultraviolet irradiation, avoiding concentration in human tissues. As efficient as hydroponics might be, large areas of soil, even open water, might be useful: oxygen production (trees), accumulators, compost, food crops, and recreation. We might call these the *arena* – not only a meeting space, but also a sandy place, recalling the term from Latin. Large spaces might combat agoraphobia (fear of open spaces) that might arise because the sky or distant horizon will bespeak little shielding overhead against outer space's hazards.

Such dreams recall enticing concepts by lunar habitat designers: building inside large, natural lunar lava tubes.[46] Underground lunar caverns exist (Figure 6.25) and collectively extend many hundreds of kilometers in length, primarily under maria. Scaling for lunar gravity and mare lava viscosity, they are wider than Earth's, typically many tens of meters. (Many with skylights are roughly 100 meters wide.) These are shielded from temperature swings, radiation, and meteoroids, and if structurally sound, could accommodate a permanent atmosphere. There should be little dust. Billions of years old, they are less prone to collapse than terrestrial structures. At 20 square meters per person, they offer potential living space for millions. Lunar lava tubes suggest spacious and secure places for humans on the Moon, but we must know more about them. Skylights into Martian lava tubes are also seen, so their use might extend later to the red planet.

We do not know the fraction of lunar lava tubes intact versus collapsed (even partially) or those too deep to sense photographically. Other ways to find them remotely are penetrating radar and locally with magnetometers, seismic arrays, and gravimeters on-site, first by robots. Work is needed on individual tubes to assess their structural integrity. How can we explore these in detail? Radar might inform us of tube conditions, with limits. One only gets strong radar return over as few as 30 radar wavelengths; a 100-meter-high tube is not easily mapped on meter scales. Local seismological arrays and magnetometers have problems seeing large, empty cavity bottoms. Intact lava tubes can be pleasures to walk in, whereas partially collapsed tubes are tortured mazes, sharp-edged and dangerous. Even largely intact lunar lava tubes may suffer damaged entrances. We have no rover for such terrain, and rockets jetting through the tube might contaminate it. We need a robot to walk over debris, maybe a quadruped on telescoping stilts (Figure 9.10).

Lava tubes imply that someday we need not think of living off Earth as cramped. Tubes are so large that one can engineer spaces sufficient to engender feelings of being outside (Figure 11.8). These may come closest to us "terraforming" the Moon. Nobody alive today will walk outside on the Moon's surface without a space suit. [Box 11.1] With lava tube habitat, lunar routine might seem at times like a walk in the park, especially for acclimated humans. Terraforming Mars or Venus might take thousands of years, perhaps too distant for a realistic human goal.

We explored the Moon four decades ago and are surely not finished. However, we have established no base, even after more than four decades. As we discussed earlier in this chapter, this is not so different from the South Pole analogy, with Apollo echoing Amundsen's generation, while we await the equivalent of a permanent South Pole station.

Box 11.1: A Breath Outside

Despite appearances, the lunar environment is not always hard vacuum, with only an exosphere of tens of tonnes mass. A thick lunar atmosphere is metastable, maintaining its configuration with insignificant change over human lifetimes, described in several works but particularly a 1974 paper ("Creation of an Artificial Lunar Atmosphere" by Richard R. Vondrak, 1974, *Nature*, 248, 657). The Moon now generates approximately 20 grams s^{-1} of gas. If this increases to approximately 60 kilograms s^{-1} (for gas of atomic mass 16), the atmosphere at its base becomes collisional (not ballistic), so below this height losses are reduced. If lunar gas is produced faster than this, atmosphere accumulates. If a thick atmosphere exists, it leaks away at approximately 60 kilograms s^{-1}. A typical comet impact, occurring approximately every 10^7 years, generates an atmosphere of approximately 10^{11} tonnes, falling to normal after approximately 30,000 years. Such an atmosphere would be approximately 1/3000 as thick as Earth's at sea level. Colliding 3,000 comets would produce a lunar atmosphere as thick as Earth's, lasting approximately 10^7 years. This is a superhuman task but would produce an effectively permanent atmosphere for all but immortals. Imagine a huge lunar oxygen plant, making several million tonnes per year, hence, a metastable breathable atmosphere lasting many human lifetimes. A breathable lunar atmosphere in 100 years implies a factory processing approximately 3×10^{12} tonnes per year, eventually consuming regolith covering 3×10^6 km^2 (or 8% of the lunar surface) – thousands of times more than Earth's most active strip mine.

Can one ever walk outside into a permanently breathable lunar atmosphere without continually producing many tonnes of gas? For an atmosphere without a confining lid – for instance, in an open-topped container – to be nearly stable, many more than several atmospheric "scale heights" must be completely enclosed by the container. The scale height indicates the vertical distance over which the atmospheric density drops by 63%, to $1/e = 0.37$ of its original value. On Earth, with our N_2 and O_2 atmosphere, the scale height at room temperature is 8.5 kilometers. On the Moon, scale height for similar gas would be 50 kilometers, inversely proportional to strength of gravity, taller than any container. We can adjust scale height inversely to the gas molecular mass. In Earth's atmosphere this is 28.8 AMU; we need a heavier molecule. The heaviest inert gas (at room temperature and sea-level pressure) is xenon at 131 AMU. The heaviest of any gas (definitely not inert) is tungsten hexafluoride (WF_6) with 300 AMU. (The slightly radioactive uranium hexafluoride with 352 AMU becomes a gas at a torrid 54°C.) A column of WF_6 at the surface and average lunar temperature has a scale height of 2.7 kilometers. To hold this gas within an open-topped container, it must be at least approximately 20 kilometers deep. Although South Pole-Aitken approaches this depth with respect to its rim, it is too wide a container. The deepest non-basin crater* on the Moon is 5–6 kilometers deep, for example, Tycho. Any hole deep enough to contain WF_6 must be constructed (or deepened) artificially.
*(Smaller than approximately 300 kilometers: *The Geological History of the Moon* by D.E. Wilhelms, 1987, U.S. Geological Survey Professional Paper, 1348.)

One cannot breath WF_6. It is highly corrosive and reacts with water and flesh. Amazingly, other fluoridated compounds (perfluorocarbons), for example, perfluorodecalin (PFD: double aromatic carbon rings populated by fluorine) do not harm human tissues, are liquid at room temperature, and can dissolve large amounts of oxygen. (One liter of PFD can dissolve 0.49 liter of O_2 gas.) This means air-breathing organisms, including humans, can breathe oxygenated PFD liquid. (I am not volunteering.) Oxygenated PFD or some equivalent might one day find use in space flight; perhaps some major engineering feat will allow one to walk out onto the open lunar surface into a lake of PFD. I would not hold my breath waiting for it.

This invites speculation of a future Moon with human populations established and what this implies. Several men and women with the right stuff make an expedition, a dozen make a base, a hundred or more and we are speaking of a settlement, although not necessarily making babies. U.S. Antarctic stations have crews of 100 or more (and from what I hear, not much procreative goings-on) and are really big bases (but not military bases). Other settlements encourage their citizens to produce Antarcticans. Argentine and Chilean Antarctic inhabitants (only on the Peninsula) are considered residents/citizens by their governments in ways encouraged to cement territorial claims, and they maintain populations of 100–200, including children, with several actually born in Antarctica (after much logistical and media fuss). Such large groups will have different needs and characteristics on the Moon. Will sponsor/supporters encourage lunar settlers to give birth to the first true selenites, and how that will be managed? We revisit this, and how large lunar settlements must be for economic activities, in Chapter 12.

There is no reason to think humans cannot procreate on the Moon. No sex has taken place there, and whether it has in outer space at all is subject to bad journalism and rumor.[47] Reasons postulated for why coitus is difficult in weightlessness do not apply to the Moon. Normal conditions in a lunar habitat should not frustrate gestation, and the primary issue is proper medical and other care during birth and infancy. One wonders if identical periods for lunar environmental variations and menstrual cycles might imply some effects. One might imagine social subtleties but nothing affecting childbirth.

Large habitations in the next 50 years may be fantasy, but these options set the boundaries for consideration. Everyday lunar life will be unique in interesting ways for inhabitants and for settlers. One-sixth gravity will influence everyday tasks. This gravity means objects tossed into space remain suspended 2.5 times longer than on Earth, opening an interval where items can be parked amid some tasks, for the nimble of hand and eye. (Something in your hand preventing you from reaching your keys? Simply toss it in the air!) Objects will be more easily blown into the air, and of course, dust will stay suspended longer. Everyday items weighing six times less can become portable on scales unfamiliar on Earth. They still take as much force to start and stop them, except for friction, which is reduced.

Are there permanent disadvantages to one-sixth gravity? People must exercise, but evidence is unclear if they suffer the potentially disabling atrophy of zero gravity, with bone loss and temporarily crippling muscle disability of long stays in orbit. This needs more research but will legislate human habitation, whether they can stay long and return to Earth. If journeyers cannot return easily without risking injury, long trips may become one-way, or never occur. They can exercise by lifting weights five times their own mass, with muscle and bone use Earth-normal. Gravity is simulated by circular motion; for inactive persons a one-g spin table with a radius of 3 meters rotating 7 times per minute, produces Earth-like pseudo-gravity. There is suspicion that astronauts suffer a reduced sense of smell, probably an effect of zero gravity, not lunar gravity.

Humans never stayed in one-sixth gravity long enough to affect bone and muscle, but 12 people slept on the Moon (or tried to). On *Apollo 11, 12,* and *14,* lunar stays were so short that astronauts stayed in their space suits. Starting on *Apollo 12* the LM furnished hammocks, so Conrad and Bean were first to sleep on the Moon, fitfully for several hours. With noise, cold, light leaks, and uncomfortable suits, sleep was uneasy. *Apollo 14* landed at a tilt, making sleep difficult. On the three-day landings of *Apollo 15–17*, remedies allowed restful sleep. Some astronauts took a mild barbiturate, and the crew stayed on a common sleep schedule. Jim Irwin of *Apollo 15* said sleeping in one-sixth gravity was much more restful than zero gravity, but he was also exhausted by his lunar workload.

Lunar astronauts seem relieved of that most common weightlessness symptom: motion sickness. Muscle pains, particularly in the back, also seem solely zero gravity, not one-sixth gravity, effects. Ignoring tragic dehydration effects on Scott and especially Irwin (Chapter 10), few lunar ailments were reported. We discussed bruising and lost fingernails caused by glove pressure. Both problems arose more from space suit design than lunar environment. Fewer lunar health problems may reflect the short cumulative stays versus much longer weightless durations.

The Moon will share other problems with orbit, even psychological ones. Mission morale suffers mundane hits such as the broken *ISS* toilet, but others can be more subtle and dangerous. Two American astronauts on *Mir* reportedly suffered stress and depression. Crew on *Skylab 4* and *Apollo 7* showed poor morale, especially dealing with Mission Control. (These six astronauts never flew again.) Cosmonauts on early space stations suffered some interpersonal crises, in worst cases compromising their mission.[48] *Soyuz 21*'s visit to space station *Salyut 5* in 1976 was curtailed suddenly by interpersonal strife caused by mental deterioration of one of the two cosmonauts, exacerbated by nitric oxide gas poisoning. The next crew vented the station's atmosphere. Before *Soyuz 11*'s tragic end, killing cosmonauts Dobrovolskiy, Volkov, and Patsayev, the mission suffered two delays from smoke inside *Salyut 1*, including an actual fire almost ending the mission. Afterwards conflict arose between Commander Dobrovolskiy and Flight Engineer Volkov, the only veteran onboard, on how to decide engineering debates.[49] The conflict subsided, but 13 days later they encountered another crisis when separating from *Salyut 1* when they could not confirm a good vacuum seal. Nobody alive knows what transpired between the three men several minutes later in their critical 40-second struggle to stop the leak robbing them of air. They died trying to control the faulty valve that killed them.

Maybe the oddest hazard encountered was the accidental spacewalk of rookie cosmonaut Yuri Romanenko of *Soyuz 26* in 1978. As his *Salyut 6* crewmate Georgi Grechko prepared for a scheduled EVA with the hatch open, Romanenko decided to peer outside, unfortunately absent his safety line. As he floated from the space station, distracted, Grechko pulled him back inside. Opinions differ as to whether his life was in danger; Romanenko went on to serve on two further Soyuz missions.

None of these factors bear only on the Moon but also on space flight in general: vacuum, cramped living space, hazards of fire, and pollution. These translate to Moon bases as well, especially early, smaller ones, particularly above the surface. These examples occur in context of defined command hierarchy rooted in Mission Control (a hierarchy that can deteriorate regardless). Mission Control set the schedule, and astronauts' careers suffered if they strayed far from this authoritarian path. I asked Bruce Melnick (who flew on Shuttle missions STS-41 and 49) how astronauts would react to isolation from Mission Control by a 20–40 minute latency time at an asteroid or Mars; he responded: "The times that we on the Shuttle are out of contact with Mission Control are almost a relief. We're trained to think for ourselves and this is our opportunity not to have people telling us what to do all of the time."

Already *ISS* astronauts can retire to individual spaces for private time and pursuits. (Despite Box 3.4, astronauts have time to play music, if not write ballads; see Figure 11.7. In the Soviet space program's early years, there were even cosmonaut ballads.) We are moving beyond the era of voice operated switch intercom (VOX) in which everyone onboard is forced to hear everyone else's business via a single

Figure 11.7. **Musica universalis**. (a) Stephen Robinson of STS-130 plays guitar in the *ISS* Cupola Module on February 18, 2010. (b) Carl E. Walz of *ISS* Expedition 4 plays keyboard in the Destiny laboratory module to (clockwise from Walz) Daniel W. Bursch, Rex J. Walheim, Jerry L. Ross, Ellen Ochoa, Lee M.E. Morin (of STS-110), and Stephen N. Frick of Expedition 4 in April 2002. Crews to *ISS* have also played saxophone, did geridoo, and flute. Video online shows *ISS* Expedition 35 Commander Chris Hadfield singing and playing guitar in May 2013. (NASA photos ISS022-E-068631 and STS110–375–032.)

Lunar Living Room

Figure 11.8. **Life in a lunar lava tube**. Several lunar lava tubes are known or strongly suspected (e.g., Figure 6.25), but likely more thread the maria. Data and theory indicate typical tube widths and heights about 100 m and lengths of tens of kilometers. They offer means to investigate pristine ancient basalts but also natural protection for humans against radiation, meteorites, and extreme temperatures. We picture a space 100 by 50 m and 200 m long, sealed by a locally produced, woven fiberglass multilayer membrane, connected by air locks to similar spaces with different functions. It serves as a long-term reservoir for dilution and decay of trace pollutants, reducing their concentration in human tissues, enhanced by pollutant-filtering vegetation (at right). The space allows for recreation via exercise, gardening, and even one-sixth gravity human powered flight. Beyond Earth it would provide a rare experience approximating a sense of outdoors without a space suit.

channel. (Few things are more irritating on a scientific expedition than suffering through the idle chat in your earphones between two of your colleagues.

What is it like to return to Earth after so long away? Even the longest stay in space (438 days by Valari Polyakov on *Mir* in 1994–1995) caused no serious psychological difficulties, although mood fluctuations occurred at the start and after the end of the mission. A six-month Antarctic stay simulating a Mars mission (ESA's White Mars in 2012) produced notable disorientation for members returning to civilization.[50]

We have not yet mastered migration off the circadian rhythm with which we were born, which is easily disrupted via strong light in the environment. In a sense it is good fortune that humans will probably live most of their time in the lunar subsurface, because there it will be easier to maintain a 24-hour cycle against the natural 29-day sunlight cycle of the surface environment.

There will be attractions to lunar living. It will bring new sports and new and exciting architecture. In places, several steps transport one from blistering heat to eternal darkness and stunning cold; imagine the industrial possibilities. In ways lunar habitat, particularly its gravity, is gentler to humans, if not natural. People may find they have longer life-spans on the Moon; eventually people may retire there to extend their years. Some diseases may largely diminish; others may replace them. People will have time for art and music (Figure 11.7). This world's rhythms will slowly penetrate; will a new fortnight culture form (in sync with paydays)? Months will have new meaning; the year and seasons losing hold (the Moon's rotational axis tilted only 1.5°). Days will be a distant, earthly concept (except at Near Side lunar night, when scenery adopts an eerie bluish cast every earthly rotation).

Recreation topside will be an attractive but rare pleasure. To hit a ball in one-sixth gravity will be truly inspirational but weighed against radiation dosage on the surface, and dangers in the vacuum. Enthusiasm will grow for building room underground for sports, and – remember – on the Moon one could fly through the air on self-powered wings, like Earth's *Gossamer Condor* but much easier. (Take that, Richard Locke!)

Topside will be a great attraction, of course. There must be viewing galleries where inhabitants can admire the glory of nighttime lunar sky and, on the Near Side, the blue, vibrant Earth. People will treasure these privileges, seeing the Universe's naked beauty and that strange, watery world. This offers great recompense for spending time underground.

Astronauts refer to reverential, even spiritual, experiences looking at Earth from the Moon.[51] Every sighted person has experienced awe in seeing the Moon, whether it be full and bathing our earthly nightscape in light, or the grace of thin crescent hanging above the western horizon, perhaps announcing Ramadan's advent, or the quarter Moon hanging startlingly in the shining blue daytime sky, its own little world, with strange blue seas the color of our firmament. Even these do not compare to the Moon seen through even a small telescope: a world of its own, with alien landscapes calling to our imagination.

Now imagine a bright blue-white, brownish-green marble many times this size and brightness hanging in a deep dark sky, changing from hour to hour, week to week, with roiling cyclones and cloud fronts moving across her, and shifting colors as continents drift by – the deep greens of central Africa and Amazonia, the sands of the Sahara, subtle changing shades of North America, Eurasia and lands of Oceania and Australia. Look on this world, forever fixed in your sky, knowing that this is your parents' birthplace, and the many generations of parents before them, the ape-men, simple mammals, early reptiles, ancient fish and simple Chordates, great-great-grandparent of us all, preceded by single-celled beings, extremophiles and primordial soup. It is the one place where everything in you and most everything you love came from, even the Moon itself, hovering before you. No, people will not easily forget the Earth for the sake of the Moon. They will always yearn to look one more time, whether in their sky, or in the flesh. This will change only because we have changed, or, hard to imagine, Earth itself has changed.

However, for those on the Moon's Far Side, for whom Earth is a distant concept never seen in ordinary life, the sky will be the darkness of a Universe at large, with worlds unvisited and inhuman forces shaping gas, dark matter/energy, and space-time itself into strange configurations, environments into which humans have never entered and never will. This is why they will be there. Staring into space, where the Moon blocks Earth and often the Sun from view, and sometimes all bright radio sources in the sky is why they are there. They may seem to live in a distant Universe, far from Earth long ago and far away, on existence's other side. Living on the Moon will change how humans think.

It seems fortunate that the lunar poles will play a central role in lunar habitation, a potential first base and source of water and organics. They will bridge Near and Far Sides to offset divergence that seems possible. (Recall the role that Washington, DC, played in uniting the young United States.) Most lunar lava tubes appear closer to the equator than poles, however, because they may be associated primarily with maria (although some appear near the maria/highlands boundary[52]). We may need to choose whether the North or South Pole will dominate.

Growing habitats will become less hierarchical and programmed, but dangers will remain. We have already had civilians onboard: tourists to the *ISS*, Space Shuttle special crew (politicians, schoolteachers). The death of Christa McAuliffe on *Challenger* suspended this development for years. (We never got professional poets in orbit, although many find work by astronauts Douglas Wheelock and Chris Hadfield inspirational.) With settlement we will need to consider children born onto the Moon and options for people who have aged to the point of retirement. These and other people will not fit in a quasi-military hierarchy.

More casual command structure on the Moon (and in space generally) will require new situational ethics, not an issue now: decisions of risk and procedure occur via Mission Control, deliberatively. A review shows that life-and-death situations arise more often than some realize. [BOX 11.2] Neither can we imagine inhabitants in everyday life maintaining attitudes of early astronauts.[53] Off-world communities (especially when latency times grow) will not be monitored so closely in real time, and situations will arise with inherent risks. (Earth-bound news media will follow this, no doubt, to the extent they can.) What is a rational response to danger, and what is a crime? How do people decide how to act and prioritize sacrifice? I recall a "hint fiction" story: "I'm sorry, but there's not enough air in here for everyone. I'll tell them you were a hero."[54] Life will be hard even without such moral ambiguity. Common sense in such environments differs from our daily expectation.

After several weeks on the Moon, its first long-term visitors will have more experience there than all today's astronauts, scientists, and science fiction writers combined. There must be new rules, some seemingly arcane today, like horse theft laws in the American Old West. Today the idea of executing someone for stealing a horse seems outrageous, but in that economy a man's life or livelihood depended on the animal. Horse theft could be crueler than murder. Even now, in areas without modern transportation and communication, problems arise, as in Mongolia more recently.[55] On the Moon equally unique rules will apply, for example, air locks must

Box 11.2: The Perilous Sky

Along with well-known, nearly fatal (*Apollo 13*) and deadly situations (*Apollo 1, Soyuz 1, Soyuz 11*, STS-51-L *Challenger*, and STS-107 *Columbia*), there are disturbingly frequent life-threatening incidents, here drawn from roughly 300 space flights by the United States and Russia/Soviet Union.

Soyuz 18A's (April 5, 1975) booster destroyed itself 5 minutes into flight, and its abort system was engaged after delay; Vasili Lazarev was injured by rapid deceleration.

Soyuz T-10A's (September 26, 1983) booster exploded on the launch pad, and its crew was saved by the launch escape system.

Gemini 8 (March 17, 1966) and *Soyuz 33* (April 12, 1979) suffered different engine malfunctions forcing their missions to be aborted – *Gemini 8* after endangering the crew with rapid rotation.

Apollo 12 (November 14, 1969) was struck twice by lightning on launch and lost power and guidance control temporarily as well as some permanent functioning of the LM. Alan Bean later suffered a concussion when struck by loose equipment on splashdown.

Mercury 9 (May 16, 1963) lost power in orbit, and reentry was controlled manually.

Soyuz 5 (January 15, 1969) reentered with its service module attached, forcing a thin portion of the heat shield to take the force of reentry; the resulting hard landing broke Boris Volynov's teeth. A similar hazard occurred on *Soyuz TMA-11* (April 19, 2008) causing neck and spinal injuries to Yi So Yeon, and other separation failures were caused by tangled wires on the first human space flight *Vostok 1* (April 1, 1961), and a faulty sensor on the first orbital flight by an American, *Mercury 6* (February 20, 1962).

Soyuz 23 (October 16, 1976) landed and sank in a frozen lake and was recovered only after 9 hours; the crew survived.

Mercury 4 (July 21, 1961) sank on splashdown, and Gus Grissom escaped by swimming.

STS-1 *Columbia* (April 1, 1981) on its first test flight suffered launch damage because of a rough Solid Rocket Booster ignition. On STS-9, *Columbia* (December 8, 1983) caught fire twice as a result of a hydrazine fuel leak.

Onboard space station *Mir* on February 23, 1997, an oxygen generator caught fire and filled the station with smoke. On June 25, 1997, the *Progress-M 34* cargo vessel rammed *Mir* and started rapid decompression forcing permanent evacuation of one of *Mir*'s science modules and damaging many of its solar panels.

On *Voskhod 2* (March 18, 1965) Alexei Leonov became stuck entering the capsule after the first human spacewalk and was forced to lower his suit pressure to dangerously low levels. The crew later initiated reentry manually and landed off course, spending the night in their capsule, amid wolves.

On STS-37, *Discovery* (April 8, 1991) astronaut Jay Apt suffered a punctured space suit glove during EVA and was saved by his finger sealing the hole. Cosmonaut Pavel Vinogradov suffered a similar glove leak on *Mir* in 1997, placing him within 5 minutes of asphyxiation danger. On STS-51, *Discovery* was punctured by a misfiring explosive bolt, punching a 13-by-3-millimeter hole in a bulkhead but not puncturing the crew cabin.

On *ISS* Expedition 36 (July 16, 2013) astronaut Luca Parmitano found his helmet interior soaking with water, obscuring his vision and hearing. His EVA was aborted, and he was helped inside by fellow EVA astronaut Chris Cassidy.

American *Apollo-Soyuz* astronauts (July 24, 1975) suffered inhalation of nitrogen tetroxide gas causing Vance Brand to lose consciousness with lesser effects on Thomas Stafford and Deke Slayton.

always be publicly accessible from outside. This is so basic it will influence air lock design: no entry refused, even by unintentional manufacture, to someone who needs to be inside to breath. Deeper entry into your habitat is another negotiation. The Outer Space Treaty (OST) of 1967 is ratified by some 100 countries and signed by 26 more, including most space-faring nations (except China, which follows it voluntarily). The OST states that nobody has authority to prevent others from going anywhere on the Moon.[56] At some point this offends expectations of privacy and private property. An effective rule will arise in practice.

Of course, many rules are unwritten, and social custom can be as influential as law. We are plied with images from Hollywood of dastardly deeds on the Old West frontier (which make for better script), but the evidence indicates the opposite prevailed. People primarily engaged in struggle with the environment need as many allies as possible. Frequent behavior is to engender trust among the neighbors; you never know who might save your life.[57] This attitude depends on proximity of agents enforcing its positive outcome and on assumptions that one's neighbors are long-term. That dynamic will likely pertain among lunar settlers. OST is vague about this, and the Moon Agreement of 1979 is not ratified and, to many people, unrealistic. OST requires inhabitants to render all possible assistance in general and particularly in accidents or distress. There must be limits to this as yet undefined; we are unprepared for such action.[58] We are progressing toward universal air locks, but more standards for sizes, threads, and specifications of couplings, containers, fluids, voltages, and so on fail in practice.[59]

New explorers and inhabitants will encounter new situations on the Moon, and anticipating these is uncertain. Humans in lunar polar environments will deal with temperatures never before experienced in nature and volatile/regolith interactions unlike anything seen before. There will be hazards. Lunar lava tubes are inviting but *luna incognita*. There is much to explore, and hazards to dodge. We need (tele)robots.

The lunar surface is hostile, sterile, and, hence, challenging. Underground conditions are comfortable and relatively safe. Reconnaissance and establishment of secure perimeters by the experts will allow the inexperienced to emigrate and homestead. Beyond this perimeter, conditions are insecure. We will need laws concerning this.

Lastly, and seemingly bizarre: is there (nonhuman) life on the Moon? Of course not, by definition: according to NASA, the Moon is sterile, by fiat. Lunar material may be brought to Earth without precaution, the only stricture being that material not be brought to the Moon that might endanger its sterility, as per Planetary Protection protocols.[60]

If this were written 30 years ago, it would allow for life on the Moon – millions of bacterial spores on Surveyor, Apollo, and Luna hardware. After *Apollo 12* astronauts clipped the camera from *Surveyor 3* (Figure 6.14) and returned it to Earth, it was found peppered with *Streptococcus mitis*, presumably dormant in lunar vacuum for 2.5 years since unsterilized *Surveyor*'s landing. Despite intense interest[61] and contemporary consensus that bacteria survived the Moon, infecting plastic foam in the camera,[62] new evidence favors bacteria originating on Earth postflight, after several opportunities for contamination (and *S. mitis* dying in –120°C lunar nights).[63]

If the Moon orbited the Sun alone, this might end the story. It orbits Earth, however – a rich and varied source of infection by species many of which can survive a trip through space to the Moon.[64] Small lunar areas store water and organics. Is there any way these could join to sustain life? This seems worthy of a thought experiment. [Box 11.3]

Box 11.3: The Natives

Earth life on the Moon? Permanently shadowed lunar surfaces are much too cold to revive extremophiles. One needs to postulate warmer places, for example, deep in the regolith. In natural Earth environments, nearly all ice samples studied are found contaminated by microorganisms ("Of Ice and Microbes" by J. Deming, 2007, *Bulletin of American Astronomical Society*, 209, 221.02; "Bacterial Activity in South Pole Snow" by Edward J. Carpenter, Senjie Lin and Douglas G. Capone, 2000, *Applied Environmental Microbiology*, 66, 4514). Down to tens of degrees below freezing, Earth ice is replete with diatoms, bacteria, viruses, and animals – often whole ecosystems, especially in contact with seawater.

Step One is liberating a vital chunk of Earth. A massive collision event is how naturally to launch from Earth at escape velocity, 11.2 km s^{-1}. Ejecta from impacts are pulverized by Earth's atmosphere, so only impacts big enough to temporarily evacuate local air propel boulders to the Moon ("Natural Transfer of Viable Microbes in Space; 1. From Mars to Earth and Earth to Mars" by Curt Mileikowsky et al., 2000, *Icarus*, 145, 391, for Earth rock launch models). Consider a large but non-evacuating impact: Ries Crater, Germany, 24 kilometers in diameter, 14.5 million years old. Average ejecta particle size decreases rapidly with crater distance: $s_{MP} = (26.7m) (r/R_D)^{-5.66}$, where s_{MP} is mean particle size, r distance from the impact, and R_D crater radius ("Bunte Breccia of the Ries: Continuous Deposits of Large Impact Craters" by Friedrich Hörz & Rolf Ostertag, 1983, *Lunar and Planetary Science Conference*, 14, 329). Impact distance depends on ejection velocity: $r \leq 2v^2/g$ (for vertical velocity v), so particle size plummets with velocity, microscopic for Earth escape. Extrapolation to large particle sizes implies few large ejecta, but large impacts violate this by removing the air. (Check atmospheric effects with "Earth Impact Effects Program" by Robert Marcus, H. Jay Melosh & Gareth Collins, 2010, http://impact.ese.ic.ac.uk/ImpactEffects/.) On the airless Moon large blocks are thrown thousands of kilometers in secondary impacts.

How big must our Earth rock be? Let us assume that it impacts the Moon with sufficient force to excavate down to freezing depth of about 10 meters. Because the rock's velocity likely barely exceeds lunar escape speed, 2.5 km s^{-1}, this sets excavation depth for given rocks: 0.1 meter diameter to reach down 0.6 meters, 1 meter to reach 5 meters, and 2.5 meters to dig down 10 meters ("Lunar Cratering" by Keith A. Holsapple: http://www.lpi.usra.edu/lunar/tools/lunarcratercalc/).

It seems possible for bacteria to be buried on the Moon at viable temperatures. What would it eat? Few carbon nutrients, unfortunately. Common at the depths likely to be excavated by the size of craters we are discussing is bacterium *Desulforudis audaxviator* ("Environmental Genomics Reveal a Single-Species Ecosystem Deep Within Earth" by D. Chivian, et al., 2008, *Science*, 322, 275), found 2.8 kilometers deep in walls of South African gold mines, infecting deep Earth that might contribute ejecta to the Moon. This bacterium needs no other organism, only inorganics (carbon from calcium carbonate, sulfate from pyrite [FeS_2], and nitrogen from smectite as well as hydrogen from water

Box 11.3: (continued)

radiolysis). Moon rocks contain pyrite (chalcopyrite [$CuFeS_2$]), but typically carbon, nitrogen, and hydrogen reside at levels of 50 parts per million or less. Water, ammonia, and methane concentrate in lunar polar regions. Water may leak from outgassing sites (Chapter 8). Can these keep bacteria alive? We do not know.

Even if nothing of Earth origin is alive on the Moon, certainly terrestrial biological material is there. This may be most compelling, scientifically, since we know little of this material from the Earth's first billion years, but some may reside on our natural satellite. Impact velocities on the Moon can be as low as 1 km s^{-1} ("Distribution of Impact Locations and Velocities of Earth Meteorites on the Moon" by John Armstrong, 2008, *Astrobiology*, 8, 4–01-O). Complex organic molecules can survive impacts into regolith even at 2 km s^{-1} ("Hypervelocity Impact Experiments in the Laboratory Relating to Lunar Astrobiology" by M.J. Burchell, J. Parnell, S.A. Bowden and I.A. Crawford, 2010, *Earth, Moon and Planets*, 107, 55).

There are strange things on the Moon: volatile/regolith mixtures, hidden lava tubes. Subsurface cavities might have loose ceilings; volatile-dominated landscapes might be unstable underfoot (especially considering the deeper deposits are warmer and might sublime away). Do we send people into this? We can expand the explored envelope with robots, preserving people from known danger. What is perilous on alien worlds? We may not know. In a situation where lives are at stake (perhaps many lives in cases of planetary protection) one must admit one's level of ignorance, what remotely possibly might occur. Surprises in science are how discoveries are made; surprises in exploration often kill people. In these situations we need separate devil's advocate Team B approaches to challenge assumptions. The last time we entered such an unknown environment was the narrow window between the Surveyor probes (1966–1967) and Apollo landings of 1969; we are out of practice (spelunking aside). We need margin not only for the worst we postulate but also for what we do not. Otherwise, we may find ourselves doomed by arrogant ignorance. There must be a rule in human planetary exploration when entering new worlds (Moon included, especially subsurface): proceed with extreme caution – that is, do not fear, but assume the worst. Do not ask just what is likely to kill us, but what conceivably could. Proceed slowly: do remote observation, some computer models, sniff what comes off the surface, do more models, send in a robot, debate the outcome, examine a loose sample (at arm's distance, maybe in a remote outpost), test it, engage in more modeling and debate, but for heaven's sake, do not take it into your system. Any mother knows that.

We should encourage exploration and exploitation of lunar resources but should institute a policing force ensuring scientific resources of the natural state of the Moon are not squandered, and we need exercise extreme caution about pre-biotic, fossil, and potentially viable matter. We need planetary protection officers (PPOs) monitoring lunar operations by private or government agents (or anyone else). With telerobotics, this should be technically easy: make sure PPOs are given all information feeds that operations personnel have. Give the officer power to question or delay entry into new areas until things are safe. Politically this will be contentious.

At least we should revisit our attitudes in exploring lunar environs with water and organics. The polar areas of the Moon should not be considered sterile by definition, despite NASA policy. It is worth reevaluating if polar and subsurface lunar activity requires different protection protocol than more familiar midlatitude surface missions. For instance, we sometimes consider crashing unpurged, unsterilized spent spacecraft into lunar polar regions. How will these affect the local sampling environment? What is the range of possible contamination? *LCROSS* was more careful about this than most. Most troublesome is that there are few formal guidelines to mandate what can be done.

Another recommendation stems from earlier observations that even a small output of gas will overwhelm natural atmospheric production (20 grams per second) raining pollution into frozen polar regions recording billions of years' history of the Moon, Sun, Solar System, and interstellar matter. Rockets exhausting sulfur compounds near the Moon should be discouraged, because this element is a prime discriminator between sources of gas interior versus exterior to the Moon. We could lose this probe forever if we pollute it with the gas from habitats or industries. If there is reason to critically examine human lunar expansion, this is the best.[65]

Some provisions of the OST seem useful in curtailing worst offenses possible from lunar exploitation; others miss the target of practical applicability and must be revised. Written in the age of Apollo and Luna, the treaty's attitude is that actions on the Moon are controlled from Earth, and little thought is given to why humans might stay there and how the treaty might empower or encourage them. Given potential ease in maintaining lunar habitation with resources we now know are available, the people living there might not be forced to listen.

If there is reason to stay on the Moon, controlling local humans from distant Earth sounds challenging, despite law and enlightened greater good.[66] There are precedents. After the Seven Years War in 1763, Britain outlawed settlement of the Ohio and the purchase of land, out of respect for their native American allies' need and reverence for their homeland. For this and like reasons, once-British settlers took up arms and their erstwhile rulers were closed out of half of the continent. Could anything on the Moon be so precious and contentious? We will discuss lunar industry and commerce in Chapter 12.

Notes

1. In Chapter 7, we find the theory advanced in 1856 by famed Dutch astronomer Peter Andreas Hansen arguing for possible life on the Moon's Far Side. Newcomb defeated the idea 12 years later, dismissing it as "mere speculation, unsupported by analogy, probability or observation." Unfortunately, Newcomb was also known later in life for skepticism of things new or speculative, e.g., saying that astronomers would have no use for telescopes much larger than those of the day.

    He was more quantitative in treating water on Mars, and prescient. In *Harper's Weekly* (July 25, 1908), he wrote, "The most careful calculation shows that if there are any considerable bodies of water on our neighbouring planet they exist in the form of ice, and can never be liquid to a depth of more than one or two inches, and that only within the torrid zone and during a few hours each day.... There is no evidence that snow like ours ever forms around the poles of Mars. It does not seem possible that any considerable fall of such snow could ever take place, nor is there any necessity of supposing actual snow or ice to account for the white caps. At a temperature vastly below any ever felt in Siberia, the smallest particles of moisture will be condensed into what we call hoar frost, and will glisten with as much whiteness as actual snow.... Thus we have a kind of Martian meteorological changes, very slight indeed and seemingly very different from those of our earth, but yet following similar lines on their small scale. For snowfall substitute frostfall; instead of feet or inches say fractions of a millimetre, and instead of storms or wind substitute little motions of an air thinner than that on the top of the Himalayas, and we shall have a general description of Martian meteorology." This is not a bad match to what we know now.

Newcomb led an interesting life. In childhood he escaped his fate as a herbalist's apprentice in Nova Scotia to teach in Maryland, then became a human "computer" at the American Nautical Almanac Office in Cambridge, Massachusetts, and soon graduated from Harvard to become a professional astronomer. With the Civil War's advent, Newcomb was appointed Professor of Mathematics for the U.S. Navy, then Director of the Nautical Almanac Office at the Naval Observatory in Washington. Cataloging and refining many of nature's fundamental constants, especially the speed of light, Newcomb showed lunar motion deviates greatly when extrapolated back to eclipses recorded by Ptolemy in the first century AD and later Arab astronomers, partially because of Earth's slowing rotation. Dr. Newcomb was honored during his life, and a navy survey ship was named for him posthumously during World War II. As *Apollo 17* approached touchdown on the last human lunar landing, it passed crater Newcomb (*The Reminiscences of an Astronomer* by S. Newcomb, 1903, Boston/New York: Houghton-Mifflin; "Obituary for Simon Newcomb," *The Times*, London, July 12, 1909).

2. *Discovery of Many Distinct Traces of Lunar Inhabitants, Especially of One of Their Colossal Buildings* by F. v. P. Gruithuisen, 1824.
3. *Impact Mechanics at Meteor Crater, Arizona* by E. M. Shoemaker, 1960, PhD thesis, Princeton University. Ralph Baldwin was also an early, significant proponent of the lunar crater impact model, presented in his book *The Face of the Moon* (1949) and earlier works. Baldwin recalls difficulty in swaying scientists from the "truth" that lunar craters formed via volcanism. Recalling a colleague referred to as "Elmer," Bardwin says:

"*The Face of the Moon* appeared in January, 1949. With nine prominent and prompt exceptions, the book was largely ignored for a few years ... I could not understand this. The observations were clear and definitive. The Moon's craters, and, of course, many terrestrial examples had to be of impact origin. Yet many otherwise excellent scientists read or listened to the evidence and remained of the same opinion, that the Moon's craters were some form of volcanism. It didn't make sense. Elmer had the same type of mentality. He 'knew' that his religious ideas were correct. He listened to scientific arguments to the contrary, but his brain refused to analyze the presented data. Anything contrary to his beliefs could not be true and therefore there was no point in his trying to interpret them. ... What Elmer had done was to show me that many people were not sufficiently flexible mentally to allow them to change their minds when new (from Baldwin's Barringer Medal Address, 2000).

4. "Great Astronomical Discoveries Lately Made by Sir John Herschel, L.L.D., F.R.S., &c, at the Cape of Good Hope" in *The Sun*, New York, August 25–29 and 31, 1835.
5. *Astronomie Populaire* by Camille Flammarion, 1880 (English edition: 1904, New York: Appleton); "Eratosthenes" (1–6) by W. H. Pickering, 1919–1925, *Popular Astronomy*, #269, 287, 312, and 317.
6. The image size θ a point source will subtend, e.g., a distant star, depends on light's wavelength λ and diameter of the telescope's primary mirror (or lens) D, given by $\theta = 0.98\, \lambda/D$. The human eye has peak sensitivity at wavelengths around 5500Å, or 0.00055 mm, so a 24-foot (7.3-meter) telescope could resolve an object larger than 0.016 arcseconds, or 4 millionths of a degree. On the Moon, R = 390,000 km away, this becomes θR = 30 m, with θ measured in radians (number of degrees times $\pi/180$). One could not easily discern smaller lunar details. In *The Sun*, the telescope is claimed to have a magnification of 42,000, which would greatly over-magnify images (of the Moon or elsewhere) given the telescope's diameter.
7. *Letters*, III, James Howell, 1645, chap. ix.
8. "Uncontrolled, *Hirondelle* goes into a glide, loses altitude, and spirals down to a nudge landing in the soft moondust of the farther shore of the Lake of Dreams. The Selenites swarm aboard, spinning dream-silk from their spinnerets. The crew are trussed within minutes. Five frail Selenites to one Terrene, the Members of the

Anglo-French Expedition to the Moon are borne into the heart of the Temple of Dreams" (*Moon Shots*, edited by Peter Crowther, 1999, New York: Daw).

9. *Apollo 17: Technical Air-to-Ground Voice Transcription*, Test Division, Apollo Spacecraft Program Office, December 1972, NASA Pub. MSC-07629.

A few more examples:

CERNAN: Whoops. Bob, I'm going to do one other thing real quick here. I've got to dust my visor off.

SCHMITT: Gene, do you want me to do that?

CERNAN: No, I can do it. I'll just do it right here. Only have to do it in a couple of places right in front of me. That didn't do much good, did it? Someone should have told me that. That just really screwed it up. Okay. Bob, you might ought to be thinking of a good way to clean that visor when I get in the cabin.

PARKER: Okay, we'll put someone on that.

ALLEN: That's got it. And I've got a LEVA-cleaning procedure which maybe you could pencil in there. It's an easy three-step procedure. And I'll go ahead and read it step-by-step here. Step number 1 is tap LEVA base to remove loose dust. Step number 2 reads: If excess dust still remains, use a towel from the LM tissue dispenser, which has been wetted with water, and gently wipe the visor from the top to the bottom; that is, in one direction. And fold this towel after each wipe to keep the contact surface clean. There's a note. 'Take care not to wet the inside that is, the concave surface of the gold visor.' And the last step is: Allow it to air-dry. And that's it on the LEVA cleaning.

SCHMITT: Some ambiguity in your statement. You want us to use a tissue or a towel on that visor cleaning?

ALLEN: Jack, they call it – they call it a towel, but it comes from the LM tissue dispenser, so I would interpret that to mean tissue.

SCHMITT: Well, you and I are thinking alike. But can you ask back there and find out?

ALLEN: Asking right now... Jack, our guess was right on the cleaning of the visors there. We're to use a tissue from the LM tissue dispenser.

10. Cernan in *Apollo 17 Technical Debriefing* by Office of Crew Training and Simulation Division, January 4, 1973, Houston: Manned Space Center, p. 20.12; Schmitt quoted by Sandra A. Wagner from February 2, 2004, Ames Research Center in *The Apollo Experience Lessons Learned for Constellation Lunar Dust Management*, September 2006, NASA/TP-2006–213726.

11. This is summarized from mission reports, technical debriefs, and science reports in Wagner, *The Apollo Experience Lessons Learned for Constellation Lunar Dust Management*, p. 63.

12. "After lunar liftoff when we were again in a 0g environment, a great quantity of dust floated free within the cabin. This dust made breathing without the helmet difficult, and enough particles were present in the cabin atmosphere to affect our vision. The use of a whisk broom prior to ingress would probably not be satisfactory in solving the dust problem, because the dust tends to rub deeper into the garment rather than to brush off" (*Apollo 12 Preliminary Science Report* by A. L. Bean, C. C. Conrad, Jr., & R. F. Gordon, 1970, NASA Pub. SP-235, 33).

13. *Apollo 12 Mission Report* by Mission Evaluation Team, March 1970, Houston: Manned Space Center, NASA Pub. MSC-01855, 6–1.

14. "Sediments of the Moon and Earth as End-Members for Comparative Planetology" by A. Basu & E. Molinaroli, 2001, *Earth, Moon and Planets*, 85, 25. The size distribution of *Apollo 11* sample 10084 is comparable in size to Sahara dust wind-blown hundreds of kilometers.

15. *LSB*, Figure 7.9. (Characterizations of lunar dust sizes below 5 microns sometimes disagree.)

16. "Is Lung Disease after Silicate Inhalation Caused by Oxidant Generation?" by A. J. Ghio, T. P. Kennedy, R. M. Schapira, A. L. Crumbliss & J. R. Hoidal, 1990, *Lancet*, 336, 967.

17. "Cardiovascular Risks from Fine Particulate Air Pollution," D. W. Dockery & P. H. Stone, 2007, *New England Journal of Medicine*, 356, 511.

18. "Augmentation of Pulmonary Reactions to Quartz Inhalation by Trace Amounts of Iron-Containing Particles" by V. Castranova, V. Vallyathan, D. M. Ramsey, J. L. McLaurin, D. Pack, S. Leonard, M. Barger, J. Y. Ma, N. S. Dalal & A. Teass, 1997, *Environmental Health Perspectives*, 105, 1319. Regarding gunsmoke smell: modern gun cartridges smell metallic on ignition although charged with guncotton (nitrocellulose) or cordite (nitroglycerine and nitrocellulose) and little metal. Cordite smoke can simulate lunar dust aroma ("Scent Study: The Science Museum" in ScentAir promotional, http://www.scentairuk.com/UK_casestudiesTheScienceMuseum.

pdf). Gunsmoke-like odor likely results from surface nanophase iron reacting in sinuses. JSC-1 Lunar regolith simulant has nearly identical composition to real regolith but with no surface nanophase iron or odor.

19. A lunar dust health issue status report: "Biological Effects of Lunar Dust Workshop," NASA Ames Research Center, May 29–31, 2005, http://lunarscience.nasa.gov/articles/biological-effects-of-lunar-dust-workshop/.
20. "Toxicity of Lunar Dust" by Dag Linnarsson et al., 2012, *Planetary and Space Science*, 74, 57.
21. "Physico-Chemical Properties of Silica in Relation to Its Toxicity" by T. Nash, A. C. Allison & J. S. Harington, 1966, *Nature*, 210, 259.
22. "Microwave Sintering of Lunar Soil: Properties, Theory and Practice" by L. A. Taylor & T. T. Meek, 2005, *Journal of Aerospace Engineering*, 16, 188; "The Lunar Dust Problem: From Liability to Asset" by L. A. Taylor, H. H. Schmitt, W. D. Carrier III & M. Nakagawa, in *1st Space Exploration Conference: Continuing the Voyage of Discovery*, 2005, American Institute of Aeronautics and Astronautics, 2510.
23. "Sintering Bricks on the Moon" by C. C. Allen, J. C. Graf & D. S. McKay, in *Engineering, Construction and Operations in Space IV*, 1994, American Society of Civil Engineering, 1220.
24. The heat flow (vector) ϕ_h is given by the differential Fourier heat conduction law $\phi_h = -\kappa \nabla T$. For our purposes think of this as a one-dimensional problem in depth z, with a dwelling consisting of a lens-shaped volume with no heat capacity but a set internal temperature (comfortable for humans) and negligible effective depth in z. In this case the heat conduction equation becomes $\phi_h = -\kappa \, \partial T/\partial z$. To adequately shield against cosmic rays one needs a regolith column density of 200 g cm^{-2}, which at bulk density of 1.5 g cm^{-3} is 135 cm deep. The density of regolith grain material (specific gravity) is 3.1 g cm^{-3}, so depths of completely compacted regolith at least 65 cm suffice. The temperature just below the surface is 253 K, while comfortable room temperature is about 293 K. The thermal conductivity k for naturally packed regolith is 0.0002 W cm^{-1} K^{-1} versus 0.001 W cm^{-1} K^{-1} for solid regolith material. These predict heat flow from natural regolith-coated buildings (with the nominal shielding depth) of $\phi_h = 6 \times 10^{-5}$ W cm^{-2} = 0.6 W m^{-2} to maximally compacted regolith with heat flow of 6×10^{-4} W cm^{-2} = 6 W m^{-2}. Because an adult human at rest generates ~100 Watts of heat, each must be assigned a surface area of 16 to 160 m^2 to radiate that heat. The Moon itself radiates 0.03 W m^{-2}, overwhelmed by human contributions.
25. "Construction of Planetary Habitation Tunnels using a Rock-Melt-Kerfing Tunnel-Boring Machine Power by a Bimodal Heat Pipe Reactor" by J. D. Blacic, M. G. Houts and T. M. Blacic, 1998, *LPI Technical Rep. 98-01*, Houston: LPI, 2.
26. *LSB*, 46, 82.
27. "HFT Events: Shallow Moonquakes?" by Y. Nakamura, 1977, *Physics of Earth and Planetary Interiors*, 14, 217.
28. *LSB*, Sec. 3.11. Earth's magnetic field partially shields the Moon several days per month.
29. "Lunar Soil as Shielding Against Space Radiation" by J. Miller, L. Taylor, C. Zeitlin, L. Heilbronn, S. Guetersloh, M. DiGiuseppe, Y. Iwata & T. Murakami, 2010, *Radiation Measurements*, 44, 163; "Shielding Strategies for Human Space" eds. J. W. Wilson, J. Miller, A. Konrad and F. A. Cucinotta, 1997, Hampton, VA: NASA Langley RC, NASA Conf. Pub. 3360.
30. "Analysis of a Lunar Base Electrostatic Radiation Shield Concept, Phase I: NIAC CP 04-01" by Charles R. Buhler, October 1, 2004, NASA Kennedy Space Center.
31. 2009 Regolith Excavation Centennial Challenge, NASA Office of Chief Technologist, http://www.nasa.gov/offices/oct/early_stage_innovation/centennial_challenges/index.html;
    NASA Kennedy Space Center Annual Lunabotics Mining Competitions, http://www.nasa.gov/offices/education/centers/kennedy/technology/lunabotics.html.
32. "Pneumatics on the Moon: Drilling and Sampling Using Gas" by K. Zacny, G. Galloway, R. Mueller, J. Craft, G. Mungas, G. Paulsen, M. Hedlund & P. Chu, in *NASA Lunar Science Institute Conference*, 2009, Ames Research Center, July 21–24; "Novel Method of Regolith Sample Return from Extraterrestrial Body Using a Puff of Gas" by K. Zacny, D. McKay, L. Beegle, T. Onstott, R. Mueller, G. Mungas, P. Chu & J. Craft, 2010, *IEEE Aerospace Conference*, #1082. The video "A Robot to Mine the Moon" by Erico Guizzo & Rosaleen Ortiz, 2008 (http://spectrum.ieee.org/video/aerospace/robotic-exploration/lunar-vacuum-cleaner) demonstrates pneumatic excavation.
33. "Principles of the Construction of Long-Functioning Lunar Settlements" by Design Bureau of General Machine Building (KBOM), December 1969; "Manned Lunar Program – Lunar Base" by Anatoly Zak, 2002, *Russian Space Web*, http://www.russianspaceweb.com/lunar_base.html.
34. H.R. 1654, "NASA Authorization Act of 2000" by Rep. Dana Rohrabacher (D-CA) and six cosponsors, introduced May 3, 1999, 106th Congress: "Sec. 128. TRANS-HAB. (a) Replacement Structure: No funds authorized by this Act shall be obligated for the definition, design, or development of an inflatable space structure to replace

any International Space Station components scheduled for launch in the Assembly Sequence released by NASA on February 22, 1999. (b) General Limitation: No funds authorized by this Act for fiscal year 2000 shall be obligated for the definition, design, or development of an inflatable space structure capable of accommodating humans in space."

35. "Private Moon Bases a Hot Idea for Space Pioneer" by Leonard David, April 14, 2010, Space.com.
36. "Study on Lunar Cement Production Using Hokkaido Anorthite and Hokkaido Space Development Activities" by T. Horiguchi, N. Saeki, T. Yoneda, T. Hoshi & T. D. Lin, 1996, *Proc. of 5th International Conference on Space*, ASCE Proceedings, 207, 86; "Behavior of Simulated Lunar Cement Mortar in Vacuum Environment" by T. Horiguchi, N. Saeki, T. Yoneda, T. Hoshi & T. D. Lin, in *Space 98*, eds. R. G. Galloway & S. L. Lokaj, 1998, Reston, Virginia: ASCE, 571; "Proposed Remote-Control, Solar-Powered Concrete Production Experiment on the Moon" by T. D. Lin, S. B. Skaar & J. J. O'Gallagher, April 1997, *Journal of Aerospace Engineering*, 104.
37. "Extraterrestrial Fiberglass Production Using Solar Energy" by Darwin Ho & Leon E. Sobon, in *Space Resources and Space Settlements*, 1979, NASA SP428, 225, proposes a 120-ton factory making 6,000 tons of fiberglass per year from 10^4 tons of regolith.
38. "Development of Lunar Ceramic Composites, Testing and Constitutive Modelling, Including Cemented Sand" by Jámos Csaba Tóth, 1994, PhD thesis, University of Arizona Civil Engineering.
39. From "Buzz Aldrin's Moon Landing Memories," interview by Nick Higham, February 25, 2009, *BBC News*.

    HIGHAM: When you arrived to the Moon and you looked out at it the phrase you used, I think, to describe it was "magnificent desolation." Is that still the way it seems to you 40 years on?

    ALDRIN: There's not a more desolate location that I have ever seen, where there is the natural, not artificial, but the natural habitat of where you are. But still I'm trying to tell people that it's not a good place to set up housekeeping.

    HIGHAM: But what about going to Mars?

    ALDRIN: Mars really is a much more habitable location, much more Earthlike than the Moon, much more potential to support active colonizing and modifying the climate. According to some people, we are the ones responsible for so drastically altering this climate here. Maybe we could get to Mars and alter its climate as easily as supposedly we seem to have done here.

40. "Pressure Modeling of Upward Flame Spread and Burning Rates over Solids in Partial Gravity" by Julie Kleinhenz, Ioan I. Feier, Sheng-Yen Hsu, James S. T'ien, Paul V. Ferkul & Kurt R. Sacksteder, 2008, *Combustion and Flame*, 154, 637.
41. "NASA Plans a Moon Garden" by Rachel Martin, December 8, 2013, *Weekend Edition*, National Public Radio: http://www.npr.org/templates/story/story.php?storyId=249570007; "NASA's Next Frontier: Plants on the Moon" by Tarun Wadhwa, November 20, 2013, *Forbes*, http://www.forbes.com/sites/tarunwadhwa/2013/11/20/nasas-next-frontier-growing-plants-on-the-moon/.
42. "Crop Productivities and Radiation Use Efficiencies for Bioregenerative Life Support" by R. M. Wheeler, C. L. Mackowiak, G. W. Stutte, N. C. Yorio, L. M. Ruffe, J. C. Sager, R. P. Prince & W. M. Knott, 2008, *Advances in Space Research*, 41, 706.
43. Some examples are arrowroot, buckwheat, carrots, chicory, clover, kelp, marigold, mint, strawberry, and many more. They are common.
44. For more information, see the Global Metallophyte Database of the International Serpentine Ecology Society (http://www.metallophytes.com).
45. "Final Report – Biological Survey of the Berkeley Pit Lake System" by Grant Mitman, 1999 November, *Mine Waste Technology Program Activity IV, Project 10; for U.S. E.P.A. and D.O.E.*, 9; see also "Remediation of Berkeley Pit Water Using Genetically Modified Extremophilic Yeast" by Andrea & Don Stierle, August 2006, *PitWatch*; "Heavy Metal-Resistant Bacteria as Extremophiles: Molecular Physiology and Biotechnological Use of Ralstonia Sp. CH34" by D. H. Nies, 2000, *Extremophiles*, 4, 77; "New Life in a Death Trap" by Edwin Dobb, December 2000, *Discover*.
46. "A Search for Intact Lava Tubes on the Moon: Possible Lunar Base Habitats" by C. R. Coombs & B. R. Hawke, in *2nd Conference on Lunar Bases and Space Activities of the 21st Century*, ed. W. W. Mendell, 1992, NASA CP-3166, 1, 219; "Utility of Lava Tubes on Other Worlds" by B. E. Walden, T. L. Billings, C. L. York, S. L. Gillett & M. V. Herbert, in *Workshop on Using In Situ Resources for Construction of Planetary Outposts*, 1998, (Houston: LPI), 16; "Lunar Lava Tube Radiation Safety Analysis" by G. de Angelis, J. W. Wilson, M. S. Clowdsley,

J. E. Nealy, D. Humes & J. M. Clem, 2002, *Journal of Radiation Research*, 43, S41. ESA conducts astronaut training with spelunking in a Sardinian cave to practice cooperative exploration and field science (CAVES: (Cooperative Adventure for Valuing and Exercising human behavior and performance Skills).

47. One married couple served together in space, on STS-47 *Endeavor* in 1992, and rumors spread about illicit goings-on aboard *TM-20/Mir* in 1994. Sex is generally discouraged to support group morale ("Do Astronauts Have Sex?" by Christopher Beam, February 7, 2007, *Slate*). Some tales should be discounted, e.g., *La Dernière Mission: Mir, l'aventure humaine* by Pierre Kohler, 2000, Phalsbourg: Calmann-Lévy.

48. *Dragonfly: NASA and the Crisis Aboard Mir* by Bryan Burrough, 1998, New York: Harper Collins, 185; "Psychological Issues Relevant to Astronaut Selection for Long-Duration Space Flight: A Review of the Literature" by Daniel L. Collins, 1985, Pub. AFHRL-TP-84–41, Brooks AFB, Texas: USAF Manpower and Personnel Division; *Red Star in Orbit* by James Oberg, 1981, New York: Random House.

49. *Salyut: The First Space Station – Tragedy and Triumph* by Grujica S. Ivanovich, 2008, Berlin: Springer Praxis, 224–226.

50. Alexander Kumar, physician at Concordia Station during White Mars, recounts: "If you look at it from the outside, it's essentially taking a small group of people who don't know one another and in fact putting them through one of the worst winters in the world. There are nine months of complete isolation in that altered daylight cycle. That's what you would undergo on long-haul space travel as well. If you were going to go to Mars, it may take eight to ten months, probably around nine, one way. Of course then you would have the problems of gravity, so of course we don't simulate zero gravity. However the isolation and the psychology far outweigh the physiology in terms of the challenges in the area of study."

"My partner flew out and sort of led me around New Zealand by the arm for ten days, you know: how to order a coffee again, and how to cross the road. It's sort of like being a child; you remember it all, but it's not all to hand. This is re-entry syndrome. That's a syndrome in name and a tribute to the collection of symptoms you undergo. You know you can be anxious; you can be irritable. You can suffer terrible concentration difficulties."

"A few people that come off the ice wander around the world locked behind hotel room doors, and then return to the ice feeling they don't fit in again. Of course, the concern is that you don't reattach; you don't reconnect. You are under a process essentially of defrosting of the mind" (*Outlook*, BBC World Service, December 12, 2012, excerpts from Alexander Kumar interview).

51. *Apollo 14*'s Edgar Mitchell: "You develop an instant global consciousness, a people orientation, an intense dissatisfaction with the state of the world, and a compulsion to do something about it. From out there on the moon, international politics look so petty. You want to grab a politician by the scruff of the neck and drag him a quarter of a million miles out and say, 'Look at that, you son of a bitch'" ("Edgar Mitchell's Strange Journey" *People*, April 8, 1974, 8, #6, 20).

52. "Lava Tubes: Potential Shelters for Habitats" by Friedrich Hörz, 1985, in *Lunar Bases and Space Activities of the 21st Century* ed. W. Mendell, Houston: LPI, 408; and "Lunar Rilles and Hawaiian Volcanic Features: Possible Analogues" by D. P. Cruikshank and C. A. Wood, 1972, *Moon*, 3, 412.

53. Gus Grissom, second American in space and first Gemini commander, who died on the launch pad of the first Apollo mission, confessed: "We're in a risky business, and we hope if anything happens to us, it will not delay the program. The conquest of space is worth the risk of life. Our God-given curiosity will force us to go there ourselves, because in the final analysis, only man can fully evaluate the Moon in terms understandable to other men" (*Risk and Space: Earth, Sea and the Stars*, eds. Steven J. Dick & Keith L. Cowling, 2005, NASA Pub. SP-4701, 190).

54. "Houston, We Have a Problem" by J. Matthew Zoss, in *Hint Fiction: An Anthology of Stories in 25 Words or Fewer* by Robert Swartwood, 2011, New York: Norton, 77.

55. Patrick Taveirne (*Han-Mongol Encounters and Missionary Endeavors: A History of Scheut in Ordos, Hetau, 1874–1911*, 2004, Leuven University, 104) cites Antoine Mostaert that horse thieves in the Ordos desert of Inner Mongolia were punished by abandoning them sans horse in the desert with their Achilles tendon cut. Horse theft could be a capital offense even in more hospitable climes, e.g., England from 1545 to 1832, more because of escalating punishment for property crimes rather than any harsh physical environment (*Confronting Animal Abuse: Law, Criminology and Human-Animal Relationships* by Piers Beirne, 2009, Maryland: Rowman & Littlefield, 41).

56. Outer Space Treaty, Article XII: "All stations, installations, equipment and space vehicles on the moon and other celestial bodies shall be open to representatives of other States Parties to the Treaty on a basis of reciprocity. Such representatives shall give reasonable advance notice of a projected visit, in order that appropriate consultations may be held and that maximum precautions may be taken to assure safety and to avoid interference with normal operations in the facility to be visited."

The United States, Russia (via the Soviet Union), the United Kingdom, France, Germany (and by extension the European Union), Japan, India and roughly 85 other countries have signed the Treaty; these particular ones and roughly 55 more have ratified it. China has not, although it has acceded to its provisions, a less binding and somewhat more nervous-prone state of affairs.

57. For more about "the Western code," see *The Old West: The Cowboys* by William H. Forbis, 1973, New York: Time-Life, 210, and opposing opinions (c.f., *It's Your Misfortune and None of My Own: A New History of the American West* by Richard White, 1993, Norman: University of Oklahoma). The issue is best documented by diaries of the era, which are diverse.

58. From OST, Article V: "States Parties to the Treaty shall regard astronauts as envoys of mankind in outer space and shall render to them all possible assistance in the event of accident, distress, or emergency landing on the territory of another State Party or on the high seas... In carrying on activities in outer space and on celestial bodies, the astronauts of one State Party shall render all possible assistance to the astronauts of other States Parties."

59. Disconnects occurred even rescuing *Apollo 13*, when CSM CO_2 scrubbers mismatched the LM and only fit with duct tape and plastic sheeting. There are still conflicts between metric and U.S. customary units. Apollo intermingled three systems, e.g., kilometers vs. statute miles vs. nautical miles. Once this multi-system relaxed, it failed catastrophically in 1999 destroying *Mars Climate Orbiter*. Expect confusion of differing space agency standards if OST Article V is ever applied.

60. "Planetary Protection Provisions for Robotic Extraterrestrial Missions" by Science Mission Directorate, Effective April 20, 2011, to April 20, 2016, NASA Pub. NPR 8020.12D.

61. In *Apollo Lunar Surface Journal*, Pete Conrad, *Apollo 12* commander is animated about this: "The thing that had the bacteria in it was the television camera. The Styrofoam in between the inner and outer shells. There's a report on that. I always thought the most significant thing that we ever found on the whole goddamn Moon was that little bacteria who came back and lived and nobody ever said shit about it."

62. "Surveyor III: Bacterium Isolated from Lunar-Retrieved TV Camera" by F. J. Mitchell & W. L. Ellis, 1971, *Proceedings of 2nd Lunar Science Conference*, 3, 2721; "Microbiological Sampling of Returned *Surveyor 3* Electrical Cabling" by M. D. Knittel, M. S. Favero, R. H. Green, 1971, *Proceedings of 2nd Lunar Science Conference*, 3, 2715; "Space Microbiology" by G. R. Taylor, 1974, *Annual Review of Microbiology*, 28, 121.

63. "A Microbe on the Moon? *Surveyor III* and Lessons Learned for Future Sample Return Missions" by John Rummel, Judith Allton & Don Morrison, in *Importance of Solar System Sample Return Missions to the Future of Planetary Science*, 2010, http://www.lpi.usra.edu/meetings/ sssr2011/presentations/rummel.pdf; an earlier summary is found in "*Apollo 12* Remembered: Lunar Germ Colony or Lab Anomaly?" in *Astrobiology Magazine*, November 21, 2004.

64. RKA/ESA's *Foton-M3* orbited Earth 11 days carrying lichen, tardigrades, and various microbial extremophiles on the craft's exterior, surviving vacuum and cosmic rays as well as indirect heat of reentry (but not solar UV) ("Tardigrades Survive Exposure to Space in Low Earth Orbit" by K. Ingemar Jönsson, Elke Rabbow, Ralph O. Schill, Mats Harms-Ringdahl & Petra Rettberg, 2009, *Current Biology*, 18, R729). Tardigrades also flew on STS-134 *Endeavor* in May 2011.

What are tardigrades? They are *polyextremophiles*: complex, differentiated multicellular organisms surviving conditions usually thought fatal, versus *extremophiles*, one-celled organisms capable of the same. Tardigrades are eight-legged, caterpillar-like, about a millimeter long, and when dormant withstand temperatures of 150°C and approaching absolute zero (hence, any lunar surface temperature). They can survive vacuum and radiation hundreds of times human fatal doses and after losing 97% of internal moisture. After *Foton-M3*'s vacuum and radiation, they laid eggs and hatched viable young. Given their hardiness, we are lucky tardigrades are not terribly active, nor find humans tasty. Bacteria, fungi, and viruses subjected directly to space conditions survived for hours or days, even hard vacuum and UV radiation by forming inactive spores, later revived. If protected from UV, spores survive vacuum for years. It seems spores can survive in amber for 30 million years ("Space Microbiology" by G. R. Taylor, 1974, *Annual Review of Microbiology*, 28, 121; "Protection of Bacterial Spores in Space, a Contribution to the Discussion on Panspermia" by G. Horneck, P. Rettberg, G. Reitz, J. Wehner, U. Eschweiler, K. Strauch, C. Panitz, V. Starke & C. Baumstark-Khan, 2001, *Journal of Origins of Life and Evolutionary Biospheres*, 31, 10.1023; "Long-Term Survival of Bacterial Spores in Space" by G. Horneck, H. Bucker & G. Reitz, 1994, *Advances in Space Research*, 14, 41; "Revival and Identification of Bacterial Spores in 25- to 40-Million-Year-Old Dominican Amber" by R. J. Cano & M. K. Borucki, 1995, *Science*, 268, 1060).

65. Outer Space Treaty, Article IX: "States Parties to the Treaty shall pursue studies of outer space, including the moon and other celestial bodies, and conduct exploration of them so as to avoid their harmful contamination and also adverse changes in the environment of the Earth resulting from the introduction of extraterrestrial matter and, where necessary, shall adopt appropriate measures for this purpose."
66. Outer Space Treaty, Article VI: "States Parties to the Treaty shall bear international responsibility for national activities in outer space, including the moon and other celestial bodies, whether such activities are carried on by governmental agencies or by non-governmental entities, and for assuring that national activities are carried out in conformity with the provisions set forth in the present Treaty. The activities of non-governmental entities in outer space, including the moon and other celestial bodies, shall require authorization and continuing supervision by the appropriate State Party to the Treaty."

Chapter 12
Lunar Power

> For I dipt into the future, far as human eye could see,
> Saw the Vision of the world, and all the wonder that would be,
> Saw the heavens fill with commerce, argosies of magic sails,
> Pilots of the purple twilight, dropping down with costly bales.
> – Alfred, Lord Tennyson, 1835, "Locksley Hall"[1]

Tennyson reminds us that despite the scope for humanity of exploration, personal fulfillment rests on immediate accomplishment: on commerce, personal opportunity, and peace versus conflict – measured on a human scale. How can outer space fit into human lives?

From the start, Columbian-era Europeans set out to explore the world for profit. Although investors often lost their stakes and crew-members their lives, both dreamed of riches, and some were rewarded. In intervening centuries ships sailed the globe to exploit and trade in natural resources: spices, whale oil, and more. In the twentieth century, exploration of Earth's polar regions and outer space was a contest for the prestige of nations and individuals more than profit. Lunar explorers need not fear unreasonably for their lives given precaution (Chapter 11), but what have investors to gain? Can distant space including the Moon be monetized for profit?

What can we gainfully import from the Moon?[2] Lunar commodities are all about opportunity costs, about being in the right place with the right substance, rather than supplying new materials or old stuff to Earth at a lesser price (with perhaps one major exception). Export costs to the Moon now reach $30,000 per kilogram, including spacecraft mass (about 30% of Apollo's costs in 2011 dollars). Importation cost from the Moon would be greater and depend on lunar resources utilization, but Earth-to-Moon transit costs tend to set the scale. Material from the Moon will incur likely net losses unless its Earth value greatly exceeds this. Precious metals are not precious enough, with gold, platinum, and rhodium about $50,000 per kilogram and mineable here.[3] Earth mines often need $200 earnings per tonne to profit. Rare earth elements, abundant within lunar KREEP (Chapter 6), are $100–$6,000 per kilogram. Synthetic elements and isotopes are made in nuclear accelerators and reactors at huge capital and energy cost. Californium-252, totally artificial, costs about $10 million per kilogram. Light helium (helium-3), made primarily in nuclear reactors for about $20 million per kilogram, is important in medicine and neutron detection screening against fissile atom bomb material but might someday fuel fusion reactors (see notes 22,23).

Earthly value for materials can be the wrong metric, because we need assets in space for remote sensing and communications. Launch costs run about $4,000 per kilogram to low Earth's orbit (LEO) and $12,000 to geosynchronous (GEO). In terms of energy, the lunar surface is much closer to GEO than Earth's is.[4] If goods on the Moon are inexpensive, sending them to GEO might undercut Earth launch. Remember this when we discuss uses and transportation costs for lunar resources.

What is available on the Moon? Lunar and terrestrial materials are similar in many ways, so most substances are also found on Earth. Terrestrial mining often exploits ores and veins from ancient action of water. Water inside Earth, deep so intensely hot but sufficiently pressurized to prevent boiling, dissolves and re-concentrates many solids. On Earth but not the Moon, magmatic "juvenile water" is renewed by ocean floor subduction. We expect little lunar aqueous activity, but it is worth discussing (see notes 39,40). Lunar material is concentrated in ways unseen on Earth. Solar wind interaction delivers substances lost from the Moon (and often Earth), especially light elements.

Water has so many applications that we unravel it first. Water allows life. It is a rocket propellant, energy storage, radiation shielding, and building material needed in ways monetizable both on the Moon and in space. Because location more than scarcity sets commodity value, discussing space economics of such a plentiful substance is valid.

Water molecules are highly stable (despite their uneven electric charge distribution), so chemicals easily forming water often do so vigorously. Liquid hydrogen (LH2) + liquid oxygen (LOX) propellant has nearly the highest speed chemical exhausts known, more than lunar alternatives, mainly because of hydrogen's small mass, which has other consequences. Its primary drawback is hydrogen's difficulty liquifying, at temperatures below 33 K (= –240°C = –400°F critical temperature, versus 155 K for LOX). Frigid as this is, some lunar locations stay even colder, in polar PSRs. Although hydrogen is storable in these sites (where it resides naturally), the main problems with practical use is their distance from hydrogen's likely utilization,[5] exacerbated by operating machinery in this cold. For LOX, sufficiently cold ambient conditions are abundant, with average surface temperatures less than 155 K within 300 kilometers of the poles.

Rather than breaking down water into propellant, one can build it up to hydrogen peroxide (H_2O_2), which dissociates explosively into water and oxygen. (Hydrogen trioxide, H_2O_3, or HOOOH, is prohibitively unstable.) H_2O_2 also serves to oxidize various fuels, but none of these propellants perform as well as their LOX equivalent. H_2O_2 dissociation makes rocket exhaust at 1.86 kilometers per second versus 4.46 for LH2/LOX (and values up to 4.7 for exotics such as LH2 + liquid fluorine).

Hydrogen peroxide's main advantages are its liquid state up to 457°C = 730 K (despite freezing at a temperatures close to water's) and its high density (1.46 grams per cubic centimeter for liquid H_2O_2 versus 0.071 for LH2 and 1.14 for LOX). The difficulty with H_2O_2 is its dangerous instability at the high purity preferred for rocketry (high test peroxide, or HTP, roughly 90% pure).[6] Many impurities break down HTP, often explosively, and decay proceeds to H_2O and O_2 even in stable conditions. HTP can explode if forced into constricted plumbing or agitated, a cause of naval accidents such as the loss of the Russian nuclear submarine *Kursk* with all 118 men after HTP exploded in a torpedo. H_2O_2 is also responsible for the March 18, 1980 explosion of a Vostok-2M rocket that killed 48 people. In spotless environs, with tanks and pipes of proper materials, HTP can remain effective longer than LH2/LOX but may never be hazard-free.

There are hydrogen-storage alternatives being studied for hydrogen-powered automobiles. Nature's smallest atom hydrogen insinuates itself into metal and organo-metallic lattices which absorb hydrogen like a sponge. Compared to gas cylinders, metal storage is compact but limited by the heavy metal atoms to 5% hydrogen uptake by mass. Carbon nanotubes are promising and may raise this. Metal-organic frameworks are solids with a lacy matrix more void than molecules, accepting more hydrogen mass, up to 10%. Storage density rises strongly at cold temperatures. Values above occur at liquid nitrogen temperature (77 K = –196°C), so good efficiency is expected from hydrogen tanks no colder than LOX. This still requires much spacecraft mass devoted to hydrogen tanks, but not unusually. (Now large tanks are needed for LH2's low density.)

Which propellant do we use? If in large tanks limiting boil-off, LH2 is suitable for missions under a month: near the Moon, to Earth (even round trip), or the initial thrust to more distant Solar System destinations.[7] HTP has slower exhaust but demands less cooling (still decaying with time) and is hypergolic (spontaneously igniting) in contact with catalysts, for example, platinum. It could provide primary propulsion for medium-length missions and maneuvering thrusters. Solid-absorbed hydrogen is long lasting at LOX temperatures, and LOX itself can last years. This is useful for the longest missions, even with dead weight in hydrogen storage solid (ten times the hydrogen's mass itself). Hydrogen and oxygen are much more powerful than HTP; solid-absorbed hydrogen wins out in many situations. LH2 + LOX works on Centaur and Delta upper stages for long missions; longer storage in orbital space propellant depots (see note 18) and the long-duration storage Integrated Cryogenic Evolved Stage (ICES) is being studied.

First consider long-distance travel on the Moon itself (no walking or driving). Unlike science fiction scenes with rocket cars gliding over the surface, hopping is best. A hopper is reusable, resembling a lunar lander, but jumping from spot to spot on the Moon (Figure 12.1). The best hop is a single bound, no intermediate stopping (or hopping).[8] To hop, blast off 45° from vertical, then after your arc's apex, fire your engine to decelerate until stopping at your destination.

One difficulty with hopping is access to fuel. To hop to the opposite point on the Moon with a 10-ton hopper (payload and propellant included) requires about 5 tons of oxygen/hydrogen and several tons on return. If this seems like much propellant, realize that a rocket from Earth into orbit consists of 93–98% propellant. In a hop from the poles (with its water) to the equator, carrying this much propellant might be possible; from Earth this would be prohibitive. Moving between successive sites far from the poles becomes increasingly problematic. Propellant pipelines

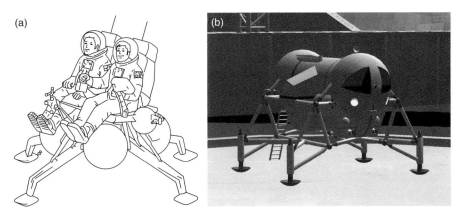

Figure 12.1. **Do the hop**. Surface-to-surface lunar hopper concepts. (a) The small Long Range Flyer proposed in 1970 as part of the Apollo Applications Program, with a 100 km round trip range across the Moon's surface, for two. (b) A rendering of a large lunar hopper designed by Andrew McSorley for the ballistic flight simulator code *Orbiter*. (Graphics from NASA and Andrew McSorley.)

over thousands of kilometers seem unlikely, so unless hoppers carry their own fuel, a tanker craft delivering propellant (at least hydrogen) is needed.

Because hydrogen fuel composes only 11–22% of propellant mass,[9] and abundant lunar oxygen is ubiquitous, a hydrogen tanker seems useful for fueling hoppers. Alternative fuel is abundant, if problematic: aluminum, 10% of the regolith's mass, is most easily made from anorthite, common in the highlands. Aluminum powder is used in many solid-fuel rockets, which need cleaning and refurbishing before any reuse. Also, aluminum propellants often contain elements rare on the Moon, for example, trimethylaluminum $Al_2(CH_3)_6$, or other oxidizers (not O_2) but often ammonium perchlorate (NH_4ClO_4).

Burning simple aluminum powder with oxygen, producing Al_2O_3, has problems: Al_2O_3 boils at 2,977°C and melts at 2,072°C, its exhaust producing a liquid/gas suspension tending to coat the engine interior. This is preferable to solid Al_2O_3 particles, which are highly abrasive, eroding the engine. Either way, engine maintenance will be messy. Development of aluminum powder/O_2 rockets is needed. Suggestions are aluminum/LOX slurries or letting oxygen flow over solid aluminum matrix, replenished after burning.[10] Because aluminum and especially oxygen are common in lunar soil, it is likely worthwhile.

A new aluminum-based fuel might work better: aluminum and water ice. (Yes, you read that correctly.) An ice and aluminum nanoparticle mixture (smaller than 100 nanometers) burns rapidly and violently, with exhaust velocity of 1.33 kilometers per second, 30% that of LH2/LOX but close to HTP's (and lunar orbital velocity).[11] Virtues of this over LH2/LOX are its eternal storage time and lesser consumption of precious hydrogen (its mass mainly aluminum). It remains solid at subsurface temperatures found anywhere on the Moon. Such rockets are not immediately reusable: once lit, they cannot be switched off, leaving a Al_2O_3 residue. Further work is needed.

Recall that along with water and H_2, *LCROSS* excavated significant carbon monoxide, hydrogen sulfide, and ammonia, so nitrogen, sulfur, and particularly carbon are available in lesser amounts, and we might consider using these in fuels. Methane is liquid as warm as 190 K = –83°C (and frozen below 91 K), compatible in temperature with LOX. Liquid methane and LOX exhaust at 3.6 kilometers per second.[12] Methane is easily synthesized from water and carbon dioxide. Since carbon monoxide at the *LCROSS* site was as abundant as water, the simplest process would probably be to oxidize CO to CO_2 and then to synthesize CH_4 in the usual way using water. The oxygen required could be extracted from lunar regolith.

We have not made a profit yet. What are plausible business scenarios? First, we are not discussing lunar real estate scams. Purveyors hold OST (Article II) does not regulate individuals, only states, against claiming lunar territory, so they claim the Moon to resell in plots to other private individuals. Such claims are void, because OST Article VI requires states to oversee and authorize such commerce. No state can authorize selling the Moon. (This is more strongly prohibited under the Moon Agreement.) More respectable is selling lunar souvenirs – not the few unfortunate cases of astronauts' items carried on missions, but literally pieces of the Moon. Lunar meteorites sell for up to $3,000 per gram. Pristine lunar samples returned by commercial spacecraft could sell for more (until market saturation), dodging the OST. At $80 million for 30 kilograms, a co-manifested lunar sample return might profit. Small telerobotic rovers rented to paying "telepilots" might profit with current hardware and are promising educational/public outreach/inspiration tools (Chapter 14). More tawdry is the idea (Chapter 8) of using PSRs to preserve, even currying "pharaonic" longings of super-rich to protect their bodies beyond mortality. There is broader utility in such capability, storing endangered DNA or other irreplaceables, exceeding a billion years, in the spirit of the Svalbard Global Seed Vault on Spitsbergen Island, arctic Norway, with 500,000 seed samples ensuring against extinction. NASA has encouraged private investment in lunar commerce by offering logistical support (but not direct funding) for lunar prospecting rovers and spacecraft, but they could provide an equally valuable service by better clarifying the legal framework for these kinds of commercial activities. This might be the most helpfully cost-effective thing NASA could do to promote private and commercial exploration of the Moon.

There is a big market in space, if not the Moon, but it can contribute. In the next decade some 1,200 satellites will launch (a 47% increase over last decade) for $178 billion. Many will go to GEO, about 235 satellites valued at $52 billion, most replacing existing satellites that break or, more often, run low on propellant.[13] Because this is primarily a question of propellant, the Moon might play a role.

Consider our proposed "cislunar tug" servicing GEO satellites otherwise drifting from their required positions. This tug would travel from Moon to Earth in three days, spend several weeks restoring GEO satellites, then return to the Moon to refuel. It does not refuel satellites, but pushes them back into place. (Alternatively, it could remain in GEO to be refilled by a tanker from the Moon.) Typical GEO satellites serve 15–20 years or more but use propellant to stay parked relative to users on the ground. In the meantime each satellite fights perturbations from the Sun, Moon, and planets forcing it from position, up to some 40 meters per second per year. Usually at life's end each satellite should boost itself to "graveyard orbit" 300 kilometers above GEO. The tug keeps satellites in place, extending their lives by conserving their propellant.

The cislunar tug's mission (Figure 12.2) comprises at least three different subsystems, all robotic. The shuttle subsystem features a lander parked most of its time on the Moon, but every few months receiving new charges of LH2/LOX or CH_4/LOX to carry into lunar orbit. Waiting in a low-maintenance frozen orbit is the tug itself, which trades empty propellant tanks for full ones from the shuttle. The shuttle lands back on the Moon using its reserve propellant, and the tug blasts toward Earth on a 3-day transfer trajectory. It settles into GEO and rendezvous with client satellites, either docking with custom fixtures on each satellite or grappling them with a claw-like device. (Eventually, when new client GEO satellites all have docking fixtures, the grappler will be detached to remain on the Moon.) After pushing several dozen satellites, the tug heads back to the Moon, LH2 largely expended, on a weak stability bound orbit putting it in lunar orbit three months later, fueled by solid-absorbed hydrogen. Soon the cycle repeats, after the robotic mining subsystem has refilled the tanks on the Moon. If abundant, methane instead of LH2 might be preferable.[14] Methane/LOX rocket motors are being developed both by NASA (at Glenn, Marshall and Johnson space centers) and SpaceX (their Raptor series). A challenge is a methane or LH2 rocket that can be restarted perhaps 100 times in space.

If regolith is loose containing water ice, simple loader and dump truck telerobots are likely best alternatives for excavating entrained water. Solid ice/regolith mixture can be harder than concrete. Machines are designed for terrestrial and lunar use to drill, fracture, and heat mixed deposits.[15] Their application depends critically on the temperature, internal versus external origin of hydration, deposition depth, associated volatile composition, and more. We discuss these momentarily.

Potential income for a cislunar tug is capped by $5 billion per year in GEO investment, 75% covering replacement satellites of which 80% die from propellant loss, meaning $3 billion per year of possible tug income. A new GEO satellite costs about $150 million, so charging $15 million per year to maintain one is a bargain, meaning the system might attract roughly 150 client satellites. Income of $2 billion per year allows for profit, depending on system costs.[16] A rough guesstimate of system cost is $10 billion initially and $100 million per year, so it should profit after five or six years. Space industry is already considering complicated options for maintaining GEO satellites from Earth-based systems.[17] When could a lunar-supplied cislunar tug reach fruition? Future space project projections and timetables are notoriously murky, but it seems this project could be operational with five years of concentrated effort.

A cislunar tug could be flexible. A mission to Mars or an asteroid might stop for propellant at the tug or its successor. We might establish propellant depots at Earth-Moon Lagrange Points L1 or L2 supplied by the tug. The tug or depot might service LEO or mid-Earth orbit (MEO). The problem with LEO and MEO is their orbits vary in energy more than GEO (all with nearly the same radii and orbital plane). With LEO and MEO markets growing several times larger than GEO, close-in, lunar-fed depots might profit.[18]

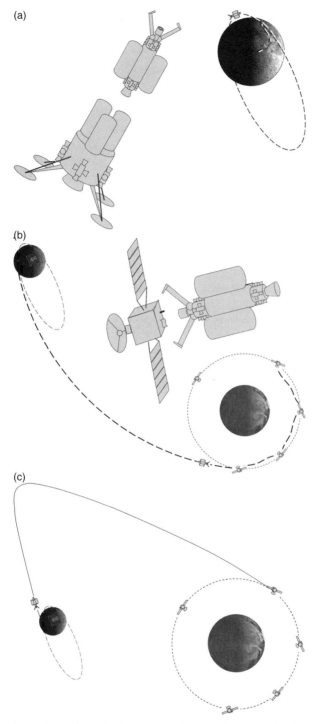

Figure 12.2. **Fly me to the moon, profitably**. A proposed system using lunar water as propellant (as liquid oxygen and hydrogen) consisting of (1) a lunar plant near one of the poles that excavates, electrolyzes, liquefies, and stores liquid oxygen and hydrogen, then loads propellant tanks onto a

Longer-range plans for lunar water use are proposed. Paul Spudis (of the Lunar and Planetary Institute, Houston, Texas) and Anthony Lavoie (NASA Marshall Space Flight Canter, Huntsville, Alabama) foresee a human and robotic lunar base generating 150 tonnes per year of water-derived propellant for use in cislunar space and beyond, including Mars. They estimate this requiring $88 billion and 14 years.[19]

A different proposal by Shackleton Energy Company would put human and robots on the Moon by 2019 for $30 billion, sending water-derived propellants to LEO from crater Shackleton at the lunar South Pole. This business failed to raise start-up funds, yet envisioned larger goals, with lunar-made solar panels beaming power to Earth to alleviate worldwide energy shortages.

Many corporations are forsaking government initiative and finding their own way to space profit. Virgin Galactic/Scaled Composites is offering short suborbital flights at $200,000 per passenger, probably by 2014. SpaceX envisions human flights separate from NASA and other agencies, using their Dragon capsule with rockets up to the Falcon Heavy (53 tonnes to LEO). Orbital Sciences, ATK, and Blue Origin are also competing for *ISS* resupply business. Bigelow Aerospace has built inflatable human space habitats but is suffering financial difficulty awaiting launch capability. More distantly, Planetary Resources, Inc., and Deep Space Industries plan to mine resources from near-Earth asteroids by the early 2020s. SpaceX's CEO and co-founder Elon Musk has even stated his intent to send people to Mars at a low cost by 2025.

To the Moon, not only are many groups competing for the Google Lunar X Prize, but also the Golden Spike Company is proposing an $8 billion system to take humans to the lunar surface for $750 million per passenger. This effort is led by Alan Stern, former NASA associate administrator, and Gerald Griffin, former Johnson Space Center director, but it is hardly a NASA enterprise. Except for lunar-specific landers and space suits, it will use other companies' technology. It has contracted for the lander design with Northrop Grumman (the original Apollo LM manufacturer).

Placing these in context, some private space efforts fail. In 2011, Shackleton Energy Company tried raising their first $1 million of $30 billion via crowd funding and collected $5,000. In 1999–2001 MirCorp succeeded in placing the first space tourist in orbit but ceased after running afoul of diplomatic/legal troubles between the United States and Russia.

Currently NASA lunar ambitions are not great, and other agencies' are not clearly greater. Opportunity promised by Golden Spike implies that others can visit the Moon without national agency support, opening many commercial options including lunar resource utilization. Let us examine the possibilities.

How do we mine water on the Moon? We considered four basic water reservoirs: rocks, the PSRs' uppermost meter, deeper regolith, and the uppermost surface, each

---

Figure 12.2. (cont.)
surface-to-orbit shuttle (*left*) that transports them to a cislunar tug (*center*) in a lunar parking orbit; (2) the tug transports itself to Earth in a fast Hohmann transfer and enters geosynchronous orbit to rapidly push satellites into their useful orbital position (and refuels others) thereby extending their useful life; (2) with most of its propellant spent except for its long duration storage, the tug slowly returns to its lunar parking orbit via a low-energy weak stability bound transfer.

requiring its own extraction technique with various advantages and difficulties. The science must be explored behind all four before mining begins in earnest.

We know of abundant water, a billion tonnes or more locked in PSRs, comprising 6% of local soil with other large volatile contributions. These are frozen, requiring a regolith-heating distillery to sublime pure water (as well as carbon monoxide, molecular hydrogen, hydrogen sulfide, and ammonia). Primary hazard of these environs is temperatures only several dozen degrees above absolute zero. Machines operate with difficulty because of brittle metals and hardened synthetics. Vacuum thermal contact is low so high-powered machines without radiators may overheat in even super-cold environs. Even machines heated while operating might later break as a result of thermal contraction when they are turned off in this frigid landscape.

If some volatiles reach polar regions not from the vacuum, but from deep interior, they accumulate down in the regolith, warmer because of temperature gradients sustained by lunar internal heat. Water ice from outgassing might form several to 10 meters deep, depending on the site's latitude. Temperatures here tend to be –100°C to –150°C, more Antarctic than PSR cryogenic deep freeze. Although Antarctica is no machinery paradise, we have experience with various transports, drills, front-end loaders, snowblowers, fork lifts, cranes, bucket loaders, graders, cement mixers, bulldozers, and even aircraft under these conditions (with preparation and precautions). Spacecraft in Earth shadow routinely enter such temperatures, as does hardware during lunar night.

We still know little of how outgassing volatiles collect and can only guess at gases involved: water and sulfur dioxide most likely, perhaps carbon monoxide or dioxide. As in PSRs, these might be excavated by trenching, except several times deeper. Unlike PSRs, the environment requires thermal protection, and solar panels are easily used nearby. Of the water-bearing environments we are discussing, this is least familiar. We are unsure its hydrogen (sensed by epithermal neutron absorption; Chapter 8) is a result of simple water ice, water trapped in clays (or other phyllosilicates), or hydrated lattices of typical regolithic minerals. These are set temperatures to which solids must be heated to extract water: about 0°C, 500°C, or 800°C, respectively.

Regolith surface bulk heating can extract loosely bound water, tested by microwave radiating ice-infused regolith simulant.[20] Ice sublimes in a vacuum at rapidly increasing rates as temperature reaches 0°C, and microwaves penetrate roughly 1 meter of regolith (depending on wavelength), tunable to the volatile depth. One can collect water vapor on a cold plate over the soil surface or a collection port manifold leading to a cold trap. This is energy intensive: in LCROSS's impact environment (–220°C, water in at least top 3 meters), to heat a square meter regolith area to 0°C 3 meters deep requires the energy from a square-meter solar panel for one week at 100% efficiency, which in reality requires months. In return this frees about 280 liters of water and similar amounts of other volatiles, mostly carbon monoxide. These might be collected at 90% efficiency.[21]

There is abundant solar power on the Moon but ways to surpass this. The muscular approach is nuclear. Many space reactors were developed since the 1960s, the largest producing 100 kilowatts.[22] Such a plant in a PSR promises several dozen kilograms of water daily (and electrolysis to LH2 and LOX requires doubling this power). Toshiba built the compact 4S nuclear plant (Super Safe, Small, and Simple, with a 1-by-2-meter core), delivering 10 megawatts for 30 years; 50 megawatts is promised. The alternative is collecting solar power and transmitting it to the PSR, requiring many kilometers of cable or microwave beams (at levels yet undemonstrated).[23]

Alternatively (in polar regions if not PSRs), ambient heat can process excavated ice-bearing regolith, implemented by deploying closed bags as greenhouses outside the excavation trench and feeding excavated regolith into them. In the trench, temperatures are close to ambient average, 150K for example, but in sun-heated bags, they reach 250 K or more, subliming water ice. This could support scientific investigation, with gas sampled and analyzed escaping each bag into plumbing that feeds the volatile processing plant, its composition measured versus its position from the trench. This process would be slower but low cost.

Large temperature differences on the Moon suggest heat engines for power, with gradients exceeding 200 K over several meters.[24] Regolith's superb thermal insulation frustrates this: areas are cold or hot, but equilibrate, reducing this gradient. Soil temperature gradients might provide some power, for example, for scientific applications or monitors, but special geometry must maintain shadowed/daylit gradients over the month. Habitation and industry needs power more muscular and consistent. For cold reservoirs (heat dumps), vacuum insulates machinery unless it has huge radiators, so almost any power generation requires large areas, even without solar panels.[25] Fabricating huge solar arrays is difficult; a simpler way to collect sunlight is often with mirrored solar concentrators.

One justification for nuclear power is continuity through 2-week lunar nights, but electrical energy is storable as LOX/hydrogen and efficiently recovered via fuel cells. The true advantage of nuclear over solar is high temperature operation requiring smaller radiators, and easier night operations than with fuel cells. The nuclear route started with NASA's 1960 Systems for Nuclear Auxiliary Power (SNAP) power supply program using heat from radioactive decay (not reactors) for up to 100 watts, and reactors up to 1 megawatt in tests and 500 W in flight. (The United States and Soviets had larger reactors, in secret.) NASA more recently worked on small reactors of 2.3 kW and eventually 50 kW, based on efficient and simple Stirling heat engines, using gaseous working fluids but liquid sodium/potassium heat transfer fluid (highly advantageous, remaining liquid over huge temperature ranges).

Having discussed power, we keep digging. Trenches can be dug telerobotically more simply than in Earth's mines today (more about this later). Lunar trenches can be more steep-sided, with regolith maintaining nearly vertical walls several meters high and arbitrary deep 40° from horizontal. A regolith trench 5 meters deep takes 190 cubic meters minimum (300 tonnes), without wall braces. Lunar machines have direct terrestrial analogs: backhoes dig 10% of their own mass per load, and (front-end) loaders 25%, determined not by gravitational force but by balance (a force ratio). Similar lunar limits apply. Lunar designs encounter reduced friction: active diggers replace blades – wheels must grip. Dump trucks on Earth carry 100%–200% of their empty mass, even more on the Moon. Trucks radically cut trips in big excavations; diggers and trucks are both necessary. From earthly experience, two 1-tonne machines could move 500 tonnes daily, one tele-operator driving both.

We overlooked utilization of water from solar wind or imbedded in rock, likely involving different extraction techniques. Solar wind only affects particles in the regolith's upper millimeters, impact-mixed over æons into the first few meters. In contrast water-bearing minerals from magma might extend far underground, or locally spread thinly on the surface, for example, by fire fountains, and might be strip-mined. Both hydration sources extend far from the poles.

Lunar hard-rock mining demands more machines: delineation drills to find ore, tunneler, digger, explosive loader, and dump truck. One might fit the mine entrance with an air lock, more resembling terrestrial mines (except lower gravity). Low gravity helps; few Earth mines exceed 2 kilometers depth because of overlying strata's weight, energy to lift rock, and high temperatures. Lunar mines would leak no water or methane (however, would still be hot at depth – but not likely hotter than Earth mines and not too hot for robots).

Lunar surface mining is a true surface operation (whereas on Earth surface mining means strip-mining – removing overburden from ore below). Solar wind products extend down several meters, so one digs only that deep. Various surface miners could harvest many commodities (water, helium-3, hydrogen, carbon, and nitrogen – all from solar wind, oxygen, and other gases). For example, the 10-tonne Mark III Lunar Volatiles Miner, a bucket-wheel excavator feeding a 700°C furnace and distillery, processes 1–2 tonnes per day, producing all these gases (except oxygen) and helium-4, methane, carbon dioxide/monoxide, enough for 750 humans, from regolith's top three meters.[26]

We reemphasize our ignorance of volatile sources and locations: from comets and asteroids frozen in PSRs, from the interior frozen deep in regolith or trapped in rock, or from the solar wind in regolith's surface. We are ill prepared to attack the Moon with a backhoe and dump truck, exploiting these resources. Several new missions, some yet unplanned, must explore these sites. However, the Moon is the one body beyond Earth that could profitably provide resources in the next decade. Mars is hundreds of times farther, and near-Earth asteroids, sometimes close, can only be visited briefly before retreating across the Solar System for years. We are far from utilizing other worlds, with notable exception of the Moon. Furthermore we can imagine installation of infrastructure for telerobotics as well as for other purposes. Machines mining one month can be used the next month to excavate lunar bases or human habitats. An active robotic lunar base might serve as the foundation and the reason for the arrival of humans on the Moon, long term (Figure 12.3).

The Moon is a scientific storehouse (Chapters 6–9), but unlike Mars (or New World 500 years ago), is unlikely a vital ecosystem vulnerable to reckless incursion. This must be checked, but for purposes of scenarios it is likely. The Moon holds material wealth and notably, energy. Fundamental components, 45% oxygen and 22% silicon, are basic constituents of simple solar cells (and another 25% in structural metals: iron, aluminum, magnesium, and abundant trace elements for electrical parts) and might supply large energy requirements for lunar operations. In some cases if developed, these resources could also benefit Earth.

Lunar Power

Figure 12.3. **Return to the Moon for adventure and profit**. Artist's conception of arrival of a human crew at a robotic base and surface mine near the lunar South Pole. The robotic base has been operating for several years, and humans have arrived for an extended stay to live off the water, LOX, and LH2 produced by the robots, to conduct science, and to upgrade, maintain, and repair the base's equipment. From left to right, one sees the Sun always close to the horizon, the newly arrived lander, its international crew of four, three solar-panel towers (turning to follow Sun), supply lander (background), a regolith-shielded electronics shelter with antenna pointed earthward, bucket-wheel excavator/processor needing repair after jamming in the regolith, LOX/LH2 distillation tanks and transfer rover, another bucket-wheel excavator in action, Earth, and another trio of solar towers on the background hill. (Composited from images from NASA, Helena Uthas and the author.)

The first output from lunar chemical industry will probably be oxygen. The simplest approach is high temperature decomposition of regolith (45% oxygen) or vapor phase pyrolysis,[27] releasing about 25% of the oxygen, at 2,000°C. One can use much lower temperatures if more discriminating in the choice of regolith mineral. Ilmenite composes up to 20% of mare basalts in some places and is reducible with hydrogen: $FeTiO_3 + H_2 \rightarrow Fe + TiO_2 + H_2O$. The water can be electrolyzed, of course, so hydrogen is conserved in net: $2\ FeTiO_3 \rightarrow 2\ Fe + 2\ TiO_2 + O_2$. Temperatures near 1,000°C are needed. In the process iron is produced, separable using magnets. Similarly one could also reduce the titanium.[28]

A larger promising realm is symbiosis between telerobotics on Earth and the Moon. Many conditions on Earth are too difficult or dangerous for humans: deep ocean; deep Earth; mountaintops, for example, high Andes mines; and local hazards. In modern mines, machines are often controlled remotely by humans via telerobotics, but the machines know their precise position.[29] The concept of "autonomation" (in Japanese *jidoka* – 自働化) describes humans and robots working together to identify errors and develop robotic ability to recognize and fix errors autonomously next time. On the Moon situations are beyond the factory floor:

changing exploration conditions require situational awareness that humans have but is augmented by artificial senses and data channels from robots. This helps on the Moon but feeds back into general robotics with likely economic spin-off. In mining and many applications, costs are mainly labor; this will reduce them (see Chapter 14).

Some lunar resources might seem abstract but might prove even more significant. The 33 kilograms of helium-3 the Mark III Lunar Volatiles Miner could collect annually has current market value of $500 million.[30] Helium-3 is vital in MRI lung research, low-temperature experimental physics, and neutron detection to find illicit nuclear material. Prices have skyrocketed recently and might continue. In a commercial fusion reactor, helium-3 could fetch more. Reacting with similar numbers of deuterium atoms, this amount would yield 5 terawatt hours of power, more than $1 billion at today's electricity prices, supplying U.S. needs for a day (assuming an unrealistic 100% electrical conversion efficiency).

Current fusion reactors force reactants together via violent implosion by laser bombardment or by holding charged particles together in intense magnetic fields. The latter approach may reach levels where helium-3 is first relevant. Today's International Thermonuclear Experimental Reactor (ITER) under construction and upcoming IGNITOR cannot achieve commercial-scale fusion. The Demonstration Power Plant (DEMO) by 2030 then PROTO by 2050 are needed before reaching commercial levels. These will burn heavy hydrogen isotopes (deuterium and tritium), not helium-3; only successor reactor generations might use helium-3. This is plausible; helium-3 fusion (^{3}He + ^{2}H → ^{4}He + ^{1}H + 18.4 MeV) makes few neutrons, whereas hydrogen fusion (^{3}H + ^{2}H → ^{4}He + n + 17.6 MeV) emits profuse neutrons.[31] Neutrons at these energies can damage reactor metal lattices, and being uncharged provide little means to contain them. Unfortunately, whereas tritium/deuterium reactions require temperatures around 300 million K, deuterium/helium-3 requires about 1 billion K – much more difficult.[32] Later reactor generations might burn boron-11/hydrogen (both common on Earth), much hotter still.

Likely sometime around 2050 the world will grow vitally interested in helium-3 for fusion reactors, with little prospect for terrestrial supply absent probably prohibitive expense. With no supply, helium-3 fusion is moribund, and fusion will languish with damaged tritium/deuterium reactors needing constant replacement/refurbishment, creating profuse radioactive waste and mocking clean fusion power's promise. Russia, China, and India express interest in mining lunar helium-3.[33] Will lunar surface mining fill the bill?[34] For a generation of reactors (20 years or more), the Moon may provide the world with clean, consistent, and abundant power without adding atmospheric carbon dioxide. This will arrive when global warming's deleterious effects are manifest, with great efforts to prevent its worsening. This seems far in the future, but solutions require decades to employ. History suggests patience: in 1492 Spain's New World economic goal was spices, then gold, but world-changing wealth ultimately derived from silver a century later.

Helium-3 power and fusion rouse varying attitudes. They are so clean we should regret not implementing them already. Scientists have built fusion reactors for five decades, since the global energy crisis was apparent. With global warming, fusion would be handy in our toolbox. Coal and petroleum can kill Earth if we rely on them more. Natural gas is limited and still greenhouse-producing. Fission is expensive, produces radioactive waste, and encourages arms proliferation. Renewables (wind, solar, hydro) are intermittent and costly, with their own environmental impacts – many severe. Commercial fusion is the sole intensive, on-demand, low-waste, non-greenhouse option, but it does not exist. Current annual global human energy use is 200 quadrillion watt hours, corresponding to lunar helium-3 from 26,000 square kilometers.[35] If helium-3 power needs 40 years of research and lunar industry to develop, it may be too little, too late, and too expensive. However, 2050 will arrive, and we will need energy; helium-3 fusion may then seem wise, if we have not dawdled.

The first hardware produced from lunar industry might be solar cells for electrical power, simple photodiodes from silicon (with the Moon low in alternatives: indium phosphide, gallium arsenide, copper indium selenide, and cadmium telluride). On Earth silicon is reduced from SiO_2 by carbon; on the Moon aluminum would suffice. Lunar SiO_2 is rare in pure form but common in former melts, for example, anorthite ($CaAl_2Si_2O_8$). The next largest mass elements are structural: lunar iron, aluminum, and titanium. Two dopants must be added to silicon: phosphorus is common (several tenths of a percent, especially in KREEP terrane). The acceptor dopant could be aluminum[36] but on Earth would be boron or gallium, less than 0.01% of lunar regolith. Auxiliary electronics need these and other elements, some common, for example, REEs. These could cut dependence on Earth imports, lowering cost.

The dream that lunar-made solar cells might someday power Earth's economy is a powerful one. Solar energy on Earth is dimmed by atmosphere, shadowing, and geometry from an incident 1,365 watts per square meter to 250 on average. To power global energy use would require some 700,000 square kilometers of solar panels, the size of France. In space it could shrink by a factor of four, on the Moon by two. Peter Glaser pioneered space-based power concepts in 1968, popularized and expanded by Gerard K. O'Neill, incorporating lunar manufacturing concepts. Congressionally mandated NASA and energy department studies in the 1970s found the venture risky and requiring more study; the incoming Reagan administration halted work. Several recent follow-ups were made by NASA.[37] David Criswell champions construction and fixture of solar arrays on the Moon, beaming economically significant energy to Earth via microwave antennae on either limb of the Near Side.[38]

It is a massively ambitious dream: assuming 10 grams per watt, the panels approach 100 million tonnes. Even if the system supplied only U.S. electrical power, it would top 10 million tonnes, several times the mass of all U.S. Navy ships. Launching this from the Moon with LH2/LOX would use roughly 20 million tonnes of propellant, 2% of known lunar water deposits. Nonetheless, it could save Earth from global warming's worst effects. Such scenarios require an established lunar colony, largely self-supporting (or at least in balanced two-way trade with Earth). The Shimizu Corporation proposes robotic (but human-tended) solar panel factories moving slowly around the lunar equator, converting regolith to solar panels, spanning the Moon with some panels always illuminated, and centering on a transmission station with a kilometers-wide microwave antenna to power Earth. The 4-by-10,920-kilometer expandable strip starts with one-seventh the length and ten times the area of the U.S. Interstate Highway System. While the bulk of the material could be derived from native lunar materials (silicon, glass, concrete, ceramics, aluminum, iron, titanium, etc.), there are probably millions of tons that might be needed from Earth, for electrical contacts, dopants, and so on.

Are there potential energy resources on smaller scales? We discussed radioactive KREEP terrane (Chapter 6); is there nuclear power potential? This radioactivity is less than it sounds. On Earth and the Moon, thorium has about three times uranium's abundance. Gamma rays reveal thorium concentrations of 10 parts per million in southern Procellarum and outer Mare Imbrium with no large area more than 13 parts per million, and around Aristarchus to 17 parts per million.[39] Thorium-rich Apollo samples reach 30 parts per million.[40] Earthly thorium often resides in thorite, $(Th,U)SO_4$, in volcanic extrusion, hydrothermal veins, or sediment. High thorium concentrations, up to 10%–20%, arise in monazite, a phosphate rich in REEs: $(Th,Ce,La,Pr,Nd,Y)PO_4$.[41] On Earth, monazite and thorite collect in sediments, being dense (given uranium and thorium's mass) and sinking when agitated in water (as opposed to precious metals concentrated by water into veins). This cannot affect lunar thorium, and deep deposits cannot rise via plate tectonics. However, with magma ocean KREEP behavior, thorium concentrates with phosphorus and REEs (and water). Thoriated lunar monazite fragments exist,[42] perhaps we can find a mother lode.

Thorium concentrations of 0.01% are not utilizable. Can we find higher abundances? Thorium gamma-ray maps only resolve details larger than 150 kilometers;[43] we need finer maps. We cannot detect thoriated mineral veins directly, but radioactive debris in regolith spread by impact over tens and hundreds of meters. This requires a special mission, at low altitude or with a high-resolution telescope – probably the former.[44] High-abundance thorium regions cited cover 400,000 square kilometers; a 100-meter resolution map is challenging.[45] Even on Earth, aerial surveys often lose out to long walks with a scintillation counter.[46]

Thorium nuclear fission is promising but commercially undeveloped, likely because of its start-up costs and military irrelevance. Any thorium mined is useable, needing no enrichment like uranium-235. Thorium-232 (the sole abundant isotope) and neutrons produce uranium-233, fissile fuel suitable for reactors. A dozen thorium test reactors were built since 1960, and India is building research thorium breeder reactors.[47] Powering lunar activities via thorium is interesting and could fuel nuclear rockets, potentially highly efficient and fast. Unfueled reactors from Earth then fueled from lunar fissiles avoid nasty possibilities launching large, radioactive payloads. Abundant lunar hydrogen might provide the optimal reaction mass, for superior high performance, efficient propulsion to explore the Solar System at large.

More determines lunar commerce than there being the will and a way; there is the law – the OST.[48] It bans countries from lunar sovereignty, but does not prevent private or commercial concerns from exploiting the Moon. It forbids anyone from excluding others from parts of the Moon, including areas being mined. The boundary separating what is owned and not is unclear: obviously if you eat plants grown in lunar soil they belong to you, but what of lunar-derived propellant, lunar-manufactured hardware, regolith shaped for your lunar base, or your roads and regolith berms? Furthermore, the treaty says who can interfere with lunar operations and to what extent. Your own national government can tell you what you can and cannot do on the Moon. If another country thinks you endanger, contaminate, or interfere with operations or the lunar environment in general, you must allow them to consult with you and permit them access to wherever on the Moon. Furthermore, persons on the Moon are mandated to "render all possible assistance to astronauts" of other countries, and inform them of any danger to their health and safety.

Exploiting lunar resources will require precedent to be established, given current law's ambiguity. Despite the OST, reasons exist to restrict access of craft to some lunar areas. Test cases will likely begin with the Google X-Prize lunar challenge. X-Prize participants receive a bonus for photographing Apollo hardware with their rovers. This may establish first precedent between OST's everyone-can-go-anywhere language and likely desired lunar commerce usage. NASA promulgates non-binding rules for how fast rovers can approach existing lunar artifacts to prevent throwing regolith on them and exclusion radii around extant Apollo and other equipment and sites to preserve their scientific and historical value.[49] Urgently, lunar laser retroreflector arrays at three Apollo and two Lunokhod sites should remain uncontaminated. Settlement of disputes under OST is virgin legal territory, considered only in the 1978 crash of Soviet radar satellite *Cosmos 954* with its nuclear power plant into Canada's Great Slave Lake. (The shire of Esperance, Western Australia, facetiously fined NASA $400 for debris strikes from *Skylab*'s reentry.) A dispute-solving mechanism is needed.[50] This will grow drastic because any lunar activity generates some exospheric pollution, which then partially transfers to the PSRs. With planned polar probes, this is urgent.

A second treaty, the Moon Agreement of 1979, has largely failed, with only 13 countries party to the agreement and four signatories, excluding the United States, Russia, China, Japan, and most of the EU (but including India and France).[51] Were the Moon Agreement in effect, it would reserve all lunar resources as "common heritage of mankind" and remove any property rights, even for commerce. It calls for international regimes to control lunar resource use, but no such action has occurred. With no treaty in effect, custom and precedent will establish international law. The spacefaring nations must write new lunar law – accept this one – or at least officially renounce it. By ignoring the treaty the United States, Russia, China, and most spacefaring nations limit their influence over it. If they signed but did not ratify it, they could work to change it.

In sum, the Moon Agreement is too restrictive were it to be ratified, discouraging lunar economic exploitation until international control is established, whereas the OST might allow too much. Although the OST prevents states from establishing lunar sovereignty, it says little about what corporations can do in their behest. This may be established in practice by the first group or two to establish lunar industry. If for no other reason this should motivate the United States to return to the Moon, to establish facts on the ground as they did in creating peaceful human presence in Antarctica in 1956 or passively by not protesting Sputnik's overflight a year later.

Maybe international consortia including the United States could not only set legalities but also combine efforts to explore the Moon and establish a lunar base, refining rules in practice.

There is faint call by likely participants for international Moon bases harnessing humankind's common heritage. Despite this, the Moon is the most popular destination discussed by International Space Exploration Coordination Group members – except the United States. Recently international lunar base proposals usually include fostering commerce. Opposed, lunar commerce advocates often disparage government control, endorsing libertarian private enterprise.[52] Government agencies have unique capabilities; we could hardly explore the Solar System without them. International collaboration succeeded in the *International Space Station* (*ISS*), including most spacefaring nations.

Not legislating ideology in lunar development, law should foster diverse effort: private, national, and international. The 1979 Moon Agreement arguably attempted imposing global socialism off-world in an age of efforts dominated by one (or two) national players; it should be rewritten. Considering humankind's common heritage, recall that most people now live in spacefaring nations. However, for centuries on the frontier, governmental-commercial partnership has succeeded; this should be fostered. If the United States reconsidered its stance to de-emphasize lunar exploration, it could lead global effort bringing commerce and international partners together in combined force.

Consider what might occur if we develop the Moon. With inexpensive lunar propellant and structures, the Moon could emerge as a gateway world to the Solar System. This would grow important, even critical. If lunar resources rise to significance in solving Earth's combined energy and global warming crises, the Moon could become politically powerful. A Moon integral to a global fusion economy has a window, perhaps around 2050, in which helium-3 is crucial, but boron-11 fusion is not yet viable. The Moon may or may not serve in providing reliable solar power.

With a reason to inhabit the Moon, will people stay long enough to belong? If individuals spend only small parts of their lives, there is no polity. Will children be born and families formed on the Moon, or will lunar society remain sterile? Economically this depends on the cost of return transportation to Earth and mandates of corporate or government control over terms of personal service. Many people would want to live much of their lives on the Moon if conditions were right, largely depending on resources used making habitation comfortable.

One model is the *ISS*, where several astronauts cycle through stays under one year before returning to their normal life. Stays on *Mir* reached a year or slightly more. On mainland Antarctica (exclusive of the Antarctic Peninsula) some stay several months, some a year, few more than a year but sometimes two – all among several thousand residents. On the Antarctic Peninsula people stay many years; children are born and mature. There is no easy, exact analogy to potential lunar communities; military bases or company towns provide interesting lessons.

How big must self-sustaining lunar communities be? A lunar lander crew or two will not suffice. The 2-year Biosphere 2 crew of eight highly trained and specialized individuals, each with unique and essential responsibilities (medicine, machine maintenance, agronomy, and waste recycling) proved overworked and insufficiently experienced (regarding farming and managing ecological balance). More recently the Mars 500 experiment put three international crews of six in succession in a small habitat (1,000 square meters) in Moscow for up to 520 days at a time. The Mars 500 experience was not as fraught as Biosphere 2, but neither did they work so hard to produce their own consumables for survival. To succeed, societies need greater experiential depth in essential skills and must allow inhabitants rest and vacations; a group of 8 multiplies by several, to about 30. In real society, not all live in their prime, ready to perform essential services. Children, very elderly, and ill will not work full days benefiting other members and require teachers, caregivers, and other providers, bringing collective minimum size to about 100. Medical teams providing ranges of care expected in developed societies number in the dozens, despite automation. Unlike Biosphere 2, built with new hardware ready for optimal performance (subject to the project's design concept), society in general needs more investment in maintenance. Inhabitants will also expand their environment. Although much could be performed robotically, it probably suffices to keep the community size only doubling, not tripling. Minimum communities grow to hundreds. Furthermore, the lunar engineering and scientific frontier demands and encourages new approaches, with many practitioners Moon dwellers, not earthlings.

A stable community needs several hundred to several thousand. This seems confirmed in many respects: the first villages held several hundred to several thousand.[53] This approximates sizes of self-contained political entities (tribes).[54] Pre-republic Roman tribes tended to several thousand, as did North American tribes before European arrival; North American nomadic tribes first held perhaps several hundred. Greenland's Viking colony started at several hundred, reaching several thousand before failing after four centuries, presumably from climate change. In contrast, English colonies in North America tried to sustain themselves for several years with about 100, and sometimes failed (as in Roanoke, and nearly in Jamestown and Plymouth). A minimum might be 300, without robotics; we would build a tribe on the Moon. In the twenty-first century, issues differ: not only is work done (tele) robotically but also cooperatively with earthlings and people communicating on both worlds via network social media. However, when community members gather, tribal dynamics will be relevant.

A community large enough to survive in isolation can also bifurcate: a problem with Biosphere 2's group of eight, or Easter Island's 15,000. Obvious centrifugal forces could pull apart a lunar society or sever it from Earth: effects as simple as Near Side versus Far Side (Chapter 11). Society on Earth cycles through introspective phases because of war or recession, and attitudes toward the Moon will vary. Conversely a prompt impact in 1969 of human lunar presence was the Earth Day effect, which allowed us to look on our Earth as an isolated, unified whole, to view politics, economics, and our place in Earth's environment differently. We must consider in advance how to minimize wasteful conflict if lunar society grows to provide vital resources for Earth. Should a lunar society be a company town? A command economy? Democracy? Humanity met such issues in recent centuries here on Earth; but few today remember the lessons directly.

Over a few generations' isolation, motherland can become Other Land. Pioneer leaders often realize intimately their dependence on the home society, even in cases such as Plymouth Colony, who in 1620 were escaping religious persecution and volatility. True political separation came, famously, a century and a half later, and throughout the Western hemisphere in 1804–1825, and thereafter. Many overseas societies are still politically associated with their home society after several centuries. In contrast Polynesian expansion created politically distinct pioneers but only in settings where commerce or communication was impractical.

Human society reinvigorates itself by the rise of the edge, people on the periphery, who incorporate strengths of central society but add their own. The succession of civilization along the Tigris-Euphrates or in India, the Kushites over the pharaohs, the Zhou over the Shang, the Yuan over the Song, the Manchu over the Ming, Rome over Athens, Britain's growth versus France, America relative to Britain/Spain is how the new replaces the old. It is rarely pretty and not always an improvement but requires a group in contact with but not dominated by the central society. It is less evident on today's Earth how this will happen. Depending on how humans move off Earth, this process may proceed anew.

If we establish a community on the Moon, politics will be important and interesting; history will continue. We should look forward to what the Moon will provide, in material resources and scientific insight and in stimulating the human spirit. Let us discuss in the next chapters the role the Moon holds for the future in the next few decades.

Notes

1. "Locksley Hall" portrays the reverie of a soldier stopping on his march as he passes his childhood home. Tennyson assuages the disappointment in his past and present by dreaming of future progress that will end all war and advance technology via united humanity:

   > "Till the war-drum throbb'd no longer, and the battle-flags were furl'd
   > In the Parliament of man, the Federation of the world.
   > There the common sense of most shall hold a fretful realm in awe,
   > And the kindly earth shall slumber, lapped in universal law."

   In 1886, aged 76, he penned "Locksley Hall Sixty Years After," bitter with lack of progress:

   > "Warless? War will die out late then. Will it ever? Late or soon?
   > Can it, till this outworn earth be dead as yon dead world the moon?
   > Dead the new astronomy calls her... On this day and at this hour,
   > In this gap between the sandhills, whence you see the Locksley tower,
   > Here we met, our latest meeting – Amy – sixty years ago –
   > She and I – the moon was falling greenish thro' a rosy glow,
   > Just above the gateway tower, and even where you see her now –
   > Here we stood and claspt each other, swore the seeming-deathless vow...
   > Dead, but how her living glory lights the hall, the dune, the grass!
   > Yet the moonlight is the sunlight, and the sun himself will pass."

2. This has been long imagined. Tsiolkovski envisioned a lunar gold mine as early as 1895 (*Dreams of Earth and Space – Grëzy o Zemle i nebe: antologiià russkogo kosmizma* by Konstantin Tsiolkovski, 1895 (1995 reprinting), St. Petersburg: Khudozh. lit-ra).

3. Dennis Wingo argues that platinum group metals are crucial to expanding a hydrogen economy and expects prices to hit lunar levels (*Moon Rush* by Dennis Wingo, 2004, Burlington, Ontario: Apogee, p. 83).
4. Effort required to reach LEO from Earth, expressed as velocity change (Δv) is 9.3 km s^{-1}, accounting for aerodynamic drag, and 3.9 km s^{-1} from LEO to GEO. In contrast, GEO to Low Lunar Orbit (LLO) is 2.0 km s^{-1}, then 1.7 km s^{-1} to the Moon. For high-performance propellant, e.g., LOX + LH2, a trip to GEO from Earth requires 10–15 times more propellant than from the Moon. LEO to LLO is 4.1 km s^{-1}, so even reaching LEO from the Moon requires about 2.5 times less propellant. Maintaining a GEO satellite on-station is a slow drain, ~0.05 km s^{-1} per year.

    Compare these to LLO to Lagrangian points (Chapter 13) around Earth and the Moon, with velocity changes of less than 1 km s^{-1}, and among the Lagrangian points themselves, several hundred m s^{-1}. Harder is LEO to Mars' vicinity at 4.3 km s^{-1}, and 1.4 km s^{-1} to low Mars orbit, and 4.1 km s^{-1} to land on Mars (often done with aerobraking). From LEO to the easiest near Earth asteroids is 3–4 km s^{-1}.
5. A rover likely requires at least a 20 km traverse from these PSRs to sunlight, where hydrogen would be used, as would a pipeline. (We discuss transportation alternatives further.)
6. HTP is made by several methods, notably using variants of organic molecule anthraquinone as a catalyst ($C_{14}H_8O_2$, three carbon hex-rings with two oxygen atoms double-bonded to the central ring). Basically, hydrogen bonds with the two oxygens to form hydroxyl groups, then added oxygen makes these two groups react, producing H_2O_2, with the two oxygen atoms unaltered. This could be implemented on the Moon, because the catalyst survives and little terrestrial input is need. H_2O_2 catalysis research continues and could improve significantly in coming years.
7. Portable LH2 tanks lose 1%–3% per day ("Development of Automotive Liquid Hydrogen Storage Systems" by G. Krainz, G. Bartlok, P. Bodner, P. Casapicola, Ch. Doeller, F. Hofmeister, E. Neubacher & A. Zieger, 2004, *Advances in Cryogenic Engineering, Cryogenic Engineering Conference*, 49, 35). The Centaur currently loses 2% per day; ICES may cut this to 0.1% per day.
8. Imagine needing to move a distance L on the Moon, in n in-line hops of distance x across the surface: $L = n\,x$. For the most efficient trajectory, a 45° hop, initial velocity perpendicular to the surface, $v_\perp$, equals velocity parallel to the surface, v_{11}; call it v. Maximum height reached on the hop is $h = {}^1/_2\,a\,t^2$, where t is time required and a the acceleration of lunar gravity, 1.622 m s^{-2}. Because $v = a\,t$, $h = {}^1/_2\,v^2/a$. For a 45° hop, it turns out $x = 4\,h$, so $x = 2\,v^2/a$. For small v, fuel usage is proportional to v, so fuel efficiency $\varepsilon = x/v \propto v \propto x^{1/2}$ ($\propto$ means "proportional to"), implying x should be maximally large. The most efficient hopping to move distance L is simply one big hop of $x = L$. Hopping halfway around the Moon corresponds to orbital velocity (1.683 km s^{-1}), as expected. (For hops this large, idealizing a flat surface breaks down, but one big hop is still most efficient.) For O_2+H_2, 500 km hops require 17% as much propellant as the spacecraft's empty mass. A hop halfway around the Moon requires about three times as much propellant. Gliding over the surface on rockets can be considered many small hops, inefficient unless the craft's forward velocity approaches orbital speed. With many lunar mountains, one cannot easily "glide" near the surface at such high speeds. Note that engineers also considered an energy-recycling hydraulic hopper that could jump 130 meters at a time. (*The Lunar Hopping Transporter* by R. Degner et al., 1971, NASA-CR-130010)
9. Water is typically 11% hydrogen by mass, but increasing hydrogen fuel ratio decreases exhaust molecular mass and subjects rocket nozzles to less oxidization, increasing their working life.
10. "Conceptual Design of Hybrid Rocket Engines Utilizing Lunar-derived Propellant" by D. Brower, W. Adams, T. Kelly, C. Ewing & T. Wiersema, 1990, 26th Joint AIAA/SAE/ASME/ASEE Propulsion Conference, 90–2114.
11. "Detailed Characterization of Al/Ice Propellants" by Timothée L. Pourpoint, Travis R. Sippel. Chris Zaseck, Tyler D. Wood, Steven F. Son, Grant A. Risha & Richard A. Yetter, 2010, 46th AIAA/ASME/SAE/ASEE Joint Propulsion Conference, Nashville, Tennessee, AIAA-2010 – 6905.
12. To make methane from available gases, perhaps best to start using oxidation $2CO + O_2 \rightarrow 2CO_2$ via catalysis ("Au-Cu Alloy Nanoparticles Confined in SBA-15 as a Highly Efficient Catalyst for CO Oxidation" by Xiaoyan Liu, Aiqin Wang, Xiaodong Wang, Chung-Yuan Mouc & Tao Zhang, 2008, *Chemical Communications*, 3187), then experimental methods for methane from CO_2 and water via nanocatalysts and light: $CO_2 + 2H_2O \rightarrow CH_4 + 2O_2$ ("High-Rate Solar Photocatalytic Conversion of CO_2 and Water Vapor to Hydrocarbon Fuels" by Oomman K. Varghese, Maggie Paulose, Thomas J. LaTempa & Craig A. Grimes, 2010, *Nano Letters*, 10, 750).
13. *Satellites to be Built and Launched by 2018, World Market Survey* by Rachel Villain, 2009, Paris: Euroconsult.
14. More can be found out about this idea in connection with our U.S. Provisional Patent.
15. Honeybee Robotics Ltd. has proposed the Icy-Soil Acquisition and Delivery System and Volatiles Extraction and Capture Systems as small prototypes for such capability http://sbir.gsfc.nasa.gov/SBIR/abstracts/11/sbir/phase1/SBIR-11-1-X1.01-8468.html.

16. The median GEO satellite mass is 5 tonne, and restoring 40 m s^{-1} will require about 45 kg of propellant per year. For 150 clients this is 7 tonnes at Earth, requiring 20 tonnes of propellant at lunar orbit and 40 tonnes at the lunar surface, exclusive of spacecraft mass (using the rocket equation)[13,16]. Including spacecraft increases this to 60 tonne (per year).

17. MacDonald, Dettwiler & Associates Ltd. (MDA) planned launching a satellite with a propellant load to dock with a GEO satellite, surgically insert itself into the satellite's propellant lines and recharge it, adding 3–5 years to its useful life. In March 2011 Intelsat set to purchase 50% of the spacecraft's propellant cargo for $280 million, delivered, to save four or five satellites by about 2015, but cancelled the deal in January 2012 because of international contracting issues. ATK and U.S. Space propose the Mission Extension Vehicle competing with MDA's, except it actually attaches to the GEO satellite instead of refueling it. It can also restore (or initially achieve) desired satellite orbit. DARPA and NASA tested a similar prototype craft in 2007 (ASTRO/NEXTSat). In 1987 NASA started work on a Flight Telerobotic Servicer to help assemble the space station. A search in 1989 for commercial customers for their servicer found none. In 1990 the project became the Robotic Satellite Servicer before work on free-flying space station support was cancelled. NASA studies further options, http://ssco.gsfc.nasa.gov/robotic_servicing_mission.html.

18. Ideas for propellant depots refilled from Earth are popular, with several proposed and more studied, potentially obviating heavy-lift boosters by spreading upper-stage propellant over several launches. MDA's system would demonstrate a propellant depot. While NASA decided in 2005 to not pursue lunar orbital depots as part of the return to the Moon, it has contracted several companies to develop depot concepts in Earth orbit and beyond (Boeing, Lockheed Martin, Ball Aerospace, and Analytical Mechanics Associates). A relevant project is DARPA's Phoenix program, to harvest functioning antennae, solar cells, and useful hardware from defunct satellites in/near GEO, with a telerobotic mission by 2015 to visit, grapple, and dismantle a GEO satellite. NASA and the Canadian Space Agency are experimenting on *ISS* with the Robotic Refueling Mission using Dextre (see Figure 10.9) to simulate operations to repair and service GEO satellites remotely.

19. "Using the Resources of the Moon to Create a Permanent, Cislunar Space Faring System" by Paul D. Spudis & Anthony R. Lavoie, American Institute of Aeronautics and Astronautics SPACE 2011 Conference, AIAA 2011–7185. See also http://www.spudislunarresources.com/Rationale.htm. Another plan is proposed by James Head and *Apollo 15*'s Dave Scott: "Astronaut Lunar Surface Exploration Destinations: New Technology to Address New and Complex Scientific Problems" by James W. Head III & David R. Scott, July 16, 2013, *Lunar Science Institute Virtual Forum*.

20. "Microwave Extraction of Water from Lunar Regolith Simulant" by Edwin Ethridge & William Kaukler, 2007, *Space Technology and Applications International Forum, AIP Conference*, 880, 830.

21. The regolith has a density of 1,660 kg m^{-3} (*LSB*, p. 475) and specific heat of 760 J kg^{-1} K^{-1} ("Specific Heats of Lunar Basalt 15555 and Soils 15301 and 60601 from 90 to 350 K" by B.S. Hemingway & R.A. Robie, 1973, *Lunar and Planetary Science Conference*, 4, 355), so heating 3 m^3 of regolith 220 K requires 830 MJoules. A square-meter solar panel collects this in 7.0 day/e, where e is the efficiency. (The Solar flux is 1,365 W m^{-2}). Water is 5.6% of the 5 tonne mass: 280 liters.

22. NASA Glenn Research Center proposed the 40 kw Fission Surface Power System, with a 1 MW option (http://www.grc.nasa.gov/WWW/TECB/fsp.htm). Marshall Space Flight Center and Los Alamos National Laboratory in 2004 made the 100 kW electric Safe Affordable Fission Engine-400 generator (400 kW thermal). Ideas for more than 1 MW include "Multi-MW Closed Cycle MHD Nuclear Space Power Via Nonequilibrium He/Xe Working Plasma" by Ron J. Litchford & Nobuhiro Harada, 2011, *Nuclear and Emerging Technologies for Space (N&ETfS) 2011*, #3349; "Trade Study of a 20 Megawatt Electric Low Specific Mass Nuclear Power System for Space Propulsion" by Wesley Deason, Regal Ferrulli, Mahima Gupta, Nic Hoifeldt, Jarred Reneau & Laura Suddert, 2011, *N&ETfS 2011*, #3289; and A Conceptual Multi-Megawatt System Based on a Tungsten CERMET Reactor" by Jonathan A. Webb & Brian J. Gross, 2011, *N&EtfS*, #3275.

23. "The History of Power Transmission by Radio Waves" by W. C. Brown, 1984, *IEEE Transactions on Microwave Theory and Techniques*, 32, 1240.

24. The maximum efficiency, ε, achievable by a heat engine is shown by Carnot cycle analysis to be $\varepsilon = 1 - T_C/T_H$, where T_C is the cold reservoir temperature and T_H the hot reservoir's. For lunar conditions, this could reach 70%, theoretically.

25. Solar cell efficiency hits 40% but more typically ~20% or ~0.3 kW m^{-2} maximum power. Heat engines typically radiate ~1 kW m^{-2} (and ~0.3 kW m^{-2}, electrical), e.g., NASA's Fission Surface Power unit built in 2006 ("A Lunar Nuclear Reactor" by Teague Soderman, 2009, http://lunarscience.nasa.gov/articles/a-lunar-nuclear-reactor/).

Lunar Power

26. The projected production (tonne year^{-1}) is H$_2$O: 109, H$_2$; 201, ^{4}He: 102, CO: 63, CO$_2$: 56, CH$_4$: 53, N$_2$: 16.5, ^{3}He: 0.033 from 1 km^2 year^{-1}, using 13 MW from solar and/or microwaves ("A Lunar Volatiles Miner" by Matthew E. Gajda, 2006 May, MS thesis, University of Wisconsin – Madison, UWFDM-1304; "A Lunar Volatiles Miner" by Matthew E. Gajda, Gerald L. Kulcinski, John F. Santarius, Gregory I. Sviatoslavsky & Igor N. Sviatoslavsky, March 7, 2006, Engineering Physics Department, University of Wisconsin – Madison, preprint. The water yield (hydroxyl, actually) will be higher given recent M^3 solar wind hydration results.

27. "Lunar Oxygen Production by Vapor Phase Pyrolysis" by Wolfgang Steurer, in *Space Manufacturing 5: Engineering with Lunar and Asteroidal Materials*, eds. Barbara Faughnan & Gregg Maryniak, 1985, New York: American Institute of Aeronautics and Astronautics, 123; "Lunar Oxygen Production by Pyrolysis" by Constance L. Senior, in *Resources of Near-Earth Space*, eds. J. Lewis & M. S. Matthews, 1993, Tucson: University of Arizona Press, 179.

28. "Production of O$_2$ on the Moon: A Lab-Top Demonstration of Ilmenite Reduction with Hydrogen" by Lawence A. Taylor, Eric A. Jerde, David S. McKay, Michael A. Gibson, Christian W. Knudsen & Hiroshi Kanamori, 1993, *Lunar and Planetary Science Conference*, 24, 1411.

29. Machines can be positioned precisely via onboard ring-laser gyroscopes to set orientation with high precision – crucial in drilling straight mine tunnels. Another system, instituted by Rio Tinto, involves triangulating with penetrating, long-wavelength radio waves from underground transmitters, using transparency of rock to long waves (Chapter 6).

30. "A Lunar Miner Design: With Emphasis on the Volatile Storage System" by Matthew E. Gajda, Gerald L. Kulcinski, Gregory I. Sviatoslavsky and Igor N. Sviatoslavsky, *Earth & Space 2006*, 188, 49.

31. Inevitably, deuterium reacts with deuterium in current tritium/deuterium mixes (^{2}H + ^{2}H → ^{3}He + n + 3.3 MeV, also ^{2}H + ^{2}H → ^{3}H + ^{1}H + 4.0 MeV), making more neutrons. Likewise, in ^{3}He+^{2}H reactors, some unwanted ^{2}H + ^{2}H → ^{3}He + n reaction occurs, making neutrons, but many fewer than ^{3}He+^{2}H (~6% of the neutrons per unit energy is anticipated for ^{3}He+^{2}H versus ^{3}H+^{2}H). Neutron production from ^{3}He+^{2}H depends strongly on temperature. At low temperatures (d $\lesssim 5 \times 10^7$ K) at least half of the reaction's energy escapes in neutrons. At 10^9 K this drops to ~1%, depending on initial ^{3}He/^{2}H ratio. (As initial ^{3}He fraction is raised, energy in neutrons drops proportionally.) One can imagine eliminating neutrons totally with ^{3}He + ^{3}He → ^{4}He + 2^{1}H + 12.9 MeV, but this requires temperatures nearly as high as for $^{11}_5$B + ^{1}H → 3 ^{4}He + 8.7 MeV.

32. One can rank fusion reaction alternatives according to ease of reaction. The Lawson criterion, with values of 1 for ^{3}H + ^{2}H, 16 for ^{3}He + ^{2}H, and 500 for $^{11}_5$B + ^{1}H, is the product of reaction density (measured for electrons) and confinement time within the reactor. This measures roughly the difficulty in initiating fusion. (Conversely, once reactions are initiated, intensity of power generation increases roughly proportionally to the Lawson criterion, as well.)

33. "Race to the Moon" by Fred Guterl, *Newsweek*, October 11, 2007.

34. An initial, compelling case for this is made in "The Moon: An Abundant Source of Clean and Safe Fusion Fuel for the 21st Century" by G. L. Kulcinski & H. H. Schmitt, in *11th International Scientific Forum on Fueling the 21st Century*, 1987, http://fti.neep.wisc.edu/pdf/fdm730.pdf

35. Annual energy use is 5×10^{20} J. Fusing 1 g of ^{3}He yields (18.4 MeV) × (6.02 × 10^{23} nuclei mole^{-1})/ (3 g mole^{-1}) = 3.7 × 10^{30} eV = 5.9 × 10^{11} J, so 850 tonne year^{-1} of ^{3}He is needed. At 33 kg km^{-2} of ^{3}He, 26,000 km^2 must be mined annually, 0.07% of the Moon's surface. There are ~1 million tonnes of ^{3}He in lunar regolith, gardened to several meters.

36. Aluminum oxidizes forming charge traps and spoiling silicon semiconductors rather than doping them ("Lifetimes in Aluminum-doped Silicon" by Jan Schmidt, Nils Thiemann, Robert Bock & Rolf Brendel, 2009, *Journal of Applied Physics*, 106, 093707). Aluminum oxidation should be avoidable in lunar vacuum.

37. "Space Solar Power Satellite Technology Development at Glenn Research Center: An Overview" by James E. Dudenhoefer & Patrick J. George, 2000, NASA Pub. TM-2000 – 210210; "Reinventing the Solar Power Satellite" by Geoffrey A. Landis, 2004, NASA Pub. TM-2004–212743.

38. "Summary of Twenty-First Century Power Needs and Supply Options" by David R. Criswell, 1998, AIP Conference Proceedings, 420, 1219; "Space/Lunar Solar Power Systems Research and Needs" by David R. Criswell & Robert D. Waldron, 1999, AIP Conference Proceedings, 458, 1513.

39. "Thorium Abundances on the Aristarchus Plateau: Insights into the Composition of Aristarchus Pyroclastic Glass Deposits" by J. J. Hagerty, D. J. Lawrence, B. R. Hawke & L. R. Gaddis, 2009, *Journal of Geophysical Research*, 114, E04002. This thorium work is more detailed than global surveys.

40. *LSB*, 337.
41. Monazite can contain rare earth elements lanthanum (La, atomic number 57), cerium (Ce, #58), praseodymium (Pr, #59), neodymium (Nd, #60), and yttrium (Y, #39), typically ~150 ppm-wt each in KREEP (*LSB*, 379). Other rare earth elements have atomic numbers 21 and 61–71. Phosphorus and lanthanum co-vary with abundance ratio P/La ≈ 31. Lanthanum's KREEP abundance is typically 115 ppm-wt, so typical phosphorus abundance is 0.35% (*LSB*, 391).
42. Thorium-bearing inclusions come from *Apollo 11*, *Luna 16*, *20*, and *24* samples, far from high-Th regions ("Lunar Monazites" by P. M. Kartashov, O. A. Bogatikov, A. V. Mokhov, A. I. Gorshkov, N. A. Ashikhmina, L. O. Magazina & E. V. Koporulina, 2006, *Doklady Earth Sciences*, 407A, 498.) Thorite was found by *Apollo 14*, near high thorium concentration sites ("The Uranium Distrubution in Lunar Soils and Rocks 12013 and 14310" by E. L. Haines, A. J. Gancarz & G. J. Wasserburg, 1972, *Abstracts of the Lunar and Planetary Science Conference*, 3, 350).
43. High-resolution telescopes suffer when resolution is wasted by insufficient signal collected to fill their maps with information. This could be attempted with coded masks, but they perform poorly with extended (not point) sources like the lunar surface ("Imaging Extended Sources with Coded Mask Telescopes: Application to the INTEGRAL IBIS/ISGRI Instrument" by Renaud Matthieu, Alexandra Gros, François Lebrun, Régis Terrier, Andréa Goldwurm, Steve Reynolds & Emrah Kalemci, 2006, *Astronomy and Astrophysics*, 456, 389). At energies of 100 keV, imaging telescopes with true reflective mirrors will be more sensitive.

    Regrettably, predicted count rates for typical thorium concentrations are low. State-of-the-art imaging high-energy X-ray telescopes, e.g., *NuSTAR*, in low Earth orbit could image the Moon with 25 km resolution, six times better than current gamma-ray maps from lunar orbit. (*NuSTAR* has a ~50 cm^2 effective area and costs $50 million, excluding launch.) Unless lunar thorium exists much more concentrated than 10 ppm, a search this way is too slow. (A broad vein of 10% thorium would be 10,000 times more concentrated and easily detectable, but from tens of kilometers over the surface, not 380,000.) The best place to find thorium is near the Moon.
44. "Global Elemental Maps of the Moon: The Lunar Prospector Gamma-Ray Spectrometer" by D. J. Lawrence, W. C. Feldman, B. L. Barraclough, A. B. Binder, R. C. Elphic, S. Maurice & D. R. Thomsen, 1998, *Science*, 281, 1484.
45. "Descriptive Models of Thorium-Rare-Earth Veins" by Mortimer H. Staatz, in *Developments in Mineral Deposit Modeling*, edited by James D. Bliss, 1992, Washington: U.S. Geological Survey, Bulletin 2004. On Earth, thorium-REE minerals (usually monazite and/or thorite) form veins up to 1.3 km long and 16 m wide, but usually much smaller. This inspires considering a high angular-resolution telescope to detect thorium hot spots. Thorium has X-ray resonant lines at 13–19 keV (L-shell α at 12.97 kev to γ at 18.98 keV) detectable with focusing-optic X-ray telescopes, but these lines are weak with 10 ppm thorium (even exciting lines with an active Sun).
46. *Geochemical Prospecting for Thorium and Uranium Deposits* by R. W. Boyle, 2013, Amsterdam: Elsevier, p. 170.
47. "Q&A: Thorium Reactor Designer Ratan Kumar Sinha" by Seema Singh, August 2008, *IEEE Spectrum*. Also see the World Nuclear Association factsheet on thorium fuel cycles, "Thorium," http://www.world-nuclear.org/info/inf62.html, and *SuperFuel: Thorium, the Green Energy Source for the Future* by Richard Martin, 2012, New York: Palgrave Macmillan.
48. This is the "Treaty on Principles Governing the Activities of States in the Exploration and Use of Outer Space, including the Moon and Other Celestial Bodies." Excerpted operative language is

    ARTICLE I: "Outer space, including the moon and other celestial bodies, shall be free for exploration and use by all States without discrimination of any kind, on a basis of equality and in accordance with international law, and there shall be free access to all areas of celestial bodies." This is analyzed in detail in "Space Settlements, Property Rights, and International Law: Could a Lunar Settlement Claim the Lunar Real Estate It Needs to Survive?" by Alan Wasser and Douglas Jobes, 2008, *Journal of Air Law and Commerce*, 73, 38.

    ARTICLE II: "Outer space, including the moon and other celestial bodies, is not subject to national appropriation by claim of sovereignty, by means of use or occupation, or by any other means."

    ARTICLE IV: "In carrying on activities in outer space and on celestial bodies, the astronauts of one State Party shall render all possible assistance to the astronauts of other States Parties. States Parties to the Treaty shall immediately inform the other States Parties to the Treaty or the Secretary-General of the United Nations of any phenomena they discover in outer space, including the moon and other celestial bodies, which could constitute a danger to the life or health of astronauts."

|ARTICLE VI:| "The activities of non-governmental entities in outer space, including the moon and other celestial bodies, shall require authorization and continuing supervision by the appropriate State Party to the Treaty."

ARTICLE XI: "States Parties to the Treaty shall pursue studies of outer space, including the moon and other celestial bodies, and conduct exploration of them so as to avoid their harmful contamination and also adverse changes in the environment of the Earth resulting from the introduction of extraterrestrial matter and, where necessary, shall adopt appropriate measures for this purpose. If a State Party to the Treaty has reason to believe that an activity or experiment planned by it or its nationals in outer space, including the moon and other celestial bodies, would cause potentially harmful interference with activities of other States Parties in the peaceful exploration and use of outer space, including the moon and other celestial bodies, it shall undertake appropriate international consultations before proceeding with any such activity or experiment. A State Party to the Treaty which has reason to believe that an activity or experiment planned by another State Party in outer space, including the moon and other celestial bodies, would cause potentially harmful interference with activities in the peaceful exploration and use of outer space, including the moon and other celestial bodies, may request consultation concerning the activity or experiment."

49. "NASA's Recommendations to Space-Faring Entities: How to Protect and Preserve the Historic and Scientific Value of U.S. Government Lunar Artifacts" by Robert M. Kelso, July 20, 2011, Houston: NASA Johnson Space Center.
50. "A New Paradigm for Arbitrating Disputes in Outer Space" by Michael Listner, January 9, 2012, *Space Review*, http://www.thespacereview.com/article/2002/1.
51. "Agreement Governing the Activities of States on the Moon and Other Celestial Bodies," http://www.unoosa.org/oosa/SpaceLaw/moon.html.
52. It is an old but rare attitude; see "Heinlein's Ghost" by Dwayne A. Day, April 16, 2007, *The Space Review*, http://www.thespacereview.com/article/851/1: "Heinlein's message of enlightened capitalists undertaking a space mission using the tools of the free market, and doing so in spite of government opposition, was virtually unheard of after the early 1950s."
53. Some examples of first, isolated villages: Jarmo, Iraq, 7,000 BCE, population of 200; Jericho, West Bank, 9,000 BCE, 1,000 people; Çatal Hüyük, Turkey, 7,500 BCE, several thousand.
54. "(Robin) Dunbar's number" refers to the cognitive limit to the number of people with whom one can maintain stable social relationships. It is quoted in the range 100–230, typically 150.

Chapter 13
Stepping Stone

> There is no easy way from the earth to the stars. (*Non est ad astra mollis e terris via.*)
> – Lucius Annæus Seneca (the Younger), ca. 50 AD[1]

> Earth is the cradle of our intelligence, but one does not live all of life in a cradle. (*Planyeta yest' kolybyel razuma, No nyelzya vietchno zhit' v kolybyeli.*)
> – Konstantin E. Tsiolkovsky, 1911[2]

The Moon, planets, and stars beckon as vast expanses of territory and potential experience and understanding beyond us. To experience it, you need to reach out. It is unlike anything you have done, and not what you evolved to do. It is true wilderness, not with predators (that we know of), but many perils. Mainly, it is oblivious. First, you must escape the clutch of the mother planet that bore you. Whether humans choose to surmount this challenge says as much about humanness as outer space. Do we choose to go? The good news is that we may have more than one chance to ascend this height. The bad news is that this may cause us to procrastinate. What is the imperative? Do we have so much time?

What are the reasons for exploring space? It is expensive, difficult, infinite, and will always be there. What would compel us? I think the answers fall into a double handful of categories, with many subcategories. Any of you could parse these divisions and generate your own set, but let us attempt to cover the range:

1. **Manifest destiny**: We belong to the species that walked out of Africa to inhabit all seven continents of Earth. We build ships to cross the largest oceans and build bases on Antarctic ice. We have traveled from the deepest ocean bottom to far above Earth's atmosphere. Humans explore. Someday we will explore outer space; let us get started. If we do not, do we lose part of our humanity?
2. **Humanity's survival on another world, or this one**: There are many agents of precedent, imagination, and the unknown that could wipe out humans and our world. We reduce the risk by not "putting all of our eggs in one basket." By this we do not necessarily mean just "living space," because acreage off Earth is extremely expensive.[3] Likewise, we may need to visit an asteroid to prevent it from striking Earth.
3. **Curiosity and scientific exploration**: Has life arisen off Earth? How were the Universe and its contents created? How do other worlds differ from ours? Why is

our planet as it is? What new natural processes occur in strange environments? What are the physical boundaries of our understanding? Major aspects of these can be addressed by space exploration.

4. **Inspiration of our youth**: Youth needs a frontier. The Universe offers a seemingly inexhaustible one. Young people want to expand and explore. The future is better for us all if we encourage them.
5. **Economic stimulus/payoff/jobs**: On a mundane level, space exploration offers a means of transfer of wealth from one group to another. Space programs offer a means of guaranteed employment for desirably trained workers.
6. **Economic benefits and spin-offs**: Technological discoveries inspired by space exploration's rigorous demands can be applied elsewhere in the economy. Utilization of outer space can reap economic benefits because of what is there. Orbital remote sensing is routine.[4]
7. **National security**: Some advantages of outer space utilization bear directly on national security and defense issues. Make use of the high ground.
8. **International cooperation/competition, instead of war**: Nations can work collectively to accomplish space goals and develop goodwill and understanding, which can be useful in other efforts. Rival nations can compete against each other via their space programs without killing each other.
9. **National/personal pride, to be first/better**: Space programs require significant competence, and the world notices feats of accomplishment in space exploration. This is a way for organizations and people to prove themselves and elevate their status.
10. **Sheer hubris**: Because it is there. Because we can do it. It is so amazing. (This closes the loop on reason 1.)

Reason 7, national security, is largely obsolete for the Moon. It is too distant for a reserve attack on Earth (given existing missile submarines) and holds no resources that we lose at national peril (except perhaps one day helium-3, depending on its prospects). Reason 5, the shifting of funds, is de facto more important than it should be, particularly in the United States (considering how funding is set by Congress). Let us not discuss this here. Reasons 6, 8, and 9 we discuss in earlier chapters. Let us look at the others (jumping between them a bit).

To summarize some fundamental points: looking forward, outer space could and probably should be part of our future and we should not shrink from it thoughtlessly. In addition, we contemporary humans often think we are in control of our largely artificial universe, but this is illusory. What is our place in existence? What formed us? What influences us that we might not understand? Many of these answers lie off world.

In this and the next chapter we consider issues of spirit, expansion, and scientific exploration. First, let us return to the tawdry issue of money. Ignore momentarily the hardware currently needed to escape Earth into space. Imagine instead some magical device teleporting you or otherwise transporting you far from Earth. All it costs is the energy equivalent of attaining escape velocity from Earth's surface. At usual electrical power utility rates, how expensive is this?

Only a few hundred dollars, typically, no more than a moderately priced airline ticket.

At such costs, we need not worry about why to go into space. People could go for whatever reason, maybe just the amazing view. Human expeditions to the Moon could be mounted for millions of dollars, not billions. Space commerce would be an easy sell.

This is impossible, correct? Now, yes, but it is worth considering. One concept comes tantalizingly close to this dream. Proposals for a "space elevator" would send passengers and freight from Earth's surface to geostationary orbit and beyond. This elaborates on Arthur C. Clarke's 1945 idea for geostationary satellites[5] but expanded to a continuous cable hanging from this satellite to the point underneath it. The idea's main impracticality is enormous lengths of self-supporting cable: 35,800 kilometers to geostationary orbit and extra length counterweighting the lower cable (Figure 13.1). The cable simply dangles in space above the Earth, orbiting one revolution per (sidereal) day, with the upper portion pulling up the lower portion. The problem is self-supporting – cables currently with highest tensile strength still fall short by a factor of ten.[6] The operational scenario would include entering the elevator capsule on Earth's surface at the

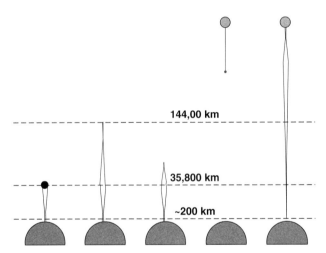

Figure 13.1. **Heavenly hoist**. Five options for space elevators from the Moon (*top*) or Earth (*bottom*). Each elevator cable is linear, width exaggerated, with distances not to scale. (1) A geosynchronous satellite (35,800 km above Earth) could lower a cable, strongest at top. The cable must be counterweighted, but a mass at 35,800 km would be huge. (2) Instead, a cable extending 108,000 km beyond geosynchronous will balance a cable to Earth without counterweight. (3) The best option might be a smaller counterweight and roughly 100,000 km cable. Carbon nanotubes at their theoretical tensile strength might permit this. (4) A lunar elevator (revolving monthly) needs a small counterweight and cable of commercially available material, about 80,000 km long. Systems must deal with micrometeoroid impacts. A similar elevator could also operate off the Far Side. (5) Because the length of the day and month differ, an elevator from Earth and the Moon is not possible, but one from the Moon to just above Earth's surface might be, with an even stronger cable. One could reach the bottom of the cable by a suborbital boost to 1,600 km/hour at perhaps 200 km altitude.

cable's bottom and riding up and out of Earth's atmosphere into space. On arrival at geostationary radius, 2.8 Earth diameters above its surface, you could step out into geosynchronous orbit. Alternatively, you could continue rising on the elevator to the counterweight, where you would be flung into space away from Earth if you jumped. At 47,000 kilometers above Earth's surface, jumping would fling you at escape velocity, to the Moon or solar orbit. The optimal design for cable and counterweight would probably put total length at about twice this, approaching 100,000 kilometers.[7]

With strong enough cable, several technical problems must be solved. We must control cable oscillations to prevent unstable moving elevator capsules, overstressed cables, or motions because of weather at the cable root. A power source, perhaps via superconducting conduits or beamed power shining on the capsule's photocells, needs development. Indeed, the capsule's transit on the cable, probably like magnetically levitated trains not using wheels, must be mastered. Finally, with the cable deployed, all satellites orbiting below the cable apex must be removed, a definitive commitment to the space elevator future. Objects in these orbits would eventually hit the cable unless active avoidance is engaged. If the cable broke, it would lay along on Earth's equator at 1,600 kilometers per hour, probably fragmenting and incinerating in the atmosphere. There are practical difficulties to be overcome, so much so that Arthur C. Clarke, when asked when a space elevator might be built, answered, "Probably about 50 years after everybody quits laughing."[8]

A space elevator for Earth would be transformative. It could turn us from an earthbound species dabbling in space flight into a truly spacefaring one. It might make space trips accessible to typical people versus less than 1,000 astronauts among billions of humans. Even with elevators only 1% efficient (with current estimates of 0.1%–2%[9]), energy costs per trip to orbit might total about $10,000.[10] One might store energy generated as objects lower themselves in Earth's gravitational field to raise objects headed skyward.

The space elevator concept is so important that we should preserve it as a long-range goal but cannot plan on it in the next 50 years, beyond this book's horizon. Limits to activity in space depend on other choke points: Earth's gravitational well, our rockets' slow exhaust speeds, human survival in space, and, telerobotically, the speed of light and latency.

The first hurdle is lifting people and cargo from the strong gravity well holding us to Earth's surface. We accelerate payloads, largely in the atmosphere, on rockets reaching at least 7.8 kilometers per second to not fall back. It is inherently violent, difficult, and wildly energy consuming.

Transport to and from airless planets is intrinsically symmetrical (going up as hard as getting down). Of course, energy is not conserved when bringing objects to Earth, evidenced by heat shields and flaming reentries. Lofting something into space, we currently partially reverse the process, still trying to save as much acceleration until we reach thin atmosphere. Accelerating in the air is wasteful, because reaching orbital velocity requires more energy (30 Megajoule per kilogram) than rising into space at zero velocity (1 Megajoule per kilogram).[11] Air resistance consumes much of the energy of a typical rocket leaving Earth.

There are alternatives, for example, getting above the air before accelerating can make the task easier, requiring smaller rocket stages, largely thrown away. One could raise the spacecraft slowly on an airplane or even a balloon. Examples include the X15 rocket plane, Pegasus air-launched rocket, and experimental rockoons (rocket balloons).[12] A space shuttle carried to high altitude via airplane before launching would exploit this.

External forces can propel spaceships from Earth. Jules Verne saw in *De la Terre à la Lune* (see note 6 in Chapter 3) that a cannon might work, then Tsiolkovsky saw the cannon projectile's acceleration would crush passengers with many thousand times Earth's surface gravity (g). Third Reich Germany built a 130-meter-long cannon, the V-3 (*Vergeltungswaffe* or "Vengeance weapon") accelerating shells at 1,000 g to 20% orbital speed. Canadian engineer Gerald Bull in the 1960s built HARP (High Altitude Research Project) for the Canadian and U.S. defense departments, firing projectiles into suborbital space. He then worked in Saddam Hussein's Iraq on Project Babylon's PC-3, 150-meter cannon, to loft projectiles into orbit. (In 1990 Bull was assassinated, allegedly by Israel's Mossad.) In 1990s America, SHARP (Super-HARP) shot projectiles into space at 90% of orbital speed. Reportedly during Operation Plumbob in mid-1957 at the Nevada Test Range, Project Thunderwell's Pascal-B test blasted a large steel plug at several times Earth's escape velocity from a vertical shaft during a nuclear explosion. It likely burned up in Earth's atmosphere.[13]

One can imagine improving the cannon, more efficient with a longer barrel and placed at high altitude, maybe inside a high mountain peak, even evacuated of air. Less violent is the mass driver or electromagnetic launcher, like a cannon with projectiles riding a crest of electrical and magnetic force (not at the speed of light) rather than an expanding gas front, as in a cannon. A lunar mass driver might be practical because it fights no atmosphere and must attain only 2.4 kilometers per second, not 11.2. However, accelerating several kilogram packages to lunar escape speed would require a driver nearly 1 kilometer long, with electromagnets, drive coils, and support structure as well as large power capacity. This implies a lunar settlement, a central idea in Gerard O'Neill's plan for human space settlements from lunar materials propelled to Earth-Moon Lagrange points L4 and L5.[14] Finally, once orbit is achieved, pure power from Earth can be sent to the spacecraft via lasers or microwaves, either powering ion engines or propelling the craft with radiation pressure.

Having left Earth, how can we travel the Solar System? Usually with rockets, but not always. Successful tests of light sails (NASA's Nanosail-D2 in low Earth's orbit and JAXA's IKAROS [Interplanetary Kite-craft Accelerated by Radiation Of the Sun] roaming the inner Solar System) use radiation pressure to navigate, like sailing ships use wind.[15] Each photon reflecting from the sail alters the craft's momentum by up to twice the photon's momentum, versus rocket exhaust changing the spacecraft's momentum by only the exhaust's momentum once. The sail accelerates without altering mass and can control its velocity by "tacking" somewhat like sailing ships, changing the sail's size and orientation.

Change in rocket speed is inherently limited by its exhaust velocity. Tsiolkovsky showed that rocket exhaust speed affects the rocket's speed exponentially.[16] Typical exhaust speeds for efficient rocket propellants, for example, oxygen and hydrogen (to water), are 4–5 kilometers per second. To escape completely from Earth's surface requires 11.2 kilometers per second. Going to Mars or Venus requires about another 5 kilometers per second and to orbit the Moon (near its surface), 1.6 kilometers per second. We want propellants with exhaust velocities above these; such rockets exist. NASA experimented with ion propulsion in the 1960–1970s, and several spacecraft (ESA's lunar *Smart-1*, JAXA's *Hayabusa* asteroid probe, NASA's interplanetary *Dawn* and *Deep Space 1*) have used ion engines with exhaust speeds of 30 kilometers per second or more, meaning they travel the Solar System without carrying most of their mass as propellant. The 1.24-tonne *Dawn* spacecraft started with 0.42 tonne of propellant to hop to Mars and then to two asteroids. Power for these comes from solar panels, so it is tiny. *Smart-1*'s ion engine took a year to push it from Earth GEO to the Moon. In American parlance, *Dawn* can accelerate from 0 to 60 miles per hour in four days. Solar-powered ion drives are insufficient for humans. They are efficient but too slow.

Nuclear-powered ion propulsion fits into efforts for nuclear rocket engines stretching over six decades. To power ion drives, NASA initiated nuclear Project Prometheus in 2003, cancelled two years later after antinuclear protests.[17] One might use a nuclear reactor to power radio-wave generators heating rocket exhaust gas (VAriable Specific Impulse Magnetoplasma Rocket [VASIMR]). NASA is testing VASIMR thrusters on the *ISS*. The most direct use of nuclear power is simply forcing gas into a hot reactor and letting superheated plasma exit as exhaust. The U.S. Atomic Energy Commission developed this concept in the 1950s, later with NASA culminating in the 75-tonne NERVA engine before cancellation in 1972. The Soviet Union had a similar program, the RD-0410. NERVA had only about twice the exhaust velocity of $H_2 + O_2$, but later designs claimed potential velocities of 20–50 kilometers per second, comparable to any needed in the Solar System. Designs are maturing for engines collapsing lithium to nuclear fusion conditions, spewing exhaust at 20 kilometers per second.[18] Project Orion (1958–1963) proposed using actual nuclear explosions to drive exhaust at roughly 10,000 kilometers per second. Exposed naked surfaces of spontaneously fissile radioisotopes (decaying violently to two roughly equal, massive nuclei) could produce exhaust of several thousand kilometers per second. Someday antimatter engines – of science fiction fame – might produce nearly lightspeed exhausts (299,793 kilometers per second). Higher performance ideas are beyond our understanding. NASA, DARPA, and other organizations discuss using antimatter or other power to energize a warp drive, expanding space-time behind and contracting it ahead of a craft to propel a bubble of space potentially faster than lightspeed, using an exotic working substance similar to dark energy. Work on these Alcubierre warp concepts have highlighted difficulties but also advanced the idea. Even the most enthusiastic protagonists foresee no such working drive in this century.[19]

Nuclear rocket fuel from Earth carries risk of a launch failure raining radioactive debris. *Galileo* and *Cassini* carried 40 kilograms of plutonium-238 between them and were targets of protest and lawsuits to prevent any small risk of radioactive dispersal. Is any nation willing to launch large radioactive payloads to power space-borne reactors – the United States, Russia, or China? This offers direct risk reduction for astronauts, allowing swift transit between planets minimizing radiation exposure for them from the Sun and Galaxy. Will this happen in our lifetimes?

The Solar System itself helps us transit between planets once we escape Earth's grasp, if we are patient. Once reaching Earth-Moon Lagrangian Point L1, where Earth's and the Moon's gravitational influence cancel, only small velocity changes send a rocket to explore other Lagrange points or to leave Earth's influence altogether (see Chapter 12). From L1, a 5-kilometers-per-second boost can send a craft to Mars or Venus, where it can fly past those planets toward others using gravitational assists, robbing slight energy from the (humongous) orbital reservoir of the planet itself and adding it to the spacecraft's motion. In this way, given enough encounters, a spacecraft can navigate the Solar System with little additional rocket thrust. It is too slow to send humans, however.

This outlines an architecture to explore the Solar System: robots and "freight" insinuating themselves between targets via Lagrangian points, gravitational assists, and ion engines or light sails. Humans are too fragile, impatient, and demanding of water, food, and oxygen for such prolonged journeys. They need rapid means to transit the Solar System, for example, nuclear engines or chemical rockets with big propellant supplies.

The Moon here is important because it offers several vital roles in navigating the Solar System. First, it holds huge potential propellant reserves derived from water, about 20 times easier in energy to access than Earth's surface. Second, it might offer the safest means of stoking nuclear rockets with radioactive fuel. Third, the jumping off points, Earth-Moon Lagrange points, are arrayed mostly near the Moon: L1 (the so-called equal attraction point between Earth and the Moon, accounting for the rotation of the whole system) and L2 (where one orbits against the summed Earth-Moon attraction once per month) are both 58,000 kilometers above the Moon – L1 over the central point of the Near Side and L2 over the Far Side. L4 and L5 are 384,000 kilometers from both Earth and the Moon, in equilateral triangles with both worlds. L3 is opposite the Moon as seen from Earth but essentially at the Moon's orbit. Orbital bases at these Lagrangian points, especially L1 and L2 (and L4 and L5 in the vision of O'Neill), might be supplied from the Moon with structures and consumables composing most of the mass but not the labor of space travel. L1 and L2 might easily provide vital Earth transit stations supported more from the Moon than Earth.

The problem of easy access to space from Earth remains unsolved, despite the enticing but forlorn hope of a space elevator. Could we build such a thing anytime soon? Yes, in a way, at the Moon. Because the Moon's mass is only 1/81 Earth's, stress on space elevator cables are much reduced. A space elevator could hang down to the Moon on conventional fibers, for example, Kevlar (famous in helmets and bulletproof vests).[20] It would reach from the lunar surface or proximity up

58,000 kilometers to L1 and then terminate at a counterweight, nearby or up to 200,000 kilometers away, depending on its mass. This counterweight might easily form a deep space station, a transit point from Earth to the Moon or farther destinations; it could be massive.

The Lunar Space Elevator Infrastructure (LSEI) cable begins as a 0.4-millimeter-wide Zylon polybenzoxazole fiber, tapering to 0.2 millimeter near the surface, totaling 42 tonnes mass (and about 6 tonnes of other structure), small enough to be emplaced with current rockets. This elevator could sustain many tonnes per year carried up or down and could even pick up and place loads at various lunar locations as an exploration vehicle. The cable's surface terminus need not be the Near Side center; it could be extended and tied in a polar region. As useful as a Near side-L1 elevator might be, a Far side-L2 cable could fling payloads into the Solar System, up to 1 kilometer per second beyond Earth escape velocity.

Cable vulnerability is a problem, its cross-section nearly microscopic, but its length larger than a planet. It exists in a dangerous environment, with meteoroids flying past at many kilometers per second. Without means to armor the cable and repair it when hit, it will soon break and fall. There must be several strands arrayed so not all are simultaneously hit.[21] Perhaps one must surround the strands with several protective sheets, similar to TransHab and Bigelow armor discussed in Chapter 11. These will increase the cable mass but likely not hugely so. Lunar space elevators are promising and should be researched, prototyped, and tested.

Orbital changes without rockets could involve space tethers. Masses can be raised and lowered from a planet, but also long, elastic tethers can exchange payloads as craft pass each other on separate orbits, using elastic force rather than rockets to accelerate. NASA intends the Comet Nucleus Sample Return Mission to fire a tethered harpoon to grab a comet sample flying by. More ideas are proposed.[22] These could test techniques for that most transformative tether, the Earth space elevator.

We have discussed many possibilities here and in previous chapters crucial to Solar System exploration and human expansion beyond Earth: propellant, building materials, solar and nuclear energy from the Moon, and possible organics, food, and oxygen from lunar sources. We have discussed transportation, from hoppers to space elevators, and lunar resources' role in this. We should remember that organics, iron, water, and some potentially lunar resources can also derive from asteroids: more on this shortly. Having discussed how to navigate space; let us return to why. Let us consider Reasons 1, 2, and 3 from our list: scientific investigation, exploration, and humanity's place on other worlds.

If we must explore, why outer space? We could explore the depths of the oceans; there is an opening Arctic frontier. We should explore these places but cannot fool ourselves about their qualities. Earth's interior is amenable only to remote sensing, except for deep drilling, and shallow caves engaging a few teams of spelunkers. Earth's oceans are huge and unexplored but dominated by instant venues for death at typical depths, in milliseconds, for any unlucky explorer encountering submarine mechanical failure.[23] This is best explored robotically. Outer space is friendlier. (You have at least a dozen seconds more, sadly insufficient for *Soyuz 11*'s poor cosmonauts, sole fatalities in space, but sufficient for several other astronauts to have found haven.) Both demand crucial supplies: energy at ocean's bottom, propellant and life support in space. Do places exist beyond Earth's surface that are friendly to humans? Even if we cannot walk outside sans space suit, where at least can we establish housekeeping? For sheer unexplored, accessible human frontier, space is the place.

Explore what? What is accessible? Where can humans survive? There are several classes of objects (smaller than stars). Gas giants (Uranus, Neptune, Saturn, and Jupiter) are uninhabitable, but ice moons orbit – several dozen in our Solar System, many with water oceans under ice layers melting via tidal interactions with their mother planet. These merge into a continuum of small objects such as Pluto/Charon, some with internal heating, to isolated icy bodies in which self-gravity is less important (comets, Centaurs, icy asteroids), to inner Solar System asteroids of metal and/or rock, some with organics. Finally, metal/rocky planets (including the Moon, for sake of discussion) in the inner Solar System have surfaces wrapped in thin exosphere/atmospheres. No human or robot will reach extrasolar planets (exoplanets) in our lifetimes. With today's technology and a trillion-dollar investment, we could hardly deliver a toothpick to an exoplanet in this century. Finding habitable exoplanets might spur technology development, someday.

Robots go many places: Venus's surface, Jupiter's atmosphere, and interstellar space. Humans are more limited, correct? Yes, but options exist that you might not expect.

The Solar System is seemingly absent human-habitable worlds. The common science fiction venue, an alien world hospitable to relatively unprotected humans, is absent our Solar System. Exposed humans on any other nearby world's surface are doomed in seconds. An exclusive factor is atmospheric pressure: only one other Solar System world has even within ten times the surface pressure of Earth – Saturn's moon Titan, with 145% of Earth's pressure at sea level, but no oxygen and surface temperatures near −180°C. Nearby planets bracket Earth's surface pressure like Goldilocks' porridge bowls: Venus with 90 times too much pressure and Mars with 160 times too little. Everywhere else is even worse.

Distant dreams exist to change this. "Terraforming" a planet might be in humanity's future but beyond the future scope of even this book. Mars is too cold, its atmosphere too tenuous, and its magnetic field too weak. We would need to import gas, mainly carbon dioxide, from comets or hydrated asteroids (roughly a hundred trillion tons, 1,000 large comets' worth). Much might be released by heating CO_2 ice trapped below Mars' surface. The resulting greenhouse effect would heat Mars, and a thicker atmosphere would persist semipermanently, even without a magnetic field's protection of the atmosphere from solar wind erosion. Venus has much more carbon dioxide and little water versus Earth. The CO_2 largely responsible for Venus's enormous greenhouse effect is trapped in rock as calcium carbonate on Earth, largely by water's action. With enough water brought to Venus (again, as comets and/or asteroids), this might sequester CO_2, leaving an atmosphere closer to Earth's in composition and density. Blocking sunlight from Venus would cool it, after monumental effort.[24] Terraforming is an issue for some future century or millennium, not the next 50 years.

Surprisingly, nearly all Earth's neighbors offer viable venues (even if not exactly backup for Earth version 1.0). Take the planet closest to the Sun, Mercury: with no atmosphere, surface temperatures on Mercury are extreme: at the equator, –175°C at night and 425°C at midday, twice the lunar temperature range. (Its gravity is 0.38 gram, the same as Mars.) How could one ever live there? First, move underground: it is still toasty down there on average. For a surface with Mercury's average reflectivity (12%), temperatures reach a fairly constant 165°C, 30°C to 40°C hotter than the Moon at its warmest. This is no problem: make surfaces overlaying your dwellings very shiny, about 82% reflective (as white as a good titanium oxide paint, less than well-aluminized mirrors). Within a decade or so, the regolith underneath will cool to about 20°C.

If life depending on good paint makes you nervous, consider moving to Mercury's poles. Within several degrees' latitude of either pole, the average temperature is about 0°C, even without paint. More valuable is what one finds nearby: water. Covering about 10,000 square kilometers of deepest crater floors near the poles are at least 10 billion tons of water ice, permanently frozen into the soil, perhaps several meters thick.[25]

Why live on Mercury? Maybe for reasons like lunar industry's: helium-3. Mercury receives 6.5 times the Moon's solar wind flux but has a magnetic field (1% Earth's strength) concentrating and likely deflecting it to the poles. Does Mercury's soil retain helium-3? This needs research. Mercury may be, beyond the Moon, helium-3's closest major source (and an excellent potential source for solar energy). In the distant future, lunar industry may prove excellent training for development of Mercury, in as good an analogy as for Mars.

More hostile than Mercury's surface is Venus's: at 470°C, the Solar system's hottest solid planetary surface. Who could live there? Likely nobody, but there is more to Venus than its surface. Venus is not hot just because of solar proximity; it absorbs less solar energy than Earth does. Although it receives twice Earth's energy, it is 2.3 times better at reflecting it. What differentiates the two planets are their atmospheric masses, with Venus's 90 times that of Earth. However, ascending through Venus's atmosphere until 89/90 remains below, conditions at that point are nearly room temperature.

More precisely, at 55 kilometers above Venus, temperatures are 310 K (37°C = 98.6°F) and pressure 45% that at Earth's sea level, the same at 5 kilometers altitude, the height of a tall mountain. Hence, after building a buoyant platform in Venus's stratosphere, one could walk outside wearing no more than a scuba outfit. It should be a full-body suit, because along with the atmosphere of mostly carbon dioxide, there are sulfuric acid mists. This outpost could even float in the atmosphere, because currents are mild, and breathable air would be buoyant. Furthermore, Venus's atmosphere has all the major elements of life (if not all the construction materials). Maybe in the far future, humans will build floating cities in Venus's stratosphere, as unnerving as that might seem.

Some speculate that early in Venus's history, when the Sun was cooler and before a global runaway greenhouse catastrophe, Venus may have sported liquid water oceans, perhaps for hundreds of millions of years.[26] We hypothesize that if conditions were right for life's origin on early Venus, it might persist in the stratosphere, or perhaps have started there. A balloon probe sensitive to life in Venus's upper atmosphere would be straightforward to plan.

Of course at proper distances from the Sun, humans can live in free space, with sufficiently substantial cocoons around them. The most involved proposal is Gerard K. O'Neill's *High Frontier* (1977), stationed at Earth-Moon Lagrangian point L5.[27] This requires generous resources derived and launched from the Moon. These stations sit above the radiation protection of Earth's magnetic field, so they must be shielded with massive absorbing layers (of lunar material). They rotate to generate artificial gravity.

Throughout the inner Solar System to Jupiter's orbit reside thousands of plausible habitats in the form of asteroids. They resemble the Moon in terms of hazards, with surfaces suffering temperature fluctuations, vacuum, meteoroid impacts, and hard radiation. Most endure impact rates more intense than Earth or the Moon; it is best to live underground. Some asteroid habitat proposals have unlimited ambition.[28] Asteroids have almost no gravity: even the largest asteroid Ceres has surface gravity only 3% of Earth's and one-sixth the Moon's. Collectively, they have water, organics, metals, and other valuables to be mined for commerce or habitation.[29] It will be hard to find the perfect asteroid with accessible sources of all commodities needed. Few asteroids are easier to access energetically than the Moon, although they have tiny escape velocities. Visiting an asteroid is a huge time commitment, with many months to get there and back. They might be easier for humans to explore once we can establish habitation to await an optimal return. There is much to study: excavating near-Earth asteroids differs depending on whether we choose solid metal bodies or a rock/dust pile. There is little known about asteroid interiors.

We should now come to Mars, at the end of the flexible path, but let us go beyond and come back. We have much to say about that world.

In terms of human habitation beyond Mars and the main asteroid belt, it seems doomed by colder temperatures as well as fatal radiation belts around Jupiter. This does not imply that life will not be found there. Saltwater oceans seem present on moons Europa and Enceledus, and perhaps Titan and Ganymede. Research has even suggested evidence of vital metabolism in Titan's atmosphere,[30] despite its –140°C surface temperature. If life has begun on any of these moons, that might be sufficient cause to establish a human outpost nearby, as we shall discuss.

Humans may never explore Jupiter itself directly, but more than 60 known moons orbit it, including Io, Europa, Ganymede, and Callisto with scales and complexity of planets themselves. The danger is Jupiter's intense radiation field, typically extending 5 million kilometers, but in times and places to 60 million, and, especially its inner 500,000 kilometers, sufficiently intense to kill humans in hours. The moon Io is subject to intense radiation (420,000 kilometers out), with Europa (670,000 kilometers) down in dosage by a factor of seven. A human could survive for several days on Ganymede (1,070,000 kilometers) but would soon receive a lifetime maximum radiation dose. Only on Callisto (1,880,000 kilometers) might humans visit a large moon and not suffer unacceptable health effects. Nonetheless, mission profiles for trips to Callisto have difficulty avoiding intense radiation belts closer to Jupiter.[31] Occasionally Callisto passes through Jupiter's extended high-

Table 13.1. *Solar System Venues for Human Habitation.*

Object	Temperature	Gravity	Area (km^2)	Justification	Notes
Mercury	< 30°C	0.38 g	6×10^6	> 77° from equator, < 1,000 km from water	Living underground; heat, radiation, and vacuum at surface
Venus	30°C	0.90 g	4.6×10^8	Pressure 45% of Earth's sea level	55 km altitude; atmosphere CO_2, some sulfuric acid
Moon	0 to 20°C	0.16 g	3.7×10^7	Within 2,500 km of water	Live above or below ground; vacuum and radiation at surface
L5 orbit	~ 20°C	0 g	–	See O'Neill (1977) (see note 27); no area constraint	Need materials (from Moon?); in vacuum; generate artificial gravity
Asteroids	−80°C to 20°C	< 0.03 g	10^7–10^{11}	Various proposed habitat designs[32]	Live above or below ground; vacuum and radiation at surface
Mars	−87°C to 10°C	0.38 g	1.4×10^8	Pressure 0.6% of Earth's sea level	Some water. Exterior atmosphere CO_2, some nitrogen; radiation
Callisto	−140°C	0.13 g	7.3×10^7	Jupiter system exploration outpost	Water and organics. Io, Europa, and Ganymede questionable: too much radiation

Notes: Beyond Earth, Venus is the only place where atmospheric pressure and temperature resemble Earth's in a stable environment. Unfortunately, this does not occur on a solid surface. The Moon and some asteroids present rare instances where the average surface temperature is Earth-like, maintained subsurface because of regolith's insulating properties. These have little atmosphere, however. Starting at Mars, temperatures are lower than Earth's, but this can be helped by artificial heating and/or building underground.

radiation tail, so long-term visitors might consider staying at Jupiter's fifth most massive moon, Hamalia (a.k.a. Jupiter VI, 170 kilometers in diameter), 11,450,000 kilometers away. NASA at times discusses human missions to Callisto in the next 50 years.

Why consider Callisto? Its typical temperature of −140°C makes it the most unpleasant place we assess (Table 13.1.) At least it hosts spectacular views, with white, brown, and blue Jupiter hanging in space, nine times the Moon's angular diameter in our sky, with Io, Europa, and Ganymede speeding past. There will be great work exploring these worlds, especially Europa with its huge saltwater ocean (probably larger than Earth's) beneath roughly 10 kilometers of water ice as hard as granite at this −170°C. Proposals are made to land a robot on Europa, engage a nuclear-powered drill to melt through this ice, and deploy a swimmer into the ocean, starting at pressure equivalent to Earth's ocean 1 kilometer deep. What might we find? With a salty ocean, abundant energy from tidal heating, likely abundant chemistry necessary for life, perhaps life's separate origin is what we will find, perhaps even complex organisms. Many scientists think Europa (or Ganymede, with saltwater ocean beneath about ten times more ice, or even Saturn's moon Titan) might be the likeliest locus beyond Earth to find life.

If we might find life swimming under Europa's surface, we want to be there, at least telerobotically. Round trip distance between Callisto and Europa is 13 light seconds, versus 90 minutes from Earth. As difficult working on Callisto might be, studying an alien ecosystem up close would make it worthwhile.

Someday humans will explore Jupiter, if not in the next 50 years. Jupiter is the monster. It is surrounded by huge, deadly radiation belts. It requires almost 30 times as much energy to simply escape Jupiter's gravitational pull as it does Earth's. Not only is it 318 times more massive than Earth, but it also contains at least a million times more deuterium than Earth and trillions of times more helium-3. If Earth ever proves insufficient for humanity's energy avarice, Jupiter awaits. It could in principle provide enough fusion power to satisfy a trillion times the humanity's current energy needs for a million years.[33] Seemingly all metals one might need are found in the Trojan and Greek asteroids in Jupiter's orbit (not to mention the asteroidal main belt nearby). Jupiter's major moons likely contain oceans with more liquid water than Earth and more than 100 times that amount as ice. Perhaps they contain life as well. There are many reasons to think the path of our long-term destiny passes through the environs of Jupiter.

Then there is the exploration of Mars – the most compelling target for many.[34] Mars approaches a full-service planet, with water, carbon, nitrogen, and enough atmosphere to help protect its surface from solar wind (if not solar ultraviolet light) as necessary for human life. We should not overstate this; atmospheric density on Mars' surface equals that 34 kilometers above Earth, almost twice the Armstrong limit and one-third way to outer space. (See note 14 in Chapter 3.) Unprotected humans would die there in seconds. However, the pressing question is whether we visit Mars first.

The key scientific question in exploring Mars is whether it currently holds life. This is also key in whether humans should land there. If we find even simple microbes on Mars, a Pandora's box of issues opens: Can they infect us or affect our health? Do microorganisms that we carry endanger them, or might they interact in unanticipated ways? Have they already, in our spacecraft on Mars? Have meteorites originating on Earth already infected Mars? If we adapt Mars to our liking, do we shut out indigenous life? Do we seek life buried deep underground or in special areas, before concluding Mars is sterile? How will we recognize Martian life when we encounter it?

We can explore Mars and even send people nearby without inhabiting and potentially contaminating its surface.[35] Mars is orbited by two small moons,

Phobos and Deimos – potential way stations to Mars. Not counting the effort to reach Mars, the moons demand less energy for a landing: Deimos 14 times less than Mars, 70 times less than Earth; Phobos, 5.5 times less than Mars. (Deimos orbits at 23,460 kilometers radius, Phobos at 9,380 kilometers, compared to Mars's surface radius of 3,400 kilometer.) However, one can land on Mars with little rocketry, after "six minutes of terror," decelerating from 3 kilometers per second to zero by fiery aerobraking in Mars's atmosphere. Leaving again takes a rocket.

Mars via Phobos is a Soviet/Russian approach. Although their missions have largely failed (*Fobos 1* and *2* in 1988–1989, and *Fobos-Grunt* in 2011), they were precursors to a clever strategy using Phobos as a space station above Mars. *Fobos 1* and *Fobos-Grunt* failed even to reach Martian orbit, and *Fobos-2* approached Phobos but failed before releasing its lander, managing to take 40 Phobos images, some of them spectacular. In the future, Russians or others might base themselves on Phobos, allowing them to monitor and control promptly any robotic craft exploring Mars's surface. Even accounting for relay satellite delays, any point on Mars is within 0.2 second of Phobos, round trip, permitting rapid telerobotic operation by cosmonauts on Phobos of their rovers, driller, and diggers on Mars – so fast that the human nervous system will barely notice any latency. Phobos itself is fascinating, probably either a captured asteroid or perhaps debris from a massive Martian impact. It appears to contain significantly hydrated minerals, potentially useful. Once thought a carbon-rich C-type carbonaceous chondrite asteroid, it is more enigmatic.[36] Closeup photographs reveal it is streaked with possible strain lines, perhaps from Martian tidal interactions (Figure 13.2).

NASA's Mars program is remarkably successful, with more than 25 years of cumulative spacecraft surface operations and more than 30 years in orbit. For these rovers, however, even lightspeed latency of 8–40 minutes is optimistic; commands from Earth are usually issued once a day. Hence, Mars Exploration Rovers average about 0.1 millimeter per second, never more than 2 millimeters per second. *Curiosity* averages 8 millimeters per second. NASA's midterm goal is to collect samples from varied locations and return them to a contained laboratory on Earth. An ultimate goal is always humans on Mars, with small contingency for intervening discovery of life. Likely if Mars were found crawling with life (maybe underground), NASA would scrutinize the Phobos option, along with an emphasis on rover artificial intelligence.

In terms of consummate science goals, then, humans in space may have compelling justification. We should not send humans directly into some vat of alien life. For many sites, for example, Europa, we could not do so even if we desired. If we find life, to study it promptly and thoroughly we need humans nearby, within light seconds. Indeed, if we discover complex, animate life, or any living system changing unexpectedly, we cannot trust preprogrammed robots to perform adequately. We would need humans to study life; this is likely to be the case for decades.

Telerobotics are essential to the central, long-term scientific goals of space exploration and why lunar robotic exploration is essential. Although the 2004 Vision for Space Exploration's lunar goals included lunar testing of human systems for use on Mars, testing telerobotics is likely just as important. If we discover alien life in this

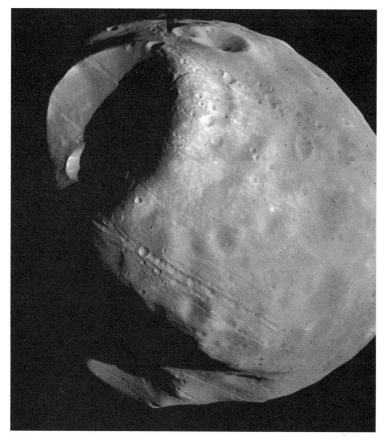

Figure 13.2. **Future base for Mars**. Phobos, 11 km diameter Martian satellite, photographed by *Viking 1* on October 19, 1978, from about 620 km away. The large crater at the upper left is Stickney. Phobos orbits about 6,000 km above Mars. Russia in particular is considering Phobos as a way station to the Martian surface. (Montage created by Dr. Edwin V. Bell, II of NSSDC/Raytheon ITSS, from NASA images F854A81, F854A82, and F854A83.)

Solar System, we will explore them with these techniques. A human outpost on Phobos is an option, as much as exploring asteroids, and easier than Mars. If we find signs of life, for example, organics in their surface ice, on Europa, Titan, Enceladus, or similar ice moons, we would need to establish an outpost on Callisto or near Saturn, an expensive proposition. Nonetheless, this could become the century's most compelling scientific question, and a healthy society would probably make the investment to answer it.

This begs the question of why Mars is our ultimate goal and why put humans on its surface. To establish humanity off Earth, there are easier ways.[37] If we seek alien life, consider the two alternatives. If Mars has life, should we set people and their attendant microorganisms into this environment? Likely not. If Mars is lifeless, do we necessarily go there first? We should finish our homework: we must explore robotically Solar System worlds in which life is plausible. We would regret first investing a trillion dollars to put humans on Mars, only to find robotically that the

Solar system's alien life flourishes on Europa or Enceladus. Exploration goals should derive from discovery more than historical ideas of these worlds' prospects. We should assess life's possibilities on both Europa/Enceladus and Mars: they are not only different – one oceanic and the other rocky – but also have diverse Earth cross-contamination histories. We might be originally Martian or Martians evolved from some meteorite reaching Earth, but likely we arose separately from any Europan life. If we find life out there, humans can follow up, in person or more likely from a light second away. If we find no life, perhaps we can inhabit these worlds ourselves, but that is not scientific justification.

Lunar colonies have been discussed technically since Tsiolkovsky.[38] If we want them as insurance against earthly global disaster, off-world society must be potentially self-sustaining, capable even of re-inhabiting Earth. As risible as 2012 campaign gaffs by U.S. presidential candidate Newt Gingrich were on the topic (although he was misquoted by most media), one can ask what lunar colonies entail. Different catastrophes could strand Earth in various states of distress: viral pandemic, nuclear war, climate change, cometary/asteroidal impact, supernovae/gamma ray burst, and so on, so the course of recovery and its difficulty is uneven. Many think we should have an option.[39] Nonetheless, Apollo required some 300,000 workers (and supporting society) to send just 12 humans to the Moon. Although a lunar colony might be replete with technically savvy workers and aerospace competence, reversing the process to restore humans to Earth would require many thousands. A colony would be monumental (but no larger than many projects undertaken recently – wars or nation building – in the trillion dollar range). Water on the Moon and habitable lunar lava tubes make such goals easier, if ambitious. On any other world it will be much more difficult, excepting perhaps Mars one distant day. Mars is also a low-pressure, regolith-covered world with trace subterranean water, with lava tubes that might best house a colony; a precursor lunar habitat program could prepare us for living on Mars.

Robert Zubrin is a strong protagonist for rapid human ascent to Mars, writing several books on the topic, recommended to interested readers.[40] Zubrin argues forcefully but often subjectively, sometimes off the mark, for example, he recently equated the Moon to Greenland as a barren way station intermediate to North America (read: Mars). A fair history of Viking Atlantic exploration proves this to be a damaging analogy for Mars.[41] Vinland failed, but Greenland was crucial and long-lived, but analogies are feeble.

Viking history teaches that outposts can fail, as could later European settlement, for example, the Roanoke colony. Those were different days; no TV audience watched the natives attack Ericson's settlement or scrutinized a starved English colonist scratching *CROATOAN* in a tidewater Carolina tree, to vanish forever. No ready help was forthcoming, and ignorance, poor judgment, or bad luck was often fatal and still could be. However, few people noticed. In contrast, one lost Mars expedition would sour public taste for space exploration for decades. Recently *Curiosity*'s Radiation Assessment Detector showed dosages in transit to Mars likely exceed human lifetime limits and might be fatal,[42] separate from the many risks associated with Mars itself.

Bring courage. Distant goals often seem foolish: had early humans leaving equatorial Africa known they would confront glaciers, bitter winters, and endless ocean, they might have stayed home. Yet without their efforts, we would be another species, maybe extinct. Explorers and colonists 500 years ago had roughly a 50-50 chance of returning home or finding a new one before dying. Today's astronauts know there is a few percent chance of perishing; still they volunteer. We must strive foremost to not increase their risk senselessly. Uncharacterized medical issues plague long missions beyond Earth's gravity and magnetic fields: radiation exposure, muscle degradation, and others (see Chapters 11 and 12).

Any biological activity on Mars is problematic for rapid human settlement, until we learn otherwise. On this I diverge with Dr. Zubrin but will not paraphrase him. Rather, I argue against deciding now on inhabiting (or terraforming) Mars. Until we know what lives there, no benefit that can derive from settling Mars justifies possible extinction of its life. Terraforming Mars would take millennia and would demand great resources from Earth.[43] Establishing if Mars is vital or sterile would take much less time and resources. In this shorter interregnum, if Earth is threatened, humans could more rapidly establish a self-sufficient lunar outpost, requiring even fewer resources.

An aggressive, anthropocentric approach to inhabiting Mars places small value on potential knowledge from studying diverse life, whereas finding alien life may grant a boon to biology and biotechnology. We would gain our first perspective on processes essential for life versus simple earthly accidents in prebiotic and early evolutionary sequences. Discovery of alien life might pull back the curtain on biotechnologies to save us if we are in peril. An evolutionarily successful species would try to learn this before barging into a new environment. It happened before: contact of Eastern and Western hemispheres 500 years ago resulted in deaths of millions of people, less from evil intent (of which there was plenty) but via unintended, unknown microbial infection. Europeans should be grateful that perhaps they gained only syphilis in the exchange.[44] However, if Earth was in immediate peril and Mars was the sole option, ethics and scientific potential would take a back seat. For lesser reasons, thoughtless destruction of Martian life is too hard a sell.

Earth's microorganisms compete with other life in rich environments, so exterior membrane surface chemistry tuned to dangerous and friendly molecules is essential. Tactics in this environment might not appear in primitive, sparse ecosystems. On Earth endosymbiosis is common – small life forms integrated into larger organisms as useful structures, especially microorganisms' organelles: mitochondria (in most eukaryotic cells), chloroplasts (plants and green algae), rhodoplasts (red algae), apicoplasts (malarial parasites), and leucoplasts (plants). These might not arise in sparse ecosystems, for example, Mars. Likewise, tactics appropriate to primitive, sparser ecosystems might be obsolete on Earth. Instead of life interacting via chemical tags on protective membranes, perhaps nutrients are attacked and ingested by less specific means: poisons or acidic/caustic reactions or thermal/mechanical disintegration.[45] In such contexts terrestrial life might not present effective surface chemistry but instead a smörgåsbord of tasty substances. Processes exist that some earthly life has superseded. Cells can combat biological agents with innate immune response rather than chemical tagging (adaptive immune system). Innate response is less efficient because it confers no long-term immunity and is considered evolutionarily older and found in plants, fungi, insects, and primitive multicellular life.

Infection need not occur on usual days to weeks time scales. Consider *prions* – inanimate protein particles duplicating by seeding energetically favored foldings of living protein once they diffuse into animal nervous systems, not alive but reproducing, unsophisticated but robust to thermal or chemical countermeasures, slow-acting but fatal (causing various transmissible spongiform encephalopathies, such as mad cow disease[46]). Imagine humans exposed to simple life forms that on Mars attack salty, aqueous concentrations of organics in some unrecognized, slow way. Imagine this agent not tuned to cells, so living both in and on cells, found on infected individuals' skin and thus transmitted. After years we might find people (and other species) thus infected, with no better warning or treatment than we did originally for prionic diseases.

Zubrin argues that if Martian life exists, it already reached Earth via Martian meteorites (at least 100 known). Not knowing what Martian microorganisms can endure, we cannot say if they survived the trip. If this happened, it likely did so during the Late Heavy Bombardment, 3.8 billion years ago, when Earth life was completely different. Whenever past extinctions involved alien pandemics, the fossil record might not reveal this.

Zubrin defines *back-contamination* as humans absorbing and spreading Martian viruses, describing it as "just plain nuts" because no such virus ever encountered humans as hosts.[47] This reasoning is circular: Martian life cannot interact with Earth life because life only interacts as does terrestrial life. Martian organisms might exist that are not viral, bacterial, or eukaryotic. We know little of alien biochemistry, with scant idea of Earth life's first generations. We should hesitate to predict how these would interact with humans or terrestrial microorganisms.

What might alien life be like? Knowing only one inhabited world, our best proxy for probability is comparing durations of life's different phases. For instance, the idea seems unlikely of finding many alien civilizations at development levels equivalent to ours, as in *Star Trek*'s science fiction universe, because our culture and technology transforms over hundreds of years compared to Earth's 4.54-billion-year age. Likewise, life consisted only of single cells or undifferentiated multicellular colonies during at least 3 billion years of Earth's existence (3.6–0.6 billion years ago). Only in the past 600 million years have differentiated, complex life forms existed: advanced plants, animals with specialized limbs, sensory organs, and so on. Likely if we find life more primitive than us on another planet, it will be undifferentiated, probably single-celled, with significant but lesser chances of having limbs, and so on.

The Universe is about 13.8 billion years old, three times older than Earth. One finds signatures of abundant heavy elements, necessary to form planets, 7 billion years older than our Solar System. The typical exoplanet is about 2 billion years older than Earth, with some 4 billion years older,[48] and most stars provide relatively stable conditions over this time. Did life arise on these exoplanets billions of years before life on Earth? Is it billions of years more advanced?

One day in mid-1950, physicist Enrico Fermi was eating lunch with Edward Teller, Herbert York, and Emil Konopinski. In those days flying saucers were frequently discussed, and the four pondered them and interstellar space travel. After pausing, Fermi reportedly looked up from lunch and asked: "Where is everybody?" The three correctly interpreted Fermi to mean there seems to be a universal existential paradox, because reasonable calculations imply that many alien civilizations have existed. If interstellar flight is possible, Fermi reasoned, some alien civilizations should have already visited Earth. So, where are they? This is the Fermi Question or Fermi Paradox. Let us examine the numbers.

Our Galaxy has about 400 billion stars. The Hubble Ultra Deep Field implies some 100 million galaxies exist. Carefully multiplying galaxies' numbers and their stellar content yields 70 sextillion = 7×10^{22} stars in the known Universe.[49] Planets likely orbit most of these stars.[50] This rivals the number of sand grains in all Earth's deserts and beaches, and the total number of cells in all 7 billion humans. Inconceivably huge numbers of planets exist in the known Universe. The idea of Earth being the only inhabited planet is untenable. Unless extremely strange forces are afoot, there is alien life in the Universe.

Many of these stars are so distant that they will die and destroy their planets before their light even reaches us. Consider instead a closer realm: our Galaxy. Our Galaxy's 400 billion stars probably altogether have at least 400 billion planets. What fraction of these can support life? Planet searches, as of 2012, have found only a few with masses close to Earth's, from thousands of likely exoplanets. This says more about our techniques than planets out there, however; most stars of nearly typical mass have planets slightly more massive than Earth (and one would guess they have Earth-like mass planets, too) (see note 48). Of these, some are discovered at the right distance from their star to be neither too hot nor too cold for life. This standard definition of *habitable zone*, describing where planets could maintain liquid water on their surface, is "surface chauvinistic." If life resides insulated subsurface, the planet can be colder (and in some cases warmer) than implied by the habitable zone. An ice layer can largely isolate a planet or moon from conditions of solar heating. If the planet holds a thick atmosphere, life can tune ambient temperature and pressure by picking altitude above the surface (like Venus). Liquid water is dependent on atmospheric pressure but requires temperatures at least 0°C unless pressures are very high and below 375°C even at highest pressures. More to the point, overlying atmospheric insulation can lead to greenhouse warming. The habitable zone concept is misleading: ice moons in exoplanetary systems, perhaps like Europa in our Solar System, might possibly support life. Of eight local planets, we have one, maybe two, maybe more that could support life. What fraction does that correspond to on average? An educated guess implies at least 60 billion worlds in our Galaxy where life could start (see note 48).

If life can start, does it? Again, we have essentially no statistical sample to support any position. Again, we could look at time elapsed. Until 3.8 billion years ago, Earth's surface, like the Moon's, was tortured by the Late Heavy Bombardment's huge impacts, some so violent as to sterilize Earth's surface. Life seems to start on Earth at least 3.5 billion years ago, maybe much sooner.[51] It seems that nearly as soon as life could start, life did start. We have no reason to suspect that habitable planets frequently go sterile. For the sake of discussion, let us assume Earth was somewhat lucky and that on average 10% of all planets that could support life actually see it take hold. That means at least 5 billion worlds have hosted life in our Galaxy.

On average how advanced does life develop on these worlds? We have essentially no idea. As we said, for 3 billion years, Earth was inhabited by single-celled organisms and undifferentiated multicellular forms such as bacterial mats. Suddenly, 550 million years ago, abundant complex life sprang into being. Most categories of life, the phyla, arise in this brief several dozen million years – the Cambrian Explosion of life's diversity and complexity.

Why did this event occur on Earth? Will it happen elsewhere? We do not know yet. Explanations divide between geological and biological evolutionary mechanisms, with interplay between the two. We can postulate oxygen's rise in Earth's atmosphere and ocean allowing larger, vigorous animals, because motion and energy depend on rapid biological oxidation and are easier for large animals when high O_2 concentration allows surface exchange in lungs or gills to supply cells deep in large bodies. An issue is that O_2 levels increased slowly because of photosynthetic algae over the 2 billion years before the Cambrian, and not so much at that time. About 850–580 million years ago, ice may have largely covered our planet as Snowball Earth. Perhaps liberation occurring with this thaw allowed the rise of animals. The 50 million years before the Cambrian may have seen the first embryos, which could sustain rapid evolution. For whatever reason, it was a special event.

Travel between solar systems requires intelligence. (If instinctive modes of interstellar space travel exist, we have not found them.) Reaching other stars implies advanced technology, something unlikely to evolve via natural selection. If animals arise on other planets, how often do they achieve intelligence? Again, we have only our Earth as an example. We do not have only our own species, however.

Humans may be Earth's most intelligent species, but have intelligent company on our planet. Other primates are intelligent, probably as much as hominids several million years ago. If we vanished, perhaps our evolutionary cousins would replace us within 1% of Earth's age, if they survived. Cetaceans such as dolphins and whales have large brains – some larger than humans' – and brain structure and components (spindle neurons) like primates, and in ways elephants. Primates, cetaceans, and elephants deviated from common ancestry some 70 million years ago, only 13% of time since the early Cambrian. In contrast, cephalopods (some able to solve puzzles, run and mazes, use tools, and communicate with diverse visual cues) are eight-limbed, boneless, and diverged from our ancestors 500 million years ago. We imagine octopi or cuttlefish – now comprehending tools and something like language – in 30 million years becoming technological, despite being as distant from human as any animal.

We reconsider "where is everybody?" This presumes life arising elsewhere in the Universe to evolve to a technological sophistication in which interstellar travel is likely. Because likely billions of planets in our Galaxy alone exist where life might arise, and because crossing the Galaxy might take only 100,000 years (or a million or even a billion, but certainly much less than 10 billion that our Galaxy has held planets), subtle reasons must explain the lacking evidence of aliens visiting Earth. What is this paradoxical reason?

We can treat this as a chain of events, each with its own probability, taking a planet capable of supporting life up to advanced life capable of traveling the Galaxy (and visiting Earth). Starting with billions of planets, some event in this sequence must have low probability – a "choke point" through which few worlds and their life forms pass.[52] Humans and Earth life have survived, and already we imagine interstellar travel in the next 1,000 years, an instant versus the Universe's age. Remember that our spacecraft are already leaving the Solar System for interstellar space. Can we identify a choke point in our past? Are we approaching a choke point in our near future, before achieving interstellar travel?

There are choke points in our past. Most species disappeared in the Permian-Triassic extinction 252 million years ago, and to lesser extent the Cretaceous-Tertiary event killing the dinosaurs 65 million years ago. Humanity itself dwindled 71,000 years ago, to several hundred or thousand people, likely a result of Indonesia's Toba supervolcanic eruption. Another choke point may be the Cambrian Explosion. Microorganisms or bacterial mats build no interstellar spaceships. For 3 billion years, two-thirds of Earth age, these were its predominant and most advanced life forms. Maybe we were lucky; maybe animals take 10 or 1,000 times longer to arise on typical planets. Maybe many planets support simple life forms but rarely products like a Cambrian Explosion's. Perhaps vanishingly little life sports limbs, sensory organs, and so on.

This poses a crucial question for Solar System exploration: Are there life forms with limbs, eyestalks, flippers, fins, and so on? These persist from Earth's event 540 million years ago. If we find crablike or fishlike creatures under Europa's or Enceladus's ice, this would suggest animal-like beings common on other worlds. This would not be the choke point. Mars is less urgent in this regard, because no evidence exists for surface animals.[53] Nonetheless we might seek fossils from its wetter past or animals underground.

If the choke point is not in our past, is it in our future? If they never visited Earth, will something prevent us from becoming one of them? This is worrisome. Does something drastic happen between now (with us building spacecraft reaching interstellar space) and the future (when we can build spacecraft to reach the stars)?

There are many possible responses to the Fermi Question. Let me categorize some logical choices in a tree diagram (Figure 13.3; here "we" refers to sentient life on Earth; "they" to advanced alien life).

Some options seem less likely, for example, Option F, that aliens reached Earth but are unaware of life here – sounds stupidly implausible but might occur. There are amazing numbers of responses to the Fermi Question, and I banish most of these to a sidebar. [Box 13.1] I list 61 general responses to the Fermi Question, so maybe before perusing this sidebar the reader should let me explain a few. (These examples are indicated by the parenthetical topics in the Box 13.1, e.g., "'Wow!' signal.")

Stepping Stone

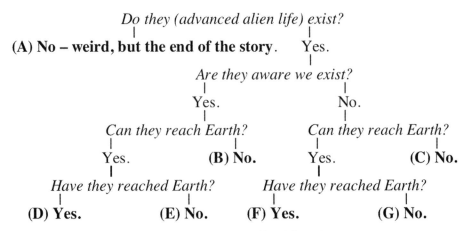

Figure 13.3. Decision tree categorizing responses to Fermi Question.

Box 13.1: Many Answers to Fermi

Responses to the Fermi Question, divided using the categories in Figure 13.3:

A) They don't exist:
   1. We are the first.
   2. We are the first in our part of the Galaxy.
   3. They never will exist.
   4. The rise of life/animals/intelligence/technology is rare.
   5. The will to explore/expand is rare.
   6. They have been replaced by artificial intelligence that does not care about life.
   7. All advanced beings die off before reaching us.
   8. Planets are much more rare than we think.
   9. Planets like Earth are required and rare.
   10. Solar systems like ours are required and rare.
   11. Our Galaxy wipes out most life found around most of its stars.
B) They cannot reach us but are aware we exist:
   12. Interstellar travel is too difficult but they have recognized our signals.
   13. They are waiting for us to answer their signal.
   14. Their thinking is so different we cannot tell they are communicating.
   15. They do not want to communicate.
C) They cannot reach us and do not know we exist:
   16. Interstellar travel and receiving our radio signals are too difficult.
   17. They are waiting for us to answer their signal.
   18. They do not want to communicate.
   19. Most life arises on ice moons and never escapes to the surface.
D) They know we exist and have been here:
   20. We are descended from aliens.
   21. They have somehow blended with us.
   22. They are meddling in human affairs.
   23. They have been here and left evidence (1908 Tunguska explosion, etc.).
   24. They last visited when we were too unsophisticated to recognize them.
   25. They visit occasionally in UFOs, etc. but leave nothing behind.

Box 13.1: (continued)

26. They have been here and our government(s) covers up this fact.
27. They prevent us from knowing about them (*Star Trek* "Prime Directive" or Zoo Hypothesis).
28. They are watching us subtly to see if they should reveal their existence to us.
29. They are preparing a surprise attack/colonization/transformation for Earth.
30. They are watching us to see if we generate useful knowledge (Interdict Scenario).
31. They are so unfamiliar/unlike what we imagine we do not see/recognize them.
32. They trick us into ignoring them.
33. We fool/rationalize ourselves into ignoring them.
34. They create a false appearance of the external Universe to deceive us (Planetarium Scenario).
35. God the Creator exists and things happen according to plan.
36. Their probes are too small/undetectable for us to have noticed.
37. There is nothing here to maintain their attention.

E) They are aware we exist and can reach Earth but have never visited:
38. They have recognized our signals but have not had time to start the trip here.
39. They are on their way but have not reached us.
40. Natural evolution at the cosmic level has led them to be shy.
41. Our Solar System has nothing that they need.
42. They are just not interested in us.
43. They find us too disturbing to deal with ("They are made out of meat!").
44. They are monitoring us from afar to see if they should reveal their existence to us.
45. They monitor us remotely in case we generate useful knowledge (Remote Interdict Scenario).
46. They are remotely preparing a surprise attack/colonization/transformation for Earth.
47. They misunderstand our signals or other information about us.
48. They are too busy doing other things.
49. They have gone somewhere else (higher dimensions, etc.).

F) They have been here but are unaware of our existence:
50. They came before life/intelligence took hold;
51. They are not looking for life/intelligence;
52. They are looking for life/intelligence but not as we know it.

G) They could reach Earth but have not and are unaware of us:
53. Interstellar exploration has gaps in the outer reaches of our Galaxy, our neighborhood;
54. Advanced beings choose not to explore the Galaxy;
55. They are signaling and await a response but we have not recognized this;
56. We heard their signal but we have not answered ("Wow!" signal);
57. We need to listen longer for their signal;
58. Advanced beings are only listening and not signaling;
59. If they find us, they will kill us (Berserkers).

AND BEYOND:
60. We simply have not conceived of what is going on ("Just because you cannot imagine it does not mean it cannot be real.").
61. The Universe is so constructed that this question makes no sense.

> **Box 13.1: (continued)**
>
> Some responses are mine; many are found in *If the Universe is Teeming with Aliens... Where is Everybody? Fifty Solutions to the Fermi Paradox and the Problem of Extraterrestrial Life* by Stephen Webb, 2002, New York: Copernicus Books. Others include *Extra-Terrestrials, Where Are They?* by Michael Hart & Ben Zuckerman, 1982, Cambridge University Press; "Where are They? I Hope the Search for Extraterrestrials Finds Nothing" by Nick Bostrom, May/June 2008, Technology Review, 72; "Cosmology, Extraterrestrial Intelligence and Resolution of the Fermi-Hart Paradox" by Paul S. Wesson, 1990, Quarterly Journal, Royal Astronomical Society, 31, 161; "The Fermi Paradox, Self-Replicating Probes and the Interstellar Transportation Bandwidth" by Keith B. Wiley, 2011, submitted to Icarus, http://lanl.arxiv.org/abs/1111.6131; "Too Damned Quiet?" by Adrian Kent, 2011, http://arxiv.org/abs/1104.0624; "A Solution to the Fermi Paradox: The Solar System, Part of a Galactic Hypercivilization?" by Beatriz Gato-Rivera, 2006, World Mystery Forum 2005, http://arxiv.org/abs/physics/0512062; "The Sustainability Solution to the Fermi Paradox" by Jacob D. Haqq-Misra & Seth D. Baum, 2009, http://arxiv.org/abs/0906.0568.

Along with discussing whether they exist and whether they know we exist comes the question of them contacting us. There are reasons why an advanced civilization might choose to not communicate. They might simply find us too disturbing. Contrarily, this provided premise for contacting humans in the film *The Day the Earth Stood Still*; we might be too disgusting, as in "They're Made Out of Meat" by Terry Bisson.[54] I doubt this applies to all advanced civilizations. More seriously, we see ideas of forces killing civilizations with which they come into contact: Berserkers (after the science fiction of Fred Saberhagen), perhaps self-replicating machines meant to seek life but mutated into destroying it, or (à la Saberhagen) defensive weapons that veer into rampaging offensive mode. Alternatively, aggressively paranoid civilizations might improve their survival chances by killing potentially competitive spacefaring species before they progress too far. A variant of answer 51 above is that they were looking for some sort of resource unrelated to humans and depleted Earth of it, whether or not they were aware of our current or future presence.

Self-replicating interstellar spacecraft are discussed in context of the Fermi Question. Although humans have several spacecraft entering interstellar space, these need millions of years to approach other stars. To avoid straining their home world with building a myriad of fast spaceships to span the Galaxy, a clever civilization might build a ship replicating itself from new planetary matter. John von Neumann invented the concept of self-replicating automata and Ronald Bracewell imagined automated interstellar probes, so we call these Bracewell–von Neumann probes. Designs for these are proposed.[55] They surpass our current capabilities but might not in a century. Perhaps civilizations make these to collect information to send home, with probes spreading exponentially over time. Even moving no faster than our current craft, such robots would cover the Galaxy in several percent of its age.

The concept of self-reproducing probes resembles life: they seek and exploit resources (energy and materials), using those for internal functions and to reproduce, maintaining defined life cycles. Like life they must carry information for their structure and operation to the next generation. That may be problematic: if that data is corrupted, probes may mutate, perhaps leading to Berzerkers (Box 3.1) or at least mission breakdown. Also, making memory devices for these data might be the hardest step in the reproductive manufacturing process. Civilizations launching these probes might modify Bracewell–von Neumann's concept by uploading data to the probes (particularly because downloading data from them is likely their reason for existing). How these data are stored, interpreted, and implemented might suffer reproductive corruption and mutation. Perhaps a safer (and economical) option is to launch the first generation of probes with thousands or millions of memory chips. If one is corrupted, at worst it produces limited numbers of mutants. If chips are small, caches of even large numbers might be shielded from radiation or other mutagens. In addition, cross-checking redundant chips could ensure hereditary quality control.

Some explain the Fermi Paradox by interstellar travel being slow and difficult, with aliens lacking motivation to make the trip. Despite this conjectured cosmic ennui, there exists good reason: avoiding extinction. In our Galaxy, some 50 billion stars have died, dooming comparable numbers of planets. An astronomically savvy civilization would have thousands or millions of years' notice of their stark choice: move or die. I imagine humanity facing this would go not gently into oblivion's night, but head for the stars. If interstellar travel is possible, and alien intelligence does encounter such conditions, surely one civilization, if not thousands, millions, or billions, survived via interstellar travel. Life grows ever adaptable: genes, epigenetics, memes, society, tools, writing, software, and so on. They would survive disaster somehow.

An alternative channel is *panspermia* in which life propagates via unconscious means to other worlds, even other planetary systems. This idea seems more plausible once one knows unpowered ballistic masses can insinuate themselves into and out of solar systems via gravitational assists, but the probability of entering a planetary system and landing on a planet's surface is low. This is exacerbated by the likelihood that most habitable worlds might be ice moons, with "earths" of habitable surfaces probably rare. We discussed earlier many places in the inner Solar System where humans could live. Unfortunately, the chances of an unguided seed finding one of these sites is minuscule if one excludes Earth – the kind of world panspermia may require.

The Fermi question's most direct answer is that they contacted us, and we heard them and did not answer. One candidate for this, the famous "Wow!" signal of August 15, 1977, was intercepted by Dr. Jerry Ehman at Ohio's Big Ear telescope searching for extraterrestrial intelligence (SETI) at radio frequency 1420.4 megahertz. It lasted at least 72 seconds but was not seen again, from a seemingly extraterrestrial source 23 degrees from our Galaxy's center, about 2 degrees from bright star chi Sagittarii. Dr. Ehman and others hesitate claiming extraterrestrial origin for this signal.

Likely success for the larger SETI program depends on how many civilizations are in radio-transmitting stages of development. This is expressed by the Drake Equation, starting with how often stars form in the Galaxy (stars/year), multiplying this by planets/star, lifetime of radio-transmitting races (in years), and a string of probability factors to yield the number of transmitting worlds. [Box 13.2] The issue is that reasonable values for these factors mean advanced civilizations arise every few hundred or thousand years. If we find other intelligence in our part of the Galaxy, it is likely to be as similar to us as cavemen or demigods, not between. As we said before, we likely do not live in a *Star Trek* universe with many alien civilizations at comparable development stages in the Galaxy. Our only such peers might be far,

---

Box 13.2: The Drake Equation

The equation was developed by Dr. Frank Drake in 1960 in considering prospects for radio SETI. He expressed the number of radio-transmitting civilizations in the Galaxy N_R:

$$N_R = R_* f_p\, n_E f_L f_I L_R$$

where R_* is the Galaxy's star formation rate (in stars/year, actually the rate several billion years ago when the planets were formed now hosting civilization), f_P is the fraction of stars with planets, n_E how many Earth-like planets (or planets capable of bearing life) per star with planets, f_L the fraction of Earth-like planets to sprout life, f_I the fraction of life-bearing planets yielding intelligence, f_C the fraction of intelligences that radio transmit, and L_R the radio lifetime of such cultures. We could also add other Drake-like chains of factors for ice moons and so on. In 1960 there was little information on most of these factors, but we have developed plausible estimates. For instance, in his 1980 series *Cosmos* (Episode 12: "Encyclopædia Galactica"), Carl Sagan makes educated guesses at these quantities, largely agreeing with estimates today: 400 billion stars in our Galaxy, 25% of stars with planets, 2 Earth-like planets per planetary system, life arising on half of suitable worlds, 10% of inhabited planets giving rise to intelligence, 10% of intelligences communicating via radio, and a century of radio life per civilization. Sagan's numbers imply $N_R \approx 10$ broadcasting civilizations, and scientific basis for these estimates has changed little (despite confirmation by many exoplanet discoveries). If we can listen 10% of the way across the Galaxy (hence approximately 1% of the Galaxy's disk), we could hear civilizations only approximately 10% of the time. Since then we have pinned down f_P and n_E much better, and perhaps in several years might estimate f_L by searching for potentially habitable planets carrying biotracers in spectra of their atmosphere and their surface: free oxygen and ozone, perhaps methane in some cases, and maybe even direct signs of life like chlorophyll. However, if there are ways to interact with civilizations longer than radio communications, the number of relevant civilizations might be much larger. As Sagan points out, if only 1% of civilizations find some way to survive times comparable to the Galaxy's lifetime, there might already millions of them, with the typical civilization billions of years old. What would be the implications of their existence? How might we be aware of them? This is reason for the Fermi Question. If any such civilizations visited the Solar System and left evidence behind, such as an artifact, this might reveal their existence. We discuss this below.

far away. *Star Trek* is enjoyable science fiction, but no model for what to expect.[56] The *Star Trek* universe arises when the age of the Galaxy is shorter by a large factor than the combined ages of all of the interstellar spacefaring civilizations in the Galaxy, meaning that they overlap significantly. In contrast, very little overlap produces what we call the Lonely Galaxy, where interstellar spacefaring civilizations are unlikely to ever encounter each other. We can also describe mathematically other conditions, such as The Old Ones, in which many civilizations survive in our Galaxy that have lived much longer than ours, or Cavemen/Demigods, largely the opposite of the *Star Trek* universe, in which the only civilizations a particular spacefaring civilization might encounter are radically more advanced or less advanced than the particular civilization.[57]

A popular response to Fermi's question is they are already here, as evidenced by unidentified flying objects. First, UFOs are not identical with alien visitors. There are conspicuous atmospheric phenomena identified only recently: sprites, blue jets, and ELVES (Emissions of Light and Very Low Frequency Perturbations from Electromagnetic Pulse Sources).[58] Ball lightning is a localized, often fast moving electrical phenomenon appearing as luminous spheroids, seemingly real but poorly explained. There are likely other such atmospheric phenomena. Until recently many such would have been considered "unidentified" and "flying," some reminiscent of early UFO reports.

Before World War II there were few claims of what today might be called UFOs, but during the war many high-altitude aviators reported small, luminous, hazy spheres ("foo fighters")[59] said to pace warplane formations non-threateningly, only to zip away suddenly or disappear. In 1946 Sweden "ghost rockets" made news, seen visually and by radar but still unexplained. Their peak during the Perseid meteor shower is suspicious, but their explanation by U.S. military intelligence as Soviet V-2 missile tests from nearby Pennemunde was not borne out. In the unfolding Cold War, authorities took these and later reports seriously.

After 1945 UFO reports exploded in frequency and notoriety, for example, the 1946 Mt. Rainier flying disks and 1947 Roswell incident, and if true, are difficult to explain conventionally. These are reported as flying "craft," seemly aerodynamic but capable of unconventional performance. I am no UFO expert, but in Chapter 9 we discuss statistically how to analyze anecdotal reports by untrained observers, for transient lunar phenomena. After TLPs made the popular news in 1957, the statistical structure of their report distribution changed radically. Sites where events were never reported suddenly became most active. Before 1957 their statistical behavior is robust, with no evidence that observers knew what others reported. After 1957 people often looked where they expected events and tended to report what they thought they should see. UFO sightings after 1945 may be unreliable for the same reason: observers are now biased. This may be as simple as people needing narrative to explain the unexplained, and alien spacecraft are compelling narrative.

Alien UFOs seemingly violate the no *Star Trek* universe idea above, which would seem implausible. Typical post-1945 UFO reports sound too much like technological products slightly more advanced than ours. One expectation is that alien technology reaching Earth might be radically advanced. It seems strange that aliens would let us observe them by accident; they would be clever enough either to hide effectively or announce their presence. Instead, these craft are too similar to what humans might build, or more precisely, what we might imagine building in the near future. In cases where aliens themselves are reported, they seem much like us, too bipedal, too human visaged, and close to how we have been shown aliens might appear. They are too much like what we imagine. They appeared only after *The War of The Worlds* and shortly after the beleaguered citizens of London and Antwerp were bombarded by supersonic rockets from outer space in 1945, after we began planning our own missions into space. New ideas, improperly understood, generate confusion. This may factor here. I do not know if most UFOs are real or if any are alien. Let us move on.

Discussion of planetary numbers, probabilities, and time scales leaves us with a strange thought. Earth is a relative latecomer to our Galaxy's planets, as presumably is life on Earth. This begs the question: If typical planetary systems are two billion years older than ours, is the most advanced life form orbiting each star typically eons older than us? Are many aliens inconceivably advanced beyond humans, by a billion years or more? Earth life a billion years ago was limited to single-celled organisms; we have no idea what billion-year advanced beings would do. Perhaps this answers the Fermi Question: advanced civilizations quickly reach levels of activities unrecognizable to us. Maybe they do not even share our space; string theory or M-theory explains physics imbedded in higher-dimensional space, six or seven dimensions beyond our experience.[60] Perhaps such advanced beings would busy themselves in this higher space. Living with us at our level might be as interesting to them as residing with bacteria in a Petri dish. We cannot "take the limits of our own field of vision to be the limits of the world" as we try to do.[61]

Either these super-aliens do not exist, do exist but are somehow unaware that we do, or know we are here. (What is awareness to a superalien?) Their existence, although perhaps intimidating, would inspire optimism. Perhaps a more interesting question for us is Does super-aliens' existence have any effect on us and our (presumed) natural world? Are there artificial aspects to existence that would not arise without these beings? It is hard to consider. Do insects crawling along your home's baseboard realize that they live in a largely artificial environment? Which of your pets do?

The Planetarium Scenario responds to Fermi by taking this to the extreme: our entire view of the Universe is generated by super-aliens hiding their existence from us. There are less drastic versions: the Zoo Hypothesis, in which super-aliens leave us alone so that we may develop unhindered (like *Star Trek*'s "Prime Directive"). In the Interdict Scenario super-aliens monitor less developed civilizations to see if they create new ideas worth adding to the cosmic compendium of knowledge. There are many alternatives.

If I were forced to guess, I would admit that I suspect that aliens exist, even super-aliens. Does their existence ever affect us? Would our past, present, or near future existence differ without them?

This is no issue mastered by religion or science. Any super-aliens presumably arose after the Big Bang, so they are not god-like creators. I neglect whether super-aliens have connection to usual human concerns with mortality. Have we any innate knowledge of such beings? Theological ideas on this extend back to at least St. Augustine (410 AD).[62] Scientific analyses of perceived divine contact precede William James' *Varieties of Religious Experience* (1902),[63] with schizophrenic or extreme bipolar patients claiming interaction with superior beings, often with paranoid result. Schizophrenic episodes of voices typically differ from perceived divine speech in being more frequent, prolonged, and often menacing.[64] Do they differ because of social influence or more organically? Scientists study perceived divine contact as normal brain operation and evolutionary adaptation[65] but see nothing like "antennae" sensing super-beings. Theology explores reasons for such perceptions, for example, revelation received empathetically, or humans made "in God's image" as in Mormonism's *Book of Moses*. Many paths to sensing the Godhead are prescribed: meditation, prayer, fasts, and icons. If super-aliens exist, we presume them universal on human scales. If religion pertains to them, we should expect diverse faiths and traditions to agree on their effects and evidence. We are far from such concurrence, even though most religions consider more human issues. Where is the universal evidence? Perhaps the signal from those who have such "antennae" has not overcome the noise from the majority of us less attuned. There may be no quick path to understanding. Meanwhile science brooks no premise of superhuman intelligences, with no method to investigate them if they exist. Instead humans seem predisposed to assign sentient causes in nature: what rustles the grass might be the breeze, but better to anticipate a stalking lion. Super-aliens might exist, but we should demur concluding they do. Finally, thinkers can over-extrapolate human impact of changes in our world view;[66] any super-aliens might play little role on human scales.

Considering super-aliens' effects seems at first antiscientific. They immediately violate Occam's Razor, being agents both complex and formed by elaborate evolution and development. Hypotheses based on such agents might be untestable, because experiment and/or specific predictions are impossible. Scientific tradition rejects supernatural explanations for natural phenomena, because local polytheistic gods were discarded by Thales in inventing the first hints of science (ca. 624–546 BC, perhaps the first true Western philosopher).[67] Nonetheless, if alien intelligence might advance beyond our own, refusing to consider its possible effects is unscientific.[68] The first scientific question, however, is how likely superior intelligence might arise given what we know about astrophysics, planetology, and biology. We do not know enough yet.

If neither science nor religion copes with this, but yet it affects us, it might be too subtle for us to recognize. If we formed under this ancient influence, we may not realize it. It can be hard to see things always there, unchanging. Our physiology may ignore or overlook their significance.

In light of our new knowledge of the huge number of planets in the Galaxy and the discussion of Fermi question possibilities earlier in the chapter, I suggest the nature of existence in our world, especially as it relates to intelligence in the Galaxy, might be divided into just a few categories. For lack of an existing term, I call these hyper-noesiological states (see note 68): (1) "Lonely Galaxy" – again, in this case we are unlikely to encounter other interstellar-capable civilizations, because few or none exist. This may be seen by some as license to exploit or do what we will with the Galaxy, because it is, in a sense, all ours. (2) "Society/clash of civilizations" – perhaps soon we will encounter alien civilizations and will need to cope with fundamentally new rules and facts, perhaps benevolent, perhaps harsh. (3) "Paranoid universe" – the aliens have already been here and have affected our existence, and may still be doing so covertly. This is more of an *X-Files* universe when taken to the extreme. The truth is not so much "out there" but already among us. In such conditions even nature and physical reality might be altered and untrustworthy. (4) "Endless landscape" – the Universe is bigger than we know, perhaps more multidimensional (à la string theory or supersymmetry), and there is so much to explore that other civilizations are too busy with that to bother much with us and our level of existence. As has been the case in the past few centuries, perhaps this places us farther from the dominant dynamics or position in our Universe, after more levels of post-Copernican expansion. Superaliens may be off exploring higher dimensions or some other reality, but they at one time existed in the Galaxy and may have left behind relic artifacts or caretaker agents to maintain their prior domain. Finally, (5) there is always the possibility that we understand so poorly the nature of reality that we cannot yet pose the question properly, and none of these options even make sense.

These hypernoesiological states are not mutually exclusive, in that one can easily imagine situations that satisfy two or more of them. I find it difficult to conjure some scenario that does not land in the first four, however. We have so little information about this that choosing one over the other seems to be largely a matter of personal preference. Which alternative one would rather believe seems to say more about the person being quizzed than the state of our world, more of a Rorschach test than a scientific issue. Scientists, for instance, seem to detest the paranoid universe, because it obviates the assumptions of physical nature needed for their work. For now, maybe we should call these hypernoesiological preferences, but wonder how we might obtain some data to distinguish among these radically different alternative realities.

We explore the Universe to understand our place in it. With Darwin this was our place among Earth's life.[69] We have a prospect for greater understanding in the context of life in the Universe. Finding this context should be our goal. It promises rewards of radically new insight in biochemistry, genetics, and even nanotechnology. What role does space exploration, and the Moon in particular, play?

Unable to psychoanalyze super-aliens, we ponder other approaches. For perspective on our future, we can outline life's past. Probes to our Solar system's major planets have found several places where life's origin seems possible.[70] Has life formed there? If so, this would eliminate life's origin as a choke point. Given Earth life's rapid emergence, this discovery might be unsurprising, albeit one of the greatest in science. How biochemical structure and processes deviate from terrestrial analogs can reveal basic processes of life. Did these worlds undergo their own Cambian Explosion? Animallike aliens would seemingly eliminate this prime candidate choke point in Earth's past, implying some surprise in our future regarding our fate as an intelligent species.

Improved hunts for exoplanets will determine the number of Earth-like worlds. Life might exist on large Earth-like planets ("super-Earths") beyond the Habitable Zone; these massive planets retaining hydrogen to supplement usual greenhouse gases to warm the planet's surface.[71] We can examine exoplanets with spectroscopy to look for indicators of life: molecular oxygen or vital compounds like chlorophyll, but some authors argue that this is impractically difficult/ambiguous. Finding ice moons, which might also harbor life, will be harder. They withhold useful signals (regular partial eclipses or velocity shifts of the central star).[72] The best way to study life on ice worlds might be missions to Jupiter's and Saturn's moons.

Does the Moon relate to these questions, besides offering a stepping stone to the planets? There is presumably no life on the Moon (and if it exists, it is likely an offshoot from Earth; see Chapter 8). The Moon is hugely significant as a repository of material unaffected by weather or volcanism for billions of years, the largest such surface in the inner Solar System except Mercury, and by far the easiest to explore. It is the best surveyed world beyond Earth in many ways, photographed to half-meter resolution over most of its area.[73]

Has the Moon possibly swept up the strangest stuff, artifacts of alien technology? As a totally random accident, this is implausible.[74] These hypothetical aliens would need some reason to be drawn to the Moon's vicinity. The Solar System is a special place, a realm of planets. If artifacts are only found within planetary systems, the density of artifacts in lunar vicinity is about a million times greater than average throughout the Galaxy, likely still insufficient to place objects on the Moon. If aliens took special interest in Earth, however, the Moon might be special vantage. The Moon is a stable, unobtrusive place wherefrom one can monitor perhaps the Solar system's only place with unobscured surface life: Earth.

If an alien artifact rests on the Moon, would we know? It need be large or a distinctive radiation source. Depending on its appearance in optical-wavelength photography for identification (for sake of argument, at LROC-NAC's 0.5 meter resolution), a minimum 3-meter size is needed.[75] This requires seeking "alien-looking" objects, poorly defined. Evidently we fail taking full advantage of existing data, because some lunar 70 archeological sites from past missions of NASA, the Soviets, and others[76] are largely unknown despite extending over 3 meters (including their impact craters). Apollo's sites are located, largely because their positions were already known, and several others were recovered anew, for example, *Lunokhod 1* (2.3-meters-long and a 1-meter shadow).[77] Most "lost" have not been found.

Alien artifacts on the Moon are old science fiction grist, particularly Arthur C. Clarke's,[78] and artifacts in the Solar System are subject to searches for extraterrestrial artifacts (SETA). Some argue that SETA is more efficient than radio SETI.[79] Where should we look for alien artifacts: the asteroid belt, the Moon, or in Earth's orbit (especially the Lagrangian points)? The first two might provide raw material for Bracewell–von Neumann probes; the latter two might attract aliens monitoring Earth life. How large must these craft be? How would they appear? What are they composed of? Would they have radiative or reflective signatures betraying them to our surveys? It is beyond our capability to guess alien techniques and intentions that would determine the answers.

A better approach is not second-guessing aliens but instituting protocols recognizing the Moon's special SETA potential. Finding one lunar artifact is a small needle in a huge haystack, but lunar mining techniques might operate in this century to improve SETA's prospects. If we surface mine for hydroxyl, helium-3, or other trace substances, we need SETA procedures to find and preserve artifacts before mining. Some resources lie subsurface, in 5–40 meters' regolith depth, as might bury artifacts. What SETA sensing methods can winnow alien material from tonnes and hectares mined? SETA might add costs, but prospects for finding these strangest of objects, perhaps the greatest scientific discovery ever to be made, cannot be dismissed. Whereas on Earth's surface these materials might be buried, weathered and destroyed in several million years, on the Moon they might last 1,000 times longer.

To summarize, why include the Moon in space exploration before 2050? There are several reasons both practical and scientific.

1. Volatiles near the lunar poles offer unique resources almost in free space (energetically). Most compelling is propellant from hydrogen in many millions or billions of tons, allowing rapid travel particularly with humans onboard, throughout the Solar System (see Chapter 8).
2. Eventually humans may build and fuel higher-powered spacecraft with nuclear fuel, and this is more easily and safely implemented on the Moon (see Chapter 12).
3. Lunar techniques share commonalities with Mars exploration, beyond those emphasized by Constellation in 2004–2010. This includes not only rockets, spacecraft, space suits, and radiation and dust strategies cited for Constellation, but technologies to search for life on Mars. Exploring lunar regolith requires deep drilling as may seeking life on Mars in subsurface hydrated regolith. Telerobotics could be essential in exploring a Mars hosting its own life, and the Moon is an ideal for developing this. Telerobotics may extend beyond Mars to exploring asteroids and moons of Jupiter, both for life and resources. Mars and the Moon probably both have habitable lava tubes (see Chapters 10 and 11).
4. Despite recent U.S. decisions, most spacefaring nations want to explore the Moon. Needing no proxy for Cold War conflict, the world's diverse peoples can explore it together (see Chapters 2–4).
5. The Moon is abundant in silicon, oxygen, iron, titanium, and other building materials, in addition to water and energy. Solar energy for manufacture is also abundant (see Chapter 12).
6. Someday, human presence in space may be revolutionized by development of space elevators. Doing this at Earth would require inventing new materials and clearing satellites from medium and low Earth orbit. These do not hinder a lunar space elevator, which could provide a practical test bed for the more ambitious Earth version, and allow us to master technical difficulties at lower cost (this chapter).
7. The Moon offers uniquely powerful venues for important frontier observations in astronomy and astroparticle physics (see Chapter 9).
8. If we ever seek artifacts from alien activity in the Solar System (SETA), the Moon is probably the best venue, not only a remarkably preservative environment over billions of years but also an obvious place for viewing the rise of life on the Sun's one planet on which life roams its surface and swims its seas, in full view (this chapter).

9. For lunar science in its own right, the Moon holds a record of its own development as well as the Sun's environment. It holds clues to its own creation along with Earth's, very different from the third planet from the Sun formed before the Moon's creation. Studying the Moon, we will learn about Earth, the Moon, and planets in general (see Chapters 1, 5–9).

Despite the diversion of this chapter's later pages, this is not a book about humanity's long-term fate. By necessity discussing humanity and the Moon deals with the next few decades. On longer scales it is anyone's guess: there are predictions that by 2050 neurologically informed meditative practices will have brought a new era of peace to humanity.[80] Others predict that by midcentury computing power will have ushered forth the singularity in the rapid rise of artificial intelligence, and humanity's fate will rest at the whim of the machines. Less drastically, other authors discuss a hybrid reality in which external physical reality is increasing supplanted by cyber existence.[81] Who knows? Before we reach any of these, however, lunar exploration and humanity in general will depend on more mundane issues such as climate, economics, politics, and demographics. We need to think about these issues as well. We also need to be analytical about that other issue: the human spirit.[82]

Notes

1. This line (#437) from Seneca's tragedy *Heracles* is spoken by Megara, daughter of Creon. This truly is a tragedy for her, for Heracles burns to death their offspring. Let us hope that space exploration's future is more uplifting than this!
2. From Tsiolkovsky's personal correspondence (in Kaluga, 1911): "Планета есть колыбель разума, но нельзя вечно жить в колыбели," which is more closely translated as "(A) planet is the cradle of the mind, but one cannot always live in a cradle."
3. Hermann Oberth, who did so much to found space exploration in Germany and beyond, had a simple idea for the goals of humanity and its exploration of outer space: "To make available for life every place where life is possible. To make inhabitable all worlds as yet uninhabitable, and all life purposeful" (*Man into Space* by Hermann Oberth, 1957, New York: Harper, 167).
4. These are even mundane, e.g., department store executives using satellite imagery of parking lots to gauge market response ("Crowded Parking Lots Hint at Black Friday Sales" by Jon Lentz, November 24, 2010, *Reuters*).
5. "Peacetime Uses for V2" by Arthur C. Clarke, February 1945, *Wireless World*, 51, 58; also "Extra-Terrestrial Relays: Can Rocket Stations Give World-wide Radio Coverage?" October 1945, *Wireless World*, 51, 305.
6. Cables of molecular matter must be dominated by low atomic weight (carbon, perhaps boron), because extra neutrons and non-valence charges only add dead weight. Cylinders of rolled-up, atom-thick graphite layers (carbon nanotubes) in principle are so strong that a continuous strand could nearly support itself from geosynchronous orbit to Earth's surface, permitting a space elevator. Unfortunately, current technology cannot make continuous strands this long. One can compensate for inadequate tensile strength by tapering the cable so lengths bearing most weight, at geosynchronous, are thickest. With no safety margin, an ideal carbon nanotube need only be tapered by 50% to support itself or tapered by several to add safety margins and excess carrying capacity. Progress is being made lengthening continuous nanotubes. New, easy methods can bond short lengths of nanotubes into ribbons and cables, but not continuously. Efficient methods are needed to bond tubes end to end at full strength. Some pertinent references include "Strong, Transparent, Multifunctional, Carbon Nanotube Sheets" by Mei Zhang, Shaoli Fang, Anvar A. Zakhidov, Sergey B. Lee, Ali E. Aliev, Christopher D. Williams, Ken R. Atkinson & Ray H. Baughman, 2005, *Science*, 309, 1215; "The Extraordinary Reinforcing Efficiency of Single-Walled Carbon Nanotubes in Oriented Poly(Vinyl Alcohol) Tapes" by Z. Wang, P. Ciselli & T Peijs, 2007, *Nanotechnology*, 18, 455709; "Fabrication of Ultralong and Electrically Uniform Single-Walled Carbon

Nanotubes on Clean Substrates" by Xueshen Wang, Qunqing Li, Jing Xie, Zhong Jin, Jinyong Wang, Yan Li, Kaili Jiang & Shoushan Fan, 2009, *Nanoletters*, 9, 3137.

7. "The Physics of the Space Elevator" by P. K. Aravind, 2007, *American Journal of Physics*, 75, 125.
8. Quoting Jerome Pearson in "Audacious and Outrageous: Space Elevators" *NASA Science News*, September 7, 2000.
9. The current estimate for beaming power efficiency to a space elevator's motor via lasers is only ~0.5%. Optimistic projections are 2%. This translates to power costs of about US$50 kg^{-1}. Elevator capital costs are estimated at $10 billion or more ("The Space Elevator: Economics and Applications" by David Raitt & Bradley Edwards, 2004, 55th International Astronautical Congress, Vancouver, Canada).
10. My informal poll of 40 typically middle-class Americans, Europeans, and Asians reveals nearly all willing to pay at least $10,000 for a one-day trip to space with no risk or great discomfort. (Some would not go at any price.) Many would pay ten times this or more. This compares to Virgin Galactic's $200,000 fare to suborbital space, somewhat irrelevant because hypothetical elevator trips would be more comfortable, higher, probably safer, and nearly weightless longer (an advantage for some, a disadvantage for others).
11. Let us compute. Accelerating 1 kg to low Earth orbit (v = 7.8 km s^{-1} = 7.8 × 10^5 cm s^{-1}) demands energy E = $\frac{1}{2}$ mv^2 = 0.5 (1,000 g) (7.8 × 10^5 cm s^{-1})^2 = 3.04 × 10^{14} erg = 30.4 MJ. Lifting 1 kg to outer space (100 km = 10^7 cm) requires E = mgh = (1,000 g) (980 cm s^{-2}) (10^7 cm) = 9.8 × 10^{12} erg = 0.98 MJ, only 3% as much energy as in achieving orbital velocity (where h is the altitude at the edge of outer space).
12. "Rockoons" by John Powell, 2010, JP Aerospace, http://www.jpaerospace.com/rockoons.html.
13. "Learning to Contain Underground Nuclear Explosions" by Robert R. Brownlee, June 2002, http://nuclearweaponarchive.org/Usa/Tests/Brownlee.html; also "The Nuclear Option" by Gregg Herken, *Air and Space*, February/March 1992, 52. One can question the plug's survival without a heat shield. The plug's dimensions were approximately 2 m diameter by 0.1 m thick. Entering at 60 km s^{-1} (the minimal speed estimated by Brownlee for the plug), an iron meteor of this size would burn up in Earth's atmosphere; it seems highly unlikely that the Thunderwell plug survived a trip through the atmosphere in the opposite direction (*Impact Cratering, A Geological Process* by H. J. Melosh, 1989, Oxford: Oxford University Press, 210.)
14. *The High Frontier: Human Colonies in Space* by Gerard K. O'Neill, 1976, Burlington, Ontario: Apogee.
15. An excellent textbook explaining light sails and other propulsion and spacecraft designs is *Deep Space Probes to the Outer Solar System and Beyond* by Gregory L. Matloff, 2005, Chichester, UK: Springer-Praxis.
16. To achieve rocket velocity V$_R$ for exhaust velocity V$_E$, one needs propellant mass (expressed in terms of payload mass m$_P$ and total rocket mass M$_R$ – composed of propellant and everything else, the latter by definition is payload), one needs a rocket with mass (mostly propellant):

    $$M_R = m_P \, e^{V_R/V_E}.$$

    This is the "rocket equation" and present-day space travel suffers under its tyranny. If exhaust velocity is 50% the desired rocket velocity, one must build a rocket that is 86% propellant by mass. To increase the rocket's speed to four times the exhaust velocity, the rocket must be 98% propellant. This is because the rocket must not only accelerate payload but also accelerate propellant later expelled as exhaust. Until propellant is used, it acts as payload.
17. "Bush Expanding Nuclear Power for Space: Threatens Planetary Ecosystem" by Bruce Gagnon, 2003, Global Network Against Weapons and Nuclear Power in Space, http://www.space4peace.org.
18. "Nuclear Propulsion through Direct Conversion of Fusion Energy: The Fusion Driven Rocket, Phase I Final Report" by John Slough, Anthony Pancotti, David Kirtley, Christopher Pihl & Michael Pfaff, September 30, 2012, NASA NAIC Grant Report NNX12AR39G.
19. "The Warp Drive: Hyper-Fast Travel within General Relativity" by Miguel Alcubierre, 1994, *Classical and Quantum Gravity*, 11, L73; "Warp Field Mechanics 101" by Harold White, NASA Tech. Report, also *Journal of Interstellar Studies*; see 100 Year Starship, http://symposium.100yss.org/symposium-proceedings. Note that while the most technologically accessible solutions to the 100-year starship are nuclear, this website goes out of its way to use the word in any prominent fashion.
20. LSEI and LADDER propose lunar elevators employing Zylon polybenzoxazole fiber, twice Kevlar's strength ("LADDER: The Development of a Prototype Lunar Space Elevator" by T. Marshall Eubanks & Michael J. Laine, 2011, *Lunar Exploration Analysis Group*, 2043). Lunar space elevator concepts may have originated with Yuri Artsutanov ("V Kosmos na Electrovoze" – "Into Space on a Train," July 31, 1960, *Komsomolskaya Pravda*). In 1972 James Cline proposed a lunar space elevator to NASA, which was rejected because of unaddressed technical

issues ("The Mooncable: A Profitable Space Transportation System" by James E. D. Cline, March 25, 1972). Jerome Pearson et al. wrote several papers starting in 1975 ("The Orbital Tower: A Spacecraft Launcher Using the Earth's Rotational Energy" by Jerome Pearson, September–October 1975, *Acta Astronautica*, 2, 785).

21. "Don't put all of your eggs in one basket" is a general principle in cable design. For a cable 0.3 mm in diameter as in LSEI and minimal 10^8 m length, a meteoroid (~0.1 mm diameter) that can break the cable impacts every 10^4 s = 3 hr. Separating the cable strands to a looser set 1 cm across, a projectile capable of severing it will hit only every 30 years, on average. Single strands will break often, so the cable must be repaired continuously (See "Lunar Microcrater Studies, Derived Meteoroid Fluxes, and Comparison with Satellite-Borne Experiments" by H. Fechtig, J. B. Hartung, K. Nagel, G. Neukum & D. Storzer, 1974, *5th Lunar Science Conference*, 3, 2463).

22. *Tethers in Space Handbook*, edited by M. L. Cosmo & E. C. Lorenzini, December 1997, NASA: Marshall SFC; "Cislunar Tether Transport System" by R. Hoyt & C. Uphoff, 1999, *35th AIAA/ASME/SAE/ASEE Joint Propulsion Conference and Exhibit*, AIAA 99–2690; "Lunar Cycler Orbits with Alternating Semi-Monthly Transfer Windows" by C. Uphoff & M. A. Crouch, *Spaceflight Mechanics 1991*, AAS 91–105, 163; "Tethers for Mars Space Operations" by Paul Penzo, *The Case for Mars II*, ed. Christopher McKay, 1985, AAS 84–174, 445.

23. Implosion of even minor components outside the crew volume can be deadly, because these can produce catastrophic shock waves. Any sea-level pressure cavity is a potential bomb. In deepest ocean a liter of un-pressurized space holds the destructive energy of 26 g of TNT.

24. Blocking Venus's cross-sectional area ($\pi r^2 = 1.15 \times 10^8$ km^2) with the thinnest known reflective sheet, e.g., aluminized Mylar a few microns thick (so 500 million tonnes) is possible on terrestrial industrial scales but 10^4 times all payloads ever sent into space ("European Space Activities in the Global Context" by Nicolas Peter, in *Yearbook on Space Policy 2007/2008: From Policies to Programmes*, eds. Kai-Uwe Schrogl, Charlotte Mathieu & Nicolas Peter, Berlin: Springer, 75). This requires factories beyond Earth, or a space elevator.

25. See "Radar Imagery of Mercury's Putative Polar Ice: 1999–2005 Arecibo Results" by John K. Harmon, Martin A. Slade & Melissa S. Rice, 2011, *Icarus*, 211, 37; "Predictions of MESSENGER Neutron Spectrometer Measurements for Mercury's North Pole Region" by David J. Lawrence, John K. Harmon, William C. Feldman, John O. Goldsten, David A. Paige, Patrick N. Peplowski, Edgar A. Rhodes, Christina M. Selby & Sean C. Solomon, 2011, *Planetary and Space Science*, 13, 1665.

26. "Searching for Evidence of Past Oceans on Venus" by David H. Grinspoon & M. A. Bullock, 2007, *Bulletin A.A.S. Division of Planetary Science*, 39, 540; "Runaway and Moist Greenhouse Atmospheres and the Evolution of Earth and Venus" by James F. Kasting, 1988, *Icarus*, 74, 472.

27. *The High Frontier: Human Colonies in Space* by Gerard K. O'Neill, 1977, New York: William Morrow & Co.

28. *Space Settlements: A Design Study* edited by R. D. Johnson & C. Holbrow, 1975, NASA Publication SP-413; *Islands in Space: The Challenge of the Planetoids* by Dandridge M. Cole & Donald W. Cox, 1964, Philadelphia: Chilton, quoting Lyndon B. Johnson, 112: "Someday, we will be able to bring an asteroid containing billions of dollars worth of critically needed metals close to earth to provide a vast source of mineral wealth for our factories." (Also *Space Resources: Breaking the Bonds of Earth* by John S. Lewis & Ruth A. Lewis, 1987, New York: Columbia University Press, 242).

29. Asteroids are rich in water or metals but rarely both. Summarizing from several sources (without explaining meteoroid classification schemes), water composes ~10% of C1 chondritic meteoroids, 5% of C2M chondrites, and 1% of C3/4 chondrites, but much less in other classes. Metals (particularly iron) compose 6% of C3/4, 4% of LL (low iron/metals), and 2% of L primitives; differentiated meteoroids can be up to 95% iron but very low in water.

30. "Possibilities for Methanogenic Life in Liquid Methane on the Surface of Titan" by C. P. McKay & H. D. Smith, 2005, *Icarus*, 178, 274; "Molecular Hydrogen in Titan's Atmosphere: Implications of the Measured Tropospheric Mole Fractions" by Darrell F. Strobel, 2010, *Icarus*, 208, 878.

31. "Radiation Analysis for Manned Missions to the Jupiter System" by G. De Angelis, M. S. Clowdsley, J. E. Nealy, R. K. Tripathi & J. W. Wilson, 2004, *Advances in Space Research*, 34, 1395.

32. Various proposals have been made for mining the asteroids and converting them to human habitat, ostensibly equivalent to up to 3,000 Earths, with a capacity for 10^{16} humans. See *Mining the Sky: Untold Riches from the Asteroids, Comets, and Planets* by John S. Lewis, 1997, New York: Perseus, and *Space Settlements: Spreading Life throughout the Solar System* 2013, ed. Ruth Globus: http://settlement.arc.nasa.gov. Many of these ideas are presented early in *Islands in Space: The Challenge of the Planetoids* by Dandridge M. Cole and Donald W. Cox, 1964, Chilton: Philadelphia.

33. Designing means to recover deuterium and helium-3 from Jupiter's atmosphere is challenging, but the energy demanded to extract, fractionate, and deliver these gases to escape velocity is less than ~10^{-5} of energy gained from efficiently fusing them. Estimates of ^{2}H and ^{3}He amounts for Jupiter are ^{2}H and ^{3}He isotopes are ~2×10^{-5} of total mass for either element, ~5×10^{22} kg. Earth's situation is more complex. Most hydrogen is probably trapped in the mantle, ~0.1% by weight ("Remote Sensing of Hydrogen in Earth's Mantle" by Shun-ichiro Karato, 2006, *Reviews in Mineralogy and Geochemistry*, 62, 343) of which ~10^{-4} is ^{2}H (or 6×10^{16} kg). The amount of ^{3}He in the Earth is poorly known but tiny, ~10^9 kg. Of course recovering these gases from deep Earth is nearly impossible. Earth's oceans hold ~40 times less ^{2}H than the mantle, and recoverable ^{3}He is even less, maybe ~10^6 kg.
34. "Mars has become a kind of mythic arena onto which we have projected our Earthly hopes and fears" (*Cosmos* by Carl Sagan, 1980, New York: Random House, 106).
35. Might we have already contaminated Mars inadvertently? *Viking*, *Pathfinder*, *MER*, *Phoenix*, and *Curiosity* landers were sterilized, nominally, as were *Mars 2* and *3*, but not completely. *Mars 2* and *3* were also loaded with thermite bombs to incinerate the landers to complete sterilization after their missions ("Rendezvous with Mars" *Time*, November 22, 1971, 65). Unfortunately, *Mars 2* crashed on Mars, and *Mars 3* landed safely but was immediately consumed by a massive dust storm (perhaps killing it with electrical discharges). Probably neither craft's thermite charges ignited, meaning that they were not completely sterilized. Neither landed in particularly hospitable locations, but earth microbes might have survived.
36. "Revisiting the Thermal Infrared Spectral Observations of Phobos" by E. Palomba, M. D'Amore, A. Zinzi, A. Maturilli, E. D'Aversa & J. Helbert, 2010, *Lunar and Planetary Science Conference*, 41, 1899.
37. By this I mean the Moon. In terms of gravitational escape, Mars is intermediate, requiring five times more energy for escape from the surface than the Moon but 4–5 times less than surface escape energy from Venus or Earth. Mars is at least 100 times farther than the Moon.
38. *Grëzy o Zemle i nebe: antologiià russkogo kosmizma* by Konstantin Tsiolkovsky, 1895 (*Dreams of Earth and Space*, 1995 reprint), St. Petersburg: Khudozh. lit-ra. Also see *Lunar Bases and Space Activities of the 21st Century* edited by W. W. Mendell, 1985, Houston: Lunar and Planetary Institute, and *The Lunar Base Handbook: An Introduction to Lunar Base Design* by Peter Eckart, 2006, New York: McGraw-Hill.
39. Consider the alternative: "In Memoriam" by Poul Anderson, 1992, *Omni*, 15, #3, 62.
40. *The Case for Mars: The Plan to Settle the Red Planet and Why We Must* by Robert Zubrin, 1996, New York: Simon & Schuster; *How to Live on Mars: A Trusty Guidebook to Surviving and Thriving on the Red Planet* by Robert Zubrin, 2008, New York: Three Rivers Press; *On to Mars: Colonizing a New World* by Robert Zubrin & Frank Crossman, 2002, Burlington, Ontario: Apogee, and several others.
41. See *The Case for Mars*, 2011 edition, p. 239. First European to see North America's mainland was not Columbus or Leif Ericson but Bjarni Herjólfsson, sailing ca. 986 AD from Iceland to visit his father in Greenland. He strayed, reaching Canadian shores before immediately returning to papa (as per the "Greenlander's Saga" of *The Icelandic Sagas*). Hearing of this misadventure, Ericson explored Canada's coast and settled some two decades later at L'Anse aux Meadows, Newfoundland. Beset by native attack and isolated from Norse civilization, that settlement died quickly, likewise many inhabitants, whereas Viking Greenland lasted 420 years, eventually failing as a result of climate change and Norse reluctance to adopt local resources (see *Collapse: How Societies Choose to Fail or Succeed* by Jared Diamond, 2006, Harmondsworth, UK: Penguin, 250). If Mars corresponds to Vinland and the Moon to Greenland, this is unflattering to Martian habitation. European settlement of the Americas succeeded 500 years later with better navigational, seafaring, and military technology (and colonial administration), whereas Greenland was long a pressure release for Icelandic society and key to Viking North American exploration.
42. "Radiation Will Make Astronauts' Trip to Mars Even Riskier" by Richard A. Kerr, 2013, *Science*, 340, 1031; "Measurements of Energetic Particle Radiation in Transit to Mars on the Mars Science Laboratory" by C. Zeitlin, et al., *Science*, 340, 1080.
43. For instance, Mars terraforming might require 1,000 years ("The Biological Terraforming of Mars: Planetary Ecosynthesis as Ecological Succession on a Global Scale" by James Graham, 2004, *Astrobiology*, 4, 168) or 100,000 years ("The Physics, Biology and Environmental Ethics of Making Mars Habitable" by Christopher P. McKay & Margarita M. Marinova, 2001, *Astrobiology*, 1, 89). Original Martian atmospheric mass has largely escaped the planet. A prime goal of terraforming is adding free atmospheric O_2. On a planet with O_2 absent, reaching oxidation equilibrium with surface and subsurface minerals will greatly exceed 50 years. Unless we pull roughly

100 times the current atmospheric mass from the rocks, we must import water and CO_2, mainly as comets. Known Martian water supplies barely suffice ("Subsurface Radar Sounding of the South Polar Layered Deposits of Mars" by Jeffrey J. Plaut, et al., 2007, *Science*, 316, 92). Cometary infall must proceed slow enough to avoid temporary runaway, so centuries are required.

44. Not all concur with syphilis's Western origin cf., "Histological Identification of Syphilis in Pre-Columbian England" by Tanya E. von Hunnius, Charlotte A. Roberts, Anthea Boylston & Shelley R. Saunders, 2006, *American Journal of Physical Anthropology*, 129, 559.
45. A terrestrial example of this is the behavior of protists engulfing and digesting bacteria by extending their cell membrane around their prey and forming a food vacuole.
46. Transmissible spongiform encephalopathies from prions include bovine spongiform encephalopathy (BSE)/mad cow disease, scrapie in sheep, and chronic wasting disease in deer, as well as Creutzfeldt-Jakob disease, kuru, and other syndromes in humans.
47. "Contamination from Mars: No Threat" by Robert Zubrin, July/August 2000, *Planetary Report*, 20, 4.
48. "An Estimate of the Age Distribution of Terrestrial Planets in the Universe: Quantifying Metallicity as a Selection Effect" by C. H. Lineweaver, 2001, *Icarus*, 151, 307; "The Galactic Habitable Zone and the Age Distribution of Complex Life in the Milky Way" by C. H. Lineweaver, Y. Fenner & B. K. Gibson, 2004, *Science*, 303, 59; "The Occurrence Rate of Small Planets around Small Stars" by Courtney D. Dressing & David Charbonneau, 2013, *Astronomical Journal*, 767, 95.
49. "Star Count: ANU Astronomer Makes Best Yet" by Simon Driver, July 13, 2003, *International Astronomical Union General Assembly XXV Press Release*, 81.
50. "The HARPS Search for Southern Extra-Solar Planets XXXIV: Occurrence, Mass Distribution and Orbital Properties of Super-Earths and Neptune-Mass Planets" by M. Mayor et al., 2011, submitted to *Astronomy and Astrophysics*, http://arxiv.org/pdf/1109.2497v1. Planets orbit most Sun-like stars. See also "Prevalence of Earth-size planets orbiting Sun-like stars" by Erik A. Petigura, Andrew W. Howard & Geoffrey W. Marcy, 2013, Proceedings of National Academy of Science, 110, 19175.
51. "Evidence for Life on Earth More Than 3850 Million Years Ago" by Heinrich D. Holland, 1997, *Science*, 275, 38.
52. The choke point concept resembles Hanson's Great Filter, except we use time delays in various life-form developments as proxy for probability ("The Great Filter – Are We Almost Past It?" by Robin Hanson, September 15, 1998, http://hanson.gmu.edu/greatfilter.html). The time-delay probability proxy is propounded by Carter ("The Anthropic Principle and Its Implication for Biological Evolution by Brandon Carter, 1983, Philosophical Transact. Royal Soc. London, 310: 347), but has been challenged by others ("Galactic Punctuated Equilibrium: How to Undermine Carter's Anthropic Argument in Astrobiology" by Milan M. Ćirković, Branislav Vukotić and Ivana Dragićević, 2007, *Astrobiology*, 9, 491 and references therein). Hanson identifies nine filter terms, one of which must be a choke point: (1) life-capable planetary system (including organics), (2) reproductive hereditary system (e.g., RNA), (3) simple (prokaryotic) single-cell life, (4) complex (archaeatic and eukaryotic) single-cell life, (5) sexual reproduction, (6) multi-cell life, (7) tool-using animals with big brains, (8) where we are now (verge of interstellar travel), and (9) colonization explosion. Hanson defines probability P_T of going from state 1 (life-capable planets) to state 9 (colonizing civilization) as the product of probability P_{12} of going from state 1 to 2 multiplied by P_{23} for passing state 2 to 3, etc. The final number of colonizing civilizations is $N_C = N_P P_T = N_P (P_{12} P_{23} P_{34} P_{45} P_{56} P_{67} P_{78} P_{89})$, where N_P is the number of life-capable planets. A P term close to zero is a choke point, implying $N_C \approx 0$. Like the Drake equation, [Box 3.2] this simple Markov Chain assumes states cannot revert backwards and ignores possible branching, either in initial state or intermediate transitions. For instance, along with life-capable planets, one might consider life-capable ice moons or evolution via sexual reproduction replaced by asexual reproduction in environments of rapid mutation (perhaps because of higher radiation). There may be multiple ways for life to achieve interstellar spacefaring ability.
53. Officially, no such evidence has been found on Mars, fossil or otherwise. There was excitement when images taken by the rover *Opportunity* using its rock abrasion tool revealed spiral-looking features ("rotelli") embedded in sedimentary rock. NASA scientists decided there was insufficient corroborating evidence to support a claim of extraterrestrial life. One can see this oneself, at the top of the groove tilted about 20 degrees clockwise from vertical, just above left of center of the photo at http://marsrover.nasa.gov/gallery/all/1/m/030/1M130859692EFF0454P2959M2M1.JPG. Be sure to view the image at full (1024 x 1024) resolution; the rotelli are at pixel x=497 and y=400 as measured from top left. See also http://marsrover.nasa.gov/gallery/press/opportunity/20040225a.html.

54. There are video and audio versions of this short story. I prefer the 2005 video directed by Stephen O'Regen, http://www.youtube.com/watch?v=gaFZTAOb7IE, and the 2011 audio version by Jonathan Mitchell, http://www.studio360.org/2011/nov/04/theyre-made-out-of-meat/.
55. "A Self-Reproducing Interstellar Probe" by Rorert A. Freitas, Jr., 1980, *Journal of British Interplanetary Society*, 33, 251. To review candidate interstellar flight technologies, see G. L. Matloff, 2005, *Deep Space Probes: To the Outer Solar System and Beyond*, Chichester, UK: Praxis, with von Neumann probes on p. 168. Costs of alternatives are treated in *Extra-Terrestrials, Where Are They?* [Box 13.1] with estimates of US$100–$1,000 billion.
56. In fairness to Gene Roddenberry, his original pitch for *Star Trek* in 1964 contained a version of the Drake equation computing the number of planets and intelligent species in the Universe.
57. One can write these conditions in terms of the number of interstellar spacefaring civilization that have existed in the Galaxy[50], N_C = the number of civilizations in the Galaxy that have achieved interstellar travel. Also let us specify t_G = the current age of the Galaxy, t_C = the typical lifetime of such civilizations, and t_t = a "technological transformation time," which is somewhat poorly defined but basically describes the rate at which a particular civilization produces inventions that its ancestors might find mysterious or even magical. For contemporary humanity I imagine this is about a century; t_t describes the average for all such civilizations now alive and contributing to N_C. One can describe several numerical regimes: (1) N_C t_C d t_G "Lonely Galaxy" – the probability of an interstellar spacefaring civilization existing in the Galaxy at any arbitrary time is not large, presumably meaning that the probability of us encountering another one is small; (2) N_C t_C t t_G and t_C p t_t (where p means "much greater than"): "The Old Ones" – the Galaxy likely contains civilizations that have attained a highly advanced state of development including interstellar travel. Depending on t_C, this might imply civilizations that have existed for billions of years; (3) t_G p N_C t_t "Cavemen or demigods" – any civilizations that we encounter will be so advanced or so backward with respect to our state of development that each civilization might not be both able to understand the technological (or societal?) development of the other; and (4) N_C t_C p t_G and N_C t_t p t_G "Star Trek universe" – we should expect to encounter eventually plural other civilizations, many of which might be close to our state of development. Obviously some of these conditions are mutually exclusive, such as the Star Trek universe versus the Lonely Galaxy, while others are compatible under some conditions.
58. Sprites (high-altitude, fast-ascending discharge from thundercloud tops) were documented in 1989, having been reported before via naked eye. Prof. David Sentman (University of Alaska at Fairbanks) named them in 1994; nomenclature of related phenomena follows this motif.
59. Even initial press reports explained these as St. Elmo's Fire (electrical discharge coronae from objects, especially sharp points, at voltages different than the surrounding air's). Others feared a new German weapon ("Science: Foo-Fighter," *Time*, January 15, 1945, 45, 70).
60. This would all be in keeping with anti-anthropocentrism's trend. First we learned that Earth is not the Universe's center, then that the Sun sits at the edge of a fairly humdrum galaxy among billions. Now we find the atoms like those composing us overwhelmed by dark matter and matter itself dominated by dark energy. Might we also find the dimensionality we inhabit to be peripheral? It would be just par for the course.
61. *Studies in Pessimism* by Arthur Schopenhauer, translated by T. Bailey Saunders, 1891, London: Swan Sonnenschein, 69.
62. Saint Augustine of Hippo seeks to understand "Blessed are the pure of heart, for they shall see God" (Matthew 5:8). In *City of God against the Pagans*, chapter 22, section 29 ("Of the Beatific Vision"), on the reality of contact with higher beings: "And now let us consider, with such ability as God may vouchsafe, how the saints shall be employed when they are clothed in immortal and spiritual bodies, and when the flesh shall live no longer in a fleshly but a spiritual fashion. And indeed, to tell the truth, I am at a loss to understand the nature of that employment, or, shall I rather say, repose and ease, for it has never come within the range of my bodily senses."
63. For instance, James cites *Natural Causes and Supernatural Seemings* by Henry Maudsley, 1886, London: Kegan Paul, Trench, 189, which draws on "The Religious Sentiment in Epileptics" by James C. Howden, 1873, *Journal of Mental Science*, 18, 482. Maudsley and Howden discuss clinical examples of purported contact with god-like supernaturals among the mentally disturbed. James argues famously that origin alone is not the measure of religious experience but that it fits into an integral system by which it must be judged. Interestingly, James quotes Maudsley: "What right have we to believe Nature is under any obligation to do her work by means of complete minds only? She may find an incomplete mind a more suitable instrument for a particular purpose. It is the work that is done, and the quality in the worker by which it was done, that is alone of moment; and it may be

no great matter from a cosmic[al] standpoint, if in other qualities of character he was singularly defective" (Maudsley, p. 256).

64. *Why We Believe What We Believe: Uncovering Our Biological Need for Meaning, Spirituality and Truth* by Andrew Newberg & Mark Robert Waldman, 2006, New York: Free Press, and *In Gods We Trust: The Evolutionary Landscape of Religion* by Scott Atran, 2002, Oxford: Oxford University Press, discuss evidence of inherent human proclivities for religious experience: magnetic resonance imaging brain scans and evolutionary biology, respectively.

65. This is detailed anthropologically and culturally in *When God Talks Back* by Tanya Luhrman, 2012, New York: Random House, especially pp. 198–200 and 227–267.

66. By this I note thinkers' tendencies to ride great insights' epiphanies to distraction. Famously, Giordano Bruno recognized the Sun as a star in 1588 (contemporary with Thomas Digges and building on Nicolas Cusanus) and visualized a universe of stars and planetary systems but met his doom when burned by the Church after populating these with infinite souls and Christs to save them. Another example is Thomas Wright in 1750 perhaps first realizing that the Milky Way is a galaxy and that galaxies fill the Universe. He went too far by postulating a spiritual superbeing in each galaxy. Five years later Kant adopted some of Wright's more credible ideas.

67. "Philosophy begins with Thales," according to Bertrand Russell in *The History of Western Philosophy*, 1946, London: George Allen & Unwin, 15.

68. We should name this. "Advanced intelligence" is already used by the artificial intelligence community. In Greek "advanced intelligence" would be *hypernoesis*, and its study, as in other life sciences, would end in "-ologys," so studying super-intelligence might be "hypernoesiology." If it is not so advanced, we already study "exobiology."

69. "The question of questions for mankind – the problem which underlies all others, and is more deeply interesting than any other – is the ascertainment of the place which Man occupies in nature and of his relations to the universe of things" ("On the Relations of Man to the Lower Animals" by Thomas Henry Huxley, 1863, in *Collected Essays*, 1894, 7, 77).

70. Mars, Europa, Enceladus, Titan, and Ganymede seem plausible. Enceladus looks misplaced on this list, because at 500 km diameter it is smaller than the other four worlds (3,140–6,780 km). However, it may not be unique. For instance, Saturn's moon Dione, at 2,250 km diameter, seems to have landforms explained by a 2–10 km ice sheet overlaying a liquid water ocean ("Determining Elastic Thickness on Dione from Flexure" by N. P. Hammond, C. B. Phillips, F. Nimmo & S. A. Kattenhorn, 2012, *Lunar and Planetary Science Conference*, 43, 2374; "Geophysical Implications of the Long-Wavelength Topography of the Saturnian Satellites" by F. Nimmo, B. G. Bills & P. C. Thomas, 2011, *Journal of Geophysical Research*, 116, E11001). There may be many oceans on ice moons with interiors heated by the tidal forces from the planet around which they orbit.

Along with 25 ice moons, many icy dwarf planets orbit the Sun, some with subsurface water oceans, perhaps even Pluto. Many have large moons (Pluto and moon Charon, Haumea with moons Hi'iaka and Namaka, Eris with Dysnomia). Most were likely blasted from their planet by giant impacts like our Moon's, heating the planet's interior. Tides from large moons may further heat the planet. Charon (11% Pluto's mass) may heat Pluto and maintain a liquid ocean to this day ("Thermal Evolution of Pluto and Implications for Despinning and Sub-Surface Oceans" by G. Robuchon & F. Nimmo, 2010, American Geophysical Union Fall Meeting, Abstract P24A-09).

71. "Hydrogen Greenhouse Planets Beyond the Habitable Zone" by Raymond Pierrehumbert & Eric Gaidos, 2011, *Astrophysical Journal Letters*, 734, L13; "Bayesian Analysis of the Astrobiological Implications of Life's Early Emergence on Earth" by David S. Spiegel & Edwin L. Turner, 2012, *Proc. of Nat'l Academy of Sciences*, 109, 395..

72. Many planets are found by movement of their star around their solar system's center of mass in reaction to the orbiting planet's gravitation. Moons have essentially no effect on this. Alternatively, the *Kepler* satellite finds exoplanets via regular eclipses of stellar light as the planets transit in front of their central stars. Moons provide much smaller, irregular signals. *Kepler* searches for moons are possible but works in progress.

73. The typical LROC NAC image is ~2 × 20 km. By late 2012, there are ~700,000 of these, totaling ~3 ' 107 km2 versus the Moon's 3.7 ' 107 km2 surface.

74. The probability of alien artifacts landing on the Moon by pure accident seems infinitesimal. Objects at rest in the local interstellar medium stream into the Solar System at ~26 km s^{-1} and would strike the Moon at roughly this velocity modulated by the ±30 km s^{-1} of the orbit of Earth and Moon around the Sun. The Moon sweeps through a cylinder in space of radius *r* about the same as the Moon's. (This impact parameter radius is only larger by a

factor $[v_{esc}^2 + v_{inf}^2]^{1/2}/v_{inf}$ where v_{inf} is the instreaming velocity above, ~26 km s^{-1}, and smaller v_{esc} is escape velocity from the lunar surface, 2.38 km s^{-1}. Thus the Moon's actual radius is a good approximation.) This volume is $V \approx (\pi r^2)\, v\, t = 6 \times 10^{15}$ km^3 year^{-1} = 6×10^{24} km^3 gigayear^{-1} = 2×10^{-16} parsec^3 gigayear^{-1}. Even if millions of objects correspond to each star, the chance of an object falling onto the Moon totally at random is negligible; more than a billion billion years would pass before this is likely.

75. The Moon's area is 3.8×10^{13} m^2, or 1.5×10^{14} resolution elements. Recognizing objects requires a minimum number of resolution elements. Idealizing a resolution element as either "bright" or "dark," the most distinct objects N that n elements can portray is $N = 2^n$. The maximum number of unique objects identifiable on the Moon is $1.5 \times 10^{14}/n$. Hence, if $N = 2^n = 1.5 \times 10^{14}/n$, $n \approx 40$, corresponding to objects ~3 meters across. This is roughly the minimal identifiable size.

76. *The International Atlas of Lunar Exploration* by Philip J. Stooke, 2007, Cambridge: Cambridge University Press.

77. "Soviet Union Lunar Rovers" by LROC group, March 18, 2010, *Missions: LRO*, http://www.nasa.gov/mission_pages/LRO/multimedia/lroimages/lroc-20100318.html.

78. Such an artifact is found in Clarke's short story "The Sentinel" (Spring 1951, *Avon Science Fiction and Fantasy Reader*, New York: Avon), which is the basis of *2001, A Space Odyssey*.

79. "Inscribed Matter as an Energy-Efficient Means of Communication with an Extraterrestrial Civilization" by Christopher Rose & Gregory Wright, 2004, *Nature*, 431, 47; "The Search for Extraterrestrial Artifacts (SETA)" by Robert A. Freitas Jr. & Francisco Valdes, 1985, *Acta Astronautica*, 12, 1027; "Searching for Alien Artifacts on the Moon" by P. C. W. Davies & R. V. Wagner, 2013, *Acta Astronautica*, 89, 261; "On the Likelihood of Non-Terrestrial Artifacts in the Solar System" by Jacob Haqq-Misra & Ravi Kumar Kopparapu, 2012, *Acta Astronautica*, 72, 15; *Deep Space Probes* by Gregory L. Matloff, 2005, Chichester, UK: Praxis, 155–184. *Journal of the British Interplanetary Society*, Vol. 51, 1998, is a special issue on SETA and future spacecraft.

80. "Brain Beat: Meditation and the Brain" by Richard J. Davidson, January 24, 2011, presented at *Brain and the Tibetan Creative Mind*, American Museum of Natural History (see http://www.amnh.org/news/tag/richard-j-davidson/).

81. "Speculations Concerning the First Ultraintelligent Machine" by I. J. Good, in *Advances in Computers*, eds. F. L. Alt & M. Rubinoff, 1965, New York: Academic Press, vol. 6, 31; "When Will Computer Hardware Match the Human Brain?" by Hans Moravec, 1998, *Journal of Evolution and Technology*, 1, 1; cf., "Tech Luminaries Address Singularity" in *IEEE Spectrum*, June 2008.

82. *Hybrid Reality: Thriving in the Emerging Human-Technology Civilization* by Parag Khanna & Ayesha Khanna, New York: TED Books (eBook).

Chapter 14
Return to Earth

> The visions we offer our children shape the future. It matters what those visions are. Often they become self-fulfilling prophecies. Dreams are maps.
>
> – Carl Sagan, 1987[1]

> I have written letters on odd pages of this book. Will you manage to get them sent? You see I am anxious for you and the boy's future. Make the boy interested in natural history if you can, it is better than games. They encourage it at some schools.
>
> – Capt. Robert Falcon Scott, near the end of his final letter to wife Kathleen Scott, March 28, 1912, a day before his death on the Ross Ice Shelf on returning from the South Pole[2]

By necessity this book concerns not just next year or 10 years, but maybe the next 20 or 40. People yet unborn will fulfill (or forsake) these efforts and decisions. What will their world be like, and who will they be? For now we depend on others' educated guesses. However, as we choose, space exploration can play important roles, in ways useful to current politics but also influencing youth central to realizing these and other dreams.

Exploration takes time, generations. It can inspire generations, one of the few strong motivations demonstrated to attract young people into science, engineering, and mathematics (over-succinctly acronymed as STEM: Science, Technology, Engineering and Mathematics). Modern economies, especially America's, require vital influx of inspired scientists, engineers, and mathematicians. These workers and innovators maintain the main seminal foundation of America's economic production (besides its agricultural base). Innovative products (and food) are what America creates from scratch that people in other nations want to buy.

On the first point, we often forget delays between exploring and settling new lands. Even for small, almost familial Viking North America, two decades passed between Herjólfsson's discovery and the L'Anse aux Meadows settlement. Columbus aggressively colonized the New World, even on his first voyage, only to see first attempts fail. More pertinently, the first permanent South Pole base in 1956–1957 followed Amundsen and Scott by 45 years, during which no explorer set foot (despite Byrd's overflight in 1929). Likewise almost two centuries transpired between first

European landfall on Australia and permanent settlements, with seven-decades delay in North America (but only half this in South America – of course already inhabited by non-Europeans in most of these cases). In this context our four-decade delay returning to the Moon is not untoward, but mortal frustration for those waiting too long, almost unbearable imagining such genius and sacrifice perhaps in vain. These people were trained to ease our return to the Moon, talent now retired.

Lunar exploration's near future depends on NASA's course and direction from the Obama Administration. This merits discussion, because Barack Obama arguably has influenced space and particularly lunar exploration more than any U.S. president since Richard Nixon.

Whatever one's opinion of his other accomplishments, Obama has left the U.S. space program suffering whiplash and confusion. In 2004–2009 we were headed to the Moon but retiring the *International Space Station*. Despite his negative 2007 presidential campaign plank, in 2009 Obama implied the Moon still had high priority. However, in 2010–2011, he dropped this course, let the Space Shuttle die, and emphasized industry leadership, not NASA, setting the pace and direction for American space. The *ISS* is intact but with no U.S. vehicle to transport astronauts there for many years. There are vague commitments, more from Congress, to a heavy-lift booster for exploration but no clear destination (see Chapter 3). Obama has promised to revisit his NASA decision in 2015, and Congress has supported NASA more than many federal programs.

Obama's thought behind canceling the Return to the Moon is unclear: "We have been there. Buzz has been there"[3] is more stating our relationship to people and events of the 1960s than the Moon, worlds beyond, and our future. It is a parochial excuse absent insight into important issues. More immediately, it is a dodge to explain away underlying political and budgetary inconvenience, missing many points.

It is useful to Obama to minimize lunar exploration's appeal. China or another country might return astronauts to the Moon first, likely seen by Americans as challenging their national prestige. Obama sees the power Sputnik symbolized in 1957, with "Sputnik moment" one favorite oratory device of his, although never referring to space. As Eisenhower learned too well, presidents cannot predict or control public reaction to such crises. By presenting the Moon as old news, Obama would blunt the impact of another nation overtaking the United States in lunar exploration. China will land nobody there before Obama's 2017 departure. He can set the space agenda, which he offers to revisit in 2015. His attitude neglects the Moon's physical reality and scientific, practical, and economic potential. Discovery of water and other lunar resources only became prominent in 2009, a year or two after Obama's advisors likely decided to end Constellation, the Return to the Moon and Vision for Space Exploration (VSE). This science policy mismatch is part the responsibility of Obama's science staff and his disinterest in the issue, but partially because lunar scientists overlook evidence against a dry, dead Moon. People recall its magnificent desolation, not its potential.

We discuss U.S. space policy decisions (Chapter 3), but what should Obama do with NASA? He has extraordinary control over it. If he wishes inspiring young people with it, he can do so.

What can he do? Another weakness in the Obama Administration negotiating stances frequently irritates his allies: unilaterally sacrificing bargaining advantage without quid pro quo, oft conceding issues like expanding offshore drilling he could use to entice Republicans to accept his energy policy goals, for example, carbon cap and trade. With NASA, politics are not bilateral (assumed for reactive devaluation, the tactic just described[4]). In the United States today regarding space, three parties – classical Republicans, Obama's Administration, and the TEA party (Taxed Enough Already) – occupy distinct positions. TEA partiers would cut federal discretionary spending (national security budgets aside), hence, NASA.[5] Standard Republicans favor strong NASA human space efforts, the Moon, Mars, and so on, perhaps because they involve defense spending constituencies, perhaps because countering Obama's space plan provides opportunity to decry Administration policy failure, or maybe they believe this to be best for the country. Arguably Obama sits between the two right-wing polarities, largely maintaining NASA's budget (for now) but privatizing human LEO access. Obama could occupy a strong position, exploiting internal opposition inconsistencies, playing TEA and classical Republicans against each other. We will see. Historically, the White House has the advantage in space policy, with NASA an executive "administration," with Congress usually more reactive, making major changes only during appropriations, with difficulty. NASA 2010–2011 budget negotiations were odd in that Congress, primarily Obama's own party, formulated such detailed counters to Administration positions.

The United States faces no urgent military issue in space now as in 1961, but there are economic ones. Later we ask why space exploration investment is needed. Space exploration fits poorly constituencies on the left or right. Few voters naturally desire more spending on space exploration but would rather not spend those funds on explicitly social programs. Contrarily, few voters desire to spend more on space as well as to shrink the overall cost of government. Traditionally space supporters thread this needle by casting space as more quasi-military, appealing to both left and right to support scientific research, or emphasizing jobs. All these describe reasons we need space. We must maintain science capability, both in research and education, to guarantee our national security, here our economic place in the world, guaranteeing continued sufficient jobs of high quality for which Americans hope.

The space program is a national effort needing multi-administration continuity, like foreign policy. A single space mission requires more than 5 years from conceptualization to completion, a larger program, employing new knowledge and capabilities, at least 10. They require specialized, highly trained professionals not easily shunted to other government programs or private industry – deep personal investments. Politicians should exercise discipline to avoid changing space goals precipitously and heedlessly. These are major national commitments possible to trash with vacillation. As discussed (Chapter 3), the United States needs a power like the once envisioned National Space Council (NSC) with persuasion over space program policy, one ensuring continuity in administration and congressional control. Advice by the National Academy of Sciences and their Decadal Surveys are, regrettably, ignored without political cost by policy makers, as was the real NSC by administrations that created, appointed, and could destroy it. Without more continuity, much U.S. space investment succeeds only in spinning wheels, not making progress. Valid complaints stick against the VSE and Constellation as conceived and executed. G.W. Bush provided little follow-up support and presided over insufficient funding profiles in the program's out years, displacing investment and sacrifice to future administrations. That said, the space program is too important to replace the Vision with no vision. President Obama called in his 2012 State of the Union address for research and innovation to advance the U.S. economy, calling for a "Sputnik moment," but offering no space role.[6] Subliminally, Obama's message is that he is too busy; he will decide later. Meanwhile we plan humans visiting an asteroid in 2025, but few believe Obama's successor's successor will uphold this. America's astronauts are lost in space.

What should constitute this vision? Why have a space program? We have a space program for several reasons: to generate high-technology jobs, to raise economic productivity by fostering new technologies, to enhance national prestige, to explore the Universe, to discover our place in the rise of life, to expand humanity's capabilities off-World, and to inspire young people to enter science, mathematics, and engineering to better accomplish these and more. Each can be advanced by the space program. In this chapter, let us address these not already analyzed.

Sometimes dismissed with flippant comments (often involving the orange drink Tang), space exploration's impact on today's economy is unquestionable. Just the initiative to develop integrated circuits for the LM guidance computer and other Apollo applications[7] probably expedited this part of the world economy by 5 years, a sector now worth $260 billion annually for the semiconductor industry alone (doubling every 4–5 years), and enabling a several trillion dollar annual business and consumer electronics industry and many trillions annually in associated services, totaling over 10% of world economic output. Accelerating this sector by a doubling time, hence, arguably doubling output in any given year, Apollo ($26 billion in 1973, inflated to $200 billion in 2010) is easily justified by this additional economic activity and productivity alone.

NASA maintains an office to promote technological spin-offs into the economy.[8] An "official" list of spin-offs (promulgated by NASA and once displayed at the Smithsonian National Air and Space Museum) consists of about a dozen items viewed by NASA as prime contributions to the U.S. economy.[9] These top developments are more direct in provenance than accelerating the integrated circuits but not as fundamental (such as the fabric roof of the Michigan Silverdome in Pontiac) and have been worth billions of dollars to the economy. The ultimate value of these and other spin-offs has provoked debate.[10] They pale before integrated circuit innovation or the NASA budget.

Are there "killer applications" for future NASA spin-offs, comparable to 1960s integrated circuits? One candidate seems obvious: robotics. By midcentury robots could be ubiquitous, and the sooner we perfect their economical operation, the sooner we benefit. Robots will operate in close cooperation with humans, so to encourage these developments in space we should foster operations close to Earth (without minutes or hours of light-travel latency). Everyday robots work in Earth gravity, so weightless development is likely misdirected. The Moon offers the best scenario for translating space-exploration robotics into earthly benefit.

In a few decades robots will clean our homes, care for our sick, and fight our wars; in fact, they already do. Most of us have seen automated floor sweepers or grass cutters. (There are about 7 million domestic robots in service, some 1 million industrial robots, and fewer military, research, underwater, and police robots.[11]) Unmanned aerial vehicles, many armed, frequent the news and already serve the armed forces of almost 40 countries, notably the United States, Israel, United Kingdom, and China. (The United States alone operates some 10,000 military robots.) Robots regularly disarm bombs and search dangerous areas. Robots will soon act as battlefield "mules" carrying up to one-half tonne, work with dolphins in submarine reconnaissance, and attack as tank-treaded, frontline shock troops.[12] Research proceeds to enable domestic servant robots to recognize and fetch objects, but their ability is still painfully halting and slow. In tasks where robots interact with people as object (medicine and warfare, primarily), software alone is not trusted to judge how to respond in life and death situations or to distinguish friend from foe. Likewise, visual pattern recognition and other sensory tasks remain often too sophisticated for artificial intelligence. In these we still rely on telerobots, where the mind controlling the action is human but "muscles" are robotic.

Lunar exploration is one environment to apply pressure on many solutions required for successful real-life robotic operation: pattern recognition, situational awareness and understanding, and object manipulation. Situational awareness and understanding are prerequisite for modeling and prediction required for field science as will be increasingly useful for lunar exploration and nearly mandatory for situations farther from Earth, as latencies remove humans from the scene. The small but significant 3-second lunar latency will encourage development of quick robotic reflexes, so it is important for robotic operation on the Earth as well.

Lunar uses could promote advances in robotic reliability and upgradeability. Consumer robots will be expensive, not disposable. Either a consumer will purchase one for many years or lease it under service contract. Either way, market pressures will make units last longer, with upgrades including hardware: sensors, data processors, motors, and so on. Buying a robot for several hundred dollars to vacuum your floors, you might expect several years' use before it breaks. At that point you might buy another or send the original for repairs. Contrarily, when owning a computer you might upgrade operating software several times before getting a new machine, in fact, often almost automatically. Robot awareness to situations will be expensive; you would not discard the whole unit every few years. The chassis at least might last a decade or more. Easier hardware (and software) upgrades will also be required on the Moon. With enough robots, such as an automated lunar base (envisioned by Japan), new robots can be made by retooling, re-sensoring and repurposing old ones, maybe like future household robots.

A reasonable but more vigorous robotic lunar exploration program might cost $1 billion annually (5% of NASA's budget), a significant but under-dominant fraction of military/domestic robotic markets.[13] An effective idea might be a private/public partnership in which robots explore the Moon, but developments quickly disseminate into industry, with design closely cognizant of industrial needs and state of the art.[14]

Many telerobotic applications exist in manufacturing and underwater exploration/production, particularly offshore oil wells (as Deepwater Horizon blowout video viewers know); these are less appropriate to lunar exploration applications. Mining might benefit from "reflexes" tuned to compensate for lunar latency. Robot manufacturers in Japan, Europe, and the United States might best apply these. China recently grew interested in telerobotics for ocean-bottom mining for polymetallic sulfide deposits for copper as the conductor for massive electrification, but this is simply "mowing the lawn" areal work.

We wonder who will be making these innovations – without some crystal ball to examine space exploration's prospects until midcentury – asking who will assume leadership and demanding investment and innovation. Who will excel? We touched on new technologies (Chapter 13) to accelerate Solar System exploration and also mentioned new organizations, for example, private consortia, for motivation and investment, and maybe more efficient means toward some goals. New lunar insight, especially regarding water, offers means to overcome space exploration's eternal problem of quick, effective propulsion though long distances and gravitational barriers. Alternative to new rocket technology, it enhances chemical propulsion, available to national or commercial agents.

Will the organizations deciding about space exploration change? Predicting moves by space agencies soon to become irrelevant is pointless: Which will survive? We discuss national space agencies' plans in Chapter 4. NASA spends as much on space exploration with more missions than all other national agencies combined. It dominates exploration beyond Earth and the Moon, with notable recent contributions by European and Japanese space agencies. Sadly, American human presence in space is now uncertain, with Russia growing dominant in human low Earth orbit (spending most of its space program funds on the *ISS*). China, Japan, and India also send spacecraft to the Moon, with NASA still a strong contender. There are two questions: Who will adopt a role of leadership, and will international cooperation or competition prevail?

A vital space program must have funding, infrastructure, engineering, and technical talent, an engineering quality control culture, and, to lead, capacity for technological innovation. Although Russia has a strong space program, its exploration beyond Earth's orbit is woefully underfunded. It plans an upcoming lunar mission and another joint mission with India, but its last lunar mission success was in 1976 and to any object beyond Earth's orbit in 1986. The next decade is likely to see Russian missions to Mars and/or Venus, but new Angara rockets and the "Rus" crew capsule will burden the budget. Russia will lead only if its space enthusiasm recovers radically (or if oil and natural gas prices once again skyrocket).

In several decades the primary change could be growing Chinese and Indian exploration. We can assume that the United States and India can grow to work together, despite India's non-aligned past. Currently the United States and China never cooperate on space, but perhaps not forever. Despite China-baiting in the United States or persistent anti-Americanism in PRC state media on economic and political issues, explicit militarism is rare. China and the United States are so economically intertwined and their militaries restrained by civilian leaders that violent prospects are minimal. China's leaders identify corruption as imperiling their political system; related threats of favoring state-owned business over private enterprise, income inequality, and environmental degradation also loom. As long as China continues progress toward individual liberties and the United States avoids the economic precipice, they both seem likely to survive and see cause to maintain general cooperation with each other over isolation. Many popular writings discuss this pairing on broad fronts, not reiterated here (Fallows, Ferguson, and Freidman, for a few), but less is said on technological innovation and space exploration leadership. We will discuss this.

World economics differ now from the Space Race's, when the United States made 40% of world gross domestic product (GDP) – now about 22%, likely surpassed in purchasing power parity (PPP) by China by 2014. In 1961 the United States enjoyed huge manufacturing and educational advantages, now diminished.

The United States versus Chinese midcentury's status is subject to many projections, usually putting the United States ahead in per capita GDP but China ahead in total GDP.[15] These predictions vary; many neglect in coming decades decelerated growth from Chinese demographic trends due to population controls since 1978 and most neglecting climate change effects. China could enter the "middle income trap" occupied by formerly rising Latin American and Middle Eastern economies and perhaps exacerbated by China's trend to return to state enterprise over expanding free market.[16] Indeed China's GDP annual growth rate has fallen several percent in recent years. China will hit human productivity decrease and radically growing social burden from the one-child generation's retirement. First, many millions of one-child sons will soon find themselves bereft of wives. China also squeezes 4.3 times the U.S. population primarily into a narrow band of arable land between rising ocean on one side (potentially drastic for cities such as Shanghai) and growing desert and shrinking water supplies on the other. Already China is importing grain after long-term drought. Longer environmental and societal changes can be severe and unpredictable, as those predicting Japan's economic rise in the 1980s may attest; yet China rises.[17]

Still, we know that leadership is not described simply by GDP, or even macroeconomics. In an age valuing innovation, individual contributions grow to paramount importance, so leadership grows granularly. The next big thing usually starts small, and growing innovation determines who influences the world.

A century ago Americans training in science may have needed independent wealth, before the GI Bill (Servicemen's Readjustment Act of 1944) and 1958 National Defense Education Act. Since Sputnik U.S. society has better portrayed science as a viable career. The Soviet Union implemented this sooner and differently. Many governments now appreciate this; studying physics, computers, or genetics depends less on personal wealth. Billions can gain access to this knowledge. Many nations decide who competes.

Mathematics and science education rankings consistently put the United States low on OECD (developed) country lists,[18] but China, typified by Hong Kong and Macao, were placed near the top (likely misrepresenting China in total). American higher-educational students have dropped from highest of any nation to roughly

25th. China has special problems: frequently decried low-end U.S. public schools ("failure factories") and concomitant social decline are matched in China with suffering by children of hundreds of millions of disenfranchised economic migrants, ethnic minorities, persons displaced from economic expansion projects (dams, energy projects, and urbanization), parents violating one-child rules, and huge numbers of people, although richer, still unable to afford a university degree path for their children. In China, approximately 60% of the population has no access to the internal social media sites and national intranet that is shaping the middle class, suggesting, absent some massive government program, the formation of a permanent "digital divide" as seen on smaller scales in the United States. The United States is a "developed nation," however.

Many Chinese anticipate their nation reoccupying a perceived rightful place in the world, the richest and most powerful until the sixteenth century. Chinese pioneered rocketry and, until the sixteenth or seventeenth century, led in space exploration's precursors: astronomy and navigation. Chinese fleets roved one-third of Earth's oceans, with Chinese astronomers the best observers.

Will the United States still lead exploration beyond low Earth orbit, or fall behind China? Will the two be sufficiently co-equal to collaborate usefully or compete meaningfully, or can China define humanity's Solar System path? This depends not just on economic power. A recent visitor to Shanghai or Beijing can be impressed by proliferating modern technology, but space exploration requires techniques beyond the commercially available; it instead requires techniques heavy in leading-edge innovation.

China now files for more patents than does any other country[19] and publishes science now at rates second only to the United States, but most such papers are far from seminal or influential, with many written to maintain quotas, even containing plagiarism.[20] In citation counts by other scientists, U.S. publications fare better (slightly behind the United Kingdom and Canada). Chinese leaders recognize basic research funding as a future priority. Publication in several premier Western journals is rewarded directly and monetarily in China, leading to distortion in motivation and performance.[21] Science functions poorly when driven by such authority and acclaim. To progress scientists must question the status quo; "scientific authority" is an oxymoron. While scientific knowledge was maintained and expanded in Asia during Europe's Middle Ages, slowly explorers challenged the established world view, a transformation unexpected but transcendent. Astronomers turned telescopes skyward and cancelled authority. They applied earthly physical laws to the heavens, and science grew to dominate. Technology helps, but freedom to question and overturn authority must be emulated.

Leading in space depends on funds, but more on technical excellence and innovation and societal motivation to explore. China is rapidly raising its students' typical aptitude in science and mathematics, although critics complain of overemphasis on rote memorization and students' reluctance to challenge their teachers' authority and reach their own conclusions. However, China wants to soon demonstrate its elevation to human intellect's forefront. This can involve human travel to the Moon. By many measures the most impressive single technological feat of any nation was Apollo, but space exploration is risky. China's leaders will not underwrite risking publicized space disasters without exhaustive temporization. Is China willing to go beyond the envelope of prior accomplishment to risk innovation, change, and objectives for which there is no precedent, where there is higher probability of overt mistakes? Additionally, in other societies small groups of innovators can make fundamental changes, under their control via their own volition, reversing fortunes of the rich and powerful. Such changes usually face opposition limited (hopefully) by rule of law. Demonstrably, David sometimes strikes Goliath with a shot squarely on the forehead. This path is closed by authority to the Chinese people, even economically. Nonetheless, space exploration is easily accomplished by large, capitalized groups like governments. A competitive, if not bold, space program seems an excellent device for China. Although we can expect them to visit the Moon, will they face true exploration unknowns? Perhaps we should expect moderate effort, not forefront expansion.

China's leaders have a curious problem. They must lift morale of enough population to keep social unrest down by encouraging capitalism and state industry that simultaneously creates economic inequality and unfairness. To maintain control they run an authoritarian state requiring unbroken continuity in leadership that suppresses other parties to survive. In this system if leaders lose control, there is no mechanism allowing their return to power from outside, assuming they survive (with rare exception from factional conflict, such as Deng, post-Mao). This is all subject to erosion from corruption and environmental decay, with oppressive inefficiencies as granular as military over-restriction of air traffic control and as basic as suppressed creativity. The world awaits to see how long they hold this balance.

Rapid economic growth in developing countries is neither correlated nor anti-correlated with democratic freedoms. The issue is whether a society that curtails freedoms can reach scientific/technological excellence. Most new avenues of wealth creation that China might pursue involve greater personal freedom, less military and party involvement, and weaker political control. Can innovation and invention be encouraged when freedom of expression is suppressed? This social experiment has been difficult in the past. Sometimes scientific creativity blossoms in an environment of expanding freedom still oppressive in some absolute sense, for example, Renaissance Italy or eighteenth- to nineteenth-century Germany. Germany (or, rather, the Holy Roman Empire and smaller states later composing Germany) had suffered seemingly endless religious strife, civil war, and invasion during the seventeenth century. Although social oppression did not lift entirely, reorganization of society into an "enlightened despotism" allowed a flowering of the German-speaking people, resulting in such personae as Bach, Mozart, Beethoven, Goethe, Kant, Leibnitz, Gauss, Planck, and, despite persistent religious/ethnic oppression, Einstein. Leibnitz, especially postmortem, was elevated to a national scientific superhero, largely to counter dismissal of German culture in the French and English-speaking world. This tool is useful: scientist as exceptional case, elevated above usual limitations imposed by society. It succeeded in Germany, until the twentieth century. China is already exploiting this for an elite entrepreneurial class, but not so much intellectual innovators, whose behavior can be more problematic.[22]

The analogy is imperfect: China today is nearly monolithic, whereas being Germanic never was. Even today with unified Germany, there is still Austria and most Swiss cantons. In the past talented Germanic artists and scientists, being more mobile than most people, could select their sponsors and politics by relocating to a different principality. In this regard China compares poorly: to disagree with the Communist Party in any prominent way is to face exile or decline. The handicap borne by Chinese intellectuals is uncertainty about when they cross the line into risking punishment by authorities. The line is unspecified by rule of law; sometimes it ebbs past your position to leave you stranded; for example, recently the Party cracked down on frivolous television programming. China might choose to loosen restrictions and let "a hundred flowers blossom,"[23] but one never knows when the reaper will mow down the flowers and enforce the monoculture. Whether and how this will occur stays ambiguous. China needs pillars, not flowers, of intellectual freedom else personal insecurity reigns; the Party must end its rounds of clampdowns. Young Communist Party of China (CPC) leaders are more social scientists who matured under Deng and Gorbachev unlike current Cold War/cultural revolution-raised technocratic engineers. China's social media networks are larger than in any other country but are state monitored, yet allow changes outside total state control limits, and are promising their expansion – perhaps.

The ill-defined forbidden zone is not just a Mao-era memory. Chinese social media (separated from the world by the "Great Firewall") is now suppressed by regulations on rumormongering of "false information," with seemingly specific, quantitative limits on how much information spread constitutes a rumor but vague definition of what is "false." (If one slightly misstates the number of victims of natural disaster, is this a rumor? Is it a rumor if one's interpretation of information disagrees with the Party line?) True or not, one cannot comment on People's Liberation Army activities, higher Party personnel, shortcomings in foreign policy, and specific hot-button controversies.

An undeniable creative explosion grows in China, in the arts and in a lesser degree the sciences, but mainly in commerce. These arose in less than one generation, so how individuals, even youth, internalize the new creativity is dubious. Liken it to lifting the 1960s societal lids that had suppressed groups and movements in the United States, still evolving 50 years later. How robust or fragile creators feel this freedom is key. Social authority may fail controlling it, but they will surely try manipulating it.

China should encourage an elite of scientific/technological excellence, isolated from many societal requirements. Deviant thought and behavior cannot be suppressed reflexively, if one wants creativity. Demanding creative people to exercise their genius according to society's demands, as laudable as it seems, usually fails. It stifles – a straightjacket in science as in art. Society rarely sees or appreciates the next new thing coming; inventors and innovators must be free to do so. Innovation is rarely formulaic; a successful innovator must also define the questions. Progress is often made by doing something nobody has ever done, and this is unquantifiable via a test. American educators in large numbers understand this; they need to instill inspiration's mysterious quality. Inspiration mixes centers of the mind so that unexpected and creative associations occur in carefully tuned ways involving reason and emotion.[24] Many experiencing this develop their own genius. We need, not just in the United States but globally, inspired problem solvers – not just a few, but millions. Innovators need the luxury of failure, which co-exists with invention. Chinese society suffers such failure poorly; success is often tightly defined and officially programmed, even in the context of the family. Much Chinese educational culture centers on students performing authority-initiated tasks: Can these young people listen to more subtle, internal voices that are more often the source of new ideas, invention, and innovation?[25] Chinese leaders are sufficiently aware of this as to allow small educational experiments.[26] Other official policies in part capitulate to the problem, relying on adapting foreign intellectual property over indigenous creativity.[27]

Recently, China gained, by many measures, its first resident Nobel laureate for work done primarily within the country.[28] To the Party's dismay, this was the Peace Prize to dissident Liu Xiaobo, imprisoned for offending social order. Regimes (mostly recent) preventing their Nobel Peace Prize laureates from attending the Oslo award ceremony soon disappeared: Nazi Germany, apartheid South Africa, the Soviet Union, the Myanmar/Burmese junta, and present-day China, with Myanmar ripe for change. In the end severe oppression fails. The impact of Liu's award has partially mitigated with 2012's Nobel literature prize to Mo Yan.

Dismissing China's rise would be naïve and historically ignorant. We ask if Chinese innovation now corresponds to nineteenth-century America's when culture was more Eurocentric, but gradually technological invention (steam ship, electric power, and telephone), then science, and then art shifted westward. Europeans then complained of American low sophistication and disrespect of intellectual property. The analogy is imperfect, with no equivalent of two world wars driving creative talent to China like artists and scientists migrating to the twentieth century United States.

Who will lead international space collaboration? China is afforded good will given its developing nation status, but this will end. China's government is still known for suppressing internal dissent and minorities, pressuring its smaller neighbors, and exploiting natural resources scruple-free. One day others may characterize China as oppressive and oppose it accordingly. This might lose to the *realpolitik* of a potent Chinese space program, but with power comes responsibility, and friction. India, Japan, Europe, or Russia could lead singularly or in collaboration by developing new enthusiasm for space (see Chapter 4).

Now America leads China in space exploration by a decade or more, especially in robotic spacecraft. If America does nothing, this could change rapidly. On the Moon, however, America could efficiently use its technological advantage for telerobotic lunar exploration, which might pave the way for effective human exploration and habitation. Simply putting astronauts on the Moon without backup to accomplish scientific, exploration, or resource goals is a stunt and may be counterproductive if in situ resource exploitation is not prepared. This applies to America, China, Russia, other nations, as well as non-national organizations. America is still specially positioned to attain this development program sensibly.

Leadership by America in space exploration can be fortuitous for both. NASA is still the leading agency; English is the lingua franca of science, business, aviation, and other activities. American mass media dominates many aspects. Americans maintain a mythos of the frontier, deep within their story of origin, although fading with each passing generation and greater urbanization. Still, the American spirit often embodies the rare combination of self-confidence and dissatisfaction with life that makes a good explorer (or revolutionary innovator). In America an innovator can fail and try again without the social stigma felt in many societies. Although America may be most advanced in space, it may be dealing poorly with challenges that our planet's future may hold.

Recent history bruised what we call the American spirit. The equation of freedom with the practice of conspicuous consumption in recent years mocks ideals and qualities of personal character forming its democracy. Rampant globalization and the 2007–2009 banking meltdown left many Americans bereft of self-confidence, in part because consumerism of previous generations was blunted, and in reaction many were willing to mortgage future personal growth. There have been worse existential crises in American history. Its capital city fell to invaders in 1777 and 1814. Its functioning or survival was fundamentally challenged in 1861, 1929, 1941, and 1973. Americans have risen from worse debacles, but this one was context for Obama's NASA decisions.

The United States must avoid a debt crisis in which lenders lose confidence in the government repaying its debts. This depends mainly on average deficit staying sufficiently low – some unspecified limit above several percent of GDP. More broadly, U.S. private and government liabilities, as a fraction of GDP, doubled from long-term levels of 400% to about 800% over the past 20 years and especially 2003–2007, whereas the net worth of the total U.S. economy remained roughly five times GDP.[29] For small national economies, confidence can collapse quickly, within days. How might this progress for a dominant economy? A prominent analysis claimed that public debt of 90% GDP, which the United States recently exceeded, is catastrophic, but reanalysis criticized this for computational errors.[30] Japan has twice the public debt/GDP ratio as the United States, although only one-third as much to foreign lenders. As long as the U.S. economy grows with low inflation, creditors are forgiving. In a world of limited resources, this accentuates needs for innovation and rising productivity.

We ponder the fate of a society's economy turning toward products and industry becoming more financial paper and bits than real products and industry. Fernand Braudel decades ago described overdependence on financial business as the "sign of autumn" marking the slow decline of empires, for example, the Venetian, Dutch, and British.[31] An immediate threat and serious debt arises from speculative investment, bringing the world economy to the brink of collapse in 2008. Growing parasitic use of money to make money has not abated. Borrowing to improve society and its technology is crowded out by investors who can find themselves trapped trying to grow wealthy via speculation. A web of hundreds of trillions of dollars in derivatives and credit default swap "insurance" threatens the economy, ready to collapse like dominos in response to significant setbacks. Slowly, the world economy backs out of this crushing burden of household, governmental, and investor overreach, and people and governments contort themselves to adjust. The financial sector struggles to avoid radical change, and many young people bank their careers on making money with money as the safe bet, rather than working toward a more resource-productive and progress-oriented society.[32] America must remember that it thrives by inventing the new and the real. The world needs young people to create wealth, not just absorb it, and to do so in a world with resources more restricted. Creators and nurturers of highest growth U.S. companies are not masters of business administration, but scientists and engineers: Bill Gates, Steve Jobs, and Sergei Brin/Larry Page. The United States needs new phalanxes of scientists, engineers, and creative thinkers, even more than economic restructuring. Other countries try to recreate Silicon Valley, but the individuals matter.[33] Rising from the 2007–2009 financial/real estate collapse, American workers and capitalists are producing a new round of creativity lifting the arts, sciences, technology, manufacturing, agriculture, and energy production; this is no gimmick.[34] For this to continue growing, Americans must be inspired to strive for needed skills.

America is also amassing a debt, figuratively, increasingly coming due, of educational neglect of younger students. Test scores for the United States relative to OECD countries and now China are low to average in the sciences and math.[35] The United States compensates in part by attracting the best foreign students to its universities and graduate schools, aided via the foresight and generosity of the 1958 National Defense Education Act. These students are discouraged from staying by U.S. immigration policy, however. The U.S. higher-education system seems the major factor in stopping rampant decline because of student losses in lower grades. Space exploration, a long-term investment as is education, might be considered part of the system attracting students to generate intellectual capital so well in the United States.

Simply increasing the number of college educated is required but insufficient for the service sector and rapidly automating manufacturing increasingly dominating developed economies. Those economies need new ideas and products or they will not be renewed. They need the creative, motivated innovators to which free societies are more conducive. Governments try to stimulate interest highlighting promising disciplines – the cyber, geno, nano, and now neuro – with mixed success. We need physical, chemical, electrical, and computer/software engineers to tackle challenges from climate change (as well as space exploration).

There are dark conversations in boardrooms of some prestigious American universities on basic research being curtailed in favor of potentially profitable disciplines such as nanotechnology or genetic engineering. This is the technological equivalent of eating our seed corn. Without Schrödinger (originator of one version of quantum mechanics) there would be no Shockley (co-inventor of the transistor). Several physicists in the mid-1920s struggling to understand fundamental quantum physics to explain basic experiments of the preceding few years developed the basis of 20% of the world economy today. Cutting basic research now means impoverishment later this century. We cannot afford descending that path. In part today's conflict was anticipated by changes in 1980s intellectual property law allowing universities to profit from developments by their science faculty.[36] Care is needed to avoid discarding the basis for long-term gains in reaping short-term profits.

Many children are naturally inventive, and there is great overlap between the basic talents of discovering reality for a scientist and for a child. If we want children more interested in science, the main essential is to let them pursue it: do not strip their self-confidence, rather convince them that as adults professional survival is possible and that their dreams can survive vagaries of economics and politics. Fail in this and many fewer enter the field. It is not simple; even if leaders appreciate this, many factors rise beyond control.

We optimistically envision growth over collapse from war or disease but anticipate challenges from environmental constraints. By the end or even midcentury, pressures from climate change will grow egregious. The challenge for humans in coming generations is reducing their environmental impact. Endemic is the paradox of equating economic health with growth and seemingly inevitably increasing natural resource exploitation, despite environmental degradation. The human population is projected to add at least 2 billion people (exceeding the current total of all continents beyond Eurasia) in the next 40 years (now 219,000 per day). Furthermore the affluence of billions could increase dramatically, including where population also increases. Americans consumed half of world economic resources 40 years ago while comprising 7% of all population. Now that has dropped to 25% of consumption and 5% of population. More people aspire to this level of affluence.

This cannot be sustained (see Figure 14.1). This is a Cerberus-like threat rearing many ugly faces. Primary is population growth. Societies cannot expect to increase both their personal wealth and population, unconstrained. Even in China, with active birth control policies starting in the mid-1970s, population grew more than 150% since the communist revolution, adding more than 800 million people to the planet. India alone will add this amount in addition. If these people all aspire to lifestyles of the 310 million Americans, our planet would choke. This is amplified by trade in wild animal parts and other simply wasteful factors. Furthermore, Americans, Europeans, and every other member of developed economies must cut resource-exploiting and greenhouse-gas producing behavior. Likewise, less greedy societies must concede that in the past few decades a new constraint arose: global warming, not well understood when developing countries became affluent.[37] Developed nations instituted environmental protection to shield their local ecosystems from obvious assaults of economic growth, but global warming only recently become manifest. It is largely invisible and slowly developing compared to environmental insults already mitigated, but its solution requires pervasive reform of the world economy's carbon-burning basis.

I, unlike some authors on resources and the Moon, do not portray our natural satellite as a cornucopia solving our problems if only we exploit it. In the mid-century future this might be achieved regarding energy production. In the near future we earthlings will run low of not just energy, but arable land, water for irrigation, industry, and human consumption, as well as minerals such as rare Earth elements, copper, and other metals. The Moon cannot be our breadbasket, and only

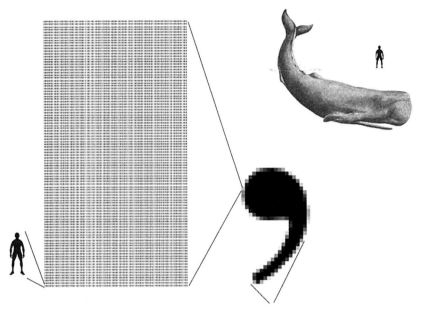

Figure 14.1. **Leviathans**. Humans versus sperm whales, the relative size and energy consumption of a typical human and an adolescent sperm whale, and an illustration of their relative number populating planet Earth. An average American uses 220,000 kcal in energy every day, the same amount as a nearly adult sperm whale, of which roughly 1 million live in the Earth's oceans. By 2050 there will be roughly 10 billion humans on Earth, or 10,000 per whale. That is 10 thousand people (about the population of Gettysburg, Pennsylvania, or Nantucket, Massachusetts) for every one of the roughly 1 million resolution pixels in this image, with roughly 600 pixels needed for a standard comma – hence, representing a city the size of metropolitan Hong Kong or Chicago. Humans, if they will use this much energy, must learn how to do so minimizing environmental impact, e.g., renewables, fusion, or fission (once issues of radioactive waste are settled).

total collapse of a mineral resource would justify lunar production for Earth export. The Moon promises resources more economically for space exploration than those from Earth, but benefits of lunar exploration for Earth itself will be subtle. The resource invigorated by lunar exploration might be ourselves. More on this later.

Early ignorance of impending global warming cannot deflect affluent societies' responsibility. Americans and others still do not discourage aspirations of "living large." In the United States 50 years ago, the common phrase "conspicuous consumption" was directed at displays of affluence by ostentatiously big cars, expensive clothes, oversized houses, and mowed expanses of lawn.[38] This concept is discouraged now as bad for business; resource exploitation is freedom of expression, many argue. In a war, selfishness and conspicuous consumption are tantamount to antipatriotic, antisocial behavior, but surely, people rationalize, we are not entering such a crisis. Many cannot accept the reality of global warming's "inconvenient truth," because doing so would call any responsible citizen to action. However, this cannot go on. If global warming reaches crisis proportions, global environmental inertia (and slow pace of rational change in the human economy) will make catastrophic suffering unavoidable. It may not end humanity but may end any environment in which we enjoy living. In facing this, the normal human reaction is to sacrifice personal aggrandizement for the common good. Freedom is not selfishness and irresponsibility to the point of collapse. Unfortunately many conspicuously consuming practices developed during the mid-twentieth century are now symbolic of success in this age of globalization. While less wasteful practices are developing, they are rarely imbued with the symbolic status of large automobiles, glass office towers, and high-caloric meat entrées. Many people worldwide, of course, reject political domination by the West but accept implicitly a Western economic symbology.

What happens when modern technological society follows a zero-growth course, as may increasingly more of us? The best model is Japan, home to 130 million people, the world's third largest economy, and arguably most advanced society technologically. It resides within modest raw resources for industry but decided a century ago to overcome this via aggressive resource acquisition overseas. This expansionist phase ended in 1945. (Japan is still one of the largest importers of raw materials.) Despite a robust technology base and rapid expansion in affluence after World War II, Japan's economy has hardly grown in two decades, maintaining essentially constant GDP since 1992, and a 4%–5% unemployment rate (overlooking major underemployment). Its population size has been flat within 1% for a decade but will decline sharply to 95 million by 2040.[39] With growth came *nikushokukei danshi* (肉食系男子) "carnivore men" or "salarymen" (サラリーマン) working hard and stereotypically drinking hard and chasing women. Productive males were offered lifetime employment; there were recognized problems of some men working themselves to death. Although Japanese lead largely comfortable lives, how do they cope with prolonged non-expansion?

Japanese do quite well, one could argue, but worrisome signs exist. One suspects that on the basis of prolonged slow growth many young people, more than in other societies, develop a sense of ennui or purposelessness. Japan's youth suicide rate is twice that of other industrialized societies and higher officially than for nearly any country (except Russia and some former Soviet republics), especially for men in their twenties and thirties.[40] School children are frequently victims of bullying, more so than in many equally affluent societies. Such issues affect young men, leading to *soushoku danshi* (草食孩男子) or "grass-eating boys," often 20- to 34-year-olds taking little interest in sex or, indeed, life.[41] The worrisome trend of *hikikomori* (ひきこもり or 引き撤もり), or "withdrawal into confinement" and implying severe social isolation, is common in Japan for males between about 14 and 26 years of age. It is no small effect, variously estimated to afflict up to 1 million Japanese or more – a significant fraction (up to about 10%) of all young men and adolescents. These young men lock themselves in their rooms for months or years, depend on parental care, and shun interaction with others except perhaps via the Internet. The phenomenon has been noticed especially since 2000, but extends to men in their thirties suffering with it since adolescence. Anecdotally, some perform bizarrely violent, antisocial crimes once leaving their lair. Introversion aside, "parasite singles" or "freeters" delay independence from their parents or have no long-term employment. Some blame this introversion on the protective relationship of Japanese parents with their sons and the growing roles of women in society (rising *nikushokukei joshi* or "carnivore women") or on the competitive pressure on students in school, but the timing of *hikikomori* suggests the importance of the no-growth economy and lack of customary stable employment. Obviously, factors exist particular to Japan but are exacerbated by the intense, zero-sum game implied by a flat economy in a highly competitive society. Many young men opt out. What corresponding trends await other societies? Most might consider themselves lucky to have only Japan's problems. I imagine these issues will lead us to seek ways to inspire more positive behavior.[42]

When societies attain affluence, population ceases exponential growth and may even decline, which is encouraging, and may be why we and the Earth's ecosystems will ultimately survive. However, the Japanese experience provokes worry about the human spirit, especially for young men (although it may include women elsewhere). Where is the adventure? What can we explore beyond the Internet? Why dream of a future promising to be much like the present, except maybe a little smaller?

The Enlightenment's program, by philosophers of the seventeenth century and politicians of the eighteenth, emphasized reason, law, and empowerment of the individual. Means toward this end was the Industrial Revolution of the eighteenth through twentieth centuries, by scientists, engineers, and industrialists. That mechanism was the harnessing of energy resources far beyond what animals or human muscle alone could provide. By flipping a switch a common citizen could energize processes that several generations before would have required armies of slaves. What freedom! That amazingly successful program has largely run its course in the current era of global limitations. What do we imagine to take its place? How do we maintain the great benefit to the common person from these centuries of progress?

In many ways expansion into space logically extrapolates the enlightenment-industrial program. The bad news has been its expense. Is there a path toward a sustainable, progressive society benefiting from space exploration?

Momentarily forget technical discussions of exploring space and its costs and consider the desirability of expansion into space in the abstract. If it were to cost little, would we necessarily want to do it? Of course many people's horizons extend little beyond their everyday life. For others any expenditure on such pursuits no matter how small detracts from their wealth and investment in the immediate world, so these distant goals waste resources, by definition. People with these inclinations will only be impressed by benefits brought straight to them, enriching their lives directly, materially. That is another topic (Chapter 12).

Fear pervades some people's reaction to space exploration: fear of the unknown but not the unimagined. Enough space monsters run through our fiction to scare people. In addition, well-known space flight perils inspire others' reluctance to let explorers risk themselves. Qualms grow for government programs with collective responsibility over explorers' fates.

Although we anticipate risk, there are always unintended consequences. For instance, alternatives to the enlightenment-industrial program exist, anathematic to expansion in space. World views foreseeing a single, world-ordered end state, be it a classless utopia, one-world caliphate, or eschatological millennium, encounter contradiction when the world is unbounded, with groups displaced by long communication delays, not easily influencing each other. Not discussing which might prevail, note that each of these three world views have, nominally, at least a billion adherents. (Their majorities do not believe such prophecy, but millions do.) Expansion in space precludes the fate for humanity envisioned by many. There can be no worldwide unity, because the world continues to expand and diversify. The changeless end state that some envision will slip their fingers, like grasping at sand. Humans on the frontier depend heavily on unbiased science, but not necessarily exclusive of other beliefs. This would not be the first time people flee to far reaches to escape religious dogma or spread it, but such vastness seems hostile to homogenization of belief. The point remains: when expansion into space looms large, some will take ideological exception, maybe to the point of extremist action.

The enlightenment-industrial program has tarnished for many people. Science and technology, providing tools to win World War II (radar, code breaking, and atomic weapons), was elevated to unprecedented preeminence in the halls of power. That grew in the atomic age and the space age, but some visions of the 1950s and 1960s were over-promised (no flying cars, atomic airplanes, automated homes, or power "too cheap to meter"). The Vietnam War demonstrated limits to technology (e.g., modern air power and surveillance to disable enemy transportation stymied by bridges built just below the surface of Vietnamese rivers). The burning Cuyahoga River and oil spilled on Santa Barbara's beaches in 1969, one oil shock in 1973, another in 1979, destruction of Three Mile Island's and Chernobyl's reactors (and now Fukushima's) cast a pall over expanding energy and the economy based on it. In this context a space program advancing the Cold War, using huge, one-shot rockets, and spearheaded by military pilots seems so 1960s. Younger people (some middle-aged by now) came to see space exploration as a phase we abandoned to these events.

Scientists are sometimes lukewarm to space exploration, particularly involving astronauts. Of course there are enthusiasts; of course many see their research tied to space, and many not working on space topics are, however, keen on keeping space exploration healthy. Historically, this is understandable: the National Defense Education Act of 1958, part of the multipronged response by Congress to the alarm sounded by *Sputnik 1*,[43] is greatly responsible for support and employment enjoyed by U.S. scientists. However, even in space science – including space astronomy and planetary and interplanetary science – a competition is perceived between human exploration and scientific research. Promoting human space exploration to scientists sometimes results in hostile stares, in part because many otherwise knowledgeable scientists have little idea of goals of NASA's and other agencies' human space flight programs and because human space flight is seen as a competitor and potential drain on other science, both in space and on the ground.

This can change. That which is passé can be renewed if it is fundamental: the frontier, the Universe beyond Earth. How can we engender new yearning for the frontier? Can we do this affordably?

First, we are no longer entreating children of the 1960s. When young people look into the sky they care little how space exploration was timed relative to geopolitics

50 years ago. Away from our cities the stars burn brightly as ever, and now our huge telescopes and the Internet can deliver images and discoveries to the palm of everyone's hand. Exploration can cost less, and more people can participate: exploring space involving people without actually launching them there. With telerobotics, participants can direct research from simple offices, with millions joining, virtually. We can do this now, like watching Deepwater Horizon befoul the Gulf of Mexico, but positively, for knowledge, once we design missions incorporating this.[44] The Moon is the best place for this.

The Moon is the only concrete destination available to us ready for wireless. With Mars 6 to 45 minutes away via round trip lightspeed, and asteroids only swooping into range occasionally, the Moon with its 3-second latency is the only "Planet Internet" that we can explore in real time, the only world from which you could conduct simple conversation with people on Earth, the only solid body where we conduct e-mail, text, video, and voice with the immediacy we grow to expect. If we want to establish meaningful contact, especially to inspire youth, this is essential. The Moon is the essential place, no farther. The Mars *Curiosity* rover has a half million Twitter followers, but this not the same thing. NASA has made a few interesting efforts,[45] but the potential is much greater.

We can incorporate this communication potential into our thinking about education and public outreach, from the beginning. Telerobotics can be incorporated from initial craft design phases to optimize the educational component. One option is to let viewers on the Internet follow exploration.[46] We could easily include an educationally oriented rover within a lander payload, as a "getaway special" for as little as several million dollars. (The rover could be small and cloned to several rovers.) Assuming useful life before overhaul/replacement of 1–2 years and 50% duty cycle, usage costs only $90 per quarter hour.[47] Millions of people would be willing to pay $90 for their 15 minutes of Moon-driving time (subject to emergency safety override protocols, of course). Any young person who desires could participate, regardless of affluence.

We see precedent for rich mixtures of science and public involvement, for example, RU 27 *Scarlet Knight* robotic submersible's Atlantic exploration by Rutgers University.[48] It finished a 7,410-kilometer autonomous mission across the ocean on December 4, 2009, completing RU 17's voyage and breaking its own 5,700-kilometer record for a free submersible over 5 months. These serve as precursors for a robotic exploration fleet. Significantly, they engaged dozens of students. More pointedly, a writer/filmmaker team worked during the mission to publicly portray its science and exploration. A similar media effort succeeded during the Balloon-borne Large Aperture Submillimeter Telescope (BLAST) mission.[49] We need Internet-based outreach programs that draw people, especially young ones, into efforts before, during, and after exploration. Corresponding productions on space exploration by NASA or other space agencies are rare (although one can find close approximations by media groups).

Without choosing sides in the debate of robots versus astronauts, harkening to von Braun or Van Allen, we have discussed various aspects (Chapters 10, 12, and 13). Almost all NASA missions are robotic at some level, although some so-called robotic missions could not occur without astronauts, for example, *Hubble* requiring more astronaut support than any spacecraft except the *ISS*, exceeding even the Apollo lunar landing missions. If we find sufficiently interesting signs of life in the Solar System, we will need astronauts, and the Moon will play a novel role in how they are perceived by us earthlings. Light travel time means our frontline explorers for alien life will need human awareness and judgment in the vicinity, even if the eyes and limbs are telerobotic. If we find life on Mars or Europa, we will propose human bases on Phobos or Callisto, despite the expense. If the aliens are interesting and we have the wherewithal, we will build them. Even if alone in the Solar System, we will eventually lift ourselves from Earth to explore and inhabit. We must learn to establish habitat on alien worlds, likely in near vacuum. The Moon is our stepping stone.

Human lunar explorers will form an elite cadre nonetheless modest in resource exploitation (versus *ISS* where all material is shipped frequently from Earth at great expense with waste dumped to eventual atmospheric reentry). On the Moon people will conserve resources even while "living off the land" whenever possible: recycle waste, grow their own food, and live successfully within limited space. This epitome of human accomplishment will offer a living example of sustainability, in contrast to the competing vision, where privatized space allows momentary play for billionaires using dozens of tonnes of rocket fuel with each space tourism junket. We sorely need better role models.

In 1969 we saw Earth from afar via Apollo's pictures and developed new appreciation for our world's fragility and uniqueness. We have not sufficiently pursued that insight. Developed nations have dawdled in advancing technologies and disciplines to save Earth's environment, and the rest of the world is only increasing an unsustainable avarice. Human nature concentrates on the most immediate threat. When the threat of global environmental degradation draws nigh, however, it will not be ameliorated quickly. There will be many years when prospects look grim, when disasters ensue, and when solutions seem not to succeed. Under such conditions humans can yield to gloom, guilt, and victimhood. At this point another perspective, another realm of activity beyond Earth will help. Over decades and centuries Earth's environment will recover, and humans must avoid despair. Maybe for the first time we all will know and feel that we live on a finite planet but in a boundless Universe. The Moon allows us the first chance to make that mental alignment.

Worldwide we must allow science and technology as means to solve environmental and economic problems. Americans particularly must admit that scientific training provides means toward better economic competitiveness. A quiet but problematic issue during the G.W. Bush years with publicly promoting NASA's science were hot-button terms *evolution* and *Big Bang* for American Christian fundamentalists, as well as discussion of global warming.[50] Science obeying politics becomes propaganda. NASA missions like *Hubble* and the *Wilkinson Microwave Anisotropy Probe* made leaps testing, constraining, and verifying the Big Bang model. Evolution controversy pertains to biology, yet galaxy evolution, a topic advanced by *Hubble* and missions like *Spitzer*, was discussed cautiously. Diffusing NASA results can be crimped by politics.

Figure 14.2. **Broad horizons**. Flight Engineer Tracy Caldwell Dyson gazing through the Cupola of the *International Space Station*. Space exploration has great, still untapped potential to inspire young people to strive beyond themselves. (NASA photo ISS024-E-014263, taken September 11, 2010, on Expedition 24.)

NASA cannot self-promote to Congress, but it spends about 1% of its budget on public outreach.[51] In a different society an agency like NASA would not measure its words singing praises of such accomplishments. The *ISS* is potentially a great source of inspiration for the people of the world (Figure 14.2); this potential is largely squandered. Urgently, NASA's outreach must be extended, not to increase public support but to increase public benefit and inspiration, in the manner described earlier.

How much should we spend? As apparent from the 1980s Soviet space program (Chapter 2), a nation can damage its economy with too much space investment. This exists nowhere now, especially in the United States. Soviet spending on *Mir*, *Energia*, *Buran*, *Polyus*, and concurrent space missions totaled roughly 2% of GDP, whereas U.S. civilian space expenditures are now 0.1% (versus the peak, 0.7% in 1966). Late Soviet space programs were mainly military; U.S. military space spending slightly outstrips civilian space.[52] The United States devotes 1/50th as much to civilian space as national security. (The latter nearly equals defense expenditures by all other nations, with 23 times the U.S. population.)

There are several arguments for cutting back on U.S. government civilian space spending: (1) federal debt is too large, and NASA is a nonessential discretionary expenditure; (2) the U.S. economy has passed through its worse shock in more than a generation and its deepest recession since World War II, and we must spread the suffering; (3) private industry should take over; or (4) we have lost interest in space exploration.

Reason 1 seems our operative policy, by happenstance over intent. Recently NASA funds have fallen, now plummeting under sequestration. Planetary science spending is slashed, and sober projections predict in several years space science budgets to be one-third lower. This stems in part from expiration of a small funding bump, salve for Obama's ending of the Vision and congressional across-the-board cuts in discretionary spending. The latter is excused by the red herring of U.S. debt reaching 90% of GDP (discussed earlier). The result is few planetary missions surviving and scientists competing 10 to 1 for research funds, untenable because peer review cannot reliably pick the top 10% of science proposals, just ones with little risk by old-hand investigators – a regressive condition.[53]

Recent authors decry this condition and argue that space exploration's benefits outweigh justifications for cuts based on debt. Neil deGrasse Tyson calls for doubling NASA's budget, and Taylor and Osborn advocate a tenfold rise.[54] These exceed what I justify, but a reliable increase is needed. A 5% rise (above precut levels) would yield an inspiring lunar program; a 20% rise ($4 billion) would expand the fore on many fronts – Mars, Europa and ice moons, asteroids, other planets, as well as space and Earth sciences, and solar and astrophysics – and maintain NASA's technology duties. With this, every year would yield rainbows of science discovery. This is only 0.025% of GDP (or 0.006%), a small fee for firing up our youth on science. We can afford this; in fact we cannot afford to fail.

Whatever your opinion of the U.S. budget and national debt, there is no doubt that money is spent voluntarily on activities less justified than space exploration. The American economy loses about $100 billion per year to cigarettes' effects.[55] Legal gambling in the United States exceeds $90 billion annually.[56] Alcohol and illicit drug spending is comparable.

Consider aspects of space exploration as educational entertainment. Many similar activities each consume of order $10 billion annually, per category of U.S. media or venue: cinema box office, live spectator sports, live theater, live music, musical recordings, video games, and DVD/Blu-ray sales.[57] The film *Avatar*, set on a moon orbiting a gas giant planet, like Jupiter's Europa, garnered the largest box office take in history: $2.8 billion, enough to fund a mission to Europa itself, ironically.[58] *Avatar*, *Star Wars*, and *Star Trek* films have grossed almost $8 billion.

Regarding suspending space exploration until better times, we argued that choosing this route we effectively abandon the effort, a long-term commitment. This only increases costs, and our scientists and engineers cannot idle with every recession. We might as well shutoff their oxygen. Ability to plan makes space exploration possible; results promised from the 20% or 5% increment above only come if it is maintained long term.

Private space travel is growing, but should not be confused with space exploration (of new worlds and conditions in the Solar System). (See note 28 in Chapter 3.) Private enterprise may accomplish great things in space exploration but not divorced from government. In the nineteenth and twentieth centuries, the U.S. government scouted the frontier, underwrote and subsidized private exploitation, and regulated its results. Historically, governmental investment has been seminal and backstopped most technological advances often credited to private entrepreneurs.[59] For demanding and technically intensive challenges like space, we cannot expect to accomplish much by doing less in the twenty-first century. Most space commerce efforts beyond LEO are counting on government clients. Governments must stay in the game. That we have lost interest in space exploration seems demonstrably erroneous, and if true, I worry for our economy and our children's spirit. I close considering this question.

Children start developing scientific patterns of thought, observation, and inference nearly from infancy,[60] applicable to science itself if they are presented with good material. School reform rarely concerns student inspiration versus teacher performance and test scores.[61] Outer space is special in capturing young people's interest in science. Give them outer space, dinosaurs, sea animals, and computers, and you stimulate most children's imagination about science of some kind. If you want to inspire physics and mechanical engineering, give them stars, planets, rockets, and astronauts. As they mature, they diversify. Despite many polls on public attitudes toward space exploration (usually positive, see Chapter 3), we rarely ask children what they think of space, or even what interests them. The data are so scant that I conducted my own poll of former children, now scientists and engineers.[62] Space, particularly Apollo, has resounding effect on most of these professionals, particularly of certain ages, so strong that an echo affects later generations: scientists who witnessed their parent's interest in space. Most of us can attest that children still dream of space travel; we need to use this constructively. We live in a time and a world soon to strike us with profound Earthly limitations, but also to reveal myriad Earth-like worlds incomprehensibly beyond our grasp. Spanning this chasm between realities is a chance to explore our own Solar System. Our children can embrace this challenge. It starts with the Moon.

If we want scientists and engineers, we need a strong space program. An inspiring, invigorating space program should be strong in lunar exploration. Technology can now combine these in new, exciting ways.

As discussed the Moon now offers a stepping stone barely imagined during Apollo. Apollo taught many lessons, but now they transform, making the Moon much more interesting and useful. These insights have not percolated to space policy makers; there is still time. The Moon will last. What may slip away is a unique opportunity using it to inspire our children to grasp solutions to challenges we face in coming decades. Give them a chance to shine; we will all benefit.

I guess that we live on the cusp of great excitement about space exploration. I bet that in a decade or so we will find strong signs of life on Mars or maybe in Europa's or Enceladus's ice. Sooner, we will find an Earth-like planet near another star, maybe with life, albeit a billion times beyond the Moon, likely beyond our grasp in this century. If we choose, we can explore and inhabit the closer world and use it to explore the Solar System starting with (tele)robots. Interest in space and its associated science and technology can step into high gear in time to enroll youth in the technological challenges that are descending. With some natural effort, we can capture this excitement and innovation.

Humanity will someday move out into the Solar System, and almost certainly it will include the Moon. The Viking clan tried to colonize North America 1,000 years ago and failed. The Ming Chinese might have sailed northeast 600 years ago and found America, but they burned their mighty exploration fleets. Eventually, almost a century later, it happened: a new world.

We can do better. In this century, we must.

Notes

1. "Dreams Are Maps: Exploration and Human Purpose" by Carl Sagan, *Planetary Report*, September/October 1992, 20–23.
2. "Natural history" here refers to observational natural sciences, particularly life sciences (biology, botany, ecology, etc.) and usually geology and meteorology, and to some extent astronomy.

    Scott's letter ("To My Widow"), displayed at the Scott Polar Research Institute, Cambridge, in 2007, continues with touching concern for their 2-year old son: "I know you will keep him out in the open air – try and make him believe in a God, it is comforting. Oh my dear my dear what dreams I have had of his future and yet oh my girl I know you will face it stoically – your portrait and the boy's will be found in my breast and the one in the little red Morocco case given by Lady Baxter – There is a piece of the Union flag I put up at the South Pole in my private kit bag together with Amundsen's black flag and other trifles – give a small piece of the Union flag to the King and a small piece to Queen Alexandra and keep the rest a poor trophy for you! – What lots and lots I could tell you of this journey. How much better it has been than lounging in comfort at home – what tales you would have for the boy but oh what a price to pay – to forfeit the sight of your dear dear face – Dear you will be good to the old mother. I write her a little line in this book. Also keep in with Ettie and the others – oh but you'll put on a strong face for the world – only don't be too proud to accept help for the boys sake – he ought to have a fine career and do something in the world. I haven't time to write to Sir Clements – tell him I thought much of him and never regretted him putting me in command of the *Discovery*." (A version of this letter highly edited by Leonard Huxley, father of Aldous and Julian, removes reference to "a God" and gratuitously inserts a rant against indolence, persisted for years in *Scott's Last Expedition, Vol. 1, Being the Journals of Captain R.F. Scott, R.N., C.V.O.*, 1914, London: Smith, Elder, 415.)

    The Scotts' son, Sir Peter Markham Scott (September 4, 1909–August 29, 1989) co-founded the World Wildlife Fund. He also contributed to the conservationist policy change by the International Whaling Commission and the signing of the Antarctic Treaty.

    Kathleen (March 27, 1878–July 25, 1947) was a noted sculptor and is portrayed in *A Great Task of Happiness: The Life of Kathleen Scott* by Louisa Young, 1995, London: Macmillan, and the play *A Father for My Son* by Jenny Coverack & Robert Edwards, 2000.
3. President Obama at Kennedy Space Center, April 15, 2010, to a NASA audience (and Aldrin).
4. "Reactive Devaluation in Negotiation and Conflict Resolution" by Lee Ross, in *Barriers to Conflict Resolution*, eds. K.J. Barrow et al., 1995, New York: Norton, 26.
5. The TEA Party in Space, http://www.teapartyinspace.org/.
6. From President Obama's State of the Union Address, January 24, 2012, as delivered:

    "The first step in winning the future is encouraging American innovation.

    None of us can predict with certainty what the next big industry will be, or where the new jobs will come from. Thirty years ago, we couldn't know that something called the Internet would lead to an economic revolution. What we can do – what America does better than anyone – is spark the creativity and imagination of our people. We are the nation that put cars in driveways and computers in offices; the nation of Edison and the Wright brothers; of Google and Facebook. In America, innovation doesn't just change our lives. It's how we make a living.

    Our free enterprise system is what drives innovation. But because it's not always profitable for companies to invest in basic research, throughout history our government has provided cutting-edge scientists and inventors with the support that they need. That's what planted the seeds for the Internet. That's what helped make possible things like computer chips and GPS.

    Just think of all the good jobs – from manufacturing to retail – that have come from those breakthroughs.

    Half a century ago, when the Soviets beat us into space with the launch of a satellite called Sputnik, we had no idea how we'd beat them to the moon. The science wasn't there yet. NASA didn't even exist. But after investing in better research and education, we didn't just surpass the Soviets; we unleashed a wave of innovation that created new industries and millions of new jobs.

This is our generation's Sputnik moment. Two years ago, I said that we needed to reach a level of research and development we haven't seen since the height of the Space Race. In a few weeks, I will be sending a budget to Congress that helps us meet that goal. We'll invest in biomedical research, information technology, and especially clean energy technology – an investment that will strengthen our security, protect our planet, and create countless new jobs for our people.

Already, we are seeing the promise of renewable energy. Robert and Gary Allen are brothers who run a small Michigan roofing company. After September 11, they volunteered their best roofers to help repair the Pentagon. But half of their factory went unused, and the recession hit them hard.

Today, with the help of a government loan, that empty space is being used to manufacture solar shingles that are being sold all across the country. In Robert's words, "We reinvented ourselves."

That's what Americans have done for over two hundred years: reinvented ourselves. And to spur on more success stories like the Allen Brothers, we've begun to reinvent our energy policy. We're not just handing out money. We're issuing a challenge. We're telling America's scientists and engineers that if they assemble teams of the best minds in their fields, and focus on the hardest problems in clean energy, we'll fund the Apollo Projects of our time."

7. See Chapter 3, pp 104–105.
8. The Office of the Chief Technologist (http://www.sti.nasa.gov/tto/) publishes the annual *Spinoff* report describing typically about 50 new technologies from NASA.
9. The NASA list of 13 items is found at http://www.sti.nasa.gov/tto/apollo.htm. A similar Smithsonian NASM list of 11 contributions include CAT/MRI digital enhancement (http://www.nasa.gov/offices/ipp/home/mythbuster/myth_mri.html), portable cool suits, kidney dialysis, cordless power tools (http://www.nasa.gov/offices/ipp/home/myth_tools.html), cardiovascular conditioner, Mylar/Al insulation barrier, vacuum metallization, water purification, hospital monitoring technology, fuel cells, and the Michigan Silverdome's fabric roof technology. Descriptions of most of these are found on the NASA list.
10. "Testimony of Robert L. Park concerning the *International Space Station* before the U.S. Senate Committee on Commerce, Science, and Transportation," October 29, 2003; "Scientists Believe ISS Is Waste of Money" by Jim Wilson, *Popular Mechanics*, December 2002; "NASA Technological Spinoff Fables" by John Pike, May 25, 1995, Foundation of American Scientists Space Policy Project e-print (http://www.fas.org/spp/eprint/jp_950525.htm).
11. "10 Stats You Should Know about Robots But Never Bothered Googling Up" by Erico Guizzo, *IEEE Spectrum*, March 21, 2008.
12. These are summarized succinctly in "War Machines: Recruiting Robots for Combat" by John Markoff, *New York Times*, November 27, 2010.
13. The service robot market in 2012 was $20 billion in sales. The largest market for robots in general is Japan, followed by South Korea, then China, the United States, and Germany, the latter three growing rapidly. The most popular service robot in sales is still the vacuuming iRobot Roomba series, with more than 6 million sold.
14. NASA runs an interagency/academic robotics alliance (http://robotics.nasa.gov/alliances.php), but the list of its industrial/private partners is tiny.
15. E.g., "China's Economic Rise – Fact and Fiction" by Albert Keidel, 2008, *Carnegie Endowment for International Peace Policy Briefing* #61, predicts a $45 trillion GDP by 2040 for China and $33 trillion for the United States, or $30,000 and $87,000 per capita, respectively; "$123,000,000,000,000: China's Estimated Economy by the Year 2040. Be Warned" by Robert Fogel, January–February 2010, *Foreign Policy*, 177, 70, predicts $123 trillion (China) and $43 trillion (United States), or $82,000 and $114,000 per capita, and 5.2% and 2.9% average growth per year, respectively. (Fogel is a Nobel laureate in economics.) Current GDP is $7 trillion (PPP of $11 trillion) and $15 trillion, respectively, per capita GDP $5,000 and $48,000, at typically 9% and 2–3% growth. A recent summary is found in "China Poised to Pass US as World's Leading Economic Power" by Chris Giles, Financial Times, April 30, 2014. Of course, there is room for skepticism about the reported numbers for China's GDP: "China's GDP Is 'Man-Made,' Unreliable: Top Leader" December 6, 2010, *Reuters*, http://www.reuters.com/assets/print?aid=USTRE6B527D20101206.
16. A worried but potentially optimistic analysis of this trend is *China 2030: Building a Modern, Harmonious and Creative High-Income Society* by staff of the World Bank and the Development Research Center of the State Council, PRC, 2012, Washington, DC: The World Bank.

17. E.g., see *When China Rules the World: The End of the Western World and the Birth of a New Global Order* by Martin Jacques, 2009, New York: Penguin Press.
18. The Organisation for Economic Co-operation and Development (OECD) Programme for International Student Assessment (PISA) compares 15-year-olds' knowledge and skills in reading, mathematical, and scientific literacy: *PISA 2012 Assessment and Analytical Framework – Mathematics, Reading, Science, Problem Solving and Financial Literacy* by Secretary-General of the OECD, 2013, Paris: OECD.
19. "China Patent Office Becomes World's Largest: WIPO" by Staff Writers, December 11, 2012, *Space Travel.com*, http://www.space-travel.com/reports/China_patent_office_becomes_worlds_largest_WIPO_999.html.
20. See "Chinese Journal Finds 31% of Submissions Plagiarized" by Yuehong Zhang, 2010, *Nature*, 467, 153; "Plagiarism Plague Hinders China's Scientific Ambition" by Louisa Lim, August 3, 2011, *National Public Radio*; also *China's Great Science Gamble*: "China Seeks Status As Scientific Superpower" August 1, 2011, and "China's Supercomputing Goal: 'Zero To Hero'" August 2, 2011, *National Public Radio*. Retractions in American science publications are rising, increasing 15 times since 2001, mainly in health sciences ("Mistakes in Scientific Studies Surge" by Gautam Naik, August 10, 2011, *Wall Street Journal*). For details, see *Retraction Watch*, http://retractionwatch.wordpress.com/.
21. "The Outflow of Academic Papers from China: Why Is It Happening and Can It Be Stemmed?" by Jufang Shao & Huiyun Shen, 2011, *Learned Publishing*, 24, 95; "Strong Medicine for China's Journals: Weak Publications Will Be 'Terminated'" by David Cyranoski, 2010, *Nature*, 467, 261.
22. A telling example is Wang Chuanfu, president of BYD Auto (Build Your Dreams), who passed up a chemistry professorship to found this successful car maker. Consider Li Na, celebrity high-ranked singles tennis player, who is allowed relatively free personal, nonpolitical expression and is held up as an example of such. In contrast, internationally renowned artist Ai Weiwei (famous for the "bird's nest" Olympic Stadium design) can air complaints against governmental authorities more strongly than average citizens but has been beaten by police and hospitalized, had his artwork bulldozed, been "disappeared" temporarily, and charged with tax evasion. Most *cognoscenti* about the case tie his first arrest more to changes in China's leadership and security environment (Jasmine Revolution, etc.) than to action by Ai. Still, the authorities could not deal with a nation of Ais. Permissive exceptions seemingly extend only to this uppermost echelon of personae. Even students at Peking University, among China's most elite, are officially counseled against being radical or unconventional ("China's Fight Against Radical Thought" by Mary Kay Magistad, March 30, 2011, *The World*, Public Radio International).
23. The Chinese Communist Party's Hundred Flowers Campaign of 1956–1957 began "Letting a hundred flowers blossom and a hundred schools of thought contend is the policy for promoting progress in the arts and the sciences and a flourishing socialist culture in our land." Within a year the campaign ended badly for intellectuals who had dared stick their necks out. I will not dwell on recent, well-known incidents. On where authorities draw the line, Ai Weiwei said, "How can you predict what's in a dictator's mind? You know if you really think about them, you are already a victim of them" ("Ai Weiwei's Artwork Travels, Despite Detainment" by Laura Sydell, May 1, 2011, *Weekend Edition Sunday*, National Public Radio).
24. I appeal to conceptual blending theory, e.g., "Conceptual Integration and Formal Expression" by Mark Tunner & Gilles Fauconnier, 1995, *Metaphor and Symbol*, 10, 183.
25. One telling case is Shi Yigong, who moved from China to the United States for his PhD, then rose to tenured professor in molecular biology at Princeton. Nonetheless, he returned to Tsinghua University in 2008. Dr. Shi relates "China does have a lot of talented scientists and engineers. People are very smart in China. But the mechanisms for people to apply their talents, for people to innovate, are not there yet. That needs to be resolved." However, Shi cannot break cultural habits in relating to his own children, as what he asks when they return home from school: "For someone who stayed in the United States for 18, 19 years, who's been really influenced by Western culture, still the question was: Did you listen to your teachers? You know, we are told to listen. We are told to accept. We are told to – not to doubt about authority. So I think that element is very hard to do away with, because that's part of our culture" ("China's Businesses Boom, But Its Brands Don't" by Renee Montagne, June 22, 2011, *Morning Edition*, National Public Radio).
26. Consider Wang Zheng, principal of Peking University High School (PUHS). When Wang came to PUHS he had all teacher's lecterns ripped from the classrooms. He asked his teachers to move among their students, engaging them. Teachers focus on process, not correct answers, and ask students their opinions: why, how, challenging their facts. ("Chinese School Defies Rigid Exam-Focused Education" by Rob Schmitz, June 9, 2011, *Marketplace*, American Public Media.)

A prime issue is annual exams, the *gaokao*. Jiang Xueqin, PUHS deputy principal, considers the National Higher Education Entrance Examination, prerequisite for entering most undergrad programs: "So, if we were to start from scratch and try to build an alternative to the gaokao, we would end up with as the only viable alternative ... the *gaokao*. That's what a lot of people tend to forget: that given the complete lack of trust in each other and in institutions, given the stifling poverty that most Chinese find themselves in, and given China's endemic corruption and inequality, the *gaokao*, for better or worse, is the fairest and most humane way to distribute China's education resources" ("The Sad Truth of China's Education" by Jiang Xueqin, June 3, 2011, *The Diplomat*).

27. From *The National Medium- and Long-Term Plan for the Development of Science and Technology (2006–2020)* by State Council of the People's Republic of China, 2006: "Despite the size of our economy, our country is not an economic power, primarily because of our weak innovative capacity." It prescribes absorbing ideas by "enhancing original innovation through co-innovation and re-innovation based on the assimilation of imported technologies." External ideas should be adapted remembering that "one should be clearly aware that the importation of technologies without emphasizing the assimilation, absorption and re-innovation is bound to weaken the nation's indigenous research and development capacity."

28. On the day of Liu's Peace Prize award (October 8, 2010), I happened on Wikipedia's "List of Nobel Laureates by Country" (http://en.wikipedia.org/wiki/List_of_Nobel_laureates_by_country), where Liu was the only laureate listed under the People's Republic of China. Over several weeks 10 names were added by online contributors, but Liu's was the only award to someone living in China for work done primarily in China. Gao Xingjian (Literature, 2000), who left China in 1997, was then the only other laureate honored primarily for work done in China.

29. "Flow of Funds Accounts of the United States Flows and Outstandings" by Board of Governors of Federal Reserve System, September 16, 2011.

30. "Growth in a Time of Debt" by Carmen M. Reinhart & Kenneth S. Rogoff, 2010, *American Economic Review*, 100, 573, is criticized by "Does High Public Debt Consistently Stifle Economic Growth? A Critique of Reinhart & Rogoff" by Thomas Herndon, Michael Ash & Robert Pollin, April 15, 2013, *Working Paper: Political Economy Research Institute*, University of Massachusetts Amherst.

31. *Civilization and Capitalism, 15th–18th Century: The Perspective of the World* by Fernand Braudel, 1979, English translation by Siân Reynolds, Berkeley: University of California Press.

32. After falling in 2008–2009, numbers of business students from the prestigious schools entering investment banking rebound to ~40% of Harvard MBAs: "Be Afraid: Harvard MBAs Infiltrating Wall St. at Near-Record Rates" by Aaron Elstein, September 13, 2011, *Crain's New York Business*.

33. In fashion now are innovation centers, in India, China, Brazil, and Russia. These depend on large numbers of participants in the right talent mix to maintain activity because most start-ups fail. Management must try many tests and commercialize new ideas and products quickly, requiring thousands of scientists and engineers, as well as many entrepreneurs with experience creating new companies. To achieve and sustain this mix, as in Silicon Valley for six decades, is challenging. There are increasing trends for centers arising in urban areas, not sub- or exurbs, requiring cities themselves and their larger societies to support innovation implicitly. (See *The Metropolitan Revolution: How Cities and Metros Are Fixing Our Broken Politics and Fragile Economy* by Bruce Katz & Jennifer Bradley, 2013, Washington: Brookings Institution Press.)

34. See, for instance "The Rise of the Creative Class – Revisited: 10th Anniversary Edition – Revised & Expanded" by Richard Florida, 2012, New York: Basic Books.

35. Shanghai, proxy for China (with Hong Kong), was first in science and math, the United States twenty-third and thirtieth, respectively ("Education: Korea and Finland top OECD's Latest PISA Survey of Education Performance" by Organisation for Economic Co-operation & Development, December 7, 2012).

36. The Bayh-Dole Patent and Trademark Law Amendments Act of 1980 gives universities control over inventions and intellectual property derived from their U.S. federally funded research.

37. This is detailed in *Consumptionomics: Asia's Role in Reshaping Capitalism and Saving the Planet* by Chandran Nair, 2011, Oxford: Infinite Ideas. Evidence of rising carbon dioxide threatening global warming was unnoticed until 1961 with Keeling's curve of Mauna Loa CO_2 levels, but greenhouse warming was foreseen in 1896 by Arrhenius relevant to oil and coal burning, and in 1824–1859 by Fourier, Pouillet, and Tyndal as basic physics. The U.S. White House described global warming's threat in 1965. Exhausting nonrenewable energy resources was foreseen sooner: "We are like tenant farmers chopping down the fence around our house for fuel when we should be using Nature's inexhaustible sources of energy – sun, wind and tide. ... I'd put my money on the sun and solar energy. What a source of power! I hope we don't have to wait until oil and coal run out before we tackle

that" (Thomas Edison quoted in 1931 in *Uncommon Friends: Life with Thomas Edison, Henry Ford, Harvey Firestone, Alexis Carrel & Charles Lindbergh* by James Newton, 1987, Boston: Houghton Mifflin Harcourt, p. 31).

38. Sustainability is difficult. Sincerely sustainably, I sold my excess property to others who can use it more efficiently, rid myself of my automobile and air conditioner, increasingly conduct business electronically rather than via plane flight, and opt for public transport whenever possible. Still, I consume more than the average human in nonrenewable resources and carbon usage. Improving this is difficult; the market for conserving approaches would be larger if more people pursued them. Huge swaths of society have no such options. We must widen this path and must all understand that the party is over.

39. *Statistical Handbook of Japan*, http:www.stat.go.jp/data/handbook/. Japanese unemployment rates are still lower than nearly any other industrialized country's.

40. *Handbook of Man in the Heisei Period (Soushoku Danshi Sedai Heisei Dansi Zukan)* by Maki Fukasawa, 2009, Tokyo: Nikkei; "The Rise of 草食系男子 Soushukusei Danshi, Masculinity and Consumption in Contemporary Japan" by Steven Chen, in *Gender, Culture and Consumer Behavior*, eds. Cele C. C. Otnes, Linda Tuncay & Tuncay Zaye, 2012, New York: Routledge, p. 285. Rather than pursue careers, *soushoku danshi* may settle for low-paying jobs, engage in traditionally feminine pursuits such as shopping and beautification, and spend more time with video games or the Internet rather than potential mates. [41] Popular media see these men as heterosexual but refer to them as *ojo*-men (お嬢マン, "girly men") and *otomen* (オトメン亿男, "young lady men").

41. See "Mental Health: Country Reports and Charts Available," World Health Organization, http://www.who.int/mental_health/prevention/suicide/country_reports/en/index.html. Even total suicide rates only in Belarus, Lithuania, Kazahkstan, South Korea, and, by some estimates, Russia are higher than in Japan. A large increase in South Korea occurred in the past decade ("Suicides Double in 10 Years to World's Highest" by Cho Jin-seo, September 9, 2010, *Korea Times*, and "S. Korea Has Top Suicide Rate Among OECD Countries: Report," Yonhap News, September 18, 2006, in *The Hankyoreh*), but this has affected older people, rising almost exponentially over age 55.

42. "Education in Contemporary Japan: Inequality and Diversity" by Kaori Okano & Motonori Tsuchiya, 1999, Cambridge: Cambridge University Press. This behavior seems rebellious against the straight jacket of conventionality. Young people, especially males, dismiss striving for a life with few surprises and no chance of reaching any distant horizon. It is boring. A novel market solution exist for young men wanting out of the *hikikomori* trap: if they find a woman willing to marry, they can avoid the public shame of their isolation by hiring impersonators playing parts of friends and family at the wedding ("Fake Guests Hired to Attend Japanese Weddings" by Danielle Demetriou, June 9, 2009, *The Telegraph*).

43. The National Defense Education Act, Advanced Research Projects Agency (which created the Internet), and the National Aeronautics and Space Act (hence, NASA) were legislated in 1958 in reaction to Sputnik.

44. Internet views of the British Petroleum "spill cam" reached more than 1 million in its first 24 hours of operation and were echoed thereafter by many other media. True telepresence may require universal high bandwidth Internet. The Next Generation Nationwide Broadband Network (NGNBN) available now in places, e.g., Singapore, should suffice.

45. The *GRAIL* spacecraft each carried several small cameras (MoonKAM), student-programmed to photograph the Moon for specific projects or pointed more at random. This marks the first NASA planetary mission carrying instruments expressly for education and public outreach.

   Chris Hadfield, commander of ISS Expedition 35, with almost 1 million Twitter followers, tweeted thousands of snapshots of Earth scenes from orbit.

46. Telerobotics with millions of viewers online looking over one's shoulder might be intimidating but tame compared to strapping oneself atop tons of high explosive to be shot into space, viewed on TV, and easier than live improvisation; it will be fresh, real, and an education to watch, with proper sidebar explanation.

47. Regarding costs, the Astrobotics Google Lunar X-Prize team offers travel to the lunar surface at $1.8 million per kg. A lunar microrover with a laptop computer's mass might cost $4 million to make and deploy, costing less if one can use a shielded central computing node (Chapter 11).

48. "Scarlet Knight Makes Landfall in Spain" by Rutgers University Office of Media Relations, December 9, 2009, *Rutgers Today*; also "Slocum Glider Energy Measurement an Simulation Infrastructure" by Hans C. Woithe, et al., 2010, in *IEEE Oceans 2010 Conference* (Sydney, Australia). Regarding outreach, see *Atlantic Crossing: A Robot's Daring Mission* by Den Seidel (producer, writer, and director: Writers House, Department of English, Rutgers University New Brunswick), 2010, which aired on PBS; also "NJ Film

Festival Dives Right in Rutgers-made 'Atlantic Crossing' Helps Kick Off Fest" by Alex Biese, January 19, 2011, *Metromix Jersey Shore*. Note that during the Scarlet mission, I never heard of her voyage or the outreach project. Hopefully, during missions by NASA, other agencies or private consortia of the public could be more engaged.

49. *BLAST!* directed by Paul Devlin, 2008, is somewhat exceptional: it stars astrophysicist Mark Devlin, who is not only BLAST principal investigator but also Paul Devlin's brother.
50. "NASA Chief Backs Openness" by Andrew C. Revkin, February 4, 2006, *New York Times*; "A Young Bush Appointee Resigns His Post at NASA" by Andrew C. Revkin, February 8, 2006, *New York Times*.
51. NASA's press office (http://www.nasa.gov/news/media/info/index.html) and a program of public outreach components associated with individual research grants each expend some $100 million/year, along with a smaller program for outreach outside NASA unassociated with grants (http://www.nasa.gov/audience/foreducators/k-4/features/F_IDEAS_Grant_Project_2006.html). NASA as a federal administration is forbidden to lobby Congress and effectively limited by Congress in its ability to solicit the public for support.
52. In FY 2005 NASA's budget was $15.6 billion versus a Department of Defense total space budget of $19.8 billion ("U.S. Military Space Programs: An Overview of Appropriations and Current Issues" by Patricia Moloney Figliola, August 7, 2006, *CRS Report for Congress*, Order Code RL33601). The National Geospatial-Intelligence Agency's budget is classified but estimated at $3 billion per year ("National Geospatial-Intelligence Agency," August 25, 2010, *Covert America, A Look at American Intelligence*, http://covertamerica.com/2010/08/25/national-geospatial-intelligence-agency/.
53. For more information on the impact of proposal oversubscription on planetary sciences, see "Planetary Community Reaction to Declining Selection Rates in Research and Analysis Programs" collected by M.V. Sykes, December 20, 2012, http://planetarypolicy.org/RA_REPORT_DEC12/community_feedback.pdf. For instance: "The impact of ~10% success rates for planetary R&A programs for me is simple. I am leaving planetary science and have accepted a job offer as a data scientist/analyst."
54. *Space Chronicles* by Neil deGrasse Tyson, 2013, New York: W.W. Norton; *A New American Space Plan* by Travis Taylor & Stephanie Osborn, 2012, Wake Forest, NC: Baen.
55. "Cigarettes Cost U.S. $7 Per Pack Sold, Study Says" *New York Times*, April 12, 2002.
56. "Industry Information: Fact Sheets: Statistics: Gaming Revenues for 2007" by American Gaming Association, http://www.americangaming.org/Industry/factsheets/statistics_detail.cfv?id=7.
57. "Communications Industry" in *Gale Encyclopedia of U.S. History*, 2008; "Blu-ray's Time Comes as DVDs Fade" by Nat Worden, December 24, 2010, *Wall Street Journal*.
58. The Juno mission to Jupiter's moons is estimated at $0.7 billion, launched on August 5, 2011, to orbit Jupiter in 2016. The Jupiter Icy Moons Orbiter (JIMO) was originally estimated at a few billion dollars, then $10 billion, then $16 billion before it was cancelled as costs of its nuclear generator ballooned. ESA plans the JUpiter ICy moon Explorer (JUICE) to launch in 2022 and orbit Europa, Ganymede, and Callisto starting in 2030, for an estimated €1.1 billion.

    Uwingu is a start-up company led by prominent space scientists and educators to create broadly popular products connecting people with the sky, with proceeds for the first ever large, private grant fund for space exploration, research, and education. Other, symbolic funding alternatives have been explored ("Scientists to Hold Bake Sale for NASA Saturday" by Clara Moskowitz, June 7, 2012, *SPACE.com*, http://www.space.com/16062-bake-sale-nasa-planetary-science.html.
59. *The Entrepreneurial State* by Mariana Mazzucato, 2011, London: Deimos.
60. *Teachers and the Uncertain American Future* by Center for Innovative Thought, College Board, July 12, 2006.
61. "Science Starts Early" by Frank C. Keil, 2011, *Science*, 331, 1022. Infants as young as five or six months seek causation in speech or events. Infants 11-month-old realize that increasing order more than disorder in a system is more likely caused by intelligent agents.
62. In August 2010 I asked 37 PhD or PhD-track scientists or engineers (admittedly not the most statistically significant sample): What portion of your original interest or motivation in the study of science would you ascribe to space exploration? Choices: (1) insignificant, (2) significant, (3) major, (4) dominant, or (5) don't know. All had an opinion, so there were no "don't knows."

    Grouping by age and nationality (16 American; 21 non-American: 3 French, 3 Hungarian, 2 Canadian, and one each from Bosnia, Czech Republic, Germany, Ireland, Italy, the Netherlands, New Zealand, Poland, Russia, South Korea, Sweden, Taiwan, and the United Kingdom), I find the following (**Mean average response** × *number of responses*):

Return to Earth

Age group	American	Non-American
21–30	**2.5** × 5	**2.7** × 7
31–50	**1.8** × 5	**1.9** × 10
51–60	**3.0** × 5	**4.0** × 2
61–70	**1.0** × 2	**1.0** × 1

Within a given age/nationality category responses are fairly consistent. I am tempted to say that those who saw Apollo tend to be strongly influenced, as were some born long thereafter. For those born at about that time or long before, the effect was weak. This may be independent of nationality, although the most "insignificant" responses come from 30- and 40-something Americans, the first cohort born after Apollo's demise.

"Sputnik babies" (now in their fifties) dominate in other space-related interests. The American Association of Variable Star Observers, a large organization of professional and largely amateur astronomers, has members mostly ages 45–65 ("The AAVSO 2011 Demographic and Background Survey" by Aaron Price & Kevin B. Paxson 2012, *Journal of AAVSO*, 40, 1). Typical subscribers to prime popular astronomy magazine *Sky and Telescope* average age 51 in 2012.

Glossary

(Terms selected from main text, not footnotes or text boxes)

7K-L1: see **L-1**.
7K-LOK: see **LOK**.

A-4: see **Aggregat-4**.
aerobraking: changing the gravitational orbit of a spacecraft around a large body by exploiting friction of the spacecraft in the large body's atmosphere.
Agena: see **Atlas**.
Aggregat-4 (A-4): the project conducted under Third Reich Germany in 1937–1945 that produced the V-2 rocket.
Almaz: a series of Soviet military Earth-orbital space stations (and several robotic satellites) flown in 1973–1992.
ALSEP: see **Apollo Lunar Surface Equipment Package**.
Altair: the proposed but now defunct Constellation spacecraft designed to carry up to four astronauts to the lunar surface and back into lunar orbit.
Andromeda Galaxy: see **M31**.
angle of repose: the angle at which an unsupported substance such as soil or regolith can stand unsupported against gravity, often depending on the height of the unsupported wall.
Apollo Applications Program: a series of proposed post-*Apollo 17* missions using Apollo and Saturn hardware and modifications, which resulted in the *Skylab* space station program.
Apollo Lunar Surface Equipment Package: about 15 various science experiments and instruments carried on at least one of the five later Apollo Moon landing missions. *Apollo 11* carried two instruments in the Early Apollo Surface Experiment Package.
Apollo program: lunar exploration project initiated by NASA in 1960 and promoted by President Kennedy to send humans to the Moon and return them to Earth. It succeeded in first placing the humans in lunar orbit in 1968 and 12 astronauts on the Moon in 1969–1972.
arcsecond (arcsec): one 3600th of a degree in angle.
Ares I: a rocket (a.k.a. Crew Launch Vehicle [CLV], or "Stick") proposed to carry crews into space as part of the Vision for Space Exploration, but cancelled in 2010.
Ares V: a large rocket (a.k.a., Cargo Launch Vehicle [CaLV]) proposed to carry cargo into space as part of the Vision for Space Exploration, but cancelled in 2010.
ARPA (Advanced Research Project Agency): see **DARPA**.
astronomical unit (AU): the average distance from Earth to the Sun, 149.6 million kilometers, about 93 million miles.

Glossary

Atlas: the first U.S. ICBM and the booster that sent the first American astronauts into Earth orbit. It has also launched many U.S. probes to the Moon and planets, topped by second-stage rockets, e.g., Agena and Centaur.

Baikonur Cosmodrome: the largest Soviet/Russian launch and space operations facility since 1956, on the Syr Darya river in Kazakhstan.

Bay of Fundy: a 290 kilometer extension of the Gulf of Maine between Nova Scotia and New Brunswick, famous for containing the world's most severe tides.

Big Whack: see **Giant Impact theory**.

breccia: a rock type formed by cementing often angular rock fragments together; on the Moon this usually occurs by melting.

British Interplanetary Society: one of the first spaceflight advocacy organizations, founded in 1933. It is well-known for Arthur C. Clarke, its former chairman, and Project Daedalus, a proposal for an interstellar spacecraft.

Bumper: a U.S. Army program from 1947–1950 launching a combination of the V-2 with the WAC-Corporal sounding rocket (the United States' first sounding rocket) as an upper stage, achieving record altitudes.

Buran: the Soviet winged space shuttle, launched into Earth orbit in 1988 without crew. Buran also refers to the accompanying shuttle program.

canali: meaning "channel," a term adopted by Giovanni Schiaparelli in 1877 to describe the apparent network of Martian surface lines (shown later to not exist)

Cassini: NASA/ESA mission launched in 1997 that flew by Venus (twice), Earth/Moon, and Jupiter to arrive in Saturn's orbit in 2004 (until the present) and carried the *Huygens* probe that landed on Saturn's large moon Titan.

Centaur: see **Atlas**.

Chadrayaan-1: an Indian Space Research Organization lunar mission that operated for ten months of its nominal 24-month life, making important discoveries including the presence of hydroxyl in the lunar soil's surface.

chondrite: a stony meteorite containing chondrules (small mineral granules).

CIA (Central Intelligence Agency): a major intelligence agency of the United States federal government.

cislunar: relating to outer space from the Earth's atmosphere extending to the Moon and lunar orbit, including the Earth-Moon Lagrangian points.

CNSA: Chinese National Space Agency, the prime space agency for exploration and military missions in China.

Columbia: 1) the first Space Shuttle launched into orbit, in April 1981, and destroyed on re-entry from its 28th mission in February 2003, 2) the Command Module of *Apollo 11*.

Command/Service Module (CSM): the Apollo spacecraft designed to carry three astronauts into space on top of a Saturn rocket and return them to Earth in the Command Module (CM).

Constellation program: several spacecraft and rockets proposed by NASA to fulfill President G.W. Bush's 2004 Vision for Space Exploration (including Ares and Earth Departure Stage rockets, and Altair and Orion spacecraft), but partially abandoned by 2010.

Crew Exploration Vehicle (CEV): see **Orion** and **Multi-Purpose Exploration Vehicle**.

Glossary

DARPA (Defense Advanced Research Project Agency): an agency of the U.S. Department of Defense charged with developing technologies for military use. DARPA was named ARPA in 1958–1972 and 1993–1996.

Deepwater Horizon: an offshore ultra-deep drilling rig that broke records for the deepest oil well before being destroyed by an explosion in April 2010 at the Macondo Prospect, off Louisiana in the Gulf of Mexico, killing 11 and creating the largest offshore oil spill in U.S. history.

Delta: a diverse rocket family in use from 1960 to present, based on the Thor ICBM used in 1958–1963. Thor and Delta rockets have launched almost 400 times, with missions to Earth orbit, the Moon, Mars, Mercury, cometary, and interplanetary destinations.

Design Bureau (Opytnoye Konstruktorskoye Buro or OKB): any of many Soviet groups (many merged or eliminated after 1991) designing high-technology systems, mainly in aerospace for military and space applications. OKB-1 was led by Sergei Korolev (1946–1966), Vasil Mishin (1966–1974), and Valentin Glushko (1974–1989), and OKB-52 by Vladimir Chelomei from 1955 to 1966 (and until 1984 under another bureau name).

DNA polymerization: synthesis of deoxyribonucleic acid from nucleotides, now via the enzyme DNA polymerase but presumably originally via other processes.

Dragon **spacecraft**: robotic spacecraft built by SpaceX, used frequently to re-supply *ISS*.

Earth Departure Stage (EDS): an upper-stage rocket on current Space Launch System designs or the defunct Ares V, intended to take spacecraft into Earth orbit and then into Earth escape trajectories.

EMU (Extravehicular Mobility Unit): NASA space suit of the Space Shuttle/*ISS* era.

Enceladus: Saturn's sixth largest moon, seen erupting with large amounts of water vapor.

Energia: (1) a powerful rocket built by the Soviet Union and launched twice in 1987–1988, and (2) a Russian aerospace corporation deriving from OKB-1, also known as OAO S.P. Korolyov Rocket and Space Corporation Energia, or RKK Energiya.

Europa: Jupiter's fourth largest moon, seen erupting water vapor and suspected to have a large subsurface water ocean.

European Space Agency (ESA): the European intergovernmental organization for exploring outer space, from 1975 to present, with its launch site in Kourou, French Guiana.

EVA (Extra-Vehicular Activity): moon walk or space "walk" outside the spacecraft.

Exploration Systems Architecture Study: a NASA-sponsored report released in November 2005 on how to implement President G.W. Bush's Vision for Space Exploration of January 2004.

F-1: the largest and highest-thrust liquid-propellant rocket engine every built. Five F-1s were used on each Saturn V rocket.

Fairchild Semiconductor International, Inc.: an independent semiconductor electronics company from 1957 to 1979 (returned to independent status in 1997) that, along with Texas Instruments Inc., invented the integrated circuit and was its initial major manufacturer.

Falcon 9: a family of moderately heavy, largely successful boosters built by SpaceX.

Far Side: the hemisphere of the Moon turned predominantly away from Earth, with longitudes of 90 to 270 degrees. It is covered primarily by highlands terrane.

feldspar: a mineral composed of aluminosilicates of potassium, sodium, and calcium, which form pale crystals.

ferromagnesian: otherwise known as "mafic" – any silicate mineral or rock rich in magnesium and/or iron.

Flexible Path: a space exploration concept proposed in 2009 and adopted for NASA by the Obama Administration in 2010 targeting various objects and locations in space rather than specializing in the Moon and Mars.

Fobos-Grunt: the failed 2011 Russian mission to Martian moon Phobos, co-manifested with Chinese Mars orbiter *Yinguo-1*, which also failed.

Freedom Space Station: a project announced by President Ronald Reagan in 1984 to build a large space station, later abandoned by NASA by 1993, but largely incorporated into the *International Space Station*.

fugacity: the real pressure of a gas supported by chemical equilibrium rather than that of the ideal gas, useful in calculating how much of the gaseous species is easily available.

gardening: the mixing and transport of regolith via impacts.

Gateway (or Exploration Gateway Platform): a concept proposed in 2011 for a station at the Earth-Moon Lagrangian Point L1 or L2 to serve as a support base for other destinations and as a servicing platform for other deep space craft.

Gemini: NASA project from 1962 to 1966 consisting of 12 missions, 10 of which carried two astronauts apiece into Earth orbit, demonstrating many techniques needed for Apollo.

Gemini 2: the second Gemini spacecraft (launched without crew on January 19, 1965) was re-launched without crew on a second mission on November 3, 1966, as part of a mockup *Manned Orbiting Laboratory* (*MOL*), a U.S.A.F. project to develop a spy space station with crew, using a modified Gemini craft, Gemini B. MOL, superseded by robotic surveillance satellites, was cancelled in 1969 without further flights.

Gemini B: see *Gemini 2*.

Genesis: a NASA robotic space mission from 2001 to 2004 that returned samples of the solar wind to Earth (despite crash landing in a Utah desert).

GEO: geosynchronous orbit, nominally 35,800 kilometers in altitude, with a period of one sidereal day.

Geosynchronous Satellite Launch Vehicle (*GSLV*): moderately heavy booster operated with mixed success by the Indian Space Research Organization.

Giant Impact theory: a theory of lunar origin in which a large object struck the proto-Earth and the Moon formed from the aggregation of resulting debris.

Google Lunar X Prize (GLXP): a prize of up to $30 million offered to a commercial or private group accomplishing certain goals with a lunar rover mission.

GRAIL (*Gravity Recovery and Interior Laboratory*): two NASA spacecraft that orbited the Moon in 2012 and mapped its gravitational field.

haptic device: an input device for humans devised to provide user feedback information for the effective operation of a remotely operated robotic device.

Hayabusa: a JAXA robotic space mission from 2003 to 2010 that returned the first samples retrieved from the surface of an asteroid (25143 Itokawa).

Glossary

heiligenschein: the enhanced reflectivity of many surfaces, particularly much of the lunar regolith, in geometries where sunlight is returned nearly 180° back toward the Sun.

Hiten: a 1990–1993 Japanese cis-lunar mission (along with its parasitic probe *Hagoramo*) that was the first spacecraft to employ ballistic lunar capture (via weak instability bound orbital transfer) and the first beyond low Earth orbit to use aerobraking to alter its orbit.

hominid: member of the taxonomic family that includes humans, great apes (gorillas, chimpanzees, bonobos, and orangutans), and also recent human ancestors. As used in chapter 1, this can also mean all species more closely related to humans than chimpanzees and bonobos.

HTP (High-Test Peroxide): high purity hydrogen peroxide (H_2O_2), which must be handled carefully to prevent violent decomposition.

Hubble Space Telescope: a 2.4-meter diameter telescope operating in Earth orbit since 1990 to observe the Universe in optical, infrared, and ultraviolet light.

ICBM: intercontinental ballistic missile.

igneous: referring to rock that has solidified from magma or lava.

IGY: see **International Geophysical Year**.

ilmenite: a black mineral, made of crystalline iron titanium oxide.

incompatible elements: elements that, due to ionic size or charge, fit poorly into the crystal lattice of the predominant mineral, and on the Earth and Moon include the rare earth elements, potassium, phosphorus, uranium, thorium, and others. As a large body of magma cools and crystallizes, these will usually be the last to separate out as solids.

International Geophysical Year (IGY): an internationally collaborative and competitive scientific project from July 1, 1957 to December 31, 1958, including exploration of Antarctica, and the Earth orbital environment with the first artificial Earth satellites.

International Space Station (*ISS*): an internationally operated station in Earth orbit, with mass of about 450 tonnes, length of 108 meters, and often carrying a crew of 6.

International Traffic in Arms Regulations (ITAR): U.S. government regulations over the export and import of potential defense-related technologies, including many involved in space exploration.

intertidal zone: the region between the extremes of high tide versus low tide along the shore of a body of fluid, such as Earth's oceans.

ISS: see *International Space Station*.

Isvestia: a prominent Soviet/Russian newspaper from 1917 to the present that was considered officially authoritative in the Soviet period.

ITAR: see **International Traffic in Arms Regulations**.

J-2: the rocket engine used on the second and third stages of the Saturn V rocket, and now being upgraded as the J-2X.

JAXA: Japan Aerospace Exploration Agency, the merged space program agency for Japan, as of January 1, 2003.

Jet Propulsion Laboratory (JPL): the research and development center founded in La Cañada Flintridge near Pasadena by the California Institute of Technology in 1936 and operated in collaboration with the U.S. Army until 1958 and with NASA since then. JPL is responsible for the creation and operation of many of NASA's robotic missions exploring the Moon and farther Solar System.

Juno: a rocket family based on the Redstone and Jupiter ICBM and used in 1958–1961 to launch many of the first U.S. satellites into Earth orbit.

Kaguya: see **SELENE**.

KGB (Komitet gosudarstvennoy bezopasnosti – Committee for State Security): the primary security agency, both internal and external, for the Soviet Union after 1954.

KREEP: an acronym (K = potassium, REE = rare earth elements, P = phosphorus) for material such as rock originating from depths of the early Moon (and other largely melted bodies, such as on Earth) where incompatible elements concentrate (including uranium, thorium, and potassium, producing radioactive heating, and also associated with water).

L-1 (or 7K-L1): a downsized version of Soyuz, designed by OKB-1 for lunar orbital flights of two cosmonauts. It flew 12 times in 1967–1970, four of these around the Moon, but never with a cosmonaut.

L1: see **Lagrangian points**.

L3: 1) the Soviet program to develop lunar landing craft, besides the N1 rocket, all designed by OKB-1. Eventually it evolved into the LK, LOK (and N1's fourth stage). 2) a metastable point between a large and small mass in circular orbit (see **Lagrangian points**).

LADEE (*Lunar Atmosphere and Dust Environment Explorer*): a NASA mission orbiting the Moon on 2013–2014 to study its atmospheric composition and dust environment.

Lagrangian points: five stable or metastable points in a system of a smaller mass orbiting a larger mass, for which a third, smallest mass can orbit in a fixed position relative to the other two masses. L1 rests between the two larger masses, L2 in line with L1 but on the other side of the smaller mass, L3 in line with L1 and L2 but on the other side of the larger mass, and L4 and L5 at the smaller mass's orbital radius but leading or following it by 60°.

Late Heavy Bombardment (LHB): an intense bombardment of the Solar system's planets by smaller bodies about 3.85 billion years ago.

latency: the delay time (here in a communications loop) due to the finite speed of light and other delays.

Launch Abort System: the rocket system atop the Orion or Multi-Purpose Crew Vehicle designed to carry the crew capsule away from a catastrophic failure of the launch vehicle, similar in function to Apollo's Launch Escape System.

lava tube: a volume evacuated from a lava field due to subsurface flow of a lava current leaving behind an empty, often elongated, chamber.

LCROSS (*Lunar CRater Observation and Sensing Satellite*): two spacecraft launched with *LRO* in 2009 that probed composition of material near the lunar South Pole by crashing into its surface.

LEO: low Earth orbit, between about 160 kilometers and 2000 kilometers altitude.

LH2: liquid hydrogen.

LHB: see **Late Heavy Bombardment**.

Life **magazine**: published 1936–2000, the highest circulation large-format, general interest magazine during much of the Space Age, featuring many articles on space topics.

Glossary

limestone: a sedimentary rock, composed primarily of calcium carbonate, $CaCO_3$.

LK: see **Lunniy Korabl**.

LK-1: a Soviet spacecraft design from OKB-52 intended for a return flyby lunar mission, cancelled in 1965 without flying.

LM: see **Lunar Module**.

LOK (or 7K-LOK, Lunniy Orbitalny Korabl – lunar orbital craft): a version of a Soviet Soyuz spacecraft modified by OKB-1 to carry two cosmonauts into lunar orbit after launch by an N1 rocket, and to support a lunar landing by the LK. Two test 7K-LOK flights occurred in 1971–1972, but never with cosmonauts or to the Moon.

Look **magazine:** published 1937–1971, during the early Space Age *Look* was second only to *Life* in terms of circulation by a large-format, general interest magazine.

LOR: see **lunar orbit rendezvous**.

LOX: liquid oxygen.

LRO: see *Lunar Reconnaissance Orbiter*.

LROC: see **Lunar Reconnaissance Orbiter Camera**.

LRV: see **Lunar Roving Vehicle**.

LSB: Lunar Sourcebook, a 1991 monograph on lunar science results primarily from the Apollo program, frequently cited herein. (See footnote 2.13)

Luna: a diverse series of Soviet robotic missions from 1959 to 1976, including the first to fly by the Moon (*Luna 1*), impact it (*Luna 2*), observe its Far Side (*Luna 3*), soft land on the Moon (*Luna 9*), orbit it (*Luna 10*), robotically return a sample to Earth (*Luna 16*), and deploy a robotic rover (*Luna 17*). As with many Soviet spacecraft programs, many Luna missions were attempted beyond those assigned a name via simple ordinal count.

Lunar Module: for the Apollo program, a two-stage craft, successfully built to carry two astronauts to the lunar surface, and then return them to lunar orbit in its upper stage.

lunar orbit rendezvous: a mission strategy for reaching the Moon in which one spacecraft stays in lunar orbit while a second lands on the Moon. The lander (or at least a portion of it) will then return to lunar orbit to rendezvous and dock with the first craft, transfer its contents, and depend on the first craft to return its crew and/or payload to Earth.

Lunar Orbiter: a series of five NASA missions in 1966–1967 that photographically surveyed the entire lunar surface and candidate Apollo landing sites, and also discovered the lumpy distribution of gravitating masses near the lunar surface.

Lunar Prospector: a 19-month NASA mission in 1998–1999 that mapped the Moon's entire surface in various magnetic and especially radioactive properties, revealing its composition.

Lunar Reconnaissance Orbiter (*LRO*): a NASA satellite in lunar polar orbit since 2009 and carrying an array seven scientific instruments.

Lunar Reconnaissance Orbiter Camera: the optical-wavelength camera on *LRO*, which can operate in a wide-field or high-resolution mode.

Lunar Resource Utilization project: see **LUNOX**.

Lunar Roving Vehicle (LRV): the two-person, four-wheeled lunar rovers used on the Apollo J missions (*Apollo 15, 16* and *17*) which roved 90 kilometers total on the Moon.

Lunar Surface Access Module (LSAM): see **Altair**.

Lunniy Korabl (Лунный корабль or "lunar ship"): a Soviet lunar landing craft from OKB-1 to carry one cosmonaut (or possibly two with re-design) to the lunar surface and back to lunar orbit. Four Lunniy Korabl orbited Earth in 1970–1971, but they never went to the Moon.

Glossary

Lunokhod: *Lunokhod 1* in 1970 and *Lunokhod 2* in 1972–1973 were the first two successful robotic lunar rovers (delivered by *Luna 17* and *21*, respectively), which operated for a total of 14 months.

LUNOX: project proposed in 1993 by NASA in response to the Space Exploration Initiative to explore the Moon at reduced costs by employing international partners and utilizing lunar resources.

M31 (or Messier 31): the Andromeda Galaxy, the nearest large galaxy to our Milky Way galaxy, about 770,000 parsecs from Earth.

magma: subsurface molten rock in a fluid or semifluid state, from which lava or igneous rock can form.

magma ocean: an expansive layer of magma, e.g., covering the Earth or Moon.

Man in the Moon: an imagined human visage on the lunar Near Side (usually with eyes formed from maria Serenitatis and Imbrium), a form of pareidolia.

Manned Orbiting Laboratory (*MOL*): see ***Gemini 2***.

mare (plural: maria): a basalt-covered plain on (or near) the lunar surface. Most maria are on the lunar Near Side hemisphere.

mascon (abbrev. for mass concentration): a high density region of rock, often associated with lunar maria.

megaregolith: the impact-fractured zone below the regolith and covering the lunar bedrock.

MEO: mid Earth orbit (below Geosynchronous and above low Earth orbit: about 2000 kilometers to 35,800 kilometers altitude).

Mercury: NASA's program from 1959 to 1963 to send a single astronaut per mission into space, two on suborbital flights in 1961 and four into Earth orbit in 1962–1963.

meteoroid: a small Solar System object capable of becoming a meteor hitting Earth's atmosphere (or a meteorite, if it reaches the ground).

micrometeorite: a space object less than 2 millimeters across that strikes the hard surface of another object.

***Mir*:** the Soviet/Russian Earth-orbital space station complex flown in 1986–2001, often with a crew of three.

Mission Control: the NASA organizational structure, based primarily at Johnson Space Center in Houston (formerly the Manned Space Center) that handles operations for human space missions. Other space agencies have other mission control facilities, while NASA has other mission control facilities for robotic missions and as backup.

Multi-Purpose Crew Vehicle (MPCV): a modification of Constellation's Orion spacecraft design intended for use with the *International Space Station*, and perhaps orbit around the Moon, Mars, and asteroids using the Space Launch System.

N1: a four-stage Soviet rocket, 105 meters tall, designed by OKB-1 to carry two cosmonauts to the Moon in the LK and 7K-LOK spacecraft. At launch it was the most powerful rocket ever built, but all four launches, in 1969–1972, failed.

NASA (National Aeronautics and Space Administration**)**: since 1958 the primary federal U.S. government organization for civilian space exploration and aeronautic technology development.

National Reconnaissance Office: a United States intelligence agency charged with building and operating spy satellites.

Glossary

National Space Council (NSC): a White House advisory council (from 1989 to 1993) to inform and advise the President of the United States on outer space issues, successor to the National Aeronautics and Space Council (NASC) of 1958 to 1973.

Near Side: the hemisphere of the Moon turned predominantly toward Earth, with longitudes of 270 to 90 degrees. It contains most of the lunar maria.

obliquity: the tilt of a body's rotation (such as the Earth's) relative the plane of its orbit around a second body (such as the Sun in the case of the Earth).

Occam's Razor: credited to William of Occam (c.1285–1349), this principle expresses the concept that the scientific explanation for an observation with the fewest assumptions is the most likely one, in demonstration of the principle of parsimony.

Office for Science and Technology Policy (**OSTP**): the office created in 1961 to advise the U.S. President on science's and technology's influence on domestic and international issues, and in particular the effects of the space program.

OKB: see **Design Bureau**.

olivine: a simple olive, gray-green, or brown mineral found in igneous rocks, $(Mg,Fe)_2SiO_2$.

OMAT: see **Optical Maturity**.

Optical Maturity (**OMAT**): a measure of aging via space weathering of regolith that measures the distance of a particular sample on an optical reflectivity-versus-color plot from its zero-age point. (See footnote 6.27.)

Orion spacecraft: the proposed Constellation spacecraft designed to carry up to six astronauts into space and later return them to Earth, now being developed as the Multi-Purpose Crew Vehicle.

Paris Match: a prominent weekly magazine published in France since 1949.

parsec (abbrev. for parallax second): a distance of 3.26 light-years = 30.9 trillion kilometers, the distance at which two objects 1 astronomical unit apart (149.6 million kilometers) perpendicular to the direction to the observer appear separated by 1 arcsecond in angle (1/3600 of a degree). The nearest star, Proxima Centauri, is 1.3 parsecs from Earth.

payload: the useful cargo, passengers, crew, instruments, and equipment carried by a rocket for purposes beyond its propulsion.

permanently shadowed region: a region on the Moon (or another body, such as Mercury) that is never illuminated by direct sunlight (and hence is often extremely cold).

Pioneer: a diverse program of NASA space probes from 1958 to 2003 exploring the planets (Venus, Jupiter, and Saturn) and the outer and inner Solar System, including the Moon. *Pioneer 4* flew by the Moon in March 1959.

PKT: see **Procellarum KREEP Terrane**.

planetary protection: the concepts and precautions in interplanetary exploration to prevent the Earth and other bodies from being contaminated by biological agents from foreign worlds.

planetesimal: in planet formation, a tiny planet-like body, which can accrete gravitationally to form a larger planetary body.

plate tectonics: the theory describing the motions in the outermost, rocky layers of Earth (and Earthlike planets) including the action of continental drift and seafloor spreading.

Glossary

Polyus (*Skif*):   a massive Soviet Earth-orbital weapons platform launched unsuccessfully in 1987 to counter the United States' Strategic Defense Initiative.

Procellarum KREEP Terrane (PKT):   Oceanus Procellarum, Mare Imbrium, and adjacent regions high in KREEP composition.

Proton: a large-payload Russian/Soviet rocket family first launched in 1966 and still in service. Protons sent various Luna and Zond missions to the Moon, and other spacecraft to Mars and Venus.

PSR: see **permanently shadowed region**.

pyroxene: a common, simple mineral made of interlocking silicate tetrahedra interspersed with other ions, in the formula $XY(Si,Al)_2O_6$ where X is calcium, sodium, magnesium, iron (Fe^{+2}), zinc, manganese, or lithium, and Y is chromium, aluminum, Fe^{+3}, magnesium, manganese, scandium, titanium, or vanadium.

R-7:   the first ICBM, developed in 1957 and modified to provide booster for the Sputnik, Vostok, Soyuz, and early Luna programs.

RAND Corporation (Research ANd Development):   a non-profit think tank from 1948 to the present providing analysis primarily to U.S. governmental defense and national security organizations.

Ranger: a NASA program from 1961 to 1965 that sent seven spacecraft to impact the Moon, three successfully, plus two craft into Earth orbit (unsuccessfully).

rare earth elements (REE):   also called the lanthanides, the elements with numbers 57 through 71 (lanthanum through lutetium) are chemically similar, with some exception for europium (number 63). Sometimes elements scandium (21) and yttrium (39) are included.

rayed crater: impacts forming craters eject debris outward in jets or "rays" which form aligned chains of secondary craters, which remain visible until obscured by later impacts and weathering.

Redstone: a rocket program during 1951–1964 producing an intermediate-range ballistic missile based on the V-2, one of the first to carry nuclear warheads, and which launched the first two American astronauts into space. Several other rocket designs used for space exploration and by the U.S. military were based on the Redstone.

REE: see **rare earth elements**.

regolith: impact-pulverized particulate soil, i.e., dust and rocks, sometimes mixed with particulates of volcanic origin. Regolith covers nearly the entire lunar surface.

Return to the Moon: part of President G.W. Bush's Vision for Space Exploration of 2004 centering on the return of astronauts to the lunar surface.

rhythmite: a sedimentary rock built up in distinct layers detectably deposited periodically during a series of individual days, months or years.

RKA: see **Russian Federal Space Agency**.

Roche limit: the smallest distance from a planet or other large object at which a smaller object such as a moon can survive fragmentation by the larger object's tidal forces, assuming the smaller object is held together by its own self-gravitation.

Roscosmos: see **Russian Federal Space Agency**.

Russian Federal Space Agency (Roscosmos or RKA):   the primary Russian civilian agency for space exploration, since 1992.

Salyut: a series of Soviet space stations orbiting Earth in 1971–1986. Salyut stations after *Salyut 1* were included in the Almaz program.

Glossary

Saturn: the rocket series including the Saturn I, IB, and V, used by NASA in 1961 to 1976 for Apollo. The IB launched crews of three astronauts into Earth orbit on *Apollo 7*, three Skylab missions and *Apollo/Soyuz Test Mission*, as did the V on *Apollo 9*. The V launched nine crews to the lunar surface or orbit, and lofted the *Skylab* space station. All three rocket designs completed other Earth-orbital missions without crews.

Search for Extra-Terrestrial Artifacts (SETA): methods and means whereby humans may find substantial evidence of visits to the Solar System or places nearby by extraterrestrial intelligences.

Search for Extra-Terrestrial Intelligence (SETI): methods and means whereby humans may find radio or optical signals as evidence of extraterrestrial intelligences.

SELENE (*SELenological and ENgineering Explorer*): a JAXA lunar satellite in polar orbit from 2007 to 2009, accompanied by two sub-satellites and carrying an array of approximately 14 scientific instruments.

Service Module (SM): the portion of Apollo (and some other spacecraft) containing supplies and equipment to support the crew in the Command Module (in the case of Apollo, see **Command/Service Module**).

SETA: see **Search for Extra-Terrestrial Artifacts**.

SETI: see **Search for Extra-Terrestrial Intelligence**.

silicates: a family of minerals, combining silicon and oxygen, usually with a metal, often containing anionic SiO_4^- in a tetrahedral geometry (described as orthosilicates).

Sintering: achieving the desired composition of a solid by allowing substances in a powder mix to diffuse together via heating.

sinus: a "bay" of basalt-covered plain, similar and adjacent to maria but usually smaller.

Skylab: the first space station flown by the United States, launched by NASA on a Saturn V on May 14, 1973, and visited by nine astronauts for 172 days total in three crews (*Skylab 2–4*) launched by Saturn 1Bs in 1973, as a post-lunar continuation of Apollo. *Skylab/Skylab 2* was the first successful space station habitation, given the death of *Soyuz 11/Salyut 1*'s crew in 1971.

SMART-1 (*Small Missions for Advanced Research in Technology*): an ESA lunar mission of 2003–2006 employing ion propulsion to transfer from Earth to lunar orbit, where it studied the Moon for 19 months.

Solid Rocket Booster (SRB): side rockets charged with solid propellant and attached to the external tank of the STS Space Shuttle (and more generally other such side boosters on other rockets).

South Pole-Aitken basin (SPA): an oval, 2500-kilometer wide impact feature on the lunar southern Far Side, perhaps the largest impact feature in the Solar System.

Soyuz: the primary spacecraft used by the Soviet Union/Russia for sending humans into Earth orbit, from 1966 to the present, completing roughly 150 missions total as of 2014.

Space Age: the period of space exploration by humankind, arguably 1957 to the present.

Space Exploration Initiative: a plan for the U.S. space program proposed by President George H.W. Bush in July 1989 but abandoned in 1993, concentrating on an Earth orbital space station, and humans on the Moon and Mars.

Space Launch System (SLS): a large rocket proposed by NASA for future space exploration.

Glossary

Space Shuttle: an Earth-orbital NASA program in 1972–2011 that launched several shuttles (*Columbia*, *Challenger*, *Discovery*, *Atlantis*, *Endeavour*, and the test vehicle *Enterprise*) and their crews of two to seven astronauts (in one case, 8) on a total of 135 missions since 1981.

Space Transportation System (STS):   see **Space Shuttle**.

SpaceX (Space Exploration Technologies Corp.):   manufacturer of Falcon rockets and Dragon spacecraft, based in California.

Spitzer: the *Spitzer Space Telescope* was launched by NASA in 2003 to explore astronomical sources of infrared light of 3 microns to 100 microns wavelength.

Sputnik: a rocket program based on the R-7, and a series of three Soviet robotic Earth satellites in 1957 and 1958, including the first two launched by humanity. The first of these is called *Sputnik 1* or simply *Sputnik*.

Star Trek: a science fiction TV series created in 1966–1969 and the ensuing film and TV franchise from 1973 to the present. The series is marked by many fictional alien species at roughly comparable states of development as human civilization on Earth.

Star Wars: (1) see **Strategic Defense Initiative**, (2) a series of feature science fiction films starting with the first (or fourth) installment in 1977 and continuing to the present.

Stardust: a NASA robotic space mission from 1999 to 2006 that returned to Earth the first samples of material retrieved directly from a comet, from the coma of comet Wild 2. The mission then morphed into *Stardust-NExT*, which flew by comet Tempel 1 and asteroid 5535 Annefrank, ending in 2011.

Strategic Defense Initiative: a U.S. military program started in 1983 to develop possible space and ground-based defenses against a ballistic missile attack on the United States.

subduction: the process in plate tectonics explaining how new ocean floor crust is created by allowing older ocean floor to disappear and be recycled below other, usually continental, crust.

Surveyor: a NASA program that from 1966 to 1968 landed seven robotic spacecraft on the Moon, five successfully.

telepresence: technologies that allow a person to experience or influence the circumstance of a distant location via the communication and translation of signals into physical effects.

telerobotics: the telepresent operation of a robot in one location by a person in another, distant location.

terraforming: the artificial and deliberate transformation of an extraterrestrial astronomical body into an environment more hospitable to human habitation there.

terrane: a region with a distinctive stratigraphy, structure, and geological history. On Earth terranes are usually bounded by faults, but this is not necessarily true on the Moon. Primary lunar terranes are the maria, highlands, the KREEP terrane, and the South Pole-Aitken basin.

Theia: the Mars-sized planetesimal theorized to have struck proto-Earth in the Moon-forming Giant Impact.

Thor: see **Delta**.

throw weight: the effective payload mass of a ballistic missile.

tide: the variation over an object of gravitational force from a second object, tending to pull the first object apart in the direction of the second object and compress it in perpendicular directions. This is seen most easily on Earth as the displacement of ocean water from the solid Earth.

Titan: a family of rockets during 1958–2005 that replaced the Atlas as the primary U.S. ICBM, launched all Earth-orbital Gemini missions, and placed many satellites in Earth orbit as well as sent probes to the Moon and all of the major outer planets.

TLP: see **Transient Lunar Phenomenon**.

Tranquility Base: the name given by the *Apollo 11* astronauts to the arrival site of their lunar module *Eagle* on July 20, 1969, as the first human landing on the Moon

Transient Lunar Phenomenon (TLP): any short-lived phenomenon seemingly associated with the lunar surface, but more specifically optical phenomena localized to small surface areas.

UFO: unidentified flying object.

urKREEP: primordial, unaltered magma-ocean KREEP material.

U.S.A.F.: United States Air Force.

U.S. Geological Survey: a bureau of the United States Department of Interior, charged with study of the geology, geography, hydrology, and biology of the U.S., and other regions of concern on the Earth and planets (via the Astrogeology Research Program).

V-2 (A-4, Aggregat-4, or Vergeltungswaffe-2): the first intermediate-range ballistic missile, first built by Germany in 1941–1945 and later by the United States and Soviet Union into the 1950s. A V-2 rocket became the first human-made object in space on October 3, 1942.

Van Allen radiation belts: two rings of intense radiation in space up to several thousand kilometers above Earth.

Vanguard: a rocket and satellite program intended to place a United States satellite into Earth orbit, which it did in 1958 (after *Sputnik 1* and *2*, and *Explorer 1*).

varved deposit: a sedimentary rock in which distinct layers are detectably deposited over periods of time, often annually.

VfR (Verein für Raumschiffahrt, League for Spaceship Travel): an amateur rocketry association from 1927–1933 based near Berlin that developed many early rockets and provided from its membership some of the leaders of the Aggregat program.

Viking: a NASA program that resulted in 1976 in the first two successful robotic landings on Mars (discounting the Soviet *Mars 3* which failed after 15 seconds on Mars in 1971).

Vision for Space Exploration: a plan announced by President George W. Bush in January 2004 for the U.S. space program, especially in Earth orbit, on the Moon and Mars. It was replaced by President Obama in June 2010.

volatiles: substances of low boiling point, usually applying to the composition of an astronomical object.

Voskhod: a Soviet program using the Voskhod spacecraft (a modified Vostok) on two Earth-orbital missions in 1964–1965. *Voskhod 1* carried the first three-person crew into space, and *Voskhod 2* two cosmonauts, resulting in the first walk in space.

Vostok: a Soviet program in 1959–1963 that carried a cosmonaut into Earth orbit on each of six missions, including Yuri Gagarin on April 12, 1961 as the first human in space.

WAC Corporal: see **Bumper**.

watt: a unit of power or rate of energy change, equal to one joule per second, about 1% the power output of a typical candle.

Glossary

weak stability bound transfer: a class of trajectories taking a spacecraft from one body to another gravitationally that allows a low-energy ballistic capture into orbit around the second body.

Wilkinson Microwave Anisotropy Probe (*WMAP*): a NASA mission from 2001 to 2010 that mapped the cosmic microwave background radiation to superlative detail in five microwave radio bands.

X37-B (or X-37): a robotic winged space shuttle operated by the U.S. Air Force, launched on several missions since 2010, some more than one year in duration.

X-ray spectrometer: an instrument sensitive to X-rays that can measure the energy of individual X-ray photons, or the wavelength of their waves. Most X-ray lines are from transitions of the most tightly bound electrons, seen in most atoms (except hydrogen, helium, and lithium). Above 20,000 electron volts in X-ray energy some transitions can involve the nucleus of the atom.

Ye-6: Soviet spacecraft and lunar landers built in 1963–1966, including the first craft to soft land on the Moon, by *Luna 9* in 1966, the 12th mission employing the Ye-6.

Zond: a diverse series of Soviet robotic missions from 1964 to 1970, for human and robotic exploration of the Moon (although no cosmonauts ever flew), and robotic exploration of Venus and Mars. Many Zond flights were test missions.

Appendix A
Von Braun et al. Space and Lunar Exploration Issues

Engineering, physical, human, and programmatic issues raised (and often answered) in von Braun et al. *Collier's* series and *Conquest of the Moon* (with key terms and concepts in bold face type):
1. Single-stage, solid-propellant rockets are impractically large. **Multistage boosters** must be strong and lightweight, burn **liquid propellants**, and survive high stresses and temperatures of launch and reentry through the atmosphere, reaching thousands of degrees.
2. They chose **hydrazine** and **nitric acid** as the **liquid oxidizer/fuel**. A more powerful mixture is liquid oxygen and hydrogen, both cryogenically cooled; the proposed duration of six weeks of the von Braun et al. lunar mission might make this impractical.
3. The stages of these **boosters might be reused** after landing in the sea or gliding back to Earth after reentry. Used engines and empty **propellant tanks can be discarded** traveling to/from the Moon to reduce weight or allowed to incinerate in the atmosphere.
4. **Craft ferrying humans and supplies in a vacuum** and toward lunar and Martian orbit need not suffer such temperatures and stresses. **Moon ship lands vertically on extended legs**.
5. **Spacecraft maneuvers controlled by machines**, not humans, are sufficiently precise. In 1955 this was envisioned as libraries of preset magnetic tape programs; now we use computers.
6. Humans must lie prone in **acceleration couches** during launch and reentry to tolerate forces of up to about six times Earth's surface gravity.
7. Humans must **communicate solely via radio** (or sight), because sound will not carry in vacuum. Earth-Moon communications suffer 2.6 seconds delay, round trip. **Spacecraft should be tracked via radio**, and their velocity determined via Doppler frequency shifts.
8. Humans must be protected at all times from vacuum by **space suits and pressurized spacecraft**. **Air locks** are required for leaving the spacecraft for the vacuum.
9. Carbon dioxide and excess water must be removed from the **air, and oxygen** replenished; they suggest avoiding nitrogen in air to discourage the "bends" or nitrogen narcosis. Water and perhaps oxygen should be recycled.
10. Each crew member will use 5 kg per day in **air, water, and food**. Food is precooked, pre-portioned, pre-sliced, heated by shortwave radio waves (read: microwaves), and served in containers not requiring cleanup or allowing leakage in weightlessness.
11. Trace gas **irritants and poisons** must be avoided or filtered out and separated from crew living areas and other commonly occupied volumes.

12. Windows and space helmets need to shield out solar **ultraviolet light**.
13. In the 1950s the danger level from **cosmic rays** was unknown. People realized that they were ignorant of magnetic fields and radiation belts that might modify exposure to radiation, and the effects of such radiation were not completely known. (See Chapter 10 regarding radiation in outer space.)
14. Spacecraft on extended missions need **shielding against meteorite impacts**, because avoidance via radar or other means is impractical. They overestimate the meteorite impact hazard; in low Earth's orbit in the twenty-first century the impact hazard is more because of our own space debris.
15. On extended lunar stays, humans **live underground** to shield against meteorites and radiation and to minimize temperature swings via natural insulation. Heaters raise natural equilibrium temperature slightly for human comfort. Lunar habitats are made from spent fuel tanks or cargo vessels.
16. Space researchers would need to understand **psychological issues** such as boredom, claustrophobia, or feelings of isolation.
17. Von Braun et al. do not consider **atrophy of muscles and bone** because of long-term effects of **zero or reduced gravity**. We discuss this in Chapter 10.
18. **Weightlessness** produces **no short-term damage** in humans, except nausea-like motion sickness in some people. Are there problems sleeping in zero gravity? Experience has shown this to be insignificant.
19. Weightlessness complicates **coordination of motions** learned on Earth. Objects have no weight, but their mass still resists acceleration/deceleration. Liquids form droplets that drift through the spacecraft. Smoke and gas do not rise via buoyancy, accumulating near their source, if absent air currents.
20. **Handholds and restraints** are needed on spacecraft exteriors to keep astronauts from floating away. Inside one needs magnets and steel surfaces. (Velcro, a common solution in space flight today, was invented in 1941 and debuted commercially in the 1960s.)
21. To avoid weightlessness' effects, a **rotating space station** could be constructed to push its occupants against the outer surface, using centripetal acceleration to simulate gravity.
22. **Large structures** cannot be lifted into orbit intact but can be **assembled in space** from smaller modules. Construction procedures must be kept simple and preplanned.
23. For **large-volume structures**, soft-sided, **inflatable** rooms can be launched, and then inflated with air to full size. To protect against meteorites, these structures should be durably shielded. (Such structures have never been built in space, but this may change soon; see Chapter 11.)
24. Spacecraft must be constructed with systems that avoid breakdown via **redundant components**, or multiple spacecraft must support each other. Repairs should be possible in flight.
25. **Excess heat** must be radiated, not convected, into space, even on the Moon. Air inside spacecraft and space suits must be temperature controlled. Temperature swings of more than 250°C are expected on the Moon, with similar temperature ranges in outer space near Earth.
26. For large vehicles, they endorse using **nuclear reactors for generating electricity**. By current standards, this raises issues of potential radioactive contamination, especially in terms of catastrophic launch failure.

27. For low power cases, they propose concave mirror **solar concentrators** heating mercury vapor (or steam) to drive turbines for electricity. Now we use silicon photovoltaic cells, more simple and robust, and less chemically reactive, as are fuel cells (in liquid hydrogen/oxygen systems).
28. Von Braun emphasizes **chemical rockets over nuclear rocket** engines (and even mentions electric ion propulsion), anticipating the actual course of development of rocket motors.[1]
29. The primary return from **lunar expeditions would be scientific**. Room should be made for astronomical, physical, geophysical, and mineralogical experiments and the researchers to perform them, as well as engineers and other crew. Wheeled or treaded crawlers will be used to explore the lunar surface.
30. **Photographs and other data** will be collected on orbit and the surface. **Seismic probes** of the lunar interior will likely require explosive charges. A satellite in **lunar orbit shows irregularities** that could probe the interior via the Moon's gravitational field.
31. There were unsettled issues of **space law**. Who can make what sort of territorial claims on the Moon or Mars? What sort of overflight "airspace" rights do nations have for objects at orbital distances above Earth? (This issue would soon prove pivotal in the Space Race.)
[1] The United States and Russia also developed nuclear rockets; several spacecraft have used ion propulsion (Chapter 12).

Appendix B
Topics in Transient Phenomena on the Moon

(Throughout this appendix we will refer to a large number of transient events reportedly observed in the vicinity of the Moon, and will refer to them as they are listed in the works by Middlehurst et al. 1968, Cameron 1978, and Cameron 2006.)[1]

Most TLP reports are visual and in recent decades originated from amateurs. Before 1900, however, many reports came from reputable, professional astronomers, even famous ones: Wilhelm Herschel in 1783–1790 with six TLPs (he also discovered Uranus, several moons, and infrared light); Edmond Halley in 1715 (Astronomer Royal, of Halley's comet fame); Edward Barnard with several TLPs in 1889–1892 (showed novae are exploding stars, discovered 17 comets and a moon of Jupiter); Ernst Tempel in 1866–1885 (discoverer of 21 comets); Johann Bode in 1788–1792 (famous celestial cartographer); George Airy in 1877 with a TLP confirmed independently (famous Astronomer Royal); Heinrich Olbers in 1821 (confirmed asteroid belt); Johannes Hevelius in 1650 (pioneering lunar topographer); Jean-Dominique Cassini in 1671–1673 (director of l'Observatoire de Paris); Camille Flammarion in 1867–1906 (founded Société Astronomique de France); William Pickering in 1891–1912 (co-founded Lowell Observatory); Johann Schröter in 1784–1792 (first noticed the phase anomaly of Venus); Friedrich von Struve in 1822 (founded Pulkovo Observatory); Francesco Bianchini in 1685–1725 (measured Earth's axis precession); and Etienne Trouvelot in 1870–1877 (noted astronomical observer). In the twentieth century noted astronomers reporting TLPs included Dinsmore Alter (in 1937–1959), Zdeněk Kopal (in 1963), and Sir Patrick Moore (in 1948–1967). Franz von Gruithuisen in 1821–1839 reported changing luminous and lunar obscured spots, yet also described the Moon inhabited and dotted by cities. (He also was first to conclude that craters result from meteorite impacts.)

Since 1900 and especially since 1957, TLPs are reported increasingly by amateurs, some of whom are not credible. Looking closely at the report catalog one notices observers producing significantly discrepant results, for example, James C. Bartlett of Baltimore in 1950–1976 reported a record 114 TLPs, 101 of them "bluish." There are only 113 reports of blue or violet TLPs, so this category arises almost entirely from Bartlett. We do not include Bartlett's events in Figure 9.2, Table 9.1, or further analysis. Nonetheless Bartlett was reputedly a careful observer and one of his non-bluish reports was confirmed retroactively as fleeting sunrise illumination of local terrain emerging from crater rim shadows.[2]

Table B.1. *Relative Frequency of Robust TLP Report Counts by Feature, Averaging Different Methods (from tables 6-9 in Crotts 2009 [1])*

Relative Frequency ± 1 σ Error	Feature
46.7% ± 3.3%	Aristarchus/Schroter's Valley
15.6% ± 1.9%	Plato
4.1% ± 1.0%	Mare Crisium
2.8% ± 0.8%	Tycho
2.1% ± 0.7%	Kepler
1.6% ± 0.6%	Grimaldi
1.4% ± 0.6%	Copernicus
(6.2% ± 1.2%	*sum of Tycho, Kepler & Copernicus*)
1.1% ± 0.5%	Alphonsus, Bessel, Cassini, Messier, Ptolemaus, or Riccioli
0.9% ± 0.5%	Eratosthenes, Gassendi, Kant, Lichtenberg, (E. of) Picard, (SW of) Pico (B), Posidonius, Proclus, Promontorium Heraclides, or South Pole
0.7% ± 0.4%	Calippus, Eudoxus, Godin, La Hire, or Theaetetus
0.5% ± 0.3%	Peak S. of Alps, Atlas, Hercules, Littrow, Macrobius, or Mare Humorum
0.2% ± 0.2%	Alpetragius, Carlini, Daniell, Hansteen, Helicon, Herschel, Humboldt, Hyginus, Manilius, Pallas, Pickering, Pierce A, or Taurus Mountains

Subjectivity is perilous in these visual reports, and we consider statistical investigations concerning whether there is consistent and/or physically meaningful information in these reports. First let us look at several notable reports, then some generalizations about their nature. Here are interesting reports.

On November 10 of 557 AD a light reportedly appeared on the thin crescent Moon. This is the oldest known TLP report (C78.1) and, unsurprisingly, is not well documented.

A similar naked-eye report comes from November (26?) 1668 by several Boston area residents (C78.9; M68.4): a bright spot between the two "horns" of the crescent Moon. Eleven naked-eye reports of lunar bright spots at various phases (more crescent) are listed, roughly one per century. This excludes several naked-eye reports from observers orbiting the Moon on Apollo.

On June 18 or 19, 1178 (Julian date), as chronicled by senior monk Gervase of Canterbury, five men or more reported "a flaming torch sprang up, spewing out, over a considerable distance fire, hot coals and sparks" between the horns of the crescent Moon.[3] This is the object of modern speculation and controversy not yet run its course (C78.5).

In 1650 Johannes Hevelius (founder of lunar topography) reported the first telescopic TLP, a transient "blood-colored mountain" (*mons porphyrie*)[4] near Aristarchus

(C78.8, M68.3). In 1671–1673 Cassini at l'Observatoire de Paris reported three TLPs that would become fairly typical: a small white cloud, a white spot, a nebulous spot, in crater Pitatus and Mare Crisium (C78.11–13; M68.6–8).

Some dramatic TLP reports are *not* typical, including the following.

Over 1783–1790, Wilhelm Herschel reported seven TLPs, sometimes supported by other (non-independent) observers. In March–April 1787 he reported a complex of events on the Moon's northwest quadrant, especially near Aristarchus. On March 13,[5] he reported three bright spots on the dark side. (Herschel had reported similar spots during a stellar occultation by the Moon in 1783.) On April 19 he reported a red glow, which he called a "volcano" near Aristarchus, and two points near craters Menelaus and Manilius on the dark side. The next night Herschel reported all three points glowing, with the Aristarchus site brighter and larger (at least 5 kilometers across), still red (M68.24–26; C78.31–33).

Since 1787 Herschel's lunar reports were dismissed, despite his transcendent observational reputation. Its legitimacy may turn on several similar reports two centuries later. On October 30, 1963, U.S. Air Force cartographers James A. Greenacre and Edward Barr were visually mapping the Moon at the Lowell Observatory 24-inch telescope near Flagstaff, Arizona. They were surprised by red spots near Aristarchus that Greenacre likened to "looking into a large, polished gem ruby." This inspired a campaign over November 1963 to observe Aristarchus. A similar but short-lived event was reported on November 11, 1963, and then a 75-minute event on November 28 by Greenacre, Barr, Flagstaff Observatory Director John S. Hall, Clyde Tombaugh (discoverer of Pluto), and several observers in Flagstaff and elsewhere in the United States (including Perkins Observatory, Ohio, confirming the red spot) and an independent report from New Jersey of uncertain timing. Both October 30 and November 28 reports correspond roughly to local lunar sunrise (M68.440–441, 443–445; C78.778, 780, 783–785). [6]

Adding notoriety to Greenacre and Barr's reports were photographs by astronomers Zdeněk Kopal and Thomas Rackham on November 1–2 and 10–11, 1963, showing a broad enhancement of red light around Aristarchus and two other craters, the second case largely confirmed independently by Patrick Moore. Kopal published his results in *Scientific American*, further increasing public interest[7] (C78.779, 782; M68.442).

At least nine photographic events appear in Cameron (1978), the earliest from 1953. Many are unpublished, with dramatic exceptions. Most notable are first TLP photographs by a professional astronomer, starting the heightened awareness of TLPs in the later twentieth century and radically changing the reports' characteristics, as described later. On October 26, 1956, Dinsmore Alter took a careful photographic sequence on Mt. Wilson Observatory's 60-inch telescope of craters Alphonsus and Arzachel, in infrared (Kodak I-N emulsion) and blue-violet light (II-O emulsion), allowing differential measurement of image properties in time and wavelength between the craters. There is fairly apparent obscuration of Alphonsus's floor not seen later or in Arzachel. It is apparent in the violet but not infrared as if some scattering cloud is present.[8] A similar (unpublished) effect in crater Purbach was photographed on April 14, 1970, by Osawa (Cameron 1978). On

Figure B.1. **Moon Flash:** Leon Stuart recorded this TLP on 1953 November 15 when he saw a bright spot on the Moon (near center, in Sinus Medii), took this photograph, but noticed that the spot disappeared before taking a second photograph. He estimates that the event lasted over 8 s. (Photo used with permission of Jerry Stuart)

January 23, 1959, Alter recorded (but never published) a photograph of a bright blue glow on Aristarchus's floor, which then turned white (M68.381; C78.653).

On November 15, 1953, radiologist and amateur astronomer Leon Stuart was observing the Moon when he noticed a bright flash near its center and managed to photograph it (Figure B.1). He did not observe the flash's start but estimated its duration at 8 to 30 seconds (although son Jerry Stuart says 8 seconds is probably too quick to take the photo.) Fifty years later planetary scientist Bonnie Buratti and student Lane Johnson noted that the flash's location on Stuart's image was near a fresh, 1.5-kilometer crater and proposed this resulting from an impact seen by Stuart. Later objections to this hypothesis are several: accurate positions of the crater and Stuart's flash image show they do not coincide, the crater exists on pre-1953 photographic plates, and the crater's surface was shown to be too old. Models show that a crater this size would cause a plasma fireball visible lasting under 8 seconds. Stuart's 1953 event appears real, but not from an impact[9] (M68.312, C78.559).

Two unpublished photographs are claimed to show red spots in craters Aristarchus and Maskelyne, respectively, with the latter report apparently confirmed

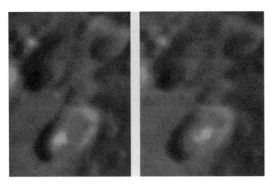

Figure B.2. **Moving on The Moon:** G. Slayton recorded these two images, about 20 minutes apart on 1981 September 5, showing a luminous spot moving (at about 30 km/hour) across the floor of Piticus. (Photograph used with permission of W. Cameron)

by separate visual observers. Two other unpublished photographs involve brightenings of Aristarchus (C78.876, 1145). Cameron (1991) presents dramatic photographs of a glowing, reddish-gray patch moving on crater Piticus's floor (by Gary Slayton of Fort Lauderdale, Florida, on September 5, 1981). See Figure B.2 (C06.152).

During a polarimetric program at l'Observatoire de Paris for lunar surface texture analysis, Dollfus (2000) recorded on December 30, 1992, a brightening in the center of crater Langrenus, and with it a rise in degree of polarization. He interpreted this as scattered light from suspended dust, presumably from gas release.[10] The phenomenon was stable for more than 6 minutes but had decreased when observed three days later. Similar polarimetric changes interpreted as suspended dust were recorded at least twice in Aristarchus, but the time scale is unclear. (C78.820)

Other kinds of permanent data of TLP exist, including spectroscopy. Most notably, Nikolai Aleksandrovich Kozyrev, intrigued by 1956 Alter's photographs, while observing the crater Alphonsus in preparation for taking spectra, on 1958 November 3 at Pulkova Observatory in the Crimea, noticed the crater's central peak turning red and varying in brightness. He began several photographic spectra of the crater's central peak and much of its diameter, some recording several absorption bands similar to molecular carbon's (C_3). Kozyrev saw another TLP on October 23, 1959, in Alphonsus and took another spectrum showing bands in emission he interpreted as C_2 and C_3. Many did not accept the molecular carbon explanation, and Kozyrev's reputation suffered. The incident marked a substantial rift between Soviet and American scientists with long-term effects[11] (C78.703,723; M68.409,423).

Several simultaneous reports by geographically distant observers of the same TLP are recorded, for example, on May 2, 1895, for 12–14 minutes on crater Plato's floor, Leo Brenner reported a streak of light, and Philip Fauth reported independently bright, parallel bands.[12] We consider Plato shortly.

The many anecdotal TLP reports are a rat's nest of selection biases and possible observer errors. Furthermore, I describe earlier many more dramatic ones; most are small spots. Rather than dwell on individual TLPs, we can glean insight from these many reports statistically.

How often do TLPs occur? This is difficult. One can see multiply reported occurrences when TLPs exceed once per month, perhaps when nearly 100% are reported. We also discuss robust TLP sites, consistently producing a fraction of TLPs, roughly 80% of total reports, peaking at several per month. Finally we discuss the gas flow available for violent outgassing and mass required, meaning up to several per month are sustainable. At what rate do TLPs occur? We are unsure, but several per month to one every few months might agree with the data.

A total of 71 reports in Middlehurst et al. (1968) include duration estimates (to within a factor of 2, excluding, e.g., Stuart's 1953 TLP). This is not a statistical sample, but indicates prolonged occurrences; binned in $\sqrt{10}$ intervals over 60–19,000 s (the longest being 18,000 s, the shortest 60 s), their distribution (with number of reports shown in italics) is 60–190 seconds, *7 reports*; 191–600 s: *9*; 601–1,900 s, *27*, 1901–6000s, *23*; and more than 6,000 s: *5*. TLPs are slow enough to allow re-inspection (albeit usually by the same observer). The median length is 15 minutes, almost certainly overestimated because shorter events are easier to miss. Assuming linear correction, the median is around 3 minutes. Nonetheless, even during longer observations, rapid internal changes are often seen. A few examples follow.

May 18, 1964, UT 03:55–05:00: Southeast of crater Ross D, "White obscuration moved 20 mph, decreased in extent. Phenomenon repeated" (C78.818, M68.458).

April 12, 1966, UT 01:05: crater Gassendi, 18 minutes long – "Abrupt flash of red settling immediately to point of red haze near NW wall. Continuous until 01:23" (C78.925, M68.536).

September 2, 1966, UT 03:16–04:18: crater Alphonsus – "A series of weak glows; Final flash observed at 04:18" (C78.971, M68.549).

Four TLPs in Middlehurst et al. (1968) are described as sudden, isolated flashes (TLPs on October 19, 1945, April 24, 1955, October 12, 1957, and September 11, 1967). None occur near known meteor showers. (April 23 is the Pi Puppid meteor peak, but these are strong only near comet 26P/Grigg-Skjellerup perihelion, in 1952 and 1957, not 1955.)

Several patrols of lunar meteoroid impacts use video cameras on Earth. (A *meteoroid* is a small object in solar orbit becoming a meteor if entering Earth's atmosphere.) This is most easily accomplished during meteor showers, when impact rates reach thousands times the average. A problem is video cameras producing artificial flashes possibly mistaken for impacts, avoided by requiring coincidence between simultaneous monitors.[13] These impacts are faint and brief, most lasting under 0.02 second. Covering more area and time, one can also wait for occasional, kilogram-range impacts, lasting 0.1 second. Most impacts have been detected by Marshall Space Flight Center's monitors – several hundred so far, the longest lasting 1 second.[14] Shower meteoroid impacts are nonuniform, tending toward lunar locations with the meteor shower radiant point overhead, but TLP loci are even more restricted. For a 1-second flash, a roughly 1-tonne impactor is needed, leaving a noticeable crater about 50 meters across. Even comets striking Jupiter, for example, Shoemaker-Levi 9, with billions-tonne masses, produced

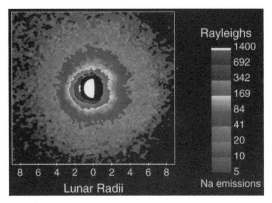

Figure B.3. **Moon Glow:** Emission from Moon's extended sodium atmosphere as observed at McDonald Observatory 1991 September 30 (Courtesy of Dr. Michael Mendillo & Boston University)

flashes of only 5–10 seconds. Few TLPs are explained by meteoroid impacts on the Moon.

Those many impacts affect the lunar atmosphere: they make it glow like a sodium vapor lamp (see Figure B.3). Meteoroid impacts erode many thousands of tonnes of lunar regolith per year.[15] Sodium is lost in this process through "desorption" – vaporization in this case by heat or solar ultraviolet irradiation, or "sputtering" – loss from particle collisions, here by solar wind. Scientists are still unsure which processes dominate.[16] Similar processes occur for potassium, and searches have been made for several other elements.[17] Solar radiation is reprocessed efficiently by these atoms' resonant scattering transitions (5893Å for sodium, 7699Å for potassium). Only tiny amounts of these atoms are found (usually less than 100 per cubic centimeter), adding negligibly to lunar atmospheric mass. Nonetheless, they surround the Moon in an optical plume to several dozen lunar radii, easily confused with the Moon's dust halo if not separated spectroscopically.

Are optical transients obviously correlated with physical processes on the Moon? Several investigator teams have correlated existing TLPs, especially their timings, with external processes. This is an uncertain procedure, depending not only on when lunar physical processes occur but also when humans are observing. Studies have looked for TLP activity's correlation with maximum tidal stress from Earth but largely failed using large report samples.[18] Early work hinted of a correlation for Aristarchus reports only, which later studies do not rule out.[19] Middlehurst et al. (1968) claim close pairings of TLP and deep moonquake loci, with epicenters always within 150 kilometers of a TLP site.

We divided structure in the TLP report catalog due to observer biases versus those due to possible physical phenomena, summarized in Crotts 2009. These reports were grouped by characteristics of the observers having presumably nothing to do with the lunar surface: their position on Earth from which they observed, the year they observed, or the season of the year (see Crotts 2009, tables 6–9). Thus we statistically "marginalize" over these observer variables and ask if the features that are always present in the distribution despite differing values of these observer variables remain the same regardless of which variables we marginalize. Amazingly, the features that remain nearly regardless of how we perform this marginalization. The resulting features include Aristarchus, Plato, Grimaldi, and the large, young Near Side impacts appearing to be robust TLP sites (correlating with radon outgassing and moonquakes), and the structure of these sites' reports breaking down with TLPs become popularly known after 1956. Those sites that show consistent behavior regardless of the location on Earth of the reporting observer or when that report was made are shown in Table B.1.

We discussed TLP history, and left the story in 1963 with Alter, Kozyrev, Kuiper, and Greenacre and Barr. TLPs were important during the Apollo era, and this should be understood. Analysis shows that observers during this period were no longer producing robust report results but rather were cognizant of what others were observing, this tale being more sociological than one of astronomical phenomena.

When the United States decided to send astronauts to the Moon, the nature of TLPs became an issue of operational mission safety. The Apollo landing site selection committee gave high priority to sites at crater Alphonsus and on the Aristarchus Plateau, in part because of reported TLP activity, although ultimately Apollo went elsewhere. Corralitos Observatory near Las Cruces, New Mexico, opened in October 1965 primarily to understand TLPs; this was sufficiently significant to be noted by President Johnson in a message to Congress. Barbara M. Middlehurst, who had worked extensively at the University of Arizona's Lunar and Planetary Laboratory with Gerard Kuiper (the lab's founder in 1960) studied TLPs extensively from 1964 to 1967. At NASA's Goddard Space Flight Center (GSFC), the lunar group including John A. O'Keefe (at GSFC from its founding in 1959) started a TLP effort, the main interest of Winifred Sawtell Cameron, arriving at GSFC in 1961.

Corralitos spent roughly 8,000 hours total looking for TLPs and was ready to follow up alerts from observers elsewhere reporting TLPs. They reported no certain TLPs (and disagreement exists on whether they rejected two TLPs from other observers).[20] There were several less sensitive patrols for TLPs, which produced several marginal detections but no dramatic events. A post-facto evaluation of nearly absent detections at Corralitos was a negative result or null.[21] Whether they should have expected to find TLPs is unclear. TLPs seem to be rare; searching the whole Near Side for many months equivalent may be necessary, hence, we propose an automated telescopic search (Chapter 9).

Heightened activity studying TLPs during the Apollo era produced no useful positive results, nonetheless some interest was maintained. At a scientific workshop in March 1976 at the Lunar Science Institute near Houston, prominent lunar scientists met to discuss and debate TLPs, lunar outgassing, and related phenomena, with contributions published in a special issue of the *Physics of Earth and Planetary Interactions* in June 1977. On September 30, 1977, ALSEP instruments on the Moon were turned off, and lunar science in general entered into a slower phase of development.

The lunar science community tends to produce encyclopedic compendia of the field every 15 years or so. The *Lunar Sourcebook: A User's Guide to the Moon* (1991), (see note 12 in Chapter 2), our most common citation in this book, discusses TLPs several times, covering more than a page, including 1 of 15 major unanswered questions about the Moon: "Lunar transient events; Present Status: controversial Earth-based observations of clouds or flashes on the Moon, impact event; Unanswered questions requiring further research/exploration: A permanent monitoring system is needed on the lunar surface. Are there gas releases from the lunar interior? What is the relation between transient events and surface radon emission?" (These derive from the Lunar Geoscience Working Group of 1986.) During the intervening 15 years, little was learned about TLPs or lunar outgassing, positive or negative, yet the next compendium, *New Views of the Moon* (2006) ignores TLPs and barely mentions outgassing, except brief mention of the Moon's sodium plume.

The study of TLPs entered a strange and unfortunate state of "pariah science" in which even its mention would bring derision and unreasoned skepticism. Why did this happen? From 1976 to 1994 was the "missing generation" of lunar science. Of course talented lunar scientists earned their college degrees and did valuable research, but no significant science about the Moon came from spacecraft missions, even robotic ones, with new data,[22] other than, fortunately, the enormous data backlog from the Apollo era. Few imagined that decades would pass before any spacecraft would explore the Moon following nearly 100 missions previously. Now new floods of data are unleashed, and new discoveries are being made at rapid pace. Unfortunately, few of these relate to TLPs.

Some scientists have studied TLP science at their peril since the early 1970s. As a career, lunar science offered a Hobson's choice regarding better understanding TLPs. Several have attempted to advance the topic and suffered discouragement. I talked with John O'Keefe about TLPs in 1979 and I was admonished to abandon them, that he would not pursue them because to do so would be a waste of the taxpayers' money. This was not NASA policy; indeed Cameron continued at Goddard until a suitable retirement age and would socialize with prominent visiting scientists (Urey, Ernst Öpik, and others) without embarrassment. However, during this time a pervasive close-mindedness settled into the field.[23]

Several attributes of TLPs are problematic in lunar science. First, many TLP reports are anecdotal, with problems discussed in Chapter 9. In the past century most TLP reports come from untrained or amateur observers, unrespected by professional scientists. There seems a third problem in that TLPs are rare and transient, with little direct apparent lunar geological effect, and lunar science is dominated by geology.

Nonscientists often see science as a monolith in which all scientists agree on how science is done. This is untrue. Different cultures arise based on tradition and in part on divergent innovation from what succeeds in particular fields. Publication procedures between disciplines, for instance, differ. In planetary science, for instance, one rarely finds journal articles with more than 30 coauthors, despite many data coming from huge group-effort space missions. In contrast, some astrophysics papers have more than 500 co-authors, and particle physics articles more than 3,000.[24] Mathematics papers usually have one or two authors.

Differences in collaboration size reflect practical considerations because of project scales, but other traditions are more subjective. Astrophysicists and particle theorists share results via preprints; journal publication frequently happens later to permanently record and assure accountability. Planetary scientists rarely use preprints but wait for the journal publication.[25] These differences go deeper: there are fundamental differences between geologists and physicists regarding data anomalies.

Geologists and physicists approach scientific anomalies differently. Physicists are taught by hard experience to focus on the anomaly, the rule's exception, almost no matter how small. The key concepts of twentieth-century physics – relativity and quantum mechanics – largely depend on this. The precession of Mercury's orbit, missing energy in nuclear decays, the extra faintness of Type Ia supernovae at high red shift, and the Lamb shift and charge/parity violation are several of many tiny effects proclaiming new physics, as often as not leading to a Nobel Prize.[26] Geology is more descriptive, concerning complex systems in which initial conditions are rarely known accurately. One cannot explain every feature; one is missing input data. Anomalies should often be ignored. This conflict is older than space exploration. Aristotle in *Physics* describes anomalies as "monstrosities," mistakes of nature, failing to attain their natural end (such as mutations), not worthy of study.[27] The concept continues through history: Pliny the Elder (23–79 AD) on monstrous human races, and Galen (131–201 AD) and Avicenna (980–1037) regarding birth defects, with traces to Mary Shelley's 1818 *Frankenstein or the Modern Prometheus*. Now medicine is interested in "monsters" as a path toward discovery, consistent with ethical considerations regarding human and animal trials. Aristotle's idea directly impacted geology regarding fossils, considered never animate freaks of nature, unworthy of study until the sixteenth century. Today concepts of geological anomalies are often exploited in searching for ore deposits and relating to unique compositional structure (also exploiting magnetic and gravitational anomalies.) Still, if minor features diverge, often the attitude is that they can be ignored.

Lunar study overlaps several scientific disciplines, those of the geologist, space scientist, and astronomers/astrophysicist. TLPs and the lunar atmosphere involve all three. Intellectual turf battles rose among these different species of scientists in Apollo's early days.[28] Strident controversy involved astrophysicist Tommy Gold and loosely packed lunar dust being hazardous (Chapter 2), an argument he ultimately lost. Some geologists saw this debate as an avoidable waste of time. (See note 13 in Chapter 2.) Gold also lost the debate with other astrophysicists as an originator of Steady State cosmology, vanquished by the Big Bang model. Gold strongly criticized NASA's Space Shuttle and was in part vindicated and derided for suggesting that pulsars are neutron stars with intense magnetic fields (which they are). Some see Gold as why the Moon should be left to the geologists.

Volatile substances' importance is quickly becoming appreciated in lunar science. The call of "extraordinary claims require extraordinary evidence" has retreated before discovery of many lunar water sources, not so drastically unlike other planets. This will correct itself faster than the mid-twentieth century plate tectonics impasse. Nonetheless sociologists of science consider these "boundary-work" issues of what is acceptable to study and how research is done set by scientists' social interactions within and between disciplines. We approach the need for rapid acceptance of volatile lunar processes so that they can be studied as vigorously as geology before. This has direct repercussion on robotic lunar exploration, as we explore in Chapter 10 about working on the Moon, much for science's sake.

Notes

1. Much of this is adapted from "Transient Lunar Phenomena: Regularity and Reality" by Arlin P. S. Crotts, 2009, *Astrophysical Journal*, 697, 1.

    Throughout Appendix B we refer to specific transient events by their listing in catalogs by Cameron 1978, etc. [see 9.36] with a coded notation, e.g., event #1165 in Cameron's 1978 list is C78.1165, the transient in Aristarchus reported by astronomers in Germany and *Apollo 11* astronauts on July 19, 1969, from Chapter 9. Rapid flashes on the surface of the Moon observed by *Apollo 16* and *17* astronauts are C78.1331 by Mattingly on April 21, 1972; C78.1352 by Schmitt on December 10, 1972; and C78.1354 by Evans on December 11, 1972 (not Gene Cernan as per Cameron 1978). Notations M68.N, C78.N, and C06.N to refer to the Nth entry from Middlehurst et al. 1968, Cameron 1978, and Cameron 2006 catalogs, respectively.

2. "Emergence of Low Relief Terrain from Shadow: An Explanation for Some TLP" by Raffaello Lena & Anthony Cook, 2004, *Journal of British Astronomical Association*, 114, 136. Winifred Cameron in introducing the 1978 TLP catalog states, "Bartlett is an assiduous, experienced observer and has been rated high although most of his observations are of bluish phenomena on Aristarchus, which may have their cause in terrestrial conditions, rather than lunar. If the user does not accept that explanation, then most of Bartlett's observations are very good."

3. From *The Historical Works of Gervase of Canterbury, Volume 1* by William Stubbs, 1879, London: Eyre & Spottiswoode, 276 (translated from Latin by Richmond Y. Hathorn, in Hartung 1976): "After sunset when the Moon had first become visible a marvelous phenomenon was witnessed by five or more men who were sitting facing the Moon. Now there was a bright new moon, and as usual in that phase its horns were tilted towards the east, and suddenly the upper horn split in two. From the midpoint of this division a flaming torch sprang up, spewing out, over a considerable distance fire, hot coals and sparks. Meanwhile the body of the Moon which was below writhed, as it were, in anxiety, and, to put it in the words of those who reported it to me and saw it with there own eyes, the Moon throbbed like a wounded snake. Afterwards it returned to its proper state. This phenomenon was repeated a dozen times or more, the flame assuming various twisting shapes at random and then returning to normal. Then after these transformations the Moon ... along its whole length took on a blackish appearance. The present writer was given this report by men who saw it with their own eyes, and are prepared to stake their honor on an oath that they have made no additions or falsification in the above narrative." Gervase also gives plausible accounts of the Mars/Jupiter conjunction on September 13, 1170 (*The Historical Works*, 221), a partial solar eclipse of September 13, 1178, (p. 21 and 279), and an apparent aurora on November 29, 1177 (p. 274).

    Hartung hypothesized that Gervase reported an asteroid impact creating crater Giordano Bruno, but several papers criticize this ("Was the Formation of a 20-km Diameter Impact Crater on the Moon Observed on June 18, 1178?" by Jack B. Hartung, 1976, *Meteoritics*, 11, 187). Crater counts superimposing ejecta from Giordano Bruno indicate a 1–20 million year crater age ("Formation Age of the Lunar Crater Giordano Bruno" by Tomokatsu Morota et al., 2010, *Meteoritic and Planetary Science*, 44, 1115). Material expected to hit Earth from such an event amounts to 10 million tonnes, with no evidence for such meteor showers ("Meteor Storm Evidence Against the Recent Formation of Lunar Crater Giordano Bruno" by Paul Withers, 2001, *Meteoritic and Planetary Science*, 36, 525; also "On the Occurrence of Giordano Bruno Ejecta on the Earth" by Jack B. Hartung, 1981, *Lunar and Planetary Science Conference*, 12, 401; "Ejecta from Lunar Impacts: Where Is It on Earth?" by D. E. Gault & P. H. Schultz, 1991, *Meteoritics*, 26, 336). The report date is disputed. ("Author's Reply" by Paul Withers, 2001, *Meteoritic and Planetary Science*, 37, 466; "The 'Lunar Event' of AD 1178: A Canterbury Tale?" by Bradley E. Schaefer; Philip M. Bagnell, 1990, *Journal of British Astronomical Association*, 100, 211. Schaefer claims Gervase's solar eclipse date is seven days in error; however, it appears correct: NASA GSFC Eclipse Web site,

Solar Eclipse of September 13, 1178, both report and post-diction on Julian September 13. He questions the AD 1177 event identified as auroral because this misses solar maximum, but determining solar activity some 53 cycles before Cycle 0 of 1755 is problematic.)

4. Porphyry is a dark red or purple rock. *Porphyritic* now independently means igneous rocks with a texture of large crystals.
5. *Selenotopographische Fragmente zur genauern Kenntniss der Mondfläche: ihrer erlittenen Veränderungen und Atmosphäre: sammt den dazu gehörigen Specialcharten und Zeichnungen* (*Selenotopographic Fragments Towards More Exact Knowledge of the Lunar Surface: Its Changes and Sustained Atmosphere Together with Accompanying Special Charts and Drawings*) by Johann Hieronymus Schröter, 1791, Lilienthal: at author's expense, 533.
6. A detailed study is found in "Revisiting the 1963 Aristarchus Events" by B. E. O'Connell & A. C. Cook, 2011, *Journal of British Astronomical Association*, 114, 136.
7. "The Luminescence of the Moon" by Zdeněk Kopal, May 1965, *Scientific American*, 212, 28.
8. "A Suspected Partial Obscuration of the Floor of Alphonsus" by Dinsmore Alter, 1957, *Publications of Astronomical Society of Pacific*, 69, 158.
9. "A Photo-Visual Observation of an Impact of a Large Meteorite on the Moon" by Leon H. Stuart, 1956, *Strolling Astronomer*, 10, 42; "Identification of the Lunar Flash of 1953 with a Fresh Crater on the Moon's Surface" by Bonnie J. Buratti & Lane L. Johnson, 2003, *Icarus*, 161, 192; "Lunar Flash Doesn't Pan Out" by J. Kelly Beatty, 2003 June, *Sky and Telescope*, 105, 24; "Optical Maturity Study of Stuart's Crater Candidate Impact" by D. T. Blewett & B. R. Hawke, 2004, *Lunar and Planetary Science Conference*, 35, 1098. See also "That Stuart Brilliant Flare and the Search for a New Lunar Crater" by Walter H. Haas, 2005, *Strolling Astronomer*, 47, 46.
10. "Langrenus: Transient Illuminations on the Moon" by Audouin Dollfus, 2000, *Icarus*, 146, 430; "First Results from Observations of the Moon by Means of a Polarimeter" by V. P. Dzhapiashvili & L. V. Ksanfomaliti, in *The Moon (IAU Symposium 14)*, eds. Z. Kopal & Z. K. Mikhailov, 1962, Waltham, MA: Academic, 463; "Some Results of Measurements of the Complete Stokes Vector for Details of the Lunar Surface (*Nekotorye Rezul'taty Izmerenii Polnom Vektora Stoksa dlia Detalei Foverkhnosti Luny*)" by Yu.N. Lipskii & M. M. Pospergelis, 1967, *Astronomicheskii Zhurnal*, 44, 410.
11. "Observation of a Volcanic Process on the Moon" by Nikolai A. Kozyrev, February 1959, *Sky and Telescope*, 18, 184; "The Kozyrev Observations of Alphonsus" by Dinsmore Alter, 1959, *Publications of Astronomical Society of Pacific*, 71, 46; "Volcanic Phenomena on the Moon" by Nikolai Kozyrev, 1963, *Nature*, 198, 979. Gerard Kuiper's skepticism and his campaign to limit effects of Kozyrev's observations are detailed in "Evaluating Soviet Lunar Science in Cold War America" by Ronald E. Doel, 1992, *Osiris*, 7, 238. Kuiper saw Kozyrev's results as counter to American geologists' majority view of lunar inactivity and studied copies of his spectrographic plates sent by *Sky and Telescope*'s editors. (According to Doel, Kuiper implied in a letter to *Sky and Telescope*'s editor that Kozyrev forged his data.) Kuiper's study was inconclusive, so he tried obtaining his own spectra at McDonald Observatory, also inconclusively. Kuiper inquired with Kozyrev's Soviet colleagues regarding his character. Urey then argued there was no good reason to doubt Kozyrev's results, further complicating interdisciplinary politics of accepting Kozyrev's work. In late 1960 Kuiper attended a scientific meeting in Leningrad and on that trip examined the original Kozyrev Alphonsus spectra for the first time. He changed his mind, saw no reason to doubt Kozyrev's basic result, and, after returning to the United States, began supporting TLP studies. (In fact Doel cites a Kuiper staff member as coining their first acronym, LTP: lunar transitory phenomena.)
12. Brenner was in Lussinpiccolo, Croatia, and Fauth in Lundstuhl, Germany, 750 km apart, so likely independent. The reports were compared in a short article two years later: "Der Lichtschein im Plato" by Johann N. Krieger, March 1897, *Sirius*, 30, 49.
13. Several surveys have seen events, e.g., five during the Leonid meteor shower, three of them observed simultaneously, by Ortiz and colleagues. ("Optical Detection of Meteoroidal Impacts on the Moon" by J. L. Ortiz, P. V. Sada, L. R. Bellot Rubio, F. J. Aceituno, J. Aceituno, P. J. Gutiérrez & U. Thiele, 2000, *Nature*, 405, 921; also "Possibility of Observing Fall of Meteorites on the Moon from a Station on the Earth" by V. A. Anoshkin, G. G., Petrov & K. L. Mench, 1978, *Astronomicheskii Vestnik*, 12, 216; "Making the Photos of Flashes on the Moon" by A. V. Arkhipov, 1991, *Zemlia i Vselennaya*, 3, 76; "Lunar Leonids: November 18th Lunar Impacts" by David Dunham, 1999, http://iota.jhuapl.edu/lunar_leonid/index.html.081299; "The First Confirmed Perseid Lunar Impact Flash" by Masahisa Yanagisawa, Kouji Ohnishi, Yuzaburo Takamura, Hiroshi Masuda, Yoshihito Sakai, Ida Miyoshi, Makoto Adachi & Masayuki Ishida, 2006, *Icarus*, 182, 489.)

14. "Rate and Distribution of Kilogram Lunar Impactors" by W. J. Cooke, R. M. Suggs, R. J. Suggs, W. R. Swift & N. P. Hollon, 2007, *Lunar and Planetary Science Conference*, 38, 1986; "The NASA Lunar Impact Monitoring Program" by Robert M. Suggs, William J. Cooke, Ronnie J. Suggs, Wesley R. Swift & Nicholas Hollon, 2008, *Earth, Moon and Planets*, 102, 293. A full sample is not yet published: "Flux of Kilogram-sized Meteoroids from Lunar Impact Monitoring" by Robert M. Suggs, W. Cooke, R. Suggs, H. McNamara, W. Swift, D. Moser & A. Diekmann, 2008, *Bulletin of American Astronomical Society DPS*, 40, 33.03, http://www.nasa.gov/centers/marshall/news/lunar/.
15. "Numerical Simulation of High-Velocity Impact Ejecta Following Falls of Comets and Asteroids onto the Moon" by N. A. Artemieva & V. V. Shuvalov, 2008, Solar System Research, 42, 329.
16. See "Photon-Stimulated Desorption as a Substantial Source of Sodium in the Lunar Atmosphere" by B. V. Yakshinskiy & T. E. Madey, 1999, *Nature*. 400, 642, and references therein.
17. "Discovery of Sodium and Potassium Vapor in the Atmosphere of the Moon" by A. E. Potter & T. H. Morgan, 1988, *Science*, 241, 675; "A Search for Far-Ultraviolet Emissions from the Lunar Atmosphere" by W. G. Fastie, P. D. Feldman, R. C. Henry, H. W. Moos, C. A. Barth, G. E. Thomas & T. M. Donahue, 1973, *Science*, 182, 710; "The *Apollo 17* Ultraviolet Spectrometer: Lunar Atmosphere Measurements Revisited" by P. D. Feldman & D. Morrison, 1991, *Geophysical Research Letters*, 18, 2105; "A Spectroscopic Survey of Metallic Species Abundances in the Lunar Atmosphere" by B. C. Flynn & S. A. Stern, 1996, *Icarus*, 124, 530; "An HST Search of Magnesium in the Lunar Atmosphere" by S. A. Stern, J. W. Parker, T. H. Morgan, B. C. Flynn, D. M. Hunten, A. L. Sprague, M. Mendillo, & M. C. Festou, 1997, *Icarus*, 127, 523.
18. "Observations of Changes on the Moon" by W. S. Cameron, 1967, *Proceedings of 5th Annual Meeting of Working Group on Extraterrestrial Research*, 47; "Comparative Analyses of Observations of Lunar Transient Phenomena'" by W. S. Cameron, 1972, *Icarus*, 16, 339; "A Survey of Lunar Transient Phenomena" by Barbara Middlehurst, 1977, *Physics of Earth and Planetary Interactions*, 14, 185; "Transient Lunar Phenomena, Deep Moonquakes and High-Frequency Teleseismic Events: Possible Connections" by Barbara M. Middlehurst, 1977, *Philosophical Transactions of Royal Society of London – A*, 285, 1327.
19. "Tidal Influences at the Lunar Crater Aristarchus" by William B. Chapman, 1967, *Journal of Geophysical Research*, 72, 6293; "Lunar Tidal Phenomena and the Lunar Rille System" by Barbara Middlehurst, in *The Moon – Proceedings of IAU Symposium #47*, 1972, Dordrecht: Reidel, 450; "Tidal Influences at the Lunar Crater Aristarchus and Transient Lunar Phenomena" by A. C. Cook, 2011, *Lunar and Planetary Science Conference*, 42, 2811.
20. "The Corralitos Lunar Transient Phenomena (LTP) Surveillance Program (1966–1972)" by J. R. Dunlap & J. A. Hynek, 1973, *Bulletin of American Astronomical Society*, 5, 37; "The Corralitos Observatory Program for the Detection of Lunar Transient Phenomena" by J. A. Hynek, J. R. Dunlap & E. M. Hendry, 1976, NASA pub. CR-147888; "A Search for Lunar Transient Phenomena: 1966–1968" by J. A. Hynek & J. R. Dunlap, 1968, *Astronomical Journal*, 73, 185.
21. The Corralitos Observatory TLP survey spent ~8,000 hours observing (10.9 months at full duty cycle), capable of covering the Near Side in 15 min (although it is unclear if it always did), hence, it should have produced some 30,000 whole-Moon epochs. What intrinsic event rate for TLPs should we assume? We might infer a rate of ~1/month. Given the distribution of observed TLP time scales, ~30% of the reports would be missed. If the Corralitos setup was equally sensitive as the typical TLP observer at large, the absence of TLPs detected by them in the untriggered survey might correspond to a –3σ fluctuation (about 0.02% Poisson, not Gaussian, random probability) in expected counts if observations were 100% efficient (which is unrealistic).

    How sensitive was the Corralitos survey? This would have quantitative implications if we knew the flux distribution function for TLPs, which we do not. Still, Corralitos was probably at least as sensitive as the typical TLP observer and probably more so, therefore, they should access at least the same event rate. The claimed sensitivity of the survey method seems improbably good: better than a 5% change in intensity in a 100Å band ("The Corralitos LTP Surveillance Program," Dunlap & Hynek)[21], so 0.5%–1% in a typical broadband characteristic of a photometric optical color employed, converted to a monochromatic, blinking contrast difference monitored by the eye. This seems several times more sensitive than the eye's threshold for detecting a constant monochromatic contrast, but the intent was apparently to improve this threshold by blinking the spot at a rate of several Hz. This may work for short durations, but the eye's response to such signals fatigues over time most significantly at rates of about 12 Hz ("Time-locked Perceptual Fading Induced by Visual Transients" by R. Kanai & Y. Kamitani, 2003, *Journal of Cognitive Neuroscience*, 15, 664.) Hynek and colleagues did their work before this effect was studied, and they may not have adjusted their procedures accordingly. The effects of fatigue should be evaluated by reconstructing the original setup of the Corralitos display equipment, which is difficult. The same should be done of Moon Blink (Chapter 9). Particularly concerning are TLP reports promptly transmitted to Corralitos Observatory during the TLP patrol for confirmation (or lack thereof). Cameron (1978) lists 25 events apparently negative (four where this is stated explicitly in terms of the data), two originating with Bartlett not included in our analysis, and two (C78.1119, C78.1150) where Cameron disagrees with Hynek and colleagues and concludes positive confirmation. This fraction of non-confirmation might lead one to conclude that many TLP reports are not objectively real, at least amid intensive campaigns like those underway when these reports were produced (April 1966 – June 1969).

22. I exclude *Hiten*, from Japan in 1990, the only twentieth-century lunar mission not from the United States or Soviet Union, a successful technology demonstration but with little lunar science instrumentation. It was first to use a low-energy transfer orbit to the Moon (using Weak Stability Boundary Theory by Edward Belbruno, Chapter 13) and showed orbits can be adjusted successfully by skimming Earth's atmosphere (aerobraking).

23. When I published my statistical study of TLPs in 2007, a lunar science popularizer posted this opinion on the Internet. For the sake of civility, I do not cite the author by name, but the quote is instructive:

    "Transient Lunar Phenomenon (TLP) have been the holy grail for some observers of the Moon. During the 18th and 19th centuries various lunar observers reported changes that they thought proved that the Moon is not dead. The flurry of international excitement over Linne in the 1860s was just the most visible of many similar incidents of exceedingly unlikely observations of change. During the last few years the Geological Lunar Research group has convincingly explained some classic TLP observations as accidents of viewing geometry, and the general trend has been to dismiss TLPs. Now a professor at a prestigious American university has reopened the question with a series of as yet unpublished research papers that conclude that ~80% of TLPs were real. He also proposes an explanation that seems unlikely in the extreme. Arlin Crotts is a professor of astronomy at Columbia University who specializes in observational cosmology. Somehow he has become captivated by TLP and has conducted sophisticated statistical studies of the classic Cameron and Middlehurst TLP catalogs. He also links the locations of TLPs with moonquakes and radon gas emissions detected by Apollo and Lunar Prospector. Crotts suggests that escaping gases might explosively loft a cloud of regolith above the surface, creating the temporary change of a TLP. This is a reasonable variant of earlier ideas. But he further speculates that if some of the gas is water vapor then it could become ice or interact with regolith particles in complex ways. Wait a minute! One of the fundamental discoveries of Apollo is that Moon rocks are dry, totally lacking water. Recent studies have suggested that there may have been extraordinarily small amounts of water in some lunar rocks, but Crotts proposes that there could be an ice layer a kilometer wide under Aristarchus and other TLP sites. This seems crazy, especially since Crotts arrives at lunar ice as a mechanism to explain a phenomenon that may not be real. It will be very interesting over the next few months to see how the professional lunar science community responds to this idea. In the meantime, Crotts and colleagues have built a specialized telescope in the Andes to search for TLPs more intensely than ever before."

    The author of this critique misunderstood several points about my papers; I do not think he read them well. However, because 2007 lunar science has largely reversed itself on the issue of lunar hydration, I am not sure many of the points made in this excerpt stand the test of time.

24. "Performance of the ATLAS Detector Using First Collision Data" by G. Aad, et al. (3,194 coauthors), *Journal of High Energy Physics*, 2010, 56; "Insights into the High-Energy Gamma-Ray Emission of Markarian 501 from Extensive Multifrequency Observations in the Fermi Era" by A. A. Abdo, et al. (453 coauthors), 2011, *Astrophysical Journal*, 727, 129. Gravitational-wave astronomy tends to particle physics customs: "Search for Gravitational Waves from Compact Binary Coalescence in LIGO and Virgo Data from S5 and VSR1" by J. Abadie, et al. (714 coauthors), 2010, *Physical Review D*, 82, 102001.

25. There is an electronic preprint server for particle physics and astrophysics (http://xxx.lanl.gov/), but the planetary science preprint archive is used mainly by extrasolar planet researchers.

26. One recalls anomalies that seemingly foretold new physics but faded: magnetic monopoles, the Pioneer 9 and 10 anomalous acceleration, etc., but that does not mean that scientists think them having been unworthy of some attention. Most physics journals make allowance for results being possibly wrong if measurement was careful and implications would benefit from open discussion.

27. From Aristotle's *Physics*: "Now surely as in intelligent action, so in nature; and as in nature, so it is in each action, if nothing interferes. Now intelligent action is for the sake of an end; therefore the nature of things also is so."

    "Now mistakes come to pass even in the operations of art: the grammarian makes a mistake in writing and the doctor pours out the wrong dose. Hence clearly mistakes are possible in the operations of nature also. If then in art there are cases in which what is rightly produced serves a purpose, and if where mistakes occur there was a purpose in what was attempted, only it was not attained, so must it be also in natural products, and monstrosities will be failures in the purposive effort."

28. See *To a Rocky Moon: A Geologist's History of Lunar Exploration* by Don E. Wilhelms, 1993, Tucson: University of Arizona Press. See particularly pages 5, 15, 20, and 26–27.

Index

Defining or introductory text for a topic indicated by **bold page numbers**. Text in footnotes indicated by *italic page numbers*. Appendix A ("Von Braun et al. Space and Lunar Exploration Issues"), Appendix B ("Topics in Transient Phenomena on the Moon") and the Glossary are not included in this index.

A9, A10 rocket designs: *115*
Abernathy, Ralph: 71
Adams, John Couch: 159
Adventures of Tom Sawyer, The: 156
aerobraking: 128
Agena rocket: 25, 27
Aggregat-4 (A-4): see V-2
Agnew, Spiro: 79
Airborne Lightweight Optical Tracking System (ALOTS): 29, 33
Aldrin, Buzz (formerly Edwin): 18, 20, 21–22, 30, 33, 34, *52–53*, 95, 97, *124*, 298, *324*, 346, *364*
aliens: see extraterrestrial life
Allen, Joe: *328*, 333, *362*
Almaz program: 51, 144
ALSEP: see Apollo Lunar Science Equipment Package
amphibole: 231–232
Amundsen, Roald: 150, 339, 432
Anders, William: 29, 31, *154*
Angara rocket series: 135, 437
　Angara 5 and 7 rockets: 97
angular size: 1–2, 158
anorthosite: **174–176**, 177, 197, 198, 201, 202, 205, *207*
Ansari X-Prize: 147
apatite: 239, 245, 247, *259*
Apollo program: 18–24, 26, 28–48, 97
　Apollo 1 (Apollo/Saturn 204): 28, 49, **69–70**, 82
　Apollo 4: 49
　Apollo 5: 31, 145
Apollo 6: 49
Apollo 7: **29**, 30, 351
Apollo 8: 23, **29–31**, 48, 49, *55*, *56*, *154*, 270
Apollo 9: **29**, 30, 31
Apollo 10: **29**, 30, **32**, 48, 270
Apollo 11: 18–24, 26, 30, **33–34**, *52–55*, 69, 71, 75, 78, 92, *119*, 175, 177, *254*, 264, 289
　descent from LM: **21–24**
　landing at Tranquility Base: 18, **19–20**, 26, 28, 44–45, 48, *52–53*, *53–54*, 104, 321
　"That's one small step for (a) man,...": 18, **21–23**, *53*, **54**
Apollo 12: 22, **33–35**, 36, 78, 189, *209*, 233, *253*, 270, 321, 357
　lightning hits at launch: 33, **75–77**, 356
Apollo 13: **35–38**, 49, 78, 82, 314
Apollo 14: **38–39**, 40, *55*, 78, 110, 176, 233, 239, 243, *253*, *254*, 270, 289, 321, *326*, *365*
Apollo 15: 23, 29, **38–40**, 41, 42, 44, 78, 175, 176, 184, 193, 195, *205*, 224, 229, 231, 233, 238, 239, *253*, *254*, *258*, 261, 270, 274, 279, 283, 289, *295*, 300
　subsatellite: 39, 43, *295*
Apollo 16: 29, 36, **40–41**, 44, 78, 101, 174, 176, 198, *205*, 212, 233, 243, *253*, *254*, *258*, 263, 270, *295*, 322, 333
　subsatellite: 43, *295*
Apollo 17: 29, **41–42**, 45, 46, 47, 78, 82, 101, 109, 176, 189, 193, 200, 201, *207*, 238, *254*, 256, *258*, 264, 270, 271, 294, 298, 314, 323, *327*, 333, *362*
Apollo 18–20 (cancelled): 38, 75, 77

Apollo program (cont.)
   astronaut science training: 35, 36, 261–262, 292, *293*, 315, *328*
   atmospheric reentry: 37, 38
   budget: 62, 69–70, 77
   Command Module (CM): 28, 37
   Command/Service Module (CSM): **29**, 31, 32, 43
   data archive: 108
   Guidance Computer (AGC): 20–21, 104–105
   H missions: 33
   hoax conspiracy theories: 108–111
   J missions: 29, **38**, **41**, 42, 76, 200, 318, 327
   lunar landing sites, selection: 27, 37–40, 176, 193–194, 263
     Descartes highlands: 40, 176
     Fra Mauro: 37–38
     Hadley Rille: 39
     Taurus-Littrow: 41, 47, 109, 184, 193–194, 200, *207*
     Tranquilitatis: 193
   Lunar Module (LM): 19–24, **29–30**, 31, 32, 34, 36, 38, 41, 43, 47–48, *54*, 62, 104
   Lunar Science Equipment Package (ALSEP): 33, 34, 38, 39, **40**, 41, 109, *120*
     ALSEP shutdown date: 261
     Central Station (and Lunar Dust Detector): 40
     Charged Particle Lunar Environment Experiment (CPLEE): 40, *254*, *295*
     Cold Cathode Ion Gauge/Gauge Experiment (CCIG/CCGE): 40, 270, 273
     Heat Flow Experiment (HFE): 39, 41, *258*, 305
     Lunar Atmospheric Composition Experiment (LACE): *253*, *254*, *256*, 270, 273, *295*
     Lunar Ejecta and Meteorites (LEAM): *294*
     Lunar Portable Magnetometer: 34, 40
     Lunar Ranging Retroreflectors (LRRR): 33, 34, 39, 40, 289
     Passive Seismic Experiment (PSE): 33, 34, 39, 40, *296*
     Radioisotope Thermoelectric Generator (RTG): 40, 311
     Solar Wind Spectrometer (SWS): *295*
     Suprathermal Ion Detector Experiment (SIDE): 34, 40, 233–234, *254*, 273
     Surface Electric Properties (SEP): 109
   Scientific Instrument Module (SIM): 43, 194
   Service Module (SM): 36, 37
Apollo Applications Program (AAP): 78–79
Apollo 13, film: *57*, 72
Apollo-Soyuz Test Project (*ASTP*, "Apollo 18"): *57*, *119*, *155*, 356
Archytas of Tarentum: 59
Ares rocket family: see Constellation
Ariane 5: 97
Aristarchus of Samos: 10–11, *16*
Aristarchus Plateau: 182, 183, 184, 185, *206*, 219, 275, 284–285, *296*, *389*
armalcolite: 177, *205*, 217
Armstrong Limit (after Harry George Armstrong): *114*
Armstrong, Neil: 18–24, 30, 33, 49, *52–53*, *54*, 298
   on *Apollo 11*: 18–24, 30, 33, 49, *52–53*, *54*, 104, 301, 305, 321
   as fighter and test pilot and on *Gemini 8*: **19–20**
asteroid
   asteroid mission proposal and Obama administration: vii, 95, 97, **100–101**, *123*
   Asteroid Return Mission (ARM) proposal: 101, 102
asteroids
   1620 Geographos: 130
   4179 Toutatis: 132
   25143 Itokawa: 43
   atmosphere: 272
   commercial mining operations proposals: 149
     Deep Space Industries: 375
     Planetary Resources, Inc.: 375
   human habitability: 402, 403, *426*
   Plymouth Rock mission proposal: 100, 101, *123*
   relative interest of NASA vs. other space agencies: 141
   Yarkovsky and YORP effects: 196
astronaut science training: 35, 36, 261–262, 292, *293*
Atlas rocket series: 24, 27, 63, 64, 87, 96
   Atlas-V: 95, 97

Index

Atomic Energy Commission, U.S.: 63
Augustine, Norman: 83, 94
    Committee on Future of U.S. Space Program (1990): 83, 94
    Review of U.S. Human Space Flight Plans (2009): 94, 95, 96, *122*

Baikonur Cosmodrome: *57*, 65, *117*, 136
Baldwin, Ralph: *361*
basalt: see lunar basalt
Bay of Pigs invasion: 65, 66
Bean, Alan: 33, 35, 36, *208*, 310, 351
Belbruno, Ed: 152, 164–165, *172*
Bell Labs; ATT Bellcom: 146, 262
Bierce, Ambrose: 331
Big Whack: see Giant Impact
Boeing Corporation: 27, 96, 100, 127, 141
Bolden, Charles: 94, 95, 96, 97, *123*
bomber airplanes
    B-29: 19
    B-47: 63
    B-52: 63
Bondarenko, Valentin: *57*
Bonestell, Chesley: 61, 62, 66
Borman, Frank: 29, 31, *53*
Bradbury, Ray: *115*
Brazilian Space Agency (Agência Espacial Brasiliera, AEB): 134, 143
breccia: 174, 176, 197, **198**, 202
Brezhnev, Leonid: 50
British Interplanetary Society: 62
Bull, Gerald: 396
Bumper rocket: 60, *114*
Buran space shuttle: 51
Burroughs, Edgar Rice: 60
Bush, George H.W.: 82, 83, 104
Bush, George W.: 84, 86, 96–97, *122*
Bush, Vannevar: 60, *115*
Bykovsky, Valeri: *56*
By Rocket into Planetary Space, book: 59

Cameron, Alistair: 165, *172*
Cameron, Winifred Sawtell: 275, *295–296*
Canadian Space Agency: 134, 143
canali: 60
Cassini, Giovanni: *17*
Cassini-Huygens mission to Saturn: 134, 143, 237, 245

catena: see crater chain
caverns, lunar: see skylights
Centaur rocket: 27, 132, 370
Central Intelligence Agency (CIA), U.S.: 25, 48, 65
Cernan, Eugene: 29, 32, 41, 45, 46, 47, 231, 261, 263, 270, 271, 298, 300, 301, 302, 303, *323–326*, 327, 333, *362*
    closing words upon departing lunar surface: 46, 47
Chaffee, Roger: 28
Chandrayaan-1 mission: 132, 235, 237–238, 244, *255*
Chawla, Kalpana: *154*
Chelomei, Vladimir: 48, 49, 50
Chinese espionage against U.S.: 127
Chinese Lunar Exploration Program (CLEP): 145
    Chang'e program
        Chang'e 1: 131–132, *156*
        Chang'e 2: 132, *152*
        Chang'e 3: 134, 145, 147
            Yutu rover: 145
            ground-penetrating radar; 145
        Chang'e 4: 144
        Chang'e 5: 144
        lunar sample return mission proposals: 144
    future lunar plans: 145
    military involvement: 144
Chinese National Space Agency (CNSA): 134, 139–140
    absence from ISECG: 141
    commercial component: 140, *154*
    cooperation with RKA: 139–140
    future as a leader in lunar exploration: 150–151
    isolation from NASA: 140, 143–144, *154*
    military component: 140, *154*
    need for Long March 5 rocket: 145
    taikonauts: 140, *154*
Clarke, Arthur C.: 62, *115*, *116*, *156*, *224*, 322, *329*, 394, *422*, *424*, *431*
Clementine, Deep Space Program Science Experiment (*DPSE*): 130–131, 181, 222, 234, *255*, 270, *294*
Clinton, William: 84, 104, 127
Cold War: 50, 51, 60, 63, 65, 83, 127
Collier's magazine: 61, 62, 63, 106, *116*
Collins, Michael: 33, *53*, 69, 108, *118*, *119*

497

Index

Comet
  Shoemaker-Levi 9: 191
  Tempel 1: 237
  Wild 2: 43
Command Module (CM): see Apollo CM
Command/Service Module (CSM): see Apollo CSM
commercialization of space: *116*, *124*, *146–149*, *368–370*, *372–376*, *378–385*, *386–387*, *388–390*
  MirCorp program for *Mir*: 375
  Moon: see lunar commercial development
Compton-Belkovich Thorium Anomaly: 219, *225*, 288
Congressional Budget Office, U.S.: 85
Conquest of Space, The, film: 63
Conquest of Space, The, books: 61, *116*
Conquest of the Moon, book: 61
Conrad, Charles "Pete": 22, 33, 35, 36, 351
Constellation space vehicle architecture: **87–90**
  Altair/Lunar Surface Access Module (LSAM): 88, 89
  Ares I rocket: 87–88, 89, 95, 97
    Ares I-X: 89
    Solid Rocket Booster (SRB): 88
  Ares V rocket: 87–88, 89, 95, 97
    Obama campaign 2007 cancellation proposal: 89
  budget: 88, 89, 94, 435
  cancellation: 96, *123*
  Earth orbit rendezvous: *122*
  Launch Escape System: 88, 90
  Multi-Purpose Crew Vehicle (MPCV)/Orion/Crew Exploration Vehicle (CEV): 89, 90, 97
  Obama Administration support and cancellation: 89–90, 94–97, 433, *455*
  technical development and challenges: 87–89
Cook expedition of 1769: *17*
Cooper, Gordon: *125*
cosmic radiation: 190, 236, 243, 289, 297, 341
Cosmic Voyage: Fantasy novella, film: 114
cosmism: *156*
crater
  central peak: 190
  crater chain: 214
    catena Davy: 191
    secondary impact chain: 192

crater count and age dating: **194–196**, *207–208*, 288–289
crater/impactor size relation: 196
cratering process: 26, **184–190**, 200, *206*, *207*
craters, named
  'Alo'i (Earth): 18
  Alphonsus: 184, 229
  Apollo: 212
  Aristarchus: 26, 180, 182, 185, 189, *206*, 264, 275, 278, 279, *296*
    age: 188, *206*, *296*
  Arzachel: *250*
  Bessel: 189
  bowl crater, anonymous: 189
  Byrgius A: *206*
  Cabeus: 242
  Camelot: 298
  Chavenet: 188
  Cone: 38, 311, 321, 322, *326–327*
  Copernicus: 27, 182, 189, 192, *206*, 213, 278
    age: 188, 192
  Craters of Moon, anonymous crater (Earth): 186
  Eratosthenes: 189, 192
  Euler: 189
  Fauth: 227–228
  Flamsteed: 27, 180
  Franz: 194
  Giordano Bruno: *54*, *206*, *254*
  Grimaldi: 264, 278, 279
  Hertzsprung: 212, 218
  Hörbiger (defunct): 227
  Ina feature: 282, *296*
  Jackson: *206*
  Kepler: *206*, 213, 278, 279
  Korolev: 212, 218
  Kreiger: 187
  Lansberg: 213
  Le Monnier: 319, *329*
  Lichtenberg: 180
  Lovelace: *210*
  Margaret Mead (Venus): 163
  McKellar: *206*
  Messier, Messier A: 191
  Ohm: *206*
  Plato: 278
  Posidonius: 187, 230
  Prinz: 185

498

Proclus: *206*
Ptolemaeus: 229
Pytheas: 182
Schiller: 191
Schrödinger: 189
Shackleton: 238
Sharanov: 212, 218
Shiva crater candidate (Earth): *172*
Shorty: 41
Spörer: 187
Sulpicius Gallus: 184
Tsiolkovskiy: 212
Tycho: 27, 41–42, 190, 213
Verdefort (Earth): 163
Wargentin: 187, *251*
   craters as TLP sites: 276–278
      correspondence to radon-222 outgassing sites: 279
   impact melt: 197
   micrometeoritic: 189, 193, 195
   morphological size progression: 189
   rays, and dating of features: 188–190, 192, *206*
Criswell, David: 270, 272, *294*, 381, *389*
Cronkite, Walter: *56*
Cuban/Caribbean Missile Crisis: 65
CubeSat program: 147
Cunningham, Walter: 29

D558–2 Skyrocket plane: 19
Darwin, Charles: 159–160
Darwin, George Howard: 159–160, *171*
Davis, Donald: 165, *172*
Dawn mission: 397
day lengthening (on Earth) due to tides: see Earth
de Bergerac, Cyrano: 58, *113*
de Morgan, Augustus: *14*
Deep Impact mission: 237, 245
Deep Space 1 mission: 128, 397
(Defense) Advanced Research Projects Agency – (D)ARPA: 65, 66
   ARPANET (.arpa): 65
Defense Reorganization Act (1958): 65
Delta rocket series: 87, 96, 370
   Delta-IV Heavy: 89, 95, 97
Deng Xiaoping: 150–151
Descartes, René: 159, *171*
Dick Tracy cartoon's Moon Valley theme: 63
Digital Acquisition Camera (DAC): 20, 22, 24
DIRECT rocket design family: 90
Disney, Walt: 61, **62–63**, 106, 211
Dragon ISS resupply spacecraft: 141, 146, 375
Drake, Frank: 417
   Drake Equation: 417, *428*, *429*
Duke, Charles: 36, 40, 44, 333
Dunthorne, Richard: 159
dust: see regolith

Earth: 2–9, 11, 13, 15–17
   age: 158, 409–410
   as seen from Moon: 110, 354–355
   atmospheric density: 272
   comparison to Mars and Venus: 8–9, 13, *16*
   composition: 176
   crust: 8–9, *16*, 167, 174–175, 185
      oldest: 193, *207*
   distance to Sun: 11, *17*
   early atmosphere: 8, *16*
   electric atmospheric phenomena: 262
   enhanced nickel-iron from Giant Impact: 167, 169
   evolution of early life: 408–409, *428*
   largest impact basin (Verdefort crater): 162
   magnetic field inconstancy: 168–169
      South Atlantic Anomaly (SAA): 168–169, *173*
   material on Moon: 204, *209*, **357–359**
   meteorites: 262, *293*
   month and Earth day lengthening due to tides: **4–7**, *15*, **158–160**, *171*, *172*
   North Pole exploration: 51, 57
   Pacific Ocean births Moon, in theory: 160
   plate tectonics' origin: 167–168, *171*
   plutons: 174, 175, 205
   quartz: 177–178
   South Pole exploration: 339–340, 349, 432
   tides from Moon: **2–7**, *15–16*
Earth orbit rendezvous (EOR): 122
economic development in 21st century and space leadership
   China: 437–442, *456–458*
   Japan: 447–448, *459*
   United States: 437–439, 441, 442–445, *458*
   world: 445–447, 448, 451–452. *458–459*
economic spin-offs and benefits: 50, 104, 435–436, *456*

educational development
    21st century education and space program(s): 455, 438–439, 441–442, 444–445, *457–458*
    space mission real-time education: 450–451, *459–460*
    National Defense Education Act (1958): 65, 66, 438
Ehricke, Krafft A.: **1**, *13*
Eisele, Donn: 29
Eisenhower, Dwight: 64, 65, 66, *117*
El-Baz, Farouk: *54*, 262–263, *293*
Enceladus: 406–407, 412, *430*
Energia rocket: 51
Ericsson, Leif: viii, 146
Europa: *17*, 402, 403–404, 405, 407, 410, 412, *430*
European Space Agency (ESA): 130, 134
    cooperation with
        CNSA: 135
        NASA: 135
        RKA: 135
    individual member nation space agencies: 134
    launch site: 135
    postponed lunar missions: 135
europium anomaly: **179–180**
Evans, Ronald: 41, 264, *325–326*
"Eve of Destruction," song: 72
Exomars 2016 mission proposal: 143
exploration, in general: 407–408
Explorer program
    Explorer 1: 64, 106
    Explorer 35: 254
    Explorer 49: 254
Exploring Expedition, U.S.: 156
extrasolar planets (exoplanets): 63, 410, 416, 421–422, *428*, *430*
extraterrestrial life: 63, *366*, 402, 404–405, 406–407, 408, 419–421, *429*
    lunar: 357–359
    superaliens and *hypernoesiology*: 419–421, *429–430*
extravehicular activity (EVA): 31, 33–35, 39, 41, 55, *56*, 142, 145, 198, **298–312**

F18 Super Hornet aircraft: 82
F9F-2 fighter plane: 19
Fairchild Semiconductor: 105

Falcon rocket series
    Falcon 1: 96
    Falcon 9: 96, 141, 148
    Falcon Heavy: 96, 97, *122*, 375
Fallaci, Oriana: 22
Fauth, Philip: 227, *251*
Federal Bureau of Investigation (FBI), U.S.: *56*, 104
feldspar: see anorthosite
Fermi, Enrico: 410
    Fermi Paradox: 410–421
Firdausī: 58, *113*
fire fountain and fire fountain glass: 40, 41, 238, 245, *256*, *257*, *259*, 261–262, **263**, 287
First Lunar Outpost (FLO): 83
First Men in the Moon, The, novel: 59, 332
First on the Moon, film: *57*
Fischer, Bobby: 50
Fisher, Osmond: 160, *171*
Flexible Path (FP): 95, 101, 102, *124*
Fobos-Grunt mission: 135, 139, 149, *153*
Forestall, James: 60
free return trajectory: 37
From the Earth to the Moon, film: 72
frontier as a concept in different societies: 151

Gagarin, Yuri: 19, 23, 44, 51, *57*, 63, 66, *117*
Galileo Galilei: 11–12, 159, 303
Galileo mission: 237
Galimov, Erik: 141–142
Garver, Lori: 94
Gateway Exploration Architecture: 101, 102
Gemini program: 28, 97, 298, *327*
    Gemini 4: 300
    Gemini 8: 19, 356
    Gemini 9: 311
    Gemini 10: 69
    Gemini 12: 300
Gemini-2/Manned Orbiting Laboratory: *75*, *119*
Genesis mission: 43
Genesis Rock: **175–176**, 261–262
geocentric model of Solar System: 12
Geosynchronous Satellite Launch Vehicle: 138–139
    planned ISRO Orbital Vehicle: 138
Gestapo: 61, 106

Index

Giant Impact: **7–8**, 160, 161, **162–170**, 288
  angular momentum of Earth-Moon and planets: 163
    transfer from Earth-Moon to Sun-Earth: 170, *173*
  chemical composition: 164, 165
  giant impacts on inner planets: 162
  impact modeling: 166–167, *172, 173*
    non-standard impacts: 170, *172, 173*
  implications for Earth: 168–169
  isotopic composition: 164, 169
    chromium: 164
    oxygen: 164, *173*
    magnesium: 164, *172, 173*
    silicon: 169, *173*
    titanium: 169
    tungsten: 169, *173*
    zinc: *173*
  mixing in gas/liquid accretion disk: 169, *173*
  pre-Impact Earth satellite hypothesis: 169
  proposed Impact debris sample mission: 170
  Theia: 163, *171*
    accretion at L4/L5: 164–165
Giffords, Gabrielle: 96
Gilbert, Grove Karl: 160
Gingrich, Newt: 100, 407
Glenn, John: 66, 68
Goddard, Robert H.: 58, **59**, 60, 106, 107, 112, *113, 114*
Godwin, Francis: 58
Gold, Thomas: *54, 251, 252, 272, 294*
Goodrich, J.T.: *156*
Google Lunar X Prize (GLXP): 147–148, 375, 383, *459*
Gorbachev, Mikhail: 51
Gordon, Bart: 96
Gordon, Richard: 33
Gott, J. Richard, III: 164–165, *172*
Gravity Recovery and Climate Experiment (GRACE) mission: 133, *152–153*
Gravity Recovery and Interior Laboratory (GRAIL) mission: 133, *153*, 220, 288, 289, *459*
Green, Charles: *17*
Griffin, Michael D.: 87, 94, *121*
Grissom, Virgil "Gus": 28, *125*, 365
Gruithuisen, Franz von Paula: 331, 332, *361*
Grumman Aircraft Engineering Corp.: 29

H missions: see Apollo H missions
HII-B rocket: 97
Haber, Heinz: 61
habitation of Moon by humans: see lunar bases and habitats
Haise, Fred: 35, 37, *55*
Hale, Edward: 58
Halley, Edmond: 159
Hansen, Peter Andreas: 211–212, *223–224*, 360
Hartmann, William: 165, *171, 172*
Hartsfield, Henry: 174
Hauri, Erik: 238, 239, *256, 257*
Hayabusa mission: 43, 128, 397
Hayden Planetarium: 61
helium-3: 142, 201
Helmholtz, Hermann von: 159
Henry the Navigator, Prince: 150
Hero of Alexandria: 59
Herrmann, Rudolph Albert: *56*
highlands: see lunar highlands
Himmler, Heinrich: 107, 227
Hiten/Hagoromo mission: 128, *152*
Hitler, Adolf: 114
Holdren, John: 92, 93, 94, 97, 143, *156*
Hörbiger, Hans: 227
Hornsby, Thomas: *17*
Hubble Space Telescope: 85, 134, 143, 291
Humphrey, Hubert: 104
Huxley, Thomas Henry: 262, *293, 430*
hydrogen bomb
  AN602 Tsar Bomba: *118*
  Castle Bravo: *118*
  explosion on Moon, proposal: 55
  first atomic and H-bomb by country: 126
  W70 and W88 warheads: 127

I Aim at the Stars, film: 63, *120*
Icaro-Menippus, story: 58
Icelandic Sagas: viii, 432, 454
igneous
  magma: see lunar magma
  processes: 8, *16*, 167, **174–184**, 185, 186, 187, 217–219, 225, 228, 238, 239–240, 242, 245, *251–252, 257, 259*
  rocks and minerals: see individual entries
ilmenite: 178, 197, 200–201, *208*, 217, 218
Incas: 1, *14*

Indian Space Research Organization (ISRO): **134–139**
   budget: 138
   cooperation with NASA: 138
   cooperation with RKA: 137–138
   public opinion on Indian space program: 138–139
inspiration of youth: 3, 50, 432, 435, 438, 439–440, 441–454
Integrated circuit development accelerated for AGC: 105, 435
Intelsat: 96
intercontinental ballistic missile (ICBM): 50, 52, 60, **63–64**, 65, 112, *115*, *118*
International Geophysical Year (IGY): 64, 140–141, 339
International Lunar Decade proposal: 149
International Lunar Network proposal: 288
international scientific cooperation
   historically: 140–141
   International Space Exploration Coordination Group (ISECG): 141, *155*
International Space Station: 11, *56*, 62, 78, 84, 95, 96, 103, *121*, *124*, **129**, 141, 137, *154*, 333
   Canadarm2: 142
   Commercial Orbital Transportation Services: 146
     Dragon capsule of SpaceX: 141, 148, 373, 375
     Liberty capsule of ATK/Lockheed Martin: 146, 375
   commercial experiments: 149
   Cupola module: 142, 352, 452
   Expeditions
     4: 352
     35: 352
     36: 352
   HTV and H-II Transfer Vehicle: 128
   Kibo module: 128
   member nations: 140, *154*
International Traffic on Arms Regulation (ITAR): 127, 145, *152*
Internet: 65, 66
Interplanetary Kitecraft Accelerated by Radiation Of the Sun (IKAROS): 128, 396
interstellar travel: 63, *116*

Io: *17*
Iran: 143, 149, *156*
Irwin, Jim: 39, 41, 42, 43, 301, 351
   dehydration & heart problems: 42, **312**, *327*, 351
isostasy: 186, 216
Israeli space program: *155*
Isvestia newspaper: 51

J missions: see Apollo J missions
James Webb Space Telescope: 101, 102, 289
Japan
   Edo period poem: *250*
   Japanese Aerospace Exploration Agency (JAXA): 128, 134
   launch sites: 128, *152*
jet airliners: *116*
Jet Propulsion Laboratory: 25, 27, 60, 126
Johnson, Lyndon B.: 65, 67, 69, 104
Joint Strike Fighter: 87
Jupiter, planet: 191, 196, 404, *460*
   Jupiter's Galilean moons: 11–12, *17*, *172*, **402–404**, *426*, *427*
Jupiter rocket: 64, 67, 106

KC 135 "Vomit Comet": *54*
Kaguya/SELENE mission: 109, **130**, 200, 248, 274, 279, 282, *295*
Kant, Immanuel: 159
Kaplan, Joseph: 61
Kármán Line: *114*, 147
Keldysh, Mstislav: 44, *55*, *155*
Kennan, George F.: 103
Kennedy, John F.: 28, 49, **66–69**, 81, 97
   Moon speech to Congress: 67, 97, *118*
   Moon speech at Rice University: 68, 97
   speech to U.N. General Assembly: *118*
Kepler, Johannes: *17*, 58, *113*, 159
Kepler Space Telescope mission: *430*
Khrushchev, Nikita: 49, 50, 63, **64**, *115*, *116*, *117*, *155*
Klushantsev, Pavel: 63
Kodak Corporation: 27
Komitet gosudarstvennoĭ bezopasnosti (KGB, Committee of State Security, U.S.S.R): 30, 48, *56*
   Apollo sabotage letters sent to KSC: *56*, 111
Komorov, Vladimir: 28, 50

Korean Aerospace Research Institute (KARI): 132, 140, 141, *155*
Korean War: 60, 62
Korolev, Sergey: 28, 48, 49, 50, 60, 63, 107
KREEP: 34–35, **176**, 170, 219
    urKREEP: 179
Kuiper, Gerard: 26
Kursk submarine: 370

lacus: 214
    Somniorum: 230–231
Lagrangian points: 95, 101, 102–103, 219, *328–329*, *387*, 396, 398
Lang, Fritz: 59, *114*
Laplace, Pierre-Simon, Marquis de: 159
Laser Interferometer Space Antenna (LISA) proposal: 143
Late Heavy Bombardment (LHB): 193, 194, 196, *207*, *208*, 221, 288–289, 409
Latham, Gary V.: 55
Laux, Dorianne: 158, *170–171*
Lehrer, Tom: 126, *152*
Leonov, Aleksei: *56–57*
Lewis and Clark expedition: 150, 151, *156*
Ley, Willy: 59, 61, 63, *116*
Life magazine: 66, 70, *118*
Lincoln, Abraham: *117*
Lindbergh, Charles: 146
Locke, Richard Adams and *The Sun* lunar hoax: 331, 332, *361*
Lockheed Corporation: 27
Lockheed Martin Space Corporation: 96, 100
Long March rocket series: 127, *156*
    Long March 3B/Intelsat 708 disaster: 127, 145, *152*, *156*
Longitude Prize: 146
Look magazine: 66
Lovell, Jim: 29, 31, 35, 37
Lucian of Samosata: 58, *113*, 311
Luna program: 23, 24–25, **26**, 28, 33, 37, 55
    Luna 1: 23
    Luna 2: 23, *55*
    Luna 3: 23, **24–25**, 26, 211, 220
    Luna 4–8: 26
    Luna 9–10: 23, 26
    Luna 11–12: 26, *254*
    Luna 13: 26
    Luna 14: 26, *254*
    Luna 15: 33, *254*
    Luna 16: 37, 42, 43
    Luna 17: 37, 319
    Luna 20: 42, 43, 176
    Luna 21: 42, *254*, 319
    Luna 23: *254*
    Luna 24: 43, 82, 131, 200, **234**, 236, 248, *254*, 261
    Luna 25 (cancelled): 43
    Luna 25 (planned): 44
    Luna missions of 1980s (cancelled): 44
    Lunokhod program: see rovers and hoppers
    trajectory for *Luna 16, 20* & *24* sample returns: 43
Luna-Resurs: 44, 131, 137–138, 143
lunar
    activity: 298–*330*
        carrying: 301
        cooperating: 323, *329–330*
        exercising: 350
        falling: *324*
        fastening: 322–323
        jumping: 301
        sleeping: 323, 350–351
        throwing: 302, 303
        walking: **298–300**, 302
    age dating: 15, 38, 41, **191–192**, 193, 252
    atmospheric composition: 38, 134, 252, **273**, *294*, *295*
        argon: 231, 242, 273–274, *295*
        detectors: 38, 270, 272
    atmospheric density: 269–270, 272, *294*, *295*
        solid bounded exosphere: 272–273
    basalt: 34, 197, 202
        vesicular basalt: 231–232, 262
    carbon: 133
    dust: see regolith
    eclipse: 10–11, 12
    gamma-ray spectrometry: *206*, *390*
    geological timeline: **192**
    GRAIL results potential: 170, 216
    gravitational field irregularity: 28, 34
    heat flow: 39
        experiment: 39
    highlands: 40, 176, 188
        Central Highlands (Near Side): 220
    impact basins: *171*, *175*, *192*, *208*
    ionized atmospheric detector: 35

lunar (cont.)
  iron: 41, 181
  laser ranging: 33, 34, 38, 39, 289
  magma: **178–184**
    europium anomaly: 179, 180
    high-aluminum: 179
    mafic versus felsic: 178
    magnesium-suite: 178–179
    titanium concentration: 178, 179–180
  magnetic field: 35, 222
    detection: 35
  mantle overturn: 178, *205–206*, 217–218
  meteorites: *55, 172*, 202, 262, 341
  North/South Pole: 221, 235, 236, 237, 238, 240–242, 244, 248, 249, *255, 256*, 379
    permanently shadowed/illuminated region: 221, 240, 241–243, *256, 257*
  optical and infrared spectrometry: **201–202**
    Christiansen infrared feature: 201
    Mg#: 202
    OMAT: see regolith (optical maturity and aging)
  phases: 1, 2, 12, *14–15*, 354
  potassium-40: 179
  power: see power production, lunar
  radar: 13, 132, 145, **200**, *208*, **234–235**, 248, *254–255*, 286, 297
    circular polarization ratio (CPR): 235, *255*
  samples: 33, 34, 38, 39, 40, 42, 43, 48, *55*, 175, 176, 190, **198**, 202
    contamination: 232, *253*
    core samples: 199–200
  seismography: 33, 34, 35, 38, 220, *226*, 267, 279–280, 288, *296*, 341
  solar wind at Moon: 35, 190, 222, 272, *294*, 341
  temperature range: 307–308, 311, 340
  thermal emission: 200
  thorium: 177, 179
  tides: **2–7**, 217
  transient phenomena: see transient lunar phenomena (TLPs)
  uranium: 177, 179
  volcanic eruptions: 178–184, 192, *206*, 218, 225
    turbulent flow: *251–252*
  water: see water, hydration and volatiles, lunar
  X-ray spectrometry: 37

Lunar Atmosphere and Dust Environmental Explorer (*LADEE*): 133–134, 248, 271
Lunar Lasercom Space Terminal (LLST): 134, *153*, 314
lunar bases and habitats: 63
  Bigelow Aerospace inflatable habitats: 149, 343, 344–345, *363–364*, 375
  Conquest of the Moon, base: 61
  Constellation concepts: 339, 342, 343
  consumables: 346–348
  First Lunar Outpost (FLO): 83
  human population size and bifurcation: 350, 355, 360, **384–386**, *391*
  lava tubes: 348–349, 353, 357, *364*
  long-term human society: 350–355, 357, *365*
  Lunar Resource Utilization (LUNOX): 83, 112, *121*, 343
  missile base: *55, 117, 118, 119*, 149
  regolith
    cement and concrete use: 202, *208, 209*, 345, *363, 364*
    shielding: 317–318, 339–345, *363*
    transport: 343–344, *363*
  Soviet Moon base design: 44, 344, *363*
lunar commercial development: *124*, 368–370, 372–376, 378–385, *386–387*, *388–390*
  Bigelow Aerospace habitats: 149
  business scenarios: 372–375
  Catalyst program: 148
  cislunar transportation costs: 368–369
  cislunar tug: 372–374, *387–388*
  commodity prices: 368–369
  GLXP competitors: 147–148
  Golden Spike corporation: 149, 375
  helium-3 production: 142, 201, 368, 378, 380–381, *389*
  Moon Agreement of 1979: 357, 383–384, *391*
  Outer Space Treaty (OST): 144, 357, 360, *365–366, 367*, 372, 382–383, *390–391*
  oxygen production: 201, 343, 344, 347, 349, 379, *389*, 423
  RKA lunar space tourism offer: 146
  Shackleton Energy Corporation: 148–149, 375
  Spudis-Lavoie proposal: 375, *388*
  thorium reactor fuel: 382, *389, 390*, 423
  Tsiolkovski 1895 gold mine concept: *387*

Index

water
  as propellant: 369, 370, *387*, 398
  related lunar propellants: 369–370, 371–372, *387*
Lunar Crater Observation and Sensing Satellite (LCROSS): 131, **132–133**, **240–241**, 242, *256–257*, 286, 372
Lunar Electric Rover (LER): 92
lunar in-situ resource utilization (ISRU): 83
  Cargo Transportation and Landing by Soft Touchdown (Catalyst): 148
  cement and concrete: 202, *208*, *209*, 345, *363*, *364*
  Shackleton Energy Company: 148–149
Lunar Jim: 3
Lunar Landing Research Vehicle (LLRV), Lunar Landing Training Vehicle, "Flying Bedstead": 19
Lunar Module (LM): see Apollo Lunar Module
lunar orbit rendezvous (LOR): 69, 103, *122*
Lunar Orbiter program: 23, **26–28**, 131
  Lunar Orbiter 1–2: 23, *254*
  Lunar Orbiter 3: 23, 28
  Lunar Orbiter 4: 23, **27**, 28, *252*, 332
  Lunar Orbiter 5: 23, **27**, 28, *252*, 283
Lunar Prospector: 131, 181, 236, 274, 279, *295*
Lunar Reconnaissance Orbiter (LRO): 11, 85, 132, 142, *256*
  Diviner: 287
  Lunar Orbiter Laser Altimeter (LOLA): 132, 289
  LRO Camera (LROC): 11, 109, 194, 197, 223, 231, *254*, 282, 320, *430*
  Lyman Alpha Mapping Project (LAMP): *256*, 273, *295*
Lunar Roving Vehicle (LRV): see rovers and hoppers
Lunar Sourcebook (*LSB*): **54**
Lunniy-1 (L1 or 7K-L1): 49
Lunniy Korabl (LK): 45, 47, 49, 304
Lunniy Korabl-1 (LK-1): 49
Lunniy Orbitalny Korabl (LOK, Moon Orbital Ship, or Soyuz 7K-LOK): 49
Lunokhod program: see rovers and hoppers

M31 galaxy in Andromeda: 2
magma ocean: 40, **174–179**, *205*, *207*, 288
Makarov, Oleg: 56

Man and the Moon, film: 62, 211
Man in Space, film: 62
Man in the Moon, pareidolic image: 1, *14*
Man in the Moone, The, novel: 58
"Man Will Conquer Space Soon", magazine articles: **61–63**, 78, *115*
Manhattan Project: 60
Mare: 25–27, 32, 34, **182–184**, 192, 195, 214
  Cognitum: 26
  Crisium: 43, 212, 213, 217, 218, 278
  Fecunditatis: 37, 43, 220
  Frigoris: 220
  Humorum: 184, 217
  Imbrium: 37, 145, 162, 176, 182, 186, 192, 193, 194, 204, *206*, 212, 213, 217, 283, 319
  Ingenii: 212
  Marginus: 25, 212
  Moscoviense: 25, 212
  Nectaris: 192, 217
  Nubium: 212, 213
  Orientale: 189, 196, *207*, 212, 217, *225*, 245, 264
  Serenitatis: 41–42, 193, 194, *208*, 212, 217, 319
  Smythii: 25, 212, 218
  Tranquilitatis: 26, 27, 32, 204, 212, **224**
  Vaporum: 184
Mariner program: 23
  Mariner 2: 25, 68
  Mariner 4: *115*
Marooned, film, and influence on *ASTP*: *119*, *155*
Mars: 60, 110
  atmosphere: 62, **115**, 272
  diameter: 162
  humans on Mars: 61, 62, 149, 353, *364*, *365*, 404–407
  impact basins: 162
  Martian life: 60, *115*, 408–409, *428*
  Martian moons as exploration bases: 404–405, 406, *427*
  robotic missions
    AAP robotic lander proposal: 78–79
    ESA 2018 lander proposal: 143, *156*
    Viking program: 79
  rotation: *171*
  temperature range: 308
Mars and Beyond, film: 62
Mars Express/Beagle 2 mission: 134, 149, *156*
Mars Project, The (*Das Marsprojekt*), book: 61, 106, *116*

505

Index

Mars Science Laboratory (*Curiosity*) rover mission: 143
Mattingly, Thomas "Ken": 40, 174, 263, 264, *293*
McCubbin, Francis: 239, *256*, *257*
McDivitt, James: 29
Medaris, Bruce: 106
media coverage: 45, 48
    American media coverage: 48, *52–53*, 56, 61, 63, 64, *115*, *119*
        American cinema: *57*, 60, *62–63*, *115*, *119*
    British media coverage: *56*
    cinema about Moon and space: 51, *57*, 59, 60, *62–63*, **72–74**
    music about Moon and space: 71, **72–74**, *120*, 352
    Russian media coverage: 45, 48, 51, 64, 65, 66
        Russian cinema: 51, *57*, 59, 63, *114*
    Western media coverage: 22, 48, *119*
Méliès, George: 59, *114*, 332
Melnick, Bruce: 352
Mercury, planet
    angular momentum: 163
    atmosphere: 272
    Caloris basin: 162, 219
    human habitability: 401, 403, *426*
    ice at poles: 235, *255*, *256*
    MESSENGER mission: *255*
Mercury program: 28
    Mercury 6 and *9*: 356
    Mercury-Redstone 3: 38
    Mercury-Redstone 4: 356
Method of Reaching Extreme Altitudes, A, book: 59
Middlehurst, Barbara M.: 275, *295*
Mikulski, Barbara: 105
Milky Way Galaxy: 2
Minuteman missile: 104–105
Mir space station: 51, 351, 352, 356, *365*
    STS missions to *Mir*: 84, 112, *121*
Miró, Joan: 3
missile gap: 64, 105
Mitchell, Edgar: 38, 39, 311, 321, *327*, *365*
Modularized Equipment Stowage System (MESA): 21, 23, 24, 35, 321, 322
Modularized Equipment Transporter (MET): 39
Montes: 214
    Agricola: 183
    Apennine: 39, 41
    Carpatus: 182
    Harbinger: 185
    "Leibnitz Mountains", unofficial: 221, *226*
    Secchi: 32
    Taurus: 41
    mons: 214
        East Massif: 45
        Marius Hills: 182, 219
        Mauna Loa (Earth): 182
        Mons Bradley: 186
        Mons Hadley: 186
        Mons Rümker: 219
        Mount Everest (Earth): 186
        "Mount Marilyn" (Secchi Theta): 32
        Olympus Mons (Mars): 182
        Sculptured Hills: 47, *207*
Moon
    angular momentum of Earth-Moon system: 163, *171*, *172*
    angular size: 1, 10, *223–224*
    Askaryan radiation: 289
    axial tilt, and lack of seasons: 163, 221, 235
    born from Pacific Ocean in theory: 160
    chemical composition: 164, 165
    contraction in radius: 133
    core: 164, 220, *226*
    crust: 174, 186, *207*, 212
    dark side: 211, 223
    differentiation: 177, *206*, 218–219
    escape velocity: 95
    Far Side: 24–25, *54–55*, 188, 200, **211–212**, 219
    formation scenarios: **158–162**
        co-formation planetesimal accretion: 160, 161, 165
        double planet: 160, 161, 163
        fission: 159–160, 161, 163, 165
        Giant Impact (see separate entry)
        plutonium breeder-induced fission: 160, *172*
        single capture: 160, 161
    gravitational field (physical geodesy): **215–217**
    mascon: 217, *225*
    gravity: 298, 301
        reduced friction: 299, 300–302, 338
    impact parameter: *224*

Index

isotopic composition: 164, 169
   deuterium: 232
   oxygen: 164, *173*, 232
   titanium: 169, *172*
Kozai mechanism: *172*
libration: *223*
Man in the Moon: 1, 212, *224*
mass: *17*, *172*
metastable atmosphere: 349
month and Earth day lengthening due to tides: see Earth
Moon as a name: 151–152
Near Side: 34, 38, *54–55*, **211**, 219–220
   Near/Far Side asymmetry: 212, 213, **214–220**, *224*
      Far Side giant collision: 219, *225*
      mantle overturn: 217–218, *225*
      Procellarum Impact Feature/Near Side megabasin: 219–220, *225*
offset in center of mass: 26, 212, *224*
orientation of orbit and rotation: 163, *172*, *223*, *224*
reflecting Earthshine: 291, *297*
selenography
   nomenclature: 213–214, *327*
   selenographic coordinates: *55*, 311
   tilt of coordinate systems: 163
shape disequilibrium: 222–223
shrinkage: 222, 223, 288
solidification: 177
unexplored continent, the Moon as an: 9, 392
   hazards experienced: 351, 356
   potential unknown conditions: 359–360, 392
Moon Agreement of 1979: 357, 383–384, *391*
Moon over Star, The, book: 92
Moore, Patrick: *54*, *295*
Musk, Elon: 107, 375

N1 rocket: 45, 47–49, 81
   L3 upper complex: 49
Napoleon: 159
National Academy of Sciences (U.S.): 64, 83, 104, 434
   National Research Council (NRC): 104, *122*
      2007 lunar exploration science goals: *153*

National Aeronautics and Space Administration (NASA) – *see also individual missions and projects*
   Ames Research Center: 170
   budget: 62, 69–70, 77, 78, 79, 83, 84, 85, 95, 103, 112, *118*, *119*, *121*, *122*, *124–125*, 134, 452–453, *460*
   China ban: 143–144
   Exploration Systems Mission Directorate: 87
   founding: 65, *459*
      National Aeronautics and Space Act (1958): 146
      encourage commercial space: 146
   future as a leader in lunar exploration: 150, 151, 437–439
   Glenn Research Center: 373, *389*
   Headquarters: 81
   international astronaut flights: *155*
   Johnson Space Center (formerly Manned Spacecraft Center): 69, 373
      Lunar Receiving Laboratory: 198
      Mission Operations Control Room: 312–313, *327*
   Kennedy Space Center (KSC, formerly Launch Operations Center): 29, 49, *56*, 68, 77, 84, 97, 99–100
      Launch Complex 34: 29
      Launch Complex 39: 29, 77, *122*
      Orbiter Processing Center: 82
      Vehicle (formerly Vertical Assembly Building (VAB)): 29, 68
   Langley Space Center: 27
   Marshall Space Flight Center: 64, 81, 87, 373
   NACA: *116*
   National Space Science Data Center: 264
   public relations: *122*
   relations with Congress: 65, 67, 70–71, 78–79, 80, 83, 85, 88, 95, 96, 97, 99, 103–104, *120*, *123*, 143, 344, *363*, 434, 451–452, *460*
   technology Challenges: 147, *156*, 307, *329*, 343–344, *363*
   Wallops Flight Facility: 133
   workers: 100, 111, *123*
National Air and Space Council (NASC): 104
National Air and Space Museum: 264
National Geographic magazine: 66
National Reconnaissance Office: 79, *120*

Index

National Space Council: 83, 104, 434–435
Nelson, Bill: 96, 105, *125*
Newcomb, Simon: 212, *224*, 311, **360–361**
Newton, Isaac: 12, 159
Nixon, Richard: 50, *52–53*, 66, 71, **74–75**, **77–78**, 81, 104
    address on New Federalism: 71

Obama, Barack
    interest in science and space: 91–93
    Occidental College/Columbia University student: 91–92
    science iconography: 92
Obama administration space policy
    announcement of new policy: 97, *122*
    campaign proposal of 2007 to cancel Ares V: 89, *121*
    future options: 102–103
    mission to asteroid: vii, 95, 97, **100–101**, *123*
    promise and failure to reestablish NSC: 104
    space and science policy decision process: 93–94
    support and cancellation of VSE: 89–90, 94–97, *122*, 433
Oberth, Hermann: 59, *114*, *424*
Oceanus Procellarum: 27, 182, 185, 200, 204, 214, 218, 219, 220
Oersted, Hans Christian: 261, *293*
Of a Fire on the Moon, book: 72, 108, *119*
Office of Science and Technology Policy (OSTP), U.S.: 95, 97
OKB (design bureau)
    OKB-1: 48, 49, 63
        RKK Energiya: 141
    OKB-52: 48
    OKB-586 (now Yuzhnoye Design Bureau): 141
O'Keefe, John: 229, *252*
O'Keefe, Sean: 84, 85, 86, 87
olivine: 2, 40, 176, **177**, 198, 201, ***205***, *206*
O'Neill, Gerard K.: 381, 396, 402, *426*
Orbital Sciences Antares/Cygnus spacecraft: 96, 375
Orteig Prize: 146
Outer Space Treaty (OST): 144, 357, 360, 365–366, 367, 372, 382–383, *390–391*

Paine, Thomas O.: 71, 79
Pal, George: 63
Paper Soldier, film: *57*
Paris Match magazine: 79, 106
Parker, Robert: 298, 302, *324*, *326*, 333–334
Partial Test Ban Treaty (Limited Test Ban Treaty, 1963): *55*
Pasteur, Louis: 261, 262, 292, *293*
Patrick, Nicholas: 142
Pegasus air-launched rocket: 396
phases of the Moon: 1, 12, 354
    human sleep and reproduction: 2, *14–15*
Phillips, Samuel C.: 107
Pickering, William H.: *294*, 331, *361*
Pioneer 1 mission: *54*
plagioclase: see anorthosite
planet formation via accretion: *172*
planetary protection: 359–360, *366*, *427*
Planetary Society space program proposals: 87, *121*, *123–124*
    attempted space missions: 149
Pluto: *172*, 272, 286, *430*
Polaris missile: 64
Politburo (U.S.S.R.): 44, 65
Polyakov, Valari: 353
Polyus (Skif-DM): 51, *57*
power production, lunar: 311, 322
    fuel cells: 311
    heat engines: 311
    nuclear power: 344
        fission reactor: 376–377
        radioisotope thermal generators: 40, 141, 311, 398
    solar photovoltaic electrical power: 344, 381–382, *384*, *388*, *389*, 423
Powers, Francis Gary: 65
Procellarum KREEP Terrane (PKT): 179–181, 219
Progress-M ISS/Mir resupply spacecraft: *121*, 356
Project A119: *55*
Project Diana lunar radar experiment: 13
Project Horizon: *117*, *118*, *119*
Proton rocket series: 48, 49, 135
    Proton-M: 97
public advocacy for space program(s): **105**, *107–108*, *125*, 449–451
public opinion of space program(s): 61, 63, **65–66**, 70, 105, 112, *116*, *117–118*, *119*, *121*, *124–125*, *449–450*, *460*
    opinion polls: 61, 105, 112, *115–116*, *117–118*, *119*, *121*, *124–125*, *460*

pyroclasts: 182, 184, 200
    dark mantle deposits: 184, 200, 201, *208*
pyroxene: 40, 176, 177, 197, **205**
    clinopyroxene: 198, 201
    orthopyroxene: 198, 201
pyroxferroite: 177, *205*
Pytheas of Massalia: 150

quartz and silicon dioxide: 177–178
Quayle, Dan: 83

R-7 rocket: 26, 63, 64
Radio Moscow: 64
Ramon, Ilan: *155*
RAND (Research ANd Development) Corporation: 63
Ranger program: 23, **24–26**
    Ranger 1–3: 25
    Ranger 4: 23, 25
    Ranger 5–6: 25
    Ranger 7: 25, 228, *251*
    Ranger 8: 25–26
    Ranger 9: 25, 228, 229
Reagan, Ronald: 83
Redstone rocket: 63, 67
regolith: **189–190**, *208*
    agglutinate: 197
    angle of repose: 321, 345–346
    bearing capacity: *54*
    composition: 176
    concrete aggregate use: 202
    density: *208*
    depth: 199–200
    dust: 20, 22–23, 26, 28, 44, 308
        abatement: 202, 304, 319, 320, 321, 322, *324*, 333–338, *362–363*
        elevated dust: 270–272, *294*
        electrical charging: 322
        heiligenschein, opposition effect/anomalous backscatter: 264, 308, **310–311**, 317, 319, 320, *329*
    impact gardening: 203, *209*
    impact glass: 197, 198
    megaregolith: 200
    odor: 333, *362–363*
    optical maturity and aging (OMAT): 190, 201, ***206–207***, 221, 281–283
    oxygen production: 83, **200–201**
    particle size distribution and particle packing: **198–200**, *208*
    radiation damage tracks: 202
    reflectivity: 310–311
    shielding uses: 317–318, 339–345, *363*
    surface rim: 190
    thickness: 200
Reiner Gamma: 214, 221–222
Return to the Moon: vii, viii, 83
    American public awareness: 85
    Constellation requirements: 88
    Obama Administration support and cancellation: 89–90, 94–97, 433
    program themes: 85–86
rhythmites: 5–6, *15*
Rich, Adrienne Cecile: *14–15*, 227, *250*
Richer, Jean: *17*
Rig Veda, hymnal: 158, *170*
Right Stuff, The
    book: 70, *125*
    film: *57*, 72
rimae: 39, 185, 214
    formation: 229–230
    Rima/Rimae
        Bode: 184
        Gruithuisen (unofficial): 204
        Hadley: 39, 41, 183, 184, 229–230
        Posidonius: 230
        Schröter's Valley (Vallis Schröteri): 181, *206*, 275, 285–286
        skylight in lava tubes: 203, 204, 247–248, 286, *287*, *296–297*, 348–349
Road to the Stars, film: 63
Robinson, Stephen K.: 142
robotics
    economic development: 436–437, *456*
    lunar exploration history 55
Roche, Edouard; Roche limit: 8, 160, 167
rocketry development: **59–60**, 67
    air launch, rockoons: 395–396, *425*
    altitude records: 59–60
    cannon launch: 396
    engines
        F1: 49, 67, 68
        J-2: 49, 68
        J2-S/J2-X: 87, 88
        "N-1" (NK-15): 48–49
        "Redstone": 67

rocketry development (cont.)
  RS-68B: 87
  Space Shuttle Main Engine (SSME, RS-25): 87
mass driver launch: 396
propulsion methods
  gunpowder rockets: 59, *114*
  ion propulsion: 397
  light sails: 396
  liquid propellants: 59
    liquid hydrogen + oxygen (LH2 + LOX): 87
  nuclear propulsion: 67, 79, 397, *425*
    VASIMR: 397
  solid rocket boosters (SRBs): 87, 88
  tether acceleration: 399
  "warp" drive concept: 397, *425*
rocket equation and velocity changes: 397, *425*, *427*
Soviet purge of rocket developers: 59
Soviet vs. U.S. draft of Aggregat rocketeers: 59–60, 106, *120*
Romney, Mitt: 100, *123*
Roosa, Stuart: 38
rovers and hoppers: 309
  ATHLETE: 309, **316**, 343
  cryogenic rover: 309, 376, 379, *387*
  hoppers: 249, **370–371**, *387*
  Lunar Electric Rover (LER): 92
  Lunar Roving Vehicle (LRV): 38, 41, 44, 263, 303, 318–319, 320–321, 334
  Lunokhod program: 23, 197, 313, **319–321**
    Lunokhod 1: 23, 37, 40, 289
    Lunokhod 2: 40, 42, 261, 270, 289, *294*, 309, 329
    Lunokhod 3 (cancelled): 44
  Mars Exploration Rovers: 309, 313, 317, 320, 329
  Mars Pathfinder Sojourner: 309
  proposed: *209*, *309*
  Yutu rover: 145
Rukavishnikov, Nikolai: *56*
Russell, Henry Norris: *172*
Russian Federal Space Agency (RKA): 103, 134, 141–143
  cooperation with
    ESA: 135, 143

ISRO: 137–138
NASA: 141–143
GLONASS: 135
launch sites: *124*, 134
mission failures: 135, *153*
planned and canceled programs (incl. lunar): 136–137, 143, *153*, *156*
space tourism program: 146
tracking network: 135
Russian Revolution, 50th anniversary of: 66
Ryan, Cornelius: *116*

Saal, Alberto: 238, 239, *256*, *257*
Sagan, Carl: *55*, 107, *119*, *427*, 432, *455*
Sagdeev, Roald: 44
Salyut program: 45, 51, *56*
  Salyut 1: *56*, 351, *365*
  Salyut 5–6: 351
  Salyut 7: 138
Sänger/Sänger-Bredt Silbervogel rocket proposal: *115*
satellite development: 60, 64, 67
  communications satellites: 68
Saturn rocket series: 28, 81, 103
  Saturn I rocket: 49, **67**, 68, *117*, *118*
  Saturn IB rocket: 29, 30, 49
  Saturn II rocket design: *117*
  Saturn V rocket: 28, 29, 30, 33, 48, 49, 62, 67, 69, 75, 88, *118*, *122*
Schmitt, Harrison "Jack": 36, 41, 45, 261, 262, 263, 264, 298, 300, 301, 302, 310, 311, *323–325*, *326*, *328*, 333, *362*
Schröter's Valley (Vallis Schröteri): 181, *206*, 275, 285–286
Schweickart, Russell "Rusty": 29
Scott, David: 19, 29, 39, 41, 42, 261–262, 300–301, 302, 303, 351
Scott, Robert Falcon: 150, 339, 432, *455*
Sea Launch (also Boeing/RKK Energiya – OKB-1): 141
Search for Extra-Terrestrial Artifacts (SETA): see space archeology
Search for Extra-Terrestrial Intelligence (SETI): 416–417, 422–423, *431*
Seneca, Lucius Annæus (the Younger): 392, *424*
Service Module (SM): see Apollo Service Module
Shakespeare, William: *14–15*

Sharma, Rakesh: 138
Shearer, Charles: 218–219, *225*
Shenzhou program: 127, **139–140**, **144–145**
   Shenzhou 7 mission and spacewalk: 145
   Tiangong 1 space lab: 144
Shepard, Alan: 38, 39, 66, 78, 311, 321, *327*
Shoemaker, Eugene: 26, *206*, *361*
Silkworm missile: 127
Sino-Soviet schism: 65, 127
Sinus: 27, 214
   Aestum: 184
   Medii: 27
Skylab program: 78, 103, *120*, 145, 342
   Skylab 4: 351
skylights: see rimae, skylights
Small Missions for Advanced Research in Technology (SMART-1): 130, 134, 397
SmallSat program: 147
Solander, Daniel: *17*
solar eclipse: 10–11
Somnium, novel: 58
South Korean robotic lunar space program: 132
South Pole-Aitken basin (SPA): 162–163, 200, *208*, 213, 218, 219, 221, 226, 245, 288
   impactor: 162–163, 219
   size: 162
Soviet Academy of Science: 64
Soviet Air Force: 66
Soviet Life magazine: *56*
Soviet space program, general: 103
   Soviet human lunar program: 26, 28, 30, 31, 33, 42, 43, **46–52**, *55*, *56*
Soviet Union: 18, 19, 24, 26, 28, 44, 45, 48, 51
   fall of: 44, 51
   Politburo: 50
Soyuz program: 28, 45, 49, 50, *56*, 64
   Soyuz 1: 28, 50
   Soyuz 5: 356
   Soyuz 11: *56*, 351
   Soyuz 18A: 356
   Soyuz 21: 351
   Soyuz 23: 356
   Soyuz 26: 351
   Soyuz 33: 356
   Soyuz T-10: 138
   Soyuz T-10A: 356
   Soyuz T-11: 138
   Soyuz TMA-5: 136

Soyuz rocket family: 135, 136
space archeology
   existing lunar sites: 158, *206*, *391*, *431*
   Apollo 17 artifacts: 109
   Search for Extra-Terrestrial Artifacts (SETA): 204–205, *210*, 422–423, *430–431*
space elevator concept: 394–395, 398–399, 423, *425–426*
space exploration, reasons for: 67, 68, **392–394**, *424*
Space Exploration Initiative (SEI): 82, *121*
Space Exploration Technologies Corp. (SpaceX): 96, 107, 146, 375 see Falcon rocket series; *Dragon* ISS resupply spacecraft
Space Launch System: 91, 97, 99
"Space Oddity," song: 72
Space Race: 50, 51, 52, 65
   current Moon race: 134, *154*
   Moon Race of 1959–1976: 48, 51, *56*, **66–70**, 105
   U.S.-U.S.S.R. space competition of 1980s: 51
space shuttle concepts and alternatives: 79, 80, *120*
   X37-B mini-shuttle program: 81
Space Shuttle/Space Transportation System (STS): 51, 82, 84, 88, 97, 103
   Atlantis: 99–100
   Challenger
      STS-51B: *154*
      STS-51–L disaster: 82, 87, 355
         "Rogers Commission": 82
         "Ride Report": 82, 83, *121*
   Columbia
      STS-1: 356
      STS-9: 356
      STS-41: 352
      STS-107 disaster: 84, 87, *154*
         Columbia Accident Investigation Board: 85
   commercial payloads: 149
   Discovery
      STS-37: 356
      STS-51: 356
      STS-114: 142
   Endeavor
      STS-49: 352
      STS-97: *154*

Space Shuttle/Space (cont.)
    STS-130: 142, *352*
    military payloads: 81, *120*
space shuttle-derived vehicle hardware: 78, 87, 88, 95, 99, 112
space station concepts: 83
    armored inflatable modules: 103, *363–364*
    Freedom: 83, 103
    'Man Will Conquer Space Soon' proposal: 61, 62, 103
    Manned Orbiting Laboratory: 75
space suit: **304–308**, **309–310**, *362*
    "backpack" (PLSS) water use: *254*, 312
    range and consumables limitations: 312, *364*
Space Task Group: 79
Spassky, Boris: 50
spinel: 198, 201
spiral development: 87
Sprites, Blue Jets and ELVES: 262, *429*
Sputnik program: 23, 25, 44, 51, 64
    Sputnik 1: 25, 51, 64, 211
        impact on American public: **65–66**, *117*, 435
    Sputnik 2: 66
Stafford, Thomas P.: 29, 32
Star Trek: 63, 89, 91, *107*, *123*, 409, 417–418, *429*
Steidle, Craig: 87
Strategic Defense Initiative/"Star Wars": 51, 83, *121*, 130, *155*
stratigraphy: 183, 186–190, 247–248, *260*
submarine launched ballistic missile (SLBM): 112, *119*, *125*
Sun: 1, 10, 243, *257*, *258*, 270
    solar activity hazard to astronauts: 101, *123*, 310, **341–342**, *427*
Superconducting Super Collider (SSC): 84
supersonic flight: 60, *115*
    supersonic transport (SST): 81
Surveyor program: 23, **26–27**, 28, 313
    Surveyor 1: 23, **27**, 270
    Surveyor 2: 23, 27
    Surveyor 3: 23, **27**, 34, 35, 36, *209*, 321
        bacterial contamination of camera: 35, *357*, *366*
    Surveyor 4: 23, 27
    Surveyor 5: 26, 27, 270
    Surveyor 6: 23, **27**, 270
    Surveyor 7: 23, **27**, 176, 270
Swigert, Leonard "Jack": 35, 37

Taylor, James: 211
Tekhnika Molodezhi (*Youth's Technology*): 3, 66
telerobotics: 262, 292, **313–320**, *328–329*; see also: rovers and hoppers, Lunokhod program
telescope, optical: 11–12, 289–290, *297*, *361*
    low-frequency radio telescope: 290–291, *297*
Telstar 1 communications satellite: 146
Temptation of Saint Anthony, The, play: *250*
Tennyson, Alfred: 368, *386*
terraforming: 400, 408, *426*, *427*
terrane: *55*
Thales Alenia Space: 127
Thales of Meletius: 420, *429*
Thomson, William (Lord Kelvin): 159
Thor missile: 64
Thule: *250*
tides
    Bay of Fundy: **4**
    lengthening of month and Earth day: **4–7**, *15*
    from Moon on Earth: **2–7**, *15–16*
    and stability of rotational axis: 7
Titan, moon of Saturn: 63, 272, 400, *426*
Titan missile: 64
Titan II: 64
titanium: 41
tranquillityite: 177, *205*
transfer trajectories/orbits: 128, *152*
    energy requirements: *387*
    weak stability boundary transfer trajectory: 128, *152*
transient lunar phenomena (TLPs): 262, 264, **275–281**, *293–294*
    associated with Aristarchus Plateau: 284–285, *296*
    correspondence of TLPs to other lunar events: 279–280, *296*
    real-time imaging monitoring: 283–284, *296*
    TLPs observed by Apollo astronauts: 264–269, 292
Trip to the Moon, A, film: 60
Triton, moon of Neptune: 272
Truman, Harry: 60
Tsien Hsue-shen (Qian Xuesen): 126, *152*
    fictional Europan mission: *156*
Tsiolkovski, Konstantin: 59, *114*, *156*, *386*, 392, 396, 397, 407, *424*, *427*
turmoil in U.S. in 1968–1969: 70–71

Tyson, Neil deGrasse: 107
Tyuratam prison camp: *57, 117*

U2 aircraft: 65
unidentified flying object (UFO): 60, 410, 418–419, *429*
United Space Alliance: 96
Urey, Harold C.: 174, **228–231**, *251, 252, 255*
U.S. Air Force: 23, 27, 60, 79
    89th Airlift Wing: *120–121*
    Vandenberg Air Force Base (California): 81
U.S. Army: 23, 106
    Fort Bliss, TX/NM: 112
    Redstone Arsenal, Huntsville: 64, 106
U.S. Geological Survey: 26, 187
U.S.–Soviet space cooperation: *57*, 69, *118, 119, 155*

V-2 (Vergeltungswaffe-2): **59–60**, 61, 106, 112, *114–115, 120*
V-3 (Vergeltungswaffe-3): 396
Van Allen, James: 105, *125, 328*
    Van Allen radiation belts: 310
Vanguard program
    Vanguard 1: 64
    Vanguard TV3: 64
Venus: 25, 110, 272
    AAP Venus human flyby proposal: 78
    angular momentum: 163, *172*
    human habitability: 401–402, 403, *426*
    largest crater (Margaret Mead): 163
    Soviet/Russian Venus program
        1970s proposals: 44
        proposed Venus 2016 program: 137
        Venera program: *55*
            Venera 5: 55
Verein für Raumschiffahrt (VfR): 59, 106
Verne, Jules: 58, 59, *114*, 151, 396
Vietnam War (with America): 52, 70, 74
Viking program: 79
Vilas, Faith: 237
Virgin Galactic/Scaled Composites suborbital flights: 127, 146, 147, 376, 453
Vision for Space Exploration (VSE): 84, 102, 112, *121*
    Constellation: see Constellation space vehicle architecture
    Exploration Systems Architecture System: 94

Obama Administration support and cancellation: 89–90, 94–97, 434, *455*
VLS-1 rocket: 143
volcanoes, volcanic: see lunar volcanic eruptions, lunar magma
Von Braun, Wernher: 18, 28, 59, 63, 64, 78, **79–81**, 103, 105, **106–107**, 111, *116, 120*, 126, *152*, 298
Von Neumann, John: 63, 415
    Bracewell-von Neumann probe concept: 415–416, *429*
Voskhod: 45, 49
Vostok program: 19, 45, 49, 64, 370
    Vostok 1: 19, 45, 356
Voyager program ("Grand Tour"): *121, 125*, 203, 313

WAC Corporal rocket: 60, *114*
Wang, Taylor Gun-Jin: 154
War of the Worlds, The, radio program: 60, 419
Ward, William: 165, *172*
water, hydration and volatiles, lunar
    alpha particle detection of radon release: 248, 274–275, *295, 296*
        alphas from daughter product polonium-210: 279, *295, 296*
        correspondence of radon release sites and TLPs: 279
    Apollo sample composition: 233, *252, 256*
    aqueous alteration of landforms, suspected: 228–229, *251–252*
    atmospheric transport of volatiles: 245
    chlorine isotopes' role in evaluating internal H: 240, 246, *256*
    cometary/meteoritic origin: 233, 239, 240, **241**, 242, 243, 245, 247, 248, *254, 257, 258, 260*
    commercialization: see lunar commercial development
    compatibility with Giant Impact model: 246–247, *259*
    cryptovolatiles: 203
    depletion: 232–233
    economic exploitation: **248–249**, *260*, 423
    H detection by epithermal neutrons: **236, 241**, 248, *257*
    H_2O in anorthosite: *256*

water, hydration and volatiles, lunar (cont.)
   H_2O in atmosphere
      Apollo 17 LACE: *252*, 273
      Moon Impact Probe (*Chandrayaan-1*): 132, 238, 256
      Suprathermal Ion Detector (SIDE) result: 233–234, *254*
   H_2O in basalts: *253*, *256*
   H_2O ice detected by radar
      Chandrayaan-1: 132, *255*
   H_2O interactions with regolith: **242–244**, *258*, 287
   isotopic composition: 232, **247**, *252–253*, *259*
   LCROSS impact: **240–241**
      excavated volatiles: 133, 241, 242, *256–257*, 372, 376
   magma ocean cumulates and hydration: 242, *257*
   moonquakes
      construction hazard: 341, *363*
      relation to lunar water content: 220–221, 246
      relation to TLPs and outgassing: 279–280, 281
   OH in lunar minerals seen with SIMS (secondary ion mass spectrometry): 237–240
   OH in regolith depths seen in IR (*Luna 24*): **234**, 236
   OH on surface seen in IR
      ground-based spectroscopy: 236–237, *255*
      Moon Mineralogy Mapper – M^3 (*Chandrayaan-1*): 132, **237–238**, 248, *255–256*, *257*, *258–259*, 287
   outgassing from lunar surface: *225*, *252*, 274–275, 280–282, *295*
   outstanding scientific issues: 247–249
   oxygen fugacity: 242, 246, *257*, **259**
   possible interstellar component: 203, 241, 242, 247–248, 256, *257*, *259*

   rust in lunar samples: 243, *253*, *254*, *258*
   sublimation in vacuum at lunar surface: 227, *251*
   variety of origins: 203, **241–242**, 245, *256*, *257*, *258*
   volatiles in permanently shadowed regions (PSRs): **241–243**, 244, 245, 248, **255**, *257*, 286, 376
Webb, James: 78, *118*, *119*
Wells, H.G.: 58, 59, *113*
Welty, Eudora: 1, *14*
Wen Ho Lee: 127
Whipple, Fred: 61
White, Edward: 28, 300
White Sands: 60, 61
"Whitey on the Moon," song: 71
Wilford, John Noble: *56*
Wilhelms, Donald: 263
Wingo, Dennis: *386*
Woman in the Moon (*Frau im Mond*), film: *114*
Worden, Al: 39, 40, 43
World War II: 59, 60
Wu-ching Tsung-yao, The, (military compendium) book: *114*

X-15 rocket airplane: 146, 396

Ye-6 lunar lander: 26
Yeager, Chuck: 60, *115*
Yinguo-1 mission: 139
Young, John W.: 29, 32, 36, 40, 44, 198, 303, 333

Zenit-3SL rocket: 141; see OKB-586
Zheng He: viii, 150, 454
zircon uranium-lead age estimation: *15*, 191, 193, *207*
Zond program: 23, 31, 48
   Zond 5: 23, 48
   Zond 6–7: 48
Zubrin, Bob: *124*, 407–408, 409, *427*, *428*

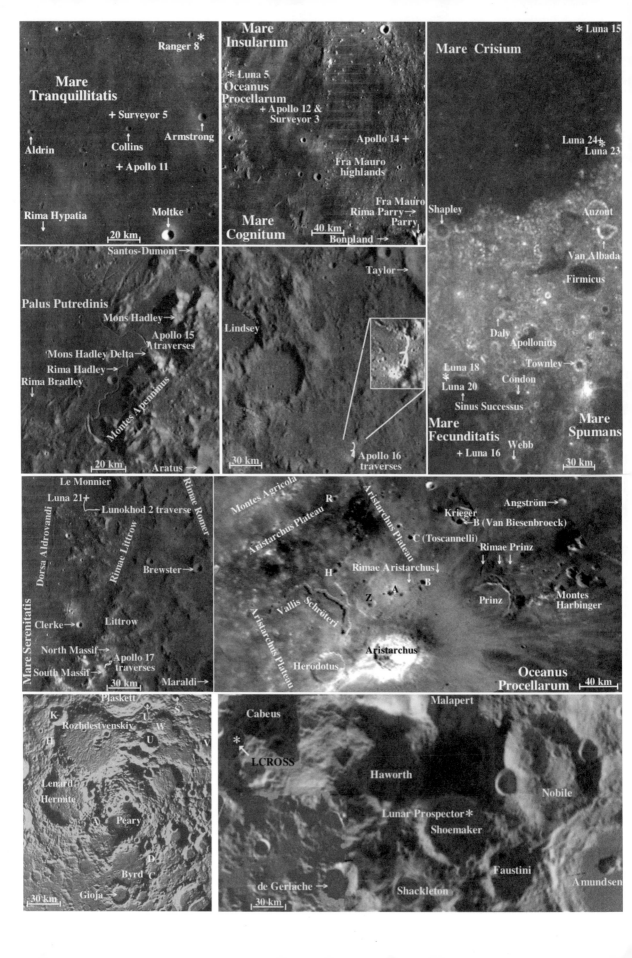